Chemical Methods in Prokaryotic Systematics

MODERN MICROBIOLOGICAL METHODS

Series Editor Michael Goodfellow, *Department of Microbiology University of Newcastle upon Tyne*

Methods in Aquatic Bacteriology (1988)
Edited by Brian Austin

0 471 91651 X

Bacterial Cell Surface Techniques (1988)
Edited by Ian Hancock and Ian Poxton

0 471 91041 4

Molecular Biological Methods for *Bacillus* (1990)
Edited by Colin R. Harwood and Simon M. Cutting

0 471 92393 1

Nucleic Acid Techniques in Bacterial Systematics (1991)
Edited by Erko Stackebrandt and Michael Goodfellow

0 471 92906 9

Chemical Methods in Prokaryotic Systematics (1994)
Edited by Michael Goodfellow and Anthony G. O'Donnell

0 471 94191 3

Chemical Methods in Prokaryotic Systematics

Edited by
Michael Goodfellow
Department of Microbiology
The University, Newcastle upon Tyne, UK

and

Anthony G. O'Donnell
Department of Agriculture and Environmental Science
The University, Newcastle upon Tyne, UK

JOHN WILEY & SONS
Chichester · New York · Brisbane · Toronto · Singapore

Copyright © 1994 John Wiley & Sons Ltd,
         Baffins Lane, Chichester,
         West Sussex PO19 1UD, England
         Telephone: National   Chichester (0243) 779777
                    International +44 243 779777

All rights reserved.

No part of this book may be reproduced by any means,
or transmitted, or translated into a machine language
without the written permission of the publisher.

Other Wiley Editorial Offices

John Wiley & Sons, Inc., 605 Third Avenue,
New York, NY 10158-0012, USA

Jacaranda Wiley Ltd, 33 Park Road, Milton,
Queensland 4064, Australia

John Wiley & Sons (Canada) Ltd, 22 Worcester Road,
Rexdale, Ontario M9W 1L1, Canada

John Wiley & Sons (SEA) Pte Ltd, 37 Jalan Pemimpin #05-04,
Block B, Union Industrial Building, Singapore 2057

Library of Congress Cataloging-in-Publication Data

Chemical methods in prokaryotic systematics / edited by Michael
   Goodfellow and Anthony G. O'Donnell.
      p. cm.—(modern microbiological methods)
   Includes bibliographical references and index.
   ISBN 0 471 94191 3
   1. Molecular microbiology. 2. Chemotaxonomy. 3. Prokaryotes—
Classification. I. Goodfellow, M. II. O'Donnell, Anthony G.
III. Series.
QR74.C48 1994
576—dc20   93-42033
                                                             CIP

British Library Cataloguing in Publication Data

A catalogue record for this book is available from the British Library

ISBN 0 471 94191 3

Typeset by Mathematical Composition Setters Ltd, Salisbury, Wiltshire
Printed in Great Britain by Bookcraft Ltd., Bath, Avon

Contents

List of Methods by Chapter	xv
Contributors	xix
Series Preface	xxi
Preface	xxiii
Abbreviations	xxv

Chapter 1: Chemosystematics: Current State and Future Prospects
    M. Goodfellow and A.G. O'Donnell   1
1.1 Introduction   1
1.2 Chemical markers   5
1.3 Whole-organism fingerprinting   13
1.4 Overview   14
References   16

Chapter 2: Analysis of Cellular Constituents from Gram-negative Bacteria
    E. Jantzen and K. Bryn   21
2.1 Introduction   21
2.2 Peptidoglycan   24
2.3 Outer-membrane proteins   25
2.4 Lipopolysaccharides   27
    2.4.1 Extraction of LPS   29
    2.4.2 Purification of LPS   34
    2.4.3 Analysis of intact LPS and lipid A   35
    2.4.4 Cleavage of LPS into lipid A and polysaccharide   36
    2.4.5 Analysis of LPS constituents   38
2.5 Extractable lipids   45
    2.5.1 Phospholipids   45
    2.5.2 Other polar lipids   45
    2.5.3 Apolar lipids   46
2.6 Polysaccharides   46
2.7 Rapid whole-cell procedures   47
    2.7.1 GLC profiles of whole-cell fatty acids   48
    2.7.2 GLC profiles of whole-cell sugars   49

		2.7.3	Combined GLC profiles of monosaccharides and fatty acids	49
		2.7.4	Constituents in clinical and environmental samples	50
	2.8	Taxonomic considerations		52
		2.8.1	Envelope chemistry and evolution	52
		2.8.2	Chemotaxonomic strategy	53
		2.8.3	Perspectives	54
References				54

Chapter 3: Analysis of Cell Wall Constituents of Gram-positive Bacteria
I.C. Hancock — 63

3.1	Introduction			63
	3.1.1	Peptidoglycan		64
	3.1.2	Anionic wall polymers		64
	3.1.3	Polysaccharides		65
	3.1.4	Proteins		65
3.2	Preparation of cell walls			66
3.3	Peptidoglycan ('murein')			68
	3.3.1	Analyses using whole bacteria		68
	3.3.2	Isolation of peptidoglycan		71
	3.3.3	Determination of amino acid composition of peptidoglycan		72
	3.3.4	Determination of peptidoglycan type		73
3.4	Anionic polymers			74
	3.4.1	Detection of teichoic acids and teichuronic acids		75
	3.4.2	Classification of teichoic acids		76
	3.4.3	Teichuronic acids		80
3.5	Polysaccharides			81
3.6	Rapid methods			82
	3.6.1	Whole-cell detection of teichoic acid components		82
References				83

Chapter 4: Analysis of Archaeal Cell Envelopes
H. König — 85

4.1	Introduction		85
4.2	Electron microscopy		85
	4.2.1	General information	85
	4.2.2	Preparation of specimens	92
4.3	Isolation and analysis		93
	4.3.1	Pseudomurein	93
	4.3.2	Methanochondroitin	98
	4.3.3	Heteropolysaccharide	100

	4.3.4	Protein S-layers	103
	4.3.5	Glycoprotein S-layers	105
	4.3.6	Surface layers with unusual stability	110
4.4	Taxonomic considerations		111
	4.4.1	Murein	112
	4.4.2	Pseudomurein	112
	4.4.3	Heteropolysaccharide	113
	4.4.4	Methanochondroitin	113
	4.4.5	S-layers	113
References			115

Chapter 5: Structural Lipids of Eubacteria
T.M. Embley and R. Wait — 121

5.1	Introduction		121
5.2	Fatty acids		121
5.3	Methods for analysing fatty acids		125
	5.3.1	Growth conditions for preparing biomass	125
	5.3.2	Extraction of fatty acids	125
	5.3.3	Thin-layer chromatography of fatty acids	127
	5.3.4	High-performance liquid chromatography of p-bromophenacyl esters of mycolic acids	131
	5.3.5	Analysis of fatty acid methyl esters by capillary gas chromatography	131
	5.3.6	Identification of fatty acids by mass spectrometry	136
5.4	Bacterial polar lipids		141
5.5	Methods for analysing polar lipids		144
	5.5.1	Extraction	144
	5.5.2	Two-dimensional thin-layer chromatography	145
	5.5.3	Mass spectrometric identification of polar lipids	148
5.6	Concluding remarks		153
References			155

Chapter 6: Analysis of Microbial Sterols and Hopanoids
G.J. Jones, P.D. Nichols and P.M. Shaw — 163

6.1	Introduction		163
6.2	Biosynthesis		164
6.3	Methods		169
	6.3.1	Extraction	169
	6.3.2	Purification	171
	6.3.3	Analysis and identification	173
6.4	Discussion		190
	6.4.1	Methods	190

	6.4.2	Taxonomic considerations	191
References			193

Chapter 7: Archaeal Lipids
M. De Rosa and A. Gambacorta — 197

7.1	Introduction		197
7.2	Isopranoid ether lipids		198
	7.2.1	Extraction of archaeal lipids	198
	7.2.2	Chromatographic analyses of extracted lipids	200
	7.2.3	Overall composition of archaeal lipids	203
	7.2.4	Degradative procedures for archaeal lipids	204
7.3	Core lipids		206
	7.3.1	Diethers	207
	7.3.2	Tetraethers	216
	7.3.3	Minor glycerol isoprenoid ether lipids	227
	7.3.4	Occurrence of diethers and tetraethers in archaeal lipids: taxonomic implications	228
7.4	Polar lipids		231
	7.4.1	Structure and distribution of polar lipid components in archaea: taxonomic implications	231
	7.4.2	Methodologies used to define polar lipid structures	242
7.5	Isoprenoid quinones		250
	7.5.1	Archaeal quinones: structural types	250
	7.5.2	Chromatographic procedures	254
	7.5.3	NMR spectroscopy	254
	7.5.4	Mass spectrometry	254
	7.5.5	Ultraviolet and visible spectroscopy	257
	7.5.6	Quinone content in archaea: taxonomic considerations	257
References			259

Chapter 8: Isoprenoid Quinones
M.D. Collins — 265

8.1	Introduction		265
8.2	Quinone structures		266
	8.2.1	Phylloquinones	266
	8.2.2	Menaquinones	267
	8.2.3	Demethylmenaquinones	272
	8.2.4	Methylmenaquinones	274
	8.2.5	Dimethylmenaquinones	276
	8.2.6	Ubiquinones	276
	8.2.7	Plastoquinones	279

		8.2.8	Sulphur-containing quinones	279
8.3	Extraction and purification			280
8.4	Chromatographic analysis			283
	8.4.1	Reverse-phase partition chromatography	|| 283
	8.4.2	Argentation chromatography	|| 286
8.5	Physico-chemical analysis			290
	8.5.1	Ultraviolet spectroscopy	|| 290
	8.5.2	Mass spectrometry	|| 294
	8.5.3	Nuclear magnetic resonance spectroscopy	|| 298
8.6	Conclusions			305
References				305

Chapter 9: Analysis of Cytochromes
R.K. Poole

9.1	Introduction		311
			311
9.2	Structure and nomenclature		312
9.3	Analytical methods		316
	9.3.1	Spectrophotometer design considerations	316
	9.3.2	Application of spectrophotometric techniques	320
	9.3.3	Techniques other than wavelength scanning and their applications	324
9.4	Assignment of visible absorption bands to cytochromes and other haemoproteins		327
	9.4.1	Overview of cytochrome types	327
	9.4.2	Non-cytochrome haemoproteins	330
	9.4.3	Interpretation of spectra	331
	9.4.4	Cytochrome quantification	335
9.5	Taxonomic considerations		337
9.6	Summary		340
References		341

Chapter 10: Analysis of Pigments: Bacteriochlorophylls
W.R. Richards

		345
10.1	Introduction	345
	10.1.1 Bacteriochlorophyll a	345
	10.1.2 Bacteriochlorophylls b and g	347
	10.1.3 Bacteriochlorophylls c, d and e	348
10.2	Analytical methods	349
	10.2.1 Isolation and estimation	349
	10.2.2 Purification by conventional methods	351
	10.2.3 Purification by HPLC	354

		10.2.4	Characterization of bacteriochlorophylls	355
		10.2.5	Characterization of the esterifying alcohols	364
10.3	Taxonomic considerations			366
		10.3.1	Purple bacteria	368
		10.3.2	Green and brown bacteria	376
		10.3.3	Heliobacteria	381
		10.3.4	Cyanobacteria	382
		10.3.5	Prochlorophytes	384
		10.3.6	The evolution of the chlorophylls of photosynthetic prokaryotes	385
References				388

Chapter 11: Analysis of Pigments: Carotenoids and Related Polyenes
K. Schmidt, A. Connor and G. Britton — 403

11.1	Introduction		403
	11.1.1	Carotenoids	403
	11.1.2	Carotenoid structure and nomenclature	403
	11.1.3	Carotenoid biosynthesis	405
11.2	Distribution and taxonomy		407
	11.2.1	Cyanobacteria	407
	11.2.2	Anoxygenic phototrophs	407
	11.2.3	Non-photosynthetic prokaryotes	408
11.3	Strategy for determining carotenoid composition		411
11.4	General methods		412
	11.4.1	Properties and handling of carotenoids	412
	11.4.2	Protection against oxygen	412
	11.4.3	Protection against light and heat	412
	11.4.4	Avoidance of acid and alkali	413
	11.4.5	Purity of solvents, adsorbents and reagents	413
11.5	Extraction and saponification of carotenoids		414
11.6	Isolation and purification: chromatography		415
	11.6.1	Choice of adsorbent	416
	11.6.2	Preliminary analysis by thin-layer chromatography	417
	11.6.3	Column chromatography	417
	11.6.4	Thin-layer chromatography	418
	11.6.5	Paper chromatography	420
	11.6.6	Purification of samples for mass spectrometry and nuclear magnetic resonance spectroscopy	420
	11.6.7	Identification by co-chromatography	421
11.7	High-performance liquid chromatography		421
	11.7.1	Reversed-phase HPLC	422
	11.7.2	Normal-phase (adsorption) HPLC	424

11.8	Ultraviolet-visible light absorption spectroscopy	426
	11.8.1 Position of the absorption maxima	426
	11.8.2 Spectral fine structure	427
	11.8.3 Geometrical isomers	433
11.9	Quantitative determination	434
	11.9.1 Spectrophotometry	434
	11.9.2 Quantitative determination by high-performance liquid chromatography	434
11.10	Diagnostic chemical reactions and formation of derivatives	436
	11.10.1 Iodine-catalysed geometrical isomerization	436
	11.10.2 Acetylation or silylation of hydroxy groups	437
	11.10.3 Reduction of carbonyl groups by sodium borohydride	437
	11.10.4 Dehydration: elimination of allylic hydroxy groups	438
	11.10.5 Carotenoid glycosides: liberation and identification of sugar residues	438
11.11	Other spectroscopic methods	439
	11.11.1 Mass spectrometry	439
	11.11.2 Nuclear magnetic resonance spectroscopy	442
	11.11.3 Circular dichroism	443
11.12	Application of carotenoid analysis in practice	443
11.13	Non-carotenoid polyenes	444
	11.13.1 Flexirubin-type pigments	444
	11.13.2 Xanthomonadins	445
	11.13.3 Properties and methods	445
References		447

Chapter 12: Determination of DNA Base Composition
J. Tamaoka — 463

12.1	Introduction	463
12.2	Methodology	463
	12.2.1 Isolation of DNA	463
	12.2.2 DNA base composition	466
12.3	Taxonomic implications	468
References		469

Chapter 13: Enzymes in Taxonomy and Diagnostic Bacteriology
A.L. James — 471

13.1	Introduction	471
13.2	Classes of enzyme of known taxonomic or identification value	473
	13.2.1 Oxidoreductases	473
	13.2.2 Transferases	474

		13.2.3 Hydrolases	474
		13.2.4 Lyases	475
		13.2.5 Isomerases	475
		13.2.6 Ligases	476
13.3	Choice of substrate, specificity of enzyme action, natural and synthetic substrates		476
13.4	Detection methods		479
		13.4.1 Visual observation	479
		13.4.2 Fluorescence techniques	480
		13.4.3 Multipoint techniques	484
13.5	Areas for development		486
		13.5.1 Hydrolytic enzymes	486
		13.5.2 Non-hydrolytic enzymes	490
References			490

Chapter 14: Analysis of Electrophoretic Whole-Organism Protein Fingerprints
B. Pot, P. Vandamme and K. Kersters — 493

14.1	Introduction		493
14.2	SDS-PAGE of whole-cell bacterial proteins		494
	14.2.1	Cultivation of bacteria	494
	14.2.2	Preparation and electrophoresis of protein samples	495
	14.2.3	Visual comparison of whole-cell protein patterns	498
	14.2.4	Computer-assisted analysis of protein electrophoregrams	500
14.3	Taxonomic considerations and applications		513
References			517

Chapter 15: Analytical Fingerprinting Methods
J.T. Magee — 523

15.1	Introduction		523
15.2	Pyrolysis techniques		524
	15.2.1	Principles	524
	15.2.2	History	524
15.3	Pyrolysis mass spectrometry		526
	15.3.1	Method	527
	15.3.2	Data analysis	533
15.4	Other pyrolysis-based techniques		535
	15.4.1	Laser pyrolysis mass spectrometry	535
	15.4.2	Pyrolysis gas-liquid chromatography-mass spectrometry	536
	15.4.3	Pyrolysis tandem mass spectrometry	538

15.5	Infrared spectrometry	538
	15.5.1 Principles	538
	15.5.2 History	539
	15.5.3 Methods	540
15.6	Ultraviolet-resonance Raman spectroscopy	543
	15.6.1 Principles	543
	15.6.2 History	544
15.7	Overview	545
References		546
Appendix: The mathematics of multivariate analyses		549
Index		555

List of Methods by Chapter

Chapter 2
2.1 Extraction of (S-type) LPS with hot aqueous phenol 32
2.2 Extraction of (R-type) LPS with phenol-chloroform-petroleum ether 33
2.3 Cleavage of LPS into PS and lipid A 37
2.4 Analysis of LPS and lipid A by TLC 38
2.5 Combined GLC analysis of sugars and fatty acids from whole cells or LPS 39
2.6 High pH anion exchange chromatography (HPAEC) profiles of LPS 44
2.7 GLC profiles of whole-cell fatty acids 48
2.8 GLC-MS analysis of the neisserial LPS marker 3-OH-12:0 in clinical samples 51

Chapter 3
3.1 Preparation of cell walls from whole bacteria 67
3.2 Determination of diaminopimelic acid isomers 69
3.3 Detection of N-glycolyl groups in peptidoglycan 69
3.4 Removal of anionic polymers and polysaccharide by treatment with hydrofluoric acid 71
3.5 Determination of N-terminal amino acids: dinitrophenylation 73
3.6 Partial acid hydrolysis 73
3.7 Measurement of teichoic acid as organic phosphorus 75
3.8 Measurement of teichuronic acid as uronic acid 76
3.9 Degradation of wall teichoic acid with hydrofluoric acid 78
3.10 Acid hydrolysis of HF degradation products 78
3.11 GC of degradation products as their alditol acetates 79
3.12 GC of degradation products as their trifluoroacetyl derivatives 80
3.13 Carboxyl-reduction of polymers containing uronic acids and aminouronic acids 81

Chapter 4
4.1 Platinum shadowing 92
4.2 Negative staining 92

4.3	Thin sections	92
4.4	Mechanical disintegration and digestion of cells	94
4.5	Hydrolysis of pseudomurein components	95
4.6	Identification and quantification of amino acids and amino sugars	97
4.7	Isolation and quantification of talosaminuronic acid	97
4.8	Identification of neutral sugars by GLC	97
4.9	Phosphate determination	98
4.10	Determination of uronic acids	99
4.11	Isolation and identification of gulosaminuronic acid	101
4.12	Isolation and quantification of sulphate	101
4.13	Isolation of S-layers and S-layer proteins	103
4.14	SDS polyacrylamide gel electrophoresis	104
4.15	Isolation of *Halobacterium* envelope components	109
4.16	Isolation of *Sulfolobus* envelope components	110
4.17	Isolation of sheaths of *Methanospirillum* and *Methanothrix*	111

Chapter 5

5.1	Preparation of fatty acid methyl esters by acid methanolysis	126
5.2	Alkaline hydrolysis for simple fatty acids	126
5.3	Alkali procedure for preparation of mycolic acid methyl esters for two-dimensional thin-layer chromatography	126
5.4	Preparation of *para*-bromophenacyl esters of mycolic acids for high-performance liquid chromatography	127
5.5	One-dimensional TLC of methanolysis products	129
5.6	Purification of fatty methyl esters using 1D TLC	129
5.7	Two-dimensional TLC of mycolic acid methyl esters	131
5.8	Catalytic hydrogenation: verification of double bonds	135
5.9	Trimethylsilylation: identifying hydroxy-substituted acids	135
5.10	Preparation of picolinyl esters	139
5.11	Integrated extraction of polar lipids and isoprenoid quinones	145
5.12	Two-dimensional TLC of polar lipid extracts	145
5.13	Spray reagents for characterizing individual components	147
5.14	Modified Bligh–Dyer extraction for FAB samples	149
5.15	Digestion of phospholipids with phospholipase A_2	150

Chapter 6

6.1	Solvent extraction of microbial lipids	169
6.2	Sterol extraction from yeasts	170
6.3	Digitonin precipitation of sterols	172
6.4	Extraction and treatment of hopanoids for chromatography	173
6.5	Determination of total sterols	174

List of Methods by Chapter　　　　　　　　　　　　　　　　xvii

Chapter 7
7.1　Lipid extraction: modified Bligh–Dyer technique　　198
7.2　Lipid extraction: modified Bligh–Dyer technique for methanogenic bacteria　　199
7.3　Lipid extraction: modified Bligh–Dyer technique for methanogenic archaea　　200
7.4　Fractionation of lipid extracts by column chromatography　　201
7.5　Acid methanolysis of core lipids for screening　　206

Chapter 8
8.1　Direct extraction of isoprenoid quinones with chloroform-methanol　　281
8.2　Purification of isoprenoid quinones by thin-layer chromatography　　282
8.3　Purification of isoprenoid quinones by silica cartridge　　282
8.4　Reverse-phase thin-layer partition chromatography of ubiquinones　　284
8.5　Reverse-phase thin-layer partition chromatography of menaquinones　　284

Chapter 11
11.1　Extraction of carotenoids from cells　　414
11.2　Saponification of carotenoid extracts　　415
11.3　TLC for preliminary analysis of extracts　　417
11.4　Column chromatography on alumina　　418
11.5　Elution of carotenoid bands following separation　　419
11.6　Paper chromatography for carotenoids　　420
11.7　Iodine-catalysed photo-isomerization for carotenoid analysis　　436
11.8　Acetylation of hydroxy groups　　437
11.9　Silylation of hydroxy groups　　437
11.10　Reduction of carbonyl groups by sodium borohydride　　438
11.11　Dehydration of allylic hydroxy groups　　438
11.12　Identification of sugars in glycosides　　438
11.13　Determination of carotenoids in lake sediments　　444

Chapter 12
12.1　Isolation of high molecular weight and double-stranded DNA　　464
12.2　Simplified method for the isolation of DNA for HPLC analysis　　466
12.3　Hydrolysis of DNA　　467

Chapter 14
14.1 Preparation of protein samples   495
14.2 Electrophoresis of protein samples   496

Contributors

Dr G. Britton *Department of Biochemistry, University of Liverpool, PO Box 147, Liverpool L69 3BX, UK*

Dr K. Bryn *National Institute for Public Health, Geitmyrsveien 75, 0462 Oslo 4, Norway*

Dr M.D. Collins *AFRC Institute of Food Research, Early Gate, Whiteknights Road, Reading RG6 2EF, UK*

Dr A.E. Connor *Department of Biochemistry, University of Liverpool, PO Box 147, Liverpool L69 3BX, UK*

Professor M. De Rosa *Istituto di Biochimica delle Macromolecole, I Facolta' di Medicina e Chirurgia, Via Costantinopoli 16, 80138 Napoli, Italy*

Dr T.M. Embley *Microbiology Group, The Natural History Museum, Cromwell Road, London SW7 5BD, UK*

Professor A. Gambacorta *Istituto per la Chimica di Molecole di Interesse Biologico, CNR, Via Toiano 6, 80072 Arco Felice, Napoli, Italy*

Professor M. Goodfellow *Department of Microbiology, The Medical School, The University, Framlington Place, Newcastle upon Tyne NE2 4HH, UK*

Dr I.C. Hancock *Department of Microbiology, The Medical School, The University, Framlington Place, Newcastle upon Tyne NE2 4HH, UK*

Dr A.L. James *Department of Chemical and Life Sciences, University of Northumbria, Newcastle upon Tyne NE7 8ST, UK*

Dr E. Jantzen *National Institute for Public Health, Geitmyrsveien 75, 0462 Oslo 4, Norway*

Dr G.J. Jones *CSIRO Division of Water Resources, Griffith Laboratory, PMB3 Griffith, NSW 2680, Australia*

Professor K. Kersters *Laboratorium voor Microbiologie, Universiteit Gent, K.L. Ledeganckstraat 35, B-9000 Gent, Belgium*

Professor Dr H. König *Abteilung für Angewandte, Mikrobiologie und Mykologie, Universität Ulm. Oberer Eselsberg, D-7900 Ulm, Germany*

Dr J.T. Magee *Department of Microbiology and Public Health Laboratory, University Hospital of Wales, Cardiff CF4 4XW, UK*

Dr P.D. Nichols *CSIRO Division of Oceanography, GPO Box 1538, Hobart, Tasmania 7001, Australia*

Dr A.G. O'Donnell *Department of Agriculture and Environmental Science, The University, Framlington Place, Newcastle upon Tyne NE2 4HH, UK*

Professor R.K. Poole *Division of Life Sciences, King's College London, Campden Hill Road, London W8 7AH, UK*

Dr P. Pot *Laboratorium voor Microbiologie, Universiteit Gent, K.L. Ledeganckstraat 35, B-9000 Gent, Belgium*

Dr W.R. Richards *Department of Chemistry, Simon Fraser University, Burnaby, British Columbia, Canada V5A 1S6*

Dr K. Schmidt *Institut für Mikrobiologie, Georg-August-Universität, Grisebachstrasse 8, D-3400 Göttingen, Germany*

Dr P.M. Shaw *Carlton & United Breweries Ltd., Analytical Development Section, Box 753F, GPO, Melbourne, Victoria 3001, Australia*

Dr J. Tamaoka *The RIKEN Institute, Hirosawa 2–1, Wako-shi, Saitama, 351-01 Japan*

Dr P. Vandamme *Laboratorium voor Microbiologie, Universiteit Gent, K.L. Ledeganckstraat 35, B-9000 Gent, Belgium*

Dr R. Wait *Centre for Applied Microbiology and Research, Porton Down, Salisbury, Wiltshire SP4 0JG, UK*

Series Preface

The science of microbiology owes its existence as well as its underlying principles to the talent and practical prowess of pioneers such as Leeuwenhoek, Pasteur, Koch and Beijerinck. Interest in microbiology has recently increased quite significantly given the exciting developments in genetics and molecular biology and the growth of microbial technology. There was a time when most microbiologists were acquainted with many of the techniques used in microbiology. It is, however, now becoming increasingly difficult for research workers to keep abreast of the bewildering range of techniques currently used in microbiological laboratories. This problem is compounded by the fact that scientists in any one field increasingly need to apply techniques developed in other scientific disciplines.

The series 'Modern Microbiological Methods' aims to identify specialist areas in microbiology and provide up-to-date methodological handbooks to aid microbiologists at the laboratory bench. The books will be directed primarily towards active research workers but will be structured so as to serve as an introduction to the methods within a speciality for graduate students and scientists entering microbiology from related disciplines. Protocols will not only be described but difficulties and limitations of techniques and questions of interpretation fully discussed.

In summary, this series of books is designed to help stimulate further developments in microbiology by promoting the use of new and updated methods. Both authors and the editor-in-chief will be grateful to hear from satisfied or dissatisfied users so that future books in the series can benefit from the informed comment of practitioners in the field.

MICHAEL GOODFELLOW

Preface

The application of chemical, molecular biological and numerical taxonomic techniques has had a profound effect on how prokaryotes, i.e. archaea and bacteria, should be classified, identified and characterized. To date, chemosystematics, like numerical systematics, has been used in both the classification and identification of prokaryotes with considerable emphasis placed on the development of rapid identification systems. In future, as techniques develop, this skew is likely to become less marked as those involved in identification move towards rapid automated chemical fingerprinting techniques such as infrared spectroscopy and pyrolysis mass spectrometry. It seems likely, therefore, that chemical methods will be increasingly used to underpin the developing phylogeny of prokaryotes. To do this successfully, it will be necessary to study the distribution of chemical markers at all levels in the taxonomic hierarchy with particular attention being paid to structural information and to the taxonomic evaluation of previously uncharacterized or poorly studied macromolecules. In concentrating on methodology, this book provides a foundation from which these glimpses of the future can be pursued.

We would like to express our sincere gratitude to all of the contributors for their stimulating and comprehensive chapters. Thanks are also due to Michael Dixon, Nicola Sawyer and Patricia Sharp of John Wiley & Sons for all of their help and encouragement during the preparation of this volume. Last, but not least, we are grateful to Julie Bottle for invaluable secretarial help.

MICHAEL GOODFELLOW
ANTHONY G. O'DONNELL
May 1993

Abbreviations

A1γ	(Gram-negative peptidoglycan)
ADMR	absorbance-detected magnetic resonance
Ara$_f$	arabinose (furanoic)
AraN	arabinosamine
Bchl	bacteriochlorophyll
Bmph	methyl bacteriopheophorbide
Bpho	bacteriopheophorbide
Bphy	bacteriopheophytin
BSTFA	N-O-bis(trimethylsilyl)trifluoracetamide
CCD	countercurrent distribution
CD	circular dichroism
CI	chemical ionization
CID	collision-induced decomposition
CM	cytoplasmic membrane
DA	discriminant analysis
DAG	diaminoglucose
DAP	diaminopimelic acid
DEPT	distortionless enhancement by polarization transfer
DGD	diglycosyl diether
DMPD	N, N-dimethyl-p-phenylene diamine
DOC-PAGE	deoxycholate-PAGE*
EI	electron impact
FAB-MS	fast atom bombardment mass spectrometry
FAME	fatty acid methyl ester
FD-MS	field desorption mass spectrometry
FTIR	Fourier transform infrared
GalA	galacturonic acid
GalN	galactosamine
GDGT	glycerol-dialkyl-glycerol tetraether
GDNT	glycerol-dialkyl-nonitol tetraether
Glc	glucose
GlcA	glucuronic acid
GLC-MS	gas-liquid chromatography-mass spectrometry
GlcN	glucosamine
GlcNA	glucosaminuronic acid
GlyP	glycerol phosphate

HPAEC	high-pH anion-exchange chromatography
HPLC	high-performance liquid chromatography
ICM	intracytoplasmic membrane
IdA	iduronic acid
IRS	infrared spectrometry
Kdo	2-keto-3-deoxyoctonate
LH	light-harvesting
LPS	lipopolysaccharide
Man	mannose
m-A$_2$pm	*meso*-diaminopimelic acid
Neu5Ac	N-acetylneuraminic acid
ODS	octadecylsilica
OMP	outer membrane protein
PAD	pulsed amperometric detection
PAGE	polyacrylamide gel electrophoresis
PCA	principal component analysis
PED	pulsed electrochemical detection
PG	phosphatidylglycerol
PGP	phosphatidylglycerol phosphate
PGS	phosphatidylglycerol sulphate
PS	polysaccharide
Py-GLCMS	pyrolysis gas-liquid chromatography-mass spectrometry
Py-MS	pyrolysis mass spectrometry
Py-MS-MS	pyrolysis tandem mass spectrometry
RC	reaction centre
Rha	thamnose
RS	Raman spectroscopy
S-DGD	sulphate diglycosyl diether
SDS	sodium dodecyl sulphate
SFC	supercritical fluid chromatography
TEA	triethanolamine
TeGD	tetraglycosyl glycolipid
TGD	triglycosyl diether
TMPD	N, N, N', N'-tetramethyl-p-phenylenediamine
TMS	trimethylsilyl
TMU	1, 1, 3, 3-tetramethylurea
UPGMA	unweighted pair group method using average linkages
UVRRS	ultraviolet resonance Raman spectrometry
Xyl	xylose

1

Chemosystematics: Current State and Future Prospects

Michael Goodfellow and Anthony G. O'Donnell
The University, Newcastle upon Tyne, UK

1.1 INTRODUCTION

It has been accepted since the earliest days of microbiology that microbial form and function are varied and diverse. The analysis of this diversity by successive generations of microbiologists has shown that organisms isolated from geographically and physico-chemically distinct environments share many common features. This has allowed microorganisms to be grouped together and has facilitated their study and exploitation. The study of microorganisms and the recognition of shared and stable attributes within groups of strains is the cornerstone of microbial systematics and as such has been fundamental to the development of microbiology as a scientific discipline.

Systematics can be defined as the scientific study of the kinds and diversity of organisms and of any and all relationships among them (Goodfellow & O'Donnell, 1993). The subject encompasses the related but quite distinct processes of classification, nomenclature and identification. Sound classification is a prerequisite for stable nomenclature and accurate identification.

Classification is the ordering of organisms into groups (taxa) to produce an orderly arrangement of strains such that a knowledge of the properties of one strain allows the properties of other members of the group to be inferred. Classification is, therefore, essential to the development of microbiology and to exploiting microbial diversity (Bull *et al.*, 1992).

Nomenclature deals with the allocation of the names used to denote taxonomic groups or categories according to internationally recognized rules and procedures (Sneath, 1992). Although it is a relatively simple task to apply the rules of nomenclature, it is important to recognize that this not

---

Chemical Methods in Prokaryotic Systematics. Edited by M. Goodfellow and A.G. O'Donnell
© 1994 John Wiley and Sons Ltd.

only involves the correct latinization of words but depends upon the thoroughness of the preceding taxonomic work. The importance of nomenclature is often underestimated, but without a reliable nomenclature the routine identification of isolates would become difficult, if not impossible.

Identification—the process of determining whether an unknown organism belongs to a previously defined group—is seen by many microbiologists as the most important objective of systematics. This quite mistaken view overlooks the interdependence of classification, nomenclature and identification and stems from the fact that systematics is a pervasive discipline and often seen as ancillary to other areas of research. This is particularly true with respect to diagnostic microbiology where the role of the 'taxonomist' is generally perceived to be the quick and reliable identification of pathogens. This view of systematics as an ancillary science has not only persisted but has significantly influenced the development of the subject with much of the research effort designed to produce rapid identification systems. However, with increased interest in the exploitation and conservation of microbial resources from natural habitats, the role of the taxonomist is changing such that rapid identification should no longer be considered the primary objective of taxonomic research.

Identification implies the existence of a classification. Classifications are not available for isolates from many natural habitats and this makes it necessary to design methodologies which will allow rapid grouping of large numbers of environmental isolates according to their shared features or properties. Such work necessitates the introduction and evaluation of rapid classification procedures. Added to this is the fact that relatively little is known about the microbial diversity of natural environments (Bull *et al.*, 1992). Indeed, estimates of species richness (Hawksworth & Mound, 1991) should be viewed as little more than hypotheses that need testing. What is clear, however, is that large numbers of, as yet, unclassified bacteria await isolation and analysis and that many of the existing methodologies will be unsuitable for the rapid characterization needed to estimate diversity and to assign conservation priorities.

Although the ultimate objective of classification should be to achieve a taxonomy which is independent of the individual types of data used, classification and identification remain markedly data dependent. As such classifications are continually being reassessed it is hardly surprising that many of the major advances in microbial systematics in recent years have come as a result of changes in the way data are collected and analysed. In prokaryotic systematics, as in other areas of microbiology, there is a continuing trend towards quantification and automation with an increasing reliance placed on analytical instrumentation. The latter is particularly true for chemosystematics, which relies heavily on analytical chemical techniques such as gas chromatography, high-performance liquid chromatography, and mass and ultraviolet spectrometry.

Given the changing role of systematics and the emergence of the 'new bacterial systematics' (Stackebrandt & Goodfellow, 1991; Goodfellow & O'Donnell, 1993) it is timely to review current developments in chemosystematics. At the time of an earlier book, *Chemical Methods in Bacterial Systematics* (Goodfellow & Minnikin, 1985), molecular systematics was a relatively new area of taxonomic research which offered, for the first time, the possibility of conducting research on microbial systematics within a phylogenetic framework. There can be no doubt that molecular systematics has fulfilled its early potential and additional effort is now needed to expand the phylogenetic framework (O'Donnell *et al.*, 1993). An important consequence of this success is that, unlike the earlier book, the nucleic acid techniques, with the exception of the determination of DNA base composition, have been considered independently of other chemosystematic approaches (Stackebrandt & Goodfellow, 1991).

It is possible to view all systematics as chemosystematics since morphology, pigmentation, serology and the biochemical properties of microorganisms reflect their chemical composition. However, given the recognition of molecular and numerical taxonomic methods as independent approaches to the classification and identification of microorganisms, it is perhaps more useful to restrict the scope of chemosystematics to the study of the distribution of specific chemical components such as lipids, wall amino acids, sugars and proteins amongst microbial taxa. This definition, however, needs to be expanded to accommodate procedures such as pyrolysis gas-chromatography, pyrolysis mass-spectrometry, infrared spectrometry and RAMAN spectroscopy which yield fingerprints or signatures of whole organisms.

Chemosystematics depends upon the chemical analysis of microbial cells and most chemotaxonomic procedures involve, to varying degrees, the extraction, fractionation, purification and resolution of target compounds. Although the introduction of new derivatization techniques and improved methodologies has led to significant advances in the extraction, fractionation and purification of cell components, the major developments in chemosystematics have come as a result of improvements in the resolution of chemical constituents, with many of the procedures now automated and capable of giving reproducible, quantitative data suitable for statistical analysis (Table 1.1). Underlying these advances in the analysis and resolution of chemical compounds have been the improvements in gas, liquid and thin-layer chromatography; these have not only provided increased and more reliable resolution but, through computer control, have made the instrumentation available to the non-specialist (Sasser, 1990; Stead *et al.*,

Table 1.1 Chemical techniques used in the classification and identification of bacteria to specific and subspecific levels.

Analysis	Taxa	Statistics	References
Complex lipids	*Mycobacterium*	None	Dobson *et al.* (1985)
Hydroxy fatty acids	*Legionella*	SIMCA principal component analysis	Jantzen *et al.* (1993)
Fatty acids	*Corynebacterium, Nocardia, Rhodococcus*	Cluster analysis	Bousfield *et al.* (1983)
Fatty acids	*Streptomyces cyaneus*	SIMCA principal component analysis	Saddler *et al.* (1987)
Fatty acids	*Amycolata, Amycolatopsis, Pseudonocardia*	Principal component analysis	Embley *et al.* (1988)
Fatty acids	*Erwinia, Pseudomonas, Xanthomonas*	Microbial identification system	Stead *et al.* (1992)
Mycolic acids	*Corynebacterium*	None	De Briel *et al.* (1992)
Pyrolysis mass spectrometry of whole organisms	*Staphylococcus epidermidis* and *S. haemolyticus*	Principal component/discriminant analysis	Freeman *et al.* (1991)
Pyrolysis mass spectrometry of whole-organisms	*Streptococcus pyogenes*	Principal component/discriminant analysis	Magee *et al.* (1991)
Whole-organism proteins	*Corynebacterium*	Correlation coefficient, cluster analysis	Jackman (1982)
Whole-organism proteins	*Xanthomonas campestris*	Correlation coefficient, cluster analysis	Vauterin *et al.* (1990)

1992) thereby allowing the microbial systematist to concentrate on taxonomic rather than technical problems.

In parallel with the technical developments in chemosystematics has come the realization that microorganisms are, in terms of their chemical composition, very diverse. Similarly the discovery of novel taxonomic groups has led to the isolation and description of new, often taxon-specific, chemical components (Kandler & König, 1985; König, 1988; De Rosa &

Gambacorta, 1988). Such studies will continue to be valuable as they provide evidence to support taxonomies based on molecular and phenotypic properties. For example, the description of many actinomycetes, which prior to the introduction of chemical procedures were regarded as fungi, now include information on lipid and wall diamino acid markers (Goodfellow, 1989; Williams et al., 1989).

Unlike numerical taxonomy, and to a limited extent molecular taxonomy, there have been relatively few proposals for a truly chemical taxonomy as chemical methods tend to be used to support classifications derived using other taxonomic methods. An advantage of this practice is that chemical data can be used to define relationships at all levels in the taxonomic hierarchy including subgeneric and in some cases subspecific rank. Unlike the qualitative analysis of key chemical markers (Goodfellow & O'Donnell, 1989), which can often be carried out following extraction using thin-layer or paper chromatography, such analyses require more analytical experience and access to suitable instrumentation. Details of chemical criteria that can be used to provide subgeneric and in some cases subspecific classification and identification are shown in Table 1.1.

1.2 CHEMICAL MARKERS

Many of the current chemotaxonomic procedures make use of the discontinuous distribution of chemical components found in the cell envelopes of prokaryotes. Bacterial cell envelopes form the interface between the organism and its environment and provide a permeability barrier enclosing the cytoplasm. This permeability barrier is complex in its chemical composition and structure and is responsible for the differential staining observed with the Gram reaction. Cell wall analyses allow most prokaryotes to be divided into two broad groups, the archaea and the bacteria (Woese, 1987; Woese et al., 1990).

Peptidoglycan, the polymer in bacterial walls responsible for structural rigidity, is common to both Gram-positive and Gram-negative bacteria. It is a heteropolymer which consists of an amino-sugar backbone cross-linked through tetrapeptide side-chains. The amino sugars form glycan strands composed of repeating units of N-acetylglucosamine and N-acetylmuramic acid in β-1,4 linkage. The tetrapeptide side-chains are linked to the carboxyl groups of muramic acid; the most usual sequence of the peptide subunit is L-alanine, D-glutamic acid, an L-diamino acid and D-alanine. In Gram-positive bacteria, the peptide subunits are usually cross-linked through interpeptide bridges composed of between one and six amino acids.

The glycan chain length varies from 10 to 65 disaccharide units. The glycan moiety shows few changes apart from variations such as *O*-acetylation or *O*-phosphorylation or the exceptional absence of peptide substituents (Schleifer, 1985; Schleifer & Seidl, 1985). Certain actinomycetes, such as those classified in the genera *Micromonospora*, *Mycobacterium* and *Nocardia*, have N-glycolated as opposed to the more common N-acetylated muramic acid (Uchida & Aida, 1977, 1979). A simple colorimetric assay is available for the determination of N-glycolyl substitution on the glycan chain of peptidoglycan (see Chapter 3).

Variation in peptidoglycan structure, first extensively studied by Cummins and Harris (1956), has been shown to be important in the systematics of Gram-positive bacteria (Schleifer & Kandler, 1972; Rogers *et al.*, 1980; Schleifer & Seidl, 1985; Komagata & Suzuki, 1987; Kodama *et al.*, 1992). Gram-negative bacteria have a remarkably uniform peptidoglycan structure with the exception of a few groups, including the spirochaetes, where L-ornithine often replaces *meso*-diaminopimelic acid at position 3 in the peptide subunit (Schleifer & Joseph, 1973). This situation is in marked contrast to the Gram-positive bacteria where considerable, taxonomically important diversity occurs in the peptide side-chain, particularly in position 3. The most widely distributed diamino acid in this position is diaminopimelic acid; L-lysine is fairly common; L-diaminobutyric acid, LL-diaminopimelic acid, and L-ornithine are less so. In some bacteria the diamino acid at position 3 is replaced by a monoamino acid such as L-alanine or L-homoserine. Where this occurs, the cross-linkage is between positions 2 and 4 of adjacent tetrapeptides (Minnikin & O'Donnell, 1984). Detection of the presence of diaminopimelic acid and identification of the isomer is one of the most useful chemotaxonomic procedures for Gram-positive bacteria, notably actinomycetes, and can be readily carried out with whole bacteria (see Chapter 3).

Simple chemical methods have been widely used to detect diagnostic diamino acids in Gram-positive bacteria, such as actinomycetes (Suzuki *et al.*, 1993). Elucidation of the primary structure of peptidoglycans requires the use of more specialized procedures that cannot readily be applied to many organisms (see Chapter 3). Nevertheless such detailed analyses are necessary to unravel the taxonomy of certain actinomycetes, such as those assigned to the genus *Arthrobacter* (Kodama *et al.*, 1992). The discontinuous distribution of wall sugars also provides valuable data for actinomycete systematics (Goodfellow, 1989). Chemical methods recommended for the detection of key wall envelope constituents are considered in Chapters 2 and 3.

The second group of prokaryotes, the archaea, are a diverse assemblage of organisms that live in extreme environments, notably in habitats that are anaerobic, have a high temperature and low pH or a very high salt concentration (König, 1988). These organisms lack peptidoglycan, though this is also true of bacteria such as *Isophaera*, *Pirella* and *Planctomyces* (König et al., 1984; Liesack et al., 1986; Giovannoni et al., 1987) and the *Mollicutes* (Razin & Freund, 1984). Archaeal heterogeneity can be exemplified by the chemical and structural diversity shown by their cell envelopes.

Most archaea are Gram-negative with a cell wall profile that consists of a single layer of protein or glycoprotein subunits associated with the cytoplasmic membrane. In contrast, Gram-positive strains have pseudomurein, methanochondroitin and acidic heteropolysaccharide walls. Cell envelope-less archaea are represented by *Thermoplasma* spp. The chemical diversity of archaeal wall polymers indicates that the common ancestor of these organisms did not have a specific wall or envelope polymer. Detailed protocols are available for the isolation and analysis of archaeal cell envelopes (see Chapter 4).

Most, possibly all, Gram positive bacteria contain anionic polymers—teichoic acids and teichuronic acids—in their walls. The former are water-soluble phosphodiester-linked polymers of glycerol phosphate, ribitol phosphate or other polyol phosphates, which may be substituted by D-alanine, sugars or amino sugars. They occur as cell wall teichoic acids covalently linked by phosphodiester bridges to muramic acid residues in the peptidoglycan, and as lipoteichoic acids associated with the plasma membrane. The wall teichoic acids show more structural diversity than lipoteichoic acids and consequently are of greater value as chemical markers and as specific surface antigens. The occurrence of different teichoic acids in the walls of Gram-positive taxa can be important in their classification, notably in the circumscription of staphylococcal species (Easmon & Goodfellow, 1990).

Less attention has been given to the teichuronic acids, anionic polysaccharides containing uronic acids or aminouronic acids and neutral sugars, linked glycosidically. The wide distribution and structural variation of anionic wall polymers make them taxonomic markers of considerable potential. At present, detailed structural determinations are not easy, but valuable information on the type and composition of teichoic and teichuronic acids can be obtained by straightforward analytical techniques (see Chapter 3).

Plasma membranes, which form the other major structural unit of bacterial cell envelopes, are based on a bilayer system composed of amphipathic polar lipids in intimate association with specific membrane proteins. Amphipathic polar lipids consist of hydrophilic head groups usually linked to two hydrophobic aliphatic chains. Phospholipids are the most common polar lipids but glycolipids and acylated ornithine amides also fall into this category. The hydrophobic interior of plasma membranes is considered to provide a location for isoprenoid quinones (Chapter 8) and for lipid-soluble pigments (Chapters 10 and 11).

A functional plasma membrane requires a balanced combination of the different structural types of amphipathic polar lipids. Such lipids interact to form a bilayer system in which the constituent fatty acid chains are interlocked and the polar head groups are exposed to aqueous environments. The physical properties of the hydrophobic core of plasma membranes are achieved to a large extent by having a suitable mix of both relatively fluid and relatively solid fatty acids esterified to the polar head group.

Due in part to the relative ease with which lipid constituents can be extracted and analysed, the variability in membrane composition has been used extensively in microbial systematics (Suzuki et al., 1993). Both polar and non-polar lipids are free lipids and as such are readily extractable into organic solvents of the correct polarity (see Chapter 5). Extracts are then simply analysed using thin-layer chromatography to give lipid patterns of diagnostic value. The most common polar lipids are phospholipids but glycolipids and amino acid amide lipids are also found. Phospholipids such as phosphatidylethanolamine, phosphatidylglycerol and diphosphatidylglycerol are widely distributed but compounds with a more restricted occurrence, and hence of greater diagnostic value, include the phosphatidylinositol mannosides of actinomycetes (Minnikin & O'Donnell, 1984), the amino acid amide lipids of certain pseudomonads (Minnikin & Abdolrahimzadeh, 1974) and aerobic, endospore-forming bacilli (Minnikin & Goodfellow, 1981).

Methods are available for the detailed characterization of polar lipids (see Chapter 5) but are generally regarded as impractical given the many analyses needed for a comprehensive taxonomic survey.

Additional taxonomic information can be obtained from a detailed analysis of the fatty acids comprising the hydrophobic core of polar lipids. Fatty acids can be defined as carboxylic acid derivatives of long-chain aliphatic molecules. In bacteria, they range in chain length from 2 to over 90 carbon molecules (Suzuki et al., 1993). Fatty acids forming the hydrophobic core of polar lipids are usually in the range C_{12} to C_{24} and have been shown to be relatively stable taxonomic characters when extracted from cells grown under carefully standardized conditions (O'Donnell, 1985, 1988; Saddler et al., 1986). Cellular fatty acids can be quantitatively analysed using gas-liquid chromatography following methylation, and numerous methods are available for their extraction and derivatization (Chapters 2 and 5). Ideally, identifications of fatty acids should be verified using additional chemical and spectroscopic techniques such as those described in Chapter 5.

A gas chromatograph equipped with a capillary column and data processor is recommended for fatty acid analysis since the reproducibility obtained makes the data suitable for subsequent numerical analysis. Numerical procedures have been used with varying success to classify and identify bacteria (O'Donnell, 1985, 1988) and it is now clear that for some groups of organisms statistical analysis of the data provides additional resolution, particularly at the subgeneric level. However, this is not the case for all taxa and like other taxonomic procedures, the effectiveness of fatty acid analysis is often taxon specific.

A promising development in recent years has been the introduction of automated, commercially available identification systems based on the chromatographic analysis of fatty acids (Sasser, 1990; Stead et al., 1992). By combining sample preparation, chromatographic analysis and data handling in one, integrated package, such systems promote standardization within and between laboratories but should not be seen as a 'stand alone' solution to all taxonomic problems. Fatty acid data are often most useful when used in combination with selected conventional tests (Wallace et al., 1988).

The outer membranes of Gram-negative bacteria and mycolic acid containing actinomycetes are structurally distinct from cytoplasmic membranes. Mycolic acids are 3-hydroxy acids with a long alkyl branch at position 2. These lipid components have been shown to vary systematically and are readily analysed using thin-layer chromatographic procedures (Dobson et al., 1985). Such routine chemical techniques provide a simple way of differentiating between the genera *Corynebacterium*, *Nocardia*, *Mycobacterium*, *Rhodococcus* and *Tsukamurella* and of classifying and identifying mycobacterial species. Protocols recommended for the extraction and analysis of mycolic acids as their *p*-bromophenacyl and methyl esters are given in Chapter 5.

The outer membrane of Gram-negative bacteria is structurally complex and is a rich source of chemical markers. It is asymmetrically assembled with certain proteins and lipopolysaccharides prevalent in the outermost layer, and phospholipids and proteins in the innermost layer, which resembles the cytoplasmic membrane (Inouye, 1979). Lipopolysaccharide is generally present in all Gram-negative bacteria and appears to be a vital structural entity of the cell surface. It consists of three covalently linked moieties: the oligosaccharide side-chain; the core oligosaccharide; and lipid A (Westphal et al., 1983). The oligosaccharide side-chain carries the determinants of O-antigenic specificity and is the most important of the three cell-surface antigens used in the serological identification of the Enterobacteriaceae at the infra-subspecific level (Kaufmann, 1966).

The core fraction of the lipopolysaccharide shows less chemical variation than that found in the O-specific chains (Wilkinson, 1977, 1988), with the sugar 2-keto-3-deoxyoctonate (KDO) forming an acid-labile linkage to lipid A in almost all lipopolysaccharides. Another common feature is the sequence of 2 or 3-L-glycero-D-manno-heptose units linked to the proximal cluster of one to three KDO moieties. Both KDO and heptose may be variably substituted by phosphate or pyrophosphate, often linked to ethanolamine.

Lipid A is generally considered to be a highly conserved molecule but recent studies show that it is structurally less homogeneous than previously believed (Mayer et al., 1989). Typically lipid A has a backbone of a β-1,6-linked D-glucosamine disaccharide with terminal phosphate or pyrophosphate groups in the 1 and 4' positions. An acetylation pattern of four to seven normal and hydroxylated fatty acids in acyloxy-, amide- and ester linkages is usual. Several taxonomically significant variations from this model have been described. The nature of the fatty acid substitutions vary between taxa (Moule & Wilkinson, 1989), with particularly complex fatty acid patterns recorded for *Legionella* (Sonesson et al., 1989; Jantzen et al., 1993). In addition, representatives of 12 genera have been shown to contain 2,3-diamino-2,3-dideoxylglucose (DAG) as a lipid A unit (Mayer et al., 1989). Lipid A may also contain non-stoichiometric amounts of other sugars such as aminoarabinose, mannose and quinovosamine (Rietschel et al., 1984; Weckesser & Mayer, 1988; Sonesson et al., 1989). Methods recommended for the extraction and purification of lipopolysaccharide, and for the analysis of its constituent parts, are described in Chapter 2. The latter also deals with whole-cell procedures used for rapid analysis of whole-cell sugars and fatty acids of Gram-negative bacteria.

Archaea cannot be distinguished from bacteria solely on the basis of the distribution of muramic acid but their membrane lipids are characterized by unusual structural features which can be considered as specific and useful taxonomic markers (De Rosa & Gambacorta, 1988). Bacterial lipids are based on ester linkages formed by condensation of alcohols and fatty acids, but those of the archaea are derivatives of isopranyl glycerol ethers formed by condensation of glycerol, or more complex polyols, with isopranoid alcohols. The presence of isopranyl ether lipids and their structural features are valuable markers in archaeal systematics. Other useful taxonomic markers include complex polar lipids based on 2,3-di-O-phytanyl-*sn*-glycerol and isoprenoid quinones. The procedures used to extract, purify and characterize archaeal isopranoid ether lipids, complex polar lipids and isoprenoid quinones are given in Chapter 7.

The plasma membrane bilayer system provides a stable matrix for the incorporation of other functional lipids that are taxonomically important. The best studied of these are the isoprenoid quinones (Suzuki et al., 1993) and carotenoid pigments (Liaaen-Jensen & Andrewes, 1985): the bacteriochlorophylls (Smith, 1991) and the cytochromes (Jones & Poole, 1985) have received less attention. Sterols and hopanoids have been even less well studied but do have potential as taxonomic markers given their structural diversity, particularly in the pattern of substitution and unsaturation in the side-chain.

Hopanoids have been found in taxonomically diverse Gram-negative and Gram-positive chemoheterotrophic bacteria and are common in cyanobacteria, purple non-sulphur bacteria and obligate methylotrophs (Rohmer et al., 1984). Sterols are generally absent from prokaryotes but are widely distributed in cyanobacteria (Paoletti et al., 1976) and mycoplasmas (Razin, 1978) and have also been found in the myxobacterium *Nannocytis exedens* (Kohl et al., 1983). The advent of high resolution capillary columns and increased availability of bench top mass spectrometers can be expected to promote interest in the nature and distribution of hopanoids and sterols amongst prokaryotes. Methods for the extraction, purification, analysis and identification of these compounds can be found in Chapter 6.

Isoprenoid quinones, which are found in the cytoplasmic membranes of most prokaryotes, play an important role in electron transport and energy generating systems (Ingledew & Poole, 1984). Some anaerobic bacteria lack isoprenoid quinones but in others these components act as coenzymes in anaerobic respiration with fumarate or nitrate functioning as terminal electron acceptors. The two most widely distributed types of isoprenoid quinones are the menaquinones or vitamin K_2 and the ubiquinones or coenzymes Q. Both types of quinoine vary in the length of the polyprenyl side-chain, with menaquinones showing marked differences in the degree of hydrogenation of the isoprene units. The significance of the structural variation is not clear but it has been suggested that in actinomycetes, where chain length and saturation can vary considerably between taxa, the differences confer the necessary physical properties for interaction with the fatty acids of the cell membrane (Minnikin & O'Donnell, 1984).

Some general rules seem to govern the distribution of isoprenoid quinones in bacteria. To date, ubiquinones have only been found in Gram-negative bacteria, whereas the isoprenoid quinones that occur in Gram-positive bacteria are predominantly menaquinones although they have also been found in Gram-negative bacteria and archaea. The taxonomic significance of isoprenoid quinones differs between bacterial groups, with some unusual isoprenoid quinones being characteristic of particular taxa. Detailed methods for the extraction, purification and analysis of isoprenoid quinones are described in Chapter 8.

Cytochromes are widely but not universally distributed in prokaryotes. Like the isoprenoid quinones, the cytochromes are also associated with the cytoplasmic membrane and involved in the respiration and energy transfer processes of the cell. Although studied more for their role in metabolism, cytochromes do show variation between taxa and are worthy of further study, particularly in relation to the emerging microbial phylogeny. Indeed, for many organisms, the description of the cytochrome pattern is an essential part of its total description. Methods used for the analysis of cytochromes are considered in Chapter 9.

Other membrane components with potential, though largely unexplored, chemotaxonomic value include lipid-soluble pigments such as the bacteriochlorophylls (Smith, 1991) and the carotenoids (Britton et al., 1993). Bacteriochlorophylls absorb light quanta of defined energy and are central to the photosynthetic apparatus of phototrophic bacteria (Oelze, 1985). The established functions of carotenoids in microorganisms include protection against photodynamic damage in both photosynthetic and non-photosynthetic organisms, indirect participation in photosynthesis by light harvesting and involvement in phototaxis (Liaaen-Jensen & Andrewes, 1985).

The anoxygenic phototrophic bacteria are a group of predominantly aquatic bacteria that are able to grow under anaerobic conditions by photosynthesis without oxygen production. Common to all species is the presence of magnesium tetrapyrrole pigments, the bacteriochlorophylls. The latter are assigned to six groups, designated *a* to *e* and *g*, on the basis of their characteristic absorption spectra in organic solvents. Detailed chemical analyses have revealed that all six components consist of a series of more than one structural form (Oelze, 1985). Structural differences have been identified with respect to both the substituents at the tetrapyrrole nucleus as well as the alcohol group esterifying the propionic acid side-chain.

Phototrophic bacteria can be assigned to five major groups on the basis of their photosynthetically active tetrapyrrole pigments, namely the anoxygenic heliobacteria, purple bacteria, and green and brown bacteria, and the oxygenic cyanobacteria and prochlorophytes. Subclasses of prokaryotes within these five groups cannot readily be distinguished by the nature of the chlorophylls they contain. Most purple bacteria, for example, have bacteriochlorophyll *a* but few produce bacteriochlorophyll *b*. Methods for the isolation, purification and characterization of bacterial chlorophylls are considered in Chapter 10, which also includes an overview of the taxonomy of phototrophic bacteria.

All phototrophic prokaryotes contain carotenoid pigments in their photosynthetic apparatus. These pigments are also common in yellow-orange coloured non-photosynthetic bacteria. Almost all carotenoids are tetraterpenes with a symmetrical C_{40} skeleton which can be either acyclic or have a six-membered ring (occasionally five-membered) at one or both ends of the molecule. The nature of these pigments and their quantitative composition provide valuable data for classification. Approximately 80 different carotenoid pigments have been found in phototrophic bacteria (Schmidt, 1978). A large variety of carotenoid structures have also been found in non-photosynthetic bacteria but the lack of detailed studies makes it difficult to recognize taxonomic trends. Cyanobacteria have several carotenoids in their photosynthetic pigment-protein complexes but the main differences between species are quantitative variations in overall carotenoid composition. Strategies and detailed protocols for determining carotenoid pigments and related polyenes are the subject of Chapter 11.

It is now well established that the mol% guanine plus cytosine content of prokaryotic DNA [% GC = 100 (G + C/A + T + G + C)] varies within the range 25 to 80% G + C between different taxa. Quick and reliable methods are available for the isolation and estimation of DNA base composition (see Chapter 12), and several generalizations can be made in the light of studies to date. Although closely related prokaryotes have similar mol% G + C values, it is important to realize that two organisms with similar mol% G + C values are not necessarily closely related; this is because the mol% G + C values do not take into account the linear arrangement of the nucleotides in the DNA. As a general guideline, bacteria with DNA differing by more than 5% G + C should not be classified in the same species and those showing differences of over 10% should not be assigned to the same genus (Goodfellow & O'Donnell, 1993).

1.3 WHOLE-ORGANISM FINGERPRINTING

All of the chemical procedures considered so far have involved the extraction, purification and characterization of specific components and as such are most valuable for providing data for the circumscription and identification of archaeal and bacterial taxa. Arguably one of the major challenges facing microbial systematists is how to classify large numbers of environmental isolates rapidly and accurately. This challenge requires the use of whole-organism fingerprinting techniques which minimize sample preparation and analysis thereby maximizing throughput. Methods available for chemical fingerprinting include rapid enzyme profiling using chromogenic and fluorogenic substrates (Manafi *et al.*, 1991) and protein gel electrophoresis (Vauterin *et al.*, 1993) and the instrument-based methods such as pyrolysis gas-chromatography, pyrolysis mass-spectrometry and RAMAN spectroscopy (Naumann *et al.*, 1991; Nelson & Sperry, 1991; Magee, 1993). Each of these approaches has specific advantages and disadvantages and all have been automated.

Rapid enzyme testing (see Chapter 13), which has its roots in diagnostic bacteriology, has considerable potential as a technique for the classification and identification of environmental isolates. Methods for the detection of specific enzymes based on the use of chromogenic and fluorogenic substrates are widely used for the classification and identification of clinically significant bacteria (Manafi *et al.*, 1991), and are increasingly being used to classify ecologically and industrially important bacteria, including streptomycetes (Goodfellow *et al.*, 1987; Kämpfer *et al.*, 1991). The development of automated systems (Hamid et al., 1993) should facilitate such studies and thereby help realize the full potential of rapid enzyme tests in bacterial systematics. Besides being rapid, an advantage of this approach is that tests for constitutive enzymes are, like the molecular methods, largely independent of growth media.

Whole-organism protein electrophoresis is a powerful, relatively simple, and cost-effective method that has been widely used in bacterial systematics (Vauterin et al., 1993). The method has the advantage that the electrophoretic mobilities of the proteins can be compared under highly standardized conditions, although pattern correction is needed to secure optimal results (Albritton et al., 1988). The electrophoretic separation of cellular proteins mainly provides information on the similarity of strains within the same species or subspecies. In many bacterial taxa excellent congruence has been obtained between results from protein electrophoretic and DNA:DNA pairing studies. The experimental and data handling procedures involved in the analysis of electrophoretic whole-organism protein fingerprints is considered in Chapter 14.

Physico-chemical techniques such as pyrolysis mass-spectrometry, infrared spectrometry and ultraviolet resonance RAMAN spectroscopy (see Chapter 15) are increasingly being used in microbial systematics. These methods have many features in common. Each involves the concurrent measurement of large numbers of characters that together reflect overall cell composition. In addition, sample and preanalysis processing are minimal; few if any specialized reagents are needed; and analysis costs are low; the equipment, though, remains expensive (Magee, 1993). However, it is conceivable that such physico-chemical procedures may provide the only practical means of coping with large numbers of microbial isolates which need to be classified for the analysis of microbial diversity in natural environments. A common feature of the whole-organism techniques is that the data produced are complex and need to be examined using data analysis routines (Blomquist et al., 1979; MacFie & Gutteridge, 1982). Fortunately, suitable statistical routines are readily available (see Chapters 14 and 15) with many of the packages suitable for use on desktop and personal computers (Bratchell et al., 1989). Neural network analysis (Chun et al., 1993) may well be a significant addition to these mathematical strategies.

1.4 OVERVIEW

The application of molecular biological techniques has had a profound effect on microbial systematics (Woese, 1992). The most spectacular development, which marked a historical turning point in microbial systematics, was the realization that related microorganisms have in their respective nucleic acids, records of the changes that have occurred since their divergence from a common ancestor over 3 billion years ago (Zuckerkandl & Pauling, 1965). Subsequent studies on conserved regions of nucleic acids, mainly ribosomal RNA, have shown prokaryotic diversity to be vast in comparison to that of eukaryotes. Further advances in microbial systematics can be anticipated as procedures for nucleic acid sequencing are simplified. Current methods have already reached the stage where it is relatively straightforward to use the polymerase chain reaction to amplify and sequence nucleic acids from small amounts of biomass (Embley & Finlay, 1993; Hutson et al., 1993).

Developments in molecular systematics should not be seen as a threat to chemotaxonomy for the two approaches are complementary. Nucleic acid sequencing data provide an evolutionary framework of the relationships amongst prokaryotes, but like other classifications need to be supported by additional evidence since the relationships between taxa can be distorted due to differences in evolutionary rates, technical problems and the type of algorithm used to generate the phylogeny (O'Donnell et al., 1993). Chemical markers are often discontinuously distributed across taxa and as such can be used to evaluate the emerging phylogeny. It is becoming increasingly evident that good congruence exists between the distribution of specific chemical markers and the relative position of species in phylogenetic trees (Goodfellow, 1989; Suzuki et al., 1993). Indeed, the integrated use of phylogenetic and phenotypic characteristics, or polyphasic taxonomy, is now seen to be necessary for the delineation of taxa at all levels from genus to kingdom (Murray et al., 1990).

Extensive chemotaxonomic analyses have revealed the inherent variability of chemical data with the same character or groups of characters being conserved in one taxon whilst varying systematically in another (Suzuki et al., 1993). For example, variation in the peptide subunit and mode of cross-linkage in peptidoglycan has been of fundamental importance in understanding the complex relationships between Gram-positive organisms but has contributed little to the systematics of Gram-negative bacteria. Similar situations exist with other chemotaxonomic markers, such as fatty acids, where it is possible, even within the same genus, to find that some species have unique profiles whereas others show no systematic variance between taxa (Saddler et al., 1987). Although this questions whether fatty acids are equally valid for all groups of organisms, it does not mean that they are of limited value since similar situations arise with other taxonomic methods, including rRNA sequencing techniques (Fox et al., 1992).

Chemical markers have been shown to be especially useful in the classification and identification of the more morphologically differentiated Gram-positive taxa such as the actinomycetes (Goodfellow, 1989; Suzuki et al.,

1993). In contrast, relatively few chemical markers have been used in the delineation of other groups, notably the proteobacteria (Murray et al., 1990). An important objective in the systematics of such taxa should be a thorough examination of representative strains from different phyletic lines to highlight appropriate chemical markers for the definition of taxa above the species level. It is encouraging that good congruence has already been found between lipid A types and the evolutionary branching points of photosynthetic bacteria (Weckesser & Mayer, 1988; Mayer et al., 1989).

To date, chemosystematics, like numerical systematics, has been used in both the classification and the identification of bacteria with considerable emphasis put on the development of rapid identification systems. In future, as the technology develops, the latter is likely to become less important as routine identification laboratories move towards rapid, automated fingerprinting techniques such as pyrolysis mass-spectrometry and infrared spectrometry. Consequently, it is to be expected that chemical methods will be used increasingly, not to identify bacteria but to underpin the developing phylogeny. To do this successfully, it may be necessary to study chemical variability between taxa at lower levels of resolution than is done at present with particular attention paid to structural information and to the evaluation of previously uncharacterized poorly studied macromolecules. This will undoubtedly require improvements in methodology and the involvement of good analytical chemists. In concentrating on methodology, this book provides a foundation from which these new and exciting developments can be pursued.

REFERENCES

Albritton, W.L., Chen, X.P.W. and Khanna, V. (1988) Comparison of whole-cell electrophoretic profiles of *Haemophilus influenzae*: Implementation of a microcomputer mainframe linked system and description of a new similarity coefficient. *Can. J. Microbiol.* **34**: 1129–34.

Blomquist, G., Johannson, E., Söderstrom, B. and Wold, S. (1979) Data analysis of pyrolysis-chromatograms by means of SIMCA pattern recognition. *J. Anal. Appl. Pyrol.* **1**: 53–65.

Bousfield, I.J., Smith, G.L., Dando, T.R. and Hobbs, G. (1983) Numerical analysis of total fatty acid profiles in the identification of coryneform, nocardioform and some other bacteria. *J. Gen. Microbiol.* **129**: 375–94.

Bratchell, N., O'Donnell, A.G. and MacFie, H.J.H. (1989) Data analysis in microbiology. In *Computers in Microbiology—A Practical Approach* (Bryant, T.N. and Wimpenny, J.W.T., eds), pp. 41–63. IRL Press: Oxford.

Britton, G. and Young, A.J. (1993) Methods for the isolation and analysis of carotenoids. In *Carotenoids in Photosynthesis* (Young, A.J. and Britton, G., eds), pp. 409–88. Chapman and Hall: London.

Bull, A.T., Goodfellow, M. and Slater, J.H. (1992) Biodiversity as a source of innovation in biotechnology. *Ann. Rev. Microbiol.* **46**: 215–52.

Chun, J., Atalan, E., Ward, A.C. and Goodfellow, M. (1993) Artificial neural network analysis of pyrolysis mass spectrometric data in the identification of *Streptomyces* strains. *FEMS Microbiol. Lett.* **107**: 321–6.

Cummins, C.S. and Harris, H. (1956) The chemical composition of the cell wall in some Gram-positive bacteria and its possible value as a taxonomic character. *J. Gen. Microbiol.* **14**: 583–600.

De Briel, D., Couderc, F., Reigel, P., Jehl, F. and Minck, R. (1992) High performance liquid chromatography of corynemycolic acids as a tool in identification of *Corynebacterium* species and related organisms. *J. Clin. Microbiol.* **30**: 1407–17.

De Rosa, M. and Gambacorta, A. (1988) The lipids of archaebacteria. *Prog. Lipid Res.* **27**: 153–75.

Dobson, G., Minnikin, D.E., Minnikin, S.M., Parlett, J.H., Goodfellow, M., Ridell, M. and Magnusson, M. (1985). Systematic analysis of complex mycobacterial lipids. In *Chemical Methods in Bacterial Systematics* (Goodfellow, M. and Minnikin, D.E., eds), pp. 237–65. Academic Press: London.

Easmon, C.S.F. and Goodfellow, M. (1990) *Staphylococcus* and *Micrococcus*. In *Topley & Wilson's Principles of Bacteriology, Virology and Immunology*, 8th edn (Parker, T.M. and Duerden, B.I. eds), pp. 161–86. Edward Arnold: London.

Embley, T.M. and Finlay, B.J. (1993) Systematic and morphological diversity of endosymbiotic methanogens in anaerobic ciliates. *Antonie van Leeuwenhoek* (in press).

Embley, T.M., O'Donnell, A.G., Rostron, J. and Goodfellow, M. (1988) Chemotaxonomy of wall type IV actinomycetes which lack mycolic acids. *J. Gen. Microbiol.* **134**: 953–60.

Fox, G.E., Wisotzkey, J.D. and Jurtshuk, J.R. (1992) How close is close: 16S rRNA sequence identity may not be sufficient to guarantee species identity. *Int. J. System. Bacteriol.* **42**: 166–170.

Freeman, R., Goodfellow, M., Ward, A.C., Hudson, S.J., Gould, F.K. and Lightfoot, N.F. (1991) Epidemiological typing of coagulase-negative staphylococci by pyrolysis mass spectrometry. *J. Med. Microbiol.* **34**: 245–8.

Giovannoni, S.J., Godchaux, W. III., Schabtach, E. and Castenholz, R.W. (1987) Cell wall and lipid composition of *Isophaera pallida*, a budding eubacterium from hot springs. *J. Bacteriol.* **169**: 2702–7.

Goodfellow, M. (1989) Suprageneric classification of actinomycetes. In: *Bergey's Manual of Systematic Bacteriology*, vol. 4 (Williams, S.T., Sharpe, M.E. and Holt, J.G., eds), pp. 2333–9. Williams & Wilkins: Baltimore.

Goodfellow, M. and Minnikin, D.E. (eds) (1985) *Chemical Methods in Bacterial Systematics*. Academic Press: London.

Goodfellow, M. and O'Donnell, A.G. (1989) Search and discovery of industrially-significant actinomycetes. In *Microbial Products: New Approaches* (Baumberg, S., Hunter, I.S. and Rhodes, P.M., eds), pp. 343–83. Cambridge University Press: Cambridge.

Goodfellow, M. and O'Donnell, A.G. (1993) Roots of bacterial systematics. In *Handbook of New Bacterial Systematics* (Goodfellow, M. and O'Donnell, A.G., eds), pp. 3–54. Academic Press: London.

Goodfellow, M., Lonsdale, C., James, A.L. and MacNamara, O.C. (1987) Rapid biochemical tests for the characterization of streptomycetes. *FEMS Microbiol. Lett.* **43**: 39–44.

Hamid, M.E., Chun, J., Magee, J.T. and Goodfellow, M. (1993) Rapid characterisation and identification of mycobacteria using fluorogenic enzyme tests. *Zbl. Bakt.* (in press).

Hawksworth, D.L. and Mound, L.A. (1991) Biodiversity databases: The crucial significance of collections. In *The Biodiversity of Microorganisms and Invertebrates* (Hawksworth, D.L., ed), pp. 17–28. Redwood Press: Melksham, UK.

Hutson, R.A., Thompson, D.E., Lawson, P.A., Schocken-Itturino, R.P., Böttger, E.C. and Collins, M.D. (1993) Genetic interrelationships of proteolytic *Clostridium botulinum* types A, B and F and other members of the *Clostridium botulinum* complex as revealed by small-subunit rRNA gene sequences. *Antonie van Leeuwenhoek* (in press).

Ingledew, W.J. and Poole, R.K. (1984) The respiratory chains of *Escherichia coli*. *Microbiol. Rev.* **48**: 222–71.

Inouye, M. (1979) *Bacterial Outer Membranes: Biogenesis and Functions*. John Wiley & Sons, New York.

Jackman, P.J.H. (1982) Classification of *Corynebacterium* species from the auxiliary skin by numerical analysis of electrophoretic protein patterns. *J. Med. Microbiol.* **15**: 485–92.

Jantzen, E., Sonesson, A., Tangen, T. and Eng, J. (1993) Hydroxy fatty acid profiles of *Legionella*: Diagnostic usefulness assessed by SIMCA principal component analysis. *J. Clin. Microbiol.* **31**: 1413–19.

Jones, C.W. and Poole, R.K. (1985) The analysis of cytochromes. *Methods Microbiol.* **18**: 285–328.

Kämpfer, P., Kroppenstedt, R.M. and Dott, W. (1991) A numerical classification of the genera *Streptomyces* and *Streptoverticillium* using miniaturized physiological tests. *J. Gen. Microbiol.* **137**: 1831–91.

Kandler, D. and König, H. (1985) Cell envelopes of archaebacteria. In *The Bacteria*, vol. VIII (Woese, C.R. and Wolfe, R.S. eds), pp. 413–57. Academic Press: London.

Kaufmann, F. (1966) *The Bacteriology of Enterobacteriaceae*. Williams & Wilkins: Baltimore.

Kodama, Y., Yamamoto, H., Amano, N. and Amachi, T. (1992) Reclassification of two strains of *Arthrobacter oxydans* and proposal of *Arthrobacter nicotinovorans* sp. nov. *Int. J. System. Bacteriol.* **42**: 234–9.

Kohl, W., Gloe, A. and Reichenbach, H. (1983) Steroids from the myxobacterium *Nannocystis exedens*. *J. Gen. Microbiol.* **129**: 1629–35.

Komagata, K. and Suzuki, K. (1987) Lipid and cell wall analysis in bacterial systematics. *Meth. Microbiol.* **19**: 161–207.

König, H. (1988) Archaeobacterial cell envelopes. *Can J. Microbiol.* **34**: 395–406.

König, E., Schlesner, H. and Hirsch, P. (1984) Cell wall studies on budding bacteria of the *Planctomyces/Pasteuria* group and a *Prosthecomicrobium* sp. *Arch. Microbiol.* **138**: 200–5.

Liaaen-Jensen, S. and Andrewes, A.G. (1985) Analysis of carotenoids and related polyene pigments. *Methods Microbiol.* **18**: 235–55.

Liesack, W., König, H. and Hirsch, P. (1986) Chemical composition of the peptidoglycan-free cell envelopes of budding bacteria of the *Pirella/Planctomyces* group. *Arch. Microbiol.* **145**: 361–6.

MacFie, H.J.H. and Gutteridge, C.S. (1982) Comparative studies on some methods for handling quantitative data generated by analytical pyrolysis. *J. Anal. Appl. Pyrol.* **4**: 175–204.

Magee, J.T. (1993) Whole-organism fingerprinting. In *Handbook of New Bacterial Systematics* (Goodfellow, M. and O'Donnell, A.G., eds), pp. 383–427. Academic Press: London.

Magee, J.T., Hindmarch, J.M. and Nichol, C.D. (1991) Typing of *Streptococcus pyogenes* by pyrolysis mass spectrometry. *J. Med. Microbiol.* **35**: 304–6.

Manafi, M., Kneifel, W. and Bascomb, S. (1991) Fluorogenic and chromogenic substrates used in bacterial diagnostics. *Microbiol. Rev.* **55**: 335–48.

Mayer, H., Ramadas-Bhat, U., Masoud, H., Radziejewska-Lebrecht, J., Widemann, C. and Krauss, J.H. (1989) Bacterial lipopolysaccharides. *Pure Appl. Chem.* **61**: 1271–82.

Minnikin, D.E. and Abdolrahimzadeh, H. (1974) The replacement of phosphatidyl-ethanolamine and acid phospholipids by an ornithine-amide lipid and a minor phosphorus-free lipid in *Pseudomonas fluorescens* NCMB 129. *FEBS Lett.* **43**: 257–60.

Minnikin, D.E. and Goodfellow, M. (1981) Lipids in the classification of *Bacillus* and related taxa. In *The Aerobic Endospore-forming Bacteria* (Berkeley, R.C.W. and Goodfellow, M., eds), pp. 59–90. Academic Press: London.

Minnikin, D.E. and O'Donnell, A.G. (1984). Actinomycete envelope lipid and peptidoglycan composition. In *The Biology of the Actinomycetes* (Goodfellow, M., Mordarski, M. and Williams, S.T., eds), pp. 337–88. Academic Press: London.

Moule, A.L. and Wilkinson, S.G. (1989) Composition of lipopolysaccharides from *Alteromonas putrefaciens* (*Shewanella putrefaciens*). *J. Gen. Microbiol.* **135**: 163–73.

Murray, R.G.E., Brenner, D.J., Colwell, R.R., De Vos, P., Goodfellow, M., Grimont, P.A.D., Pfennig, N., Stackebrandt, E. and Zavarzin, G.A. (1990) Report of the ad hoc committee on approaches to taxonomy within the proteobacteria. *Int. J. System. Bacteriol.* **40**: 213–15.

Naumann, D., Helm, D., Labischinski, H. and Griesbrecht, P. (1991) The characterization of microorganisms by Fourier-transform infrared spectroscopy. In *Modern Techniques for Rapid Microbiological Analysis* (Nelson, W. H., ed), pp. 43–96. VCH Publishers: New York.

Nelson, W.H. and Sperry, J.F. (1991) UV resonance Raman spectroscopic detection and identification of bacteria and other microorganisms. In *Modern Methods for Rapid Microbiological Analysis* (Nelson, W.H., ed), pp. 43–96. VCH Publishers: New York.

Nurminen, M., Rietschel, E.T. and Brade, H. (1985) Chemical characterization of *Chlamydia trachomatis* lipopolysaccharide. *Infect. Immun.* **48**: 573–5.

O'Donnell, A.G. (1985) Numerical analysis of chemotaxonomic data. In *Computer-Assisted Bacterial Systematics* (Goodfellow, M., Jones, D. and Priest, F.G., eds), pp. 403–14. Academic Press: London.

O'Donnell, A.G. (1988) Recognition of novel actinomycetes. In *Actinomycetes in Biotechnology* (Goodfellow, M., Williams, S.T. and Mordarski, M., eds), pp. 69–88. Academic Press: London.

O'Donnell, A.G., Embley, T.M. and Goodfellow, M. (1993) Future of bacterial systematics. In *Handbook of New Bacterial Systematics* (Goodfellow, M. and O'Donnell, A.G., eds), pp. 513–24 Academic Press: London.

Oelze, J. (1985) Analysis of bacteriochlorophylls. *Methods Microbiol.* **18**: 257–84.

Paoletti, C., Pushparaj, B., Florenzano, G., Capella, P. and Lercker, G. (1976) Unsaponifiable matter of green and blue-green algal lipids as a factor of biochemical differentiation of their biomasses. II. Terpenic alcohol and sterol fractions. *Lipids II*: 266–71.

Razin, S. (1978) The mycoplasmas. *Microbiol. Rev.* **42**: 414–70.

Razin, S. and Freund, E.A. (1984) The mycoplasmas. In *Bergey's Manual of Systematic Bacteriology*, vol. 1 (Krieg, N.R. and Holt, J.G., eds), pp. 740–1. Williams & Wilkins: London.

Rietschel, E.T., Wollenweber, H-W., Brade, H., Zahringer, U., Linder, B., Seydel, U., Bradaczek, H., Barnickel, G., Labischinski, H. and Griesbrecht, P. (1984) Structure and composition of the lipid A component of lipopolysaccharides. In *Handbook of Endotoxin* (Rietschel, E.T., ed), pp. 187–220. Elsevier: Amsterdam.

Rogers, H.J., Perkins, H.R. and Ward, J.B. (1980) *Microbial Cell Walls and Membranes*. Chapman and Hall: London.

Rohmer, M., Bouvier-Nave, P. and Ourisson, G. (1984) Distribution of hopanoid triterpenes in prokaryotes. *J. Gen. Microbiol.* **130**: 1137–50.

Saddler, G.S., Goodfellow, M., Minnikin, D.E. and O'Donnell, A.G. (1986) Influence of the growth cycle on the fatty acid and menaquinone composition of *Streptomyces cyaneus* NCIB 9616. *J. Appl. Bacteriol.* **60**: 51–6.

Saddler, G.S., O'Donnell, A.G., Goodfellow, M. and Minnikin, D.E. (1987) SIMCA pattern recognition in the analysis of streptomycete fatty acids. *J. Gen. Microbiol.* **133**: 1137–47.

Sasser, M. (1990) Identification of bacteria through fatty acid analysis. In *Methods in Phytobacteriology* (Klement, Z., Rudolph, K. and Sands, D.C. eds), pp. 199–203. Académiai Kiadó: Budapest.

Schleifer, K.H. (1985) Analysis of the chemical composition and primary structure of murein. *Meth. Microbiol.* **18**: 123–56.

Schleifer, K.H. and Joseph, R. (1973) A directly cross-linked L-ornithine containing peptidoglycan in cell walls of *Spirochaeta stenostrepta*. *FEBS Lett.* **36**: 83–6.

Schleifer, K.H. and Kandler, O. (1972) Peptidoglycan types of bacterial cell walls and their taxonomic implications. *Bacteriol. Rev* **34**: 407–77.

Schleifer, K.H. and Seidl, P.H. (1985) Chemical composition and structure of murein. In *Chemical Methods in Bacterial Systematics* (Goodfellow, M. and Minnikin, D.E., eds), pp. 201–19. Academic Press: London.

Schmidt, K. (1978) Biosynthesis of carotenoids. In *The Photosynthetic Bacteria* (Clayton, R.K. and Sistrom, W.R., eds), pp. 729–49. Plenum Press: New York.

Smith, K.M. (1991) The structure and biosynthesis of bacteriochlorophylls. In *New Comprehensive Biochemistry of Biosynthesis of Tetrapyrroles*, vol. 19 (Jordan, P.M., ed), pp. 237–55. Elsevier Amsterdam.

Sneath, P.H.A. (ed) (1992) *International Code of Nomenclature of Bacteria: 1990 Revision*. American Society for Microbiology: Washington.

Sonesson, A., Jantzen, E., Bryn, K., Larsson, L. and Eng, J. (1989) Chemical composition of a lipopolysaccharide from *Legionella pneumophila*. *Arch. Microbiol.* **153**: 72–8.

Stackebrandt, E. and Goodfellow, M. (eds) (1991) *Nucleic Acid Techniques in Bacterial Systematics*. John Wiley & Sons: Chichester.

Stead, D.E., Sellwood, J.E., Wilson, J. and Viney, I. (1992) Evaluation of a commercial microbial identification system based on fatty acid profiles for rapid, accurate identification of plant pathogenic bacteria. *J. Appl. Bacteriol.* **72**: 315–21.

Suzuki, K., Goodfellow, M. and O'Donnell, A.G. (1993) Cell envelopes and classification. In *Handbook of New Bacterial Systematics* (Goodfellow, M. and O'Donnell, A.G., eds), pp. 195–249. Academic Press: London.

Uchida, K. and Aida, K. (1977) Acyl type of bacterial cell wall: Its simple identification by colorimetric method. *J. Gen. Appl. Microbiol.* **23**: 249–60.

Uchida, K. and Aida, K. (1979) Taxonomic significance of cell-wall type in *Corynebacterium-Mycobacterium-Nocardia* group by a glycolate test. *J. Gen. Appl. Microbiol.* **25**: 169–83.
Vauterin, L., Vantomme, R., Pot, B., Hoste, B., Swings, J. and Kersters, K. (1990) Taxonomic analysis of *Xanthomonas campestris* pv. *begoniae* and *X. campestris* pv. *pelargonii* by means of phytopathological, phenotypic, protein electrophoretic and DNA hybridization methods. *Syst. Appl. Microbiol.* **13**: 166–76.
Vauterin, L., Swings, J. and Kersters, K. (1993) Protein electrophoresis and classification. In *Handbook of New Bacterial Systematics* (Goodfellow, M. and O'Donnell, A.G., eds), pp. 251–80. Academic Press: London.
Wallace, P.L., Hollis, D.G., Weaver, R.E. and Moss, C.W. (1988) Cellular fatty acid composition of *Kingella* species, *Cardiobacterium hominis*, and *Eikennella corrodens*. *J. Clin. Microbiol.* **26**: 1592–4.
Weckesser, J. and Mayer, H. (1988) Different lipid A types in lipopolysaccharides of phototropic and related non-phototropic bacteria. *FEMS Microbiol Rev.* **54**: 143–54.
Westphal, O., Lüderitz, O. and Bister, R. (1952) Uber die Extraktion van Bakterien mit Phenol/Wasser. *Zeitschr. Naturforsch.* **7b**: 148–55.
Westphal, O.W., Jann, K. and Himmelspach, K. (1983). Chemistry and immunochemistry of bacterial lipopolysaccharides as cell wall antigens and endotoxins. *Prog. Allergy* **33**: 9–39.
Wilkinson, S.G. (1977) Composition and structure of bacterial lipopolysaccharides. In *Surface Carbohydrates of the Prokaryotic Cell* (Sutherland, I.W., ed), pp. 97–175. Academic Press: London.
Wilkinson, S.G. (1988) Gram-negative bacteria. In *Microbial Lipids*, vol. 1 (Ratledge, C. and Wilkinson, S.G., eds), 299–488. Academic Press: London.
Williams, S.T., Sharpe, M.E. and Holt, J.G. (eds) (1989) *Bergey's Manual of Systematic Bacteriology*, vol. 4. Williams & Wilkins: Baltimore.
Woese, C.R. (1987) Bacterial evolution. *Microbial Rev.* **51**: 221–71.
Woese, C.R., (1992) Prokaryotic systematics: The evolution of a science. In *The Prokaryotes* 2nd edn. (Balows, A., Trüper, H.G., Dworkin, M., Harder W. and Schleifer, K.H., eds), pp. 3–18. Springer-Verlag: New York.
Woese, C.R., Kandler, O. and Wheelis, M.L. (1990). Towards a natural system of organisms: Proposal for the domains Archaea, Bacteria and Eucarya. *Proc. Ntl. Acad. Sci. USA* **87**: 4576–9.
Zuckerkandl, E. and Pauling, L. (1965) Molecules as documents of evolutionary history. *J. Theor. Biol.* **8**: 357–66.

2
Analysis of Cellular Constituents from Gram-negative Bacteria

Erik Jantzen and Klaus Bryn
National Institute of Public Health, Oslo, Norway

2.1 INTRODUCTION

With a few exceptions, Gram-negative bacteria possess a cell envelope with certain fundamental ultrastructural features in common. This unique biological construction consists primarily of a distinct outer membrane, an inner cytoplasmic membrane, and an interjecting thin layer of peptidoglycan (Lugtenberg & van Alphen, 1983; Nikaido & Vaara, 1985). Additional external layers, termed capsules and usually of polysaccharide nature, may also be present (Jann & Jann, 1977; Kenne & Lindberg, 1983). A schematic representation of the Gram-negative cell envelope is given in Figure 2.1, with the potential value of the main envelope constituents for bacterial systematics indicated in Table 2.1.

The inner (cytoplasmic) membrane is a unit membrane consisting mainly of phospholipids and proteins. Whereas the fatty acid constituents of phospholipids are well established in the taxonomy of Gram-negative bacteria (see Section 2.7 and Chapter 5), the membrane proteins have only been considered as part of whole-organism sodium dodecyl sulphate (SDS) polyacrylamide gel electrophoresis (PAGE) patterns (see Chapter 14). Also located within the cytoplasmic membrane are the isoprenoid quinone lipids involved in electron transport (see Chapter 8).

The unique bacterial structure that overlies the cytoplasmic membrane, the peptidoglycan, consists of a polymeric network of glycan strands cross-linked by short peptide side-chains. This molecule appears to be conserved in almost all Gram-negative bacteria. In some taxa, e.g. *Escherichia coli*, the peptidoglycan is covalently linked to low molecular weight lipoproteins (Braun, 1973) and possibly to certain other proteins (Benz, 1988).

The outer-membrane bilayer is asymmetrically assembled. In the outer-

---

Chemical Methods in Prokaryotic Systematics. Edited by M. Goodfellow and A.G. O'Donnell
© 1994 John Wiley and Sons Ltd.

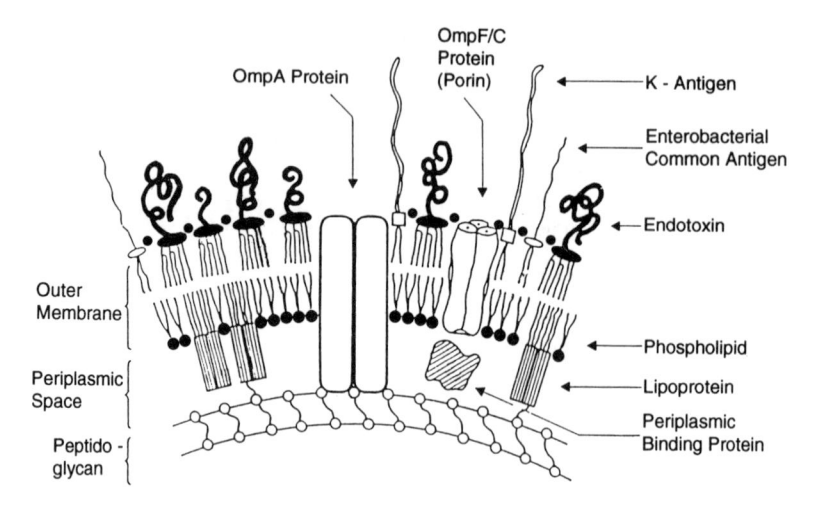

Figure 2.1 Schematic diagram of a section of the Gram-negative bacterial cell wall. From Rietschel *et al.* (1988) with permission.

most layer, certain proteins and lipopolysaccharides (LPS) dominate whilst the innermost layer is composed of proteins and phospholipids and thereby resembles the cytoplasmic membrane (Inouye, 1979).

Outer membrane protein (OMP) composition is usually relatively simple. Analysis using SDS-PAGE shows a limited number of bands that represent major proteins. These may comprise the important pore-forming entities named porins, the low molecular weight (7.2 kDa) Braun's lipoproteins, OmpA, and some minor proteins (Benz, 1988). Porins are essential for facilitating the passage of hydrophilic nutrients through defined pores in the outer membrane (Hancock, 1987). OmpA and lipoprotein have no pore function but seem to be essential for the integrity of the outer membrane (Di Rienzo *et al.*, 1978). Minor proteins apparently have important functions for the uptake of specific nutrients, and under special starvation conditions some of these may become major SDS-PAGE bands (Benz, 1988). Related bacteria generally have similar PAGE patterns which can be subject to detailed immunoblotting, i.e. the use of specific antibodies after electrophoretic transfer of OMPs to a membrane (Towbin *et al.*, 1979).

The unique lipopolysaccharides are an obligate class of major antigens of Gram-negative bacteria, located asymmetrically on the outside of the outer membrane with the lipid part serving as an anchor in the hydrophobic phase. LPS exhibit a variety of biological ('endotoxic') effects including fever, shock and even death (for review see Rietschel *et al.*, 1988). Their general antigenic properties have been utilized in diagnostic work for a long time, e.g. the well-known heat-stable O-antigens of

2. Analysis of Gram-negative Cell Constituents

Table 2.1 Envelope constituents of Gram-negative bacteria and their potential in chemotaxonomy.

Constituents	Main methods*	Specific information
Capsular polysaccharides	Hydrolysis, GLC, GLC-MS, HPLC, immunoblot	Sugar profiles, epitopes
Protein S-layers	SDS-PAGE	Protein patterns
Waxes, lipid layers	GLC, HPLC, TLC	Lipids and/or their fatty acid, fatty alcohol constituents
Outer-membrane proteins	SDS-PAGE, sequencing, immunoblot	Protein patterns, amino acid sequence data
Lipopolysaccharides	Hydrolysis, SDS/DOC-PAGE, immunoblot, GLC	Band patterns, epitopes, sugar and fatty acid profiles
Lipid A-part	Hydrolysis, GLC, TLC, HPLC	Sugar and fatty acid profiles
Polysaccharide-part	Hydrolysis, GLC, TLC, HPLC	Sugar profiles
Phospholipids	TLC, HPLC	Lipid patterns
Fatty acids	Hydrolysis, GLC	Fatty acid profiles
Lipoproteins	SDS-PAGE, sequencing	Protein patterns, amino acid sequence data
Peptidoglycan	Hydrolysis, GLC, HPLC, electrophoresis	Peptide and amino acid profiles, sequence data
Cytoplasmic membrane	SDS-PAGE	
Proteins		Protein patterns
Lipids	Hydrolysis, TLC, GLC	Patterns of lipids, fatty acids

* The first analytical step involves extraction with aqueous or organic solvents, and further purification, e.g. by chromatograpy on columns, is either necessary or useful.

enterobacteria. Detailed structural analyses of LPS from a variety of Gram-negative bacteria have confirmed the structural basis for these serological features (Lüderitz et al., 1971). Thus, LPS contain both fatty acids and sugar structures of considerable potential in bacterial systematics (Jantzen & Bryn, 1985; Weckesser & Mayer, 1988; Wilkinson, 1988; Mayer et al., 1989).

Capsular polysaccharides (PS) often surround Gram-negative organisms and hence contribute significantly to the chemical and immunological character of the bacterial surface, including the organism's resistance to the immune defence system (Lee, 1987). Though PS appear, at least in part, to be anchored in the hydrophobic phase of the outer membrane by a small lipid unit (Gotschlich et al., 1981; Schmidt & Jann, 1982), they are more loosely attached than LPS. Capsular polysaccharides are important candidates for vaccines and are also of interest in the more fundamental studies of bacterial infections (Lee, 1987). Mainly for these reasons, many laboratories have contributed to the comprehensive list of structurally characterized bacterial PS (Kenne & Lindberg, 1983; Lindberg, 1990).

The biosynthesis of PS is more dependent on cultural conditions than LPS; PS may even disappear *in vitro* after growth on artificial media (Masson *et al.*, 1982), or *in vivo* as recorded during an epidemic outbreak of group B meningococcal disease (Kristiansen *et al.*, 1986). Such variability could represent a problem for taxonomic applications, but PS are generally considered stable characteristics of strains and species. Examples of clinical importance include PS of urinary tract-infective strains of *E. coli* (Jann & Jann, 1987) and the group-specific PS of *Haemophilus influenzae* and *Neisseria meningitidis* (Jennings, 1983).

The value of chemical data on Gram-negative cell envelopes is currently difficult to assess fully in bacterial systematics. However, the situation has become easier to understand given the publication of several reviews (Lechevalier, 1977; Jantzen & Bryn, 1985; Jann & Jann, 1987; Mayer *et al.*, 1989). The comprehensive book by Ratledge and Wilkinson (1988) on microbial lipids is especially helpful. Albeit slowly, the impact of chemical data from cell envelopes is becoming more evident in the circumscription of Gram-negative taxa.

In this chapter we shall focus mainly on chemical methods applied to the analysis of outer membrane constituents. Peptidoglycan is considered in more detail in Chapter 3. Likewise, methods applied to isoprenoid quinones and cytochromes (Chapters 8 and 9), lipids (Chapter 5), pigments (Chapters 10 and 11) and proteins (Chapter 14) are dealt with in detail elsewhere. The reader is also referred to the comprehensive book on bacterial cell surface techniques (Hancock & Poxton, 1988a).

2.2 PEPTIDOGLYCAN

Compared to the thick, multilayered peptidoglycan of most of their Gram-positive counterparts (Schleifer & Seidl, 1985), Gram-negative organisms share a relatively simple, directly cross-linked monomolecular layered peptidoglycan structure. In this highly conserved structure (A1τ *sensu* Schleifer & Kandler, 1972; Chapter 3), the diamino acid in position 3 of the stem peptide is *meso*-diaminopimelic acid (m-A$_2$pm), with exceptions mostly occurring amongst anaerobic organisms. Thus, the taxonomic potential appears limited, but can be of some value for certain bacterial groups. For instance, the peptidoglycans of spirochaetes often contain L-ornithine instead of m-A$_2$pm (Schleifer & Joseph, 1973), or may, as found in *Treponema pallidum*, use glycine as the interpeptide bridge (Umemoto *et al.*,

1981). In several fusobacteria (*Fusobacterium gonidiaformans*, *F. necrophorum*, *F. nucleatum* and *F. russi*) the m-A$_2$pm is replaced by its sulphur analogue m-lanthionine (Kato *et al.*, 1979; Vasstrand *et al.*, 1979). *Fusobacterium mortiferum* contains both lanthionine and m-A$_2$pm (Vasstrand *et al.*, 1982).

Gram-negative peptidoglycan can be isolated directly from whole cells or from previously isolated murein-outer membrane complexes (Hancock & Poxton, 1988b). Improved methods for analysing peptide subunits of peptidoglycan are available (Garcia-Bustos *et al.*, 1987; Glauner, 1988); the use of muramidase and high performance liquid chromatography (HPLC) enabled about 80 different peptides from *E. coli* peptidoglycan to be distinguished (Glauner, 1988). HPLC patterns of this kind also have considerable potential in chemotaxonomic studies of Gram-negative bacteria.

2.3 OUTER-MEMBRANE PROTEINS

Cell envelope proteins belong to three main groups:

(a) Proteinaceous appendages, i.e. fimbriae (or pili) and flagellae.
(b) Enzymes localized on the surface of the cell, or in the periplasmic space.
(c) Outer membrane proteins (OMPs; Lugtenberg & van Alphen, 1983; Benz, 1988); these have structural functions and may serve as porins (Hancock, 1987).

The amounts of individual enzymes are low, hence their contribution to PAGE patterns is limited (see Chapter 14). Whole-cell extracts can therefore be boiled in sodium dodecyl sulphate (SDS) and analysed directly (Lambert, 1988) without previous isolation of the OMP fraction. Such analyses generally reveal OMP patterns comprising four major bands (Figure 2.2). Minor bands in such OMP profiles are often due to cytoplasmic contamination or to flagellar or fimbrial proteins, which by specific procedures (Lambert, 1988) may contribute to the overall profile.

The protein patterns of cell envelopes hitherto utilized in taxonomy or epidemiology are largely restricted to the OMPs. After solubilization with detergent these are visualized by PAGE, followed by staining with Coomassie Brilliant Blue or a silver reagent. Alternatively, the OMPs are electrophoretically transferred to a membrane where immunological detection may give specific additional information (immunoblot; Towbin *et al.*, 1979).

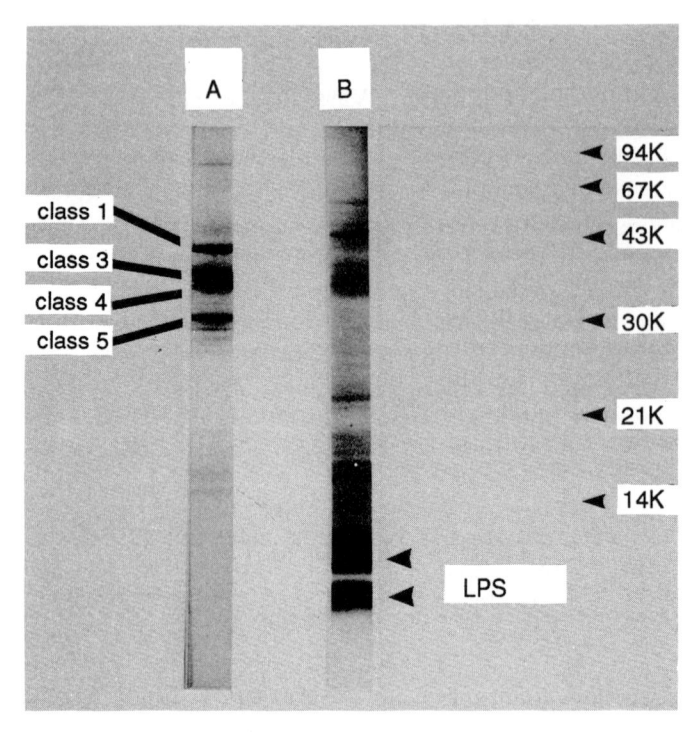

Figure 2.2 Outer membrane proteins (OMPs) of a meningococcal vesicle vaccine, separated by SDS-PAGE. Proteins of the major OMP classes and LPS are indicated. From Fredriksen et al. (1991) with permission.

A basic problem in using protein patterns in systematics is the tendency for the expression of certain proteins to be lost upon subcultivation or growth on artificial media. This loss of expression may be permanent due to loss of the plasmids harbouring the genes involved. Strict standardization of the procedures, including the growth conditions, can help to reduce this problem.

The taxonomic potential of envelope proteins or OMP patterns has not been fully explored, but results have been obtained for several taxa of central interest in medicine, including the Enterobacteriaceae (Lugtenberg et al., 1975; Tokunaga et al., 1979; Lakey et al., 1985), Bordetella (Armstrong et al., 1986), Haemophilus (Vachon et al., 1985) and Neisseria (Young et al., 1983). However, the lack of a generally accepted OMP nomenclature still makes comparative intergeneric studies problematic.

2.4 LIPOPOLYSACCHARIDES

The outer-membrane lipopolysaccharides (LPS) of Gram-negative bacteria are chemically complex molecules consisting of distinct polysaccharide (PS) and lipid (lipid A) parts (Figure 2.3). Most of their many biological effects are assigned to lipid A, whereas the PS part essentially confers antigenicity (Rietschel et al., 1988).

The structure of the PS part seems to belong to one of three main categories (Figure 2.3). The complete 'classical' LPS, as studied most extensively for the Enterobacteriaceae, has two distinct PS parts: a 'core' oligosaccharide comprising some 10 sugar moieties, and a polysaccharide chain of repeating units, e.g. pentasaccharides (O-specific antigen). Such a complete LPS with O-chain, core, and lipid A is termed the S-form (smooth). In contrast an R (rough) form LPS, usually found in mutants, lacks the O-chain and may also have an incomplete core part.

The innermost part of the core has certain features well conserved among taxa, with the sugar 2-keto-3-deoxyoctonate (Kdo) forming an acid-labile linkage to lipid A in almost all LPS. Another common feature is the sequence of two or three L-glycero-D-manno-heptose units linked to the proximal cluster of one to three Kdo moieties. Furthermore, both Kdo and heptose may be variably substituted by phosphate or pyrophosphate, often linked to ethanolamine. A third distinct type of PS is found in the LPS of taxa such as *Bordetella*, *Haemophilus* and *Neisseria*. This PS type lacks a distinct O-part and is considerably less conserved among taxa than the core of S- and R-forms. Lipopolysaccharide of this type is often termed lipo-oligosaccharide (LOS), as suggested by Schneider et al. (1984).

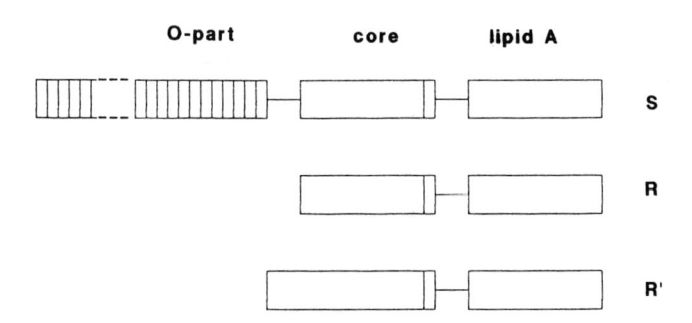

Figure 2.3 Schematic diagram of lipopolysaccharides (LPS). S = S-form or 'smooth' LPS; R = R-form or 'rough' (mutant) LPS; R' = R-like LPS. The lipid A, inner core, core and O-parts are less well conserved in that order.

However, an expert panel from five LPS research centres recommended the use of LPS also for these short-chain structures (Hitchcock *et al.*, 1986).

Lipid A is generally considered to be a highly conserved structure among Gram-negative taxa. However, recent chemotaxonomic studies covering a phylogenetically more diverse range of organisms have revealed that the lipid A structure is less homogeneous than previously believed (Mayer *et al.*, 1989). Typically lipid A has a backbone of a β-1,6-linked glucosamine disaccharide with terminal phosphate or pyrophosphate groups in the 1 and 4' positions. An acylation pattern of four to seven normal and hydroxylated fatty acids in ester, amide and acyloxy linkages is usual (Figure 2.4). However, several deviations from this model have been described (Table 2.2; Mayer *et al.*, 1989). About 25 species belonging to 12 different genera have been shown to contain 2,3-diamino-2,3-dideoxyglucose (DAG) as a lipid A unit (Mayer *et al.*, 1989), including the important pathogen *Brucella abortus* (Moreno *et al.*, 1990). In addition to these DAG-containing variants, lipid A may also contain non-stoichiometric amounts of other (amino)sugars (aminoarabinose, quinovosamine, mannose); some examples of these are given in Table 2.2.

The fatty acid substitution pattern is another notable taxonomic feature of lipid A. Usually it consists of a limited number of saturated, straight-

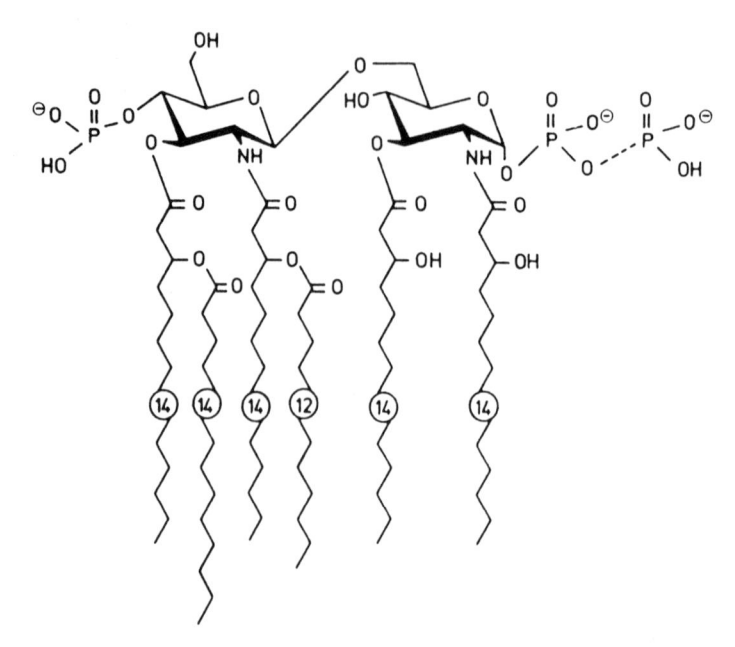

Figure 2.4 Proposed chemical structure of lipid A of *E. coli*. From Rietschel *et al.* (1984) with permission.

chained or mono-methyl-branched (*iso* or *anteiso*) acids and at least one type of fatty acid that is hydroxylated at the 3-position, or infrequently at the 2-position. Recently, considerably more complex patterns have been described, consisting of up to 23 different 3-OH-fatty acids (Nurminen *et al.*, 1985; Moule & Wilkinson, 1989; Sonesson *et al.*, 1989b). Exceptionally complex LPS fatty acid patterns were observed for certain species of *Legionella*, which also include 2,3-di-OH fatty acids (Sonesson *et al.*, 1989b) and a pattern of long-chain (C_{28}–C_{30}) oxo- and dioic acids (Moll *et al.*, 1992; Sonesson *et al.*, 1993). The use of OH-fatty acid patterns for differentiation of *Legionella* species has recently been described (Jantzen *et al.*, 1993). Other long-chain fatty acids (e.g. *n*-2-hydroxylated; C_{26}–C_{30}) have been demonstrated as LPS constituents of species of Rhizobiaceae and *Brucella* (Bhat *et al.*, 1991). Interestingly, Weckesser and Mayer (1988) have noted a correlation between phylogenetic position (as measured by 16S rRNA homology) and the chemical nature of the lipid A variants.

An LPS preparation most often occurs as a microheterogeneous population of molecules differing chemically and immunologically, especially due to variation in the phosphorylation of Kdo and heptose, in the fatty acylation pattern of lipid A (Jiao *et al.*, 1989) (see below), and in PS chain length, e.g. the polymerization degree of O-chains of the S-form LPS (Palva & Mäkelä) or incompleteness of the R-like PS chain (Michon *et al.*, 1990). The heterogeneity, which may render difficulties in detailed structural analysis, can be demonstrated by complex SDS-PAGE (or TLC) patterns of LPS preparations (see below).

2.4.1 Extraction of LPS

Numerous methods for the extraction of LPS have been proposed. Generally, LPS of different bacterial species vary in hydrophobicity due to variation in PS chain length, degree of acylation, or in the amount of phosphate and apolar sugars. These factors enhance differences in their extractability with organic solvents. Extraction losses of LPS molecules of marginal polarities must therefore be considered and it is worth noting that the efficiency of several extraction methods depends heavily upon the previous purification steps.

2.4.1.a Extraction of LPS by hot aqueous phenol

The procedure most widely used to extract LPS remains the 'hot aqueous phenol method' of Westphal *et al.* (1952) (Method 2.1). Non-covalent forces e.g. to membrane proteins, are disrupted using the one-phase system of 45% phenol at 68 °C, and after phase separation by cooling, the protein-containing phenol layer can be removed by centrifugation. Usually LPS

Table 2.2 Lipid A characteristics of representative Gram-negative species.

Species	16S rRNA[a]	Back bone sugar[b]	Fatty acids 3-OH-FA	Fatty acids FA	Additional (unusual) constituents FA	Additional (unusual) constituents Sugars	References
Rhodopila globiformis	α-1	DAG	14,18,19	16,18			Pietsch et al. (1990)
Brucella abortus		DAG, GlcN	14,16	14,16,18	27-OH-28		Moreno et al. (1990)
Pseudomonas diminuta	α-2	DAG	12,13 14,16	14,14:1	5,9-diOH14		Kasai et al. (1987) Arata et al. (1987)
Rhizobium trifolii		GlcN	14,16		27-OH-28	GlcNA	Hollingsworth & Lill-Elghanian (1989); Hollingsworth & Carlson (1989)
Rhodobacter capsulatus	α-3	GlcN	10,14	12:1	3-oxo-14		Krauss et al. (1988)
Rhodocyclus gelatinosus	β-1	GlcN	10	12,14			Mayer & Weckesser (1984)
Rhodocyclus tenuis	β-2	GlcN	10	12,16		Ara_f GlcN AraN	Weckesser & Mayer (1988)

Species	Phylogenetic group[a]	Sugar[b]	Fatty acids		Other	Reference
Neisseria gonorrhoeae	β-3	GlcN	12,14	12,14		Takayama et al. (1986)
Chromatium vinosum	γ-1	GlcN, DAG	14	12	Man	Weckesser & Mayer (1988)
Legionella pneumophila[c]	γ-2	?	i14,a15,18,i20, 20,21,i22,22	i14,a15, i16,16, a17,18,20	2,3-diOHi14, 2,3-diOHn14, 27-oxo-28, Dioic-27	Sonesson et al. (1989b) Moll et al. (1992)
Escherichia coli	γ-3	GlcN	14	12,14,15	AraN	Rietschel et al. (1984)
Pseudomonas aeruginosa	γ-3	GlcN	10,12	12	2-OH-12	Kulshin et al. (1991)

[a] Phylogenetic groups (evolutionary branching points) established by 16Sr RNA sequences (Woese, 1987).
[b] Sugar constituent of the lipid A backbone structure (Mayer & Weckesser, 1984; Rietschel et al., 1984).
[c] Lipid A of this species has a very complex fatty acid pattern of 19 3-OH-FA and 8 FA; only the most abundant are listed.
Abbreviations: Sugars: DAG, diaminoglucose; GlcN, glucosamine; GlcNA, glucosaminuronic acid; Man, mannose; Ara, arabinose (furanoic); AraN, arabinosamine; QuiN, quinosamine; GlyP, glycerol phosphate. Fatty acids (FA): *m*-OH-*n*, hydroxylated or oxygenated (oxo) FA (*m*, position of OH- or oxo-groups; *n*, number of C atoms); a simple number, carbon atoms in a straight-chain non-hydroxylated FA; prefix i, methylbranch in ω-1 position; α in ω-2 position.

is recovered in the top water phase, but less hydrophilic LPS molecules may have increased affinity for the phenol phase. Thus, both layers may have to be collected to recover adequate amounts, e.g. of meningococcal LPS (Jennings & Johnson, 1983). Two chemically and immunologically distinct forms of *Yersinia enterocolitica* LPS have been recovered from the two phases (Caroff *et al.*, 1984). Unlike other extraction methods, LPS extracted by phenol has been shown to be resistant to enzymatic degradation during prolonged dialysis (Leive & Morrison, 1972).

Method 2.1 Extraction of (S-type) LPS with hot aqueous phenol (modified from Westphal *et al.*, 1952)

---

1. An aqueous suspension of 10–50 mg bacteria (dry weight; e.g. growth from two agar plates) is transferred to a (round bottom) glass centrifuge tube.
2. Water is added to a volume of 1.5 ml and the suspension heated to 68 °C. A solution (1.5 ml) of 90% phenol (90 g + 11 ml water) prewarmed to 68 °C, is added and the suspension stirred with a magnet at 68 °C for 15 min.
3. Phase separation is obtained by cooling below 10 °C in an ice bath, and centrifugation (20 min) preferably at low temperature or in precooled holders. The magnet may be left in the tube during centrifugation.
4. The top (aqueous) phase is carefully pipetted off, leaving the opaque interphase behind.
5. Another 1.5 ml hot water is added and steps 2 to 4 repeated.
6. The combined water phase is dialysed overnight against tap water, followed by distilled water until the smell of phenol disappears.
7. The LPS solution (contaminated with RNA and PS) is freed from insoluble material by centrifugation. If LPS is not to be purified further by ultracentrifugation (steps 8 and 9), it is freeze-dried for storage.
8. The LPS solution is concentrated (approximately 5 times) on a rotary evaporator, and any insoluble deposits removed by centrifugation.
9. Ultracentrifugation (3 h; $100\ 000 \times g$) sediments the clear gelatinous LPS. If the pellet has an opaque central base part of insoluble material, this is left behind when carefully suspending the LPS in a small volume of distilled water.
10. At the expense of yield, step 9 may be repeated 1 to 2 times to obtain a purer product. The LPS supernatant is freeze-dried and stored at room temperature.
11. As an *alternative* to steps 6 and 7, the aqueous phase may be precipitated at −20 °C using four volumes of ethanol in the presence of 1/20 volumes 5 M NaCl, and the collected sediment washed with 75% (v/v) ethanol before freeze-drying. This yields a less pure product.

Note: The toxic properties of phenol must be considered during handling and disposal. Use gloves, and immediately rinse contaminated skin with water.

2.4.1.b Extraction of rough LPS with phenol/chloroform/petroleum ether

Galanos *et al.* (1969) introduced a method (Method 2.2) which significantly raises the yield of rough (i.e. R-form and R-like) LPS. Briefly, LPS is extracted into a mixture of phenol (90%), chloroform and petroleum ether (PCP). The latter two solvents are removed by evaporation and the LPS precipitated out of the phenol by the dropwise addition of water. Alternatively the LPS can be precipitated using two volumes of ethanol (Erwin & Munford, 1989). When using the PCP extraction method devised for less polar LPS, one should be aware of possible contamination with phospholipids (see below). The PCP method has also proven useful in the purification of S-form LPS that has been obtained by previous extraction using the hot aqueous phenol procedure.

Method 2.2 Extraction of (R-type) LPS with phenol-chloroform-petroleum ether (modified from Galanos et al., 1969).

Extraction solvent: phenol (90 g + 11 ml water)/chloroform/petroleum ether (2:5:8). The petroleum ether (bp 40–60 °C) may be substituted by hexane. If cloudy, the solution is cleared by adding a small amount of solid phenol.

1. One part (weight) of dry bacteria is transferred to a (round bottom) glass centrifuge tube.
2. About 5 vols extraction solvent is added, and the suspension (below 20 °C) is vortexed for 2 min.
3. After centrifugation (15 min) the supernatant is filtered through a coarse filter paper into a glass centrifuge tube which can be fitted to a rotary evaporator.
4. Steps 2 to 3 are repeated 1 to 2 times, and the supernatants combined.
5. Petroleum ether and chloroform are removed by rotary evaporation. If solidification occurs, this is counteracted by adding a minimal amount of water.
6. LPS is precipitated from the remaining phenol solution by careful addition (dropwise) of distilled water. LPS is collected by centrifugation, the phenol removed using a Pasteur pipette and the tube walls cleaned with paper tissue.
7. Alternatively LPS may be precipitated with 2 vols ethanol.
8. LPS washed three times with small volumes of 75% (v/v) ethanol, followed by acetone and dried under vacuum or by N_2. Purification can be performed with ultracentrifugation as described in Method 2.1.

2.4.1.c Other extraction methods

Besides the important phenol-containing extraction agents, a number of other chemicals which may act as solvents, detergents or chelators have been introduced for the solubilization and extraction of LPS. Butanol is considered to be an especially mild agent, as butanol : water partition does not disrupt *Escherichia coli* cells (Leive & Morrison, 1972). The use of Tris-EDTA, which removes the calcium ions stabilizing the outer membrane, has proven to be more effective in *E. coli* than in *Salmonella*, with increased yields for cell walls rich in phosphorus (Leive & Morrison, 1972; Wilkinson et al., 1973). The effect of detergents, sodium dodecyl sulphate (SDS; Darveau & Hancock, 1983) or deoxycholate (Wu et al., 1987), has been shown to be positive both for the yields obtained and for the water solubility of the final product.

2.4.2 Purification of LPS

The analysis of LPS for taxonomic work may be performed on preparations purified to variable degrees, depending on the specific problem and the detection methods. Thus, for determinations of LPS fatty acids only lipid contamination need be considered whilst for sugar analyses, PS are of primary concern. Demonstration of microheterogeneity by SDS- or deoxycholate-(DOC-)PAGE (see below) or TLC appears to be unaffected by the presence of proteins, when specific silver stains are used. It is useful to monitor contaminants throughout the purification scheme because of batch-to-batch variations that may influence the outcome of a given purification step.

2.4.2.a *Ultracentrifugation*

Crude LPS preparations, e.g. aqueous phenol extracts, are, after dialysis against distilled water, often purified by ultracentrifugation (three times at 100 000 × g (Westphal *et al.*, 1952). Co-extracted contaminants such as capsular PS and nucleic acids generally sediment less (in *Neisseria* 5 to 10% in one run; unpublished) than LPS and marked purification of an LPS preparation can be achieved by ultracentrifugation. However, many LPS sediment only moderately (*Bordetella* 5%, *Neisseria* 60%; unpublished), resulting in extensive loss of material and making the addition of calcium ions necessary (Borneleit *et al.*, 1989).

2.4.2.b *Enzymatic treatment*

Enzymatic hydrolysis of proteins and nucleic acids during preparation of LPS may have two positive effects: more LPS is solubilized and extracted, and contaminating macromolecules are removed. Several authors recommend treatment with proteinase K, DNAse and RNAse at one or more stages of the purification procedure although this obviously has to be optimized for each kind of LPS. The low molecular weight enzymic products may be removed easily by ultracentrifugation, dialysis, column chromatography or centrifugation, and can be monitored colorimetrically or by UV (see below).

2.4.2.c *Fractionation on columns*

Due to the hydrophobic character of lipid A, large LPS aggregates (molecular weights of multi-millions) form spontaneously in the absence of detergents. Although direct column chromatographic purification of LPS is

2. Analysis of Gram-negative Cell Constituents

possible, the addition of detergents during chromatography is generally useful in separating the LPS monomers (MW approx. 3–30 kDa) from other material (Tsai et al., 1989).

2.4.2.d Electrodialysis

After extraction, phosphate and carboxyl groups of LPS are associated, in unknown proportions, with both inorganic (Na^+, K^+, Ca^{2+}, Mg^{2+}) and organic (various amines) cations. These cations strongly influence the solubility and aggregation of LPS, and thereby its biological properties. Galanos and Lüderitz (1975) introduced the removal of the LPS-associated cations by electrodialysis, followed by neutralization with a selected base to convert LPS into a defined salt form. Electrodialysis is recommended especially for biological experiments. The triethylamine salt commonly used gives a more readily water-soluble form.

2.4.2.e Check for impurities

Both DNA and RNA (predominant) may be detected by their UV absorption maximum at 260 nm. Alternatively, RNA can be readily detected (by GLC) by the content of ribose, a sugar which only rarely occurs as a true LPS constituent. DNA may be determined fluorimetrically (Downs & Wilfinger, 1983).

Proteins may be determined with the Folin–Ciocalteu reagent (Lowry et al., 1951), but the possible interference due to detergents or other material must be controlled. For certain more immunological purposes the absence of outer membrane proteins is important, and this can be checked using SDS/DOC-PAGE or immunoblotting (see below).

Capsular polysaccharides known to be present in the bacterium under study should be monitored when required, using standard chromatographic methods, colorimetry or immunological techniques as described elsewhere.

Phospholipids are most easily revealed by GLC detection of unsaturated fatty acids, which occur almost exclusively in extractable lipids (Jantzen & Bryn, 1985; Wilkinson, 1988), or can be detected as intact lipids on silicic acid TLC after specific phospholipid extraction (Christie, 1982). However, the latter method requires relatively large amounts of material.

2.4.3 Analysis of intact LPS and lipid A

2.4.3.a SDS-PAGE

When S-form LPS is analysed by SDS-PAGE (Jann et al., 1975), or DOC-PAGE (Komuro & Galanos, 1988), a ladder-like pattern of bands is observed for S-form LPS. Distances between adjacent bands correspond to molecular weight differences of one oligosaccharide repeating unit of the O-chain. In the low molecular weight region PAGE bands of S-form LPS correspond to the small number of bands of R-form LPS, which lack O-chains. R-like LPS show only a small number of PAGE bands in the low molecular weight region, approximately of similar molecular weights as R-form LPS bands. When bacteria have been cultivated at high growth rates, obtained either at higher temperatures or in rich media, the LPS band patterns may be shifted towards lower molecular weight LPS. As has been reported both for S-form LPS (Acker et al., 1980) and for R-like LPS (Tsai et al., 1983), this suggests an incomplete synthesis of LPS under these conditions.

Detection of LPS can be carried out in several ways following SDS- (or DOC-) PAGE. Staining can either be done directly on fixed and washed gels using Coomassie Brilliant Blue, or by the more sensitive and LPS-specific silver staining (Tsai & Frasch, 1982) or by both in sequence. Alternatively, electrophoretic blotting of LPS onto nitrocellulose membranes allows antibody detection of specific epitopes (Schneider et al., 1984; Pyle & Schill, 1985). Direct 'staining' in SDS-PAGE gels using labelled lectins or antibodies (Burridge, 1978) may also prove useful for LPS.

2.4.3.b Analysis of LPS and lipid A by TLC

Chromatographic separation of complex and large molecules such as LPS is an inherently difficult task. Firstly, both the PS and the lipid A part are often structurally microheterogeneous. Secondly, the large difference in hydrophobicity between the PS and the lipid A part presents a challenge in devising a TLC system which takes the chemical nature of both parts into account. Nevertheless, the pattern of TLC spots and their specific colours can be used to characterize LPS preparations (Buttke & Ingram, 1975; Chen et al., 1975). More recently Samu et al. (1988) have demonstrated a distinct separation between S- and R-forms of LPS and have shown that different R-form LPS can be separated.

Lipid A preparations (see below) are considerably more homogenous in their hydrophobic properties than whole LPS, and hence TLC (and probably also HPLC) provides excellent separation of the individual molecular forms in a microheterogeneous lipid A preparation (Mattsby-Baltzer & Alving, 1984; Samu et al., 1988; Jiao et al., 1989).

2.4.4 Cleavage of LPS into lipid A and polysaccharide

Lipid A is almost always linked by a ketosidic linkage to a Kdo moiety of the PS. This linkage is very acid-labile and can be cleaved under mild conditions, e.g. 0.1 M HCl or 1% acetic acid at 100 °C for 1–3 h (Lüderitz et al., 1971). However, after only 30 min of hydrolysis, the 1-phosphate is split off, resulting in monophosphorylated lipid A. In addition, even some 10% of the O-ester-linked fatty acids are lost (Jiao et al., 1989); milder less acidic conditions for hydrolysis have recently been introduced to minimize such degradation. Brade and Brade (1985) recommended a sodium acetate buffer at pH 4.4, and Jiao et al. (1989) a calcium acetate buffer at pH 3.5. Both procedures provide the intact diphosphorylated form of lipid A. The sodium acetate buffer method also may render ketosidic Kdo-Kdo linkages in PS uncleaved. Separation of lipid A and PS is obtained either by spontaneous precipitation of lipid A (especially for monophosphoryl-lipid A), precipitation with ethanol, extraction, or by column chromatography.

Method 2.3 Cleavage of LPS into PS and lipid A (modified from Lüderitz et al., 1971; Brade & Brade, 1985).

2.3.A Cleavage into PS and monophosphoryl-lipid A

1. Lipopolysaccharide (0.01–1 mg) is transferred to a glass tube (with Teflon-lined screw cap) and 0.3–1 ml of HCl (0.1 M) or acetic acid (1%) is added.
2. The tube is heated (1–2 h at 100 °C) and monophosphoryl-lipid A is collected by centrifugation. The yield may be increased by precipitation with 4 vols ethanol. The sedimented lipid A is washed twice with ethanol, once with water, and dried. Lipid A may be solubilized in triethylamine, preferably with sonication.
3. Polysaccharide in the supernatant is freed from contaminating lipid A by extraction with 1 vol chloroform/methanol (4 : 1 v/v). The bottom organic phase is removed by a Pasteur pipette, and organic solvent remaining in the aqueous PS phase by N_2.
4. The last traces of lipid A may then be removed from PS by solid phase extraction. A commercial reversed phase (C_{18}) cartridge is prewetted with methanol and rinsed with water (4 × 1 ml) before application of the PS solution. Elution of PS is completed with 1 ml water.

2.3.B Cleavage of LPS into PS and diphosphoryl-lipid A

This is a more gentle method leaving the 1-phosphate of lipid A intact, and reduces the loss of ester-linked fatty acids.

1. Heating is performed as in Method 2.3A, with the substitution of HCl or acetic acid with sodium acetate buffer (20 or 100 mM, pH 4.4).
2. As precipitation of diphosphoryl-lipid A may be limited, addition of 4 vol ethanol (see A above) may be required.

Method 2.4 Analysis of LPS and lipid A by TLC (modified from Samu et al., 1988; Jiao et al., 1989).

2.4.A Lipopolysaccharide

1. Lipopolysaccharide (e.g. 3 µl of 1 mg/ml in water) is applied to a TLC plate (aluminium sheets with silica gel).
2. The plate is developed twice (with hot air drying between runs) using a mixture of isopropanol/water/chloroform/conc. NH$_4$OH/triethylamine (120 : 60 : 16 : 2 : 1 by volume).
3. The plate is carefully dried (hot air), dipped into a staining solution (e.g. 0.1% orcinol in water/methanol/sulphuric acid; 16 : 4 : 1 by volume). Excess staining solution is removed (paper tissue), the plate is dried (hot air) and LPS detected by carefully heating on a hot plate (125 °C).

2.4.B Lipid A

1. Lipid A (e.g. 3 µl of 1 mg/ml in triethylamine) is applied to a TLC plate (aluminium sheet with silica gel).
2. Development is performed using the same solvent as for LPS (see Method 2.4.A above) or chloroform/methanol/conc. NH$_4$OH/water (100 : 50 : 5 : 8 by volume).
3. Lipid A is detected by dipping into, or spraying with, 5% methanolic sulphuric acid and charring at 170 °C.

2.4.5 Analysis of LPS constituents

2.4.5.a Combined analysis of LPS sugars and fatty acids

A simple one-step procedure for analysis of the majority of LPS constituents has been described (Bryn & Jantzen, 1982, 1986). The procedure is a one-vial method and comprises the following four steps:

(a) Heating LPS (or LPS-containing material; 10–1000 µg) with dry methanol/HCl.
(b) Removal of HCl and methanol with N$_2$.
(c) Derivatization of the hydroxy- and amino-groups using trifluoroacetic anhydride (TFAA).
(d) GLC on an apolar capillary column.

Whereas glycosidic and ester linkages are quantitatively cleaved by 2 M HCl in methanol at 85 °C overnight, more accurate determination of strongly bonded (amide-linked, N-acetylhexosamine-linked) constituents requires 4 M HCl at 100 °C overnight. Except for the 1-linked phosphate of lipid A, phosphate linkages are stable even at these conditions, but can be quantitatively split by aqueous hydrofluoric acid (HF) at 4 °C.

To avoid degradation of labile sugars, milder conditions may prove necessary. Thus, Kdo which shows 60% degradation into anhydro- and other compounds at 85 °C overnight, is quantitatively liberated and largely undegraded following methanolysis (2 M HCl) at 60 °C for 2 h (Bryn & Jantzen, 1982, 1986). The even more labile dideoxyhexoses may require a lower temperature (37 °C) to be stable. Generally a GLC profile of LPS constituents obtained by this method (Figure 2.5) contains the sugar peaks (as trifluoroacetyl (TFA)-methylglycosides) in the first part of the chromatogram, followed by the fatty acids (as methyl esters and TFA-hydroxy methyl esters).

Methanolysis yields more than one peak for each monosaccharide due to alpha and beta forms of pyranose and furanose rings. This results in a complex sugar pattern in the chromatograms (Figure 2.5). Nevertheless, for each sugar there is a remarkably reproducible peak configuration which can be used for peak identification. During evaporation of the methanolic HCl a partial loss may occur of the more volatile non-hydroxylated fatty acid methyl esters, especially those with 14 C-atoms or less.

The attack on the stationary phase of the GLC column by the derivatization reagent TFAA, or its hydrolytic product trifluoroacetic acid, can reduce its useful lifetime. This may, however, be substantially increased (to 300–400 or more injections; unpublished) by reducing the TFAA concentration from 50 to 10% before injection onto the chromatograph (see below). For storage in capillaries at −20 °C, the samples should be kept in 50% TFAA.

Method 2.5 Combined GLC analysis of sugars and fatty acids from whole-cells or LPS (modified from Jantzen et al., 1978; Bryn & Jantzen, 1982).

The GLC profiles essentially reveal (unphosphorylated) sugars quantitatively, and fatty acid methyl esters (FAME; semiquantitatively). FAMEs (especially $<C_{14}$) show evaporation losses, and amide-linked hydroxylated FAMES are released only to approximately 70% yield.

1. Dry material (e.g. a loopful of cell material from an agar plate) in a Teflon-lined screw-cap cone vial (1 ml capacity) is heated overnight at 85 °C in 300 µl of methanolic HCl (2 M, anhydrous, cautiously made by dropwise addition of 15 ml acetylchloride to methanol; final volume 90 ml). Inspect for leakage and re-tighten cap after 30 min.
2. Concentrate the reaction mixture to approximately 1/2 volume using N_2, and remove possible deposits by centrifugation. Complete drying with N_2, add 0.3 ml methanol and repeat drying.
3. Add to the dried sample 50 µl of TFAA; (50% in acetonitrile), heat with hair dryer for 2 min. Concentrate just to dryness with N_2, and redissolve in 30 µl 10% TFAA/acetonitrile.
4. Analyse the sample by GLC (25 m × 0.2 mm fused-silica coated with SE-30, methylsilicone) using a temperature gradient of 80 °C (2 min) → (8 °/min) → 280 °C.
5. Use the profile on a fingerprint basis, or identify peaks by mass spectrometry/ retention data and apply response factors (Bryn & Jantzen, 1982).

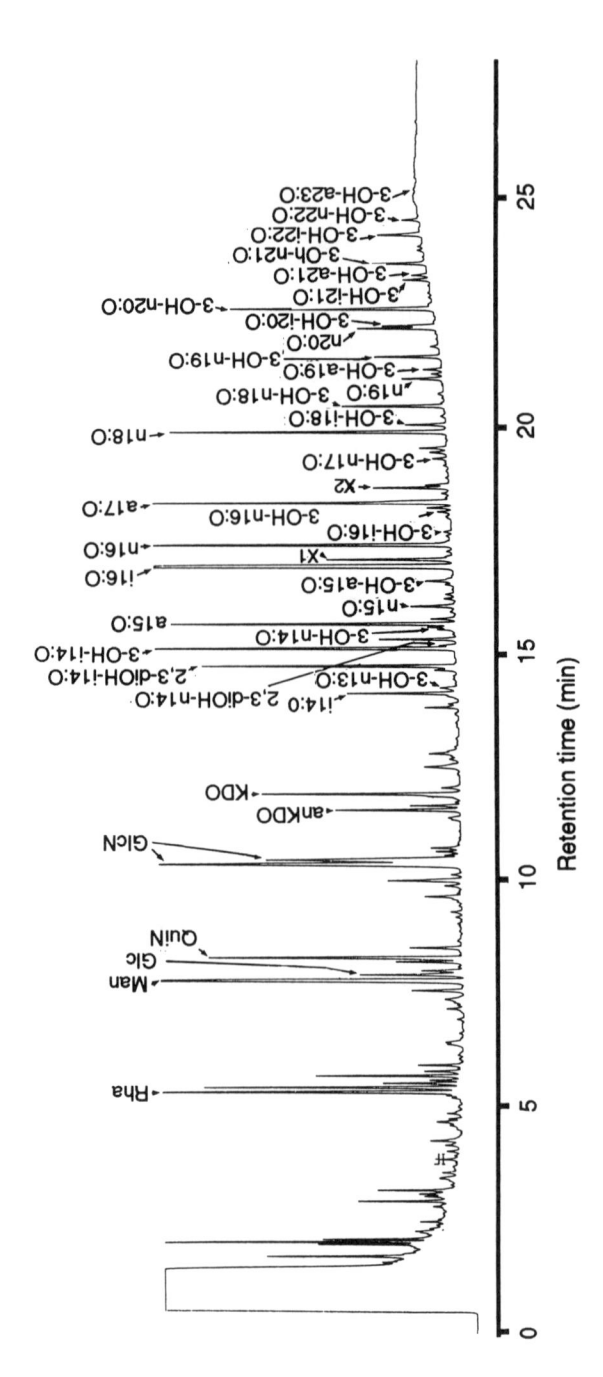

Figure 2.5 GLC profile of *L. pneumophila* LPS after methanolysis, trifluoroacetylation and analysis on a 25 m SE-30 fused-silica capillary column. Temperature programmed from 80 to 280 °C at 8 °C/min. Rha, rhamnose; Man, mannose; Glc, glucose; GlcN, glucosamine; KDO, methylester/methylketoside of 2-keto-3-deoxyoctonic acid; anKDO, anhydro form of KDO; n19:0, nonadecanoic acid (internal standard). From Sonesson *et al.* (1989b) with permission, © Springer-Verlag 1989.

2.4.5.b Analysis of the PS sugars

Analysis of the ('degraded') polysaccharide (PS) after splitting off the lipid A, follows the general scheme for analysis of polysaccharides (see below). The reader is also referred to more comprehensive works on carbohydrate analysis (Aspinall, 1983; Mayer et al., 1985; Chaplin & Kennedy, 1986). For qualitative and quantitative analysis of the monosaccharides, the most common approach involves hydrolysis under two different sets of conditions followed by the well established alditol acetate method for derivatization and analysis using GLC or GLC-MS (Patouraux-Promé & Promé, 1984; Gilbert et al., 1988). More recently, the special HPLC technique of high-pH anion-exchange chromatography (HPAEC) with pulsed amperometric- (PAD) or pulsed electrochemical (PED) detection, has been shown to be an attractive alternative for routine analysis of (underivatized) mono- and oligosaccharides (Lee, 1990), including bacterial saccharides (Phillips et al., 1990). Due to its high resolution and sensitivity, this technique should also be well suited for diagnostic purposes (Sonesson & Jantzen, 1992). Thus, as seen in Figure 2.6, HPAEC profiles give distinct peaks of N-acetylneuraminic acid (Neu5Ac), Kdo and LPS oligosaccharides.

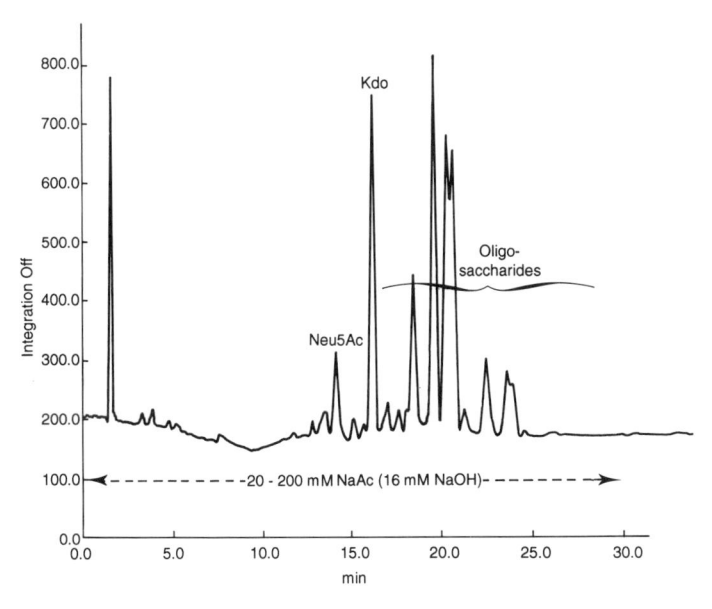

Figure 2.6 HPAEC-PED profile of meningococcal LPS. Lipid A was removed by hydrolysis (0.1 M HCl; 100 °C, 2 h) and adsorption on a reverse phase (C_{18}) column. Neu5Ac, N-acetylneuraminic acid; Kdo, 2-keto-3-deoxyoctonic acid.

We have for several years applied an alternative or supplementary technique based on methanolysis and TFA-derivatization (see above). Here we will focus only on the use of this technique for GLC and GLC-MS analyses of certain characteristic LPS constituents, namely Kdo, L- and D-glycero-D-*manno*heptoses, deoxysugars and 3,6-dideoxyhexoses.

Kdo is a sugar that is acid-labile and often occurs phosphorylated in LPS. Optimal conditions of acidic release are essential to minimize degradation, a problem in the colorimetric determination of Kdo, e.g. using thiobarbituric acid (TBA) assays. By methanolysis (60 °C for 2 h) and GLC these problems can largely be avoided as Kdo is liberated quantitatively from virtually all linkages (except phosphate, see below) with minimal degradation. When Kdo is linked to hexosamine, as in the meningococcal PS29E, its release may require prolonged methanolysis (6 h), which can be undertaken without substantial degradation (Bryn & Jantzen, 1986).

Phosphorylated Kdo is not easily determined by GLC, and very low levels are obtained using colorimetric assays due to phosphorylation of Kdo in the 5-position. These quantification problems can be overcome by

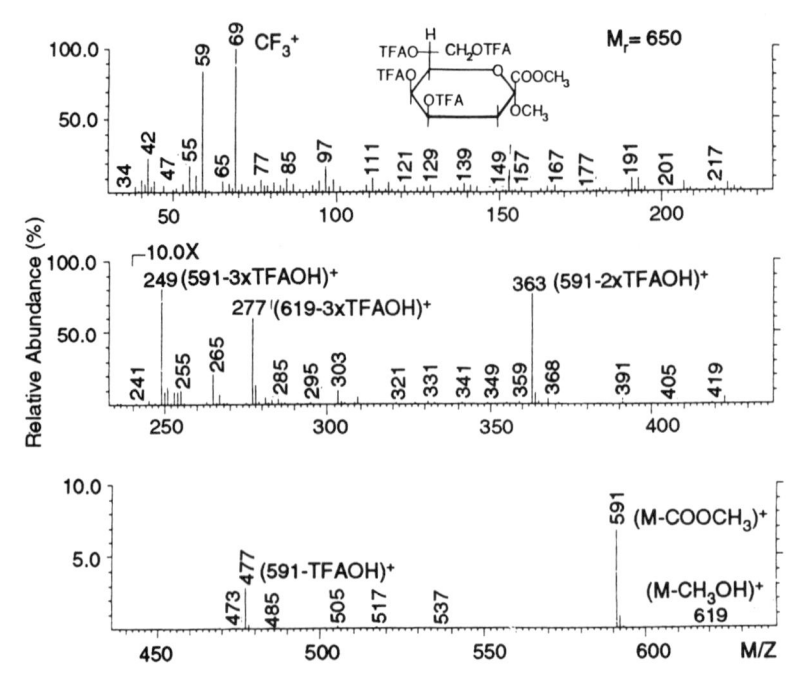

Figure 2.7 Mass spectrum of Kdo. The sugar was released from LPS by methanolysis (2 M HCl, 85 °C, 18 h) trifluoroacetylated and analysed by ion-trap GLC-MS (25 m SE-30 column) in electron-impact (70 kV) mode. From Bryn & Jantzen (1986) with permission.

2. Analysis of Gram-negative Cell Constituents

the removal of phosphate groups using aqueous HF (4 °C for 48 h) (Caroff et al., 1987a,b; Sonesson et al., 1989a). The identity of Kdo may be verified by GLC-MS after methanolysis and TFA derivatization. The major Kdo peak gives a characteristic pattern (Figure 2.7) with a prominent fragment of m/z 591 (M-59; loss of carboxylester) followed by serial losses of one and two TFAOH groups (m/z 477 and 363).

Heptoses are stable towards both hydrolysis and methanolysis and when unsubstituted by phosphate their GLC analysis does not represent a problem. When phosphorylated, however, total heptose may be determined by GLC only after HF-mediated removal of phosphate (Sonesson et al., 1989a) or by HPLC. By far the most common heptose is L-glycero-D-*manno*-heptose, which can be accompanied by its biosynthetic precursor, the D-glycero isomer.

6-Deoxyhexoses (fucose, rhamnose) are frequent constituents of LPS. They are easily liberated and determined by methanolysis followed by GLC and can be readily verified by MS.

3, 6-Dideoxyhexoses (abequose, colitose, tyvelose) occur in the O-chain of some enterobacterial LPS. These sugars are extremely acid labile. A GLC determination of 3,6-dideoxyhexoses may be performed following methanolysis at 37 °C (Bryn & Jantzen, 1982).

In addition several other rare sugars of diagnostic interest have been described (Lindberg, 1990).

2.4.5.c Analysis of the lipid A constituents

For analysis of *total fatty acids* (amide- and ester-linked) strong acidic conditions (e.g. 4 M HCl in methanol, 100 °C for 18 h) are required to ensure complete cleavage of the amide-linked acids. Alkaline conditions generate a beta-elimination from 3-acyloxyacyl residues resulting in significant amounts of alpha-beta-unsaturated fatty acids as artefacts. Furthermore, alkaline conditions cleave amide-linkages incompletely (Jantzen et al., 1978). *Ester-linked fatty acids* can be selectively released by treatment with methanolic NaOH (Koeltzow & Conrad, 1971), by NaOCH$_3$ (Wollenweber & Rietschel, 1990), or preferably by a one-step transmethylation using 2 M HCl in methanol at 60 °C for 2 h (Bryn & Jantzen, 1982). After de-*O*-acetylation and removal of the fatty acids, the remaining *amide-linked fatty acids* can be released as described above for total fatty acids. The GLC conditions used to analyse fatty acids are given in Method 2.7.

Aminosugars The backbone structure of lipid A is most commonly a disaccharide of two glucosamine (GlcN) units. For LPS in some taxa, one unit of dideoxydiaminoglucose (DAG) has the same function (see above). In addition, some lipid A structures contain an (unacylated) aminosugar linked to the non-reducing end of the GlcN disaccharide (Rietschel et al., 1988). Generally, the linkages between aminosugars are strong and require relatively strong conditions for complete cleavage, e.g. 4 M aqueous HCl, 100 °C for 18 h (Yokata et al., 1987). As described above, the monomeric aminosugars can, after liberation, be determined by GLC as alditolacetates (Yokata et al., 1987) or as TFA methylglycosides (Bryn & Jantzen, 1982).

2.4.5.d Analysis of other LPS constituents

Phosphate occurs in LPS in either acid-labile (1-position of glucosamine in lipid A) or acid-stable linkages. Labile phosphate is released by brief hydrolysis (1 M HCl at 100 °C for 7 min), whereas release of stable phosphate requires complete oxidative ashing of the organic material (Leloir & Cardini, 1957). The released phosphate can then be determined colorimetrically (Chen et al., 1975; Hancock & Poxton, 1988). *Phosphate-linked components* can, in some cases, be analysed directly or after removal of phosphate by phosphatase (Brade, 1985; Zamze et al., 1987), or by HF at 4 °C for 48 h (Caroff et al., 1987a; Sonesson et al., 1989a). The recently described HPAEC-PAD/PED technique (see above) may prove useful for the analysis of phosphorylated sugars, although the sensitivity apparently is significantly reduced (unpublished observations). *Acetyl* groups may be specifically removed by 0.05 M NaOH at 25 °C for 3–4 h and determined by GLC (Fromme & Beilharz, 1978). *Ethanolamine* may be determined by GLC (unpublished) or after TLC or electrophoresis, on the basis of the free amino groups, using stains such as ninhydrin.

Method 2.6 High pH anion exchange chromatography (HPAEC) profiles of LPS (modified from Philips et al., 1990).

1. Lipopolysaccharide (50 µg) is freed from its lipid A part (Method 2.3). Glucuronic acid (5 µg) may be added as internal standard before hydrolysis.
2. The aqueous PS solution is freeze-dried, dissolved in ultrapure water (200 µl), and filtered (0.45 µm; with centrifugation).
3. Filtered PS is freeze-dried, dissolved in ultrapure water (100 µl), and 25 µl is injected onto the HPAEC-PAD/PED instrument (20 µl loop).
4. The following gradient is used for released acidic monomers (Kdo and Neu5Ac) and oligo- or polysaccharides: 20 to 400 mM sodium acetate in 16 mM NaOH during 1 h.

2.5 EXTRACTABLE LIPIDS

The lipid composition of Gram-negative bacteria has been extensively analysed (see Wilkinson, 1988) and it is well documented that environmental factors can influence this both qualitatively and quantitatively (see Rose, 1989). Thus, lipid profiles are relatively sensitive to environmental factors and certain 'unexplainable' differences between closely related strains may occur. However, by standardizing growth conditions the reproducibility can be increased significantly. There can be no doubt that lipid profiles have provided valuable chemotaxonomic information (Asselineau, 1966; Shaw, 1974; Lechevalier, 1977; Wilkinson, 1988; Lechevalier & Lechevalier, 1988).

2.5.1 Phospholipids

Phospholipids are, together with proteins, vital membrane constituents and they are generally present in high proportions (Wilkinson, 1988). Comparative phospholipid data are available for numerous Gram-negative taxa (Lechevalier & Lechevalier, 1988; Wilkinson, 1988). Phosphatidylethanolamine is most common among Gram-negative bacteria, followed by the widely distributed phosphatidylglycerol and diphosphatidylglycerol; less frequently occurring are phosphatidylinositol and phosphatidylcholine.

2.5.2 Other polar lipids

Other polar lipids such as glycolipids, sphingolipids and sulphonolipids have a limited distribution among Gram-negative organisms, but this can in fact enhance their chemotaxonomic potential (Wilkinson, 1988). Gram-negative glycolipids include acylated sugars, glycosyldiacylglycerols and phosphoglycolipids, all of which provide specific chemotaxonomic markers (Shaw, 1974; Wilkinson, 1988). Recently, a group of interesting surface-active lipids produced by *Serratia rubidaea*, with 3-hydroxylated fatty acids as main fatty acid substituents, have been described (Matsuyama *et al.*, 1990). As noted by Wilkinson (1988), the occurrence of sphingolipids seems so far to be limited to a few taxa (e.g. *Bacteroides*, some species of *Flavobacterium/Sphingobacterium* and *Pseudomonas paucimobilis*). On the other hand, lipoamino-acids of the ornithine amide type are more widely distributed (Kawai & Moribayashi, 1982; Wilkinson, 1988). Members of the *Cytophaga-Flexibacter* group contain large quantities of unusual sulphonolipids (*N*-fatty acyl 2-amino-3-hydroxyheptadecane-1-sulphonic acids) in their cell envelope; these lipids appear essential for the characteristic gliding motility of these taxa (Godchaux & Leadbetter, 1988).

2.5.3 Apolar lipids

Apolar lipids (other than isoprenoid quinone lipids involved in electron transport, see Chapter 8, and pigments, see Chapters 10 and 11) have been reported as constituents of only a few Gram-negative taxa. For instance, wax esters were found to distinguish species belonging to the genera *Branhamella*, *Moraxella* and *Acinetobacter* (Bryn et al., 1977). Relevant chemotaxonomic methods for individual lipid classes are given in Chapter 8.

2.6 POLYSACCHARIDES

Gram-negative bacteria are frequently surrounded by polysaccharides, and a wide range of different structures are known (Kenne & Lindberg, 1983; Lindberg, 1990). Most attention has inevitably been paid to the PS of species readily cultivated in the laboratory and to organisms known to have a role in human infection. Polysaccharide structures may be homopolymers of a single sugar or they may consist of repeating oligosaccharide units. They may be linear polymers, or branched polymers composed of two to five different sugars and additional moieties, e.g. *O*-acetyl, pyruvic acid, glycerol, ribitol and phosphodiesters (Kenne & Lindberg, 1983).

Polysaccharides have for a long time played an important role in bacterial systematics, mainly as the basis of serology. Thus, they constitute more than 100 different K antigens of the Enterobacteriaceae, and define the many clinically important serogroups of *Haemophilus* and *Neisseria* (Jennings, 1983; Jann & Jann, 1987). Usually the chemical structures of PS have been established a long time after the introduction of serology for clinical use. Classical chemical procedures, and increasingly nuclear magnetic resonance (Perlin & Casu, 1982; Koerner et al., 1987) and mass spectrometry (Lönngren & Svensson, 1974; Reinhold, 1987) have revealed (and revised) detailed information on PS structures.

A high proportion of the PS of Gram-negative bacteria are acidic, consisting of a few constituents such as those found in the group-specific PS of *N. meningitidis* (Jennings, 1983). These are linear polymers of one or of two different sugars with the acidic character conferred by phosphate (e.g. serogroups A, X and Z), *N*-acetyl-neuraminic acid (Neu5Ac; groups B, C, Y and W135) or Kdo (group 29E) (Jennings, 1983). In *E. coli* certain K antigens are also simple linear acidic PS whilst others contain additional neutral sugars linked together in more complex branched structures (Jann & Jann, 1987). Similar structures are also found in *Klebsiella* (Jennings, 1983).

Negatively charged PS may be prepared in a highly purified form in relatively few steps by binding to quaternary ammonium detergents (Cetavlon) which enables their separation from the cell. The PS-detergent complexes are then solubilized by a solution of CaCl₂, and the PS recovered following ethanol precipitation. Recent studies of *E. coli* K1 PS (colominic acid) and several other bacterial PS preparations by a modified PAGE technique (Pelkonen *et al.*, 1988) revealed that PS preparations may contain molecules of wide-ranging molecular weights. Such an approach also facilitates the use of antibodies for specific detection. Similarly, oligosaccharides can be detected by specific antibodies after chromatographic development on amino-bonded high-performance TLC plates (Magnani, 1987).

Chemical relatedness among the polysaccharides of bacteria does not necessarily reflect taxonomic relatedness. Indeed, taxonomically distant organisms can have identical PS. Thus, chemically and immunologically similar homopolymers of 2,8-linked Neu5Ac are found in *N. meningitidis* group B (Jennings, 1983), *E. coli* K1 (Jann & Jann, 1987) as well as in strains of *Moraxella nonliquefaciens* (Bøvre *et al.*, 1983). Intergeneric similarities are also found for other Gram-negative species, e.g. for PS of *E. coli*, *Klebsiella* and *Shigella* strains (Jann & Jann, 1987). The possible biological importance of such similarity in surface chemistry of unrelated bacteria has not been fully explored, but may become of use in studies of bacterial infections or as an aid to vaccine production.

Several techniques exist for the extraction and purification of native PS, the analysis of their chemical composition and for the determination of their detailed structures. These procedures are amply described elsewhere (e.g. Aspinall, 1983; Chaplin & Kennedy, 1986; Hancock & Poxton, 1988a). The emphasis below is founded on a few simple techniques based on whole cells and adapted for practical taxonomic use.

2.7 RAPID WHOLE-CELL METHODS

Chromatographic profiles based on bacterial whole-cells are already well established and as such represent a realistic means of directly utilizing the taxonomic information obtainable from the chemical composition of the cell envelope. Two main types of analysis are evident. The first of these provides a rapid characterization of bacteria after brief cultivation. Using this approach the complex chromatographic profiles obtained by hydrolysis (or methanolysis) and derivatization of whole cells serve as 'chemical fingerprints'. The second approach is a taxonomically more complex challenge and involves the direct detection and identification of bacterial cells in clinical or environmental samples. The use of this latter approach necessitates the detection of one or more bacterial marker substances against a large background signal. In spite of the inherent practical problems, such a direct approach is of obvious potential in routine bacterial diagnostic work.

The successful establishment of such methods in diagnostic microbiology has as a prerequisite a reliable and reproducible database derived from carefully performed chemotaxonomic studies. These must, in the first instance, be based on purified bacterial compounds isolated from strains genetically representative of the taxa in question.

2.7.1 GLC profiles of whole-cell fatty acids

Most of the taxonomic data on bacterial lipids relate to the fatty acid composition of the organisms (see Chapter 5). Unfortunately, however, several publications are incomplete with respect to the firmly bound amide-linked constituents such as the 3-hydroxy-fatty acids of LPS. These are taxonomic entities of high information content that require strong acidic conditions for liberation. Therefore they are likely to be totally or partly missed when alkaline methanolysis or hydrolysis are used alone for fatty acid liberation. Cyclopropane-substituted fatty acids, on the other hand, are degraded under acidic conditions (Jantzen et al., 1978), so that depending on the taxon under study, both acidic and alkaline conditions may have to be used; this can be effectively achieved using a recently introduced single tube method (Mayberry & Lane, 1993).

Gram-negative bacteria contain one or more of the following categories of fatty acids: straight-chain, monounsaturated, cyclopropane substituted, 2- and 3-hydroxylated and methyl-branched (*iso* and *anteiso*) fatty acids. Diunsaturated fatty acids reported to be present in certain species are likely to have been taken up from the growth medium. Two substitutions on the same fatty acid (e.g. methyl-branched hydroxy fatty acids) also occur but are less common. Usually only a few of the main categories of fatty acids predominate within a taxon, and a simple recording of fatty acid types can often be of value as a preliminary step in diagnosis (Jantzen & Bryn, 1985). For detailed taxonomic studies, numerical analysis of the fatty acid composition can prove useful and may highlight important taxonomic relationships (see Chapter 5; O'Donnell, 1985; Jantzen et al., 1987, 1993).

Method 2.7 GLC profiles of whole-cell fatty acids (modified from Jantzen & Bryn, 1985).

1. Bacterial cells (e.g. two or three loopfuls directly from the agar plate, or about 500 μg freeze-dried material) in a Teflon-lined screw-cap tube are heated overnight at 85 °C in methanolic 2 M HCl (1 ml; see Method 2.6). Inspect for leakage and re-tighten cap after 30 min.

2. Concentrate the sample to approximately 1/2 volume by using N_2 to reduce the HCl concentration.
3. Add two vols half-saturated NaCl and three vols hexane. Vortex and separate the phases by centrifugation.
4. The (top) hexane phase is transferred to another tube, and the hexane extraction is repeated.
5. The combined hexane phases are taken just to dryness with N_2.
6. Add TFAA (50% in acetonitrile; 250 µl), heat for 2 min with a hair dryer, concentrate just to dryness with N_2 and dissolve in 30 µl TFAA (10% in acetonitrile).
7. Analyse the sample by capillary GLC (25 × 0.2 mm fused-silica coated with SE-30, methylsilicone) using a temperature gradient of 70 °C (2 min) → (20 °/min) → 130 °C → (8 °/min) → 280 °C.

2.7.2 GLC profiles of whole-cell sugars

Compared to the GLC profiles of fatty acids, the corresponding profiles of whole-cell sugars have only been used to a limited extent in systematics. An important reason for this might be that the well-established serological techniques are based on polysaccharides which provide more specific information than that obtained from total sugar composition. However, direct chemical analysis of sugar composition may be especially useful for differentiating within taxonomic groups without an established serology and where other diagnostic tests are limited. For example, the two major pathogens of the genus *Legionella*, *L. micdadei* and *L. pneumophila*, are readily differentiated by their methylpentose and aminodideoxyhexose content (Walla et al., 1984) as well as the occurrence of the branched-chain octose, yersinose A, in *L. micdadei* (Sonesson & Jantzen, 1992). Sugar profiles are also useful in differentiating between certain species of *Moraxella* and *Neisseria* (Jantzen et al., 1976).

2.7.3 Combined GLC profiles of monosaccharides and fatty acids

When the technique of methanolysis followed by TFA-derivatization and GLC (see above) is applied to Gram-negative whole-cell preparations, the profiles mainly comprise peaks of methylglycosides and fatty acid methyl esters (Jantzen et al., 1978). The former arise from RNA, capsular PS and LPS, whereas the fatty acid peaks originate primarily from LPS and extractable lipids. Such profiles are rapidly produced since no preliminary extractions are involved. They also have the advantage of combining the taxonomic information from two different classes of compounds, i.e. sugars and fatty acids. Although the profiles can be difficult to interpret due to their complexity, they may still be valuable at a 'fingerprint' level.

2.7.4 Constituents in clinical and environmental samples

Direct detection of bacterial cells or fragments in mammalian tissue or other complex biological material has been shown to be possible using capillary GLC coupled to a mass spectrometer or other sensitive detectors. This approach relies on the knowledge of specific bacterial markers which can be analysed against a chemically very complex, primarily non-bacterial, background. For detection of Gram-negative bacteria, the 3-hydroxy fatty acids originating from LPS are of definite potential. Thus, LPS has been used for monitoring microbial contamination in industrial fermentation processes (Elmroth et al., 1990, 1992), industrial products (Maitra et al., 1986) and in environmental samples (Sonesson et al., 1988). Recently, the amount of LPS in sera was determined directly from patients suffering from systemic meningococcal disease using a GLC-MS assay, with 3-OH-12:0 as a specific marker (Figure 2.8; Brandtzaeg et al., 1992). A detection level for LPS of 1 pg was indicated, and for five patients the very high values in the Limulus assay corresponded to the amount of LPS found by GLC-MS. However, a low but distinct background level of 3-OH-fatty acids of unknown origin in human plasma may complicate the interpretation of measurements of more moderate LPS levels.

Several sugars and amino acids that do not occur in mammalian tissue, e.g. D-alanine, diaminopimelic acid, heptoses, Kdo, muramic acid, and L-rhamnose, have also been suggested as bacterial markers (Sonesson

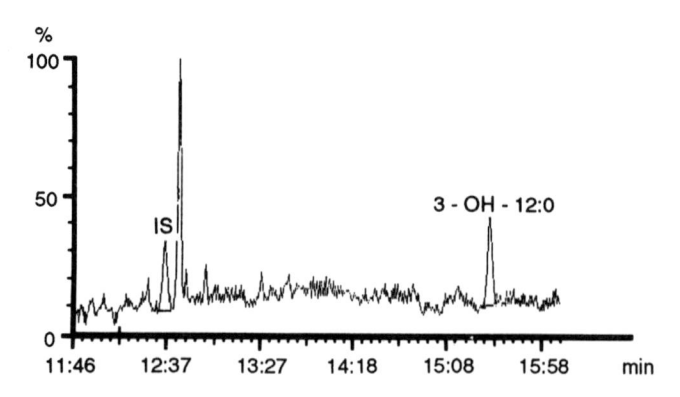

Figure 2.8 GLC-MS profile of plasma from a patient with fulminant meningococcal septicaemia. Ester-linked 3-OH-12:0 was released from LPS by methanolysis (2 M HCl, 85 °C, 2 h) extracted into hexane and enriched by solid phase extraction. Trimethylsilyl ethers were analysed by GLC (25 m SE-30 column)-MS (ion-trap; electron-impact; selected ion: m/z 287 (M-15)). From Brandtzaeg et al. (1992) by copyright permission of the American Society for Clinical Investigation.

2. Analysis of Gram-negative Cell Constituents

et al., 1988; Larsson et al., 1989; Morgan et al., 1989; Elmroth et al., 1992), but their true value in clinical diagnosis awaits further investigations.

As mentioned above, a major problem in the direct *in situ* analysis of bacterial markers is the high background of chemically related structures. Usually it is necessary to do a clean-up consisting of several steps, to eliminate or reduce this background noise. Strong hydrolytic conditions should be avoided when liberating carbohydrate or ester-linked fatty acids. Mild conditions leave proteins, peptidoglycan and DNA relatively intact and uncleaved, and they are thus easily removed by centrifugation. Commercial solid-phase extraction columns are efficient in removing interfering substances on the basis of differences in polarity or charge (Christie, 1987). Furthermore, derivatization of hydroxyl groups by a halogen-containing reagent followed by GLC in combination with mass spectrometry (in negative ion detection mode) is apparently an effective way of amplifying signals of particular bacterial constituents (Sonesson et al., 1988). More recently, however, Mielniczuk et al. (1992) showed that the difference in sensitivity between 3-OH-fatty acid methyl esters (3-OH-FAMEs) detected as trimethylsilyl-derivative (in electron impact mode) and pentafluorobenzoyl-derivative (chemical ionization/negative ion mode) was almost negligible, and the simpler trimethylsilyl-derivatization procedure was recommended.

Method 2.8 GLC-MS analysis of the neisserial LPS marker 3-OH-12:0 in clinical samples (modified from Brandtzaeg et al., 1992).

1. Plasma (200 μl) is freeze-dried (1 h) in a 1 ml glass vial fitted with Teflon-lined screw cap. To the residue is added 20 μl of 3-OH-10:0 (0.5 μg/ml in methanol) as internal standard, and 400 μl of 2 M HCl in dry methanol. The mixture is heated at 85 °C for 2 h (to cleave O-ester linkages).
2. The volume is reduced to approximately half by N_2 and the liberated FAMEs extracted twice with hexane (3 vols) after addition of 2 vols of LPS-free water, half-saturated with NaCl.
3. The combined hexane phase is concentrated to approximately 0.2 ml and then applied to a commercial silica extraction cartridge prewashed with 2 vols each of hexane, methylene chloride, and hexane/methylene chloride (1:1).
4. Non-hydroxylated FAMEs are first removed by 3 ml of hexane-methylene chloride (1:1).
5. Hydroxylated FAMEs are then eluted by 3 ml of dry diethyl ether.
6. The solvent is removed by N_2, and the trimethylsilyl (TMS) derivative is formed by the addition of 50 μl of the following reagent: bis-trimethylsilyl-trifluoroacetamide/acetonitrile/pyridine (50:40:10), and heating at 85 °C for 30 min before injection (2 μl) into the GLC-MS instrument.
7. GLC-MS instrument: Quadrupole or Ion-Trap with a (25 m × 0.2 mm) fused-silica SE-30 column, used with electron-impact ionization at 70 eV, and in selective ion mode in the mass range 170–350.

8. Calculations are based on the peaks obtained using the abundant (M-15)$^+$ ions of the TMS-derivatized 3-OH-12:0 and 3-OH-10:0 (internal standard) for detection. A standard curve is constructed on the basis of 3-OH-12:0 in concentrations of 100, 50, 25, and 5 µg/l, respectively, with 50 µg/l of 3-OH-10:0 as internal standard.

Note: The molecular mass of meningococcal LPS varies within certain limits (3.2–5.1 kDa) and a value of 4.0 kDa may be used for calculation of the amount of neisserial LPS in the sample.

2.8 TAXONOMIC CONSIDERATIONS

2.8.1 Envelope chemistry and evolution

The extensive group, the 'Gram-negative bacteria', encompasses a genetically very heterogeneous collection of organisms that occupy almost all kinds of environmental niches. Common to virtually all such organisms is the general cell envelope construction with an outer membrane and a thin peptidoglycan layer (Figure 2.1). Hence, this biologically unique cell assembly has remained remarkably stable throughout evolution.

On the molecular level the peptidoglycan structure also has proved extraordinarily resistant to evolutionary pressure, with only a few known exceptions to the general and simple peptidoglycan structure of Gram-negative organisms represented by *E. coli*. Hence, peptidoglycan structure plays a relatively anonymous role in the taxonomy of Gram-negative bacteria, although it has been suggested that more precise analytical techniques might revise this perception (Glauner, 1988).

The outer membrane constituent LPS is normally present in all Gram-negative bacteria and appears to be a vital structural entity of the cell surface. Its lipid A part, which has been most extensively studied for *Salmonella* LPS, has also been shown to be structurally well conserved. Recently, however, studies of lipid A structures from organisms genetically more distant to *Salmonella* have revealed that lipid A constitutes an interesting chemical entity for use in systematics. This is mainly due to its apparent ability to differentiate between certain phylogenetically defined taxa (see Table 2.2; Mayer *et al.*, 1989).

The more exposed structures of the outer membrane, such as OMPs, the PS part of LPS, and capsular PS, apparently have other roles in the process of bacterial survival. Adaptation to rapid changes in the environment is clearly important, and the bacteria may do this by modulation of the chemical composition of their surface structures. As a result these structures may not necessarily be distributed according to phylogeny, and intraspecific variations may also occur. However, as noted previously, by

2. Analysis of Gram-negative Cell Constituents

carefully standardizing the growth conditions, this group of structures can also provide useful markers for bacterial identification.

The major breakthroughs in genetic techniques have influenced bacterial systematics in recent years. Nucleic acid hybridization and 16S rRNA analyses contribute significantly to prokaryotic systematics. Studies of 16S rRNA sequences even reveal evolutionary trends (Woese, 1987), and enable variations in envelope composition to be compared directly with the phylogeny of the organisms. For example, Mayer, Weckesser and their co-workers (Weckesser & Mayer, 1988; Mayer et al., 1989) have examined the LPS composition of several photosynthetic and other bacteria and demonstrated congruence between lipid A types and the evolutionary branching points established by 16S rRNA sequences (Woese, 1987).

With DNA-RNA hybridization as the central technique, De Ley and co-workers have studied several taxa with unclear systematics (Willems et al., 1989, 1990). Several of these comprehensive studies included chemical data, e.g. protein profiles and cellular fatty acid composition. Together these data provided an improved classification of many problematic taxa such as the *Pseudomonas*-affiliated and misnamed *Pseudomonas* organisms. These studies also confirmed that fatty acid and protein patterns correlated well with relationships at a genetic level.

2.8.2 Chemotaxonomic strategy

The analysis of a single chemical constituent can often be of limited taxonomic value and the experienced chemotaxonomist must carefully judge which analyses are most relevant to the bacterial group under investigation. As a general rule, however, the success of any chemotaxonomic procedure almost always depends on factors related to the chemical methodology and to the chemical and biological stability of the constituents being analysed. Ideally one should know the markers in detail with respect to their chemistry, biosynthesis and function, and if possible their quantitative distribution between and within taxa. Possible effects of variation in cell chemistry due to growth conditions must always be considered. Usually these prerequisites are only partially fulfilled, but nevertheless unique and useful taxonomic information can often be obtained. Some important considerations for evaluating chemotaxonomic methods and the results obtained are listed below.

(a) The chemical procedure should be reliable and reproducible, i.e. not liable to fail because of chemical instability, complicated procedures, variable release or degradation of the compound during analysis.
(b) The bacterial constituent should occur consistently in the taxon under

consideration. Alternatively, its biological variation should be established.

(c) The analysed bacterial compounds should be sufficiently good markers to differentiate the taxon from other relevant taxa. This may require the application of several taxonomic methods.

(d) For direct chemical analysis of biological samples (without culturing of bacteria) the sensitivity and signal/noise ratio of the procedure should be carefully considered.

(e) Interference from taxonomically unrelated bacterial constitutents should be absent (or known). Such interference may lead to false positive reactions due to chemical similarity, or to false negative results arising from factors suppressing the chemical reaction.

(f) As for non-chemical procedures, variation due to factors such as strain differences, mutational changes, growth characteristics, and choice of growth media should not be overlooked. Careful standardization is of advantage in reducing the influence of environmental factors.

(g) Quantitative differences in marker composition may be difficult to interpret reliably for some constituents, and careful consideration of relevant neighbouring taxa is of special importance. Automated data acquisition may greatly improve the handling of such variability.

2.8.3 Perspectives

Future taxonomic work, such as screening of strains for specific constituents, identification of bacterial isolates, or creation of new taxonomic units including those of higher ranks such as genera and families, will undoubtedly be based on, and strengthened by, molecular systematic studies based on nucleic acid data. Nevertheless, chemical analysis of envelope constituents will continue to provide simpler alternatives for classification and identification.

Until simple and reliable molecular genetic procedures of practical use are available for systematics there will remain a need to improve chemical procedures, to map chemical variation between taxa, to simplify data retrieval and acquisition analysis, as well as to improve interlaboratory communication. These measures will also inevitably improve taxonomic methodology in all areas.

REFERENCES

Acker, G., Wartenberg, K. and Knapp, W. (1980) Zuckerzusammensetzung des Lipopolysaccharids und Feinstruktur der äusseren Membran (Zellwand) bei *Yersinia enterocolitica*. Zbl. Bakt. Hyg., I. Abt. Orig. A **247**: 229–40.

2. Analysis of Gram-negative Cell Constituents

Arata, S., Hirayama, T., Kasai, N., Itoh, T. and Ohsawa, A. (1989) Isolation of 9-hydroxy-delta-tetradecalactone from lipid A of *Pseudomonas diminuta* and *Pseudomonas vesicularis*. *FEMS Microbiol. Lett.* **60**: 219–22.

Armstrong, S.K., Parr., T.R. Jr., Parker, D.C. and Hancock, R.E.W. (1986) *Bordetella pertussis* major outer membrane porin protein forms small, anion-selective channels in lipid bilayer membranes. *J. Bacteriol.* **166**: 212–6.

Aspinal, G.O. (1983) *The Polysaccharides*. Academic Press: New York.

Asselineau, J. (1966) *The Bacterial Lipids*. Hermann: Paris.

Benz, R. (1988) Structure and function of porins from Gram-negative bacteria. *Ann. Rev. Microbiol.* **42**: 359–93.

Bhat, U.R., Carlson, R.W., Busch, M. and Mayer, H. (1991) Distribution and phylogenetic significance of 27-hydroxyoctacosanoic acid in lipopolysaccharides from bacteria belonging to the alpha-2 subgroup of *Proteobacteria*. *Intern. J. Syst. Bacteriol.* **41**: 213–7.

Borneleit, P., Blechschmidt, B., Blasig, R., Franke, P., Günther, P. and Kleber, H.-P. (1989) Preparation of R-type lipopolysaccharides of *Acinetobacter calcoaceticus* by EDTA-salt extraction. *Current Microbiol.* **19**: 77–81.

Bøvre, K., Bryn, K., Closs, O., Hagen, N. and Frøholm, L.O. (1983) Surface polysaccharide of *Moraxella nonliquefaciens* identical to *Neisseria meningitidis* group B capsular polysaccharide. *NIPH Annals (Oslo)* **6**: 65–73.

Brade, H. (1985) Occurrence of 2-keto-deoxyoctonic acid 5-phosphate in lipopolysaccharides of *Vibrio cholerae* Ogawa and Inaba. *J. Bacterial.* **161**: 795–8.

Brade, L. and Brade, H. (1985) Characterization of two different antibody specificities recognizing distinct antigenic determinants in free lipid A of *Escherichia coli*. *Infect. Immun.* **48**: 776–81.

Brandtzaeg, P., Bryn, K., Kierulf, P., Øvstebø, R., Namork, E., Aase, B. and Jantzen, E. (1992) Meningococcal endotoxin in lethal septic shock plasma studied by gas chromatography, mass spectrometry, ultracentrifugation, and electron microscopy. *J. Clin. Invest.* **89**: 816–23.

Braun, V. (1973) Molecular organization of the rigid layer and the cell wall of *Escherichia coli*. *J. Infect. Dis.* **128**: S9–S16.

Bryn, K. and Jantzen, E. (1982) Analysis of lipopolysaccharides by methanolysis, trifluoroacetylation, and gas chromatography on a fused-silica capillary column. *J. Chromatogr.* **240**: 405–13.

Bryn, K. and Jantzen, E. (1986) Quantification of 2-keto-3-deoxy-octonate in (lipo)polysaccharides by methanolytic release, trifluoroacetylation, and capillary gas chromatography. *J. Chromatogr.* **370**: 103–12.

Bryn, K., Jantzen, E. and Bøvre, K. (1977) Occurrence and patterns of waxes in *Neisseriaceae*. *J. Gen. Microbiol.* **102**: 33–43.

Burridge, K. (1978) Direct identification of specific glycoproteins and antigens in sodium dodecyl sulfate gels. In *Methods in Enzymology*, vol. 50 (Ginsburg, V., ed.), pp. 54–64. Academic Press: New York.

Buttke, T.M. and Ingram, L.O. (1975) Comparison of lipopolysaccharides from *Agmenellum quadruplicatum* to *Escherichia coli* and *Salmonella typhimurium* by using thin-layer chromatography. *J. Bacteriol.* **124**: 1566–70.

Caroff, M., Bundle, D.R. and Perry, M.B. (1984) Structure of the O-chain of the phenol-phase soluble cellular lipopolysaccharide of *Yersinia enterocolitica* serotype O:9. *Eur. J. Biochem.* **139**: 195–200.

Caroff, M., Lebbar, S. and Szabo, L. (1987a) Detection of 3-deoxy-2-octulosonic acid in thiobarbiturate-negative endotoxins. *Carbohydr. Res.* **161**: 4–7.

Caroff, M., Lebbar, S. and Szabo, L. (1987b) Do endotoxins devoid of 3-deoxy-D-*manno*-2-octulosonic acid exist? *Biochem. Biophys. Res. Comm.* **143**: 845–7.

Chaplin, M.F. and Kennedy, J.F. (eds) (1986) *Carbohydrate Analysis: A Practical Approach*. IRL Press: Oxford.

Chen, C.-L., Chang, C.-M., Nowotny, A.M. and Nowotny, A. (1975) Rapid biological and chemical analyses of bacterial endotoxins separated by preparative thin-layer chromatography. *Anal. Biochem.* **63**: 183–94.

Christie, W.W. (1982) *Lipid Analysis: Isolation, Separation, Identification and Structural Analysis of Lipids*. Pergamon Press: Oxford.

Christie, W.W. (1987) *High-Performance Liquid Chromatography and Lipids*. Pergamon Press: Oxford.
Darveau, R.P. and Hancock, R.E.W. (1983) Procedure for isolation of bacterial lipopolysaccharides from both smooth and rough *Pseudomonas aeruginosa* and *Salmonella typhimurium* strains. *J. Bacteriol.* **155**: 831–8.
Di Rienzo, J.M., Nakamura, K. and Inouye, M. (1978) The outer membrane proteins of Gram-negative bacteria: biosynthesis, assembly and functions. *Ann. Rev. Biochem.* **47**: 481–532.
Downs, T.R. and Wilfinger, W.W. (1983) Fluorometric quantification of DNA in cells and tissue. *Anal. Biochem.* **131**: 538–47.
Elmroth, I., Valeur, A., Odham, G. and Larsson, L. (1990) Detection of microbial contamination in fermentation processes. *Biotechnol. Bioeng.* **35**: 787–92.
Elmroth, I., Sundin, P., Valeur, A., Larsson, L. and Odham, G. (1992) Evaluation of chromatographic methods for the detection of bacterial contamination in biotechnical processes. *J. Microbiol. Methods* **15**: 215–28.
Erwin, A.L. and Munford, R.S. (1989) Comparison of lipopolysaccharides from Brazilian purpuric fever isolates and conjunctivitis isolates of *Haemophilus influenzae* biogroup *aegyptius*. Brazilian Purpuric Fever Study Group. *J. Clin. Microbiol.* **27**: 762–7.
Fredriksen, J.H., Rosenqvist, E., Wedege, E., Bryn, K., Bjune, G., Frøholm, L.O., Lindbak, A.-K., Møgster, B., Namork, E., Rye, U., Stabbetorp, G., Winsnes, R., Aase, B. and Closs, O. (1991) Production, characterization and control of MenB-vaccine 'Folkehelsa': an outer membrane vesicle vaccine against group B meningococcal disease. *NIPH Annals* **14**: 67–80.
Fromme, I. and Beilharz, H. (1978) Gas chromatographic assay of total and O-acetyl groups in bacterial lipopolysaccharides. *Anal. Biochem.* **84**: 347–53.
Galanos, C. and Lüderitz, O. (1975) Electrodialysis of lipopolysaccharides and their conversion to uniform salt forms. *Eur. J. Biochem* **54**: 603–10.
Galanos, C., Lüderitz, O. and Westphal, O. (1969) A new method for the extraction of R lipopolysaccharides. *Eur. J. Biochem.* **9**: 245–9.
Garcia-Bustos, J.F., Chait, B.T. and Tomasz, A. (1987) Structure of the peptide network of pneumococcal peptidoglycan. *J. Biol. Chem.* **262**: 15400–5.
Gilbert, J., Harrison, J., Parks, C. and Fox, A. (1988) Analysis of the amino acid and sugar composition of streptococcal cell walls by gas chromatography-mass spectrometry. *J. Chromatogr.* **441**: 323–33.
Glauner, B. (1988) Separation and quantification of muropeptides with high-performance liquid chromatography. *Anal. Biochem.* **172**: 451–64.
Godchaux, W. III and Leadbetter, E.R. (1988) Sulfonolipids are localized in the outer membrane of the gliding bacterium *Cytophaga johnsonae*. *Arch. Microbiol.* **150**: 42–7.
Gotschlich, E.C., Fraser, B.A., Nishimura, O., Robbins, J.B. and Liu, T.Y. (1981) Lipid on capsular polysaccharides of Gram-negative bacteria. *J. Biol. Chem.* **256**: 8915–21.
Hancock, I. and Poxton, I. (eds) (1988a) *Bacterial Surface Techniques*. John Wiley & Sons: Chichester.
Hancock, I. and Poxton, I. (1988b) Separation and purification of surface components. In *Bacterial Surface Techniques* (Hancock, I. and Poxton, I., eds), pp. 67–135. John Wiley & Sons: Chichester.
Hancock, R.E.W. (1987) Role of porins in outer membrane permeability. *J. Bacteriol.* **169**: 929–33.
Hitchcock, P.J., Leive, L., Mäkelä, H., Rietschel, E.T., Strittmatter, W. and Morrison, D.C. (1986) Lipopolysaccharide nomenclature—past, present, and future. *J. Bacteriol.* **166**: 699–705.
Hollingsworth, R.I. and Carlson, R.W. (1989) 27-Hydroxyoctacosanoic acid is a major structural fatty acyl component of the lipopolysaccharide of *Rhizobium trifolii* ANU 843. *J. Biol. Chem.* **264**: 9300–3.
Hollingsworth, R.I. and Lill-Elghanian, D.A. (1989) Isolation and characterization of the unusual lipopolysaccharide component 2-amino-2-deoxy-2-N-(27-hydroxyoctacosanoyl)-3-O-(3-hydroxytetradecanol)-gluco-hexuronic acid and its de-O-acylation product from the free lipid A of *Rhizobium trifolii* ANU 843. *J. Biol. Chem.* **264**: 14039–42.
Inouye, M. (1979) *Bacterial Outer Membranes: Biogenesis and Functions*. John Wiley & Sons: New York.

Jann, B., Reske, K. and Jann, K. (1975) Heterogeneity of lipopolysaccharides. Analysis of polysaccharide chain lengths by sodium dodecylsulfate-polyacrylamide gel electrophoresis. *Eur. J. Biochem.* **60**: 239–46.

Jann, K. and Jann, B. (1977) Bacterial polysaccharide antigens. In *Surface Polysaccharides of the Procaryotic Cell* (Sutherland, I. ed), pp. 247–87. Academic Press: London.

Jann, K. and Jann, B. (1987) Polysaccharide antigens of *Escherichia coli*. *Rev. Infect. Dis.* **9**: S517–26.

Jantzen, E. and Bryn, K. (1985) Whole-cell and lipopolysaccharide fatty acids and sugars of Gram-negative bacteria. In *Chemical Methods in Bacterial Systematics* (Goodfellow, M. and Minnikin, D. eds, pp. 145–71. Academic Press: London.

Jantzen, E., Bryn, K. and Bøvre, K. (1976) Monosaccharide patterns of *Neisseriaceae*. *Acta Path. Microbial. Scand.* **84B**: 177–88.

Jantzen, E., Bryn, K., Hagen, N., Bergan, T. and Bøvre, K. (1978) Fatty acid and monosaccharides of *Neisseriaceae* in relation to established taxonomy. *NIPH Annals (Oslo)* **1**: 59–71.

Jantzen, E., Kvalheim, O.M., Hauge, T.A., Hagen, N. and Bøvre, K. (1987) Grouping of bacteria by SIMCA pattern recognition on gas chromatographic lipid data: patterns among *Moraxella* and rod-shaped *Neisseria*. *Syst. Appl. Microbiol.* **9**: 142–50.

Jantzen, E., Sonesson, A., Tangen, T. and Eng, J. (1993) Hydroxy-fatty acid profiles of *Legionella*: diagnostic usefulness assessed by SIMCA principal component analysis. *J. Clin. Microbiol.* **13**: 1413–19.

Jennings, H.J. (1983) Capsular polysaccharides as human vaccines. *Adv. in Carbohydr. Chem.* **41**: 155–208.

Jennings, H.J. and Johnson, K.G. (1983) The structure of an R-type oligosaccharide core obtained from some lipopolysaccharides of *Neisseria meningitidis*. *Carbohydr. Res.* **121**: 233–41.

Jiao, B.H., Freudenberg, M. and Galanos, C. (1989) Characterization of the lipid A component of genuine smooth-form lipopolysaccharide. *Eur. J. Biochem.* **180**: 515–8.

Kasai, N., Arata, S., Jun-ichi, M., Akiyama, Y., Tanaka, C., Egawa, K. and Tanaka, S. (1987) *Pseudomonas diminuta* LPS with a new endotoxic lipid A structure. *Biochem. Biophys. Res. Comm.* **142**: 972–8.

Kato, K., Umemoto, T., Fukuhara, H., Sagawa, H. and Kotani, S. (1979) Variation in dibasic amino acid in the cell wall peptidoglycan of *Fusobacterium nucleatum*. *Current Microbiol.* **3**: 147–52.

Kawai, J. and Moribayashi, A. (1982) Characteristic lipids of *Bordetella pertussis*: simple fatty acid composition, hydroxy fatty acids, and an ornithine-containing lipid. *J. Bacteriol.* **511**: 996–1005.

Kenne, L. and Lindberg, B. (1983) Bacterial polysaccharides. In *The Polysaccharides*, vol. 2 (Aspinall, G.O., ed), pp. 287–363. Academic Press: New York.

Koeltzow, D.E. and Conrad, H.E. (1971) Structural heterogeneity in the lipopolysaccharide of *Aerobacter aerogenes* NCTC 243. *Biochem.* **10**: 214–24.

Koerner, T.A.W., Prestegard, J.H. and Yu, R.K. (1987) Oligosaccharide structure by two-dimensional proton nuclear magnetic resonance spectroscopy. In *Methods in Enzymology*, vol. 138 (Ginsburg, V., ed), pp. 38–59. Academic Press: Orlando.

Komuro, T. and Galanos, C. (1988) Analysis of *Salmonella* lipopolysaccharides by sodium deoxycholate-polyacrylamide gel electrophoresis. *J. Chromatogr.* **450**: 381–7.

Krauss, J.H., Reuter, G., Schauer, R., Weckesser, J. and Mayer, H. (1988) Sialic acid-containing lipopolysaccharides in purple nonsulfur bacteria. *Arch. Microbiol.* **150**: 584–9.

Kristiansen, B.-E., Sørensen, B., Bjorvatn, B., Falk, E., Fosse, E., Bryn, K., Frøholm, L.O., Gaustad, P. and Bøvre, K. (1986) An outbreak of group B meningococcal disease: tracing the causative strain of *Neisseria meningitidis* by DNA fingerprinting. *J. Clin. Microbiol.* **23**: 764–7.

Kulshin, V.A., Zähringer, U., Lindner, B., Jäger, K.E., Dmitriev, B.A. and Rietschel, E.T. (1991) Structural characterization of *Pseudomonas aeruginosa* wild-type and rough mutant lipopolysaccharides. *Eur. J. Biochem.* **198**: 697–704.

Lakey, J.H., Watts, J.P. and Lea, E.J.A. (1985) Characterization of channels induced in planar bilayer membranes by detergent solubilized *Escherichia coli* porins. *Biochim. Biophys. Acta* **817**: 208–16.

Lambert, P.A. (1988) Isolation and purification of outer membrane proteins from Gram-negative bacteria. In *Bacterial Cell Surface Techniques* (Hancock, I. and Poxton, I., eds), pp. 110–25. John Wiley & Sons: Chichester.

Larsson, L., Sonesson, A., Jantzen, E. and Bryn, K. (1989) Detection and identification of bacteria using gas chromatographic analysis of cellular fatty acids. In *Rapid Methods and Automation in Microbiology and Immunology* (Balows, A., Tilton, R.C. and Turano, A., eds), pp. 389–95. Brixia Academic Press: Brescia, Italy.

Lechevalier, H. and Lechevalier, M.P. (1988) Chemotaxonomic use of lipids—an overview. In *Microbial Lipids*, vol. I (Ratledge, C. and Wilkinson, S.G., eds), pp. 869–902. Academic Press: London.

Lechevalier, M.P. (1977) Lipids in bacterial taxonomy—a taxonomist's viewpoint. *Crit. Rev. Microbiol.* **5**: 109–210.

Lee, C.-J. (1987) Bacterial capsular polysaccharides—biochemistry, immunity and vaccine. *Molec. Immunol.* **24**: 1005–19.

Lee, Y.C. (1990) High-performance anion-exchange chromatography for carbohydrate analysis. *Anal. Biochem.* **189**: 151–62.

Leive, L. and Morrison, D.C. (1972) Isolation of lipopolysaccharides from bacteria. In *Methods in Enzymology*, vol. 28 (Ginsburg, V., ed). pp. 254–62. Academic Press: New York.

Leloir, L.F. and Cardini, C.E. (1957) Characterization of phosphorus compounds by acid lability. In *Methods in Enzymology*, vol. 3 (Colowick, S.P. and Kaplan, N.O., eds), pp. 840–50. Academic Press: New York.

Lindberg, B. (1990) Components of bacterial polysaccharides. *Adv. Carbohydr.* **48**: 279–318.

Lönngren, J. and Svensson, S. (1974) Mass spectrometry in structural studies of natural carbohydrates: *Adv. Carbohydr. Chem.* **29**: 41–106.

Lowry, O.H., Rosebrough, A.L., Farr, A.L. and Randall, R.J. (1951) Protein measurement with the Folin phenol reagent. *J. Biol. Chem.* **193**: 265–75.

Lüderitz, O., Westphal, O., Staub, A.M. and Nikaido, N. (1971) Isolation and chemical and immunological characterization of bacterial lipopolysaccharides. In *Microbial Toxins*, vol. 4 (Weinbaum, G., Kadis, S. and Ajl, S.J., eds), pp. 145–233. Academic Press: London.

Lugtenberg, B. and van Alphen, L. (1983) Molecular architecture and functioning of the outer membrane of *Escherichia coli* and other Gram-negative bacteria. *Biochim. Biophys. Acta* **737**: 51–115.

Lugtenberg, B., Meyers, J., Peters, R., Van der Hoek, P. and Van Alphen, L. (1975) Electrophoretic resolution of the 'major outer membrane protein' of *Escherichia coli* K-12 into four bands. *FEBS Lett.* **58**: 254–8.

Magnani, J.L. (1987) Immunostaining free oligosaccharides directly on thin-layer chromatography. In *Methods in Enzymology*, vol. 138 (Ginsburg, V., ed), pp. 208–12. Academic Press: Orlando.

Maitra, S.K., Nachum, R. and Pearson, F.C. (1986) Establishment of β-hydroxy fatty acids as chemical marker molecules for bacterial endotoxin by gas chromatography-mass spectrometry. *Appl. Environm. Microbiol.* **52**: 510–14.

Masson, L., Holbein, B.E. and Ashton, F.E. (1982) Virulence linked to polysaccharide production in serogroup B *Neisseria meningitidis*. *FEMS Microbiol. Lett.* **13**: 187–90.

Matsuyama, T., Kaneda, K., Ishizuka, I., Toida, T. and Yano, I. (1990) Surface-active novel glycolipid and linked 3-hydroxy fatty acids produced by *Serratia rubidaea*. *J. Bacteriol.* **172**: 3015–22.

Mattsby-Baltzer, I. and Alving, C.R. (1984) Lipid A fractions analysed by a technique involving thin-layer chromatography and enzyme-linked immunosorbent assay. *Eur. J. Biochem.* **138**: 333–7.

Mayberry, W.R. and Lane, J.R. (1993) Sequential alkaline saponification/acid hydrolysis/esterfication: a one-tube method with enhanced recovery of both cyclopropane and hydroxylated fatty acids. *J. Microbiol. Methods* **18**: 21–32.

Mayer, H. and Weckesser, J. (1984) Unusual lipid A's: structure, taxonomic relevance and potential value for endotoxin research. In *Handbook of Endotoxin*, vol. 1 (Rietschel, E.T., ed), pp. 221–47. Elsevier Science Publishers: Amsterdam.

Mayer, H., Tharanathan, R.N. and Weckesser, J. (1985) Analysis of lipopolysaccharides of Gram-negative bacteria. In *Methods in Microbiology*, vol. 18 (Gottschalk, G., ed), pp. 157–207. Academic Press: London.

2. Analysis of Gram-negative Cell Constituents

Mayer, H., Ramadas Bhat, U., Masoud, H., Radziejewska-Lebrecht, J., Widemann, C. and Krauss, J.H. (1989) Bacterial lipopolysaccharides. *Pure and Appl. Chem.* **61**: 1271–82.

Michon, F., Beurret, M., Gamian, A., Brisson, J.-R. and Jennings, H.J. (1990) Structure of the L5 lipopolysaccharide core oligosaccharides of *Neisseria meningitidis*. *J. Biol. Chem.* **265**: 7243–47.

Mielniczuk, Z., Alugupalli, S., Mielniczuk, E. and Larsson, L. (1992) Gas chromatography-mass spectrometry of lipopolysaccharide 3-hydroxy fatty acids: comparison of pentafluorobenzoyl and trimethylsilyl methyl ester derivatives. *J. Chromatogr.* **623**: 115–22.

Moll, H., Sonesson, A., Jantzen E., Marre, R. and Zähringer, U. (1992) Identification of 27-oxo-octacosanoic acid and heptacosane-1,27-dioic acid in *Legionella pneumophila*. *FEMS Microbiol. Lett.* **97**: 1–6.

Moreno, E., Stackebrandt, E., Dorsch, M., Wolters, J., Busch, M. and Mayer, H. (1990) *Brucella abortus* 16S rRNA and Lipid A reveal a phylogenetic relationship with members of the alpha-2 subdivision of the class *Proteobacteria*. *J. Bacteriol.* **172**: 3569–76.

Morgan, S.L., Fox, A. and Gilbert, J. (1989) Profiling, structural characterization, and trace detection of chemical markers for microorganisms by gas chromatography-mass spectrometry. *J. Microbiol. Methods* **9**: 57–69.

Moule, A.L. and Wilkinson, S.G. (1989) Composition of lipopolysaccharides from *Alteromonas putrefaciens* (*Shewanella putrefaciens*). *J. Gen. Microbiol.* **135**: 163–73.

Nikaido, H. and Vaara, M. (1985) Molecular basis of bacterial outer membrane permeability. *Microbiol. Rev.* **49**: 1–32.

Nurminen, M., Rietschel, E.T. and Brade, H. (1985) Chemical characterization of *Chlamydia trachomatis* lipopolysaccharide. *Infect. Immun.* **48**: 573–5.

O'Donnell, A.G. (1985) Numerical analysis of chemotaxonomic data. In *Computer-Assisted Bacterial Systematics* (Goodfellow, M., Jones, D. and Priest, F.G., eds), pp. 403–14. Academic Press: London.

Palva, E.T. and Mäkelä, P.H. (1980) Lipopolysaccharide heterogeneity in *Salmonella typhimurium* analysed by sodium dodecyl sulfate/polyacrylamide gel electrophoresis. *Eur. J. Biochem.* **107**: 137–43.

Patouraux-Promé, D. and Promé, J.-C. (1984) Carbohydrates. In *Gas Chromatography Mass Spectrometry Applications in Microbiology* (Odham, G., Larsson, L. and Mårdh, P.-E., eds) pp. 105–56. Plenum Press: New York.

Pelkonen, S., Häyrinen, J. and Finne, J. (1988) Polyacrylamide gel electrophoresis of the capsular polysaccharides of *Escherichia coli* K1 and other bacteria. *J. Bacteriol.* **170**: 2646–53.

Perlin, A.S. and Casu, B. (1982) Spectroscopy methods. In *The Polysaccharides*, vol. 1 (Aspinal, G.O., ed), pp. 133-93. Academic Press: London.

Phillips, N.J., John, C.M., Reinders, L.G. and Gibson, B.W. (1990) Structural models for the cell surface lipooligosaccharides of *Neisseria gonorrhoeae* and *Haemophilus influenzae*. *Biomed. Environm. Mass Spectrom.* **19**: 731–45.

Pietsch, K., Weckesser, J., Fischer, U. and Mayer, H. (1990) The lipopolysaccharides of *Rhodospirillum rubrum*, *Rhodospirillum molischianum* and *Rhodopila globiformis*. *Arch. Microbiol.* **154**: 433–7.

Pyle, S.W. and Schill, W.B. (1985) Rapid serological analysis of bacterial LPS by electrotransfer to nitrocellulose. *J. Immunol. Meth.* **85**: 371–82.

Ratledge, C. and Wilkinson. S.G. (1988) *Microbial Lipids*, vol. 1. Academic Press: London.

Reinhold, N. (1987) Direct chemical ionization mass spectrometry. In *Methods in Enzymology*, vol. 138 (Ginsburg, V., ed), pp. 59–84. Academic Press: Orlando.

Rietschel, E.T., Wollenweber, H.-W., Brade, H., Zähringer, U., Lindner, B., Seydel, U., Bradaczek, H., Barnickel, G., Labischinski, H. and Giesbrecht, P. (1984) Structure and conformation of the lipid A component of lipopolysaccharides. In *Handbook of Endotoxin*, vol. 1 (Rietschel, E.T., ed), pp. 187–220. Elsevier: Amsterdam.

Rietschel, E.Th., Brade, L., Schade, U., Seydel, U., Zähringer, U., Kusumoto, S. and Brade, H. (1988) Bacterial endotoxins: properties and structure of biologically active domains. In *Surface Structures of Microorganisms and Their Interaction with Mammalian Host* (Schrinner, E., Richmond, M.H., Seibert, G. and Schwartz, U., eds), pp. 1–41. VCH Publishers: Weinheim.

Rose, A.H. (1989) Influence of the environment on microbial lipid composition. In *Microbial Lipids*, vol. 2 (Ratledge, C. and Wilkinson, S.G., eds), pp. 255–78. Academic Press: London.

Samu, J., Kovats, E., Nguyen, V., Keler, T. and Nowotny, A. (1988) Thin-layer chromatography of endotoxins, their derivatives and contaminants. *J. Chromatogr.* **435**: 167–183.

Schleifer, K.H. and Joseph, R. (1973) A directly cross-linked L-ornithine-containing peptidoglycan in cell walls of *Spirochaeta stenostrepta*. *FEBS Lett.* **36**: 83–6.

Schleifer, K.H. and Kandler, O. (1972) The peptidoglycan types of bacterial cell walls and their taxonomic implications. *Bacteriol. Rev.* **36**: 407–77.

Schleifer, K.H. and Seidl, P.H. (1985) Chemical composition and structure of murein. In *Chemical Methods in Bacterial Systematics* (Goodfellow, M. and Minnikin, D.E., eds), pp. 201–19. Academic Press: London.

Schmidt, M.A. and Jann, K. (1982) Phospholipid substitution of capsular (K) polysaccharide antigens from *Escherichia coli* causing extraintestinal infections. *FEMS Microbiol. Lett.* **14**: 501–10.

Schneider, H., Hale, T.L., Zollinger, W.S., Seid, R.C. Jr., Hammack, C.A. and Griffiss, J.M. (1984) Heterogeneity of molecular size and antigenic expression of lipooligosaccharides of individual strains of *Neisseria gonorrhoeae* and *Neisseria meningitidis*. *Infect. Immun.* **45**: 544–9.

Shaw, N. (1974) Lipid composition as a guide to the classification of bacteria. *Adv. Appl. Microbiol.* **17**: 63–108.

Sonesson, A. and Jantzen, E. (1992) The branched-chain octose yersinose A is a lipopolysaccharide constituent of *Legionella micdadei* and *Legionella maceachernii*. *J. Microbiol. Methods* **15**: 241–8.

Sonesson, A., Larsson, L., Fox, A., Westerdahl, G. and Odham, G. (1988) Determination of environmental levels of peptidoglycan and lipopolysaccharide using gas chromatography with negative-ion chemical-ionization mass spectrometry utilizing bacterial amino acids and hydroxy fatty acids as biomarkers. *J. Chromatogr.* **431**: 1–15.

Sonesson, A., Bryn, K., Jantzen, E. and Larsson, L. (1989a) Gas chromatographic determination of (phosphorylated) 2-keto-3-deoxyoctonic acid, heptoses and glucosamine in bacterial lipopolysaccharides after treatment with hydrofluoric acid, methanolysis and trifluoroacetylation. *J. Chromatogr.* **487**: 1–7.

Sonesson, A., Jantzen, E., Bryn, K., Larsson, L. and Eng, J. (1989b) Chemical composition of a lipopolysaccharide from *Legionella pneumophila*. *Arch. Microbiol.* **153**: 72–8.

Sonesson, A., Moll, H., Jantzen E. and Zähringer, U. (1993) Long-chain alpha-hydroxy-(ω-1)-oxo fatty acids and α-hydroxy-1, ω-dioic fatty acids are lipid constituents of *Legionella jordanis*, *Legionnella maceachernii* and *Legionella micdadei*. *FEMS Microbiol. Lett.* **106**: 315–20.

Takayama, K., Qureshi, N., Hyver, K., Honovich, J., Cotter, R.J., Mascagni, P. and Schneider, H. (1986) Characterization of a structural series of lipid A obtained from the lipopolysaccharides of *Neisseria gonorrhoeae*. *J. Biol. Chem.* **261**: 10624–31.

Tokunaga, H., Tokunaga, M.Y. and Nakae, T. (1979) Characterization of porins from the outer membrane of *Salmonella typhimurium*. *Eur. J. Biochem.* **95**: 433–47.

Towbin, H., Staehelin, T. and Gordon, J. (1979) Electrophoretic transfer of proteins from polyacrylamide gels to nitrocellulose sheets: procedure and some applications. *Proc. Natl. Acad. Sci. USA* **76**: 4350–5.

Tsai, C.-M. and Frasch, C.E. (1982) A sensitive silver stain for detecting lipopolysaccharides in polyacrylamide gels. *Anal. Biochem.* **119**: 115–9.

Tsai, C.-M., Boykins, R. and Frasch, C.E. (1983) Heterogeneity and variation among *Neisseria meningitidis* lipopolysaccharides. *J. Bacteriol.* **155**: 498–504.

Tsai, C.-M., Frasch, C.E., Rivera, E. and Hochstein, H.D. (1989) Measurements of lipopolysaccharide (endotoxin) in meningococcal protein and polysaccharide preparations for vaccine usage. *J. Biol. Standard.* **17**: 249–58.

Umemoto, T., Ota, T., Sagawa, H., Kato, K., Takada, H., Tsujimoto, M., Kawasaki, A., Ogawa, T., Harada, K. and Kotani, S. (1981) Chemical and biological properties of a peptidoglycan isolated from *Treponema pallidum* Vazan. *Infect. Immun.* **31**: 767–74.

Vachon, V., Lyew, D.J. and Coulton, J.W. (1985) Transmembrane permeability channels across the outer membrane of *Haemophilus influenzae* type b. *J. Bacteriol.* **162**: 918–24.

Vasstrand, E.N., Hofstad, T., Endresen, C. and Jensen, H.B. (1979) Demonstration of lanthionine as a natural constituent of the peptidoglycan of *Fusobacterium nucleatum* Fev1. *Infect. Immun.* **25**: 775–80.

Vasstrand, E.N., Jensen, H.B., Miron, T. and Hofstad, T. (1982) Composition of peptidoglycan in *Bacteroidaceae*: determination and distribution of lanthionine. *Infect. Immun.* **36**: 114–22.

Walla, M.D., Lau, P., Morgan, S.L., Fox, A. and Brown, A. (1984) Capillary gas chromatography-mass spectrometry of carbohydrate components of legionellae and other bacteria. *J. Chromatogr.* **288**: 399–413.

Weckesser, J. and Mayer, H. (1988) Different lipid A types in lipopolysaccharides of phototropic and related non-phototropic bacteria. *FEMS Microbiol. Rev.* **54**: 143–54.

Westphal, O., Lüderitz, O. and Bister, R. (1952) Über die Extraktion von Bakterien mit Phenol/Wasser. *Zeitschr. Naturforsch.* **7b**: 148–55.

Wilkinson, S.G. (1988) The gram-negative bacteria. In *Microbial Lipids*, vol. 1 (Ratledge, C. and Wilkinson, S.G., eds), pp. 299–488. Academic Press: New York.

Wilkinson, S.G., Galbraith, L. and Lightfoot, G.A. (1973) Cell walls, lipids, and lipopolysaccharides of *Pseudomonas* species. *Eur. J. Biochem.* **33**: 158–74.

Willems, A., Busse, J., Goor, M., Pot, B., Falsen, E., Jantzen, E., Hoste, B., Gillis, M., Kesters, K., Auling, G. and De Ley, J. (1989) *Hydrogenophaga*, a new genus of hydrogen-oxidizing bacteria that includes *Hydrogenophaga flava* comb. nov. (formerly *Pseudomonas flava*), *Hydrogenophaga palleronii* (formerly *Pseudomonas paleronii*), *Hydrogenophaga pseudoflava* (formerly *Pseudomonas pseudoflava* and "*Pseudomonas carboxydoflava*"), and *Hydrogenophaga taeniospiralis* (formerly *Pseudomonas taeniospiralis*). *Int. J. System. Bacteriol.* **39**: 319–33.

Willems, A., Falsen, E., Pot, B., Jantzen, E., Hoste, B., Vandamme, P., Gillis, M., Kersters, K. and De Ley, J. (1990) *Acidovorax*, a new genus for *Pseudomonas facilis*, *Pseudomonas delafieldii*, EF group 13, EF group 16, and several clinical isolates, with the species *Acidovorax facilis* comb. nov., *Acidovorax delafieldii* comb. nov. and *Acidovorax temperans* sp. nov. *Int. J. System. Bacteriol.* **40**: 384–98.

Woese, C.R. (1987) Bacterial evolution. *Microbiol. Rev.* **51**: 221–71.

Wollenweber, H.-W. and Rietschel, E.T. (1990) Analysis of lipopolysaccharide (lipid A) fatty acids. *J. Microbiol. Methods* **11**: 195–211.

Wu, L., Tsai, C.-M. and Frasch, C.E. (1987) A method for purification of bacterial R-type lipopolysaccharides (lipooligosaccharides). *Anal. Biochem.* **160**: 281–9.

Yokota, A., Rodriguez, M., Yamada, Y., Imai, K., Borowiak, D. and Mayer, H. (1987) Lipopolysaccharides of *Thiobacillus* species containing lipid A with 2,3-diamino-2,3-dideoxyglucose. *Arch. Microbiol.* **149**: 106–11.

Young, J.D.E., Blake, M., Mauro, A. and Cohn, Z.A. (1983) Properties of the major outer membrane protein from *Neisseria gonorrhoeae* incorporated into model lipid membranes. *Proc. Natl. Acad. Sci. USA*, **80**: 3831–5.

Zamze, S.E., Ferguson, A.J., Moxon, E.R., Dwek, R.A. and Rademacher, T.W. (1987) Identification of phosphorylated 3-deoxy-manno-octulosonic acid as a component of *Haemophilus influenzae* lipopolysaccharide. *Biochem. J.* **245**: 583–7.

3
Analysis of Cell Wall Constituents of Gram-positive Bacteria

Ian. C. Hancock
The Medical School, Newcastle upon Tyne, UK

3.1 INTRODUCTION

The cell walls of Gram-positive bacteria exhibit great diversity and therefore offer much potential grist to the taxonomist's mill. Many of the techniques described in the previous chapter in relation to Gram-negative bacterial walls are still appropriate for analysis of individual components. The unique problems posed by Gram-positive bacteria relate to the very different chemical architecture of their walls, in which most of the components are covalently linked to one another. Thus the whole wall is the minimum structure that can be isolated without resorting to chemical degradation of the bacteria. Analysis of individual components requires either examination of complex whole-wall hydrolysates or selective chemical degradation followed by, often difficult, separation techniques to obtain the components in pure form.

In Gram-positive bacteria, the wall frequently represents more than 20% of the cell dry weight. Peptidoglycan constitutes between 40 and 80% of the wall weight, while the remainder is made up largely of other macromolecules covalently linked either directly to peptidoglycan or to one another. Apart from the wall lipids of some actinomycetes, dealt with in Chapter 5, the predominant wall components are anionic polymers such as teichoic acids and acidic polysaccharides, and proteins. The anionic polymers are invariably linked covalently to peptidoglycan. Proteins may be covalently linked, as in the case of protein A of *Staphylococcus aureus*, but native walls also contain many strongly adsorbed proteins, such as autolytic enzymes, in relatively small amounts.

For analysis of the covalently linked components, isolation of the cell wall as an insoluble product, free of cytoplasmic and membrane material

---

Chemical Methods in Prokaryotic Systematics. Edited by M. Goodfellow and A.G. O'Donnell
© 1994 John Wiley and Sons Ltd.

provides an important initial purification stage that is likely to be essential for high-quality results. The procedures for carrying out this initial step depend very much on what subsequent analysis is to be undertaken, and it is necessary to consider the properties of the individual components in devising an appropriate method. An important consideration is that the same degradation products may arise from very different wall polymers. For example, sugars in an acid hydrolysate could originate in polysaccharide or teichoic acid, or both. Degradation procedures should therefore be as selective as possible.

3.1.1 Peptidoglycan

Although there is considerable inter-species variation in the detailed structure of peptidoglycan, its chemical architecture remains constant. β1-4-linked disaccharides of N-acetylglucosamine and N-acetylmuramic acid form glycan chains up to 100 units long, which are covalently cross-linked in three dimensions by oligopeptides interconnecting the 3-O-lactoyl groups of muramic acid residues in different glycan chains. This structure is very stable. To hydrolyse the glycosidic links in the glycan chains efficiently, 4 M HCl at 100 °C for 4 h is required while up to 16 h under the same conditions is needed to degrade the oligopeptides to their constituent amino acids, by which time considerable degradation of the aminosugars of the glycan has occurred. A consequence of this stability is that peptidoglycan remains insoluble and intact under a variety of conditions that can be used to solubilize and remove other cellular constituents. Thus extraction of whole bacteria with hot detergents, organic solvents or phenol, and treatment with proteases and nucleases, can all be used to obtain material greatly enriched in cell wall, from which individual components can subsequently be isolated in relatively pure form.

The value of peptidoglycan structure in taxonomy has been widely recognized, though its exploitation outside a few specialist laboratories has been limited by the lack of really simple rapid methods.

3.1.2 Anionic wall polymers

Two types of anionic macromolecules, teichoic acids and teichuronic acids, may occur in Gram-positive walls, where they can constitute up to 60% of the weight. Teichoic acids are short polymer chains in which alditols (glycerol, ribitol or, more rarely, mannitol) and/or sugars (particularly *N*-acetylhexosamines) are linked by phosphodiesters. Teichuronic acids are anionic polysaccharides containing uronic acids or aminouronic acids and neutral sugars, linked glycosidically. Each type of polymer has regular, repeating unit structure, terminating in a sugar-1-phosphate through which it is linked by a phosphate ester bond to N-acetylmuramic acid in peptidoglycan. Because of the acid lability of the sugar-1-phosphate linkage, these polymers can be detached from peptidoglycan, and hence solubilized from the cell wall, by treatment at pH 2, for a few minutes at 100 °C, without significant degradation of peptidoglycan or, with the exception of a few structural types of teichoic acid that contain an N-acetylhexosamine-1-phosphate repeating unit, of the anionic polymers. The special 'linkage unit' that attaches teichoic acids to peptidoglycan is also alkali-labile, permitting selective alkaline extraction of teichoic acids.

Anionic polymers can also be detached from peptidoglycan, and teichoic acids selectively degraded, without damaging other components in the wall by treatment with concentrated hydrofluoric acid, which selectively cleaves phosphate esters.

The wide distribution and great structural variation of anionic wall polymers make them excellent candidates for taxonomic studies. While detailed structural determinations are not easy, a considerable amount of information on type and composition can be obtained by straightforward analytical techniques.

3.1.3 Polysaccharides

Although a variety of polysaccharides other than the anionic teichuronic acids have been described, there has been little detailed study of their mode of attachment to the cell wall. Some appear to be linked through a sugar-1-phosphate, like teichuronic acids, and can be removed by dilute acid. Thus sugars detected in a dilute acid extract of a cell wall cannot necessarily be assumed to have arisen from an anionic polymer. However, this form of linkage of polysaccharide may not be universal. There is evidence for direct glycosidic attachment to peptidoglycan in a few cases. To obtain useful information on polysaccharide distribution and composition it is necessary to separate the polymers from other sugar-containing components of the wall. Not enough is known at the moment to permit the design of a routine method for accomplishing this. Within a particular group of bacteria in which the presence of polysaccharide is recognized by the detection of large amounts of sugars in the cell wall, it may be possible to determine selective extraction or degradation techniques. Under those circumstances polysaccharides could be valuable chemotaxonomic markers.

3.1.4 Proteins

Wall-associated proteins occur in a wide range of Gram-positive bacteria. They have received some attention in staphylococci and streptococci

(Russell, 1988; Kehoe, 1993), but they appear not to have been exploited for taxonomic purposes, except for a few specific types such as the M-proteins of Group A streptococci and the Fc receptor proteins of *Staphylococcus aureus* and Group C and G streptococci.

Proteins can be analysed by SDS-polyacrylamide gel electrophoresis (PAGE) of SDS extracts of washed cell walls prepared by mechanical breakage of the bacteria. Further proteins may be released by enzymic solubilization of the cell wall. Interpretation of data obtained in this way is difficult, however. Some of the proteins found non-covalently attached to the wall may not be genuine wall components. The large negative charge on the cell wall may lead to binding of extraneous proteins, while the bulk and limited porosity of the wall may trap substantial amounts of proteins ultimately destined for export (Harwood *et al.*, 1990). Moreover, even well-washed walls may retain bound fragments of membrane. Thus, without clear evidence for a permanent surface location for a particular protein, its detection in detergent extracts of cell walls is not a reliable characteristic. Proteins released from SDS-purified cell walls by treatment with lytic enzymes can be regarded as true wall components, but the wide spectrum of sensitivity to a variety of wall-lytic enzymes makes the establishment of a standard method for their extraction difficult. The methods available, and some applications, are described by Russell (1988). Proteins that can only be extracted in this way are probably covalently linked to peptidoglycan in the native cell wall. Consequently, they will be released by lytic enzymes still covalently linked to solubilized fragments of the cell wall. Unless peptidoglycan degradation has proceeded to completion, therefore, there is a risk of heterogeneity in the protein complex, leading to poor resolution, or even multiple peaks, on PAGE gels.

While extreme care is thus needed in designing experimental protocols, the small amount of published information indicates that reliable, reproducible results are possible. Once genuine wall proteins have been identified, rapid screening by immunochemical techniques becomes possible. Wall proteins therefore represent a valuable, poorly exploited resource for the taxonomist.

3.2 PREPARATION OF CELL WALLS

Although not always necessary, preparation of a cell wall fraction generally facilitates clean analysis of covalently linked wall components. Further simple selective degradation can then be used to isolate peptidoglycan, anionic wall polymers, neutral polysaccharides and wall-bound proteins. Isolation of absolutely pure walls requires mechanical cell disruption. This is seldom feasible when screening large numbers of strains, but for a wide

3. Analysis of Gram-positive Cell Walls

variety of species treatment of whole bacteria with detergent provides acceptable material for analysis, which can be further purified on a small scale by treatment with proteases and nucleases. A boiling solution of sodium dodecylsulphate (SDS) is the reagent of choice for this technique, since it has high solubility in water, and excellent protein and lipid-solubilizing properties, without causing detectable chemical damage to the wall structure. A detailed protocol is given below (Method 3.1). With some strains it may be possible to obtain clean walls without the protease and nuclease steps. It should be recognized, however, that walls containing highly cross-linked peptidoglycan act as molecular sieves and may block the complete extraction of intracellular proteins and nucleic acids unless these have been degraded enzymically. If enzyme treatment is required, great care must be taken that all the SDS has been removed from the samples beforehand. This can conveniently be monitored by the dye-absorption procedure of Hayashi (1975). SDS readily precipitates from solution at low temperatures and is then extremely difficult to redissolve. The temperature should preferably never fall below 15 °C until all SDS has been removed.

An important consideration in any wall analysis work is the avoidance of wall autolysis. Autolysis may be activated by a variety of conditions, including cold shock, change of pH, anaerobiosis, and perturbation of protein–wall interactions by many reagents, including detergents. It is therefore essential that bacteria are harvested as rapidly as possible, at a low temperature, and either used immediately or rapidly frozen. Detergent treatment must be carried out at 100 °C and the bacteria should be added to already boiling reagent so that it remains at that temperature throughout. The neutral protease from *Streptomyces griseus*, ('Pronase E') is frequently used in wall purification because of its broad specificity. Some preparations, however, contain wall-lytic enzymes, and these have to be inactivated at 60 °C before use.

Method 3.1 Preparation of cell walls from whole bacteria.

3.1.A Treatment with sodium dodecyl sulphate (SDS)

1. Freshly harvested or thawed bacteria, or lyophilized material is suspended in ice-cold water (2–4 mg dry wt/ml) and added gradually to an equal volume of boiling SDS solution (BDH-Merck, 'specially pure') (8% w/v) with continuous stirring on a hotplate to maintain boiling.
2. Boiling and mixing is continued for 30 min, with addition of water to maintain approximately constant volume.
3. After cooling to room temperature, insoluble crude cell wall is recovered by centrifugation (30 000 × g, 15 min).
4. The SDS treatment is repeated at least once more on the insoluble material.

Preliminary experiments may be necessary to determine how many treatments are required to remove all extractable UV-absorbing material.
5. The final insoluble pellet is washed by resuspension and centrifugation 5 times with hot water.

3.1.B Removal of nucleic acids and protein

Traces of nucleic acid and protein remaining in the preparation can be removed enzymically.
1. The wall is resuspended in 0.05 M-Tris-HCl, pH 7.0 containing 1 mM $MgCl_2$, 0.2 mM dithiothreitol.
2. RNase A (5 Kunitz units/ml) and DNase 1 (50 Kunitz units/ml) are added and the mixture is incubated at 37 °C for 3 h.
3. The walls are recovered by centrifugation, washed three times in the Tris buffer, and resuspended in the same buffer.
4. Pronase (protease type XIV, Sigma Chemical Co.), previously heated as a 10 mg/ml solution in Tris buffer for 2 h at 60 °C, is added to a final concentration of 100 µg/ml and the mixture incubated at 60 °C for 1 h.
5. The pure walls are recovered by centrifugation at 48 000 × g for 15 min, washed four times with water, and lyophilized.

3.3 PEPTIDOGLYCAN ('MUREIN')

3.3.1 Analyses using whole bacteria

Since, with a few exceptions, the degradation products of Gram-positive cell walls are not uniquely cell wall components, there is limited scope for whole-cell analysis in this field. Thus, although whole-cell sugar analysis has been used for defining 'cell wall' chemotypes, for example among the actinomycetes (Lechevalier & Lechevalier, 1970), it cannot be assumed *a priori* that the sugars originate in the wall. Glycolipids, carbohydrate storage products and firmly, but non-covalently attached capsular polysaccharides may all contribute to the cellular sugar composition. Komagata and Suzuki (1987) describe a procedure for whole-cell sugar analysis which may be useful for comparison of a relatively homogeneous group of well-understood strains.

There are two cases in which unique cell wall components can be detected conveniently in whole cells. Detection of the presence of diaminopimelic acid (DAP) and identification of the isomer is one of the most useful chemotaxonomic procedures for Gram-positive bacteria, including actinomycetes, and can be carried out conveniently with whole bacteria. Similarly, detection of N-glycolyl substituents on the glycan chain of peptidoglycan is a useful procedure for the classification of actinomycetes which can be carried out without prior cell wall isolation. Detailed protocols for these techniques are given below (Methods 3.2 and 3.3).

Method 3.2 Determination of diaminopimelic acid isomers (based on Staneck & Roberts, 1974).

1. Dried bacteria (up to 5 mg) are hydrolysed in 1 ml 4 M HCl at 100 °C for 16 h, in a screw-capped reaction vial or tube. These conditions are usually sufficient to hydrolyse the peptidoglycan completely. Sonesson *et al.* (1988) have shown that heating in 6 M HCl at 150 °C for 6 h under nitrogen is a faster alternative.
2. Insoluble material is removed by filtration through a glass-fibre filter (e.g. Whatman GF/F) and combined with a further 1 ml of water passed through the filter.
3. The sample is then dried, either by rotary evaporation or under vacuum in a desiccator over NaOH pellets. It is essential to redissolve the dried material in water and dry again at least once more, to remove all traces of acid. The final residue should be dissolved in 0.1 ml water per mg of starting material, for analysis.

Thin-layer chromatography

1. Diaminopimelic acid (DAP) isomers are separated by thin-layer chromatography on 20 cm × 20 cm cellulose plates (e.g. Merck no. 5716) using the developing solvent methanol-water-6 M HCl-pyridine (80 : 26 : 4 : 10 by volume). It is advisable to line the tank with filter paper and allow equilibration with the solvent for several hours before use.
2. Samples (3 µl) are applied to the baseline of the chromatogram, together with standards of authentic DL-DAP (Sigma Chemical Co.) and bacterial hydrolysates containing the *meso-* and LL-isomers of DAP (e.g. from *Bacillus* and *Streptomyces* respectively), placed on either side of the experimental samples.
3. Development is carried out until the solvent front is 1 cm from the top of the plate (about 3.5 h).
4. The dried plate is sprayed with ninhydrin (0.2% w/v in water-saturated butan-1-ol) and heated at 100 °C for 5 min to visualize the amino acids.

DAP isomers run slower than other amino acids and give characteristic green spots with the LL isomer running slightly faster than DL, *meso* (R_Fs approximately 0.29 and 0.24 respectively). 3-Hydroxy-DAP, reported to be present with *meso*-DAP in, for example, *Actinoplanes*, has an R_F of about 0.20 under the same conditions.

The isomers of DAP can be separated as their heptafluorobutyryl isobutyl derivatives by GC on a fused silica capillary column coated with a chiral stationary phase such as Chirasil-L-Val (Chromopack) (Sonesson *et al.*, 1988).

Method 3.3 Detection of N-glycolyl groups in peptidoglycan (from Uchida & Aida, 1977).

Some actinomycetes contain peptidoglycan in which some, or all, of the muramic acid residues are N-acylated with glycolyl groups instead of the more usual acetyl. The presence of glycolyl muramic acid can be readily determined using whole bacteria.

1. Dry bacteria (10 mg) are hydrolysed with 0.1 ml 6 M HCl at 100 °C for 2 h in a screw-capped reaction vial or tube.
2. The hydrolysate is added to a column (0.5 × 5 cm) of Dowex 1 × 8 (Analytical

Grade) in the acetate form, and the column is subsequently washed with 2 ml water and 1 ml 0.5 M HCl. The glycolic acid is then eluted in a further 2 ml 0.5 M HCl.

3. 0.1 ml samples of the eluate are incubated with 2 ml freshly prepared 2,7-dihydroxynaphthalene solution (0.02% w/v in conc. H_2SO_4) at 100 °C for 10 min in a screw-capped tube.

4. The development of a red-purple colour indicates the presence of glycolic acid, which can be quantified spectrophotometrically, by measuring absorbance at 530 nm against a reagent blank. A standard curve should be constructed using samples of sodium glycolate solution (up to 10 mM) eluted from the ion-exchange column in the same way as the bacterial hydrolysates.

The presence of diaminopimelic acid isomers and N-glycolyl substitution in peptidoglycan can be determined using hydrolysates of whole bacteria. However, more detailed examination of the peptide composition of peptidoglycan, which provides extremely valuable chemotaxonomic data (Schleifer & Seidl, 1985), requires purified peptidoglycan, from which all other peptide and protein components have been rigorously removed. In most cases cell wall prepared by treatment of whole bacteria with detergent as described above (Method 3.1) is a satisfactory starting point for isolation of peptidoglycan, though in rare cases where the preparation cannot be freed of protein by that procedure it may be necessary to disrupt the bacteria by sonication before detergent treatment, as described by Komagata and Suzuki (1987). The residue from lipid extraction of whole bacteria is another potential source of peptidoglycan.

The chemical stability of peptidoglycan allows the extraction of other cell wall components under relatively mild conditions, leaving peptidoglycan as an insoluble residue, recoverable by centrifugation. Thus covalently linked anionic polymers can be removed by dilute acid hydrolysis, and covalently linked protein by protease treatment. It should be recognized, however, that these procedures leave chemical residues at the attachment points on the peptidoglycan. Acid extraction of an anionic polymer or a polysaccharide leaves a phosphate group esterified at C6 of the muramic acid to which the polymer was attached, while protease removal of a protein covalently linked to the peptide moiety of the peptidoglycan may leave a residue of terminal amino acids of the protein. The advantage of starting the procedure with purified cell walls is that the wall components can be quantified and hence the quantitative significance of these residues in the purified peptidoglycan can be assessed.

The provenly most valuable information to be obtained from detailed peptidoglycan analysis is the type and composition of the peptide cross-link between adjacent glycan chains in the polymer network. This may vary from a direct peptide bond between amino acids in the cross-linked

3. Analysis of Gram-positive Cell Walls

glycopeptides to the presence of an additional peptide chain of several amino acids, intervening between the linked glycopeptides. The different types of cross-link have been classified by Kandler, Schleifer and colleagues, and their taxonomic significance widely discussed (Schleifer & Kandler, 1972; Schleifer & Seidl, 1985). Growth conditions can affect the detailed amino acid composition of cross-linking peptide chains, though not, usually, the type of cross-link (Schleifer et al., 1976). Recent detailed HPLC analysis of individual glycopeptides derived by muramidase treatment of peptidoglycan has revealed that a small proportion of glycopeptides may have a cross-link type different from the predominant one in the wall. In *E. coli*, for example, some glycopeptides are linked by a peptide bond between two diaminopimelic acid residues, rather than by the predominant diaminopimelic acid-D-alanine cross-link (Glauner & Schwartz, 1988). The taxonomic significance of these minor cross-link types has not yet been examined.

Another variable in peptidoglycan structure that has received little attention is the degree of N and O-acetylation of the aminosugars of the glycan chain. Partial de-N-acetylation of both N-acetylglucosamine and N-acetylmuramic acid occurs, for example, in *Bacillus*, and appears to vary from species to species (Zipperle et al., 1984). Other species, such as *Staphylococcus*, exhibit acetylation of the hydroxyl group at C6 of some of the amino sugars (Tipper & Strominger, 1966). These modifications are one of the principal causes of resistance to lysozyme.

3.3.2 Isolation of peptidoglycan

In many cases, particularly when the cell wall does not contain substantial amounts of protein, it is possible for rapid screening purposes to prepare peptidoglycan from whole bacteria by treatment with hot trichloroacetic acid followed by treatment with trypsin (Schleifer & Kandler, 1972). The preparation is, however, likely to be contaminated with protein and care is needed during subsequent analyses to distinguish between peptidoglycan and protein amino acids.

For more critical work, peptidoglycan is prepared from cell walls purified as described above (Method 3.1).

Method 3.4 Removal of anionic polymers and polysaccharide by treatment with hydrofluoric acid.

---

1. Purified cell walls (1–10 mg) (Method 3.1) are suspended in 0.5 ml hydrofluoric acid (48% w/v, Analytical Grade, BDH-Merck) in 15 ml polypropylene centrifuge tubes.
2. The mixture is incubated at 2 °C for 4 h, then neutralized by addition of the

appropriate volume of 2 M KOH while the tube is immersed in a −70 °C freezing bath.
3. The insoluble peptidoglycan is recovered by centrifugation (48 000 × g, 15 min) and washed three times with water.

Other methods of purification using acid treatment may give incomplete removal of polymers and, as described above, leave residual phosphate groups on the peptidoglycan. They are summarized in Schleifer and Seidl (1985).

3.3.3 Determination of amino acid composition of peptidoglycan

Peptidoglycan is hydrolysed in a screw-capped reaction tube in 4 M HCl at 100 °C for 16 h. Peptidoglycans containing the amino acids threonine and isoasparagine may be more resistant to acid hydrolysis, and in this case complete hydrolysis requires 6 M HCl at 120 °C. In both cases, hydrolysis under vacuum or nitrogen improves recovery. Acid is removed as described in Method 3.2 above.

Qualitative analysis of the amino acids is conveniently carried out by two-dimensional cellulose thin-layer chromatography. Elution in the first dimension is carried out using isopropanol-acetic acid-water (75 : 10 : 15 by vol). After thorough drying in air at room temperature, the plate is run in the second dimension in 2-methylpyridine-25% NH$_4$OH-water (70 : 2 : 28 by vol). Amino acids are detected with ninhydrin as described in Method 3.2. The procedure also separates and detects the aminosugars glucosamine and muramic acid. An example of a chromatogram is shown in Schleifer and Seidl (1985). A mixture of standard cell wall amino acids should be chromatographed on a separate plate at the same time. In order to distinguish unequivocally between the diamino acids DAP, lysine and ornithine, a sample should also be chromatographed as described for DAP in Method 3.2.

Quantitative analysis can be carried out on an automatic amino acid analyser, or by reverse-phase HPLC of the dabsylated amino acids (Chang et al., 1983). The amino acids are dabsylated by incubation in 0.2 ml 0.2 M sodium hydrogen carbonate with 0.2 ml dabsyl chloride solution (0.65 mg/ml in acetone) in screw-capped reaction vials at 70 °C for 15 min with occasional shaking. Samples of the reaction mixture are separated on a C18 reverse-phase column using a gradient of 20 to 70% acetonitrile in 0.045 M sodium acetate buffer, pH 4.1 over 25 min. Elution is monitored at 420 nm. The compounds elute in the order: dabsyl chloride, muramic acid, glutamic acid, glucosamine, alanine, DAP and ammonia.

3.3.4 Determination of peptidoglycan type

The determination of peptidoglycan type involves peptide analysis that does not lend itself to rapid screening of large numbers of samples. For a detailed account, see Schleifer and Kandler (1972) and Schleifer and Seidl (1985). Two relatively simple procedures, however, distinguish different peptidoglycan types sufficiently clearly for comparative purposes and are described below (Methods 3.5 and 3.6). Comparison of the results of these procedures with the data presented by Schleifer and Kandler (1972) allows a preliminary identification of type to be made.

Method 3.5 Determination of N-terminal amino acids: dinitrophenylation (from Takebe, 1965).

Some peptidoglycan peptides are not cross-linked. These have a free N-terminus that can be labelled using fluorodinitrobenzene, permitting subsequent hydrolysis and identification of the N-terminal amino acid.

1. Peptidoglycan (about 1 mg) is suspended in 1 ml 1% (w/v) sodium hydrogen carbonate and mixed with 2 ml 2.5% (v/v) fluorodinitrobenzene in ethanol.
2. The mixture is incubated in the dark at 37 °C for 6 h, then the peptidoglycan is washed with 67% (by vol) ethanol, and repeatedly with water.
3. The peptidoglycan is hydrolysed as described above for amino acid analysis (see Section 3.3.3) and the products are examined by silica-gel thin-layer chromatography using the solvent chloroform-methanol-acetic acid (95 : 5 : 1 by vol). Authentic dinitrophenylated amino acids and amino sugars should be prepared and run in parallel for identification purposes.

Method 3.6 Partial acid hydrolysis.

Different types of peptidoglycan yield distinctive small peptide products on partial acid hydrolysis. Two-dimensional chromatography of these produces 'fingerprints' characteristic of the peptidoglycan types. These have been studied extensively, and described in detail by Schleifer and Kandler (1972).

1. Peptidoglycan (about 5 mg) is hydrolysed in 4 M HCl at 100 °C for periods ranging from 15 to 60 minutes. It is suggested that two samples, hydrolysed for 20 and 60 minutes, are compared initially.
2. Acid is removed as previously described in Method 3.2.
3. Two-dimensional paper chromatography on, for example, Whatman no. 1 paper, in the solvents already described for amino acid analysis by thin-layer chromatography (Section 3.3.3) is carried out on the whole sample. In order to achieve adequate separation of the peptides of low mobility, it is necessary to elute the chromatogram in each dimension several times, drying between runs. The number of elutions will depend on the size of paper sheet that can be used—the larger the better. The separation can also be carried out on cellulose thin-layer plates as described for amino acid analysis, but a large number of repeat elutions is required to obtain the necessary separation, and this can cause deterioration of the edges of the thin layer.

4. The chromatograms are stained with ninhydrin. The colour of spots during development can be a valuable distinguishing feature.

3.4 ANIONIC POLYMERS

Most, possibly all, Gram-positive bacteria contain anionic polymers—teichoic acids or teichuronic acids—in their cell walls. Although a wide range of these has been characterized (for the most recent review see Archibald *et al.*, 1993) they have been used surprisingly rarely as chemotaxonomic characters (Endl *et al.*, 1983; Uchikawa *et al.*, 1986). The polymers are readily detected and quantified in whole cell walls by estimation of phosphorus and uronic acid, as described below. The teichoic acids are susceptible to more detailed structural identification because they can be selectively degraded, by concentrated hydrofluoric acid, to dephosphorylated repeating units that can be analysed by GC or HPLC. Since under appropriate conditions only phosphoester bonds are cleaved in this procedure, it is very specific and can be carried out directly on purified cell walls.

The teichoic acids fall into four main classes: (I) poly(alditolphosphate); (II) poly(glycosylalditolphosphate); (III) poly(sugar-1-phosphate-alditolphosphate); and (IV) poly(sugar-1-phosphate). In classes I, II and III the alditol may be glycerol (G), ribitol (R) or mannitol (M), and may be substituted with a sugar, of which glucose, galactose, *N*-acetylglucosamine and *N*-acetylgalactosamine are predominant. Thus, the wall teichoic acid of *Staphylococcus aureus* could be classified as I : R : *N*-acetylglucosamine. In types II, III and IV the same range of sugars occurs, with hexoses predominant in type II and *N*-acetylhexosamines predominant in types III

Table 3.1 Distribution of teichoic acid types amongst genera of Gram-positive bacteria.

Type I, Glycerol:	*Actinomadura, Arthrobacter, Bacillus, Lactobacillus, Staphylococcus, Streptomyces*
Type I, Ribitol:	*Bacillus, Lactobacillus, Listeria, Staphylococcus, Streptococcus, Streptomyces*
Type I, Mannitol:	*Brevibacterium*
Type II, Glycerol:	*Actinomadura, Actinoplanes, Bacillus, Lactobacillus, Streptomyces*
Type II, Ribitol:	*Brevibacterium, Streptococcus*
Type III:	'*Micrococcus*', *Staphylococcus*
Type IV:	*Bacillus*, '*Micrococcus*', *Staphylococcus, Streptococcus*

3. Analysis of Gram-positive Cell Walls

and IV. Additional structural variation occurs in the anomeric configuration of the sugar, in the position of attachment of the sugar on the alditol, and in the presence of esterified amino acids. These features are, however, less amenable to routine analysis and are not dealt with here. Procedures for assignment to the basic structural groups are described below. The known distribution of the four types of teichoic acid amongst Gram-positive genera is shown in Table 3.1. For more detailed information see Archibald et al. (1993).

Pyrolysis mass spectrometry of cell walls and whole cells of *Bacillus* has shown the technique's potential for recognizing some features of teichoic acids and teichuronic acids (Boon et al., 1981). It offers considerable possibilities for rapid analysis of these polymers in the future.

3.4.1 Detection of teichoic acids and teichuronic acids

Teichoic acid in the cell wall is readily detected and measured as organic phosphorus. The only other sources of phosphorus in purified wall are the attachment points of polysaccharides and teichuronic acids linked to peptidoglycan by phosphate esters. These sources contribute very little to the weight of the wall, whereas teichoic acid phosphorus would be expected to constitute between 0.5% and 5% of the weight of the wall. The detection of uronic acids in the wall similarly indicates the presence of a teichuronic acid—it is convenient in this context to regard a polysaccharide containing uronic acid residues as a teichuronic acid. Phosphorus and uronic acids can be detected in microgram quantities by straightforward spectrophotometric techniques, without the necessity of extracting and purifying the individual polymers.

Method 3.7 Measurement of teichoic acid as organic phosphorus.

The procedure involves the conversion of organic phosphorus to inorganic orthophosphate, followed by the measurement of inorganic phosphate by the method of Chen et al. (1956). The conversion to inorganic phosphate can be carried out by digestion either with a mixture of perchloric acid and sulphuric acid (Chen et al., 1956) or with ethanolic magnesium nitrate (Ames, 1966). The former procedure gives more reliable quantitative results, but the latter is safer, and more convenient. Both procedures are given.

1. Prepare the digestion reagents as follows: (A) conc. H_2SO_4-60% perchloric acid, 3 : 2 by volume, or (B) magnesium nitrate 10% w/v in absolute ethanol. All reagents should be analytical grade. These can be stored satisfactorily.
2. Prepare the inorganic phosphate reagent. This is unstable and must be prepared fresh from stock solutions immediately before use, in scrupulously clean, phosphate-free glassware. Note that many laboratory cleaning agents contain phosphate. The stock solutions are: 3 M sulphuric acid; 2.5% (w/v) ammonium

molybdate; and freshly prepared 3.3% (w/v) ascorbic acid. Working reagent (C) is prepared by mixing these three in the ratio 1:1:3 by volume.
3. (a) Cell wall samples (0.1–1 mg, containing up to 10 μg phosphorus) in 0.1 ml water are completely dried in borosilicate glass tubes on an electric heater. 1 ml digestion reagent A is added and the solution is refluxed on the same heater for 20 min. *Alternatively* (b) the cell wall suspensions are heated, without prior drying, with 0.1 ml digestion reagent B until the samples are dry and no brown nitrogen dioxide is evolved. 0.3 ml 0.05 M HCl is added and again the samples are boiled to dryness.
4. The digested samples are thoroughly mixed with 3.9 ml water and 4 ml reagent C, and incubated at 37 °C for 90 min.
5. Inorganic phosphate produces a blue colour, and a linear absorbance response at 820 nm is obtained in the range 0–10 μg phosphorus. For quantitative work the assay should be calibrated using potassium dihydrogen phosphate subjected to the same digestion procedure as the experimental samples.

Method 3.8 Measurement of teichuronic acid as uronic acid.

Uronic acids in cell walls are detected by the procedure of Blumenkrantz and Asboe-Hansen (1973). The procedure is not satisfactory for aminouronic acids.

1. Prepare the reagents. Two reagents are required, both of which can be stored at 4 °C in the dark for several weeks: (A) 0.0125 M disodium tetraborate in conc. H_2SO_4; (B) 0.15% (w/v) 3-phenylphenol (*m*-phenylphenol; 3-hydroxybiphenyl) in 0.5% (w/v) aqueous sodium hydroxide.
2. Cell wall samples (0.1–1 mg) in 0.6 ml water are vigorously mixed with 3.6 ml reagent A and incubated in a boiling water bath for 5 min, then rapidly cooled in ice to room temperature.
3. Reagent B (0.06 ml) is added and mixed thoroughly.
4. The absorbance is measured at 520 nm against a reagent blank exactly 5 min after addition of the reagent. Glucuronic acid or glucuronolactone is a convenient standard. Glucuronic acid (10 μg) gives an absorbance of about 0.25.

3.4.2 Classification of teichoic acids

Although a few genera, such as *Micrococcus*, appear to lack teichoic acids, these polymers are too widely distributed for their occurrence alone to be very useful taxonomically. It is necessary to determine the type of teichoic acid present, as discussed above. This is best done in two stages—identification of the alditols and/or sugars present, followed by studies of the way these units are attached to one another.

A convenient procedure is to selectively degrade teichoic acid in the cell wall with hydrofluoric acid and to analyse the low molecular weight products before and after acid hydrolysis. Glycerol, ribitol, anhydroribitol or mannitol in the acid hydrolysate confirms the presence of a glycerol, ribitol or mannitol teichoic acid. The absence of one of these products from walls shown by phosphorus analysis to contain teichoic acid suggests that a teichoic acid of type IV is present. The sugars in the acid hydrolysate may derive from wall polymers other than teichoic acid, that were linked to the cell wall by phosphate esters. Quantitation of sugars and comparison with the amount of wall phosphorus may indicate the presence of excess sugars, derived from polysaccharide.

Analysis of the HF degradation products without further acid hydrolysis gives more information about type of teichoic acid (Table 3.2) and provides a useful 'fingerprint' for comparative purposes. The formation of equimolar amounts of free alditol and free sugar suggests the presence of a type III teichoic acid, while sugar alone indicates type IV. Glycosylalditols are formed from types I and II; type I, but not type II, will also yield some free alditol if not all the alditols are glycosylated. The alditol glycosides may be tentatively identified chromatographically by comparison with glycosides derived by HF treatment of teichoic acids of known structure. An example of this, and useful chromatographic data, is given in Fiedler *et al.* (1981).

Where analysis suggests the presence of a teichoic acid of type III or type IV, advantage can be taken of the acid lability of the sugar-1-phosphate linkage to obtain the repeating unit of the polymer—either alditol-phosphate-sugar (type III) or sugar-phosphate (type IV). These anionic products are not suitable for analysis by GC, but can be examined by paper or thin-layer chromatography (Hancock & Poxton, 1988).

Analysis of other teichoic acid degradation products can be carried out either by direct ion-exchange HPLC, using a refractometric or, preferably, pulsed amperometric detection, by GC of peracetylated derivatives, or by GC of the more volatile trifluoroacetyl derivatives. In all cases, both alditols and sugars can be analysed, but the trifluoroacetyl derivatives are the most appropriate for glycosylalditols produced by HF degradation.

Table 3.2 Products of HF cleavage of teichoic acids of different classes.

Class	Major products	
	Before acid hydrolysis	After acid hydrolysis
I. Glycerol	Glycerol, glycosylglycerol	Glycerol, sugar
I. Ribitol	Ribitol, glycosylribitol	Anhydroribitol, sugar
II. Glycerol	Glycosylglycerol	Glycerol, sugar
II. Ribitol	Glycosylribitol	Anhydroribitol, sugar
III. Glycerol	Glycerol, sugar	Glycerol, sugar
III. Ribitol	Ribitol, sugar	Anhydroribitol, sugar
IV	Sugar	Sugar

Method 3.9 Degradation of wall teichoic acid with hydrofluoric acid.

1. Dry cell wall (up to 10 mg) is suspended in 0.25 ml hydrofluoric acid (analytical grade, 48%) in a 10–15 ml capacity polythene or polypropylene tube and incubated at 0 °C for 48 h.
2. The HF is removed under vacuum in a plastic desiccator over NaOH pellets, with a sodium fluoride trap between the desiccator and the pump.
3. The dried material is suspended in 1 ml water and centrifuged at 30 000 × g for 15 min at 2 °C to remove insoluble material.
4. The supernatant is applied to a column (1 ml) of Dowex 2 (×8) in the bicarbonate form, eluted in 2 ml water and dried under vacuum.

Method 3.10 Acid hydrolysis of HF degradation products.

1. Some of the product from the above process is hydrolysed in 0.5 ml 1 M-H$_2$SO$_4$ in a screw-capped reaction tube under nitrogen at 100 °C for 3 h. If necessary for quantitative measurements, an internal standard (such as arabinose for neutral sugars or N-methylglucosamine for aminosugars) is added.
2. Acid is removed by addition of 2.5 ml 20% N,N-dioctylmethylamine in chloroform followed by vigorous mixing. On standing, the mixture separates into two phases and the upper, aqueous layer is retained.
3. Traces of organic solvents are removed by passing the aqueous phase through a small column (1 ml) of Bond-Elut previously rinsed with 2 ml methanol and 2 ml water.
4. The aqueous phase is washed through with a further 1 ml water and the combined eluate is dried under vacuum at 30 °C (Whitton *et al.* (1985).

Glycosides of N-acetylaminosugars may not be completely hydrolysed by this procedure. If de-N-acetylation precedes hydrolysis of the glycosidic bond in a particular molecule, the resulting positively charged amino group renders the glycoside resistant to further hydrolysis. For quantitative measurements when aminosugars are present, therefore, hydrolysis should also be carried out in 2 M H$_2$SO$_4$ and care should be taken that the sample is rapidly heated to 100 °C, for example by immersion in a boiling water bath rather than a dry block incubator or oven.

3.4.2.a HPLC of degradation products

A wide range of HPLC columns for the separation of sugars is available from manufacturers, and each requires its own elution protocol. However, detection is difficult with standard HPLC instruments. Pulsed amperometric detection (PAD) affords excellent sensitivity and selectivity, but requires a strongly alkaline eluant or post-column addition of alkali due to the requirement for ionization of the species to be detected. The technique is described in detail by Hardy (1989). The CarboPac MA1 column (Dionex Corporation, Sunnyvale, California, USA), permits separation of alditols,

3. Analysis of Gram-positive Cell Walls

sugars and alditol glycosides, in conjunction with PAD, by elution with sodium hydroxide.

Himanen (1992) has described another HPLC method satisfactory for the measurement of glycerol in teichoic acids.

3.4.2.b GC of degradation products

The most reliable method for the analysis of sugars by GC is as their volatile fully O- and N-acetylated alditol derivatives: the sugars are converted to alditols by reduction with sodium borohydride, and then peracetylated with acetic anhydride. This derivatization technique has the advantage that, unlike other procedures, it only produces one chromatography peak per sugar. It is equally applicable to analysis of alditols and glycosylalditols from teichoic acids, with the proviso that the sugars ribose and mannose will give the same derivatives as the alditols ribitol and mannitol. In the case of ribitol this is not too serious, as acid hydrolysis also produces anhydroribitol, which is not formed from ribose-containing compounds such as RNA, and is readily detected by the same chromatographic methods. Anhydroribitol can be prepared by heating ribitol in 2 M H_2SO_4 for 16 h at 100 °C, under nitrogen, and removing the acid as described in Method 3.10.

The original method of Sawardeker et al. (1967) for preparation of alditol acetates has been modified in various ways to reduce the number of manual manipulation steps in the process. The method recommended here is that of Whitton et al. (1985). This procedure is suitable for all neutral sugars, alditols and aminosugars found in teichoic acids.

Method 3.11 GC of degradation products as their alditol acetates (from Whitton et al., 1985).

---

1. The products of HF and sulphuric acid degradation described above are dried under vacuum and redissolved in 0.25 ml water.
2. Sodium borohydride solution (60 µl; 10 mg/ml in N-methylimidazole) is added and the mixture is incubated in a reaction vial for 90 min at 37 °C.
3. Excess borohydride is then destroyed by addition of 20 µl glacial acetic acid.
4. Acetic anhydride (0.6 ml) is added and the mixture incubated again at 37 °C for 45 min.
5. The sample is cooled on ice and neutralized by the slow addition of concentrated ammonia solution.
6. Chloroform (1 ml) is added and the mixture shaken vigorously.
7. After separation the organic phase is applied to a 1 ml Bond-Elut column and eluted in 2 ml chloroform.
8. The eluate is passed a further time through a similar column and again eluted in 2 ml chloroform.
9. The sample is dried, and redissolved in chloroform for GC.

For successful separation of aminosugar and alditolglycoside derivatives, high GC temperatures are required, and polar stationary phases of high thermal stability must be employed. Wide bore columns and fused silica capillary columns with stationary phases such as SP-2340, OV-225, Silar 10C and SP-2380 can be used successfully, and the manufacturer's guidelines for alditol acetate separation should be followed. Integrated peak areas may need correction for individual sugars and alditols when used for quantitative measurements.

Method 3.12 GC of degradation products as their trifluoroacetyl derivatives.

This procedure is based on that of Endl *et al.* (1983), where detailed chromatographic data can be found. Derivatization is carried out without prior borohydride reduction.

1. The dry, acid-free products from HF degradation, or acid hydrolysis, are heated in 0.2 ml trifluoroacetic acid-trifluoroacetic anhydride (1:50 by volume) at 100 °C for 20 h in a reaction vial with a Teflon-lined cap. The long reaction time is required for full acylation of glycosides.
2. After cooling, samples can be analysed directly by GC without further treatment. GC has been carried out successfully using 3% QF1 stationary phase, with temperature programming from 100 °C to 230 °C at 5°/min.

The procedure separates α- and β-glycosides of glycerol and ribitol, as well as the alditols themselves. It is not, however, very satisfactory for complex mixtures of free sugars, since these give multiple peaks.

3.4.3 Teichuronic acids

Analysis of teichuronic acids beyond the stage of identification and quantification of the sugar constituents has only been carried out in a few cases (Hase & Matsushima, 1979; Lifely *et al.*, 1980; Yoneyama *et al.*, 1984; Aono & Uramoto, 1986) and is probably beyond the scope of most chemotaxonomic studies. Surveys of the presence of hexuronic acid-containing teichuronic acids (White, 1977) and of aminouronic acid-containing polymers (Yoneyama *et al.*, 1982) in *Bacillus* show the potential of even a simple search for the presence of uronic acids. However, more detailed identification of uronic acids and aminouronic acids is difficult because few authentic standard compounds are available. In addition, aminouronic acids are extremely unstable in acid and decompose to non-carbohydrate products. If the presence of a teichuronic acid in the cell wall is revealed by uronic acid analysis, then it can be valuable to carry out reduction of the uronic acid to the corresponding aldose, which can usually be identified more readily. This can be carried out on the teichuronic acid polymer, extracted from the cell wall, or on the whole cell wall. Sugar analysis of hydrolysates before and after reduction reveals the sugar derived from the uronic acid.

3. Analysis of Gram-positive Cell Walls

Acid hydrolysis and sugar analysis are carried out as described for teichoic acid.

Method 3.13 Carboxyl-reduction of polymers containing uronic acids and aminouronic acids (Yoneyama et al., 1982).

This procedure can be carried out with purified cell walls, without prior isolation of the teichuronic acid.

1. An aqueous suspension of cell wall (about 10 mg in 5 ml) is mixed with 200 μmol solid 1-ethyl-3-(3-dimethylaminopropyl)-carbodiimide.
2. The pH of the suspension is adjusted to 4.8 with HCl.
3. The suspension is stirred for 2 h at room temperature, with further addition of HCl, if necessary, to maintain the pH at 4.8.
4. 3 ml 2 M sodium borohydride (freshly prepared) is added slowly, with stirring, and mixing is continued overnight.
5. The cell wall is recovered by centrifugation, and the reduction procedure repeated.
6. The resulting reduced cell wall is recovered by centrifugation, washed with water, and lyophilized.

3.5 POLYSACCHARIDES

Chemical and physiological studies have revealed the presence of polysaccharides containing neutral sugars in a wide range of Gram-positive bacteria, but there has been no systematic study of their structures and distribution. In some cases uronic acids are also present and the polymers might be alternatively described as teichuronic acids. Analysis of polysaccharide structure has been revolutionized by the use of sophisticated NMR and mass spectrometry techniques (Bundle, 1988), but these require substantial quantities of highly purified material and are not currently suitable for systematic screening of bacteria. Probably the only simple approach available at the moment is to identify and quantify cell wall sugars. This approach does not, however, distinguish the presence of more than one polysaccharide species, and is complicated by the presence of sugar-containing anionic polymers, whose sugar composition would have to be analysed separately. Where very distinctive sugars are present, whole-cell sugar analysis may be appropriate. Saddler et al. (1991), for example, were able to use this procedure to search for madurose (3-O-methyl-D-galactose) in whole-cell hydrolysates of actinomycetes, by GC of alditol acetates.

Following acid hydrolysis of the polysaccharide, sugars can be analysed by several well-established procedures employing GC or HPLC, as already described for sugars derived from anionic wall polymers. The gas chromatographic analysis of alditol acetate derivatives should be the method of first choice. This has been considerably simplified in recent years by the introduction of methylimidazole as the catalyst for peracetylation. In the original procedure laborious removal of borate, formed from the borohydride used in the reduction step, was necessary to prevent inhibition of acetylation. This is not required when methylimidazole is used, and the procedure is satisfactory for a wide range of sugars, alditols and aminosugars (see Method 3.11). However, Whitton et al. (1985) found that very little acetylation of muramitol, derived from muramic acid of peptidoglycan, occurred in the methylimidazole-catalysed procedure. Since in some quantitative wall analyses measurement of muramic acid serves as a useful normalization process for multiple samples, an alternative method of acetylation is required in these cases, and the use of sodium acetate, as described by Whitton et al. (1985) is recommended.

3.6 RAPID METHODS

As described in Section 3.1, the walls of a single strain of a Gram-positive bacterium may contain a variety of polymers containing common constituents, such as sugars and amino acids; moreover, the constituents of some wall polymers are also found in other cellular components. Consequently, rapid, whole-cell analyses have limited value for taxonomic studies based on identifiable cell wall components, and there are few cases in which such studies have been attempted, except for the detection of diaminopimelic acid and N-glycolyl groups in peptidoglycan (see Section 3.3.1), and madurose from actinomycetes (see Section 3.5). There are, however, some approaches that warrant exploration.

3.6.1 Whole-cell detection of teichoic acid components

The detection of glycerol and ribitol in whole bacteria provides little information because glycerol can derive from a wide range of cellular components—lipids, lipoteichoic acid, metabolic intermediates, for example—while ribitol can arise from ribose in RNA if the standard alditol acetate GC method is employed for analysis. However, bacteria containing entirely unsubstituted teichoic acids are rare. In most cases, type I teichoic acids are at least partially glycosylated and therefore give rise to glycosyl alditols on treatment with HF. Under carefully controlled conditions, HF will only cleave phosphate esters, and in addition the products of teichoic acid degradation may be unique, and therefore readily recognizable in chromatograms.

It therefore appears feasible that HF treatment of whole bacteria, followed, for example, by GC of the trifluoroacetyl derivatives, could provide an initial screening procedure of considerable value, though the possibility of glycosyl glycerol arising from lipoteichoic acid and glycolipids would have to be considered A preliminary treatment with hot 50% phenol would provide considerable initial purification by solubilizing both these contaminants.

REFERENCES

Ames, B.N. (1966) Assay of inorganic phosphate, total phosphate and phosphatases. *Meth. Enzymol.* **8**: 116–18.
Aono, R. and Uramoto, M. (1986) Presence of fucosamine in teichuronic acid of the alkalophilic *Bacillus* strain C-125. *Biochem. J.* **233**: 291–4.
Archibald, A.R., Hancock, I.C. and Harwood, C.R. (1993) Cell wall structure, synthesis and turnover. In *Bacillus subtilis and other Gram-positive Bacteria: Biochemistry, Physiology and Molecular Genetics* (Hoch, J., ed), ch. III. American Society for Microbiology: Washington.
Blumenkrantz, N. and Asboe-Hansen, G. (1973) New method for the quantitative determination of uronic acids. *Analyt. Biochem.* **54**: 484–9.
Boon, J.J., De Boer, W.R., Kruyssen, F.J. and Wouters, J.T.M. (1981) Pyrolysis mass spectrometry of whole cells, cell walls and isolated cell wall polymers of *Bacillus subtilis* var. *niger* WM. *J. Gen. Microbiol.* **122**: 119–27.
Bundle, D.R. (1988) Structural analysis of polysaccharides by high-resolution ^{1}H and ^{13}C NMR. In *Bacterial Cell Surface Techniques* (Hancock, I.C. and Poxton, I.R., eds), pp. 146–58. Wiley: Chichester.
Chang, Y.-Y., Knecht, R. and Braun, D.G. (1983) Aminoacid analysis in the picomole range by pre-column derivatisation and high performance liquid chromatography. *Meth. Enzymol.* **91**: 41–8.
Chen, P.S., Toribara, T.Y. and Warner, H. (1956) Microdetermination of phosphorus. *Analyt. Chem.* **28**: 1756–61.
Endl, J., Seidle, H.P., Fiedler, F. and Schleifer, K.H. (1983) Chemical composition and structure of cell wall teichoic acids of staphylococci. *Arch. Microbiol.* **135**: 215–23.
Fiedler, F., Schaffler, M.J. and Stackebrandt, E. (1981) Biochemical and nucleic acid hybridisation studies on *Brevibacterium linens* and related strains. *Arch. Microbiol.* **129**: 85–93.
Glauner, B. and Schwartz, U. (1988) Investigation of murein structure and metabolism by HPLC. In *Bacterial Cell Surface Techniques* (Hancock, I.C. and Poxton, I.R., eds), pp. 158–70. Wiley: Chichester.
Hancock, I.C. and Poxton, I.R. (eds) (1988) *Bacterial Cell Surface Techniques*. Wiley: Chichester.
Hardy, M.R. (1989) Monosaccharide analysis of glycoconjugates by high performance anion-exchange chromatography with pulsed amperometric detection. *Meth. Enzymol.* **179**: 76–85.
Harwood, C.R., Coxon, R.D. and Hancock, I.C. (1990) The *Bacillus* cell envelope and secretion. In *Molecular Biological Methods for Bacillus* (Harwood, C.R. and Cutting, S.M., eds), Ch. 8. Wiley: Chichester.
Hase, S. and Matsushima, Y. (1979) The structure of the branching point between acidic polysaccharide and peptidoglycan in *Micrococcus lysodeikticus* cell wall. *J. Biochem.* **81**: 1181–6.
Hayashi, K. (1975) A rapid determination of sodium dodecyl sulphate with methylene blue. *Analyt. Biochem.* **67**: 503–6.
Himanen, J.-P. (1992) Determination of glycerol in bacterial cell wall teichoic acid by high performance liquid chromatography. *J. Chromatog.* **607**: 1–6.
Kehoe, M.A. (1993) Cell-wall-associated proteins in Gram-positive bacteria. In *New Comprehensive Biochemistry* (Ghuysen, J.-M. and Hackenbeck, R., eds), Ch. 11. Elsevier: Amsterdam.
Komagata, K. and Suzuki, K.-I. (1987) Lipid and cell wall analysis in bacterial systematics. *Meth. Microbiol.* **19**: 161–207.

Lechevalier, M.P. and Lechevalier, H.A. (1970) Chemical composition as a criterion in the classification of aerobic actinomycetes. *Int. J. Syst. Bacteriol.* **20**: 435–43.

Lifely, M.R., Tarelli, E. and Baddiley, J. (1980) The teichuronic acid from the wall of *Bacillus licheniformis* ATCC 9945. *Biochem. J.* **191**: 305–18.

Russell, R.R.B. (1988) Isolation and purification of protein linked to the cell wall in Gram-positive bacteria. In *Bacterial Cell Surface Techniques* (Hancock, I.C. and Poxton, I.R., eds), pp. 104–10. Wiley: Chichester.

Saddler, G.S., Tavecchia, P., Lociuro, S., Zanol, M., Colombo, L. and Selva, E. (1991) Analysis of madurose and other actinomycete whole cell sugars by gas chromatography. *J. Microbiol. Meth.* **14**: 185–91.

Sawardeker, J.S., Sloneker, J.H. and Jeanes, A. (1967) Quantitative determination of monosaccharides as their alditol acetates by gas liquid chromatography. *Analyt. Chem.* **37**: 1602–4.

Schleifer, K.H. and Kandler, O. (1972) The peptidoglycan types of the bacterial cell walls and their taxonomic implications. *Bacteriol. Rev.* **36**: 407–77.

Schleifer, K.H. and Seidl, P.H. (1985) Chemical composition and structure of murein. In *Chemical Methods in Bacterial Systematics* (Goodfellow, M. and Minnikin, D., eds), pp. 201–19. Academic Press: London.

Schleifer, K.H., Hammes, W.P. and Kandler, O. (1976) Effect of endogenous and exogenous factors on the primary structures of bacterial peptidoglycan. *Adv. Microbial Physiol.* **13**: 245–92.

Sonesson, A., Larsson, L., Fox, A., Westerdahl, G. and Odham, G. (1988) Determination of environmental levels of peptidoglycan and lipopolysaccharide using gas chromatography with negative-ion chemical-ionisation mass spectrometry utilising bacterial amino acids and hydroxy fatty acids as biomarkers. *J. Chromatog.* **431**: 1–15.

Staneck, J.L. and Roberts, G.D. (1974) Simplified approach to the identification of aerobic actinomycetes by thin-layer chromatography. *Appl. Microbiol.* **28**: 226–31.

Takebe, I. (1965) Extent of cross-linkage in the murein sacculus of *Escherichia coli* B cell wall. *Biochim. Biophys. Acta* **101**: 124–6.

Tipper, D.J. and Strominger, J.L. (1966) Isolation of 4-O-β-N-acetylmuramyl-N-acetylglucosamine and 4-O-β-N,6-O-diacetylmuramyl-N-acetylglucosamine and the structure of the cell wall polysaccharide of *Staphylococcus aureus*. *Biochem. Biophys. Res. Commun.* **22**: 48–56.

Uchida, K. and Aida, K (1977) Acyl type of bacterial cell wall: its simple identification by colorimetric method. *J. Gen. Appl. Microbiol.* **23**: 249–60.

Uchikawa, K., Sekikawa, I. and Azuma, I. (1986) Structural studies on teichoic acids in cell walls of several serotypes of *Listeria monocytogenes*. *J. Biochem.* **99**: 315–29.

White, P.J. (1977) A survey for the presence of teichuronic acid in the walls of *Bacillus megaterium* and *Bacillus cereus*. *J. Gen. Microbiol.* **102**: 435–9.

Whitton, R.S., Lau, P., Margan,S.L., Gilbert, J. and Fox, A. (1985) Modification of the alditol acetate method for analysis of muramic acid and other neutral and aminosugars by capillary gas chromatography-mass spectrometry with selected ion monitoring. *J. Chromatog.* **347**: 109–20.

Yoneyama, T., Koike, Y., Araki, Y., Arakawa, H., Yokohama,K., Sasaki, Y., Kawamura, T., Ito, E. and Takao, S. (1982) Distribution of mannosamine and mannosaminuronic acid among cell walls of *Bacillus* species. *J. Bacteriol.* **149**: 15–21.

Yoneyama, T., Araki, Y. and Ito, E. (1984) The primary structure of teichuronic acid in *Bacillus subtilis* AHU 1031. *Eur. J. Biochem.* **141**: 83–9.

Zipperle, G.F., Ezzell, J.W. and Doyle, R.J. (1984) Glucosamine substitution and muramidase susceptibility in *Bacillus anthracis*. *Can. J. Microbiol.* **30**: 553–9.

4
Analysis of Archaeal Cell Envelopes

Helmut König
Universität Ulm, Federal Republic of Germany

4.1 INTRODUCTION

The archaea (neé archaebacteria) represent a third domain separate from the domains bacteria and eucarya (Woese, 1987; Woese *et al.*, 1990). Significant differences to the cell walls of bacteria were established prior to the publication of this new phylogenetic concept when the cell envelopes of *Halobacterium* (Houwink, 1956; Mescher & Strominger, 1974), *Halococcus* (Brown & Cho, 1970; Steber & Schleifer, 1975), *Sulfolobus* (Weiss, 1974) and of the methanogenic bacteria (Kandler & Hippe, 1977) were analysed. Once the taxonomic significance of cell envelope structure was appreciated, phenotypic characteristics derived from studies on wall envelopes were used to distinguish archaea from bacteria (Kandler, 1979). It soon became clear that archaea did not have a common cell wall polymer but exhibited heterogeneous cell envelopes which lacked the typical cell wall component of bacteria, the murein (Kandler & König, 1985; König, 1988). Cell wall-less organisms like *Thermoplasma acidophilum* (Darland *et al.*, 1970) are also found among the archaea.

4.2 ELECTRON MICROSCOPY

4.2.1 General information

The fact that the archaea are a heterogeneous group is reflected by the great variety of cell envelope structures they show (Table 4.1; Figure 4.1; Kandler & König, 1985; König, 1988, Sleytr *et al.*, 1986; Jain *et al.*, 1988; Lechner & Wieland, 1989). Archaea have not developed a common cell wall polymer like the bacterial murein; the common profile of Gram-negative bacteria, which consists of a thin murein layer and an outer

---

Chemical Methods in Prokaryotic Systematics. Edited by M. Goodfellow and A.G. O'Donnell
© 1994 John Wiley and Sons Ltd.

Table 4.1 Systematics and chemical composition of the cell envelopes of archaea.*

Order	Family	Genus	Species	Cell envelopes A	Cell envelopes B	MM[a] (kDa)
I. Methanobacteriales	I. Methanobacteriaceae	1. Methanobrevibacter	M. arboriphilus	5	PM	–
			M. ruminantium	5	PM	–
			M. smithii	5	PM	–
		2. Methanobacterium	M. alcaliphilum	5	o	
			M. bryantii	5	PM	–
			M. formicicum	5	PM	–
			M. ivanovii	5	o	
			M. thermoaggregans	5	o	
			M. thermoalcaliphilum	5	o	
			M. thermoautotrophicum	5	PM	–
			M. thermoformicicum	5	o	
			M. uliginosum	5	PM	–
			M. wolfei	5	PM	–
		3. Methanosphaera	M. cuniculi	5	PM	–
			M. stadtmanae	5	PM	–
	II. Methanothermaceae	1. Methanothermus	M. fervidus	6	PM, GSL	92
			M. sociabilis	6	PM, GSL	89
II. Methanococcales	I. Methanococcaceae	1. Methanococcus	M. aeolicus	1	PSL	64
			M. deltae	1	o	
			M. halophilus	1	o	
			M. jannaschii	1	PSL	90
			M. maripaludis	1	o	
			M. thermolithotrophicus	1	PSL	83
			M. vannielii	1	PSL	60
			M. voltae	1	PSL	76

III. Methanomicrobiales	I. Methanomicrobiaceae	1. *Methanomicrobium*	*M. mobile*	1	PSL	175
			M. paynteri	1	o	
		2. *Methanogenium*	*M. aggregans*[d]	1	GSL	92
			M. bourgense[e]	1	GSL	101
			M. cariaci	1	PSL	125/117
			M. frittonii	1	GSL	106
			M. marisnigri[f]	1	GSL	143/138
			M. olentangyi[g]	1	GSL	132
			M. tationis	1	GSL	120
			M. thermophilicum[h]	1	GSL	130
		3. *Methanospirillum*	*M. hungatei*	2	PS[b]	o
	II. Methanoplanaceae	1. *Methanoplanus*	*M. endosymbiosus*	1	GSL	110
			M. limicola	1	GSL	143
	III. Methanosarcinaceae	1. *Methanosarcina*	*M. acetivorans*	1	SL	
			M. barkeri	4	MC	
			M. frisia	4	o	
			M. mazei	4	MC	
			M. thermophila	4	MC	
			M. vacuolata	4	o	
		2. *Methanolobus*	*M. siciliae*	1	GSL	180
			M. tindarius	1	GSL	156
			M. vulcani	1	GSL	116
		3. *Methanococcoides*	*M. methylutens*	1	GSL	159
		4. *Methanothrix*	*M. concilii*[i]	2	PS[b]	
			M. soehngenii	2	GPS	
			M. thermoacetophila[j]	2	o	
		5. *Methanohalophilus*	*M. mahi*	1	o	
			M. zhilinae	1	o	
IV. Methanocorpusculaceae	1. *Methanocorpusculum*		*M. parvum*	1	o	

(Continued)

Table 4.1 (Continued)

Order	Family	Genus	Species	Cell envelopes A B	MM[a] (kDa)
IV. Halobacteriales	V.	1. Methanohalobium	M. evestigatus	1 o	
	I. Halobacteriaceae	1. Halobacterium	H. cutirubrum	1 GSL	200
			H. denitrificans	1 o	
			H. halobium	1 GSL	200
			H. saccharovorum	1 GSL	220
			H. salinarium	1 GSL	200
			H. sodomense	1 o	
			H. trapanicum	1 o	
		2. Haloarcula	H. hispanica	1 o	
			H. vallismoritis	1 o	
		3. Haloferax	H. gibbonsii	1 o	
			H. mediterranei	1 o	
			H. volcanii	1 SL	185
		4. Halococcus	H. morrhuae	5 HP	
		5. Natronobacterium	N. gregoryi	1 o	
			N. magadii	1 o	
			N. pharaonis	1 o	
		6. Natronococcus	N. occultus	1 o	
		7. Walsby's square bacterium		1 GSL	
V. Thermoplasmales	I. Thermoplasmaceae	1. Thermoplasma	T. acidophilum	3 WE[c]	152
			T. vulcanium	3 o	
VI. Archaeoglobales	I. Archaeoglobaceae	1. Archaeoglobus	A. fulgidus	1 GSL	

VII. Thermococcales	I. Thermococcaceae	1. Thermococcus	T. celer	1	o	
		2. Pyrococcus	P. furiosus	1	GSL	138
			P. woesei	1	o	
VIII. Sulfolobales	I. Sulfolobaceae	1. Sulfolobus	S. acidocaldarius	1	GSL	140
			S. solfataricus	1	SL	122
		2. Acidianus	A. brierleyi	1	SL	116
			A. infernus	1	o	
		3. Desulfurolobus	D. ambivalens	1	o	
IX. Thermoproteales	I. Thermoprotaceae	1. Thermoproteus	T. neutrophilus	1	SL[b]	
			T. tenax	1	SL[b]	
		2. Pyrobaculum	P. islandicum	1	o	
			P. organotrophum	1	o	
		3. Thermofilum	T. librum	1	o	
			T. pendens	1	o	
	II. Desulfurococcaceae	1. Desulfurococcus	D. mobilis	1	o	
			D. mucosus	1	o	
			D. saccharovorans	1	o	
	III. Staphylothermaceae	1. Staphylothermus	S. marinus	1	o	
	IV. Pyrodictiaceae	1. Pyrodictium	P. brockii	1	GSL	
			P. occultum	1	GSL	172
	V. Thermodiscaceae	1. Thermodiscus	T. maritimus	1	GSL	

* Data from: Kandler and König (1985), Sleytr et al. (1986), Jain et al. (1988), König (1988); Lechner and Wieland (1989) and Zellner (1989). *Abbrevialtions:* A, cell envelope profile (cf. Figure 4.1); B, chemical composition of the cell envelopes; GPS, glycoproteinaceous sheath; GSL, glycoprotein S-layer; HP, heteropolysaccharide; MC, methanochondroitin; MM, molecular mass; PM, pseudomurein; PS, proteinaceous sheath (carbohydrates present); PSL protein S-layer; SL, S-layer (glycosylation not proven); WE, without cell envelope; o, not determined; – not found.

[a] Apparent molecular masses of S-layer (glyco-)proteins were determined by SDS-polyacrylamide gel electrophoresis; [b] carbohydrates present; [c] a mannose-rich glycoprotein present in the cytoplasmic membrane; [d] renamed *Methanoculleus aggregans*; [e] renamed *Methanoculleus bourgense*; [f] renamed *Methanoculleus marisnigri*; [g] renamed *Methanoculleus olentangii*; [h] renamed *Methanoculleus thermophilicus*; [i] renamed *Methanosaeta concillii*; [j] renamed *Methanosaeta thermoacetophila*.

GRAM REACTION	PROFILE	REPRESENTATIVE GENERA
Negative	1. ●●●●●●●●●● SL	Halobacterium Methanococcus Thermoproteus
Negative	2. ●● ●● ●● ●● PS ●●●●●●●●●● SL ———— CM	Methanospirillum
Negative	3. ▮▮▮▮▮▮▮▮▮▮ CM	Thermoplasma
Positive or negative	4. ▬▬▬▬▬▬ MC ●●●●●●●●●● SL ———— CM	Methanosarcina
Positive	5. ▬▬▬▬▬▬ PM,HP ———— CM	Halococcus Methanobacterium
Positive	6. ●●●●●●●●●● SL ▬▬▬▬▬▬ PM ———— CM	Methanothermus

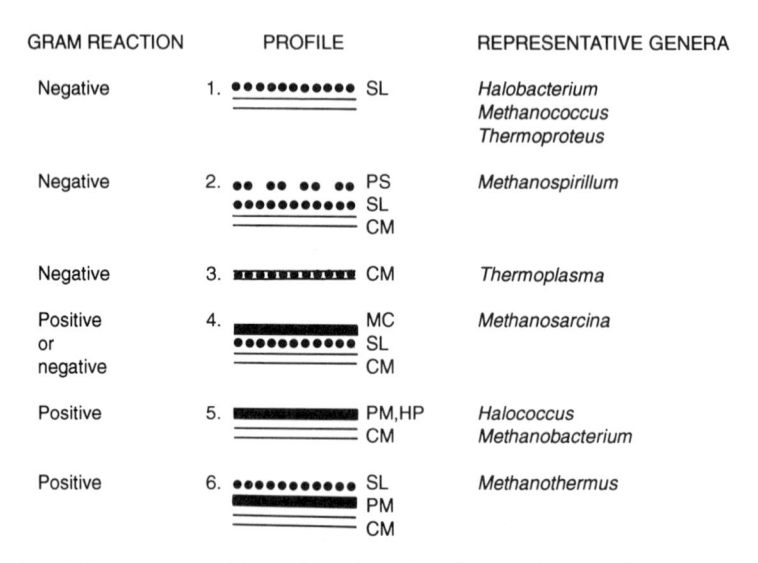

Figure 4.1 Schematic profiles of archaeal cell envelopes. CM, cytoplasmic membrane; HP, heteropolysaccharide; PM, pseudomurein; PS, proteinaceous sheath; SL, surface layer (S-layer); MC, methanochondroitin. *Methanosarcina* may lack one of the two layers. It is still not known whether the layer (named 'SL') surrounding the individual cells in *Methanospirillum* (and *Methanothrix*) strains is composed of proteinaceous subunits.

membrane, is also absent. Morphologically the cell envelope profiles of the Gram-positive members of both prokaryotic domains are similar. The Methanobacteriales (Figures 4.1 and 4.2a,b) exhibit only one electron-dense layer, 15–20 nm in width, apart from members of the genus *Methanothermus* which have two layers, the pseudomurein and S-layers.

Most archaea are Gram-negative. The most common cell wall profile of these organisms is represented by a single layer (S-layer) of protein or glycoprotein subunits closely associated with the cytoplasmic membrane (Figures 4.1 and 4.2e,g). The filamentous species of the genera *Methanospirillum* and *Methanothrix* show individual cells within the filaments which are surrounded by a layer (Figures 4.1 and 4.2f) which has not been defined chemically; the outer envelope of the filament is formed by a tubular sheath

Figure 4.2 (*opposite*) Electron micrographs of archaea. (a) *Methanobacterium uliginosum*, thin sections, bar = 1 µm; (b) *Methanothermus fervidus*, thin sections of dividing cells, bar = 0.2 µm; (c) *Methanosarcina* sp., thin sections, bar = 2 µm; (d) *Halococcus morrhuae*, thin section, bar = 0.5 µm; (e) *Pyrodictum occultum*, thin sections, bar = 0.5 µm; (f) *Methanothrix soehngenii*, isolated sheath, platinum shadowing, bar = 0.5 µm; (g) *Thermoproteus tenax*, isolated S-layers, platinum shadowing, bar = 0.25 µm; (h) *Thermoplasma acidophilum*, thin sections, bar = 0.5 µm.

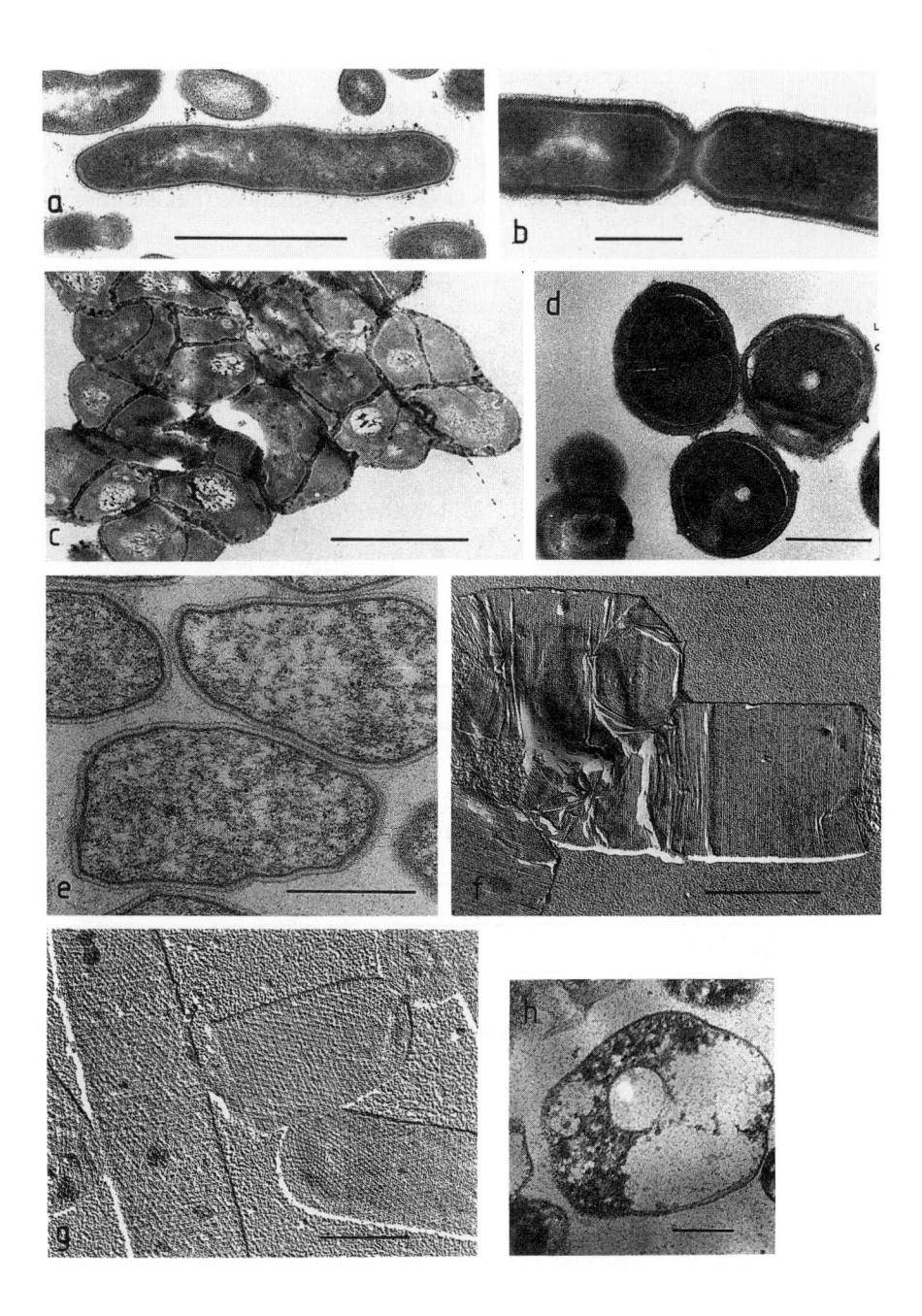

4. Analysis of Archaeal Cell Envelopes 91

composed of hoop-like rings. Special image processing techniques have been used to determine the three-dimensional structure of several S-layers (Taylor et al., 1982; Sleytr & Messner, 1983; Stewart et al., 1985; Shaw et al., 1985, Beveridge et al., 1986a; Messner et al., 1986; Patel et al., 1986; Baumeister & Engelhardt, 1987; Wildhaber et al., 1987; Wildhaber & Baumeister, 1987; Kessel et al., 1988). The three-dimensional structure of pseudomurein sacculi has been revealed by X-ray analysis (Labischinski et al., 1980; Formanek 1985).

4.2.2 Preparation of specimens

The methods described below are taken from König & Stetter (1982).

Method 4.1 Platinum shadowing.

1. Parlodion grids are dipped onto a drop of fresh archaeal culture and washed (3×) with water or basal salt solution (growth medium without organic constituents).
2. The cells are then fixed in a solution of glutaraldehyde (20 g/l) for 3 min at room temperature and washed (3×).
3. The grids are shadow-casted (Edwards vacuum coater) with a platinum-iridium alloy at an angle of 7° followed by carbon coating of the parlodion film for stabilization.

Method 4.2 Negative staining.

Instead of shadow-casting, the specimens can be negatively stained by dipping the grids onto a drop of aqueous uranyl acetate (20 g/l) for 5 min. The grids are then air dried.

Method 4.3 Thin sections.

1. Archaeal cells or cell envelopes are fixed in a solution of glutaraldehyde (20 g/l) in water or basal salt solution for 2 h, washed (3×) and then placed in a solution of OsO_4 (10 g/l) in water or basal salt solution for 45 min.
2. After washing (3×) the cells are dehydrated with ethanol (70%, 3×; 80%; 90%; 96%; 100%, 2×; each for 10 min; propylene oxide, 2× for 15 min).
3. The specimens are incubated for 16 h at room temperature in a mixture of propylene oxide and Durcupan (1:1) followed by embedding in Durcupan ACM (Fluka; 24 h, 37 °C; 36 h, 60 °C).
4. Thin sections are prepared with an ultramicrotome (Reichert; LKB) using a diamond knife (Diatome).
5. The sections are then contrasted with lead citrate (26.6 g $Pb(NO_3)_2$/l; 32.2 g trisodium citrate/l) for 5 min, washed, in uranyl acetate (20 g/l) in methanol/water (7:3) for 5 min, followed by hydration, which is achieved

by dipping the grids in aqueous ethanol (70%, 2×; 50%; 30%; 20%) and water.
6. Finally, the sections are contrasted once again with lead citrate for 3 min.

4.3 ISOLATION AND ANALYSIS

4.3.1 Pseudomurein

4.3.1.a General information

Pseudomurein is the characteristic cell wall polymer of members of the order Methanobacteriales (König *et al.*, 1982) and *Methanopyrus* (Kurr *et al.*, 1991). This polymer forms approximately 15 nm-thick cell wall sacculi (Figure 4.2a,b) which surround the cells. The density of the pseudomurein has been determined as $1.39-1.46$ g/cm^3 and the unit cell has a dimension of $4.5 \times 10 \times 21-22.4$ Å (Labischinski *et al.*, 1980; Formanek, 1985).

The pseudomurein is composed of glycan strands cross-linked by peptide subunits (Figure 4.3). The glycan strands consist of alternating $\beta(1 \rightarrow 3)$-linked *N*-acetyl-D-glucosamine or *N*-acetyl-D-galactosamine and *N*-acetyl-L-talosaminuronic acid residues. The peptide moiety is composed of a set of the three L-amino acids: glutamic acid, alanine, and lysine (Table 4.2). In some genera alanine can be replaced by threonine or serine, and ornithine can occur in addition to lysine. Biosynthesis of the pseudomurein proceeds via a UDP-activated disaccharide and UDP-activated peptides; undecaprenol acts as lipid carrier (Hartmann *et al.*, 1989; Hartmann & König 1989; König *et al.*, 1989). Nothing is known about the decomposition of pseudomurein-containing envelopes in nature.

The pseudomurein sacculi have been partially removed in some species by adding elevated concentrations of amino acids (König, 1985). Recently, an autolytic activity was described for *Methanobacterium wolfei* (König *et al.*, 1985; Kiener *et al.*, 1987); it is based on the production of a peptidase that splits the peptide bond between alanine and lysine. This enzyme has been used to produce protoplasts for both membrane studies (Mountfort *et al.*, 1986) and for the isolation of nucleic acids (Kiener *et al.*, 1987).

Pseudomurein shows antigenicity. The glycan components glucosamine, galactosamine, and talosaminuronic acid and the C-terminal dipeptide γ-glutamyl-alanine are the antigenic determinants (Conway de Macario *et al.*, 1983). Biosynthesis of pseudomurein is resistant to antibiotics acting against the synthesis of the cytoplasmic precursors, such as β-lactam antibiotics (Hammes *et al.*, 1979), and to lytic enzymes (Kandler & König, 1978).

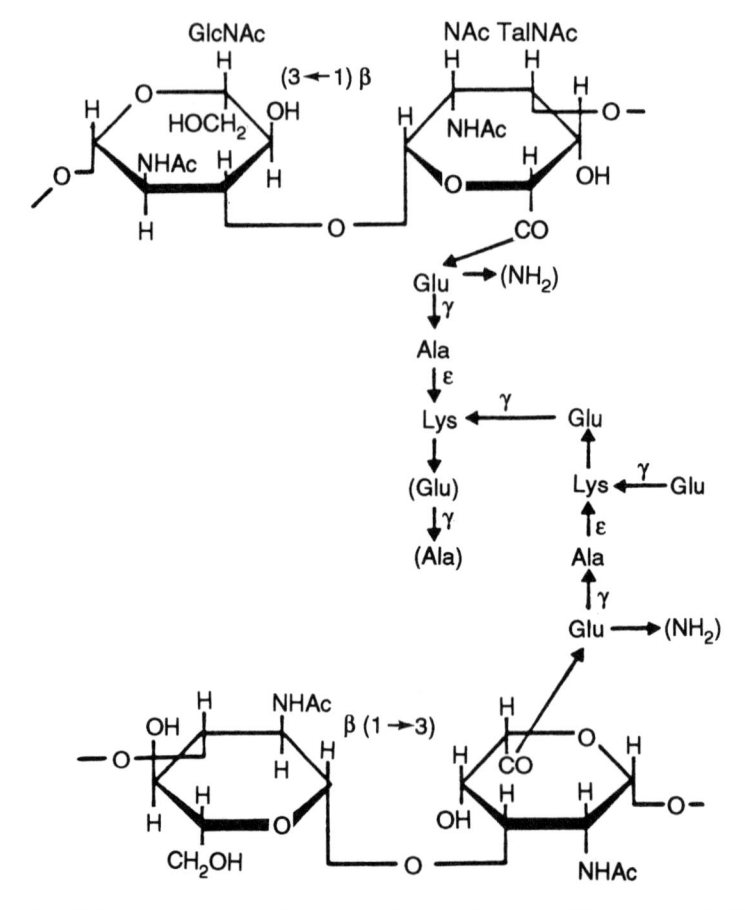

Figure 4.3 Primary structure of the pseudomurein. From König et al. (1982) with permission.

4.3.1.b Isolation

Pseudomurein sacculi of all members of the order Methanobacteriales can be isolated by mechanical disintegration and digestion with trypsin as usually applied to Gram-positive bacteria (Schleifer & Kandler, 1972; Kandler & König, 1978; Schleifer, 1985).

Method 4.4 Mechanical disintegration and digestion of cells.

---

1. Harvested cells are disrupted either with glass beads or by ultrasonication.

2. A cell paste is then prepared in water and mixed with about two volumes of glass beads (diameter 0.18 mm).
3. The cells and glass beads are shaken in a special cell mill (Vibriogen cell mill Vi4; Bühler) for 20 to 30 min. The disruption of the cells is checked by phase-contrast microscopy.
4. The beads are then removed from the cells by filtering through a coarse-grade sintered filter (grade 1; Schott); the glass beads are cleaned by washing with concentrated hydrochloric acid and distilled water.
5. The crude cell wall suspension is centrifuged (18 000 rpm, Sorvall rotor SS34) for 20 min.
6. The pellet is suspended in potassium phosphate buffer (0.05 M, pH 7.8) and incubated with trypsin (0.5 mg/ml) at 37 °C for about 16 h; bacterial growth is prevented by adding toluene (~1 ml/100 ml).
7. The suspension is then centrifuged and the pellet washed with water (3–4×). Wet cell walls can be kept at −20 °C or at room temperature after freeze-drying (Edwards).

Instead of using a cell mill, archaea can be disrupted by ultrasonication (Sonifier B-12; Branson Sonic Power Company); cells are suspended in distilled water and sonicated for 30–60 min at 60 W. Instead of using trypsin, crude walls can be purified by heating a cell wall suspension in a solution of sodium dodecyl sulphate (2–10%) in a boiling water bath for 30 min. Cell wall polysaccharides can be partly extracted by heating cell walls (5 mg/ml) in hot formamide (170 °C) for 30 min (Schleifer et al., 1968).

4.3.1.c Analysis of constituents

The chemical composition of trypsin-purified cell walls of members of the order Methanobacteriales is shown in Table 4.2 (Kandler & König, 1978; König et al., 1982). The walls consist of pseudomurein, neutral polysaccharides and inorganic salts, but teichoic acids, lipoteichoic acids and teichuronic acids, all typical constituents of the cell walls of Gram-positive bacteria, are absent.

Method 4.5 Hydrolysis of pseudomurein components.

1. Amino acid and amino sugar residues of the pseudomurein are released as monomers by hydrolysis with hydrochloric acid (4 M; 0.1 ml/mg) for 16 h at 100 °C.
2. Talosaminuronic acid is qualitatively determined after a short period of hydrolysis (4 M HCl, 30 min, 100 °C); this component is nearly completely destroyed during prolonged acid hydrolysis (4 h).
3. Excess hydrochloric acid is removed in a stream of warm nitrogen (40 °C).
4. Neutral sugars are released with 2 M HCl (0.1 ml/mg) for 2 h at 100 °C.

Table 4.2 Molar ratios of the components of the cell walls of the order Methanobacteriales.

Species	Amino acids Ser	Ala	Thr	Glu	Lys	Orn	NH$_3$	GalN	N-acetyl amino sugars GlcN	TalNA[a]	Neutral sugars Gal	Glc	Man	Rha	Phosphate (%DW)
Methanobrevibacter															
M. arboriphilus DH1	–	1.52	–	2.13	1.00	–	1.36	0.67	–	(1.00)	–	–	0.02	0.02	4.54
M. arboriphilus AZ	–	1.17	–	2.32	1.00	–	3.92	0.91	–	(1.00)	o	o	o	o	o
M. smithii PS	–	1.42	–	2.04	1.00	0.40	0.81	0.44	0.56	(1.00)	0.11	0.14	–	0.09	1.00
M. ruminantium M1[b]	–	–	0.90	1.85	1.00	–	0.85	0.80	0.60	(1.00)	0.29	0.02	0.02	0.21	4.65
M. ruminantium M1[b]	–	0.70	0.40	1.80	1.00	–	1.00	0.80	0.50	(1.00)	o	o	o	o	o
Methanobacterium															
M. bryantii M.o.H.	–	1.39	–	2.39	1.00	–	1.12	0.45	1.25	(1.00)	0.84	0.09	1.18	+	0.20
M. bryantii M.o.H.G.	–	1.32	–	2.21	1.00	–	0.84	–	0.84	(1.00)	0.53	0.26	0.26	+	o
M. formicicum MF	–	1.51	–	2.11	1.00	–	1.53	1.00	1.16	(1.00)	0.76	0.27	0.37	0.62	0.64
M. thermoautotrophicum ΔH	–	1.20	–	2.27	1.00	–	0.61	0.16	1.18	(1.00)	0.02	0.13	–	+	0.28
M. thermoautotrophicum Marburg	–	1.32	–	2.37	1.00	–	0.77	0.87	0.23	(1.00)	o	o	o	o	o
M. wolfei	–	1.65	–	3.82	1.00	–	0.87	1.54	0.39	(1.00)	o	o	o	o	o
M. uliginosum P2 St.	–	1.35	–	1.74	1.00	–	0.89	0.16	0.75	(1.00)	1.08	0.29	0.71	0.17	o
Methanopyrus															
M. kandleri AV19	–	1.51	–	2.33	1.00	–	o	0.91	–	(1.00)	o	o	o	o	o
Methanothermus															
M. fervidus V24S	–	1.47	–	2.23	1.00	–	1.64	0.35	0.49	(1.00)	–	0.01	–	o	o
M. sociabilis KF1-F1	–	1.14	–	2.51	1.00	–	0.96	0.36	1.30	(1.00)	o	o	o	o	o
Methanosphaera															
M. stadtmanae MCB-3	0.77	0.26	–	1.94	1.00	–	1.56	0.32	0.59	(1.00)	0.04	0.24	–	–	o

Data modified from König et al. (1982) and Kurr et al. (1991).
[a] The molar ratio of N-acetyltalosaminuronic acid was estimated from the analysis of partial acid hydrolysates since it is completely destroyed in total hydrolysates.
[b] Two different batches of cells.
–, not found; +, trace; o, not determined.
Abbreviations: Ala, alanine; DW, dry weight; Gal, galactose; GalN, galactosamine; Glc, glucose; GlcN, glucosamine; Glu, glutamic acid; Lys, lysine; Man, mannose; Orn, ornithine; Rha, rhamnose; Ser, serine; TalNA, talosaminuronic acid; Thr, threonine.

4. Analysis of Archaeal Cell Envelopes

Method 4.6 Identification and quantification of amino acids and amino sugars.

Identification and quantification of the cell wall amino acids (glutamic acid, alanine, lysine, threonine, serine, ornithine) and the amino sugars (galactosamine, glucosamine) is carried out using an amino acid analyser (LC 5000, Biotronic) programmed for 'Bacterial Hydrolysates' (Biotronik).

Method 4.7 Isolation and quantification of talosaminuronic acid.

The acid and lactone forms of talosaminuronic acid can also be identified by using the amino acid analyser. However, talosaminuronic acid should be isolated first by thin-layer chromatography (König & Kandler, 1979) in order to avoid interference with other oligomers produced after partial acid hydrolysis (4 M HCl, 30 min 100 °C).

1. The components of the partial acid hydrolysate of the cell walls are applied to cellulose thin-layer plates (Merck) and a mixture of α-picoline: $NH_3(25\%)$: H_2O (70 : 2 : 28) is used as the solvent system.
2. The ninhydrin-positive band that migrates faster than glucosamine ($R_F = 0.78$; lactonized form) is eluted with water and incubated with 1 M NH_3 for 4 h at room temperature to effect delactonization.
3. It is then re-chromatographed in the same system.
4. The ninhydrin-positive band (free acid) that migrates half the distance of glucosamine ($R_F = 0.33$) is eluted with 0.01 M HCl.
5. The lactonized $R_{GlcN} = 1.19$, Liquimat III; Kontron; König & Kandler, 1979) and delactonized forms ($R_{GlcN} = 0.92$) give two separate peaks when examined using an amino acid analyser.

Method 4.8 Identification of neutral sugars by GLC.

Neutral sugars can be identified as their alditol derivatives by gas-liquid chromatography using a method modified from Albersheim et al. (1967).

1. The dried hydrolysate of 3 mg of cell walls is dissolved in 1 ml 0.5 M NH_3.
2. After addition of 10 mg $NaBH_4$ the mixture is incubated for 2 h at room temperature.
3. The excess of $NaBH_4$ is destroyed by adding a drop of acetic acid.
4. Water is removed in a stream of nitrogen.
5. Methanol (1 ml) is added (4×) and the sample is dried with nitrogen.
6. The residue is dissolved in 1 ml of a mixture of pyridine/acetic anhydride (1 : 1) and heated for 30 min at 100 °C; the solvent is then removed with a stream of nitrogen.
7. The residue is dissolved in 100 µl acetic acid ethyl ester and 1 µl of this preparation is injected into the gas chromatograph, Durabond 1701 capillary column: 25 m; ICT Laboratories: 120 °C isothermally for 2 min → (20 °C/min) → 240 °C isothermal for 20 min.

Method 4.9 Phosphate determination (after Chen et al., 1956).

1. Cell walls (5 mg) are suspended in 1 ml conc. H_2SO_4 and ashed by heating in a Bunsen flame until the colour turns black.
2. The mixture is then treated with 0.2 ml H_2O_2 and heated until a clear solution is obtained.
3. The latter is brought to a volume of 15 ml with water.
4. Prepare the reagent by mixing 10 ml 3 M H_2SO_4, 20 ml H_2O, 10 ml ammonium molybdate (2.5%) and 10 ml ascorbate (10%)
5. An aliquot of the ashed sample is mixed with 4 ml H_2O and 4 ml reagent and incubated for 2 h at 37 °C.
6. The extinction is measured at 680 nm.
Measuring range: 0.02 to 0.26 μmol PO_4^{3-}.

4.3.1.d Determination of the pseudomurein type

In bacteria more than 100 different murein modifications have been identified which differ in the primary structure of the peptide subunit (Schleifer & Kandler, 1972; Schleifer & Seidl, 1984; Schleifer, 1985). In the case of the pseudomurein only one primary structure has been found. However, alanine can be replaced by either serine or threonine, and ornithine can occur in addition to lysine (Table 4.2). In contrast to murein, glucosamine can be replaced by galactosamine. For taxonomic purposes it is sufficient to determine the chemical composition of the pseudomurein (Table 4.2; König et al., 1982). Consequently, the primary structure of the peptide subunit of the pseudomurein has not to be determined by analysing overlapping oligopeptides (Schleifer & Kandler, 1972; König & Kandler, 1979; Schleifer, 1985).

4.3.2 Methanochondroitin

4.3.2.a General information

Cells of *Methanosarcina* strains often form large cuboid aggregates. Most of the cells have an outer cell wall layer that is up to 200 nm thick (Figure 4.2c). This layer consists of N-acetylglucosamine, glucuronic or galacturonic acid, glucose, and small amounts of mannose (Table 4.3; Kreisl & Kandler, 1986). The shape-maintaining component is a fibrillar nonsulphated polymer composed of uronic acid and N-acetylgalactosamine residues in a molar ratio of 1 : 2 (Figure 4.4). This polymer forms a compact or sometimes loose matrix reminiscent of components of eukaryotic connective tissue. It was, therefore, named 'methanochondroitin' (Kreisl & Kandler, 1986).

4. Analysis of Archaeal Cell Envelopes

Table 4.3 Chemical composition (molar ratios) of isolated cell walls of *Methanosarcina* stains.*

DSM 800	D-Glc	D-GalN	D-GlcA	D-GalA	Acetate	$PO_4^{3(-)}$	$SO_4^{2(-)}$
DSM 800	0.24	1.88	1.00	–	2.31	0.21	0.04
DSM 804	0.36	2.57	1.00	–	2.63	0.20	0.02
Bryant G1	0.29	2.71	0.25	1.00	2.78	0.08	0.03
DSM 1232	0.13	1.86	1.00	–	2.17	0.08	0.02
DSM 1825	0.03	2.37	1.00	–	2.47	0.02	0.02
DSM 2053	0.07	6.37	1.00	1.00	6.37	0.32	0.05
DSM 1538	0.32	1.98	1.00	–	o	o	o

* Modified from Kreisl and Kandler (1986).
o, not determined.

Abbreviations: D-GalA, D-galacturonic acid; D-GalN, D-galactosamine; D-Glc, D-glucose; D-GlcA, D-glucuronic acid.

$$[\rightarrow 4)\text{-}\beta\text{-D-GlcA-}(1\rightarrow 3)\text{-D-GalNAC-}(1\rightarrow 3 \text{ or } 4)\text{-D-GalNAc-}(1\rightarrow]_n$$

Figure 4.4 Structure of the cell wall polymer (methanochondroitin) of *Methanosarcina*. From Kreisl and Kandler (1986) with permission.

4.3.2.b Isolation and analysis

Methanochondroitin sacculi can be isolated using the methods described for pseudomurein (see Method 4.4).

Neutral and amino sugars can be determined as described for pseudomurein (see Methods 4.5, 4.6 and 4.8).

Uronic acids can be determined as their alditol pentafluorpropionylates after methanolysis (Wieland *et al.*, 1983).

Method 4.10 Determination of uronic acids (after Wieland *et al.*, 1983).

1. Cell walls (3 mg) are suspended in 1 ml 0.5 N methanolic HCl (Serva) and hydrolysed at 80 °C for 20 h.
2. The methanolic HCl is removed in a stream of nitrogen.
3. Dichloromethane (100 μl) and pentafluoropropionic anhydride (100 μl) are added to the residue and the preparation is heated for 30 min at 100 °C.
4. The solvent is removed in a stream of nitrogen and the residue is dissolved in 100 μl methylene chloride.
5. 1 μl of the preparation is injected onto the column. The pentofluorpropionyl derivatives are detected as described in Method 4.8.

4.3.3 Heteropolysaccharide

4.3.3.a General information

The Gram-positive cells of *Halococcus morrhuae* are surrounded by an electron-dense cell wall layer that is about 50 nm thick (Figure 4.2d; Brown & Cho, 1970). The rigid cell wall sacculi prevent the lysis of the cocci in media at low ionic strength. The cell wall polymer is composed of a complex heteropolysaccharide consisting of *N*-acetylglucosamine, *N*-acetylgalactosamine, glucose, galactose, mannose, glucuronic acid, galacturonic acid, acetate, sulphate, glycine and gulosaminuronic acid (Table 4.4; Reistadt, 1972, 1974, 1975; Steber & Schleifer, 1975). The carbohydrate monomers are arranged in three domains (Figure 4.5; Schleifer *et al.*, 1982), which may be partly linked by *N*-glycyl-glucosaminyl bridges (Steber & Schleifer, 1979).

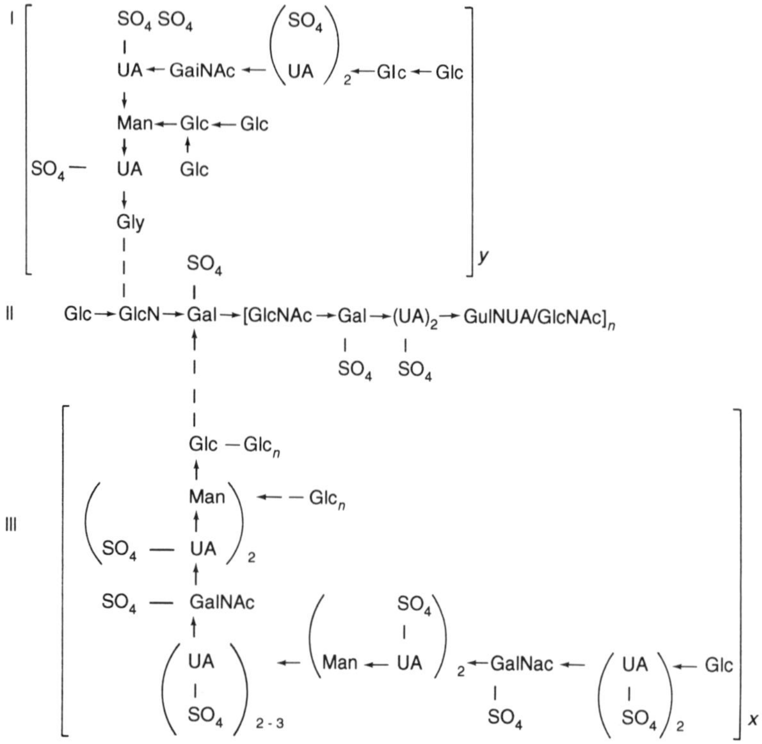

$$\text{I} \begin{bmatrix} & SO_4\ SO_4 & \begin{pmatrix} SO_4 \\ | \\ UA \end{pmatrix}_2 \leftarrow Glc \leftarrow Glc \\ & | & | \\ & UA \leftarrow GalNAc \leftarrow \\ & \downarrow \\ & Man \leftarrow Glc \leftarrow Glc \\ & \downarrow \quad\quad \uparrow \\ SO_4 - & UA \quad\quad Glc \\ & \downarrow \\ & Gly \\ & | \\ & | \quad SO_4 \\ & | \quad\quad | \end{bmatrix}_y$$

$$\text{II} \quad Glc \rightarrow GlcN \rightarrow Gal \rightarrow [GlcNAc \rightarrow Gal \rightarrow (UA)_2 \rightarrow GulNUA/GlcNAc]_n$$

$$\text{III} \begin{bmatrix} & Glc - Glc_n \\ & \uparrow \\ & \begin{pmatrix} Man \\ \uparrow \\ SO_4 - UA \end{pmatrix}_2 \leftarrow - Glc_n \\ & \uparrow \\ & SO_4 - GalNAc \\ & \uparrow \\ & \begin{pmatrix} UA \\ | \\ SO_4 \end{pmatrix}_{2-3} \leftarrow \begin{pmatrix} Man \leftarrow UA \end{pmatrix}_2 \leftarrow GalNac \leftarrow \begin{pmatrix} SO_4 \\ | \\ SO_4 \end{pmatrix} \begin{pmatrix} UA \\ | \\ SO_4 \end{pmatrix}_2 \leftarrow Glc \end{bmatrix}_x$$

Figure 4.5 Structure of the heteropolysaccharide of *Halococcus morrhuae*. Modified from Schleifer *et al.* (1982) with permission.

4. Analysis of Archaeal Cell Envelopes

Table 4.4 Chemical composition of purified (trypsin-treated) cell walls of three *Halococcus morrhuae* strains.

Components	CCM 859	CCM 537	CCM 889
Glucose	0.44	0.39	0.30
Mannose	0.35	0.21	0.30
Galactose	0.27	0.18	0.17
Glucosamine	0.38	0.21	0.35
Galactosamine	0.20	0.08	0.24
Gulosaminuronic acid	0.11	0.09	0.10
Uronic acids	0.67	0.24	0.63
Acetate	0.62	o	o
Glycine	0.10	0.13	0.1
Sulphate	1.47	0.80	0.18

Data modified from Schleifer *et al.* (1982). Values are μmol/mg cell wall.
o, not determined.

4.3.3.b Isolation and analysis

Heteropolysaccharide sacculi can be isolated as described for pseudomurein (Method 4.4).

Amino acids (Methods 4.5 and 4.6), amino sugars (Methods 4.5 and 4.6), neutral sugars (Methods 4.5 and 4.8) and uronic acids (Method 4.10) can be identified as described above. Isolation and analysis of gulosaminuronic acid and of ester-bound sulphate employs the following procedures.

Method 4.11 Isolation and identification of gulosaminuronic acid.

1. Release gulosaminuronic acid from the heteropolysaccharide by partial acid hydrolysis (4 N HCl, 100 °C, 30 min).
2. Subject the hydrolysate to paper chromatography and run for 60 h in a mixture of pyridine/acetic acid ethyl ester/acetic acid/H$_2$O (5:5:1:3; v/v) as the solvent system. The band of $R_{GlcN} = 0.39$ contains gulosaminuronic acid (Steber, 1976).
3. Gulosaminuronic acid can be further identified by using an amino acid analyser ($R_{GlcN} = 0.95$; Liquimat III, Kontron; Steber, 1976; König & Kandler, 1979).

Method 4.12 Isolation and quantification of sulphate.

Reagent

The reagent (Dogston & Price, 1962) is prepared in the following manner:

1. 2 g gelatin are dissolved in 400 ml H$_2$O at 70 °C and the solution is kept for 16 h at 40 °C.
2. Barium chloride (2 g) is then added, the mixture stirred for 3 h and then kept for 24 h at 4 °C.

3. The solution is centrifuged to get a weak turbidity in the supernatant.
4. Barium chloride (2 g) is dissolved in the supernatant and the pH adjusted to 2.5 with 2 N HCl.

Measuring range: 0.01 to 0.20 μmol SO_4^{2-}.

Preparation of the sample

Ester bound sulphate is released by treatment with 4 N HCl (0.1 ml/mg cell wall) at 100 °C for 4 h. The acid is removed in a stream of air and the residue is redissolved in 0.5 ml H_2O.

Procedure

The sample is mixed with 1 ml reagent and the extinction point measured at 366 nm.

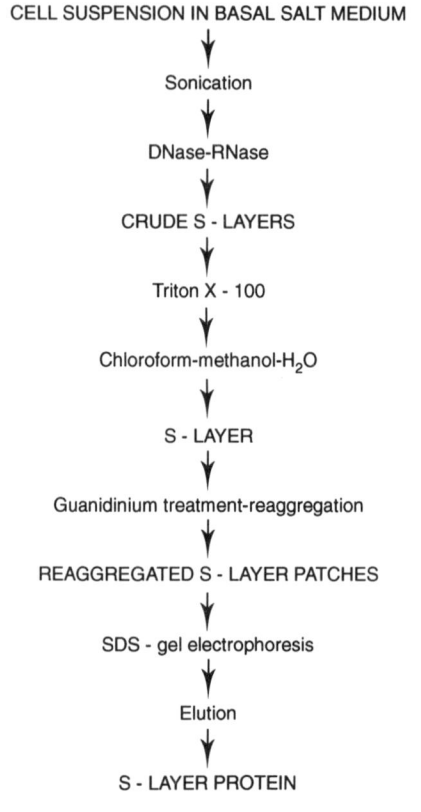

```
CELL SUSPENSION IN BASAL SALT MEDIUM
                ↓
            Sonication
                ↓
           DNase-RNase
                ↓
         CRUDE S - LAYERS
                ↓
           Triton X - 100
                ↓
      Chloroform-methanol-H₂O
                ↓
             S - LAYER
                ↓
   Guanidinium treatment-reaggregation
                ↓
    REAGGREGATED S - LAYER PATCHES
                ↓
        SDS - gel electrophoresis
                ↓
              Elution
                ↓
          S - LAYER PROTEIN
```

Figure 4.6 Procedure used to isolate S-layers and S-layer proteins of methanococci.

4. Analysis of Archaeal Cell Envelopes

4.3.4 Protein S-layers

4.3.4.a General information

Methanococci form a second order of methanogens (Balch et al., 1979). The cells only have an S-layer outside the cytoplasmic membranes; the S-layer is composed of nonglycosylated protein subunits in a hexagonal arrangement. The S-layers vary according to species. They are heterogeneous with respect to their molecular masses (Table 4.1; Koval & Jarrel, 1987; Nußer & König, 1987) and antigenicity (Conway de Macario et al., 1984). Proteinaceous nonglycosylated surface layers are probably primitive cell envelope structures as the information for the formation of the stable crystalline layers is exclusively preserved in the amino acid sequence. Outside the cells other 'activated' compounds are not required for the biosynthesis of the cell envelopes as with glycoproteins or peptidoglycans.

4.3.4.b Isolation and analysis

The procedure for isolating the S-layers of methanococci is outlined in Figure 4.6 and described in Method 4.13 below.

Method 4.13 Isolation of S-layers and S-layer proteins.

1. Cells are suspended in a basal salt solution (medium 3 (Balch et al., 1979) without organic components).
2. The cells are disrupted by sonication (Sonifier B-12; Branson Sonic Power Company; 20 s, 60 W) followed by incubation with DNase I (1 mg/l; Serva) and RNase A (1 mg/l; Boehringer) for 30 min at 37 °C (crude S-layer).
3. The suspension is then incubated with Triton X-100 (0.5%, 60 °C, 30 min; Sleytr & Thorne, 1974) and washed by centrifugation in the basal salt solution.
4. The pellet is extracted in the basal salt solution with an equal volume of chloroform/methanol/H_2O (1:2:0.8; 1 × overnight, 2 × 2 h; 37 °C; modified from Bligh & Dyer, 1959).
5. Prior to centrifugation of the chloroform extract an equal volume of chloroform and basal salt solution (1:1) is added.
6. The preparation is centrifuged and the material in the interphase is withdrawn (S-layer) then solubilized in guanidinium chloride (5 M; Sleytr & Thorne, 1974) to isolate the S-layer protein.
7. The preparation is dialysed for 4 h against the basal salt solution at room temperature.
8. The preparation is centrifuged (20 000 rpm, Sorvall rotor SS 34).
9. The proteins of the reaggregated S-layer patches are separated by SDS-gel electrophoresis (Method 4.14 below).
10. The protein bands from the gel are eluted by cutting the gel into pieces and incubating them with ammonium acetate buffer (0.5 M; 2% SDS, 0.1 mM EDTA; 60 °C, 16 h; modified from Maxam & Gilbert, 1977) or by electroelution.
11. The eluted proteins are precipitated by adding two volumes of acetone at 20 °C (S-layer protein).

Protein determination
1. The sample (0.5 ml) is added to 0.5 ml reagent A and incubated for 10 min at room temperature.
2. Reagent B (0.25 ml) is added and the mixture is incubated for 30 min at room temperature.
3. The extinction is determined at 750 nm (Peterson, 1972).

Reagents and solutions
Mixture of $CuSO_4$ (0.1%), potassium-sodium tartrate (0.2%) and Na_2CO_3 (10%) (final concentrations in parentheses).
10% SDS
0.8 M NaOH
Folin and Ciocalteu's phenol reagent (Merck)

Reagents
(A) Mixture of equal volumes of solutions 1, 2, 3 and H_2O
(B) Mixture of solution 4 and H_2O (1:1; v/v).

Method 4.14 SDS polyacrylamide gel electrophoresis (from Laemmli, 1970).

Stacking gel (3%)
0.125 M Tris/HCl buffer, pH 6.8
3% Acrylamide solution (taken from the stock solution)
0.1% SDS (w/v)
0.3% Ammonium peroxydisulphate (w/v)
0.025% Tetramethylethylenediamine (v/v)

Separation gel (10%):
0.375 M Tris/HCl buffer, pH 8.8.
10% Acrylamide (taken from the stock solution)
0.1% SDS (w/v)
0.3% Ammonium peroxydisulphate (w/v)
0.025% Tetramethylethylenediamine (v/v)

Acrylamide stock solution (30%; w/v):
Acrylamide, 30 g
N,N'-methylene-bis-acrylamide 0.8 g
100 ml H_2O

Sample buffer:
SDS, 2 g
Mercaptoethanol 5 ml
Glycerol 10 ml
Bromophenol blue (0.01 %) 10 ml
0.065 M Tris/HCl buffer pH 6.8 to 100 ml

Electrode buffer:
0.025 M Tris/HCl, pH 8.3
0.1% SDS (w/v)
0.192 M Glycine

Sample preparation:
The (glyco-)proteins are dissolved in the sample buffer by heating in boiling water for 10 min. About 10 μg protein in 10 to 20 μl are applied onto the gel.

4. Analysis of Archaeal Cell Envelopes

Electrophoresis:
Electrophoresis is carried out at a current intensity of 3 to 10 mA at 4 °C until the bromophenol blue marker reaches the bottom of the gel.

Staining of proteins

The gel is stained at 60 °C for 30 min and destained at 60 °C.

Staining solution:
Coomassie Blue R	2.2 g
Ethanol	400 ml
Acetic acid	80 ml
H$_2$O	400 ml

Destaining solution:
Ethanol	5 ml
Acetic acid	7 ml
H$_2$O	88 ml

Staining of glycoproteins

Staining for glycoproteins is performed using the periodate—Schiff reagent (Segrest & Jackson, 1972).
1. The gel is fixed in a solution of 40% ethanol, 5% acetic acid and 55% H$_2$O for 3 h.
2. Transfer to a solution of 0.7% periodic acid in 5% acetic acid for 3 h at room temperature and then to a solution of 0.2% sodium disulphite in 5% acetic acid for 3 h.
3. The gel is finally stained with the periodate-Schiff reagent for 16 h at 4 °C when carbohydrate-containing bands become red.

Periodate-Schiff reagent:
1. 2.5 g basic fuchsin is dissolved in H$_2$O for 30 min and 50 ml 1 N HCl and 4.25 g Na$_2$S$_5$O$_5$ is added.
2. The mixture stirred for 3 h.
3. Two spoons of charcoal are then added and the preparation is stirred for 30 min.
4. The reagent is kept at 4 °C and filtered before use.

4.3.5 Glycoprotein S-layers

4.3.5.a Methanothermus

Crystalline glycoprotein surface layers associated with the pseudomurein have been found only in extremely thermophilic members of the genus *Methanothermus* (Figure 4.2b; Stetter *et al.*, 1981; Nußer *et al.*, 1988).

The S-layer glycoprotein can be solubilized by extracting whole cells with trichloroacetic acid (4%, 30 min, room temperature). The extracts are nearly free of protein impurities. Coextracted low molecular weight compounds are removed by reversed-phase chromatography (ProRPC HR 5/10; Pharmacia) using aqueous formic acid (10%; Heukeshoven & Dernick, 1985) with subsequently increasing concentrations (0–85%) of methanol and isopropanol as eluants. The glycoprotein elutes at an

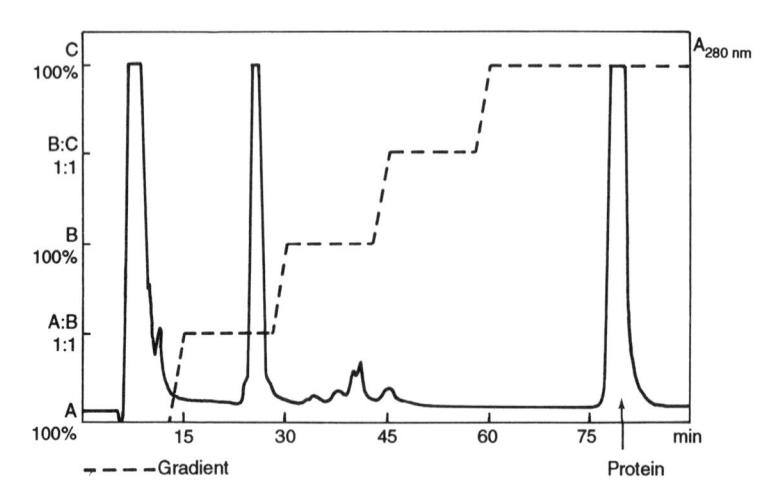

Figure 4.7 Elution profile of the trichloroacetic acid extract of whole cells of *Methanothermus fervidus* separated on a reversed-phase column. Eluent A: 10% formic acid; eluent B: 85% methanol in A (v/v); eluent C: 85% isopropanol in A (v/v).

isopropanol concentration of 85% (Figure 4.7). In the case of *M. fervidus* the glycoprotein gives two bands on SDS-polyacrylamide gels, with molecular masses of 92 and 60 kDa; with *M. sociabilis* a band of 89 kDa is apparent (Table 4.5). The two bands with different molecular masses belong to one protein species (Nußer *et al.*, 1988).

The molecular masses (Table 4.1) of the S-layer glycoproteins together with other archaeal features can be used as taxonomic markers. Glycoprotein (Methods 4.13 and 4.14) and the amino acids (Methods 4.5 and 4.6) are analysed as described above. The carbohydrate moiety of the S-layer glycoproteins of *Methanothermus fervidus* and *M. sociabilis* contains only neutral and amino sugars, which are determined by gas-liquid chromatography and the amino acid analyser (Methods 4.8 and 4.6), gas-liquid chromatography and the amino acid analyser (Methods 4.8 and 4.6), respectively.

4.3.5.b Halobacterium

Halobacterium salinarium (synonymous with *Halobacterium halobium*) produces a surface layer that is composed of glycoprotein subunits. The surface layer glycoprotein was the first glycosylated protein detected in prokaryotes (Mescher & Strominger, 1974, 1976). Analysis of the amino acid sequence and the carbohydrate moiety revealed a molecular mass of ~116 kDa (Lechner & Sumper, 1987), while on SDS-polyacrylamide gels a molecular mass of 200 kDa was determined (Mescher & Strominger, 1976).

Table 4.5 Carbohydrate composition of some S-layers, sheaths and S-layer glycoproteins.

Species[a]	Man	Gal	Glc	3-O-Me-Glc[b]	GlcN	GalN	Xyl	Rha	Rib	Ara	GalA	3-O-Me-GalA	GlcA	IdA
Halobacterium halobium (GP)	−	+	+	−	+	+	−	−	−	−	+	+	+	+
Methanothermus fervidus V24S (GP)	+	+	−	+	+	+	−	−	−	−	−	−	−	−
Methanothermus sociabilis KF1–F1 (GP)	+	+	−	+	+	+	−	−	−	−	−	−	−	−
Methanolobus tindarius Tindari3 (GP)	−	−	+	−	−	−	+	−	−	−	−	−	−	−
Methanospirillum hungatei JF1 (Sh)	+	+	+	−	−	−	−	−	+	−	−	−	−	−
Methanospirillum hungatei GP1 (Sh)	+	+	+	−	−	−	−	+	+	−	−	−	−	−
Methanothrix concilii GP6 (Sh)[c]	+	+	+	−	−	−	−	+	+	−	−	−	−	−
Methanothrix soehngenii Opfikon (Sh)	+	−	+	−	+	−	−	−	−	−	−	−	−	−
Sulfolobus acidocaldarius (SL)	+	+	+	−	+	−	−	+	+	−	−	−	−	−
Thermoproteus tenac Kra1 (SL)	+	−	+	−	+	−	−	+	−	+	−	−	−	−

[a] SL, S-layer; Sh, sheath; GP, glycoprotein.
[b] Mixture of 3-O-MeGlc and 3-O-MeMan.
[c] Renamed Methanosaeta concilii.

Abbreviations: Ara, arabinose; Gal, galactose; GalA, galacturonic acid; GalN, galactosamine; Glc, glucose; GlcA, glucuronic acid; GlcN, glucosamine; IdA, iduronic acid; Man, mannose; Rha, rhamnose; Rib, ribose; Xyl, xylose.

The carbohydrate moiety is composed of three different glycopeptides that have novel linkage types (Figure 4.8; Wieland et al., 1980, 1981, 1982, 1983, 1986; Paul et al., 1986):

(a) A sulphated glycosaminoglycan of high molecular weight consisting of N-acetylglucosamine, N-acetylgalactosamine, galacturonic acid, galactose, and 3-O-methyl galacturonic acid. This heterosaccharide occurs once per protein molecule and it is connected via N-acetylgalactosamine to an asparagine residue.
(b) Sulphated glycans of low molecular weight consisting of glucose, glucuronic acid, and iduronic acid the latter are linked via glucosyl-asparaginyl residues to the peptide moiety.
(c) Disaccharides, composed of glucose and galactose, which are connected to threonine residues.

Biosynthesis studies (Sumper 1987; Wieland, 1989) revealed that for the glycosaminoglycan, dolichyl pyrophosphate serves as lipid carrier. Dolichyl monophosphate is the lipid carrier for the second sulphated oligosaccharide; Asn-Ala-Ser serves as acceptor peptide for the glycosaminoglycan. A transient methylation of a glucose residue is required for the transfer of dolichyl oligosaccharides onto the protein (Lechner et al.,

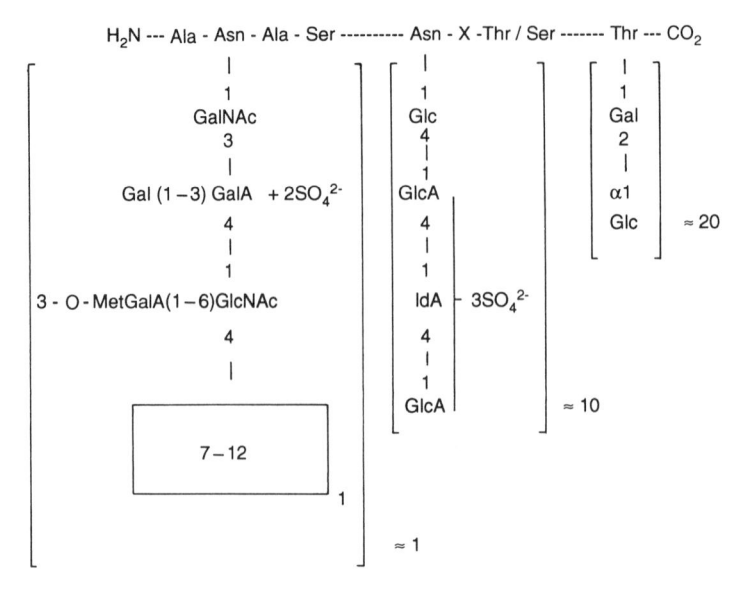

Figure 4.8 Glycan structures of the glycoprotein from *Halobacterium halobium*. Modified from Lechner and Wieland (1989) with permission.

4. Analysis of Archaeal Cell Envelopes

1985a,b, 1986). The shape of *Halobacterium salinarium* is maintained by its surface layer glycoprotein. This is demonstrated by the generation of spheroid cells in the presence of bacitracin; the latter inhibits the generation of the lipid monophosphate carrier (Wieland *et al.*, 1982). As with methanochondroitin, the halobacterial S-layer protein is reminiscent of eukaryotic components of the connective tissue, in particular proteoglycans.

Method 4.15 Isolation of *Halobacterium* envelope components (from Mescher & Strominger, 1974, 1976).

---

1. The harvested cells are washed twice in basal salt solution that has the same composition as the growth medium (w/vol): 25% NaCl, 1.0% MgSO$_4$·7H$_2$O, 0.02% CaCl$_2$·2H$_2$O and 0.5% KCl. Washing and all subsequent operations are done at 0 °C.
2. The washed wet cell pellet is suspended in three volumes of basal salt solution and two volumes of acid-washed, 120 μm glass beads and ground for 5 min in a Mini-mill (Gifford-Wood Co.).
3. The broken cell suspension is then centrifuged at 4500 × g for 10 min to remove glass beads, debris and whole cells, and the supernatant solution is centrifuged at 100 000 × g for 40 min.
4. The resulting envelope pellet is then washed twice by resuspending in 50 volumes of the basal salt solution followed by centrifugation at 100 000 × g for 40 min.
5. Lipids are extracted by suspending the envelopes in 10 ml of a mixture of chloroform/methanol (2:1) per gram of wet cell envelopes and mixing with a Vortex mixer.
6. The suspension is allowed to stand for 15 min at room temperature with occasional mixing and is then centrifuged. The pellet is re-extracted twice in the same manner.
7. Extraction with phenol is carried out for 30 min at room temperature using an equal volume of 0.05 M Tris/HCl buffer (pH 7.2) containing 0.5% NaCl and buffer-saturated phenol.
8. The glycoprotein is recovered in the aqueous phase. It is further purified by chromatography on DEAE-cellulose following dialysis of the aqueous phase against distilled water. The dialysed aqueous phase (100 mg protein/100 ml) is applied to a DE52 column (2.0 × 17 cm) equilibrated with 0.05 M Tris/HCl buffer, pH 7.2. Fractions of 10 ml are collected.
9. Beginning at fraction 12 the column is eluted with a linear gradient consisting of 500 ml of 0.05 M Tris/HCl pH 7.2 in the mixing chamber and 500 ml of 1 M NaCl in 0.05 M Tris/HCl, pH 7.2 in the reservoir. Fractions 40 to 48 contain the glycoprotein.

Analysis of the glycoprotein (Methods 4.13 and 4.14), amino acids and carbohydrates (Methods 4.5, 4.6 and 4.10), and sulphate (Method 4.12) is performed as described above.

4.3.5.c Sulfolobus

The S-layer of *Sulfolobus acidocaldarius* is composed of a glycoprotein with an apparent molecular mass of 170 kDa or 140 kDa (two modifications; Michel *et al.*, 1980). The polar amino acids serine and threonine predominate as opposed to the high amount of acidic amino acids in *Halobacterium halobium* and of asparagine in *Methanothermus fervidus*. The glycoprotein is hexagonally arranged (P6 symmetry) with a unit cell dimension of 22 nm (Taylor *et al.*, 1982; Deatherage *et al.*, 1983; Baumeister & Engelhardt, 1987). Only 30% of the 10 nm thick S-layer is occupied by protein; the latter gives rise to a rather spongy structure.

Method 4.16 Isolation of *Sulfolobus* envelope components (after Michel *et al.*, 1980).

---

1. The broth culture is neutralized with solid NaHCO$_3$ and the cells spun down.
2. The cell pellet is suspended in 10 mM HEPES buffer (containing 2 mM EDTA) at pH 7 and 0.15% SDS is added to the preparations.
3. DNase is then added.
4. Following digestion of the DNA the SDS concentration is brought to 2% and the suspension stirred overnight.
5. After centrifugation (36 000 × g, 30 min), the upper white layer of the pellet is suspended in a 2% aqueous SDS solution and incubated at 60 °C for 1 h.
6. The cell envelopes are washed (3×) by centrifugation, then disintegrated by heating to 60 °C in 100 mM phosphate buffer at pH 9.
7. The solubilized cell envelope glycoprotein is purified by molecular sieve chromatography on Sepharose 6B with 0.1 M potassium phosphate buffer at pH 7.

Analysis of the glycoprotein (Methods 4.13 and 4.14), amino acids and carbohydrates (Methods 4.5 and 4.6) is performed as described above.

4.3.6 Surface layers with unusual stability

The sheaths or S-layers, respectively, of the two methanogenic genera, *Methanospirillum* and *Methanothrix*, and of members of the anaerobic extremely thermophilic sulphur-metabolizing genera *Thermoproteus* and *Thermofilum*, are highly resistant to detergents and proteases (Kandler & König, 1978; König & Stetter, 1986). The (glyco-)protein subunits may be partly covalently linked. Only in some cases can single protein species be released. Consequently, only the isolation of the sheaths and S-layers is described.

4.3.6.a Methanospirillum and Methanothrix

The cells of *Methanospirillum* and *Methanothrix* occur in chains that form long filaments. The outer envelope of the filaments is formed by tubular sheaths composed of hoop-like rings (Zeikus & Bowen, 1975; Shaw et al., 1985; Stewart et al., 1985; Beveridge et al., 1986a,b; Patel et al., 1986). The sheath is mainly composed of protein and carbohydrate (Kandler & König, 1978, 1985; Sprott & McKellar, 1980); a covalent linkage between the two has been demonstrated for *Methanothrix* (Pellerin et al., 1990).

The sheath exhibits unusual resistance against chemical agents and lytic enzymes (Kandler & König, 1978; Beveridge et al., 1985). Treatment with dithiothreitol releases spheroplasts, but the sheath remains intact (Sprott et al., 1979). The sheaths of *Methanothrix concilii* were disintegrated into individual hoops and soluble glue peptides using mercaptoethanol. Additional incubation with mercaptoethanol and sodium dodecyl sulphate solubilizes the hoops (Beveridge et al., 1986b; Sprott et al., 1986).

Method 4.17 Isolation of sheaths of *Methanospirillum* and *Methanothrix*.

---

1. Cells suspended in distilled water are sonicated for 15 min at 50 to 70 W (Sonifier B12, Branson Sonic Power Company).
2. The suspension is then incubated with DNase (1 mg/l) and RNase A (1 mg/l) for 30 min at 37 °C.
3. After centrifugation, the pellet is suspended in an SDS solution (20 g/l), heated in a boiling water bath for 30 min and washed (3×) with water in an Eppendorf centrifuge.

Analysis of the amino acids and carbohydrates is performed as described above (see Methods 4.5, 4.6, 4.8 and 4.10).

4.3.6.b Thermoproteus and Thermofilum

S-layers with unusual stability against chemical agents have been isolated from *Thermoproteus tenax* and *Thermofilum pendens* (Kandler & König, 1985; König & Stetter, 1986). The subunits of *Thermoproteus tenax* are hexagonally arranged (P6 symmetry) with a centre distance of 30 nm (Messner et al., 1986; Wildhaber & Baumeister, 1987).

The S-layers can be isolated as described for *Methanospirillum* and *Methanothrix* (Method 4.17).

Analysis of amino acids and carbohydrates is performed as described above (Methods 4.5, 4.6, 4.8 and 4.10).

4.4 TAXONOMIC CONSIDERATIONS

Archaea generally have similar morphologies and lack easily determined physiological features. The cell envelopes of these organisms have not been investigated as intensively as those of bacteria, but it is evident that constituents of cell envelope structures may be useful taxonomic markers.

In the following section the determinative features of archaeal cell envelopes are summarized; these markers allow the identification of some archaeal groups, or even strains in some cases.

4.4.1 Murein

Archaea lack murein, but this is also true for members of some bacterial genera with cell envelopes, e.g. *Planctomyces* and *Pirella* (König et al., 1984; Liesack et al., 1986) and the cell wall-less mycoplasmas (Razin & Freund, 1984). Consequently, the lack of muramic acid alone cannot be taken as a marker for distinguishing archaea and bacteria. Other typical archaeal features, such as the presence of phytanyl ether lipids (see Chapter 7) or ADP-ribosylation of elongation factor EF2 (Kessel & Klink, 1982), should also be determined.

4.4.2 Pseudomurein

Pseudomurein is present in all members of the methanogenic order Methanobacteriales and *Methanopyrus* spp. and hence it is a useful taxonomic marker. Methanogens with pseudomurein are Gram-positive. Some methanogens can be recognized at the species level (Table 4.2) owing to modifications in chemical composition; such modifications are mainly found in species living in the digestive tract, such as *Methanobrevibacter* and *Methanosphaera* species. Some taxonomically useful modifications are given below.

(a) *Methanobacterium bryantii* strain M.o.H.-G. and *Methanobacterium thermoautotrophicum* strain ΔH: these strains have ~ 1 mol glucosamine per mole of lysine, but only traces of galactosamine. In contrast to *Methanobacterium bryantii* strain M.o.H.-G., *Methanobacterium bryantii* strain M.o.H. contains both galactosamine and glucosamine.
(b) *Methanobacteruim thermoautotrophicum* strain Marburg: the amino sugars galactosamine and glucosamine yield a molar ratio of 1 against lysine. This is also true for *Methanobrevibacter smithii* PS. The presence of galactosamine is a difference between the two strains *Methanobacterium thermoautotrophicum* Marburg and ΔH. In the JW strains of *Methanobacterium thermoautotrophicum* the two amino sugars exceed a molar ratio of 1.
(c) *Methanobrevibacter arboriphilus* strain AZ and DH1: the glycan strand only contains galactosamine.
(d) *Methanobrevibacter ruminantium* strain M1: alanine is either partially or totally replaced by threonine. It is evident that galactosamine is the main sugar of the glycan strand as glucosamine can be released by

hydrofluoric acid at 0 °C indicating a linkage via phosphodiester bonds. Higher amounts of phosphate (4.5%) are present.
(e) *Methanobrevibacter smithii* strain PS: ornithine occurs in addition to lysine.
(f) *Methanosphaera stadtmanae*: alanine is replaced by serine.
(g) *Methanothermus* species: a double-layered cell envelope profile (Figures 4.1 and 4.2b), which is composed of pseudomurein and S-layer, is characteristic for *Methanothermus*.
(h) *Methanopyrus kandleri* strain AV19: contains only galactosamine; ornithine occurs in addition to lysine.

4.4.3 Heteropolysaccharide

Halococcus morrhuae is characterized by the presence of cell walls consisting of a single polymer composed of a complex spectrum of carbohydrates. In contrast to methanochondroitin, this polymer consists of a more extended spectrum of carbohydrates which are sulphated (Table 4.4).

4.4.4 Methanochondroitin

Only members of the methanogenic genus *Methanosarcina* have cell walls composed of methanochondroitin with the exception of *Methanosarcina acetivorans* and *Methanosarcina frisia*, which have only a single S-layer. Strains G1 and DSM 2053 have galacturonic acid in addition to glucuronic acid, while the other strains investigated only have glucuronic acid in their cell walls. Methanogens with methanochondroitin cell walls are Gram-positive.

4.4.5 S-layers

The chemistry of S-layer (glyco-)proteins has rarely been studied in detail, but in many cases the apparent molecular weights have been determined and found to give data of some taxonomic value (Table 4.1).

4.4.5.a Protein S-layers

Single S-layers composed of nonglycosylated protein subunits are found in all members of the methanogenic order Methanococcales, which contains the single genus *Methanococcus*. The different species of *Methanococcus* can be distinguished by the molecular masses of their S-layer proteins (Table 4.1). Nonglycosylated surface layer proteins also occur in other archaeal species, e.g. *Methanogenium cariaci*.

4.4.5.b Glycoprotein S-layers

Halobacterium halobium: the S-layer protein of this organism has an apparent molecular mass on SDS-polyacrylamide gels of 200 kDa. Characteristic components not found in other S-layer glycoproteins are, for example, iduronic acid, 3-O-methylgalacturonic acid and sulphated sugar derivatives. The S-layer is solubilized in the absence of high salt (>12%); other S-layers normally have the tendency to form aggregates in water.

Methanothermus spp.: the S-layer glycoproteins of *Methanothermus fervidus* and *Methanothermus sociabilis* have apparent molecular masses of 92 and 60 kDa (two bands), and 89 kDa, respectively (Table 4.1). These species also contain characteristic glycoprotein sugars, namely 3-O-methylmannose/3-O-methylglucose. The latter have not been found in other intact S-layer glycoproteins. The solubility of these organisms in 4% trichloroacetic acid is also unusual.

Sulfolobus acidocaldarius: the S-layer of *Sulfolobus acidocaldarius* shows elevated resistance against detergents. It is not disintegrated at 60 °C in the presence of SDS (2%), but can be solubilized at 100 °C or at 60 °C in phosphate buffer at pH 9. In contrast the S-layer of *Sulfolobus solfataricus* is solubilized at room temperature in a solution of 2% SDS.

4.4.5.c SDS-resistant layers

Sheaths or S-layers which are not disintegrated in a solution of SDS (2–10%) at 100 °C are only found in the methanogenic genera *Methanospirillum* and *Methanothrix* or in the extreme sulphur-metabolizing genera *Thermofilum* and *Thermoproteus*. The sheaths of methanogens can be distinguished morphologically from the S-layers of sulphur metabolizers, as they show a striated texture. *Methanospirillum hungatei* strains GP1 and JF1 show differences in pentose content (Table 4.5). Strain JF1 possesses arabinose and GP1 ribose. *Methanothrix soehngenii* has only two sugars in its sheath, while *Methanothrix concilii* has a spectrum of five sugars (Table 4.5).

4.4.5.d Cell envelope-less bacteria

Cell envelope-less bacteria are represented by *Thermoplasma* spp. The latter can be identified in ultrathin sections due to their characteristic cell profile (Figures 4.1 and 4.2).

REFERENCES

Albersheim, P., Nevins, D.J., English, P.D. and Karr, A. (1967) A method for the analysis of sugars in plant cell-wall polysaccharides by gas-liquid chromatography. *Carboh. Res.* **5**: 340–5.

Balch, W.E., Fox, G.E., Magrum, L.J., Woese, C.R. and Wolfe, R.S. (1979) Methanogens: Reevaluation of a unique biological group. *Microbiol. Rev.* **43**: 260–96.

Baumeister, W. and Engelhardt, H. (1987) Three-dimensional structure of bacterial surface layers. In *Electron Microscopy of Proteins*, vol. 6 (Harris, J.R. and Horneeds, R.W., eds), pp. 109–54. Academic Press: London.

Beveridge, T.J., Stewart, M., Doyle, R.J. and Sprott, G.D. (1985) Unusual stability of the *Methanospirillum hungatei* sheath. *J. Bacteriol.* **162**: 728–37.

Beveridge, T.J., Patel, G.B., Harris, B.J. and Sprott, G.D. (1986a) The ultrastructure of *Methanothrix concilii*, a mesophilic aceticlastic methanogen. *Can. J. Microbiol.* **32**: 703–10.

Beveridge, T.J., Harris, B.J., Patel, G.B. and Sprott, G.D. (1986b) Cell division and filament splitting in *Methanothrix concilii*. *Can. J. Microbiol.* **32**: 779–86.

Bligh, E.G. and Dyer, W.J. (1959) A rapid method of total lipid extraction and purification. *Can. J. Biochem. Physiol.* **37**: 911.

Brown, A.D. and Cho, K.J. (1970) The walls of the extremely halophilic cocci. Gram-positive bacteria lacking muramic acid. *J. Gen. Microbiol.* **62**: 267–70.

Chen, P.S., Toribara, T.Y. and Warner, H. (1956) Microdetermination of phosphorus. *Anal. Chem.* **28**: 1756–8.

Conway de Macario, E., Macario, A.J.L., Magarinos, M.C., König, H. and Kandler, O. (1983) Dissecting the antigenic mosaic of the archaebacterium *Methanobacterium thermoautotrophicum* by monoclonal antibodies of defined molecular specificity. *Proc. Natl. Acad. Sci. USA*. **80**: 6346–50.

Conway de Macario, E., König, H., Macario, A.J.L. and Kandler, O. (1984) Six antigenic determinants in the surface layer of the archaebacterium *Methanococcus vannielii* revealed by monoclonal antibodies. *J. Immunol.* **132**: 883–7.

Darland, G., Brock, T.D., Samsonoff, W. and Conti, S.F. (1970) A thermophilic, acidophilic mycoplasma isolated from a coal refuse pile. *Science* **170**: 1416–18.

Deatherage, J.F., Taylor, K.A. and Amor, L.A. (1983) Three-dimensional arrangement of cell wall protein of *Sulfolobus acidocaldarius*. *J. Mol. Biol.* **167**: 823–52.

Dogston, K.S. and Price, R.G. (1962) A note on the determination of the ester sulfate content of sulphated polysaccharides. *Biochem. J.* **84**: 106–10.

Formanek, H. (1985) Three-dimensional models of the carbohydrate moieties of murein and pseudomurein. *Z. Naturforsch. C* **40**: 555–61.

Hammes, W.P., Winter, J. and Kandler, O. (1979) The sensitivity of the pseudomurein-containing genus *Methanobacterium* to inhibitors of murein synthesis. *Arch. Microbiol.* **123**: 275–9.

Hartmann, E. and König, H. (1989) Isolation of lipid activated pseudomurein precursors from *Methanobacterium thermoautotrophicum*. *Arch. Microbiol.* **153**: 444–7.

Hartmann, E., König, H., Kandler O. and Hammes, W. (1989) Isolation of nucleotide activated peptide precursors from *Methanobacterium thermoautotrophicum*. *FEMS Microbiol. Lett.* **61**: 323–8.

Heukeshoven, J. and Dernick, R. (1985) Characterization of a solvent system for separation of water-insoluble poliovirus proteins by reversed-phase high performance liquid chromatography. *J. Chromatogr.* **326**: 91–101.

Houwink, A.L. (1956) Flagella, gas vacuoles and cell-wall structure in *Halobacterium halobium*; An electron microscope study. *J. Gen. Microbiol.* **15**: 146–50.

Jain, M.K., Bhatnagar, L. and Zeikus, J.G. (1988) A taxonomic overview of methanogens. *Indian. J. Microbial.* **28**: 143–77.

Kandler, O. (1979) Zellwandstrukturen bel Methanbakterien. Zur Evolution der Prokaryonten. *Naturwissenschaften* **66**: 95–105.

Kandler, O. and Hippe, H. (1977) Lack of peptidoglycan in the cell walls of *Methanosarcina barkeri*. *Arch. Microbiol.* **113**: 57–60.

Kandler, O. and König, H. (1978) Chemical composition of the peptidoglycan-free cell walls of methanogenic bacteria. *Arch. Microbiol.* **118**: 141–52.
Kandler, O. and König, H. (1985). Cell envelopes of archaebacteria. In *The Bacteria* (Wolfe, R.S. and Woese, R.C., eds), pp. 413–57. Academic Press: New York.
Kessel, M. and Klink, F. (1982) Identification and comparison of eighteen archaebacteria by means of the diphtheria toxin reaction. *Zbl. Bakt. Hyg., I. Abt. Orig.* **C3**: 140–8.
Kessel, M., Wildhaber, I., Cohen, S. and Baumeister, W. (1988) Three-dimensional structure of the regular surface glycoprotein layer of *Halobacterium volcanii* from the Dead Sea. *EMBO J.* **7**: 1549–54.
Kiener, A., König, H., Winter, J. and Leisinger, Th. (1987) Purification and use of *Methanobacterium wolfei* pseudomurein endopeptidase for lysis of *Methanobacterium thermoautotrophicum*. *J. Bacteriol.* **169**: 1010–6.
König, H. (1985) Influence of amino acids on growth and cell wall composition of *Methanobacteriales*. *J. Gen. Microbiol.* **131**: 3271–5.
König, H. (1988) Archaeobacterial cell envelopes. *Can. J. Microbiol.* **34**: 395–406.
König, H. and Kandler, O. (1979) 2-Amino-2-deoxytaluronic acid and 2-amino-2-deoxyglucose from the pseudomurein of *Methanobacterium thermoautotrophicum* possess the L- and D-configurations, respectively. *Hoppe Seyler's Z. Physiol. Chem.* **361**: 981–3.
König, H. and Stetter, K.O. (1982) Isolation and characterization of *Methanolobus tindarius*, sp. nov., A coccoid methanogen growing only on methanol and methylamines. *Zbl. Bakt. Hyg., I. Abt. Orig.* **C3**: 478–90.
König, H. and Stetter, K.O. (1986) Studies on archaebacterial S-layers. *Syst. Appl. Microbiol.* **7**: 300–9.
König, H., Kralik, R. and Kandler, O. (1982) Structure and modifications of pseudomurein in *Methanobacteriales*. *Zbl. Bakt. Hyg., I. Abt. Orig.* **C3**: 179–91.
König, H., Schlesner, H. and Hirsch, P. (1984) Cell wall studies on budding bacteria of the *Planctomyces/Pasteuria* group and on a *Prosthecomicrobium* sp. *Arch. Microbiol.* **138**: 200–5.
König, H., Semmler, R., Lerp, C. and Winter, J. (1985) Evidence for the occurrence of autolytic enzymes in *Methanobacterium wolfei*. *Arch. Microbiol.* **141**: 177–80.
König, H., Kandler, O. and Hammes, W. (1989) Biosynthesis of pseudomurein: Isolation of putative precursors from *Methanobacterium thermoautotrophicum*. *Can. J. Microbiol.* **35**: 176–81.
Koval, S.F. and Jarrel, K.F. (1987) Ultrastructure and biochemistry of the cell wall of *Methanococcus voltae*. *J. Bacteriol.* **169**: 1298–306.
Kreisl, P. and Kandler, O. (1986) Chemical structure of the cell wall polymer of *Methanosarcina*. *Syst. Appl. Microbiol.* **7**: 293–9.
Kurr, M., Huber, R., König, H., Jannasch, H.W., Fricke, H., Tricone, A., Kristjansson, J.K. and Stetter, K.O. (1991) *Methanopyrus kandleri*, gen. and sp. nov. represents a novel group of hyperthermophilic methanogens, growing at 110 °C. *Arch. Microbiol.* **156**: 239–47.
Labischinski, H., Barnickel, G., Leps, B., Bradaczek, H. and Giesbrecht, P. (1980). Initial data for the comparison of murein and pseudomurein conformations. *Arch. Microbiol.* **127**: 195–201.
Laemmli, U.K. (1970) Cleavage of structural proteins during the assembly of the head of bacteriophage T$_4$. *Nature* **227**: 680–5.
Lechner, J. and Sumper, M. (1987) The primary structure of a procaryotic glycoprotein. Cloning and sequencing of the cell surface glycoprotein gene of halobacteria. *J. Biol. Chem.* **262**: 9724–9.
Lechner, J. and Wieland, F. (1989) Structure and biosynthesis of procaryotic glycoproteins. *Ann. Rev. Biochem.* **58**: 173–94.
Lechner, J., Wieland, F. and Sumper, M. (1985a) Biosynthesis of sulfated saccharides *N*-glycosidically linked to the protein via glucose. *J. Biol. Chem.* **260**: 860–6.
Lechner, J., Wieland, F. and Sumper, M. (1985b) Transient methylation of dolichyl oligosaccharides is an obligatory step in halobacterial sulfated glycoprotein biosynthesis. *J. Biol. Chem.* **260**: 8984–9.
Lechner, J., Wieland, F. and Sumper, M. (1986) Sulfated dolicholphosphate oligosaccharides are transiently methylated during biosynthesis of halobacterial glycoproteins. *Syst. Appl. Microbiol.* **7**: 286–92.

Liesack, W., König, H. and Hirsch, P. (1986) Chemical composition of the peptidoglycan-free cell envelopes of budding bacteria of the *Pirella/Planctomyces* group. *Arch. Microbiol.* **145**: 361–6.
Maxam, A.M. and Gilbert, W. (1977) A new method for sequencing DNA. *Proc. Natl. Acad. Sci. USA.* **74**: 560–54.
Mescher, M.F. and Strominger, J.L. (1974) Protein and carbohydrate composition of the cell envelope of *Halobacterium salinarium*. *J. Bacteriol.* **120**: 945–54.
Mescher, M.F. and Strominger, J.L. (1976) Purification and characterization of a procaryotic glycoprotein from cell envelope of *Halobacterium salinarium*. *J. Biol. Chem.* **251**: 2005–14.
Messner, P., Pum, D., Sára, M. and Stetter, K.O. (1986) Ultrastructure of the cell envelope of the archaebacteria *Thermoproteus tenax* and *Thermoproteus neutrophilus*. *J. Bacteriol.* **166**: 1046–54.
Michel, H., Neugebauer, D.C. and Osterhelt, D. (1980) The 2-d crystalline cell wall of *Sulfolobus acidocaldarius*: Structure, solubilization, and reassembly. In *Electron Microscopy at Molecular Dimensions* (Baumeister, W. and Vogell, W., eds), pp. 27–35. Springer-Verlag: Berlin.
Mountfort, D.O., Mörschel, E., Beimborn, D.B. and Schöheit, P. (1986) Methanogenesis and ATP synthesis in a protoplast system of *Methanobacterium thermoautotrophicum*. *J. Bacteriol.* **168**: 892–900.
Nußer, E. and König, H. (1987) S-layer studies on three species of *Methanococcus* living at different temperatures. *Can. J. Microbiol.* **33**: 256–61.
Nußer, E., Hartmann, E., Allmeier, H., König, H., Paul, G. and Stetter, K.O. (1988) A glycoprotein surface layer covers the pseudomurein sacculus of the extreme thermophile *Methanothermus fervidus*. In *Proceedings of the Second International Workshop on S-layers in Procaryotes*, Vienna (Sleytr, U.B., Messner, P., Pum, D. and Sàra, M., eds), pp. 21–25. Springer Verlag: Berlin.
Patel, G.B., Sprott, G.D., Humphrey, R.W. and Beveridge, T.J. (1986) Comparative analyses of the sheath structures of *Methanothrix concilii* GP6 and *Methanospirillum hungatei* GP1 and JF1. *Can. J. Microbiol.* **32**: 623–31.
Paul, G., Lottspeich, F. and Wieland, F. (1986) Asparaginyl-N-acetylgalactosamine: Linkage unit of halobacterial glycosamino-glycan. *J. Biol. Chem.* **261**: 1020–4.
Pellerin, P., Fournet, B. and Debeire, P. (1990) Evidence for the glycoprotein nature of the cell sheath of Methanosaeta-like cells in the culture of *Methanothrix soehngenii* strain FE. *Can. J. Microbiol.* **36**: 631–6.
Peterson, G. (1972). A simplification of the protein assay method of Lowry *et al.* which is more generally applicable. *Anal. Biochem.* **83**: 346–356.
Razin, S. and Freund, E.A. (1984) The mycoplasmas. In *Bergey's Manual of Systematic Bacteriology*, vol. 1 (Krieg, N.R. and Holt, J.G., eds), pp. 740–1. Williams & Wilkins: London.
Reistadt, R. (1972) Cell wall of an extremely halophilic coccus. Investigation of ninhydrin-positive compounds. *Arch. Microbiol.* **82**: 24–30.
Reistadt, R. (1974) 2-Amino-2-deoxyguluronic acid: A constituent of the cell wall of *Halococcus* sp., strain 24. *Carbohydr. Res.* **36**: 420–3.
Reistadt, R. (1975) Aminosugar and amino acid constituents of the cell wall of the extremely halophilic cocci. *Arch. Microbiol.* **102**: 71–3.
Schleifer, K.H. (1985) Analysis of the chemical composition and primary structure of murein. In *Methods in Microbiology* (Gottschalk, G. ed), pp. 123–56. Academic Press: New York.
Schleifer, K.H. and Kandler, O. (1972). Zur chemischen Zusammensetzung der Zellwand der Streptokokken. I. Die Aminosäuresequenz der Mureins von *Str. thermophilus* und *Str. faecalis*. *Arch. Microbiol.* **57**: 335–64.
Schleifer, K.H. and Seidl, P.H. (1984) Chemical composition and structure of murein. In *Chemical Methods in Bacterial Systematics* (Goodfellow, M. and Minnikin, D.E., eds), pp. 201–19. Academic Press: New York.
Schleifer, K.H., Plapp, R. and Kandler, O. (1968) Zur chemischen Zusammensetzung der Zellwand der Streptokokken. III. Die Aminosäuresequenz eines glycinhaltigen Mureins aus *Peptostreptococcus evolutus* (Prevot) Smith. *Arch. Microbiol.* **61**: 292–301.

Schleifer, K.H., Steber, J. and Mayer, H. (1982) Chemical composition and structure of the cell wall of *Halococcus morrhuae*. *Zbl. Bakt. Hyg., I. Abt. Orig.* **C3**: 171–8.

Segrest, J.P. and Jackson, R.L. (1972) Molecular weight determination of glycoproteins by polyacrylamide gel electrophoresis in sodium dodecyl sulfate. In *Methods in Enzymology*, vol. XXVIII (Ginsburg, V. ed), pp. 54–63. Academic Press: New York.

Shaw, P.J., Hills, G.J., Henwood, J.A., Harris, J.E. and Archer, D.B. (1985) Three-dimensional architecture of the cell sheath and septa of *Methanospirillum hungatei*. *J. Bacteriol.* **161**: 750–7.

Sleytr, U.B. and Messner, P. (1983) Crystalline surface layers on bacteria. *Ann. Rev. Microbiol.* **37**: 11–9.

Sleytr, U.B. and Thorne, K.J.I. (1974) Chemical characterization of the regularly arranged surface layers of *Clostridium thermosaccharolyticum* and *Clostridium thermohydrosulfuricum*. *J. Bact.* **126**: 377–83.

Sleytr, U.B., Messner, P., Sàra, M. and Pum, D (1986) Crystalline envelope layers in archaebacteria. *System. Appl. Microbiol.* **7**: 310–3.

Sprott, G.D. and McKellar, R.C. (1980) Composition and properties of the cell wall of *Methanospirillum hungatei*. *Can. J. Microbiol.* **26**: 115–20.

Sprott, G.D., Colvin, J.R. and McKellar, R.C. (1979) Spheroplasts from *Methanospirillum hungatei* formed upon treatment with dithiothreitol. *Can. J. Microbiol.* **25**: 730–8.

Sprott, G.D., Beveridge, T.J., Patel, G.B. and Ferrante, G. (1986) Sheath disassembly in *Methanospirillum hungatei* strain GP1. *Can. J. Microbiol.* **32**: 847–54.

Steber, J. (1976) Untersuchungen zur chemischen Zusammensetzung und Struktur der Zellwand von *Halococcus morrhuae*. PhD dissertation: University of Munich.

Steber, J. and Schleifer, K.H. (1975) *Halococcus morrhuae*: A sulfated heteropolysaccharide as the structural component of the bacterial wall. *Arch. Microbiol.* **105**: 173–7.

Steber, J. and Schleifer, K.H. (1979) N-Glycyl-glucosamine, a novel constituent in the cell wall of *Halococcus morrhuae*. *Arch. Microbiol.* **123**: 209–12.

Stetter, K.O., Thomm, M., Winter, J., Wildgruber, G., Huber, H., Zillig, W., Janékovic, D., König, H., Palm, P. and Wunder, S. (1981) *Methanothermus fervidus* sp. nov., a novel extremely thermophilic methanogen isolated from an Icelandic hot spring. *Zbl. Bakt. I. Hyg., I. Abt. Orig.* **C2**: 166–78.

Stewart, M., Beveridge, T.L. and Sprott, G.D. (1985) Crystalline order to high resolution in the sheath of *Methanospirillum hungatei*: A cross beta structure. *J. Mol. Biol.* **183**: 509–15.

Sumper, M. (1987) Halobacterial glycoprotein biosynthesis. *Biochim. Biophys. Acta* **906**: 69–79.

Taylor, K.A., Deatherage, J.F. and Amos, L.A. (1982) Structure of the S-layer of *Sulfolobus acidocaldarius*. *Nature* **299**: 840–2.

Weiss, L.R. (1974). Subunit cell wall of *Sulfolobus acidocaldarius*. *J. Bacteriol.* **118**: 275–84.

Wieland, F. (1989) The cell surfaces of halobacteria. In *Halophilic Bacteria*, vol. II. (Rodriguez-Valera, F., ed), pp. 55–65. CRC Press: Boca Raton.

Wieland, F., Dompert, W., Bernhardt, G. and Sumper, M. (1980) Halobacterial glycoprotein saccharides contain covalently linked sulphate. *FEBS Lett.* **120**: 110–14.

Wieland, F., Lechner, J., Bernhardt, G. and Sumper, M. (1981) Sulphation of a repetitive saccharide in halobacterial cell wall glycoprotein. *FEBS Lett.* **132**: 319–23.

Wieland, F., Lechner, J. and Sumper, M. (1982) The cell wall glycoprotein of halobacteria: Structural, functional and biosynthetic aspects. *Zbl. Bakt. Hyg., I. Abt. Orig.* **C3**: 161–70.

Wieland, F., Heitzer, R. and Schaefer, W. (1983) Asparaginylglucose: A novel type of carbohydrate linkage. *Proc. Natl. Acad. Sci. USA* **80**: 5470–4.

Wieland, F., Lechner, J. and Sumper, M. (1986) Iduronic acid: Constituent of sulphated dolichylphosphate oligosaccharides in halobacteria. *FEBS Lett.* **195**: 77–81.

Wildhaber, J. and Baumeister, W. (1987) The cell envelope of *Thermoproteus tenax*: Three-dimensional structure of the surface layer and its role in shape maintenance. *EMBO J.* **6**: 7475–80.

Wildhaber, I., Santarius, U. and Baumeister, W. (1987) Three-dimensional structure of the surface protein of *Desulfurococcus mobilis*. *J. Bact.* **169**: 5563–8.

Woese, C.R. (1987) Bacterial evolution. *Microbiol. Rev.* **51**: 221–71.

Woese, C. R., Kandler, O. and Wheelis, M.L. (1990) Towards a natural system of organisms: Proposal for the domains Archaea, Bacteria and Eucarya. *Proc. Natl. Acad. Sci. USA* **87**: 4576–9.

Zeikus, J.G. and Bowen, V.G. (1975) Fine structure of *Methanospirillum hungatei. J. Bacteriol.* **121**: 373–80.

Zellner, G. (1989) *Physiologisch biochemische Untersuchungen der Methanogenese anhand bekannter und neu isolierter Mikroorganismen.* Ph.D thesis, University of Regensburg.

5
Structural Lipids of Eubacteria

T. Martin Embley[1] and Robin Wait[2]
[1]*Natural History Museum, London, UK*
[2]*Centre for Applied Microbiology and Research, Porton Down, Salisbury, UK*

5.1 INTRODUCTION

The analysis of the structural lipids of bacteria is one of the most useful phenotypic approaches for their classification and identification. The fatty acids and polar lipids which form the lipid bilayer of bacterial membranes are among the most valuable for this purpose. These compounds are relatively easy to determine but display considerable qualitative and quantitative variation which when analysed carefully has excellent discriminative value. The present chapter aims to provide a selection of the most useful methods for fatty acid and polar lipid analysis that are applicable in taxonomic studies. For further information on lipid analysis the specialist textbooks by Christie (1982, 1987, 1989), Kates (1985, 1990), Kuksis (1978, 1987) and Marinetti (1976) are recommended.

5.2 FATTY ACIDS

The fatty acids most commonly found in eubacteria possess between 12 and 20 carbon atoms and are relatively simple in structure (Table 5.1). Straight-chain saturated, monounsaturated and monomethylbranched fatty acids generally account for the majority of the components encountered. The occurrence and relative amount of each component reflects the biosynthetic capabilities of the microorganism and the conditions under which it was grown (Fulco, 1983; Harwood & Russell, 1984; Rose, 1989; Schweizer, 1989). The carbon chain length, the location of methyl groups (e.g. *iso-* or *anteiso-*) and the positions of double bonds (e.g. vaccenic or oleic) all have a taxonomic utility.

Different bacteria can have totally different fatty acids. Some fatty acids

Table 5.1 Structures of fatty acids commonly found in eubacteria.

Type	Example	Abbreviation
Straight chain		
Saturated:	Hexadecanoic	16:0
Unsaturated:		
Oleic series	cis-9-octadecenoic	18:1(9)
Vaccenic series	cis-11-octadecenoic	18:1(11)
Methylbranched:		
iso- Series	14-methylpentadecanoic	i-16:0
anteiso- Series	14-methylhexadecanoic	a-17:0
'Tuberculostearic'	10-methyloctadecanoic	10-Me-18:0
Cyclopropane:		
'Dihydrosterulic'	cis-9,10-methyleneoctadecanoic	18:0cy
'Lactobacillic'	cis-11,12-methyleneoctadecanoic	18:0cy
Cyclohexane:	13-cyclohexyltridecanoic	
Hydroxy:		
2-hydroxy	2-hydroxytetradecanoic	2-OH-14:0
3-hydroxy	3-hydroxyhexadecanoic	3-OH-16:0
2,3-dihydroxy	2,3-dihydroxy-12-methyltridecanoic	2,3-diOH-i-14:0

have a restricted distribution and may be diagnostic for particular groups. Hydroxy fatty acids are important constituents of the lipid A of Gram-negative bacteria (Jantzen & Bryn, 1985; Weckesser & Mayer, 1988) but they are also found in sphingolipids (Miyagawa *et al.*, 1979) and may occur as minor components in some Gram-positive organisms including actinomycetes (Kroppenstedt, 1985). Many actinomycetes contain 10-methyl branched fatty acids, such as 10-methyloctadecanoic acid ('tuberculostearic acid') and its homologues (Kroppenstedt, 1985; Embley *et al.*, 1987), and bacteria such as *Campylobacter* and *Lactobacillus* contain cyclopropane fatty acids (e.g. Wait & Hudson, 1985; Wilkinson, 1988; O'Leary & Wilkinson, 1988). Cyclohexyl and cycloheptyl fatty acids are characteristic components of some acidothermophilic bacilli (De Rosa *et al.*, 1971; Oshima & Ariga, 1975; Deinhard *et al.*, 1987).

In addition to simple long-chain fatty acids such as those shown in Table 5.1, members of the actinomycete genera *Corynebacterium*, *Gordona*, *Mycobacterium*, *Nocardia*, *Rhodococcus* and *Tsukamurella* contain mycolic acids (Minnikin & Goodfellow, 1980; Dobson *et al.*, 1985; Collins *et al.*, 1988; Stackebrandt *et al.*, 1988; Minnikin, 1993). These are high molecular weight (20–90 carbons), long-chain, 3-hydroxy fatty acids with an alkyl branch at position 2 (Figure 5.1a). Most mycolic acids are esterified to the arabinogalactan, which is present in the cell envelope. Examples of the structures which are most commonly encountered are illustrated in Figure 5.1b. Species within a genus usually have the same simple fatty acids, but there may be significant differences in the relative amounts of individual components. In recent years, the application of computerized pattern recognition techniques to capillary chromatograms of fatty acids has become a powerful tool for the rapid identification of bacteria (Miller & Berger, 1985; Larsson *et al.*, 1989; Sasser, 1990; Janse, 1991; Bernard *et al.*, 1991; Welch, 1991; Stead *et al.*, 1992). A variety of statistical methods can be used to analyse the quantitative data (Wold & Sjostrom, 1977; O'Donnell, 1985; Saddler *et al.*, 1987; Eerola & Lehtonen, 1988; Ninet *et al.*, 1992).

The distribution and taxonomic significance of bacterial fatty acids has been the subject of a number of reviews and the reader is referred to these

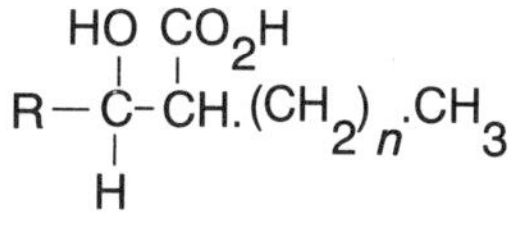

$$R - \underset{\underset{H}{|}}{\overset{\overset{HO}{|}}{C}} - CH.(CH_2)_n.CH_3$$

Figure 5.1a General structure of mycolic acids. The group R may be up to 59 carbons in length and can include various functional groups (see Figure 5.1b). The value of n varies between 5 and 23.

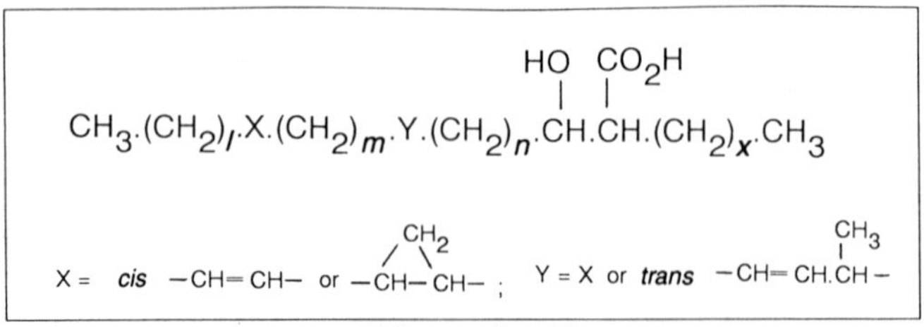

$$CH_3 \cdot (CH_2)_l \cdot X \cdot (CH_2)_m \cdot Y \cdot (CH_2)_n \cdot \underset{\underset{HO}{|}}{CH} \cdot \underset{\underset{CO_2H}{|}}{CH} \cdot (CH_2)_x \cdot CH_3$$

$$X = cis \;\; -CH=CH- \;\; \text{or} \;\; -\overset{\overset{\displaystyle CH_2}{\diagup \;\; \diagdown}}{CH}-CH- \;\; ; \quad Y = X \;\; \text{or} \;\; trans \;\; -CH=\overset{\overset{\displaystyle CH_3}{|}}{CH} \cdot CH-$$

Alpha-mycolic acid

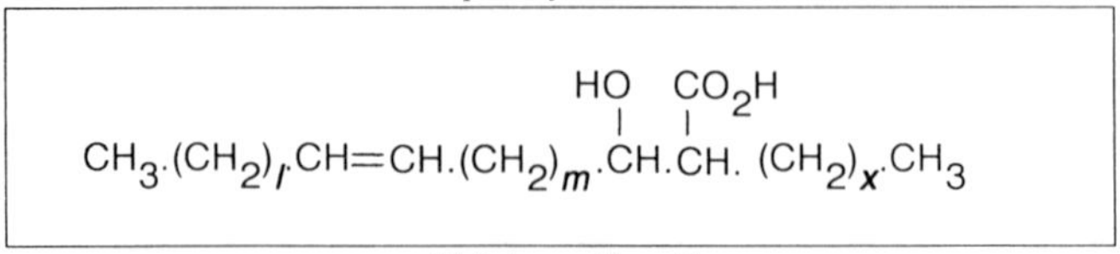

$$CH_3 \cdot (CH_2)_l \cdot CH=CH \cdot (CH_2)_m \cdot \underset{\underset{HO}{|}}{CH} \cdot \underset{\underset{CO_2H}{|}}{CH} \cdot (CH_2)_x \cdot CH_3$$

Alpha'-mycolic acid

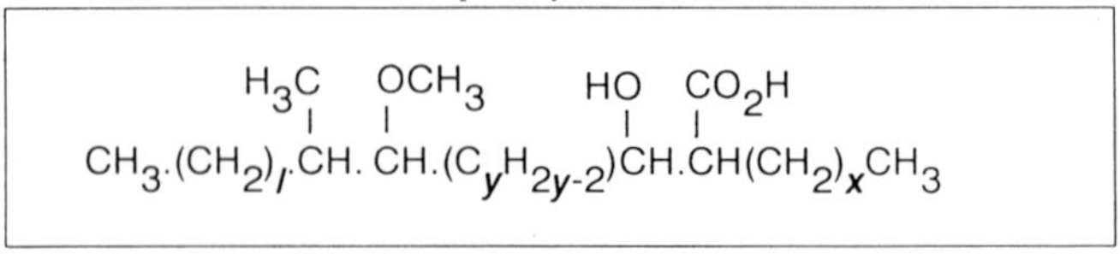

$$CH_3 \cdot (CH_2)_l \cdot \underset{\underset{H_3C}{|}}{CH} \cdot \underset{\underset{OCH_3}{|}}{CH} \cdot (C_yH_{2y-2}) \underset{\underset{HO}{|}}{CH} \cdot \underset{\underset{CO_2H}{|}}{CH} (CH_2)_x CH_3$$

Methoxymycolic acid

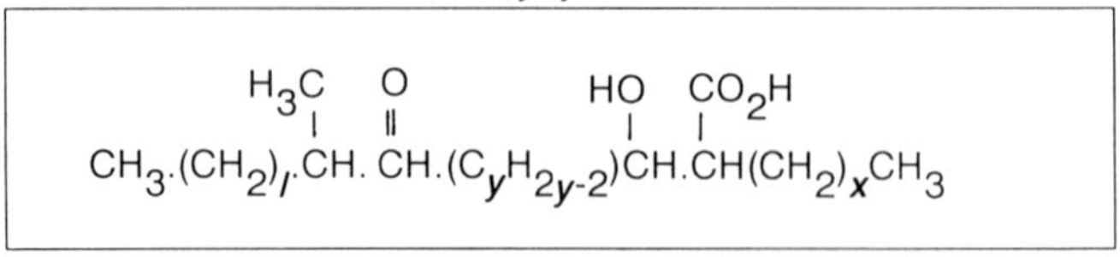

$$CH_3 \cdot (CH_2)_l \cdot \underset{\underset{H_3C}{|}}{CH} \cdot \overset{\overset{\displaystyle O}{\|}}{C} \cdot (C_yH_{2y-2}) \underset{\underset{HO}{|}}{CH} \cdot \underset{\underset{CO_2H}{|}}{CH} (CH_2)_x CH_3$$

Ketomycolic acid

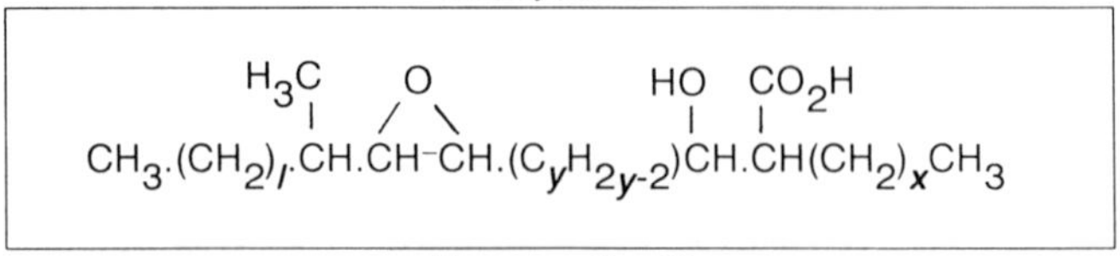

$$CH_3 \cdot (CH_2)_l \cdot \underset{\underset{H_3C}{|}}{CH} \cdot \overset{\overset{\displaystyle O}{\diagup \;\; \diagdown}}{CH-CH} \cdot (C_yH_{2y-2}) \underset{\underset{HO}{|}}{CH} \cdot \underset{\underset{CO_2H}{|}}{CH} (CH_2)_x CH_3$$

Epoxymycolic acid

$$HO_2C \cdot (CH_2)_l \cdot CH=CH \cdot (CH_2)_m \cdot \underset{\underset{HO}{|}}{CH} \cdot \underset{\underset{CO_2H}{|}}{CH} \cdot (CH_2)_x \cdot CH_3$$

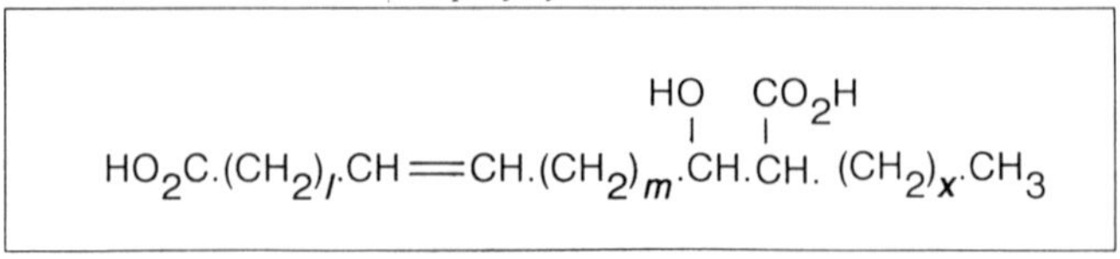

ω - Carboxymycolic acid

Figure 5.1b The structures of some frequently encountered mycolic acid types. The subscripts l, m, n, x, and y have values in the range 11–35 carbons. The unit (C_yH_{2y-2}) incorporates either *cis* double bonds, or *trans* double bonds or cyclopropane rings with an adjacent methyl branch in the middle of the chain, as is shown in the structure of the alpha mycolate. Modified from Minnikin (1993) with permission.

for more information on particular taxa (Kroppenstedt, 1985; Lechevalier & Lechevalier, 1988; O'Leary & Wilkinson, 1988; Wilkinson, 1988; Welch, 1991; Suzuki et al., 1993).

5.3 METHODS FOR ANALYSING FATTY ACIDS

5.3.1 Growth conditions for preparing biomass

There is considerable evidence that fatty acid composition is markedly affected by growth temperature and medium composition (McElhaney, 1976; Kroppenstedt & Kutzner, 1978; Saddler et al., 1987; Rose, 1989). Care should therefore be taken to ensure that bacteria for comparative studies are grown under standardized conditions and, where possible, in chemically defined media. As far as is possible media should omit material containing fatty acids, such as Tweens and serum. Material for the acid methanolysis procedure should be lyophilized after harvest as methylation may not go to completion in the presence of water. Growth from plate cultures can be used for the alkali procedure, providing that colonies are taken from the third quadrant of a streak plate as detailed in Miller and Berger (1985) and Sasser (1990).

5.3.2 Extraction of fatty acids

In the intact bacterial cell the bulk of the fatty acids occur as constituents of more complex molecules such as phospholipids, glycolipids and lipid A. It is therefore necessary to liberate them by means of a hydrolytic or solvolytic step, prior to chemical derivatization and analysis by a procedure such as gas chromatography. There are two basic approaches (and many variations) for the extraction of fatty acids from bacteria. Alkaline hydrolysis (saponification), followed by transesterification to form methyl esters, is the commonest method (Miller, 1982; Miller & Berger, 1985). An alkali-based procedure is also normally used for the 2-dimensional analysis of mycolic acids as it preserves some of the more labile components (Dobson et al., 1985; Minnikin, 1988). Acid methanolysis (Minnikin et al., 1980) has been widely used for the analysis of fatty acids from Gram-positive taxa. Both techniques give broadly comparable data but in taxonomic studies it is important to use a consistent method.

Method 5.1 Preparation of fatty acid methyl esters by acid methanolysis (from Minnikin et al., 1980).

1. To approximately 10 mg of lyophilized biomass in a tube with a polytetrafluoroethylene (PTFE) cap insert, add 3 ml of dry methanol-toluene-sulphuric acid (30:15:1, v/v) and incubate at 50 °C overnight.
2. Cool to room temperature and add 2 ml of hexane, mix by shaking and centrifuge at low speed to break the emulsion.
3. Remove the upper hexane layer and repeat step 2 using 1 ml of hexane; pool the upper layers.
4. Prepare a column of ammonium hydrogen carbonate in a Pasteur pipette as follows: push a small plug of clean glass wool or non-absorbent cotton wool as far as the constriction in a short Pasteur pipette; add ammonium hydrogen carbonate to form a 1 cm column; wash through twice with 1 ml of diethyl ether.
5. Add the pooled hexane supernatants to the top of the column and collect the eluent in a clean vial. Dry the eluent in a stream of nitrogen and redissolve in 100 μl of hexane for subsequent analysis.

Note: Care must be taken to avoid the degradation of cyclopropane fatty acids (Lambert & Moss, 1983) when using the acid-based procedure. The problem is much reduced by performing the incubation at 50 °C.

Method 5.2 Alkaline hydrolysis for simple fatty acids (after Miller & Berger, 1985; Sasser, 1990).

1. Take a 4 mm loopful of growth from the third quadrant of an overnight bacterial plate culture and coat the bottom of a PTFE-capped 8.5 ml tube with the biomass.
2. Add 1 ml of 15% (w/v) NaOH in 50% (v/v) aqueous methanol, vortex and heat at 100 °C in a boiling bath for 5 min, vortex again, continue heating for a further 25 min and cool to room temperature.
3. Add 2 ml of 6 M HCl in 50% (v/v) aqueous methanol, vortex, heat for 10 min at 80 °C in a water bath and cool rapidly on ice water.
4. Add 1.25 ml of hexane/methyl tert-butyl ether (1:1, v/v), mix for 10 min, remove and discard bottom phase.
5. Add 3 ml of 1.2% (w/v) aqueous NaOH, mix for 5 min, remove upper phase and transfer to a clean vial for direct analysis or for drying under a stream of nitrogen.

Method 5.3 Alkali procedure for preparation of mycolic acid methyl esters for two-dimensional thin-layer chromatography (after Minnikin, 1988, 1993).

1. Place 50 mg of dried biomass in an 8.5 ml tube with a PTFE cap.
2. Add 2 ml of 15% (v/v) aqueous tetrabutylammonium hydroxide (TBAH) and heat at 100 °C overnight.

5. Structural Lipids of Eubacteria

3. Allow the tube to cool and add 2 ml of water, mix, centrifuge at low speed and transfer the supernatant to a clean tube.
4. Add 25 μl of a solution of iodomethane in dichloromethane (2 ml) and mix on a tube rotator for 1 h.
5. Discard the upper aqueous layer and wash the lower organic layer with 2 ml of 20% aqueous HCl, followed by 2 ml of water. Dry down the organic phase under a stream of nitrogen. Redissolve in 50 μl of hexane for TLC analysis.

Note: This procedure gives almost identical extracts to those obtained with the procedure of Dobson et al. (1985).

Method 5.4 Preparation of *para*-bromophenacyl esters of mycolic acids for high-performance liquid chromatography (after Durst *et al.*, 1975; Butler & Kilburn, 1990; Butler *et al.*, 1991, 1992; the following procedure was kindly supplied by W.R. Butler).

1. Take a loopful of cells from solid growth media and saponify in a 13 mm tube with a PTFE cap by adding 2 ml of saponification reagent (25% w/v potassium hydroxide in methanol : water, 1 : 1 v/v) and autoclaving for 1 h at 121 °C.
2. After cooling, the mixture is acidified by adding 1.5 ml of a 50% (v/v) mixture of concentrated HCl and water (acid solution 1). Add 2 ml of chloroform and mix thoroughly. Remove the lower chloroform layer to a clean 13 × 100 mm screw-capped tube with a PTFE cap and dry down.
3. Add 1 ml of a solution of potassium bicarbonate (2% in methanol : water, 1 : 1 v/v) and after mixing, dry down once again.
4. Add 1 ml of chloroform and 50 μl of *p*-bromophenacyl-8 reagent (Pierce Chemical Co., Rockford, Illinois, USA) and vortex briefly, heat at 85 °C for 20 min and cool.
5. Add 1 ml of acid solution 2 (acid solution 1 diluted with methanol to 50% v/v). Remove the lower chloroform layer and evaporate to dryness. Redissolve in 50 μl of chloroform for HPLC analysis.

5.3.3 Thin-layer chromatography of fatty acids

Simple one-dimensional thin-layer chromatography can be used to analyse the crude methanolysates in order to monitor the progress of the reaction and the presence of fatty acid methyl esters, mycolic acids or hydroxy fatty acids (Minnikin *et al.*, 1980; Kroppenstedt, 1985). It can also be used to purify fatty acids prior to gas-liquid chromatography and thus avoid contaminating the capillary column. One-dimensional systems may also give some indication of the type of mycolic acid present (Minnikin, 1993) and hence the generic assignment of the organism. Mycolic acid-containing taxa other than mycobacteria give single spots corresponding to simple alpha-mycolates and these may have characteristic mobilities in some solvent systems (Minnikin *et al.*, 1980; Minnikin, 1993). For example, members of the genus *Corynebacterium* normally give single spots with R_F

a Petrol-acetone 95:5 x1

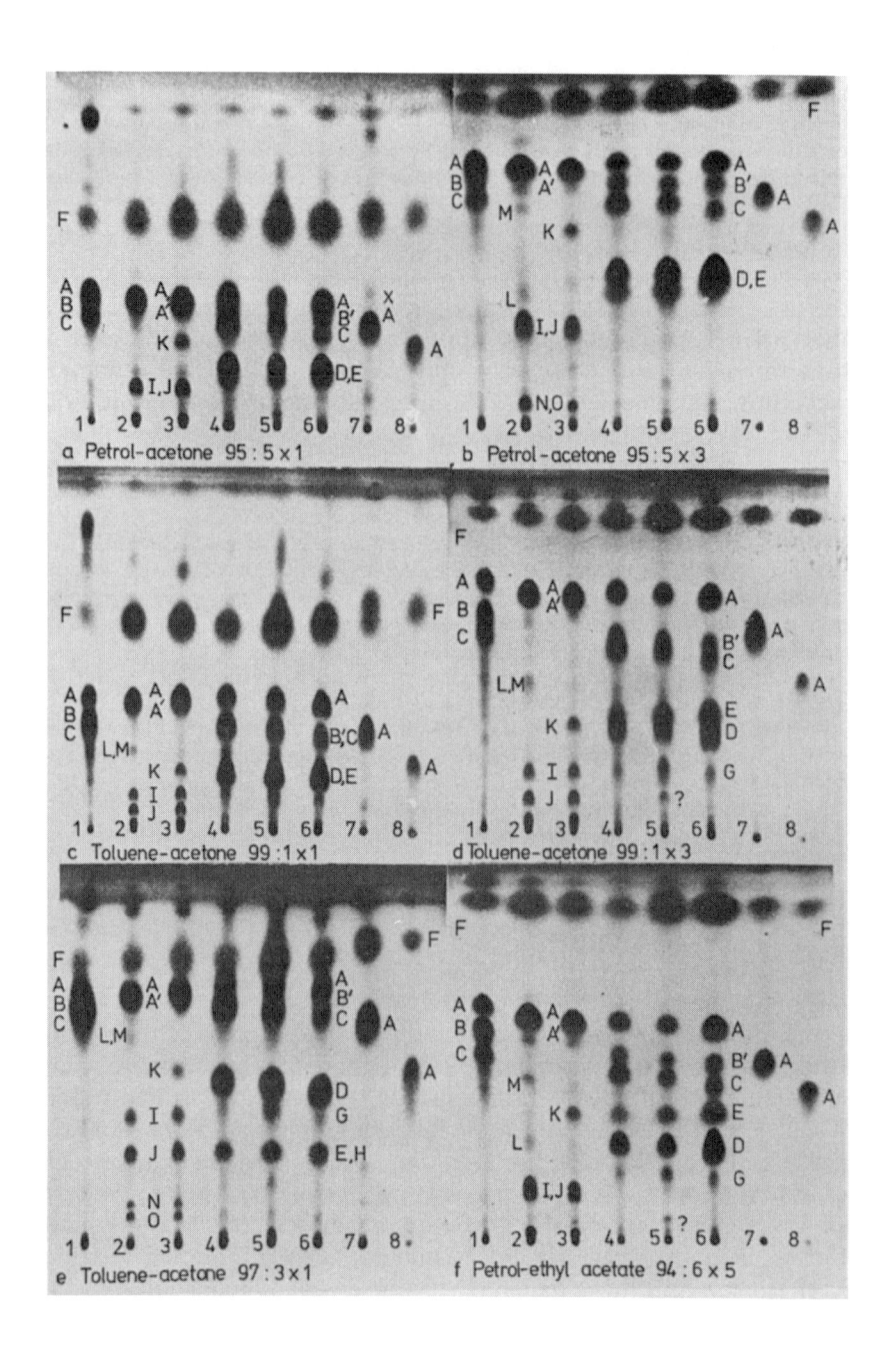

b Petrol-acetone 95:5 x3

c Toluene-acetone 99:1 x1

d Toluene-acetone 99:1 x3

e Toluene-acetone 97:3 x1

f Petrol-ethyl acetate 94:6 x5

values of 0.2 to 0.25; strains of *Gordona* and *Nocardia* produce single spots with R_F values in the range 0.3 to 0.31 (Minnikin, 1993). Strains of *Tsukamurella* may give one (*T. wratislaviensis*) or two spots (*T. paurometabola*) and their R_F values may vary (Hamid *et al.*, 1993; Minnikin, 1993). Members of the genus *Mycobacterium* may have mixtures of mycolic acids which differ in both their length and type of substitution. Single 1D developing systems are not able to resolve these but Minnikin *et al.* (1980) have systematically investigated the use of successive developments in different solvents (Figure 5.2). Two-dimensional separations are more useful and diagnostic patterns (Figure 5.3) can be obtained for some mycobacteria using this approach (Minnikin *et al.*, 1980, 1984a,b; Dobson *et al.*, 1985; Minnikin, 1988, 1993).

Method 5.5 One-dimensional TLC of methanolysis products (from Minnikin *et al.*, 1980).

1. Dissolve the dried material in 100 μl of hexane.
2. Spot ~5 μl of the sample onto an aluminium-backed thin-layer plate (Merck silica gel 60F$_{254}$ cat. 5554) about 1 cm from the bottom of the plate.
3. Develop with hexane-diethyl ether (85:15, v/v) or light petroleum ether (bp 60–80 °C)-acetone (95:5, v/v) until the solvent front is about 1 cm from the top of the plate. Dry the plate in a fume cupboard.
4. Spray the plate with a solution of 10% (w/v) molybdophosphoric acid in 95% (v/v) aqueous ethanol (MPA reagent) and heat in an oven at 120 °C for 10 to 15 min.

Lipids are revealed as black spots on a green background (see Figure 5.2). Figure 5.2 shows the effects of different solvent systems on the separation behaviour of mycolic acids and other long-chain compounds (Minnikin *et al.*, 1980).

Method 5.6 Purification of fatty acid methyl esters using 1D TLC.

1. Apply the extract (dissolved in ~50 μl of hexane) to the bottom of a plastic-backed, silica-gel plate (Merck 5735) as a narrow band about 1 cm long.
2. Develop using the same solvent system as used for the analytical chromatography.

Figure 5.2 (*opposite*) Use of different solvent systems for single dimensional TLC of whole-organism methanolysates from: 1, *Mycobacterium tuberculosis*; 2, *M. smegmatis*; 3, *M. fortuitum*; 4, *M. avium*; 5, *M. johnei*; 6, *M. phlei*; 7, *Nocardia asteroides*; 8, *Corynebacterium diphtheriae*. A, A' = methyl mycolates devoid of oxygen functions other than the 3-hydroxy ester system; B = methoxymycolate; B' = unknown component from 4,5,6; C = ketomycolate; D = ω-carboxy mycolate methyl ester; E = 2-eicosanol and homologies; F = non-hydroxylated fatty acid methyl esters; G = unknown from 4,5,6; H = unknown from 6; I,J = unknowns from 2,3. From Minnikin *et al.* (1980) with permission.

3. Spray the developed plate with a solution of 0.01% rhodamine 6G in 95% (v/v) aqueous ethanol and allow to dry in a fume cupboard.
4. Visualize the separated components under shortwave UV light. Fatty acids have an R_F value of about 0.7 to 0.8. Mark the location of the band with a soft graphite pencil.
5. Scrape the silica off the plate and place it in a small vial with 1 ml of diethyl ether, shake intermittently for 5 min to extract the fatty acid methyl esters from the gel, remove the supernatant, and repeat the extraction.
6. Pool the extracts and pass through a small column of neutral aluminium oxide prepared in a Pasteur pipette, and prewashed with 1 ml of diethyl ether. Wash the column with 1 ml of diethyl ether and add this to the pooled eluents. Dry under a stream of nitrogen at 37 °C. Redissolve in 100 µl of hexane for GC analysis.

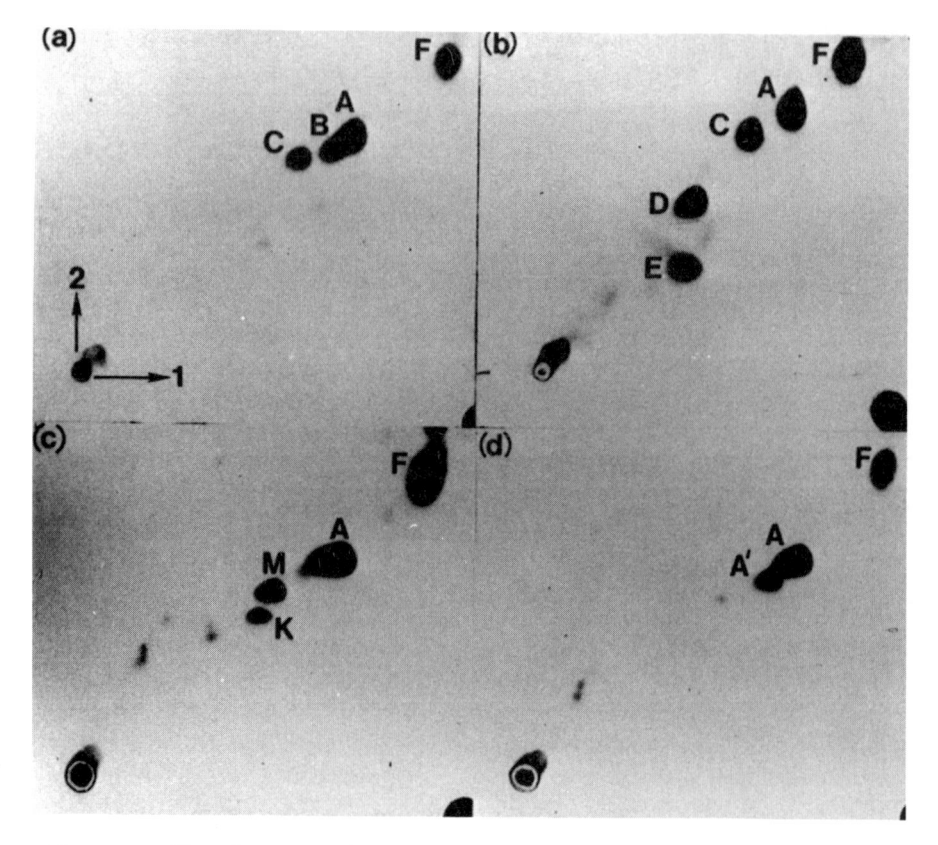

Figure 5.3 Two-dimensional TLC of mycolic acid methyl esters from: (a) *Mycobacterium tuberculosis*; (b) *M. avium*; (c) *M. fortuitum* and (d) *M. chelonei*. A, α-mycolate; A', α'-mycolate; B, methoxymycolate; C, ketomycolate; D, ω-carboxymycolate; E, eicosanol and homologues; F, fatty acid methyl esters; K, unidentified mycolate; M, epoxymycolate. From Dobson *et al.* (1985) with permission.

Method 5.7 Two-dimensional TLC of mycolic acid methyl esters (Minnikin et al., 1980; Dobson et al., 1985).

1. Dissolve the sample in 100 µl of hexane, spot 5 µl of extract on the bottom left corner of a piece (6 × 6 cm) of Merck 5554 aluminium TLC sheet.
2. Develop in the first direction (×3) with petroleum ether : acetone (95 : 5, v/v) drying between each run. Develop in the second direction using toluene : acetone (97 : 3, v/v). Visualize using MPA reagent as for 1D TLC.

Figure 5.3 shows typical 2D patterns from *Mycobacterium tuberculosis*, *M. avium*, *M. chelonei* and *M. fortuitum* (Dobson et al., 1985).

5.3.4 High-performance liquid chromatography of *p*-bromophenacyl esters of mycolic acids

HPLC of mycolic acids is more sensitive and is capable of better resolution than thin-layer chromatography. Under defined growth conditions it can also be used to yield quantitative information. Reverse-phase chromatography gives good separation of *p*-bromophenacyl esters of mycolic acids, and appropriate columns, detectors and solvent gradient profiles are described in the papers by Butler and colleagues (Butler & Kilburn, 1988; Butler et al., 1987, 1991, 1992; De Briel et al., 1992). An example of a typical profile is shown in Figure 5.4.

5.3.5 Analysis of fatty acid methyl esters by capillary gas chromatography

Open tubular (capillary) columns have largely superseded packed columns for the separation of bacterial fatty acid methyl esters. We have found the most useful general purpose column to be a 25 m (0.25 mm i.d.) fused silica open tubular column with a thin (0.2–0.3 µm) chemically bonded non-polar, methylsilicone phase such as OV-1 or equivalent. This combines stability with extremely good resolution of the common fatty acids. A complementary choice is a column coated with a polar cyanopropyl siloxane phase (e.g. BPX 70 from SGE Ltd.). Fatty acids exhibit different retention behaviour on the two phases and this can often facilitate the identification of components. For example, monounsaturated fatty acids elute after saturated acids on polar phases, whereas this order is reversed on non-polar columns. Polar columns generally provide superior resolution of isometric unsaturated fatty acid methyl esters (FAMEs). The highest chromatographic efficiencies are obtained when hydrogen is used as carrier gas. Helium is only slightly inferior in performance and its use avoids the risk of explosion. To fully exploit the resolving power of capillary columns the sample must be applied as a narrow band. This is commonly achieved using an inlet splitter (Grob, 1988). The disadvantage of this technique is that the split may not be linear across the entire boiling point range of the sample, which can lead to unreliable quantification of FAMEs with 20 or more carbon atoms. Discrimination can also result from poor injection technique causing the loss of volatile components. The best technique is the so called 'hot needle' method. The sample is withdrawn into the syringe barrel (about 1 cm), the needle is then placed in the injection port

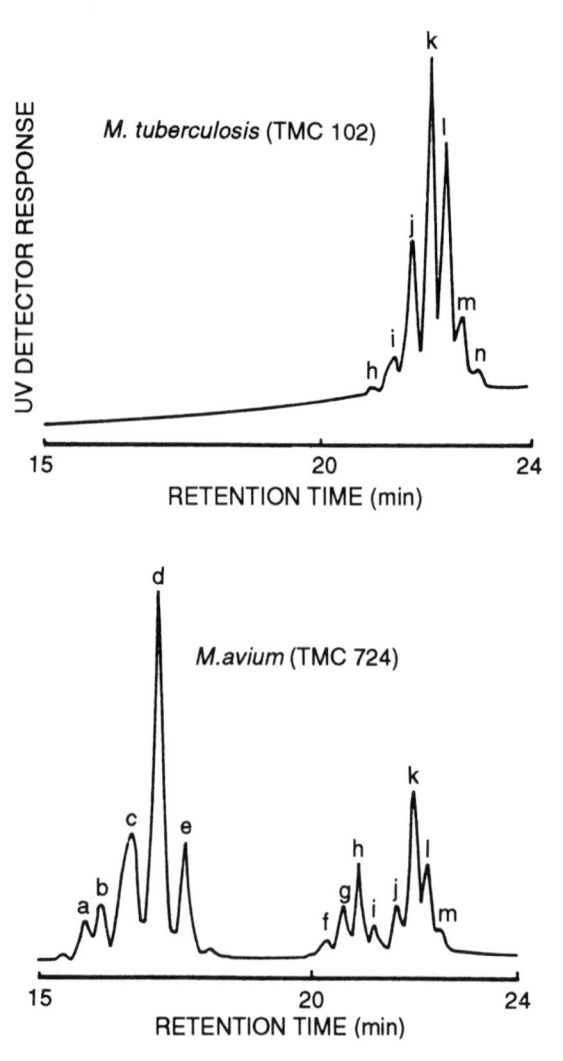

Figure 5.4 High performance liquid chromatograms of *p*-bromophenacyl esters from saponified whole-cells of *Mycobacterium tuberculosis* and *M. avium*. Abbreviations merely refer to order of elution. From Butler & Kilburn (1988) with permission.

and allowed to come to temperature before the plunger is rapidly depressed leading to flash vaporization of the sample. Figure 5.5 shows the results of typical analyses obtained using sample splitting and the hot needle method. When the amount of sample is limited or if more reliable quantitation is required then splitless (Grob, 1988) or cold on-column injection (Grob, 1992) are the methods of choice.

Fatty acids can be identified in a number of ways. The first step is to compare the retention behaviour of the sample components with that of known fatty acid methyl esters. Standard mixtures containing representatives of the common bacterial fatty acids are commercially available (e.g. bacterial acid methyl ester mix cat. 4–5436, Supelco, Belefonte, Pennsylvania, USA). More unusual fatty acids can be extracted from bacteria of known lipid composition, e.g. cyclopropane fatty acids can be extracted from *Lactobacillus rhamnosus* DSM 20021 and 10-methyloctadecanoic acid (tuberculostearic acid) from *Nocardia asteroides* DSM 43757 (Kroppenstedt, 1985). The calculation of retention times relative to an internal standard provides a correction for variation in chromatographic behaviour which may occur between runs. Fatty acids may also be identified from their equivalent chain lengths either calculated (Miwa, 1963) or determined by a graphical method (Kroppenstedt, 1985). However, retention data can only rigorously prove non-identity as chemically distinct species may exhibit identical retention times. Ideally, identifications should be verified by additional chemical and spectroscopic techniques as described below.

Method 5.8 Catalytic hydrogenation: verification of double bonds.

The presence of double bonds can be verified by catalytic hydrogenation to form the corresponding saturated analogue. However, positional isomers, *cis/trans* isomers, and polyunsaturated acids with the same carbon skeleton will all be converted to the same saturated product. Most cyclopropane rings are unchanged by hydrogenation under mild conditions.

1. Transfer the sample to a 5 ml flat-bottomed vial, and evaporate the solvent under nitrogen.
2. Redissolve the sample in 2 ml of chloroform : methanol (2 : 1, v/v), and add approximately 5 mg of 5% (w/v) palladium on charcoal (BDH Ltd., UK) together with a small PTFE-coated stirrer bar. Seal the vial with a PTFE-faced rubber septum.
3. Pierce the septum with two hypodermic needles, flush the vial with hydrogen, and place on a magnetic stirrer block at room temperature for 15 min.
4. Repeat step 3 twice, reflushing the vial with hydrogen each time.
5. Open the vial and remove the catalyst by filtration through a plug of glass wool in a Pasteur pipette; wash the residue on the glass wool with 2 ml of chloroform : methanol. Pool the filtrate and washings and evaporate to dryness. Redissolve the sample in the original solvent and repeat gas chromatography using identical chromatographic conditions as previously.

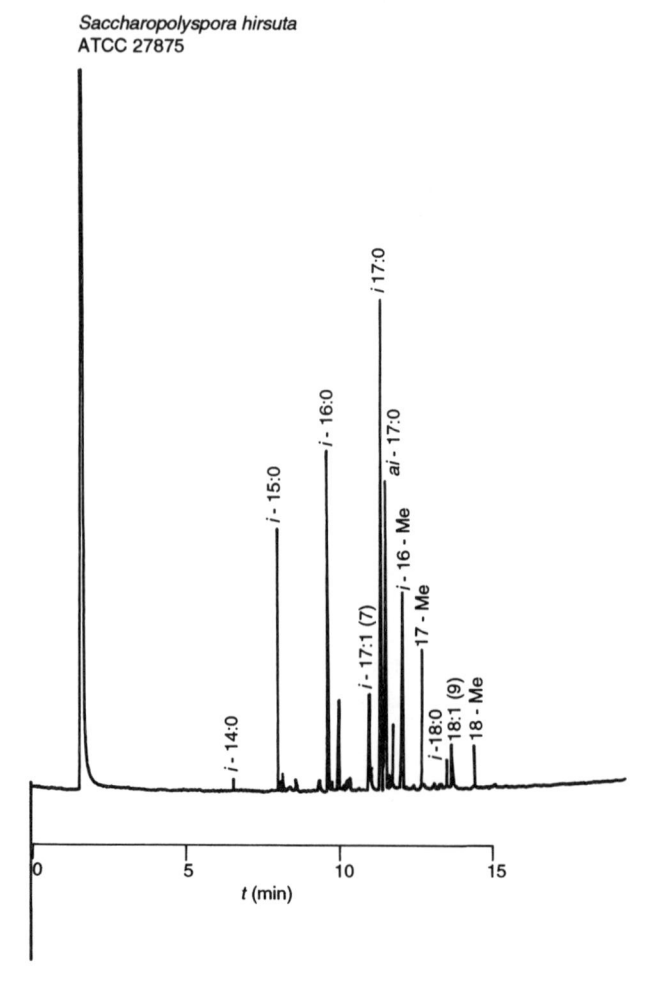

Figure 5.5 Capillary gas chromatograms showing fatty acids from *Saccharopolyspora hirsuta* ATCC 27875 and a *Pseudomonas* sp (*opposite*). The column was a 25 m × 0.2 mm 5% phenol methyl silicone fused silica capillary column (Ultra-2, Hewlett Packard Ltd., HP 19091B-102); the operating parameters were: injector temperature 250 °C, detector temperature 300 °C, temperature programme 170 °C to 270 °C at 5 °C per minute, 50 ml of H_2 per minute onto the column with a 50 : 1 split. Abbreviations are explained by the following examples: 16 : 0, hexadecanoic acid; *i*-16 : 0, 14-methylpentadecanoic acid; *ai*-15 : 0, 12-methyltetradecanoic acid; 18 : 1(9), *cis*-9-octadecanoic acid; 18-Me, 10-methyloctadecanoic acid; *i*-16-Me, 10,15-dimethylhexadecanoic acid; 2OH or 3OH, hydroxy groups substituted at positions 2 or 3 (from the carboxyl end), respectively.

5. Structural Lipids of Eubacteria

Pseudomonas sp.

Peaks labeled: 10:0 (3OH), 12:0 (2OH), 12:0, 12:0 (3OH), 16:1 (9), 16:0, 18:1 (9), 18:0

t (min)

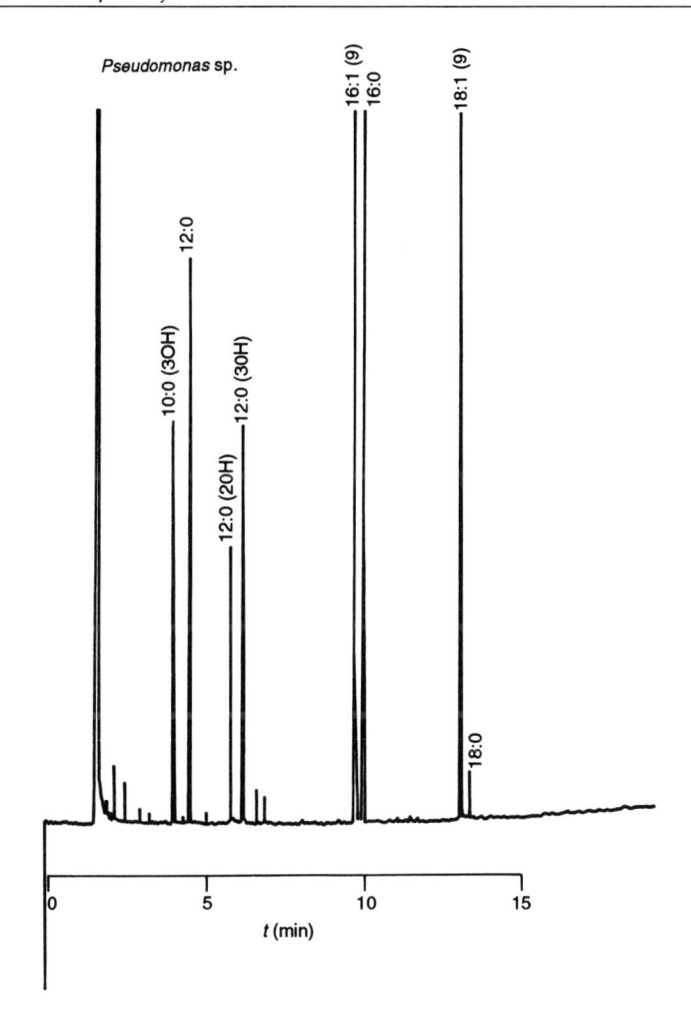

Method 5.9 Trimethylsilylation: identifying hydroxy-substituted acids.

The conversion of free hydroxyl groups to trimethylsilyl (TMS) ethers is a useful means of identifying hydroxy-substituted acids in mixtures, since derivatization alters their retention behaviour. Furthermore, derivatization often improves chromatography by eliminating peak tailing, and reducing losses due to polar interactions. Retention times on non-polar phases are generally increased by trimethylsilylation. It should be noted, however, that the abolition of polar interactions can result in a loss of resolution; for example, the TMS ethers of 2- and

3-hydroxy isomers may co-chromatograph on some stationary phases which fully resolve the corresponding free acids. Trimethylsilyl ethers are prone to hydrolysis and should be analysed soon after preparation. The reaction mixture may be directly injected if using an inlet splitter, but for splitless and on-column injections it is preferable to remove the reagents and redissolve the sample in hexane to minimize contamination of the detector with SiO_2 deposits.

1. Place a dry sample (up to 10 mg) of FAME in a screw-capped Wheaton vial and dissolve in 0.5 ml of N-O-bis(trimethylsilyl)trifluoracetamide (BSTFA).
2. Incubate at 70 °C for 15 min, cool and remove excess reagent in a stream of dry nitrogen, or vacuum centrifuge.
3. Redissolve the sample in an appropriate volume of hexane and analyse by GC or GC-MS.

5.3.6 Identification of fatty acids by mass spectrometry

The most powerful method for on-line identification of the peaks in a gas chromatogram is to use a mass spectrometer as the detector. Bombardment of the sample molecules with a beam of electrons produces a mixture of characteristic molecular ions and fragments that together define the mass and structure of each component. For some compounds, e.g. hydroxylated fatty acids, the molecular ion is weak or absent, making it difficult to establish the chain length. In such cases a 'soft ionization' method such as chemical ionization, which generates molecular ions by cation attachment, should be used.

5.3.6.a Conditions for gas chromatography-mass spectrometry of fatty acids

Chromatographic conditions matching those described above may be used but some differences in retention behaviour are sometimes observed with directly interfaced columns terminating in the mass spectrometer vacuum. Chemically bonded stationary phases with low column bleed are strongly recommended for GC-MS. Spectra are usually recorded at an ionization energy of 70 eV which facilitates comparison with library and literature spectra. Lower ionization energies (e.g. 20 eV) are sometimes used as a means of reducing excessive fragmentation, though with some sacrifice of sensitivity.

5.3.6.b Mass spectra of fatty acid methyl esters

Fatty acid methyl esters were among the first biological molecules to be systematically studied by mass spectrometry (Ryhage & Stenhagen, 1960) and their fragmentation behaviour has been extensively reviewed (Odham & Stenhagen, 1972; McCloskey, 1970; Odham, 1980; Asselineau & Asselineau, 1984; Jensen & Gross 1987; Kuksis & Myher 1989). Saturated

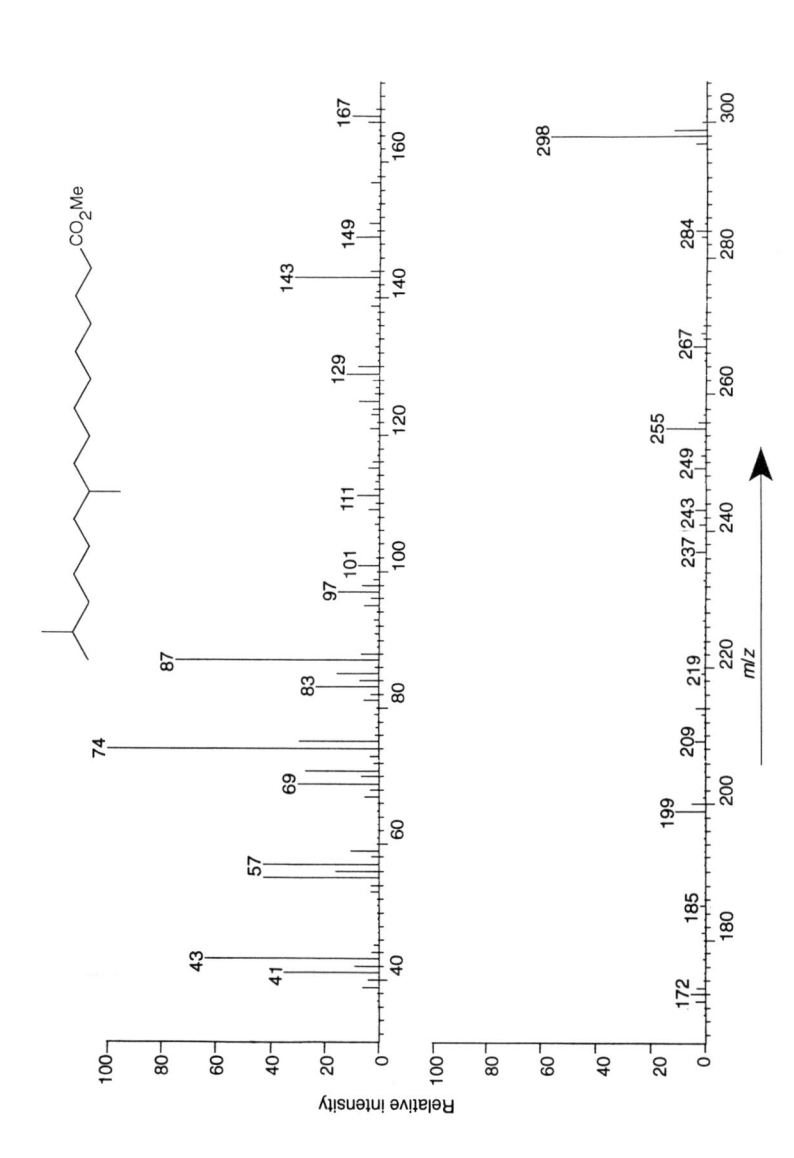

Figure 5.6 Electron impact mass spectrum of the methyl ester of 10,15-dimethylhexadecanoic acid. The overall fragmentation is typical of a saturated FAME. The position of the methyl branch in position 10 is indicated by the two ions m/z 167 and 149, which have no counterpart in the straight chain analogue. The *iso*-branch has very little effect on the fragmentation pattern of the methyl ester.

FAMEs (Figure 5.6) are characterized by molecular ions [M]˙ (typically about 20% of the intensity of the base peak), and a series of weak, regularly spaced fragments (14 mass units apart) arising from cleavages of the carbon skeleton. Elimination of H_2CH_3 and CH_2CH_2CH from the carboxyl terminal produces ions at m/z M-29 and M-43. The peak m/z 74 is characteristic of saturated FAMEs. Since it retains carbon 2, substituents in this position are easily recognized by the mass shift they induce. For example in the 2,4-dimethyltetradecanoic acid found in *Mycobacterium kansasii* the corresponding ion is found at m/z 88 (Julák et al., 1980). Alkyl substituents on carbons 3 and 4 are likewise recognizable by the mass shifts of the M-29 and M-43 ions. Branches near the centre of the alkyl chain induce characteristic changes in the fragmentation pattern, because cleavage adjacent to the branch produces a stable secondary carbocation. In the spectrum of the 10-methyl branched acid in (Figure 5.6) this is the ion m/z 199, which then undergoes sequential losses of methanol (32) and water (18), producing fragments of m/z 167 and m/z 149 which have no counterpart in the straight-chain analogue. Terminal methyl substituents are more difficult to identify from the spectra of methyl esters. *Anteiso-* branches are recognizable because the fragment M-29 is more intense than M-31, whereas the converse is true in *iso-*branched and straight-chain acids. *Iso-*branching in particular has little effect on the fragmentation of methyl esters, so *iso-*branched acids are best identified from capillary retention data, or from the mass spectra of derivatives such as picolinyl esters (Harvey, 1982). Another useful approach is reduction to the corresponding fatty alcohol (Karlsson et al., 1973) or alkane (Simon et al., 1990), both of which produce distinctive spectra.

Lipopolysaccharides (LPS) of Gram-negative organisms commonly contain 2- and 3-hydroxylated fatty acids. *Iso-*branched 3-hydroxy acids are components of the sphingolipids of *Porphyromonas gingivalis* and some related species, while 2,3-dihydroxy fatty acids have been identified in both cell and LPS hydrolysates of several *Legionella* species (Mayberry, 1981, 1984; Benson et al., 1989, 1991; Thacker et al., 1992). The presence of a hydroxyl substituent provides an alternative site for ionization enabling additional fragmentation pathways. Most of the characteristic fragmentations of the alkyl chain are still observed, though at reduced intensity. In 2-hydroxy acids a characteristic cleavage between carbons 1 and 2 produces an intense ion at m/z M-59. If the substituent is located on carbon 3, or some other position more remote from the ester group, the molecular ion is weak or absent, and the major fragmentation process is cleavage on the alkyl side of the hydroxylated carbon atom. In the case of a 3-hydroxy acid this produces a characteristic ion m/z 103 (shifted to m/z 175 in the corresponding TMS ether) which is generally the most intense peak in the spectrum. It is advantageous to analyse hydroxy acids as their TMS ethers since the chain length is deducible from an abundant ion at M-15.

Unsaturation or the presence of carbocyclic rings reduces the mass of the molecular ion by two mass units per double bond compared to the corresponding saturated acid. Unsaturated FAMEs undergo more extensive fragmentation than their saturated analogues, but the position of the double bond is not usually deducible from their spectra. The spectra of cyclopropane-containing acids are virtually identical to their unsaturated isomers, presumably because ring opening occurs on ionization, followed by processes similar to the double bond migration and indiscriminate fragmentation of unsaturated FAMEs. Fatty acids containing cyclohexane rings have been found in some acid-tolerant thermophiles (Oshima & Ariga, 1975). The molecular ions of the methyl esters of these compounds are two mass units lower in mass than the corresponding straight-chain saturated acid, but the overall appearance of the spectra is similar to saturated acids, except for an ion m/z 83, representing cleavage through the cyclopropane ring (Ohya et al., 1986). The size and location of the ring can be confirmed from the spectra of the picolinyl derivatives (Wait et al., 1989). Methods for double bond and cyclopropane ring location have been reviewed (Minnikin, 1978; Schmitz & Klein, 1986; Harvey, 1992).

5.3.6.c Picolinyl esters

Picolinyl esters (Harvey, 1982, 1984, 1992) offer advantages over methyl esters for the identification of terminal branches, and the localization of double bonds and cyclopropane rings in fatty acids. The method below (Method 5.10) gives better recoveries of unsaturated acids than Harvey's original procedure.

Method 5.10 Preparation of picolinyl esters (Christie & Stefanov, 1987).

1. Dissolve the unesterified fatty acids (in a screw-capped test-tube with a PTFE-faced septum) in 250 µl of trifluoroacetic anhydride (TFAA) at 50 °C for 30 min. Remove the TFAA by evaporation in a stream of nitrogen.
2. Add 100 µl of a solution containing 10 mg 3-hydroxymethyl pyridine and 2 mg of 4-dimethylaminopyridine (Fluka) in dichloromethane to the sample tube, seal the tube and maintain at room temperature for 3 h.
3. Evaporate the solvent under nitrogen, and dissolve the residue in 4 ml of hexane; wash with three successive 2 ml portions of distilled water.
4. Use a final wash with 1 ml of a 0.3 M solution of potassium hydroxide to remove any residual free fatty acids, remove the hexane layer containing the picolinyl esters and evaporate to dryness.

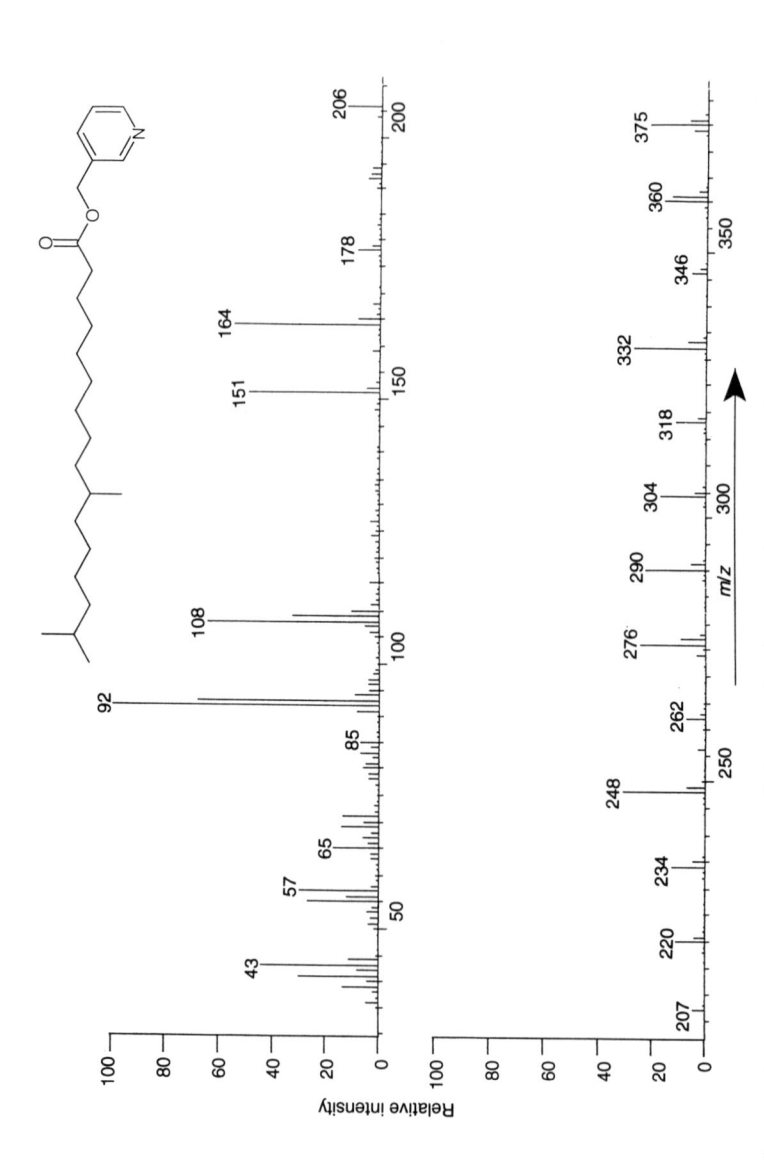

Figure 5.7 Electron impact mass spectrum of the picolinyl ester of 10,15-dimethylhexadecanoic acid. The locations of the methyl branches are indicated by discontinuities in the regular series of alkyl chain cleavage ions, in this case at m/z 346 and 262.

5. Structural Lipids of Eubacteria

The elution temperatures of picolinyl esters are about 50 °C higher than those of the corresponding methyl esters. Either polar or non-polar phases can be used for their separation, but non-polar phases are preferred on account of their superior thermal stability. Peak tailing is frequently observed because of the ring nitrogen, and some loss of resolution is to be expected compared to methyl esters. It may be advantageous to use split rather than splitless injection, since run times are shorter due to higher initial oven temperatures.

The mass spectra of picolinyl esters are characterized by intense molecular ions, and fragments at m/z 92, 108, 151 and 164, the base peak usually being m/z 92 (or sometimes m/z 108). Above m/z 164 the spectra are dominated by a series of regularly spaced ions, which represent cleavage without rearrangement at each carbon of the alkyl chain. The positions of double bonds, carbocyclic rings and alkyl branches can be deduced from the masses of these ions. In the case of straight-chain saturated acids, the spacing of these ions is a uniform 14 mass units. If a double bond is present, the molecular ion, and all the fragments containing the site of unsaturation will be reduced in mass by two mass units; however, all the ions originating from cleavages on the ester side of the unsaturated carbon will be the same mass as in the corresponding saturated compound (i.e. $164 + (14)_n$). Thus the separation between the two ions resulting from cleavage on either side of the double bond will be 26 mass units, not 28. Cyclopropane rings are located with particular ease, since the alkyl chain fragments at the site of the ring. This cleavage involves the breakage of two bonds so the resulting odd electron ion has an odd mass number, in contrast to most of the other ions in the spectrum. Cyclopropane rings in positions 9,10 and 11,12 are characterized by ions of m/z 247 and m/z 275 respectively.

The presence of alkyl branching results in an interruption in the sequence of alkyl cleavage ions from which the position of the branch may be deduced. *Iso-* and *anteiso-* branched acids have diminished intensity of the ions at M-29 and M-43 respectively, while reduced intensity at m/z 262 is indicative of a 10-methyl branch. In the example in Figure 5.7 methyl branches are present in both the *iso-* position (weak ion at m/z 346) and position 10 (reduced intensity at m/z 262).

5.4 BACTERIAL POLAR LIPIDS

The polar lipids of bacteria are a heterogeneous group of molecules which have amphipathic properties because the lipid is covalently linked to a hydrophilic component. The most commonly occurring polar lipids in eubacterial membranes are glycerophospholipids, in which fatty acids are esterified to positions *sn* 1 and 2 of the glycerol, and the polar function is linked to position 3 by a phosphodiester linkage. The glycerophospholipids are formally derivatives of phosphatidic acid (1,2-diacyl-*sn*-glycerol-3-phosphate), as shown in Figure 5.8. The group linked to the phosphate may be an amino alcohol (e.g. ethanolamine), an amino acid (e.g. L-serine), a polyol (e.g. glycerol) (Figure 5.8), a polyol derivative (e.g. desphosphatidylglycerol or phosphatidyl-*O*-aminoacyl glycerol) or an inositol derivative (Figure 5.9). The fatty acids on positions 1 and 2 of the glycerol are extremely heterogeneous, though unsaturated acids when present are usually located predominantly (but not exclusively) on position 2. Thus, while most bacteria contain relatively few different phospholipid types (Lechevalier *et al.*, 1977; Minnikin & Goodfellow, 1981; Harwood & Russell, 1984: Kroppenstedt, 1985), the heterogeneity of fatty acid substitution produces many discrete molecular species resulting in mixtures of formidable complexity.

Phosphatidylglycerol and diphosphatidylglycerols (cardiolipin; Figure 5.9) are widely distributed among the eubacteria and are of rather limited taxonomic use. Phosphatidylethanolamine is more restricted in its occurrence (most Gram-negatives, but only some Gram-positive bacteria such as *Bacillus* spp. and actinomycetes). Mono- and dimethylphosphatidyl ethanolamine, and phosphatidylcholines are relatively uncommon in prokaryotes. Phosphatidylcholines are one of the major constituents of the cell membranes of the Legionellaceae (Finnerty *et al.*, 1979; Mauchline *et al.*, 1992) and they also occur in some actinomycetes (Lechevalier *et al.*,

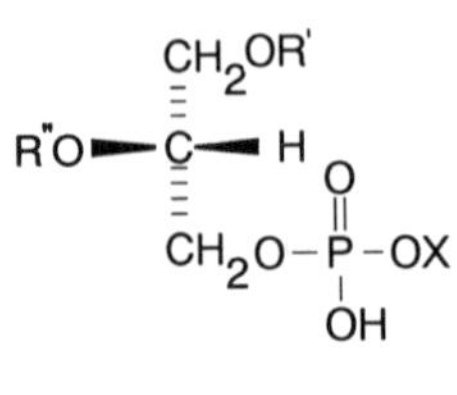

Phospholipid	Head group-X
Phosphatidic acid	$-H$
Phosphatidylserine	$-CH_2CH(NH_2)COOH$
Phosphatidylethanolamine	$-CH_2CH_2NH_2$
Phosphatidyl-*N*-monomethylethanolamine	$-CH_2CH_2NHCH_3$
Phosphatidyl-*N,N*-dimethylethanolamine	$-CH_2CH_2NH(CH_3)_2$
Phosphatidylcholine	$-CH_2CH_2NH(CH_3)_3$
Phosphatidylglycerol	$-CH_2CH(OH)CH_2OH$

Figure 5.8 Structures of some common glycerophospholipids.

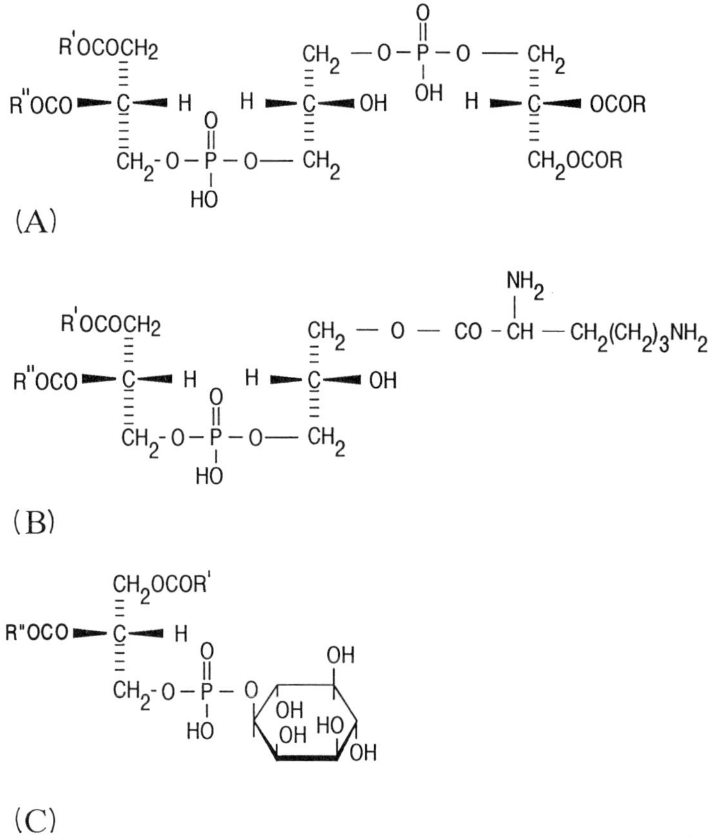

5. Structural Lipids of Eubacteria

Figure 5.9 Chemical structures of: (A) 1,3-bis-(3-*sn*-phosphatidyl)glycerol (cardiolipin); (B) 3-*sn*-phosphatidyl-1'-(3'-*O*-lysyl)-*sn*-glycerol; (C) 3-*sn*-phosphatidyl-*sn*-1'-*myo*-inositol.

1977; Kroppenstedt, 1985; Embley *et al.*, 1988). Phosphatidylinositols are likewise relatively rare. Phosphatidyl-*O*-aminoacyl glycerols, in which an amino acid (usually lysine, ornithine, arginine or alanine) is linked to the *sn*-3' position of phosphatidylglycerol, are found in some Gram-positive organisms. *Listeria monocytogenes*, for example, contains 3-*sn*-phosphatidyl-1'-(3'-*O*-lysyl)-*sn*-glycerol (Figure 5.9).

Sometimes one or both the fatty acyl chains are replaced by an ether-linked group. Compounds in which there is an alk-1-enyl group in position 1 and an acyl group in position 2 are known as plasmalogens, and are very common in anaerobic organisms. Long-chain dimethyl acetals derived from the degradation of plasmalogens (Gray, 1976) are occasionally observed in acid methanolysates of anaerobic bacteria (Jantzen & Hofstad, 1981).

Sphingophospholipids, in which a fatty acid N-acylates a long-chain base and a phosphate group is attached to the 1-hydroxyl group of the base, are occasionally encountered, for example in some *Bacteroides* and *Porphyromonas* species (Miyagawa *et al.*, 1979).

The other major group of eubacterial polar lipids are the glycolipids. The simplest of these are acylglycoses in which a mono- or oligosaccharide is esterified by a long-chain fatty acid, for example the acylated trehaloses found in *Corynebacterium, Mycobacterium, Nocardia* and other actinomycetes (Brennan, 1988). Glycosyl diacyl glycerols in which mono-, di- or (sometimes) oligosaccharides are attached to position *sn*-3 of glycerol are very widely distributed particularly in Gram-positive organisms. The monosaccharides most commonly encountered are galactose, glucose and mannose (Harwood & Russell, 1984). Bacterial glycolipids, particularly those of mycobacteria, have been comprehensively reviewed (Brennan, 1988; Goren, 1990; Kates, 1990; Puzo, 1990).

In taxonomic studies, polar lipids have largely been analysed by one- or two-dimensional thin-layer chromatography. This approach has been particularly useful for Gram-positive taxa with a high GC content in their DNA, where individual genera often display characteristic patterns (e.g. Minnikin *et al.*, 1977, 1979; Collins *et al.*, 1980, 1982; Embley *et al.*, 1988). The separated components may be tentatively identified by comparison with standards and by using spray reagents which are specific for different functional groups. The analysis of polar lipids by high-performance liquid chromatography has potential for taxonomic investigations but it has yet to see significant use. Nevertheless, solvent systems and columns are available which will separate the common bacterial polar lipids (e.g. Patton *et al.*, 1982; Christie, 1987; Fallon *et al.*, 1988). The most sophisticated and powerful method for the analysis of polar lipids is fast atom bombardment mass spectrometry (FAB-MS).

5.5 METHODS FOR ANALYSING POLAR LIPIDS

5.5.1 Extraction

The classic method of polar lipid extraction is the modified Folch procedure devised by Bligh and Dyer (1959). This utilizes a monophasic mixture of chloroform, methanol and water for the extraction. The addition of more chloroform and water forces a phase separation. The lower, mainly chloroform layer contains the polar lipids whereas non-lipid components remain in the upper aqueous phase. Minnikin *et al.* (1984c) introduced a modified

5. Structural Lipids of Eubacteria

procedure in which an initial extraction with hexane removes non-polar compounds such as isoprenoid quinones. In this way menaquinones and polar lipids can be extracted from a single sample of biomass.

Method 5.11 Integrated extraction of polar lipids and isoprenoid quinones (from Minnikin *et al.*, 1984c).

1. Place approximately 50 mg of dried biomass into an 8.5 ml tube with a PTFE-lined cap.
2. Add 2 ml of aqueous methanol (10 ml of 0.3% aqueous NaCl added to 100 ml of methanol) followed by 2 ml of hexane and shake in a tube rotator for 15 min.
3. Centrifuge at low speed to break the emulsion and remove the upper layer to a small vial (this contains non-polar lipids).
4. Repeat the extraction with a further 1 ml of hexane and add the second supernatant to the first. The pooled supernatants contain menaquinones which can be further purified and analysed.
5. Heat the remaining biomass and aqueous methanol in a boiling bath for 5 min and cool to 37 °C in a water bath.
6. Add 2.3 ml of chloroform-methanol-aqueous 0.3% (w/v) NaCl (90:100:30, v/v) to the biomass and lower layer and mix on a tube rotator for 1 h.
7. Centrifuge at low speed to pellet biomass, remove supernatant (monolayer) to a clean tube.
8. Add 0.75 ml of chloroform-methanol-0.3% (w/v) NaCl (50:100:40, v/v) to the cell residue and mix for 30 min.
9. Centrifuge to pellet cells, remove supernatant and add to previous supernatant; repeat step 8.
10. Add 1.3 ml of chloroform and 1.3 ml of aqueous NaCl to the pooled supernatants and mix thoroughly; centrifuge; remove top layer and dry lower layer (contains polar lipids) using a flow of nitrogen while heating at 37 °C.

5.5.2 Two-dimensional thin-layer chromatography

Thin layer chromatography can be used to determine simple two-dimensional patterns of polar lipids which may be characteristic of individual taxa and can be used to rapidly screen large numbers of organisms. Figure 5.10 shows some examples of the pattern which can be obtained.

Method 5.12 Two-dimensional TLC of polar lipid extracts (from Minnikin *et al.*, 1977).

1. Dissolve the dried polar lipids in 50 μl of chloroform-methanol (2:1, v/v). Spot 5 μl to the bottom of a 6 × 6 sheet of thin-layer plate coated with silica gel (Merck 5554) as shown in Figure 5.10.
2. Develop with chloroform-methanol-water (65:25:3.8, v/v) in the first direction and dry for at least 30 min in a fume cupboard.

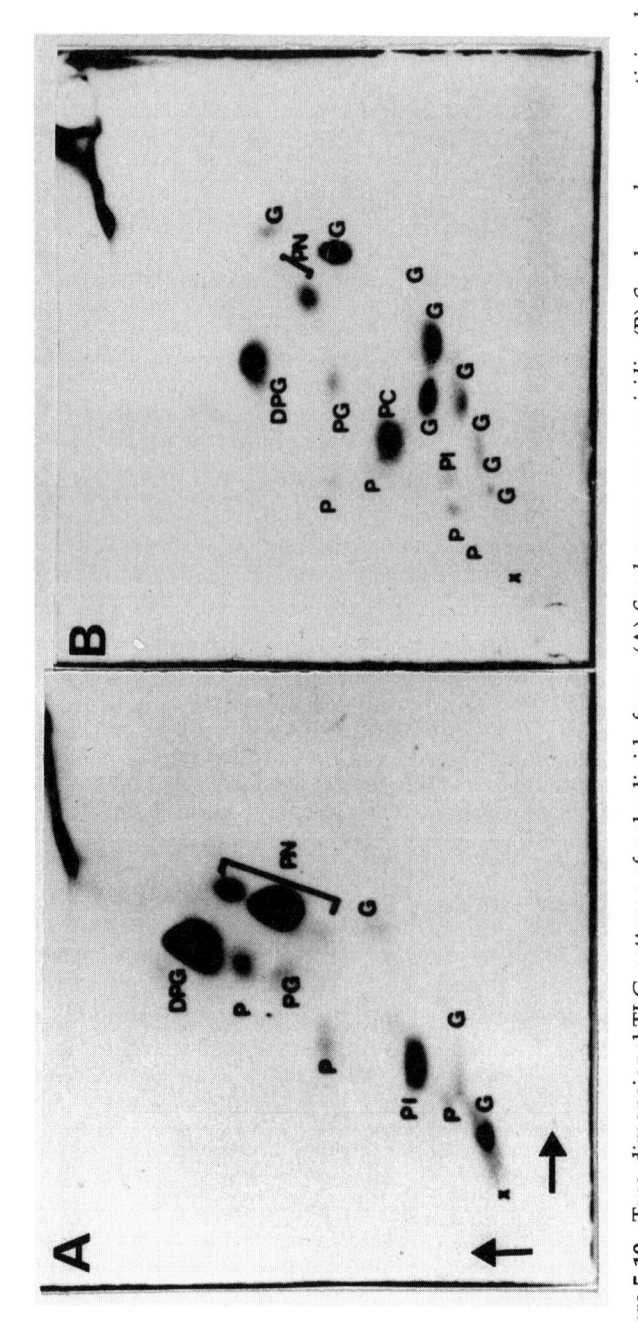

Figure 5.10 Two-dimensional TLC patterns of polar lipids from: (A) *Saccharopolyspora rectivirgula*. (B) *Saccharomonospora viridis*. DPG, diphosphatidylglycerol; G, unidentified glycolipid; P, unidentified phospholipid; PC, phosphatidylcholine; PG, phosphatidylglycerol; PI, phosphatidylinositol; PN, phosphatidylethanolamine and derivatives. Arrow to the right indicates direction of development in the first solvent system. **x** marks the point of application.

5. Structural Lipids of Eubacteria

3. Develop in the second direction using chloroform-methanol-acetic acid-water (40:7.5:6:1.8, v/v). Dry the plates in a fume cupboard.

Method 5.13 Spray reagents for characterizing individual components.

Molybdophosphoric acid for all lipids (Gunstone & Jacobsberg, 1972)
1. Dissolve 10% (w/v) molybdophosphoric acid in (95%, v/v) ethanol.
2. Spray plate and heat at 150 °C for at least 10 min. Lipids show as dark spots on a light-green background.

Ninhydrin reagent for lipids containing free amino groups (Consden & Gordon, 1948)
1. Dissolve ninhydrin (0.2 %, w/v) in water saturated butan-1-ol.
2. Lightly spray the TLC plate and heat at 100 °C for 5 min to reveal lipids which contain amino groups as pink spots. Mark the spots with a soft pencil as they tend to fade on storage. The same plate can be used for the detection of lipid phosphorus using Zinzadze reagent.

Zinzadze reagent for phosphorus-containing lipids (Dittmer & Lester, 1964)
1. Add molybdenum trioxide (40.11 g) to 1 l of 25 N H_2SO_4 and boil gently in a fume cupboard until all the residue dissolves (solution A).
2. Add powdered molybdenum (1.5 g) to 500 ml of solution A, and boil the mixture gently for 15 min and leave to cool (solution B).
3. Mix equal volumes of solutions A and B and dilute with two volumes of distilled water to make a working solution.
4. Spray the plate very lightly at room temperature; lipids containing phosphorus show as blue spots.

Periodate–Schiff reagent (Shaw, 1968)

Lipids containing vicinal hydroxyl groups, e.g. phosphatidylglycerol, phosphatidylinositol and glycolipids, give colour reactions with this reagent.
1. Spray the plate with a solution of sodium metaperiodate (1%, w/v) until saturated and leave at room temperature to oxidize for 10 min.
2. Decolorize with sulphur dioxide gas to remove excess periodate.
3. Spray very lightly with Schiff's reagent (Sigma 395–2–016) before treating again with sulphur dioxide until the first colours start to appear.
4. Leave in a fume cupboard for several hours for characteristic colours to develop.

α-Naphthol reagent (Jacin & Mishkin, 1965)
1. Dissolve 15 g of α-naphthol in 100 ml of 95% aqueous ethanol.
2. Mix 10.5 ml of this solution with H_2SO_4 (6.5 ml), 40.5 ml of 95% ethanol, and water (4 ml) to make a working solution.
3. Spray the plate lightly and heat at 100 °C for 10 min. Glycolipids appear as brown spots.

5.5.3 Mass spectrometric identification of polar lipids

Two-dimensional TLC is a simple and inexpensive method which provides a very good qualitative picture of polar lipid composition. However, since the available detection reagents are not completely specific, recourse must be made to other techniques for definitive identification. Methods for the detailed characterization of polar lipids are well established (Christie, 1982; Kates, 1985; Kuksis, 1987) but are mostly impractical for the large number of analyses required for taxonomic studies.

Underivatized phospholipids are too polar for GC and GC-MS but are amenable to analysis by soft ionization mass spectrometric techniques (Jensen & Gross, 1988), including chemical ionization (Crawford & Plattner, 1983), thermospray (Kim & Salem, 1987), plasmaspray (Odham et al., 1988) and fast atom bombardment (FAB). Although plasmaspray/ thermospray allows on-line liquid chromatography MS of mixtures of bacterial polar lipids (Odham et al., 1988), most of the published studies of bacterial phospholipids have employed FAB (Heller et al., 1987, 1988; Platt et al., 1988; Frederickson et al., 1989; Pramanik et al., 1990; Cole & Enke 1991). In FAB experiments (Barber et al., 1981) the sample, dissolved in a liquid of low vapour pressure such as glycerol or 3-nitrobenzyl alcohol (the matrix), is bombarded by a beam of energetic neutral particles, usually xenon atoms, which sputter ionized sample molecules directly into the mass spectrometer vacuum.

FAB spectra are characterized by abundant molecular ions resulting from proton or other cation attachment and by weaker fragment ions which are usually structurally informative. The spectra also contain peaks originating from ionization of the liquid matrix, which may sometimes obscure sample-derived ions. The most useful matrix for FAB-MS of lipids is 3-nitrobenzyl alcohol. A 5:1 mixture of dithiothreitol and dithioerythritol is also quite successful, but is more prone to form adducts with sample molecules, particularly when unfractionated mixtures are analysed. The addition of about 5% of crown ether (5-crown-15) may sometimes improve the data by scavenging residual alkali metal ions.

To perform an experiment, 1–2 μl of matrix is applied to the probe tip, and a similar volume of sample solution is added and thoroughly mixed. Generally 1–5 μg of sample are ample, and it may be possible to use considerably less, depending on the sensitivity of the instrument. The FAB technique is very unforgiving of even trace amounts of ionic and surface-active contaminants, which in the worst case may suppress the sample signal altogether. Fortunately most polar lipids can be adequately desalted by washing a chloroform solution of the sample with de-ionized water. The following modified Bligh–Dyer procedure (Method 5.14), which avoids the use of NaCl solution, is recommended for the extraction of polar lipids

5. Structural Lipids of Eubacteria

from bacterial cells and for the clean-up of salt-contaminated samples prior to analysis.

Method 5.14 Modified Bligh–Dyer extraction for FAB samples.

1. Remove a loopful (approx. 5 mg wet weight) of cells from the surface of an agar plate and place in a 5 ml screw-capped test-tube. There is no need to freeze dry, though lyophilized material can also be extracted.
2. Vortex the bacteria for 2 min with 2 ml of a mixture of chloroform : methanol : water (1 : 2 : 0.8, v/v).
3. Add 0.6 ml of chloroform and mix for 30 s, then add 0.6 ml of de-ionized water, also with 30 s vortexing.
4. Allow the tube to stand for a few minutes until separation into two phases occurs. If an emulsion forms, phase separation may be encouraged by gentle centrifugation.
5. Transfer the lower (mainly chloroform) layer to another tube with a Pasteur pipette, taking care not to disturb the cell debris, which concentrates at the interface.
6. Wash the chloroform twice with an equal volume of distilled water, remove to a conical bottomed vial, and evaporate in a stream of nitrogen.
7. Dissolve the residue in 50–100 μl of chloroform : methanol (1 : 1, v/v), and mix 1 μl with matrix on the tip of the FAB probe.

5.5.3.a FAB spectra of polar lipids

Fast atom bombardment-mass spectrometry (FAB-MS) can be applied to both purified phospholipids, and to unfractionated mixtures, though additional techniques such as tandem mass spectrometry may be necessary for the interpretation of the latter experiments (Jensen et al., 1986, 1987; Muster & Budzikiewicz, 1987, 1988). The general features of the spectra of phosphatidyl-glycerols, -serines, -inositols and -ethanolamines are similar (Jensen & Gross, 1988). In the positive-ion mode intense $[M^+H]^+$ ions are produced, which fragment by cleavage of the CH_2–O bond of the phosphate group with charge retention on the glycerol backbone ion (Fenwick et al., 1983; Chen et al., 1990). The mass of this ion thus reveals the total number of carbon atoms and degree of unsaturation of the two fatty acid substituents. The mass of the eliminated fragment is diagnostic of the polar head group; 141 is characteristic of phosphatidylethanolamine, 155 of N-methyl-phosphatidylethanolamine, 172 of phosphatidylglycerol and so on.

Analogous processes are observed in the spectra of some sphingophospholipids such as the ceramide-1-phosphoethanolamines found in *Bacteroides* and *Porphyromonas* species. The FAB spectra of phosphatidylcholines are characterized by intense $[M + H]^+$ ions and cleavage of the CH_2–O–P bond, but charge is retained on the choline phosphate moiety giving rise to a characteristic fragment at m/z 184 rather than a glycerol backbone ion as in other phospholipids.

The negative ion spectra of phospholipids frequently provide complementary information (Munster et al., 1986). Compounds such as phosphatidylglycerols and phosphatidic acids, which do not produce molecular ions in the positive mode, yield abundant deprotonated molecular anions $[M-H]^-$. Phosphatidylcholines however do not usually produce an $[M-H]^-$ ion, $(M-CH_3)^-$ generally being the highest mass signal produced. Negative ion spectra provide more information about the nature and disposition of substituents because the major fragmentations in the negative mode involve loss of the fatty acids rather than the polar group (Jensen et al., 1986, 1987).

Thus, FAB spectra of glycerophospholipids define both the polar headgroup and the combined number of carbon atoms and double bonds in the fatty acid substituents, though not their individual identities or relative positions on the glycerol backbone. The nature and location of the fatty acids may be deduced by using phospholipase A_2 to specifically remove the substituent from position 2 of the phospholipid, followed by identification of the hydrolysis product by FAB-MS. In the case of stearoyl/oleoyl phosphatidylcholine, for example, the $[M^+H]^+$ signal at m/z 788 shifts to m/z 524, corresponding to the $[M^+H]^+$ of a *lyso*-phosphatidylcholine bearing a saturated C18 substituent, which demonstrates that the unsaturated substituent is on position 2. A protocol for phospholipase digestion which is compatible with FAB-MS and which can be applied to either pure or mixed phospholipids is given below (Method 5.15).

Method 5.15 Digestion of phospholipids with phospholipase A_2.

1. Prepare borate buffer pH 8.9 by dissolving 0.476 g of sodium borate ($Na_2B_4O_7.10H_2O$) in 50 ml of distilled water and 7.1 ml of 0.1 M HCl. Bring the volume to 100 ml with more distilled water and adjust the pH if necessary.
2. Dissolve 250 units of purified *Crotalus atrox* phospholipase A_2 (Sigma P-3770) and 1.2 mg of calcium chloride in 12.5 ml of the buffer. This stock solution is stable for several months at $-40\,°C$.
3. Dissolve the sample in 3 ml of diethyl ether/methanol (98:2, v/v), and add 0.5 ml of the enzyme stock solution (approx. 10 units). If the phospholipid to be digested has been purified by TLC or HPLC then less enzyme (1–5 units) is needed. Set up a tube containing a standard phospholipid as a check on the enzyme activity.
4. Incubate overnight at 37 °C with constant shaking.
5. Remove the solvent in a stream of nitrogen, recover the digestion products by Bligh–Dyer extraction of the residue (Method 5.14) and analyse by FAB-MS.

Note: As a complementary experiment the fatty acids liberated from position 2 should be purified by TLC, methylated and analysed by GC and GC-MS.

5.5.3.b Analysis of mixtures of bacterial polar lipids

The FAB technique may be used for the direct analysis of solvent-disrupted suspensions of bacterial cells (Heller et al., 1987) though better data are usually obtained from cell-free solutions of phospholipid mixtures. It must be stressed, however, that the results are qualitative since the intensity of the signals in the spectra reflect the relative ease of desorption rather than the solution concentration of the various phospholipid species.

Figure 5.11 shows the positive- and negative-ion FAB spectra of a Bligh–Dyer extract of *Campylobacter jejuni*. Signals in the region m/z 664 to m/z 744 of the positive-ion spectrum correspond to the $[M+H]^+$ ions of phosphatidylethanolamines of varying degrees of methylation, all differing in their fatty acid substitution pattern. The cluster of signals between m/z 523 and m/z 603 are attributable to the glycerol backbone fragments of phosphatidylethanolamines and phosphatidylglycerols. In the negative-ion spectrum additional signals at m/z 747 and 773 correspond to the deprotonated molecular ions ($[M-H]^-$) of phosphatidylglycerols. Although these spectra are complex, most of the signals can be assigned by some supplementary mass spectrometric techniques, of which the most useful are collisional activation, tandem mass spectrometry and linked scanning. A scan at constant B/E records only the decomposition products of a selected precursor ion. Thus, in a complex mixture the $[M+H]^+$ of each component is selected as precursor ion in turn, and a full spectrum is obtained from each, without interference from the rest of the mixture. This experiment fails when two or more discrete molecular species have the same mass (as is common in mixtures of isometric phospholipids). Molecular ions to which several molecular species contribute may be deconvoluted by means of a constant neutral loss scan (CNL). This experiment records only species that decompose by the elimination of a selected fragment. Since neutral loss of the polar head group of phospholipids is a facile process, a CNL scan for the appropriate mass will selectively detect different classes of phospholipid (Heller et al., 1988; Cole & Enke, 1991). For example, a constant neutral loss scan for 141 will selectively record the molecular ions of all phosphatidylethanolamines present in a mixture. Figure 5.12 shows the phosphatidylethanolamine composition of *Campylobacter jejuni* obtained by means of a constant neutral loss scan using the same phospholipid extract as shown in Figure 5.11. The use of the enzymatic digestion strategy described above in conjunction with constant neutral loss and other linked scanning experiments enables the determination of both phospholipid type and fatty acid substitution patterns in mixtures of bacterial polar lipids (Wait, 1992).

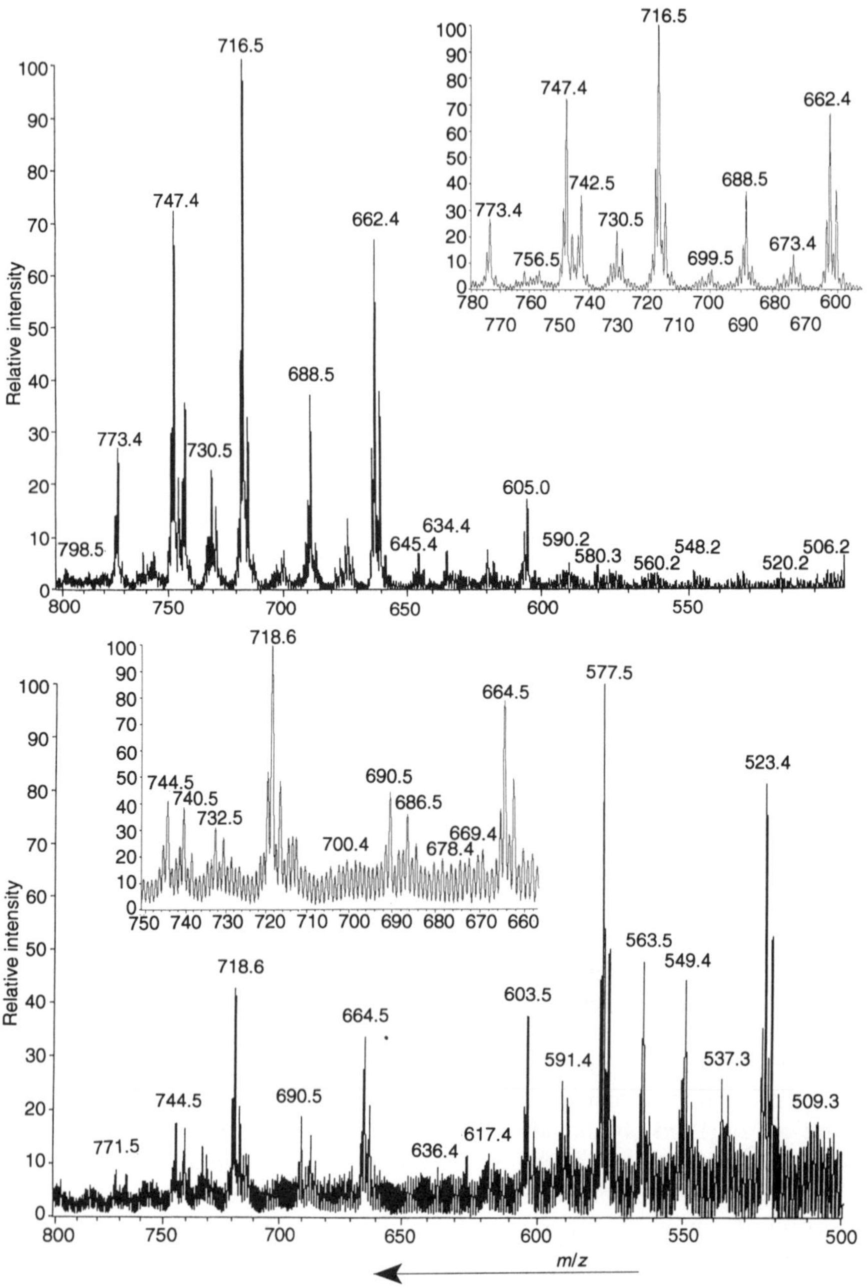
716.5
100
90
80
70
60
50
40
30
20
10
0
Relative intensity
747.4
773.4
730.5
688.5
662.4
605.0
798.5
645.4
634.4
590.2
580.3
560.2
548.2
520.2
506.2
800
750
700
650
600
550
742.5
756.5
699.5
673.4
780
760
740
720
700
680
600
770
750
730
710
690
670
718.6
664.5
744.5
740.5
732.5
690.5
686.5
700.4
678.4
669.4
660
577.5
523.4
563.5
549.4
603.5
591.4
537.3
509.3
771.5
636.4
617.4
500
m/z

5.6 CONCLUDING REMARKS

Over the past 25 years, the results of fatty acid and polar lipid analyses have been instrumental in clarifying the taxonomy of a large number of bacterial genera. For example, the systematic application of lipid, whole-cell and cell wall analyses to Gram-positive taxa resulted in a clear appreciation of 'genus level' groups by the early 1980s (Schleifer & Kandler, 1972; Lechevalier, 1977; Lechevalier et al., 1977; Minnikin et al., 1979; Minnikin & Goodfellow, 1980, 1981; Collins & Jones, 1981). The existence of these chemically defined groups enabled the diversity of Gram-positive taxa to be effectively and economically sampled using the more expensive and technically difficult nucleic acid pairing and sequencing methods (Döpfer et al., 1982; Stackebrandt et al., 1980, 1981; Stackebrandt & Woese, 1981). Although the results of these analyses produced some surprises at the family level and above, the groups circumscribed by chemical features were remarkably accurate.

Subsequently, the development of the polymerase chain reaction has made it simple and relatively cheap to sequence rRNA genes for the classification and identification of microorganisms even if they cannot be cultivated (Giovannoni et al., 1990). Nucleic acid probe technology is rapidly improving and it is now a simple procedure to obtain a probe for any microorganism once a sequence is available. Under carefully controlled conditions probes give unambiguous answers, they are very sensitive, and they can be used directly on environmental material. It is thus timely to consider if lipid analyses still have a role to play in the classification and identification of microorganisms. On purely scientific grounds the answer must be yes. The polyphasic approach to taxonomy, i.e. one that uses genotypic and phenotypic measures, provides the most stable and useful classification for microorganisms. Moreover, phenotypic data reveal what

---

Figure 5.11 (*opposite*) (Lower trace) Mass assigned profile plot of the positive-ion fast-atom bombardment spectrum of the mixture of polar lipids obtained by Bligh–Dyer extraction of *Campylobacter jejuni* cells. The inset shows an expansion of the molecular ion region. The signals m/z 744, 718, 690 and 664 are attributable to the molecular ions of phosphatidylethanolamine species substituted 18:1/18:1, 16:0/18:1, 16:0/16:1 and 14:0/16:0 respectively. The peaks between m/z 603 and m/z 523 are fragment ions originating from the elimination of the polar head group from phosphatidylethanolamines and phosphatidylglycerols.
(Upper trace) Negative-ion FAB spectrum of the same sample. Deprotonated molecular ion signals (M-H)⁻ of the phosphatidylethanolamines are present at m/z 742, 716, 688 and 662. The signals at m/z 773 and 747 correspond to deprotonated molecules of phosphatidylglycerol, substituted 18:1/18:1 and 16:0/18:1 respectively. The molecular ions of these species are not observed in the positive-ion experiment.

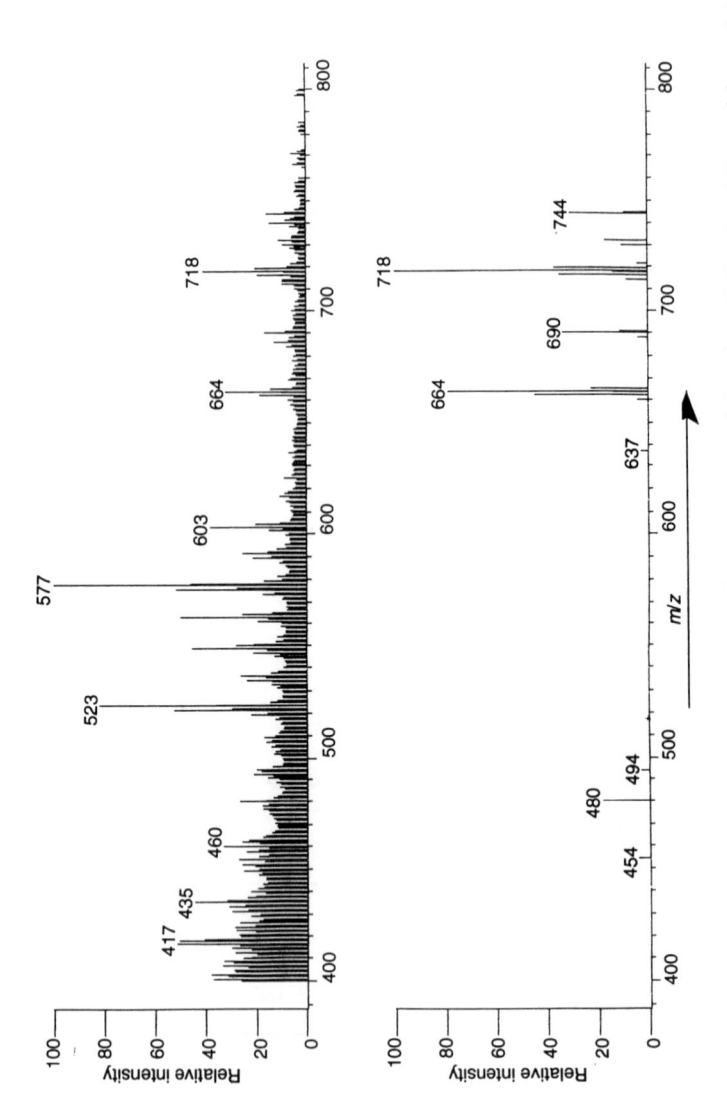

Figure 5.12 (Lower trace) Centroided plot of a constant neutral loss spectrum of the phospholipids of *Campylobacter jejuni*. The experiment was set up to record only the molecular ions of those species which decomposed by loss of 141 mass units (i.e. the mass of the phosphoethanolamine group). Thus phosphatidylethanolamines are selectively recorded without interference from the chemical background intrinsic to the FAB experiment (cf. the conventional FAB spectrum in the upper trace).

a microorganism can do and they provide interesting and biologically relevant information. Thus, while only molecular data can be used to construct a genealogy for microorganisms, phenotypic data are essential for describing the groups so defined and for aiding the process of demarcation between groups. Lipid data have an excellent track record in taxonomic studies and should be included as part of this process. Lipids also have a role to play in identification, particularly for routine analysis of large numbers of different bacteria where it would be too expensive to produce probes or sequence genes. For example, fatty acid fingerprinting is a cheap (once a gas chromatograph and computer have been purchased), fast and reliable method for identifying bacteria to strain and species level (e.g. Sasser, 1990; Welch, 1991; Stead et al., 1992).

REFERENCES

Asselineau, C. and Asselineau, J. (1984) Fatty acids and complex lipids. In *Gas Chromatography/Mass Spectrometry: Applications in Microbiology* (Odham, G., Larsson, L. and Mårdh, P.A., eds), pp. 57–103. Plenum Press: New York.
Bårber, M., Bordoli, R.S., Sedgwick, R.D. and Tyler, A.N. (1981) Fast atom bombardment of solids as an ion source in mass spectrometry. *Nature* 293: 270–5.
Benson, R.F., Thacker, W.L., Walters, R.P., Quinlivan, P.A., Mayberry, W.R., Brenner, D.J. and Wilkinson, H.W. (1989) *Legionella quinlivanii* sp. nov. isolated from water. *Curr. Microbiol.* 18: 195–7.
Benson, R.F., Thacker, W.L., Lanser, J.A., Sangster, N., Mayberry, W.R. and Brenner, D.J. (1991) *Legionella adelaidensis*, a new species isolated from cooling tower water. *J. Clin. Microbiol.* 29: 1004–6.
Bernard, K.A., Bellefeuille, M. and Ewan, P. (1991) Cellular fatty acid composition as an adjunct to the identification of asporogenous aerobic Gram-positive rods. *J. Clin. Microbiol.* 29: 83–9.
Bligh, E.G. and Dyer, W.J. (1959) A rapid method of total lipid extraction and purification. *Can. J. Biochem. Physiol.* 37: 911–7.
Brennan, P. (1988) *Mycobacterium* and other actinomycetes. In *Microbial Lipids* (Ratledge, C. and Wilkinson, S.G., eds), pp. 203–98. Academic Press: London.
Butler, W.R. and Kilburn, J.O. (1988) Identification of major slowly growing pathogenic mycobacteria and *Mycobacterium gordonae* by high performance liquid chromatography of their mycolic acids. *J. Clin. Microbiol.* 26: 50–3.
Butler, W.R., Kilburn, J.O. and Kubica, G.P. (1987) High performance liquid chromatography of mycolic acids as an aid in laboratory identification of *Rhodococcus* and *Nocardia* species. *J. Clin. Microbiol.* 25: 2126–31.
Butler, W.R., Jost, K.C. and Kilburn, J.O. (1991) Identification of mycobacteria by high performance liquid chromatography. *J. Clin. Microbiol.* 29: 2468–72.
Butler, W.R., Thibert, L. and Kilburn, J.O. (1992) Identification of *Mycobacterium avium* complex strains and some similar species by high performance liquid chromatography. *J. Clin. Microbiol.* 30: 2698–704.
Chen, S., Kirschner, G. and Traldi, P. (1990) Positive ion fast atom bombardment massspectrometric analysis of the molecular species of glycerophosphatidylserine. *Anal. Biochem.* 191: 100–5.
Christie, W.W. (1982) *Lipid Analysis*. Pergamon Press: Oxford.
Christie, W.W. (1987) *High Performance Liquid Chromatography and Lipids: A Practical Guide*. Pergamon Press: Oxford.

Christie, W.W. (1989) *Gas Chromatography and Lipids: A Practical Guide*. The Oily Press: Ayr, Scotland.
Christie, W.W. and Stefanov, K. (1987) Separation of picolinyl derivatives by high-performance liquid chromatography for identification by mass spectrometry. *J. Chromatogr.* **392**: 259–65.
Cole, M.J. and Enke, C.G. (1991) Direct determination of phospholipid structures in microorganisms by fast atom bombardment triple quadruple mass spectrometer. *Anal. Chem.* **63**: 1032–8.
Collins, M.D. and Jones, D. (1981) Distribution of isoprenoid quinone structural types and their taxonomic implications. *Microbiol. Rev.* **45**: 316–54.
Collins, M.D., Goodfellow, M. and Minnikin, D.E.M. (1980) Fatty acid, isoprenoid quinone and polar lipid composition in the classification of *Curtobacterium* and related taxa. *J. Gen. Microbiol.* **118**: 29–37.
Collins, M.D., Goodfellow, M. and Minnikin, D.E.M. (1982) Polar lipid composition in the classification of *Arthrobacter* and *Microbacterium*. *FEMS Microbiol. Lett.* **15**: 299–302.
Collins, M.D., Smida, J., Dorsch, M. and Stackebrandt, E. (1988) *Tsukamurella* gen. nov. harboring *Corynebacterium paurometabolum* and *Rhodococcus aurantiacus*. *Int. J. Syst. Bacteriol.* **38**: 385–91.
Consden, R. and Gordon, A.H. (1948) The effect of salt on partition chromatograms. *Nature* **162**: 180–1.
Crawford C.G. and Plattner, R.D. (1983) Ammonia chemical ionization mass spectrometry of intact diacylphosphatidylcholine. *J. Lipid Res.* **24**: 456–60.
De Briel, D., Couderc, F., Riegel, P., Jehl, F. and Minck, R. (1992) High performance liquid chromatography of corynemycolic acids as a tool in identification of *Corynebacterium* species and related organisms. *J. Clin. Microbiol.* **30**: 1407–17.
Deinhard, G., Saar, J., Krischke, W. and Poralla, K. (1987) *Bacillus cycloheptanicus* sp. nov., a new thermoacidophile containing ω-cycloheptane fatty acids. *System. Appl. Microbiol.* **10**: 68–73.
De Rosa, M., Gambacorta, A., Minale, L. and Bu'lock, J.D. (1971) Cyclohexane fatty acids from a thermophilic bacterium. *Chem. Comm.* 1334.
Dittmer, J.C.F. and Lester, R.L. (1964) A simple specific spray for the detection of phospholipids on thin-layer chromatograms. *J. Lipid Res.* **5**: 126–7.
Dobson, G., Minnikin, D.E., Minnikin, S.M., Parlett, J.H. and Goodfellow, M. (1985) Systematic analysis of complex mycobacterial lipids. In *Chemical Methods in Bacterial Systematics* (Goodfellow, M. and Minnikin, D.E., eds), pp. 237–266. Academic Press: London.
Döpfer, H., Stackebrandt, E. and Fiedler, F. (1982) Nucleic acid hybridisation on *Microbacterium*, *Curtobacterium*, *Agromyces* and related taxa. *J. Gen. Microbiol.* **128**: 1697–708.
Durst, H.D., Milano, M., Kitka, E.J., Connelly, S.A. and Grushka, E. (1975) Phenacyl esters of fatty acids via crown ether catalysts for enhanced ultraviolet detection in liquid chromatography. *Anal. Chem.* **47**: 1797–801.
Eerola, E. and Lehtonen, O.P. (1988) Optimal data processing procedure for automatic bacterial identification by gas-liquid chromatography of cellular fatty acids. *J. Clin. Microbiol.* **26**: 1745–53.
Embley, T.M., Wait, R., Dobson, G. and Goodfellow, M. (1987) Fatty acid composition in the classification of *Saccharopolyspora hirsuta*. *FEMS Microbiol. Lett.* **41**: 131–5.
Embley, T.M., O'Donnell, A.G., Rostron, J. and Goodfellow, M. (1988) Chemotaxonomy of wall IV actinomycetes which lack mycolic acids. *J. Gen. Microbiol.* **134**: 953–60.
Fallon, A., Booth, R.F.G. and Bell, L.D. (1988) *Laboratory Techniques in Biochemistry and Molecular Biology: The Application of HPLC in Biochemistry*. Elsevier: Oxford.
Fenwick, G.R., Eagles, J. and Self, R. (1983) Fast atom bombardment mass spectrometry of intact phospholipids and related compounds. *Biomed. Mass. Spectrom.* **10**: 382–6.
Finnerty, W.R., Makula, R.A. and Feeley, J.C. (1979) Cellular lipids of the Legionnaires' disease bacterium. *Ann. Intern. Med.* **90**: 631–4.
Frederickson, H.L., de Leeuw, J.W., Tas, A.C., van der Greef, J., LaVos, G.F. and Boon, J.J. (1989) Fast atom bombardment (tandem) mass spectrometric analysis of intact polar ether lipids extractable from the extremely halophilic archaebacterium *Halobacterium cutirubrum*. *Biomed. Environ. Mass Spectrom.* **18**: 96–105.

Fulco, A.J. (1983) Fatty acid metabolism in bacteria. *Prog. Lipid Res.* **22**: 133–60.
Giovannoni, S.J., Britschgi, T.B., Moyer, C.L. and Field, K.G. (1990) Genetic diversity in the Sargasso Sea bacterioplankton. *Nature* **345**: 60–3.
Goren, M.B. (1990) Mycobacterial fatty acid esters of sugars and sulfosugars. In *Handbook of Lipid Research*, vol. 6 (Kates, M., ed), pp. 363–461. Plenum Press: New York.
Gray, G.M. (1976) Gas chromatography of the long chain aldehydes. In *Lipid Chromatographic Analysis*, vol. 3 (Marinetti, G.V., ed), pp. 897–923. M. Dekker: New York.
Grob, K. (1988) *Classical Split and Splitless Injection in Capillary GC*. Hüthig Publishing: New York.
Grob, K. (1992) *On-column Injection in Capillary Gas Chromatography*. Hüthig Publishing: New York.
Gunstone, F.D. and Jacobsberg, F.R. (1972) Fatty acids, part 35: The preparation and properties of a complete series of methyl epoxyoctadecanoates. *Chem. Phys. Lipids* **9**: 26–64.
Hamid, M.E., Minnikin, D.E. and Goodfellow, M. (1993) A simple chemical test to distinguish mycobacteria from other mycolic acid-containing actinomycetes. *J. Gen. Microbiol.* 2203–2213.
Harvey, D.J. (1982) Picolinyl esters as derivatives for the structural determination of long chain branched and unsaturated fatty acids. *Biomed. Mass Spectrom.* **9**: 33–8.
Harvey, D.J. (1984) Picolinyl derivatives for the characterisation of cyclopropane fatty acids by mass spectrometry. *Biomed. Mass Spectrom.* **11**: 187–92.
Harvey, D.J. (1992) Mass spectrometry of picolinyl and other nitrogen-containing derivatives of lipids. In *Advances in Lipid Methodology-1* (Christie, W.W., ed.), pp. 19–80. The Oily Press: Ayr, Scotland.
Harwood, J.L. and Russell, N.J. (1984) *Lipids in Plants and Microbes*. George Allen & Unwin: London.
Heller, D.N., Cotter, R.J., Fenselau, C. and Uy, O.M. (1987) Profiling of bacteria by fast atom bombardment mass spectrometry. *Anal. Chem.* **59**: 2806–9.
Heller, D.N., Murphy, C.M., Cotter, R.J., Fenselau, C. and Uy, O.M. (1988) Constant neutral loss scanning for the characterization of bacterial phospholipids desorbed by fast atom bombardment. *Anal. Chem.* **60**: 2787–91.
Jacin, H. and Mishkin, A.R. (1965) Separation of carbohydrates on borate impregnated silica gel G plates. *J. Chromatogr.* **18**: 170–3.
Janse, J.D. (1991) Pathovar discrimination within *Pseudomonas syringae* subsp. *savastanoi* using whole cell fatty acids and pathogenicity as criteria. *System. Appl. Microb.* **14**: 79–84.
Jantzen, E. and Bryn, K. (1985) Whole cell and lipopolysaccharide fatty acids and sugars of Gram-negative bacteria. In *Chemical Methods in Bacterial Systematics* (Goodfellow, M. and Minnikin, D.E., eds), pp. 145–72. Academic Press: London.
Jantzen, E. and Hofstad, T. (1981) Fatty acids of *Fusobacterium* species: Taxonomic implications. *J. Gen. Microbiol.* **123**: 163–71.
Jensen, N.J. and Gross, M.L. (1987) Mass spectrometry methods for structure determination and analysis of fatty acids. *Mass Spectrom. Rev.* **6**: 497–536.
Jensen, N.J. and Gross, M.L. (1988) A comparison of mass spectrometry methods for structural determination and analysis of phospholipids. *Mass Spectrom. Rev.* **7**: 41–69.
Jensen, N.J., Tomer, K.B. and Gross, M.L. (1986) Fast atom bombardment and tandem mass spectrometry of phosphatidylserine and phosphatidylcholine. *Lipids* **21**: 580–7.
Jensen, N.J., Tomer, K.B. and Gross, M.L. (1987) FAB MS/MS for phosphatidylinositol, -glycerol, -ethanolamine and other complex phospholipids. *Lipids* **22**: 480–9.
Julák, J., Turecek, F. and Miková, Z. (1980) Identification of characteristic branched-chain fatty acids of *Mycobacterium kansasii* and *gordonae* by gas-chromatography-mass spectrometry. *J. Chromatogr.* **190**: 183–7.
Karlsson, K.A., Samuelsson, B.E. and Steen, G.O. (1973) Improved identification of monomethyl paraffin chain branching (close to the methyl end) of long chain compounds by gas chromatography and mass spectrometry. *Chem. Phys. Lipids* **11**: 17–38.
Kates, M. (1985) *Techniques of Lipidology*. Elsevier: Amsterdam.
Kates, M. (1990) Glyco-, phosphoglyco- and sulphoglycoglycerolipids of bacteria. In *Handbook of Lipid Research*, vol. 6 (Kates, M., ed), pp. 1–122. Plenum Press: New York.

Kim, H.-Y. and Salem, N. (1987) Application of thermospray high-performance liquid chromatography/mass spectrometry for the determination of phospholipids and related compounds. *Anal. Chem.* **59**: 722-6.

Kroppenstedt, R.M. (1985) Fatty acids and menaquinones of actinomycetes and related organisms. In *Chemical Methods in Bacterial Systematics* (Goodfellow, M. and Minnikin, D.E., eds), pp. 173-200. Academic Press: London.

Kroppenstedt, R.M. and Kutzner, H.J. (1978) Biochemical taxonomy of some problem actinomycetes. *Zbl. Bact. Parasit. Infeckt. Hyg. (Abt. 1) Suppl.* **6**: 125-33.

Kuksis, A. (1978) *Handbook of Lipid Research, Volume 3, Fatty Acids and Glycerides*. Plenum Press: New York.

Kuksis, A. (1987) *Chromatography of Lipids in Biomedical Research and Clinical Diagnosis*. (Journal of Chromatography library vol. 37). Elsevier: Amsterdam.

Kuksis, A. and Myher, J.J. (1989) Lipids. In *Mass Spectrometry* (Lawson, A.M., ed), pp. 265-351. W. de Gruyter: Berlin.

Lambert, M.A. and Moss, C.W. (1983) Comparison of the effects of acid and base hydrolyses on hydroxy and cyclopropane fatty acids in bacteria. *J. Clin. Microbiol.* **17**: 1370-7.

Larsson, L., Sonesson, A., Jantzen, E. and Bryn, K. (1989) Detection and identification of bacteria using gas-chromatographic analysis of cellular fatty acids. In *Rapid Methods and Automation in Microbiology and Immunology* (Balows, A., ed), pp. 389-95. Academic Press: Brixia.

Lechevalier, M.P. (1977) Lipids in bacterial taxonomy—a taxonomist's viewpoint. *Crit. Rev. Microbiol.* **5**: 109-210.

Lechevalier, H. and Lechevalier, M.P. (1988) Chemotaxonomic use of lipids—An overview. In *Microbial Lipids*, vol. 1 (Ratledge C. and Wilkinson, S.G.E., eds), pp. 869-902. Academic Press: London.

Lechevalier, M.P., De Biévre, C. and Lechevalier, H. (1977) Chemotaxonomy of aerobic actinomycetes: Phospholipid composition. *Biochem. System. Ecol.* **5**: 249-60.

McCloskey, J.A. (1970) Mass spectrometry of fatty acid derivatives. In *Topics In Lipid Chemistry*, vol. 1 (Gunstone, F.D., ed), pp. 369-440. Logos Press: London.

McElhaney, R.N. (1976) The biological significance of alterations in the fatty acid composition of microorganism membrane lipids in response to changes in environmental temperature. In *Extreme Environments: Mechanisms of Microbial Adaptation* (Heinrich, M.R., ed), pp. 255-81. Academic Press: New York.

Marinetti, G.V. (1976) *Lipid Chromatographic Analysis*, vols. 1, 2 and 3, 2nd edn. M. Dekker: New York.

Mauchline, W.S., Araujo, R., Wait, R., Dowsett, A.B., Dennis, P.J. and Keevil, C.W. (1992) Physiology and morphology of *Legionella pneumophila* in continuous culture at low oxygen concentration. *J. Gen. Microbiol.* **138**: 2371-80.

Mayberry, W.R. (1981) Dihydroxy and monohydroxy fatty acids in *Legionella pneumophila*. *J. Bacteriol.* **147**: 373-81.

Mayberry, W.R. (1984) Monohydroxy and dihydroxy fatty acids in *Legionella* species. *Int. J. Syst. Bacteriol.* **34**: 321-6.

Miller, L.T. (1982) Single derivatisation method for routine analysis of bacterial whole cell fatty acid methyl esters, including hydroxy acids. *J. Clin. Microbiol.* **16**: 584-6.

Miller, L.T. and Berger, T. (1985) Bacteria identification by gas chromatography of whole cell fatty acids. Hewlett-Packard Application Note 228-241.

Minnikin, D.E. (1978) Location of double bonds and cyclopropane rings in fatty acids by mass spectrometry. *Chem. Phys. Lipids* **21**: 313-47.

Minnikin, D.E.M. (1988) Isolation and purification of mycobacterial wall lipids. In *Bacterial Cell Surface Techniques* (Hancock, I.C. and Poxton, I.R., eds), pp. 125-35. Wiley: Chichester.

Minnikin, D.E.M. (1993) Mycolic acids. In *CRC Handbook of Chromatography: Lipids III.* (Mukherjee, K.D. and Weber, N., eds). CRC Press: Boca Raton (in press).

Minnikin, D.E. and Goodfellow, M. (1980) Lipid composition in the classification and identification of acid fast bacteria. In *Microbial Classification and Identification* (Goodfellow, M. and Board, R.G., eds), pp. 189-256. Academic Press: London.

Minnikin, D.E. and Goodfellow, M. (1981) Lipids in the classification of *Bacillus* and related taxa. In *The Aerobic Endospore-forming Bacteria* (Berkeley, R.C.W. and Goodfellow, M., eds), pp. 59–90. Academic Press: London.
Minnikin, D.E., Patel, P.V., Alshamaony, L. and Goodfellow, M. (1977) Polar lipid composition in the classification of *Nocardia* and related bacteria. *Int. J. System. Bacteriol.* **27**: 104–17.
Minnikin, D.E., Collins, M.D. and Goodfellow, M. (1979) Fatty acid and polar lipid composition in the classification of *Cellulomonas, oerskovia* and related taxa. *J. Appl. Bacteriol.* **47**: 87–95.
Minnikin, D.E., Hutchinson, I.G., Caldicott, A.B. and Goodfellow, M. (1980) Thin-layer chromatography of methanolysates of mycolic acid containing bacteria. *J. Chromatogr.* **188**: 221–33.
Minnikin, D.E., Minnikin, S.M., Hutchinson, I.G., Goodfellow, M. and Grange, J.M. (1984a) Mycolic acid patterns of representative strains of *Mycobacterium fortuitum*, *Mycobacterium peregrinum* and *Mycobacterium smegmatis*. *J. Gen. Microbiol.* **130**: 363–7.
Minnikin, D.E., Minnikin, S.M., Parlett, J.H., Goodfellow, M. and Magnusson, M. (1984b) Mycolic acid patterns of some species of *Mycobacterium*. *Arch. Microbiol.* **139**: 225–31.
Minnikin, D.E., O'Donnell, A.G., Goodfellow, M., Alderson, G., Athalye, M., Schaal, A. and Parlett, J.H. (1984c) An integrated procedure for the extraction of bacterial isoprenoid quinones and polar lipids. *J. Microbiol. Meth.* **2**: 233–41.
Miwa, T.K. (1963) Identification of peaks in gas liquid chromatography. *J. Amer. Oil Chem. Soc.* **40**: 309–13.
Miyagawa, E., Azuma, R., Suto, T. and Yano, I. (1979) Occurrence of free ceramides in *Bacteroides fragilis* NCTC 9343. *J. Biochem. (Tokyo)* **86**: 311–20.
Munster, H. and Budzikiewicz, H. (1987) Structural analysis of phospholipids by fast atom bombardment/collision activation with a tandem mass spectrometer. *Rapid Commun. Mass. Spectrom.* **1**: 126–8.
Munster, H. and Budzikiewicz, H. (1988) Structural and mixture analysis of glycerophosphoric acid derivatives by fast atom bombardment. *Biol. Chem. Hoppe-Seyler* **369**: 303–8.
Munster, H., Stein, J. and Budzikiewicz, H. (1986) Structural analysis of underivatized phospholipids by negative ion fast atom bombardment mass spectrometry. *Biomed. Environ. Mass. Spectrom.* **13**: 423–7.
Ninet, B., Traitler, H., Aeschlimann, J.-M., Horman, I., Hartmann, D. and Bille, J. (1992) Quantitative analysis of cellular fatty acids (CFA) composition of the seven species of *Listeria*. *System. Appl. Microbiol.* **15**: 76–81.
Odham, G. (1980) Fatty acids. In *Biochemical Applications of Mass Spectrometry; First Supplementary Volume* (Waller, G.R. and Dermer, O.C., eds), pp. 153–171. Wiley: New York.
Odham, G. and Stenhagen, E. (1972) Fatty acids. In *Biochemical Applications of Mass Spectrometry* (Waller, G.R., ed), pp. 211–28. Wiley: New York.
Odham, G., Valeur, A., Michelsen, P., Aronsson, E. and McDowall, M. (1988) Highly sensitive determination and characterization of intact cellular ester-linked phospholipids using liquid chromatography-plasma spray mass spectrometry. *J. Chromatogr.* **434**: 331–41.
O'Donnell, A.G. (1985) Numerical analysis of chemical data. In *Computer-assisted Bacterial Systematics* (Goodfellow, M., Jones, D. and Priest, F.G., eds), pp. 403–14. Academic Press: London.
Ohya, H., Komai, Y. and Yamaguchi, M. (1986) Zinc tolerance of an isolated bacterium containing co-cyclohexyl fatty acid. *FEMS Microbiol. Lett.* **34**: 247–60.
O'Leary, W.M. and Wilkinson, S.G. (1988) Gram-positive bacteria. In *Microbial Lipids*, vol. 1 (Ratledge, C. and Wilkinson, S.G., eds), pp. 117–202. Academic Press: London.
Oshima, M. and Ariga, T. (1975) ω-Cyclohexyl fatty acids in acidophilic thermophilic bacteria. *J. Biol. Chem.* **250**: 6963–8.
Patton, G.M., Fasulo, J.M. and Robins, S.J. (1982) Separation of phospholipids and individual molecular species of phospholipids by high performance liquid chromatography. *J. Lipid Res.* **23**: 190–5.
Platt, J.A., Uy, O.M., Heller, D.N., Cotter, R.J. and Fenselau, C. (1988) Computer based linear regression analysis of resorption mass spectra of microorganisms. *Anal. Chem.* **60**: 1415–9.

Pramanik, B.N., Zechman, J.M., Das, P.R. and Bartner, P.L. (1990) Bacterial phospholipid analysis by fast atom bombardment mass spectrometry. *Biomed. Environ. Mass Spectrom.* **19**: 164–70.

Puzo, G. (1990) The carbohydrate and lipid containing cell wall of mycobacteria, phenolic glycolipids; structure and immunological properties. *Crit. Rev. Microbiol.* **17**: 305–27.

Rose, A.H. (1989) Influence of the environment on lipid composition. In *Microbial Lipids*, vol. 2 (Ratledge, C. and Wilkinson, S.G., eds), pp. 255–78. Academic Press: London.

Ryhage, R. and Stenhagen, E. (1960) Mass spectrometry in lipid research. *J. Lipid Res.* **1**: 361–90.

Saddler, G.S., O'Donnell, A.G., Goodfellow, M. and Minnikin, D.E. (1987) SIMCA pattern recognition in the analysis of streptomycete fatty acids. *J. Gen. Microbiol.* **133**: 1137–47.

Sasser, M. (1990) Identification of bacteria through fatty acid analysis. In *Methods in Phytobacteriology* (Klement, Z., Rudolph, K. and Sands, D.C., eds), pp. 199–203. Académiai Kiadó: Budapest.

Schleifer, K.-H. and Kandler, O. (1972) Peptidoglycan types of bacterial cell walls and their taxonomic implications. *Bacteriol. Rev.* **36**: 407–477.

Schmitz, B. and Klein R.A. (1986) Mass spectrometric localization of carbon-carbon double bonds: a critical review of recent methods. *Chem. Phys. Lipids.* **39**: 285–311.

Schweizer, E. (1989) Biosynthesis of fatty acids and related compounds. In *Microbial Lipids* vol. 2 (Ratledge, C. and Wilkinson, S.G., eds), pp. 3–50. Academic Press: London.

Shaw, N. (1968) The detection of lipids on thin layer chromatograms using the Periodate–Schiff reagent. *Biochim. Biophys. Acta* **164**: 407–77.

Simon, E., Kern, W. and Spiteller, G. (1990) Localization of the branch in monomethyl branched fatty acids. *Biomed. Environ. Mass Spectrom.* **19**: 129–36.

Stackebrandt, E. and Woese, C.R. (1981) Towards a phylogeny of actinomycetes and related organisms. *Curr. Microbiol.* **5**: 131–6.

Stackebrandt, E., Lewis, B.J. and Woese, C.R. (1980) The phylogenetic structure of the coryneform group of bacteria. *Zbl. Bakt. Abt. Orig.* **C1**: 137–149.

Stackebrandt, E., Wunner-Fussl, B., Fowler, V.J. and Schleifer, K.H. (1981) Deoxyribonucleic acid homologies and ribosomal ribonucleic acid similarities among spore forming members of the order Actinomycetales. *Int. J. Syst. Bacteriol.* **31**: 420–31.

Stackebrandt, E., Smida, J. and Collins, M.D. (1988) Evidence of phylogenetic heterogeneity within the genus *Rhodococcus*: Revival of the genus *Gordona* (Tsukamura). *J. Gen. Appl. Microbiol.* **34**: 341–8.

Stead, D.E., Sellwood, J.E., Wilson, J. and Viney, I. (1992) Evaluation of a commercial microbial identification system based on fatty acid profiles for rapid, accurate identification of plant pathogenic bacteria. *J. Appl. Bacteriol.* **72**: 315–21.

Suzuki, K., Goodfellow, M. and O'Donnell, A.G. (1993) Cell envelopes and classification. In *Handbook of New Bacterial Systematics* (Goodfellow, M. and O'Donnell, A.G., eds), pp. 195–250. Academic Press: London.

Thacker, W.L., Dyke, J.W., Benson, R.F., Havlichek, D.H. Jr, Robinson-Dunn, B., Stiefel, H., Schneider, W., Moss, C.W., Mayberry, W.R. and Brenner, D.J. (1992) *Legionella lansingensis* sp. nov. isolated from a patient with pneumonia and underlying chronic lymphocytic leukemia. *J. Clin. Microbiol.* **30**: 2398–401.

Wait, R. (1992) The use of FAB-MS of cellular lipids for the characterisation of medically important bacteria. In *Mass Spectrometry in the Biological Sciences: A Tutorial* (Gross, M.L., ed), pp. 427–41. Kluwer Academic Publishers: Dordrecht.

Wait, R. and Hudson, M.J. (1985) The use of picolinyl esters for the characterization of microbial lipids: Application to the unsaturated and cyclopropane fatty acids of *Campylobacter* species. *Lett. Appl. Microbiol.* **1**: 95–9.

Wait, R., Hudson, M.J. and Embley, T.M. (1989) Mass spectra of the picolinyl esters of some unusual microbial lipids. *Adv. Mass Spectrom.* **11**: 1520–1.

Weckesser, J. and Mayer, H. (1988) Different lipid A types in lipopolysaccharides of phototrophic and related non-phototrophic bacteria. *FEMS Microbiol. Rev.* **54**: 143–54.

Welch, D.F. (1991) Applications of cellular fatty acid analysis. *Clin. Microbiol. Rev.* **4**: 422–38.

Wilkinson, S.G. (1988) Gram-negative bacteria. In *Microbial Lipids*, vol. 1 (Ratledge, C. and Wilkinson, S.G., eds), pp. 299–488. Academic Press: London.

Wold, S. and Sjostrom, M.J. (1977) SIMCA: A method for analysing chemical data in terms of similarity and analogy. In *Chemometrics: Theory and Application* (Kowalski, B., ed), pp. 243–82. ACS Symposium Series No. 52. American Chemical Society: Washington DC.

6

Analysis of Microbial Sterols and Hopanoids

Gary J. Jones[1], Peter D. Nichols[2] and Philip M. Shaw[3]
[1]CSIRO Division of Water Resources, PMB3 Griffith, New South Wales, Australia
[2]CSIRO Division of Oceanography, Hobart, Tasmania, Australia
[3]Carlton and United Breweries Ltd., Melbourne, Victoria, Australia*

6.1 INTRODUCTION

Triterpenoids, which are formed by the cyclization of the C_{30} isoprenoid hydrocarbon squalene, are found throughout the bacteria and eukarya but appear to be absent from archaea. As triterpenoid biosynthesis is generally anaerobic in bacteria and aerobic in eukarya, the major class of triterpenoids in bacteria—the hopanoids—is different to that in eukarya—the sterols (Figure 6.1). There is a further bifurcation in the pathway of sterol biosynthesis in that photosynthetic organisms (algae, plants) form primarily cycloartenol as the first cyclization product, whereas non-photosynthetic organisms (animals, fungi) form mostly lanosterol. There are exceptions to this general pattern of triterpenoid distribution, with hopanoids being found in some eukarya (e.g. ferns and lichens) and sterols being found in certain methanotrophic bacteria and in cyanobacteria. The hopanoid content of bacteria (0.2–2 mg/g dry weight) is generally of the same order of magnitude as the sterol content of eukaryotic cells, whilst cyanobacteria tend to be at least an order of magnitude lower in sterol content than eukaryotic microorganisms.

Sterols and hopanoids display considerable structural diversity, particularly in the pattern of substitution and unsaturation in the side-chain, making them good candidates as chemotaxonomic markers. In hopanoids, hydroxyl and amino group substitution of the side-chain occurs, whereas in sterols the substitution is by methyl and ethyl groups. In the bacteriohopanepolyols tetra-hydroxyl substitution of the side-chain is most

---

*Present address: NAL Environmental Consultants.

common, although pentols are generally found as minor components (Rohmer et al., 1984). Sterols may be methyl or ethyl substituted in algae but are usually only methyl substituted in fungi. Sterols also exhibit a range of unsaturation (C_5, C_7, C_8) and methyl-substitution (C_4, C_{14}) patterns on the polycyclic nucleus. Some examples of common bacterial hopanoids and common sterol side-chain and ring structures are given in Figures 6.2 and 6.3. Sterols and hopanoids occur in free and complexed forms and the relative proportions of the different forms may vary significantly between taxa. These forms include fatty acid esters and glycosides for sterols (Weete, 1976), and aminoacyl (Neunlist et al., 1985) and glycoside (e.g. N-acetylglucosamine) (Langworthy et al., 1976) derivatives for hopanoids.

6.2 BIOSYNTHESIS

The enzymes responsible for the cyclization of squalene to hopanoids in bacteria, and squalene to sterols in microeukaryotes are quite distinct. Hopanoids are formed by the direct cyclization of squalene by the enzyme squalene cyclase (Figure 6.1). In sterol synthesis, squalene is first oxygenated by squalene epoxidase to form (3S)-squalene epoxide ((3S)-2,3-oxidosqualene), which may then be cyclized by one of two related enzymes: 2,3-oxidosqualene:lanosterol cyclase or 2,3-oxidosqualene:cycloartenol cyclase. Lanosterol and cycloartenol undergo a series of further transformations to yield the stable sterol end-products found in microbial cells. These reactions are:

(a) opening of the cyclopropane ring (cycloartenol);
(b) saturation at C_{24};
(c) alkylation of side-chain;
(d) desaturation at C_{22};
(e) demethylation at C_4 and C_{14}.

Ring opening normally occurs after alkylation but the sequence of these reactions is variable (Goodwin, 1981; Heftman, 1983). Some microbial taxa lack the enzymes involved in one or more of these steps and they therefore accumulate sterols that are substrates for the missing enzyme(s). For example, methanotrophic bacteria appear to lack C_4-demethylase and they thus accumulate 4-methyl and 4,4'-dimethyl sterols (Jahnke & Nichols, 1986). More detailed discussions of sterol biosynthesis are provided by Heftman (1983) and Goodwin (1981).

Comparatively less is known about the biosynthesis of bacterial hopanoids. De novo biosynthesis has been conclusively demonstrated in bacteria

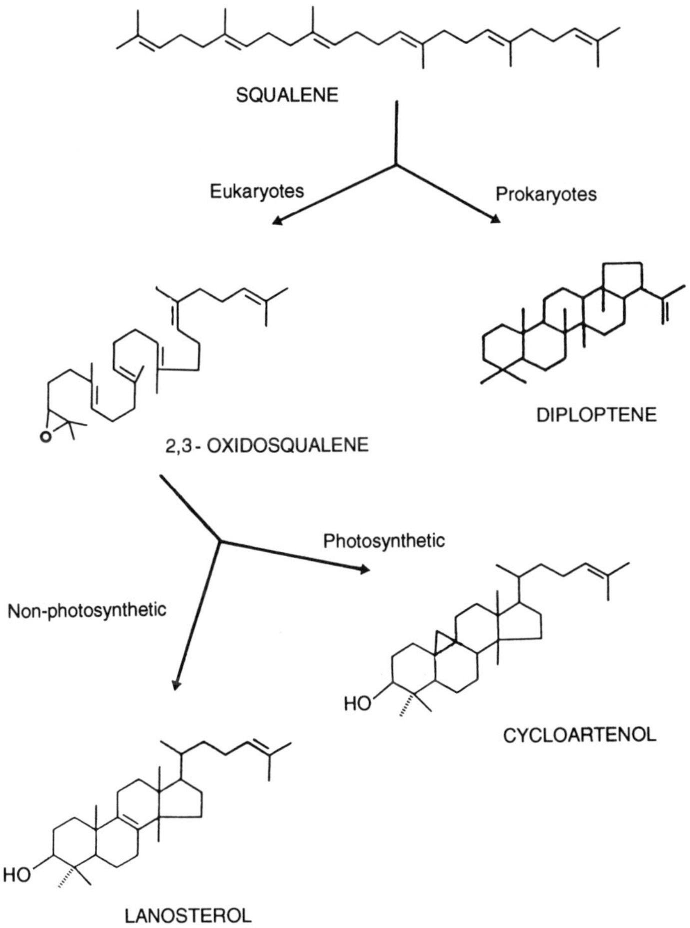

6. Analysis of Sterols and Hopanoids

Figure 6.1 Pathway of bacterial hopanoid biosynthesis and sterol biosynthesis by photosynthetic and non-photosynthetic eukaryotes.

(Rohmer et al., 1984). Although hopanepolyols and sterols are hydroxylated, the hydroxyl groups of the hopanepolyols do not arise from oxygen as they do in the sterols. Recent work by Neunlist et al. (1988) suggests that the hydroxylated side-chain is formed from the ribose group of adenosine, following loss of adenine from an adenosylhopane precursor. Hopanoids and sterols are structurally similar molecules and both act as reinforcers in the cell membrane. It has been postulated that hopanoids are the phylogenetic precursors of the oxidatively synthesized sterols (Rohmer et al., 1979). Rohmer et al. (1984), drawing partly upon some earlier suggestions by Nes and Nes (1980), have listed a number of reasons why hopanoid biosynthesis can be considered to be more primitive than sterol biosynthesis:

(a) The cyclization substrate in hopanoid biosynthesis, squalene, is simpler than that in sterol biosynthesis, (3S)-squalene epoxide.
(b) Squalene cyclization occurs in the all pre-chair conformation which is thermodynamically more favourable than that required for the formation of lanosterol, which is partly in a pre-boat conformation. The formation of cycloartenol is proposed to occur with squalene in the chair-boat-chair conformation (Goodwin, 1981).
(c) Unlike hopanoid biosynthesis, sterol biosynthesis requires further oxidative degradation and rearrangement of the cyclization product,
(d) Squalene cyclases are not highly substrate specific, with both enantiomers of squalene epoxide, as well as squalene being cyclized, whereas (3S)-squalene epoxide cyclase is stereospecific and is active only with (3S)-squalene epoxide.
(e) Hopanoid biosynthesis is completely anaerobic and is therefore compatible with an ancient prebiotic atmosphere.

Hopanoids (Figure 6.2) have been detected in only half of the bacterial species examined, whereas sterols (Figure 6.3) have been found in virtually all eukaryotic organisms. Whilst there is good evidence to suggest a

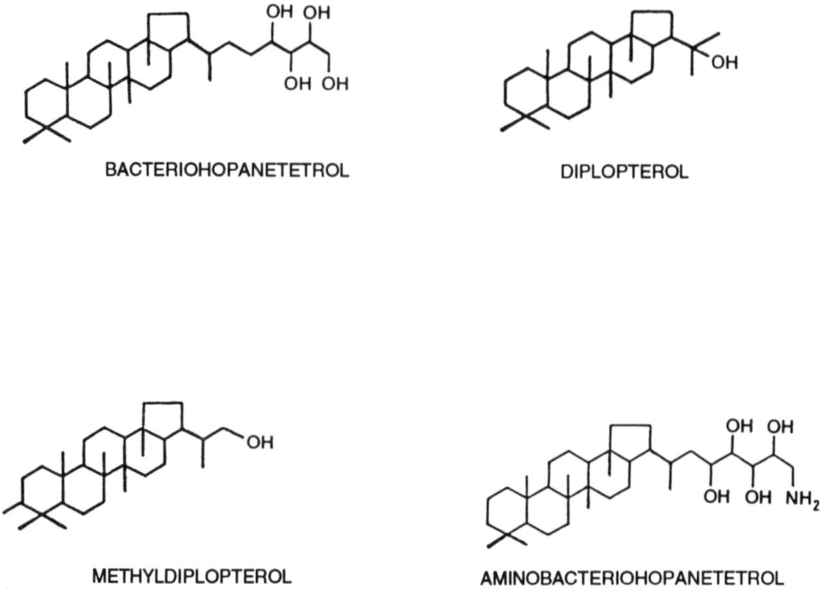

Figure 6.2 Structures of some common bacterial hopanoids.

6. Analysis of Sterols and Hopanoids

I

II

III

IV

V

VI

VII

VIII

IX

X

XI

XII

XIII

(a)

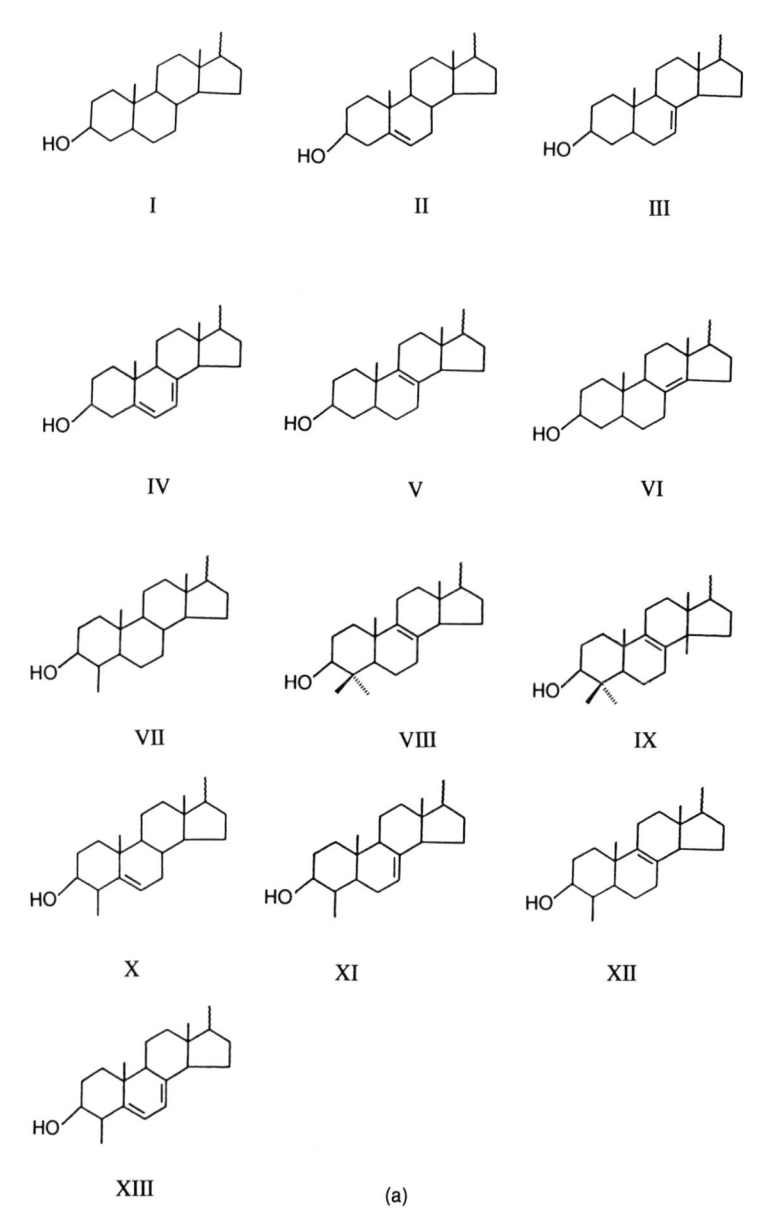

Figure 6.3a *(For legend, see overleaf)*

(b)

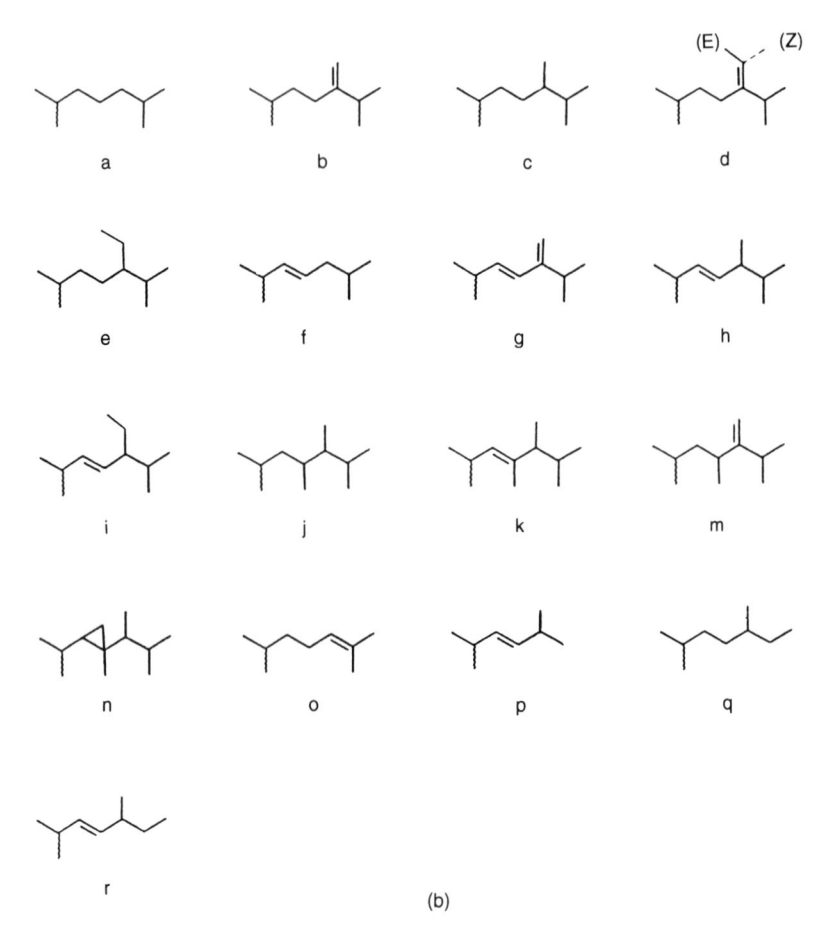

Figure 6.3 Structures of common sterol polycyclic nuclei (a, *previous page*) and side-chains (b). Letters and numerals below each structure correspond to the structural codes in Table 6.2. Note: except for structure d in (b), the stereochemistry of C_{24} and C_4 methyl and ethyl substitution, C_{22} double bond, and C_5 hydrogen is not represented. All other stereochemistry is as for lanosterol.

phylogenetic relationship between sterols and hopanoids, the distribution of hopanoids amongst the bacteria, algae, fungi, higher plants and animals does not totally support this contention (Nes & Nes, 1980). Thus, hopanoids are found in some, but not all bacteria, they are rare in eukaryotic algae, frequently found in higher plants, yet absent in animals. Given that cyanobacteria contain both sterols and hopanoids, it is possible that the two pathways are very primitive and that one or other of these pathways has for functional reasons been subsequently selected in different groups of organisms.

6.3 METHODS

6.3.1 Extraction

6.3.1.a Algae, bacteria and related material

Algal cells or particulate matter from natural waters can be harvested using pre-cleaned glass fibre (GF/F or GC/C) filters or by centrifugation. For particulate matter samples, it may be necessary to exclude larger material, which may be derived from animal or other non-algal sources, using a mesh pre-filter. Bacteria are usually harvested by centrifugation (10 000 × g.) Immediate extraction of harvested material is the preferred choice for subsequent analysis of sterols, hopanoids and other lipid components. If this is not practical, samples should be stored frozen at temperatures below 0 °C.

Extraction of lipids, including sterols and hopanoids, is usually performed in one of two ways: (a) extraction using organic solvents and (b) direct sample saponification. Many variations in the extraction conditions have been used (e.g. the combination and volume of solvents, number of extractions, extraction time, the addition of dilute acid to the extraction mixture, aiding the extraction with ultrasonication or other means, the concentration of saponification reagent) with individual laboratories generally having their preferred method. This can lead to major quantitative differences in sterol analyses between laboratories.

For the solvent extraction method, one of the most frequently used systems is the Bligh and Dyer (1959) procedure using $CHCl_3/MeOH/H_2O$ (see also White *et al.*, 1979; Guckert *et al.*, 1985).

Method 6.1 Solvent extraction of microbial lipids (after Bligh & Dyer, 1959).

1. Extract sample with a single-phase solvent system ($CHCl_3$-MeOH-H_2O, 1:2:0.8, v/v/v) for a minimum of 4 h.
2. Add $CHCl_3$-H_2O (final solvent ratio $CHCl_3$-MeOH-H_2O, 1:1:0.9, v/v/v to form two phases. The initial and final solvent ratios are selected to achieve maximum lipid recovery in the lower $CHCl_3$ layer.
3. Reduce the lipid phase using a rotary evaporator.
4. Transfer with $CHCl_3$ to a glass vial or other container and dry under a stream of nitrogen gas prior to workup or storage (at −20 °C).

Details of general considerations for lipid extraction are provided by Kates (1975).

Saponification, the second general method for obtaining algal lipid and sterol material, involves treatment of samples directly with methanolic KOH (typically 5%, w/v) at elevated temperature or under reflux. After cooling and the addition of water, saponified neutral lipids including sterols are extracted into an organic solvent, typically a mixture of hexane with petroleum ether, diethyl ether or CHCl$_3$.

Some sterol and hopanoid derivatives are water-soluble even after base saponification and, consequently, are not extracted by non-polar solvents. Hydrolysis and subsequent extraction of the free sterols or hopanoids can be achieved by initial treatment with acid, or other treatments known to cleave particular linkages, e.g. glycosides.

Superficial fluid extraction offers potential for future studies of prokaryote and eukaryote-derived triterpenoids. As few data are available, this technique will not be discussed in this review.

6.3.1.b Yeasts

Yeasts deserve special mention due to the difficulty in breaking yeast cell membranes and thus efficiently extracting lipid components. The variable lipid yields reported for yeasts in the literature must, in part, be due to the different extraction efficiencies of the methods used. Any extraction regime should be tested for completeness and selectivity for sterols before application to a series of measurements. This applies not only to yeasts but also to the extraction of all materials and although obvious, is often overlooked.

If the sample is to be analysed some time after collection, the yeast should be killed to prevent further metabolism and possible changes to the sterols. Direct sampling into 100% ethanol or methanol to give a final alcohol concentration of 50% (v/v) has been shown to be effective in killing yeasts without extracting sterols (Shaw, unpublished data). If inactivation of lipolytic enzymes is required, the yeast can be treated with hot ethanol which is later combined with the solvent extract.

The procedures for extracting sterols from yeasts differ from those for other organisms in the need to disrupt the membrane before extracting with the chosen solvents (see Method 6.2).

Method 6.2. Sterol extraction from yeasts (after Calderbank *et al.*, 1984).

---

1. Mix 500 mg (dry weight equivalent) of cells with methanol (10 ml), a suitable antioxidant (e.g. butylated hydroxy-toluene, 15 mg), an internal standard (if used) and a lipase inhibitor (e.g. *p*-chloromercuribenzoate, 1 mM) in a Braun bottle.
2. Add glass beads (35 g, 0.45–0.55 mm diameter) and disrupt the sample in a Braun homogenizer (B. Braun, Melsungen, FRG) for six periods of 15 s at 4000 rpm using CO$_2$ cooling. The Braun homogenizer has been specified here as to the knowledge of the authors it is the only one which effectively disrupts yeasts.
3. After disruption, addition of a less polar solvent may be made and the sample treated in the same manner as for any other microbial extraction.

6. Analysis of Sterols and Hopanoids

Free and esterified sterols may also be isolated from cell-free extracts of yeast by the method of Parks and Stromberg (1978). This is particularly relevant for enzymatic studies. Diatomaceous earth is added to the yeast extract (1.0 ml, prepared using standard methods) after any incubation step required for enzyme studies. Protein is caused to adhere to the diatomaceous earth by addition of acetone (5 ml) while vortexing the suspension, and the solvent is recovered by decanting after centrifugation (500 g). The diatomaceous earth is then extracted with $CHCl_3$-MeOH (4:1, v/v, 10 ml) and the solvent collected as before. The residue is finally extracted with diethyl ether (5 ml) and the three extracts combined and dried. The sterols may then be analysed directly or separated from other lipid components into free and esterified components using chromatographic techniques.

An alternative approach involves the direct saponification of yeast cells in hot (usually refluxing) methanolic KOH (e.g. Breivik & Owades, 1957). Simply heating the sample in methanolic KOH is in some cases sufficient to liberate the free and esterified sterols, but this needs to be tested for the particular samples under investigation.

Sterol yields higher than those obtainable through direct saponification of yeasts are possible with acid pre-treatment (Gonzalez & Parkes, 1977). This involves resuspending a cell pack (10 ml of culture spun at $500 \times g$ for 15 min) in HCl (0.1 M, 10 ml) and heating at 100 °C for 20 min. The residue is then saponified as usual. The acid pre-treatment is reported to increase the amount of extracted sterols by a factor of four without altering their distribution. This technique may be applicable to other organisms for increasing sterol yield.

6.3.1.c Soils and sediments

Soils and sediments can be extracted with organic solvents or by direct saponification in the same manner as for the samples described above. Sediments can be extracted either wet or after freeze-drying. If wet sediments are extracted, water is not added in the first stage of the Bligh and Dyer (1959) extraction (Method 6.1). Care may be needed in checking the recovery of lipid from certain soil types, particularly with respect to the method chosen for lipid extraction.

6.3.2 Purification

6.3.2.a Sterols

After lipids have been isolated by either the solvent extraction or saponification procedures, sterols can be separated using one of a number of methods: digitonin precipitation, thin-layer chromatography (TLC), column chromatography and high-performance liquid chromatography (HPLC).

Method 6.3 Digitonin precipitation of sterols.

1. Add excess of a 1% digitonin solution in 90% ethanol to the extract dissolved in ethanol-CH$_2$Cl$_2$ (3:1, v/v) (Boon et al., 1979).
2. Stir for 2 h at room temperature.
3. Centrifuge the mixture and wash the precipitate with ethanol.
4. Reflux the precipitate in pyridine for 30 min; the digitonin is then precipitated from the reaction mixture by addition of diethyl ether.
5. The pyridine-diethyl ether solution is next treated with 2 M HCl, and the ether layer containing the freed 3β-OH sterols is separated and concentrated.

Thin layer and column chromatographic methods using silica gel are the most commonly employed techniques of sterol purification. Details on these methods are provided by Kates (1975). Separation of 4-methyl and 4-desmethyl sterols has been obtained using either TLC (Kieselgel 60; toluene-ethylacetate, 9:1, v/v; de Leeuw et al., 1983) or column chromatography (Gagosian et al., 1983a,b). With the column technique of Gagosian et al. (1983a,b) (7 g silica gel deactivated with 5% distilled water w/w), 14 fractions were collected by elution with mixtures of hexane, toluene and ethylacetate. Fraction VII (15% ethylacetate in hexane) contained the 4-methyl sterols and fraction VIII (20% ethylacetate in hexane) contained the 4-desmethyl sterols.

Methods for the HPLC separation of simple lipids (including sterols), and of both simple and complex lipid classes have been elegantly reviewed by Christie (1987). High-performance liquid chromatography has not been widely used to date for the separation of sterols from other lipids in microbial and related samples, although such procedures offer many advantages such as ease of sample loading and fraction collection, and automation. An isocratic HPLC system has been routinely used for the separation of simple lipid classes, including sterols, obtained from saponification of total lipid extracts (Spherisorb 5 μm CN-bonded column, 25 cm × 4.6 mm i.d.; solvent = hexane-propanol, 97:3 v/v, 1 ml/min flow rate) (Nichols, 1983; Nichols et al., 1985). Total run time is less than 10 min with this system and separation of sterols from hopanoids is also achieved.

In studies where lipid has been extracted from yeasts, algae and some environmental samples by base saponification, purification of sterols is not often performed. This is due to the high concentration of sterols relative to other lipid fractions or to the fact that other lipid components, even if in high concentration, do not interfere with sterol analysis and quantitation.

6.3.2.b Hopanoids

After extraction of bacterial cellular lipids using $CHCl_3/MeOH$, extracts can be treated according to a number of procedures (e.g. Neunlist & Rohmer, 1985a–c; Rohmer et al., 1980, 1984). A preferred treatment (Method 6.4) results in the conversion of bacteriohopanepolyols to simpler products that can be readily analysed by gas chromatography (Rohmer et al., 1984).

Method 6.4 Extraction and treatment of hopanoids for chromatography.

1. Stir the crude $CHCl_3$-MeOH extract at room temperature for 1 h with a solution of H_5IO_6 (300 mg) in tetrahydrofuran-water (3 ml, 8:1, v/v).
2. Add water (10 ml), and extract the lipids three times with petroleum ether (10 ml).
3. Dry the solution over anhydrous Na_2SO_4 and evaporate to dryness.
4. Reduce the residue by stirring for 1 h at room temperature with an excess of $NaBH_4$ (100 ml) in ethanol (3 ml).
5. After addition of a solution of KH_2PO_4 (15 ml, 100 mM), the hopanoids are extracted as previously described with petroleum ether.

The reaction mixture obtained after the H_5IO_6 treatment can be separated by TLC using a double development with dichloromethane on Merck HF 254 (0.25 mm) silica plates into hydrocarbon brands containing diplotene and squalene ($R_F = 0.79$), diplopterol ($R_F = 0.26$), 22S-bacteriohopane derivatives ($R_F = 0.19$) and 22R-bacteriohopane derivatives ($R_F = 0.15$) (Rohmer et al., 1984). After spraying with a 0.1% alcoholic solution of berberin chlorhydrate, bands are visualized under UV light (366 nm) and scraped off; the hopanoids are recovered from the silica gel using dichloromethane. The primary alcohols liberated by the H_5IO_6 treatment are acetylated and the acetates purified by TLC (cyclohexane-ethyl acetate, 90:10, $R_F = 0.52$) (Rohmer et al., 1984).

An alternative treatment used particularly for the analysis of aminohopanoid derivatives, involves direct acetylation of the crude $CHCl_3$-MeOH extract (see below for details of acetylation procedures) (Neunlist et al., 1985). Excess reagent is removed under reduced pressure, and the residue separated by TLC using $CHCl_3$-H_2O (95:5, v/v) as eluent. The hopanoid polyacetates can be further purified by reverse-phase HPLC (Neunlist & Rohmer, 1985a–c).

6.3.3 Analysis and identification

6.3.3.a Total sterols

Total sterols can be determined spectrophotometrically after reaction of an isolated sterol fraction (Courchaine et al., 1959; Zlatkis et al., 1963).

Method 6.5 Determination of total sterols.

1. Up to 0.3 mg of sterol fraction is dried under a stream of nitrogen gas, and acetic acid (glacial, 6.0 ml) is added with mixing.
2. Acidic ferric chloride (2.5% $FeCl_3.6H_2O$ (w/v) in 85% orthophosphoric acid (v/v) made up to 50 ml with sulphuric acid (conc); 4 ml) is added and the mixture vortexed, cooled, then allowed to stand at room temperature for 10 min.
3. The concentration is then determined from the absorbance at 550 nm compared with a standard curve determined using pure sterol and subtraction of reagent bank.

6.3.3.b $\Delta^{5,7}$ Unsaturated sterols

The total $\Delta^{5,7}$ unsaturated sterols can be determined spectrophotometrically in a sterol fraction (or total lipid extract if the sample is found to contain no interferences) by measurement of its absorption at 282 nm (Breivik & Owades, 1957). The sample is dissolved in hexane, and the concentration determined by comparing the absorbance at 282 nm with a calibration curve prepared from different concentrations of a standard $\Delta^{5,7}$ sterol. Distinction between ergosterol and $\Delta^{24(28)}$ ergosterol can be made by subtracting the contribution of the latter calculated from the absorbance measured at 230 nm. However, this is in a region of significant solvent absorption, and the results are prone to large errors for low concentrations of ergosterol.

6.3.3.c HPLC analysis of sterols

A major use of HPLC has been in the analysis of ergosterol, primarily because of the strong absorbance at 282 nm of the $\Delta^{5,7}$ unsaturated sterols (e.g. Newell et al., 1988; Zill et al., 1988). Both reversed-phase and silica stationary phases have been used to separate ergosterol from contaminants, which may interfere with the spectroscopic method described above. Mixtures of free sterols have been separated on a range of columns (e.g. Colin et al., 1979) including silica, C_{18}, $AgNO_3$ and pyrocarbon. Ultraviolet detection is commonly used (~250 nm as a typical wavelength), although refractive index detection can also be used for clean fractions. The use of HPLC to perform enantiomeric separations is discussed below.

Individual separations can be obtained readily for almost any pair of sterols by judicious choice of solvent but generally there is insufficient resolution to separate the complex mixture of sterols found in most organisms and in environmental samples.

6.3.3.d Derivatization for GC and GC-MS analysis

A number of simply formed sterol and hopanoid derivatives have been used for GC and GC-MS analysis. These reactions are often performed in test tubes or glass GC vials. Hopanoid and sterol acetates are typically formed by treatment with acetic anhydride-pyridine (1:1, v/v, room temperature overnight, or 100 °C for 2 h) (Copius-Peereboom, 1965; Itoh et al., 1982). Excess reagents are removed under a stream of nitrogen.

Trimethylsilyl ethers (TMS ethers) are probably the most commonly used sterol derivatives in GC analysis. Sterol-TMS ethers readily produce diagnostic mass spectrometric fragmentation ions (Knight, 1967; Brooks et al., 1968; de Leeuw et al., 1983), making them particularly suitable for GC-MS analysis. Sterol-TMS ethers can now be readily formed by treatment with one of a number of reagents. Early workers prepared derivatives by treatment of the sterols in dry pyridine with hexamethyldisilazane and trimethylchlorosilane at room temperature (Wells & Makita, 1986; Brooks et al., 1968). More recently sterol-TMS ethers have been formed using bis-(trimethylsilyl)trifluoroacetamide (BSTFA) or bis-(trimethylsilyl)acetamide (BSA) (reaction conditions typically: 30 μl BSTFA 100 °C/1 h; 30 μl of CHCl$_3$ can be added if material has not fully dissolved; Edlund et al., 1985; Nichols et al., 1987). Excess reagents are removed under a stream of nitrogen. The shelf life of the TMS ethers is limited, and the derivatized sterols should be analysed promptly by GC or GC-MS. It is best to re-derivatize a sample if the sterol-TMS ethers have been stored for any length of time (greater than 24–48 h). A number of studies have analysed sterols as their corresponding methyl ethers (e.g. Clayton, 1962).

TMS ether derivatives can be formed with simple hopanols, but derivatization of bacteriohopanepolyols is difficult due to steric hindrance. Even some simple hopanols (e.g. diplopterol) may be difficult to silylate because of steric constraints, thus derivatization efficiency should always be carefully checked.

6.3.3.e GC and GC-MS analysis

To date gas chromatography has been the most commonly used method for the separation and identification of sterols and hopanoids. The GC operating conditions for these compounds are generally similar and will not be considered separately here. Whilst most GC analyses of sterols have been accomplished using only one column, typically a non-polar column, it has been previously stated that at least three columns are necessary to even tentatively identify a sterol by GC (Patterson, 1971a). This criterion still holds if sterol identification is based on GC analysis alone. Most early studies were performed using low-resolution packed columns, but sterol analyses are now undertaken using high-resolution fused-silica capillary columns which provide much better separation of components than those used in the early 1970s. However, even with today's high-resolution columns, at least two phases are considered necessary to identify a component if mass spectrometric detection is not used. Representative gas chromatograms together with GC conditions, illustrating the type of resolution obtainable using modern 'state of the art' capillary columns are shown in Figures 6.4 to 6.8.

The chromatograms chosen represent some of the diverse applications encountered in the analysis of sterols. The oven temperature heating rate may be varied depending on whether or not the analyst is interested in components eluting prior to the sterol region. Peak numbers refer to GC and GC-MS data provided in Tables 6.1, 6.2 and 6.3. The unicellular marine alga, *Chattonella antiqua*, contains an unusually high abundance of 24-ethylcholesterol (Figure 6.4; Nichols et al., 1987). Until recently, it was generally believed that few marine algae contained significant amounts of

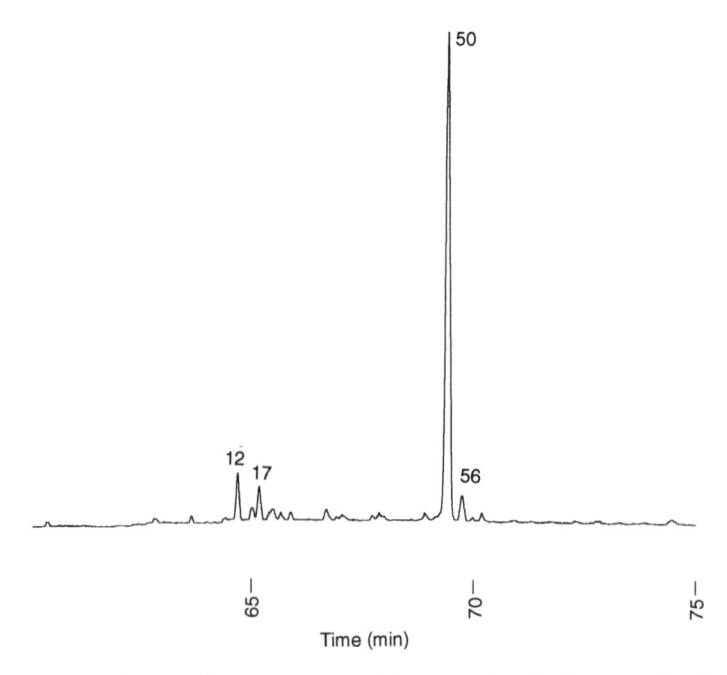

Figure 6.4 Partial gas chromatogram (60–75 min) of the sterols from the unicellular red tide flagellate *Chattonella antiqua* (Raphidophyceae). Instrument: HP5890 gas chromatograph fitted with a split/splitless injector (sample injected in splitless mode) and a flame ionization detector. Column: fused silica capillary, 50 m × 0.2 mm i.d. cross-linked methyl silicone (HP-1). Oven temperature: 50 °C for 1 min, 30 °C/min to 150 °C, then 4 °C/min to 300 °C. The carrier gas was hydrogen. Peak numbers as in Table 6.2.

6. Analysis of Sterols and Hopanoids

Table 6.1 Molecular ions of sterol TMS-ether derivatives.

Carbon number	Molecular ion (m/z)			
	Number of double bonds			
	0	1	2	3
26	446	444	442	440
27	460	458	456	454
28	474	472	470	468
29	488	486	484	482
30	502	500	498	496

this sterol, and that a high relative abundance of this sterol in either marine sediments or particulate samples was usually indicative of input from terrestrial sources. The finding of 24-ethylcholesterol in an increasing number of marine algae (Volkman, 1986; Jones *et al.*, 1987; Nichols *et al.*, 1987) indicates that care needs to be taken with environmental samples, through the analysis of other lipid classes, before a solely terrestrial source of this sterol can be supported.

The separation of a range of 4-methyl and 4-desmethyl sterols found in the alga FCRG 51 is demonstrated in Figure 6.5. The occurrence of

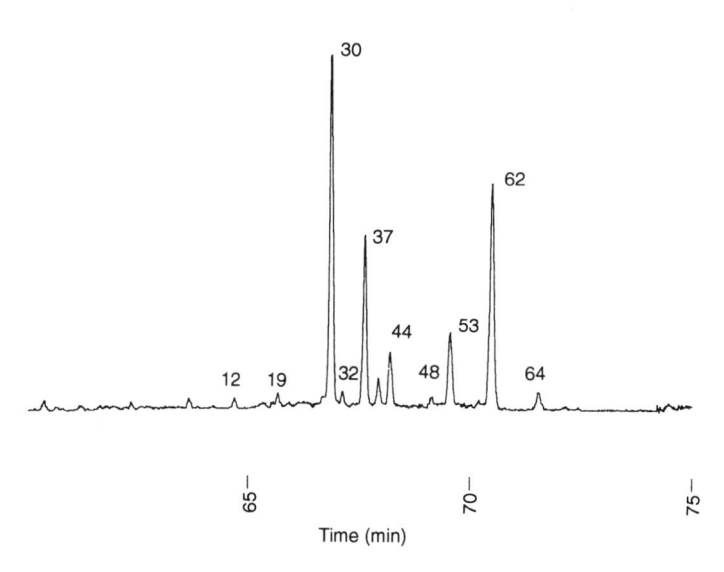

Figure 6.5 Partial gas chromatogram (60–75 min) of the sterols from the unicellular alga FCRG 51 (Scripps Institution of Oceanography, clonal designation PY-37). Gas chromatographic conditions as for Figure 6.4. Peak numbers as in Table 6.2.

Table 6.2 Mass spectral and gas chromatographic retention data for sterol TMS-ether derivatives.

Peak number[b]	Sterol	Trivial name[a]	Structure[c]	RRT[d]	M+	Base peak	Other major ion fragments (m/z)
1	24-norcholesta-5,22E-dien-3β-ol	24-nordehydrocholesterol	IIp	0.50	442	97	353, 313, 255, 215, 130
2	24-nor-5α-cholest-22E-en-3β-ol	24-nordehydrocholestanol	Ip	0.53	444	75	406, 391, 374, 345, 257
3	24-nor-4-methyl-5α-cholest-22E-en-3β-ol		VIIp	0.80	458	97	388, 359, 271
4	5β-cholestan-3β-ol	coprostanol	Ia	0.83	460	370	445, 403, 355, 257, 215, 75
5	27-nor-24-methylcholest-5,22E-dien-3β-ol	occelasterol	IIr	0.83	456	111	441, 366, 351, 327, 255, 129
6	cholesta-5,22Z-dien-3β-ol	cis-22-dehydrocholesterol	IIf	0.83	456	111	441, 366, 351, 327, 255, 129
7	27-nor-24-methyl-5α-cholest-22E-en-3β-ol	patinosterol	Ir	0.87	458	257	374, 359, 345, 75
8	5α-cholest-22Z-en-3β-ol	cis-22-dehydrocholestanol	If	0.87	458	257	374, 359, 345, 75
9	cholesta-5,22E-dien-3β-ol	trans-22-dehydrocholesterol	IIf	0.90	456	111	441, 366, 351, 327, 255, 129
10	5α-cholest-22E-en-3β-ol	trans-22-dehydrocholestanol	If	0.93	458	257	374, 359, 345, 75
11	cholesta-5,7,22E-trien-3β-ol		IVf	0.97	454	454	439, 355, 341, 253
12	cholest-5-en-3β-ol	cholesterol	IIa	1.00	458	129	443, 368, 353, 329, 247
13	5α-cholest-8(14)-en-3β-ol		VIa	1.01	458	75	443, 368, 353, 255, 229, 213
14	5α-cholestan-3β-ol	cholestanol	Ia	1.03	460	75	445, 403, 355, 215, 107
15	27-nor-24-methyl-5α-cholest-3β-ol		Iq	1.06	460	75	445, 355, 215, 95
16	cholesta-5,7-dien-3β-ol	7-dehydrocholesterol	IVa	1.07	456	75	441, 360, 351, 211, 198, 129
17	5α-cholest-8-en-3β-ol	24-dihydrozymosterol	Va	1.07	458	75	443, 368, 353, 255, 229, 213
18	cholesta-5,24-dien-3β-ol	desmosterol	IIo	1.09	456	129	441, 366, 343, 327, 253
19	24-methylcholesta-5,22E-dien-3β-ol	brassicasterol/crinosterol	IIh	1.12	470	69	455, 380, 365, 340, 255, 255, 129
20	24-methyl-5α-cholest-22E-en-3β-ol	brassicastanol/crinostanol	Ih	1.15	472	69	457, 374, 345, 257, 109

21	27-nor-4,24-dimethyl-5α-cholest-22E-en-3β-ol		VIIr	1.15	472	69	388, 373, 359, 271
22	5α-cholest-7-en-3β-ol	lathosterol	IIIa	1.17	458	75	443, 368, 353, 255, 229, 213
23	5α-cholesta-8,24-dien-3β-ol	zymosterol	Vo	1.17	456	69	441, 366, 351, 229, 213, 107
24	24-methylcholesta-5,7,22E,24(28)-tetraen-3β-ol[e]		IVg	1.18	466	361	376, 251, 73
25	4-methyl-5α-cholest-22E-en-3β-ol		VIIf	1.20	472	69	388, 373, 359, 271, 95
26	4-methyl-5α-cholest-8-en-3β-ol		XIIa	1.21	472	75	457, 382, 367, 269, 243, 227
27	24-methylcholesta-5,7,22E-trien-3β-ol	ergosterol	IVh	1.22	468	69	378, 363, 337, 253, 131
28	24-methyl-5α-cholesta-7,22E-dien-3β-ol	stellasterol	IIIh	1.26	470	75	455, 343, 255, 229, 213
29	4-methyl-5α-cholest-7-en-3β-ol	lophenol	XIa	1.28	472	75	457, 382, 367, 269, 243, 227
30	24-methylcholesta-5,24(28)-dien-3β-ol	24-methylenecholesterol	IIb	1.28	470	129	455, 386, 365, 341, 296, 257, 253
31	24-methyl-5α-cholest-8(14)-en-3β-ol	ergost-8(14)-enol	VIc	1.28	472	75	457, 378, 367, 343, 255, 229, 213
32	24-methyl-5α-cholest-24(28)-en-3β-ol	24-methylenecholestanol	Ib	1.31	472	75	457, 388, 345, 255, 215
33	24-methylcholest-5-en-3β-ol[f]	campesterol/dihydrobrassicasterol	IIc	1.31	472	129	382, 367, 343, 255
34	4-methyl-5α-cholestan-3β-ol	4-methylcholestanol	VIIa	1.32	474	75	459, 384, 369, 345, 229
35	24-methyl-5α-cholestan-3β-ol	campestanol/dihydrobrassicastanol	Ic	1.35	474	215	459, 384, 369, 215, 147
36	24-ethyl-5β-cholestan-3β-ol[f]	24-ethylcoprostanol	Ie	1.35	488	398	473, 383, 343, 257, 215, 75
37	23,24-dimethylcholesta-5,22E-dien-3β-ol		IIk	1.39	484	69	469, 394, 372, 343, 255, 139, 83
38	24-methyl-5α-cholesta-7,24(28)-dien-3β-ol	24-methylenelophenol	IIIb	1.40	470	343	455, 386, 365, 253, 213, 131, 75
39	4,24-dimethyl-5α-cholest-22E-en-3β-ol		VIIh	1.41	486	69	388, 359, 271, 247, 229, 125

(*Continued*)

Table 6.2 (*Continued*)

Peak number[b]	Sterol	Trivial name[a]	Structure[c]	RRT[d]	M⁺	Base peak	Other major ion fragments (m/z)
40	23,24-dimethyl-5α-cholesta-22E-en-3β-ol		Ik	1.42	486	69	374, 345, 257, 97, 83
41	24-ethylcholesta-5,22E-dien-3β-ol	stigmasterol/poriferasterol	IIi	1.42	484	83	394, 379, 355, 255, 129
42	24-methyl-5α-cholest-7-en-3β-ol	ergost-7-enol	IIIc	1.46	472	75	457, 378, 367, 343, 255, 229, 213
43	24-ethyl-5α-cholesta-22E-en-3β-ol	stigmastanol/poriferastanol	Ii	1.47	486	83	374, 353, 345, 257
44	4,24-dimethyl-5α-cholesta-22E-en-3β-ol		VIIh	1.47	486	69	471, 388, 373, 359, 271
45	23,24-dimethyl-5α-cholesta-7,22E-dien-3β-ol		IIIk	1.51	484	69	469, 372, 343, 255, 229, 213, 75
46	24-ethyl-5α-cholesta-7,22E-dien-3β-ol	chondrillasterol/spinasterol	IIIi	1.56	484	55	469, 372, 343, 255, 229, 213, 75
47	23,24-dimethylcholest-5-en-3β-ol		IIj	1.60	486	129	471, 396, 381, 357, 255, 245
48	4,24-dimethylcholesta-5,7-dien-3β-ol		XIIIc	1.60	484	69	393, 379, 355, 267, 241, 227, 159
49	4,4,14-trimethyl-5α-cholesta-8,24-dien-3β-ol	lanosterol	IXo	1.60	498	393	483, 109, 69
50	24-ethylcholest-5-en-3β-ol	sitosterol/clionosterol	IIe	1.63	486	129	471, 396, 381, 357, 255
51	24-ethylcholesta-5,24(28)E-dien-3β-ol	fucosterol	IId	1.63	484	386	463, 371, 355, 296, 281, 257
52	23,24-dimethyl-5α-cholestan-3β-ol		Ij	1.63	488	75	473, 305, 215
53	4,24-dimethyl-5α-cholest-24(28)-en-3β-ol		VIIb	1.65	486	402	471, 387, 359, 269, 229, 75
54	24-ethyl-5α-cholestan-3β-ol	sitostanol	Ie	1.67	488	75	473, 398, 383, 305, 230, 215

#	Systematic name	Trivial name	Code	RRT			
55	24-ethyl-5α-cholest-24(28)E-en-3β-ol	fucostanol	Id	1.67	486	69	471, 388, 373, 297, 283, 229, 215
56	24-ethylcholesta-5,24(28)Z-dien-3β-ol	isofucosterol	IId	1.67	484	69	463, 371, 355, 296, 281, 257
57	4,4-dimethyl-5α-cholesta-8,24-dien-3β-ol		VIIIo	1.68	484	386	469, 394, 379, 135
58	4,24-dimethyl-5α-cholestan-3β-ol		VIIc	1.69	488	75	473, 398, 383, 359, 229, 130
59	24-ethyl-5α-cholest-24(28)Z-en-3β-ol	isofucostanol	Id	1.70	486	69	471, 388, 373, 297, 283, 229, 215
60	24-ethyl-5α-cholest-7-en-3β-ol	stigmast-7-enol	IIIe	1.86	486	75	471, 396, 381, 255, 229, 213
61	4,23,24-trimethylcholesta-5,22E-dien-3β-ol	dehydrodinosterol	Xk	1.74	498	139	388, 367, 359, 129, 69
62	4,23,24-trimethyl-5α-cholest-22E-en-3β-ol	dinosterol	VIIk	1.79	500	69	485, 388, 367, 359, 283, 271
63	4-methyl-24-ethyl-5α-cholest-22E-en-3β-ol		VIIi	1.81	500	83	388, 367, 359, 283, 271
64	4,23,24-trimethyl-5α-cholest-24(28)E-en-3β-ol		VIIm	1.93	500	57	485, 387, 373, 359, 297, 283, 269
65	4,23,24-trimethyl-5α-cholest-7-en-3β-ol		XIj	1.96	500	57	485, 410, 395, 269, 243, 227, 147
66	4,23,24-trimethyl-5α-cholestan-3β-ol	dinostanol	VIIj	2.08	502	57	487, 412, 397, 373, 261, 229, 130
67	22,23-methylene-23,24-dimethyl-cholest-5-en-3β-ol	gorgosterol	Xn	2.19	498	129	483, 408, 400, 386, 343, 337, 255

[a] Trivial name for sterols where such names are commonly used. Where two names are given, they refer to the C24(α) and (β) epimers respectively.
[b] Peak numbers refer to those cited in Figures 6.4 to 6.14.
[c] See Figure 6.3.
[d] RRT, relative retention time on non-polar (methyl silicone or BP-1) fused silica capillary columns. Cholesterol = 1.00; 24-ethylcholesterol = 1.63.
[e] Tentative identification.
[f] 24-Ethylcoprostanol (peak 36) and 24-methylcholesterol (campesterol) (peak 33) whilst separated in this study, co-elute on non-polar columns, relative elution times of these, and other, sterols may be affected by oven heating rate.

4-methyl sterols in this alga, in particular 4,23,24-trimethyl-5α-cholest-22E-en-3β-ol (dinosterol, sterol 62; Table 6.2) was used to suggest that the organisms should be assigned to the Dinophyceae (Nichols *et al.*, 1983). The separation and high proportion of 5β-stanols (coprostanol and 24-ethyl-5β-cholestanol, sterols and 36; Table 6.2) in sediment collected from a sewage-contaminated freshwater environment is shown in Figure 6.7. Coprostanol has been widely used as a 'biomarker' for sewage contamination (e.g. Wells & Makita, 1962; Hatcher & McGillivary 1979; Vivian 1986). The sterol distribution of a brewing yeast is illustrated in Figure 6.8. A

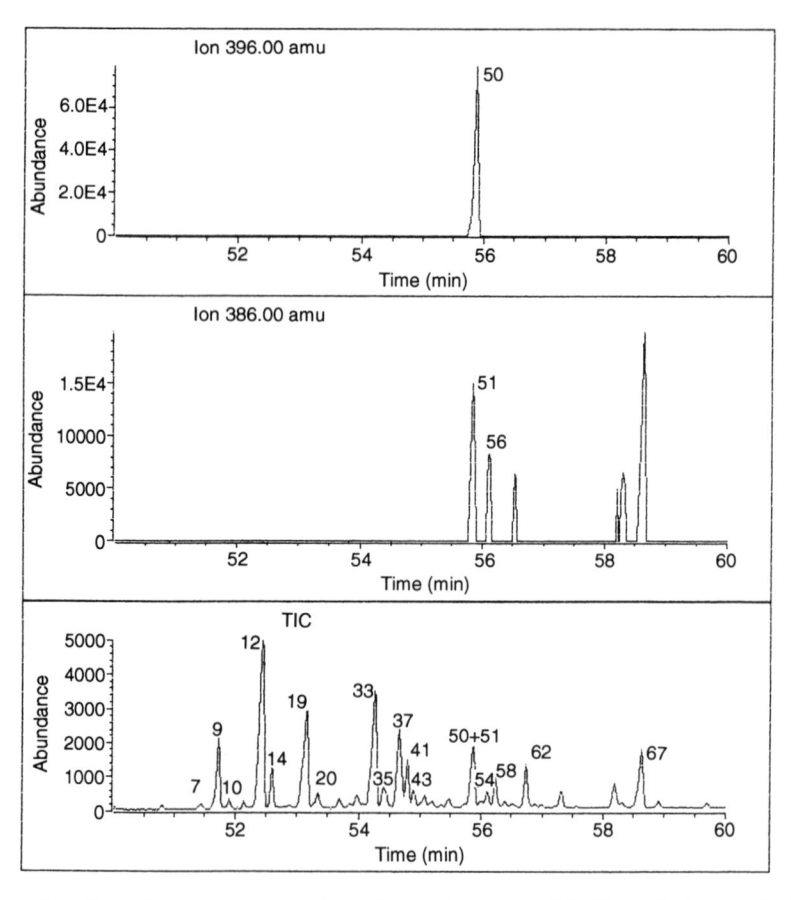

Figure 6.6 Partial reconstructed ion chromatograms (50–60 min) from GC-MS analysis of sterols in suspended particulate matter from a tropical coral reef lagoon. Upper trace: m/z 396, 24-ethylcholest-5-en-3β-ol (sterol 50). Middle trace: m/z 386, fucosterol and isofucosterol (sterols 51 and 56). Lower trace: total ion chromatogram. Gas chromatographic conditions are as for Figure 6.4 with the exception that helium was the carrier gas. Peak numbers as in Table 6.2.

6. Analysis of Sterols and Hopanoids

number of components remain only partially identified. In the brewing industry, sterol profiles can be used to monitor yeast metabolism and the pathways of oxygen utilization during fermentation.

Patterson (1971a) determined relative retention times (RRTs) for 92 sterols and related compounds on four different columns; these data provided an excellent bench-mark for other workers studying sterols. A second large set of RRT data was reported by Itoh *et al.* (1982). The calculation of RRTs for individual sterols enables a preliminary tentative identification to be made prior to positive identification of sterols. The formula for calculating the RRT for any sterol is:

$$RRT_x = 1.00 + [0.63(RT_x - RT_{chol})]/(RT_{sit} - RT_{chol})$$

where RRT_x = relative retention time of sterol x
RT_x = retention time of sterol x
RT_{chol} = retention time of cholesterol (cholest-5-en-3β-ol)
RT_{sit} = retention time of sitosterol (24-ethylcholest-5-en-3β-ol)

This formula will interpolate RRT values between 1.00 and 1.63, and can be used to extrapolate from RRTs between 0.6 and 1.00, and 1.63 and 2.4.

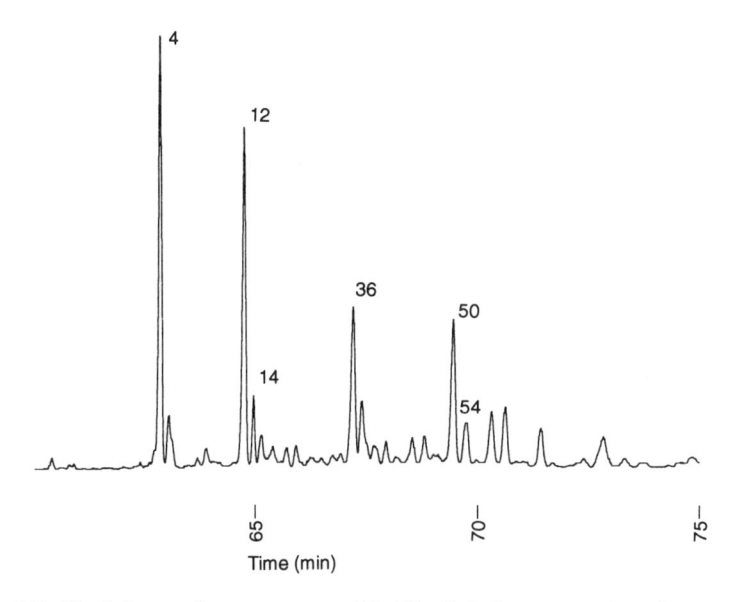

Figure 6.7 Partial gas chromatogram (60–75 min) demonstrating the separation and high proportion of 5β-stanols (coprostanol and 24-ethyl-5β-cholestanol, sterols 4 and 36) in the sterols of sediment collected from a sewage-contaminated freshwater environment. Gas chromatographic conditions as for Figure 6.4. Peak numbers as in Table 6.2.

If cholesterol and/or 24-ethylcholesterol are not present in the sample, other sterols with known RRT values may be substituted and the values 1.00 and 1.63 can be replaced by RRT$_{sterol\ 1}$ and (RRT$_{sterol\ 2}$ − RRT$_{sterol\ 1}$). The components chosen should have as large a retention time difference as possible to minimize extrapolation errors.

Calibration of the flame ionisation detector is an important step which is neglected in many studies of sterols and hopanoids. This omission often leads to significant errors in absolute and relative concentrations of individual components. Ideally the detector should be calibrated for each component to be measured and the linearity of the detector response determined over the concentration range expected. This is usually impossible due to the paucity of standards, and an estimate of the response factor based on the response for structurally similar components usually gives reasonable results. It must be noted, however, that even structurally similar compounds may have response factors differing by up to a factor of two.

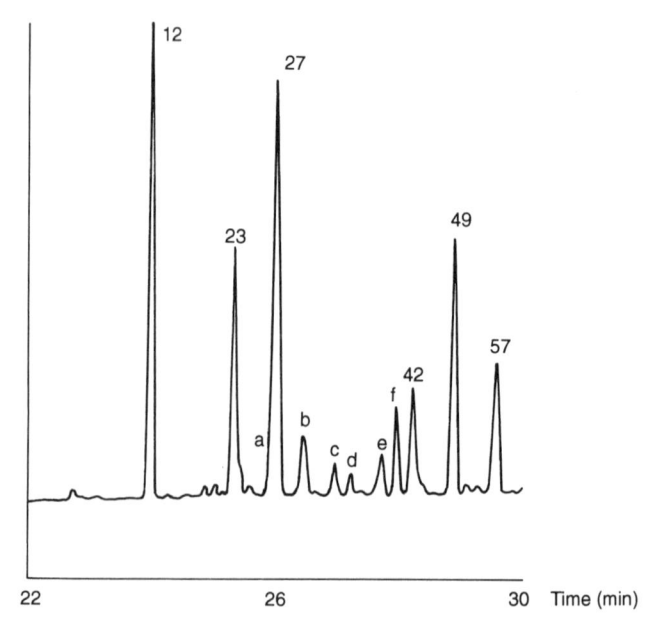

Figure 6.8 Partial gas chromatogram (22–30 min) of a yeast total sterols fraction obtained following saponification of the whole cells. Gas chromatographic conditions as for Figure 6.4 with the exception that the oven temperature program was 200 °C (initial temperature, 0 min), 10 °C/min to 280 °C, 1.3 °C/min to 296 °C, 6 °C/min to 315 °C (7 min). Peak numbers as in Table 6.2. Other peaks partially identified by GC-MS are: a, 28:4 (C$_{28}$ with four double bonds), RRT 1.18; b, 28:4, RRT 1.31; c, 28:2, RRT 1.35; d, 28:1, RRT 1.37; e, 28:2, RRT 1.44; f, 28:2, RRT 1.48.

6. Analysis of Sterols and Hopanoids

Positive identification of sterols and hopanoids is best achieved by coinjection with standards (if available) and, more importantly, by GC-MS analysis. The increasing access to GC-MS systems, largely brought about by the advent of low-cost bench-top units over the last decade, has enabled sterol structures to be confirmed without the use of multiple GC columns. Literature mass spectral data are often not available or not readily accessible for many sterols and hopanoids. There are few reports which

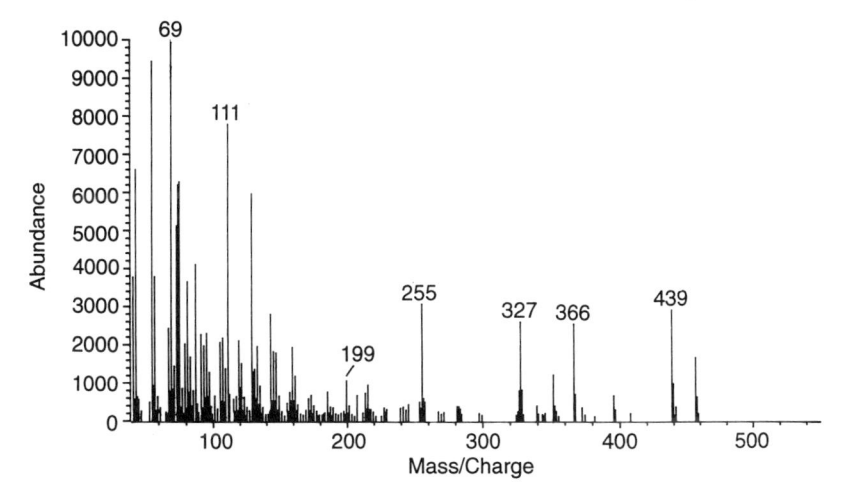

Figure 6.9 Mass spectrum of *trans*-22-dehydrocholesterol (sterol 9).

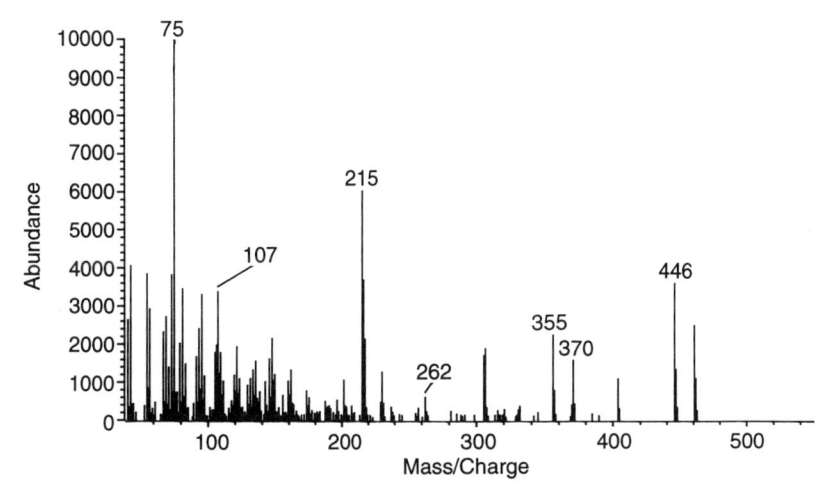

Figure 6.10 Mass spectrum of 5α-cholestanol (sterol 14).

document both GC retention time and GC-MS fragmentation data together. We have provided mass spectral data on the molecular ions and a number of characteristic ion fragments of sterol TMS ethers in Tables 6.1 and 6.3. Similarly, mass spectral and gas chromatographic data obtained for a range of sterols (as TMS ethers) are presented in Table 6.2. These data provide reference information, not always readily obtainable,

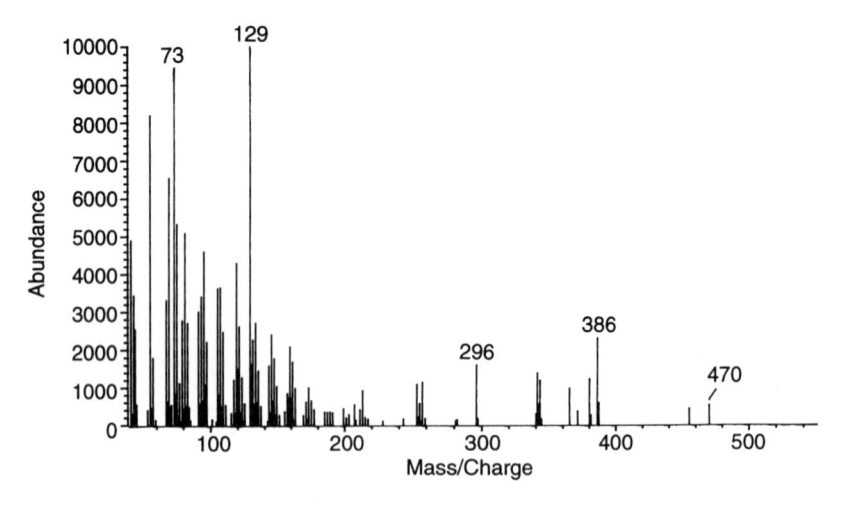

Figure 6.11 Mass spectrum of 24-methylenecholesterol (sterol 30).

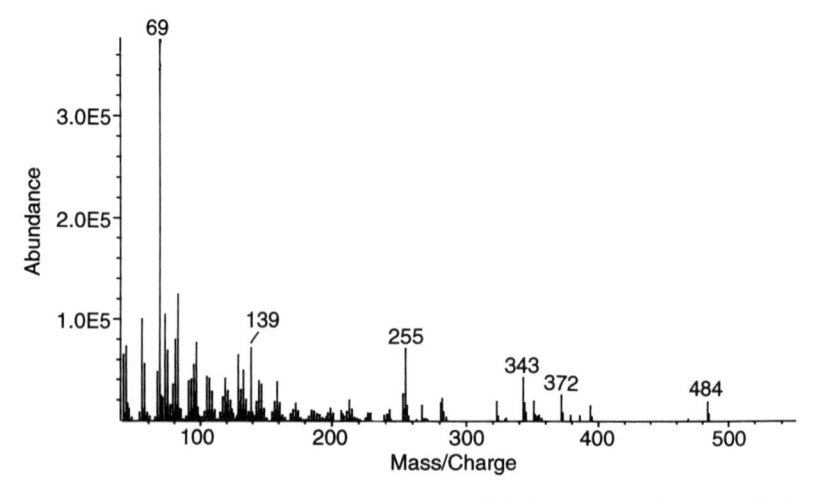

Figure 6.12 Mass spectrum of 23,24-dimethylcholesta-5,22E-diene-3β-ol (sterol 37).

6. Analysis of Sterols and Hopanoids

for sterols commonly encountered in studies of algae, bacteria, yeast and environmental samples. Although variations in GC and GC-MS operating conditions can sometimes cause large changes in mass spectral ion intensities, these data should be useful to workers in a number of fields. Similarly, GC RRT data may also fluctuate slightly due to overloading of individual compounds and other factors such as changes in the column performance

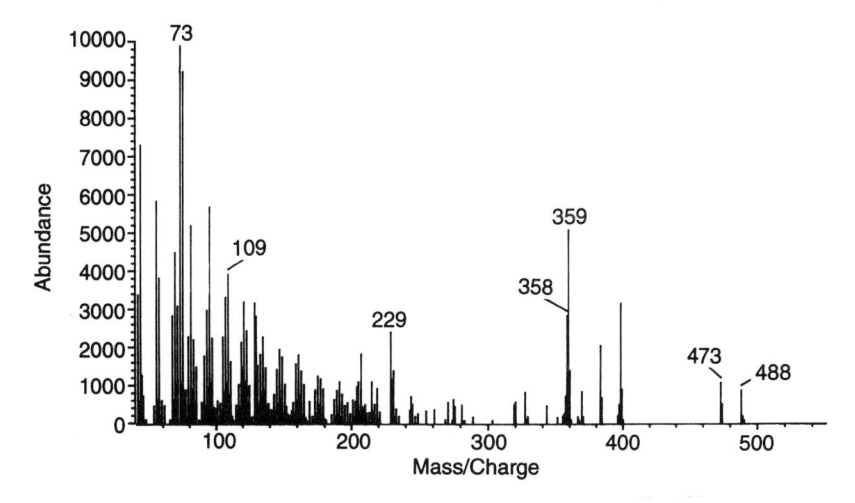

Figure 6.13 Mass spectrum of 4,24-dimethyl-5α-cholestan-3β-ol (sterol 58).

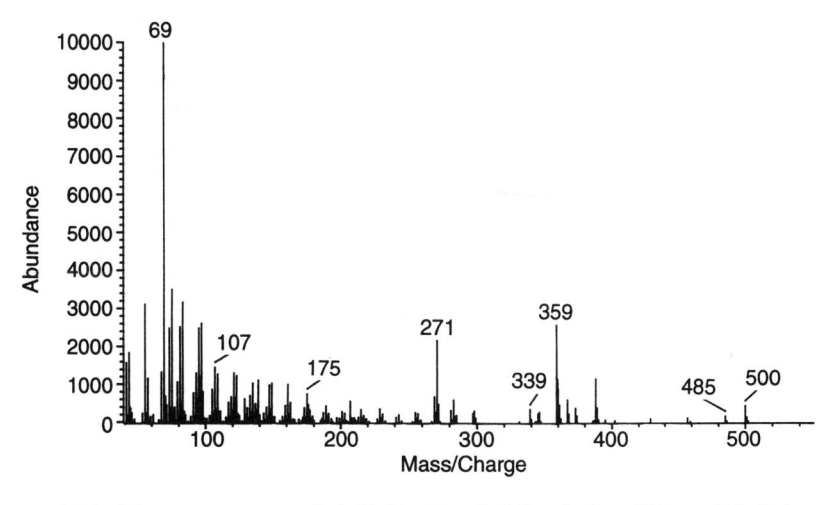

Figure 6.14 Mass spectrum of 4,23,24-trimethyl-5α-cholest-22E-en-3β-ol (sterol 62).

Table 6.3 Characteristic ion fragments for sterol TMS-ether derivatives and hopanoids.

Ion fragment(s)	Source of fragmentation
Sterols	
129, M-129	Δ^5 sterol
213, 229, 255	Δ^7 sterol
345, 229	$\Delta^{8(9)}$ and $\Delta^{8(14)}$ sterol
215, 305/306, 75	5α-stanol
227, 243, 269	C4-methyl-Δ^7-sterol
229	C4-methyl
	C4-dimethyl
372	Δ^{22} sterol
M-90 (strong), 215, 75	5β-stanol
Steroidal hydrocarbons	
217	Steranes
215, 257	Sterene, one double bond
213, 255	Sterene, two double bonds
211, 253	Sterene, three double bonds
Hopanoids	
191	Hopanes
205	Methylhopanes

brought about by column aging. The calculation of RRT data for individual sterols will nonetheless provide useful information allowing possible structure assignment prior to GC-MS analysis. In a number of cases calculation of RRT can be used to distinguish sterols which have identical mass spectra. For example, fucosterol and isofucosterol show similar major fragmentation ions, but can be distinguished by their differing GC RRTs. Similarly, cholest-7-en-3β-ol, cholest-8(9)-en-3β-ol and cholest-8(14)-en-3β-ol all show similar mass spectra, but can also be distinguished using GC retention data.

In many studies of algal-derived sterols or environmental samples of complex origin, coelution of sterols will probably occur. In such instances GC-MS selected ion monitoring (SIM) can be used to distinguish and, if data for standards are available, quantitate coeluting sterols. As an example, the separation of fucosterol (sterol 51) and 24-ethyl cholesterol (sterol 50) is shown in Figure 6.6 (GC-MS conditions; Nichols et al., 1988). Organic geochemists have made particular use of this method for the identification of sterols and steroidal-derived compounds in recent and ancient sediments.

6.3.3.f Separation of C24-sterol epimers

Many terrestrial and marine-derived sterols contain an alkyl group at the C_{24} position (e.g. sitosterol, 24R/α; clionosterol, 24S/β). In many instances they occur as C_{24} epimeric mixtures. The separation, assignment of stereochemistry and quantitation of these epimers is therefore important in biochemical, geochemical and taxonomic studies.

The separation of the C_{24} epimers of sterols remains a challenge in the analysis of sterols. Nuclear magnetic resonance spectroscopy can be used for the assignment of C_{24} stereochemistry (Thompson et al., 1980; Chiu & Patterson, 1981), but it cannot be applied to the analysis of natural mixtures of sterols. Other methods used for assignment of C_{24} configuration include X-ray crystallography (Ling et al., 1970), differences in melting points (Gershengorn et al., 1968), and optical rotation measurements (Bernstein & Wallis, 1938; Lenfant et al., 1970). These techniques typically require a few milligrams of the isolated sterol, therefore their application to studies of small quantities of complex sterol mixtures is limited.

The separation of C_{24} epimers of steranes (Maxwell et al., 1980) and sterols (Thompson et al., 1981) by GC using glass capillary columns has been reported. The very long columns and long analysis times used (e.g. 100 m glass column coated with DEGS/PEGS, 3:1 w/w; approx. 10–12 h GC run time) and the reproduction of columns that efficiently separate C_{24} epimers have not been satisfactory, preventing the routine use of these methods.

Reverse-phase HPLC procedures that achieve separation of C_{24} epimers have also been developed. Separation of a number of epimeric pairs of C_{24} alkyl sterols containing a $C_{22(23)}$ double bond was obtained by reverse-phase HPLC on a Whatman Partisil M9 10/50 ODS-2 column (Bohlin et al., 1981). Under these conditions, however, the corresponding sterols having a saturated side-chain were not separated. More recently, separation of C_{24} epimeric sterol benzoates using a reversed-phase HPLC (TSK-Gel ODS-120A column; hexane/2-propanol/acetonitrile 1:3:16, v/v, and other solvent systems; 240 nm detection) has been achieved (Ikekawa et al., 1989). The HPLC procedures currently appear to be the most useful for routine separation of the C_{24} epimers. A preliminary separation of complex sterol mixtures into various fractions by argentation chromatography and conventional capillary GC (or GC-MS) analysis of the fractions may be necessary in certain instances prior to use of the HPLC procedures.

The development and availability of suitable GC columns and methodology for routine determination of C_{24} stereochemistry would provide a more practicable alternative, due to the greater resolution generally obtainable for sterol analysis by GC compared to HPLC.

6.4 DISCUSSION

6.4.1 Methods

A few general methodology issues need to be addressed, particularly relating to standardization of culture conditions. The major problem with the use of sterol analysis as a taxonomic tool is that there are many factors which may contribute both qualitatively and quantitatively to the sterol composition determined for an organism. These include light regime, incubation temperature and time of harvest, as well as other parameters which have not yet been examined. Within any one laboratory this should not be a major problem, but it makes comparison with data from other laboratories difficult. Indeed, many of the earlier studies on sterol composition of algae and fungi were conducted under poorly defined or, sometimes, unspecified culture conditions and without the aid of mass spectral analysis. Comparison of present work with that reported in the earlier literature must be undertaken with this caveat in mind.

There is a need, especially in taxonomic studies, for a standardization of culture conditions and harvest times between laboratories. We are not attempting to set down what these conditions should be; however, it is hoped that recommendations can be formulated in the future through the agency of one or more of the relevant international bodies. These conditions might need to be tailored for the different chemotypic groups of algae, bacteria and fungi. A better understanding of the effects of a wide range of environmental parameters is also required before reasonable recommendations can be made. The anticipated guidelines should enable more meaningful and accurate interpretation of sterol/hopanoid profiles in geochemical studies. Standardization of extraction, derivatization and analytical procedures would be of similar benefit.

An area that requires further methodological development is the analysis of C_{24} epimers of sterols. A number of techniques have been used to date (see Section 6.3.3.f) but none has been used routinely for the analysis of small quantities of sterol mixtures. Should appropriate GC methods become available, the utility of sterols in taxonomic and other studies will be further enhanced. Chiral phase GC capillary columns are presently available but, to our knowledge, are not routinely used for sterol analysis.

Provision of a standard reference sterol and hopanoid material, as can be obtained for trace metal analysis, would facilitate better interpretation, quantification and production of reproducible triterpenoid profiles. It is worthwhile to stress again the importance of checking the percentage recovery of sterols and hopanoids from microbial samples. There are many extraction protocols available and it is always prudent to confirm the extraction efficiency of a method by comparison with other techniques.

6.4.2 Taxonomic considerations

Although sterols and hopanoids are not, at present, extensively used as taxonomic markers, they have the potential to assume a significant role in microbial systematics. Certainly, if one compares the distribution of fatty acids (a widely used group of taxonomic markers) with that of sterols and hopanoids it is evident that the latter are more discontinuously distributed between taxa. The main barrier to extensive use of these compounds historically has been an instrumental one—sterol and hopanoid determination generally requires GC-MS analysis whereas fatty acid analysis can usually be achieved using simpler GC systems. This obstacle has been greatly reduced in recent years by the advent of high-resolution capillary columns and the introduction of comparatively inexpensive bench-top mass spectrometers.

The application of multivariate statistical methods should also extend the utility of triterpenoids as taxonomic markers, as has been the case for fatty acids. For example, in recent times analyses of the fatty acids derived from the whole cells of more than 8000 strains of bacteria have been collected and used in a fully automated bacterial identification system that is commercially available (Miller & Berger, 1985). One of the major advantages of multivariate statistical methods is that minor components have equal weighting in the overall analysis procedure. To date, many researchers (ourselves included) have considered only the major sterols and hopanoids of an organism, yet it is quite possible that extensive taxonomic information can be obtained from the abundances and variations in the minor components. This is most likely to be true at the genus or species level where variations in major components are not always significant.

A detailed description of sterol and hopanoid distribution in bacteria, fungi and algae will not be attempted here as excellent reviews by Volkman (1986), Goodwin (1981), Goad (1978) and Weete (1976) for sterols and Rohmer et al. (1984) for hopanoids already cover this subject.

There are probably insufficient data available on bacterial hopanoids at present to discuss the taxonomic implications of their distribution. At best it can be stated that they have been found in all purple nonsulphur bacteria and almost all cyanobacteria and obligate methylotrophs examined, and appear to be absent from archaea and purple sulphur bacteria. Some mention, however, can be made of algal sterol taxonomy as it demonstrates the possible application of triterpenoids in the study of microbial systematics.

The major algal groups generally contain one or two principal sterols (Table 6.4), although there are exceptions to this characterization. In some cases it is possible to assign an alga to a taxonomic group based on the presence of only one sterol. In particular, dinoflagellates (with a few

Table 6.4 Major sterols of the algae and cyanobacteria.*

Order/Group	Major sterols
Chlorophyceae	$C_{28}\Delta^5$, $C_{29}\Delta^5$, $\Delta^{5,22}$
Prymnesiophyceae	$C_{27}\Delta^5$, $C_{28}\Delta^{5,22}$, $C_{29}\Delta^{5,22}$
Chrysophyceae	$C_{29}\Delta^{5,22}$
Cryptophyceae	$C_{28}\Delta^{5,22}$
Bacillariophyceae	$C_{27}\Delta^5$, $C_{28}\Delta^5$, $\Delta^{5,22}$, $C_{29}\Delta^{24(28)}$
Phaeophyceae	$C_{29}\Delta^{5,24(28)}$
Rhodophyceae	$C_{27}\Delta^5$, $\Delta^{5,22}$
Dinophyceae	$C_{27}\Delta^5$, $C_{30}\Delta^{22}$
Euglenophyceae	$C_{28}\Delta^{5,7,22}$, $C_{29}\Delta^5$
Prasinophyceae	$C_{28}\Delta^5$, $\Delta^{5,24(28)}$, $C_{29}\Delta^{5,24(28)}$
Raphidophyceae	$C_{29}\Delta^5$
Cyanobacteria	$C_{27}\Delta^5$, $C_{29}\Delta^5$

* This table should be interpreted with caution as only a few representatives of some of these orders have been analysed. The nomenclature used represents the number of carbon atoms and the position of double bonds in the molecule.

notable exceptions) contain the unusual 4-methylsterol dinosterol (4,23,24-trimethyl-5α-cholest-22E-en-3β-ol) which has only been found in this algal group. Dinosterol has been used in chemotaxonomic (Jones et al., 1983) and geochemical studies (e.g. Mackenzie et al., 1982) as a marker for dinoflagellates or organic matter of dinoflagellate origin. Sterols have also been used to distinguish algae at the genus level, a good example of which is the study of the genus *Chlorella* by Patterson (1971b). Patterson found that the genus could be divided into three subgroups based on the presence of either Δ^5, Δ^7 or $\Delta^{5,7}$ sterols as the major sterol components.

Triterpenoid phylogenetics has been extensively described by Nes and Nes (1980) in their monograph on lipid evolution. These authors rightly highlight the importance of biosynthetic pathways in systematic and phylogenetic studies. Most microbial sterols, for example, can be synthesized by more that one route in different, and occasionally in the same, organisms. Thus, important taxonomic information may sometimes be lost by examination of only the sterol end-products. Of course, the study of biosynthetic pathways is often complex and most taxonomists are precluded by time or funding constraints from such detailed examination. In the study of difficult taxa, however, such an option may be worth consideration.

At present, the main use of sterols and hopanoids as markers is not in taxonomic analysis but in geochemical studies. As sterols and hopanoids (or their easily recognizable diagenetic products) are very stable molecules, they can provide a great deal of information about the biological source of

6. Analysis of Sterols and Hopanoids

organic matter in both recent and ancient sediments. Furthermore, the state of oil maturation can be determined from the extent of chemical transformation of component sterols and hopanoids (see Mackenzie et al. (1982) for an excellent review of this subject). Ultimately, geochemical studies are limited by the availability of taxonomic information concerning the sterol and hopanoid composition of microorganisms. Expansion of the use of sterols and hopanoids as marker compounds will not only be of aid to microbial systematics but will also provide an important database for organic geochemical studies, many of which are of major economic significance.

REFERENCES

Bernstein, S. and Wallis, E.S. (1938) The structure of β-sitosterol, and its preparation from stigmasterol. *Org. Chem.* **2**: 341–5.
Bligh, E.G. and Dyer, W.J. (1959) A rapid method of total lipid extraction and purification. *Can. J. Biochem. Physiol.* **37**: 911–7.
Bohlin, L., Sjostrand, U.S., Djerassi, C. and Sullivan, B.W. (1981) Minor and trace sterols in marine invertebrates. Part 20. 3β-Hydroxymethyl-A-nor-patinosterol and 3β-Hydroxymethyl-A-nor-dinosterol. Two new sterols with modified nucleus and side-chain from the sponge *Teichaxinella morchella*. *J. Chem. Soc. Perkin Trans.* **1**: 1023–8.
Boon, J.J., Rijpstra, W.I.C., De Lange, F. and De Leeuw, J.W. (1979) Black Sea sterol—a molecular fossil for dinoflagellate blooms. *Nature* **277**: 125–7.
Breivik, O.N. and Owades, J.L. (1957) Spectrophotometric semimicro-determination of ergosterol in yeast. *Agric. Food Chem.* **5**: 360–3.
Brooks, C.J.W., Horning, E.C. and Young, J.S. (1968) Characterization of sterols by gas-chromatography-mass spectrometry of the trimethylsilyl ethers. *Lipids* **3**: 391–402.
Calderbank, J., Keenan, M.J.H., Rose, A.H. and Holman, G.D. (1984) Accumulation of amino acids by *Saccharomyces cerevisiae* Y185 with phospholipids enriched in different fatty-acyl residues: a statistical analysis of data. *J. Gen. Microbiol.* **130**: 2817–24.
Chiu, P.L. and Patterson, G.W. (1981) Quantitative estimation of C-24 epimeric sterol mixtures by 220 MHz nuclear magnetic resonance spectroscopy. *Lipids* **16**: 203–6.
Christie, W.W. (1987) *HPLC and Lipids. A Practical Guide*. Pergamon Press: Oxford.
Clayton, R.B. (1962) Gas-liquid chromatography of sterol methyl ethers and some correlations between molecular structures and retention data. *Biochem.* **1**: 357–366.
Colin, H., Guichon, G. and Siouffi, A. (1979) Comparison of various systems for the separation of free sterols by high performance liquid chromatography. *Anal. Chem.* **51**: 1662–6.
Copius-Peereboom, J.W. (1965) Gas chromatography of phytosterol mixtures. Part I. Identification of phytosterols. *J. Gas Chromatogr.* **3**: 325.
Courchaine, A.J., Miller, W.H. and Stein, D.B. (1959) Rapid semi-micro procedure for estimating free and total cholesterol. *Clin. Chem.* **5**: 609–14.
De Leeuw, J.W. , Rijpstra, W.I.C., Schenck, P.A. and Volkman, J.K. (1983) Free, esterified and residual bound sterols in Black Sea Unit I sediments. *Geochim. Cosmochim. Acta* **47**: 455–65.
Edlund, A., Nichols, P.D., Roffey, R. and White, D.C. (1985) Extractable and lipopolysaccharide fatty acid and hydroxy acid profiles from *Desulfovibrio* species. *J. Lipid Res.* **26**: 982–8.
Gagosian, R.B., Nigrelli, G.E. and Volkman, J.K. (1983a) Vertical transport and transformation of biogenic organic compounds from a sediment trap experiment off the coast of Peru. In *Coastal Upwelling, Its Sediment Record* (Suess, E. and Thiede J., eds), Part A, pp. 241–72.
Gagosian, R.B., Volkman, J.K. and Nigrelli, G.E. (1983b) The use of sediment traps to determine sterol sources in coastal sediments off Peru. In *Advances in Organic Geochemistry 1981* (Bjoroy, M. et al., eds), pp. 369–79.

Gershengorn, M.C., Smith, A.R.H., Goulston, G., Goad, L.J., Goodwin, T.W. and Haines, T.H. (1968) The sterols of *Ochromonas danica* and *Ochromonas malhamensis*. *Biochem.* **7**: 1698–1706.

Goad, L.J. (1978) The sterols of marine invertebrates: composition, biosynthesis and metabolites. In *Marine Natural Products* (Scheuer, P.J., ed), pp. 75–172. Academic Press: Sydney.

Gonzalez, R.A. and Parkes, L.W. (1977) Acid-labilization of sterols for extraction from yeast. *Biochim. Biophys. Acta* **489**: 507–9.

Goodwin, T.W. (1974) Sterols. In *Algal Physiology and Biochemistry* (Stewart, W.D.P., ed) pp. 266–80. Blackwell Scientific Publications: London.

Goodwin, T.W. (1981) Biosynthesis of sterols. In *Biochemistry of Plants*, vol. 4 (Stumpf, P.K., ed), pp. 485–507. Academic Press: New York.

Guckert, J.B., Antworth, C.P., Nichols, P.D. and White, D.C. (1985) Phospholipid, ester-linked fatty acid profiles as reproducible assays for changes in prokaryotic community structure of estuarine sediments. *FEMS Microbiol. Ecol.* **31**: 147–58.

Hatcher, P.G. and McGillivary, P.A. (1979) Sewage contamination in the New York Bight. Coprostanol as an indicator. *Environ. Sci. Technol.* **13**: 1225–9.

Heftman, E. (1983) Biogenesis of steroids in Solanaceae. *Phytochem.* **22**: 1843–60.

Ikekawa, N., Fujimoto, Y., Kadota, S. and Kikuchi, T. (1989) Effective separation of sterol C-24 epimers. *J. Chromatogr.* **468**: 91–8.

Itoh, T., Tani, H., Fukushima, K., Tamura, T. and Matsumoto, T. (1982). Structure-retention relationship of sterols and triterpene alcohols in gas chromatography on a glass capillary column. *J. Chromatogr.* **234**: 65–76.

Jahnke, L.L. and Nichols, P.D. (1986) Methyl sterol and cyclopropane fatty acid composition of *Methylcoccus capsulatus* grown at low oxygen tensions. *J. Bacteriol.* **167**: 238–42.

Jones, G.J., Nichols, P.D. and Johns, R.B. (1983) The lipid composition of *Thoracosphaera heimii*: evidence for inclusion in the Dinophyceae. *J. Phycol.* **19**: 416–20.

Jones, G.J., Nichols, P.D., Johns, R.B. and Smith, J.D. (1987) The effect of mercury and cadmium on the fatty acid and sterol composition of the marine diatom *Asterionella glacialis*. *Phytochem.* **26**: 1343–8.

Kates, M. (1975) *Vol. 3. Laboratory Techniques in Biochemistry and Molecular Biology. II. Techniques of Lipidology: Isolation, Analysis and Identification of Lipids*. North Holland Publishing Company: Amsterdam.

Knight, B.A. (1967) Identification of plant sterols using combined GLC/mass spectrometry. *J. Gas Chromatogr.* **5**: 273–82.

Langworthy, T.A., Mayberry, W.R. and Smith, P.F. (1976) A sulfonolipid and novel glycosimidylglycolipids from the extreme thermoacidophile *Bacillus acidocaldarius*. *Biochim. Biophys. Acta* **431**: 550–69.

Lenfant, M., Lecompte, M.F. and Farrugia, G. (1970) Identification des sterols de *Physarum polycephalum*. *Phytochem.* **9**: 2529–35.

Ling, N.C., Hale, R.L. and Djerassi, C. (1970) The structure and absolute configuration of the marine sterol gorgosterol. *J. Am. Chem. Soc.* **92**: 5281–2.

Mackenzie, A.S., Brassell, S.C., Eglinton, G. and Maxwell, J.R. (1982) Chemical fossils: The geological fate of steroids. *Science* **217**: 491–504.

Maxwell, J.R., Mackenzie, A.S. and Volkman, J.K. (1980) Configuration at C-24 in steranes and sterols. *Nature* **286**: 694–7.

Miller, L. and Berger, T. (1985) Bacteria identification by gas chromatography of whole cell fatty acids. *Hewlett-Packard Application Note* 228–41.

Nes, W.R. and Nes, W.D. (1980) *Lipids In Evolution*. Plenum Press: New York.

Neunlist, S. and Rohmer, M. (1985a). Novel hopanoids from the methylotrophic bacteria *Methylococcus capsulatus* and *Methylomonas methanica*: (22S)-35-aminobacteriohopane-30,31,32,33,34-pentol and (22S)-35-amino-3β-methylbacteriohopane-30,31,32,33,34-pentol. *Biochem. J.* **231**: 635–9.

Neunlist, S. and Rohmer, M. (1985b) A novel hopanoid, 30-(5'-adenosyl)hopane, from the purple non-sulphur bacterium *Rhodopseudomonas acidophila*, with possible DNA interactions. *Biochem. J.* **228**: 769–71.

Neunlist, S. and Rohmer, M. (1985c) The hopanoids of '*Methylosinus trichosporium*': Aminobacteriohopanetriol and aminobacteriohopanetetrol. *J. Gen. Microbiol.* **131**: 1363–7.

Neunlist, S., Holst, O. and Rohmer, M. (1985) The hopanoids of the purple non-sulphur bacterium *Rhodomicrobium vannielli*: an aminotriol and its aminoacyl derivatives, N-tryptophanyl and N-ornithinyl aminotriol. *Eur. J. Biochem.* **147**: 561–8.

Neunlist, S., Bissert. P. and Rohmer, M. (1988) The hopanoids of the purple non-sulphur bacteria *Rhodopseudomonas palustris* and *Rhodopseudomonas acidophila* and the absolute configuration of bacteriohopanetetrol. *Eur. J. Biochem.* **171**: 245–52.

Newell, S.Y., Arsuffi, T.L. and Fallen, R.D. (1988) Fundamental procedures for determining ergosterol content of decaying plant material by liquid chromatography. *Appl. Environ. Microbiol.* **54**: 1876–9.

Nichols, P.D. (1983) *Biological Markers in the Marine Environment*. PhD Thesis, University of Melbourne, 349 pp. University Microfilms International: Ann Arbor.

Nichols, P.D., Volkman, J.K. and Johns, R.B. (1983) Sterols and fatty acids of the marine unicellular alga, FCRG51. *Phytochem.* **22**: 1447–52.

Nichols, P.D., Klumpp, D.W. and Johns, R.B. (1985) Lipid components of the epiphyte material, suspended particulate matter and cultured bacteria from a seagrass, *Posidonia australis*, community as indicators of carbon source. *Comp. Biochem. Physiol.* **80B**: 315–25.

Nichols, P.D., Volkman, J.K., Hallegraeff, G.M. and Blackburn, S.I. (1987) Sterols and fatty acids of the red tide flagellates *Heterosigma akashiwo* and *Chattonella antiqua* (Raphidophyceae). *Phytochem.* **26**: 2537–41.

Nichols, P.D., Volkman, J.K., Palmisano, A.C., Smith G.A. and White, D.C. (1988) Occurrence of an isoprenoid C25 diunsaturated alkene and high neutral lipid content in Antarctic sea-ice diatom communities. *J. Phycol.* **24**: 90–6.

Parks, L.W. and Stromberg, V.K. (1978) Measurement *in vitro* of the esterification of yeast sterols. *Lipids* **13**: 29–33.

Patterson, G.W. (1971a) Relationship between structure and retention time of sterols in gas chromatography. *Anal. Chem.* **43**: 1165–70.

Patterson, G.W. (1971b) The distribution of sterols in algae. *Lipids* **6**: 120–7.

Rohmer, M., Bouvier, P. and Ourisson, G. (1979) Molecular evolution of biomembranes: Structural equivalents and phylogenetic precursors of sterols. *Proc. Natl. Acad. Sci. USA* **76**: 847–51.

Rohmer, M., Anding, C. and Ourisson, G. (1980) Non-specific biosynthesis of triterpenes by a cell-free system from *Acetobacter pasteurianum*. *Eur. J. Biochem.* **112**: 541–7.

Rohmer, M., Bouvier-Nave, P. and Ourisson, G. (1984) Distribution of hopanoid triterpenes in prokaryotes. *J. Gen. Microbiol.* **130**: 1137–50.

Thompson, R.H., Patterson, G.W., Dutky, S.R., Svoboda, J.A. and Kaplanis, J.N. (1980) Techniques for the isolation and identification of steroids in insects and algae. *Lipids* **15**: 719–33.

Thompson, R.H., Patterson, G.W., Thompson, M.J. and Slover, H.T. (1981) Separation of pairs of C-24 epimeric sterols by glass capillary gas liquid chromatography. *Lipids* **16**: 694–9.

Vivian, C.M.G. (1986) Tracers of sewage sludge in the marine environment: a review. *Sci. Tot. Environ.* **53**: 5–40.

Volkman, J.K. (1986) A review of sterol markers for marine and terrigenous organic matter. *Org. Geochem.* **9**: 83–99.

Weete, J.D. (1976) Algal and fungal waxes. In *Chemistry and Biochemistry of Natural Waxes* (Kolattukudy, P.E. ed.), pp. 349–418. Elsevier: Amsterdam.

Wells, W.W. and Makita, M. (1962) The quantitative analysis of fecal neutral sterols by gas-liquid chromatography. *Anal. Biochem.* **4**: 204–12.

White, D.C., Davis, W.M., Nickels, J.S., King, J.D. and Bobbie, R.J. (1979) Determination of the sedimentary microbial biomass by extractable lipid phosphate. *Oecologia* **40**: 51–62.

Zill, G., Englehardt, G. and Wallnofer, P.R. (1988) Determination of ergosterol as a measure of fungal growth using Si 60 HPLC. *Z. Lebansm. Unters. Forsch.* **187**: 246–9.

Zlatkis, A., Zak, B. and Boyle, A.J. (1963) A new method for the direct determination of serum cholesterol. *J. Lab. Clin. Med.* **41**: 486–92.

7
Archaeal Lipids

Mario De Rosa[1] and Agata Gambacorta[2]

[1]*Istituto di Biochimica delle Macromolecole, Napoli, Italia*
[2]*Istituto per la Chimica di Molecole di Interesse Biologico, Arco Felice, Napoli, Italia*

7.1 INTRODUCTION

Living organisms are assumed to stem from the early divergence of a putative progenitor into three primary domains: archaea, bacteria and eukarya. The archaea have only been recognized as a separate group in recent years, although various species belonging to this domain have been well studied for some time and described as eubacteria. The archaea form a phylogenetically coherent group of microorganisms that differ from bacteria and eukarya in several important genetic and molecular respects (Woese, 1987).

The archaea encompass a collection of disparate phenotypes which thrive in environments that would normally kill most known organisms. In fact, they are confined to a few ecological niches, notably habitats that are anaerobic, have a very high temperature and low pH or a very high salt concentration. The initial archaeal phylogeny was structured essentially along three phenotypic lines: halophiles, which require habitats saturated in sodium chloride; methanogens, which live in niches with very low redox potential; and thermophiles, confined to habitats with temperatures between 50 °C and 110 °C. Although archaeal taxonomy is complex, all recent new isolates essentially conform to one of these original phenotypic types (Woese, 1987; Juez, 1988; König, 1988). The plasma membrane plays a key role in the ability of archaea to survive in the face of extreme environmental stress. Consequently, extensive studies have been undertaken to characterize the structural identity and biogenetic origin of their membrane lipids.

A comparison of the molecular components of archaea with those of bacteria and eukarya highlights deep chemical differences in lipid composition, especially in the context of other components whose essential chemical features appear to have been preserved. In fact, the structure of archaeal membrane lipids remains one of the major distinctions between these and all other organisms.

All the hitherto identified membrane lipids of archaea are characterized by unusual structural features which can be considered as specific and useful taxonomic markers. Unlike bacterial and eukaryotic lipids, which are based on ester linkages formed by condensation of alcohols and fatty acids, archaeal lipids are mainly based on isopranyl glycerol ethers. These molecules are formed by condensation of glycerol (or more complex polyols) with isopranoid alcohols consisting of 20, 25 or 40 carbon atoms. It is also worth noting that the C2 configuration of glycerol in archaeal lipids is opposite to the conventional diglyceride configuration. In fact, glycerol ethers of archaea contain a 2,3-di-O-sn-glycerol moiety, while naturally occurring glycerophosphatides or diacyl glycerols are known to have an sn-1,2-glycerol stereochemistry (Jones et al., 1987; De Rosa & Gambacorta, 1988; Kamekura & Kates, 1988).

Although isopranoids with specific functions occur in the lipid membrane of most cells, the ether lipids considered here are the only compounds that provide major structural components of the membrane in which they occur (De Rosa & Gambacorta, 1988). The uniqueness of archaeal lipids also suggests important differences in their topology within the membrane structure. It is interesting to speculate how lipids with such different chemical composition can perform similar functions and why chemical variability is required for sustaining life (De Rosa et al., 1983c; Woese & Wolfe, 1985).

This chapter surveys the methodologies used to purify and characterize isopranoid ether lipids and quinones of archaea with attention focused on taxonomic considerations.

7.2 ISOPRANOID ETHER LIPIDS

7.2.1 Extraction of archaeal lipids

Total lipids are generally extracted from washed cells of archaea using modifications of the method of Bligh and Dyer (1959).

Method 7.1 Lipid extraction: modified Bligh–Dyer technique (after Collins et al., 1980).

1. Freeze-dried biomass (1 g) is stirred for a 1 min in 25 ml of aqueous methanol-NaCl (10 ml; 0.3% v/w, aqueous NaCl in 100 ml of methanol) then heated to 100 °C for 5 min.

2. After cooling, 8 ml of aqueous NaCl (0.3%, w/v) and 12 ml of chloroform are added to the suspension.
3. The mixture is stirred for 2–3 h at room temperature and cell debris removed by centrifugation prior to extraction with 20 ml of chloroform-methanol-aqueous NaCl (5:10:4, v/v) with stirring for 15 min.
4. The mixture is then centrifuged as before and the first and second supernatants pooled.
5. The pooled supernatant is treated with 7.5 ml aqueous NaCl and 18 ml of chloroform and, after shaking, the two layers are separated by centrifugation.
6. The lower lipid-containing layer is collected and evaporated to dryness in a stream of oxygen-free nitrogen.

Increases in the amount of lipid phosphorus extracted can be achieved by lysing cells at 10,000 psi prior to the extraction procedure.

Method 7.2 Lipid extraction: modified Bligh–Dyer technique for methanogenic archaea (from Mancuso *et al.*, 1985).

1. Lyophilized biomass (1 g) is sonicated, in test-tubes fitted with Teflon-lined screw-caps, with 200 ml of methanol for 30 min.
2. 100 ml of chloroform and 80 ml of 50 mM phosphate buffer (pH 7.4) are then added and the sample extracted at room temperature for 18 h.
3. The preparation is then centrifuged and the supernatant transferred to a 1 l separation funnel when sufficient chloroform and buffer are added to give a final ratio for the chloroform-methanol-buffer of 2:2:1.8 by volume.
4. After separation of the two phases, the lower chloroform layer is collected through a Whatman 2V filter into a screw-cap test-tube and dried under a dry stream of nitrogen below 40 °C.
5. The methanol-water phase is evaporated to dryness and the residue refluxed with hot chloroform-methanol (1:1, v/v) and filtered in order to test whether the methanol-water phase contains additional lipids that are difficult to extract (Kramer *et al.*, 1987).
6. The filtrate is then treated with 0.9 parts of water (chloroform-methanol-water, 1:1:0.9, v/v), and the chloroform phase examined for lipids by thin-layer chromatography (TLC).

A further modification of the Bligh–Dyer method has been reported for the extraction of lipids from methanogenic archaea (Nishihara & Koga, 1987). These authors demonstrated that the replacement of water or buffer in the extraction solvent by 5% trichloroacetic acid resulted in a sixfold increase in the amount of lipid extracted. The use of HCl (2 M) as an alternative to trichloroacetic acid is also effective, but prolonged extraction with HCl-containing solvent causes the degradation of some phosphoglycolipids.

Method 7.3 Lipid extraction: modified Bligh–Dyer technique for methanogenic archaea (from Nishihara & Koga, 1987).

1. Wet biomass (1 g) is suspended in 30 ml of water followed by successive additions of 150 ml of methanol, 75 ml of chloroform, and 30 ml of 10% aqueous trichloroacetic acid solution.
2. The mixture is stirred at room temperature for 2 h before the addition of chloroform (75 ml) and water (75 ml).
3. After separation into two phases by low-speed centrifugation, the lower chloroform phase (1 vol) is washed with 1.9 vols of methanol-water (1:0.8, v/v), to remove trichloroacetic acid. Neither the phosphate nor the sugar content in the chloroform layer decrease during washing.

The extraction of lipids from halophilic archaea is carried out using the method of Bligh and Dyer, as modified by Kates (1986). In this procedure biomass is suspended in 25% NaCl (w/w) to bring the water content of the suspension to 80% (w/w), and the lipids extracted by mixing one volume of the suspension with two volumes of methanol and one volume of chloroform. The mixture is then either centrifuged or filtered, prior to being diluted with one volume each of chloroform and water to make two phases. The lower chloroform phase, which contains the total lipids and pigments, is recovered and brought to dryness under vacuum.

As a further alternative to the Bligh and Dyer method, some workers (De Rosa et al., 1976a) have obtained lipids from lyophilized cells of archaea by continuous extraction in Soxhlet using solvents of increasing polarity such as chloroform, methanol and water.

The yield of extracted lipids from archaea strongly depends on culture age, growth conditions and extraction procedure. The yield of extracted lipids from different archaea is shown in Table 7.1. In general, yields range from 2 to 10% of dried cells.

7.2.2 Chromatographic analyses of extracted lipids

Total lipid extracts can be resolved into neutral and polar lipid fractions by precipitation with acetone under cold conditions. Acetone precipitation yields an acetone-insoluble tan-coloured precipitate of polar lipids that accounts for about 80 to 90% of the total lipids by weight and an acetone-soluble 'neutral lipid' fraction that forms about 10 to 20% of the total lipids.

Alternatively, lipid extracts can be fractionated into neutral lipids, glycolipids and acidic lipids by column chromatography on silica gel and DEAE-cellulose (Langworthy, 1977a).

Method 7.4 Fractionation of lipid extracts by column chromatography.

1. Lipids from 1–2 g lyophilized biomass are fractionated on columns (2 × 8 cm) of silicic acid (Unisil, 100 to 200 mesh). Neutral lipids are eluted with 250 ml of chloroform followed by 250 ml of methanol for elution of the combined glycolipid and acidic lipid fraction.
2. The methanol eluate is taken up in chloroform-methanol-water (60 : 30 : 4.5, v/v) and applied to a DEAE-cellulose column (2.5 × 22 cm) in the acetate form.
3. Glycolipids are eluted with 500 ml of chloroform-methanol (7 : 3, v/v) followed by chloroform-methanol-ammonia (70 : 30 : 2, v/v) containing 0.4% ammonium acetate to separate the acidic lipids.

Table 7.1 Yields of extracted lipids in different types of archaea.

Organism	Lipid content (% of cell dry weight)	References
Desulfurococcus mobilis	2.5	Lanzotti *et al.* (1987)
Desulfurolobus ambivalens[a]	8.0	Trincone *et al.* (1989)
Halobacterium cutirubrum	2.2–3.5	
H. halobium	2.6–4.1	
H. marismortui[b]	9.3	Kamekura & Kates (1988)
H. salinarium	3.6	
Haloferax mediterranei[b]	3.3–11	
Strain 54R	6.6	Lanzotti *et al.* (1989a)
Methanobacterium thermoautotrophicum	5.6	Nishihara & Koga (1987)
Methanococcus jannaschii	3.2	Comita *et al.* (1984)
M. voltae	5.1	Ferrante *et al.* (1986)
Methanospirillum hungatei GP1	5.5	Kushwaha *et al.* (1981)
Methanothrix concilii GP6	10.0	Ferrante *et al.* (1988b)
Natronobacterium pharaonis	6.8	
N. magadii	7.0	De Rosa *et al.* (1988)
N. gregorii	9.2	
Natronococcus occultus	7.2	
Strain SP-8	8.0	
Pyrococcus woesei	1.5	De Rosa *et al.* (1983a)
Pyrococcus isolate ANI	3.5	Lanzotti *et al.* (1989c)
Sulfolobus acidocaldarius	2.5	Langworthy *et al.* (1974)
S. solfataricus	8.0	De Rosa *et al.* (1983c)
Thermococcus celer	1.3	De Rosa *et al.* (1987)
Thermoplasma acidophilum	3.0	Langworthy *et al.* (1972)
Thermoproteus tenax	6.0	Thurl & Schafer (1988)

[a] The lipid content is the same irrespective of whether organisms are grown under aerobic or anaerobic conditions.
[b] The lipid content for these microorganisms is reported as a percentage of the cell protein.

A modification of the elution solvents applied to the DEAE-cellulose column was reported by Thurl and Schafer (1988). These workers eluted glycolipids with chloroform-methanol (7:3, v/v); pigments were eluted using chloroform-acetic acid (3:1, v/v) followed by a methanol wash for acetic acid removal; for less polar phospholipids they used chloroform-methanol-33% ammonium hydroxide-ammonium acetate (70:30:2:0.4, v/v/v/w), and for more polar phospholipids they used chloroform-methanol-water-potassium hydrogenphthalate (60:30:4.5:0.18, v/v/v/w) adjusted to pH 2.5 with concentrated HCl. The eluates were dried and partitioned between the upper and lower phases from chloroform-methanol-water (2:1:0.6, v/v) to remove salts from the phospholipids.

The final purification of fractions is often achieved by repeated preparative TLC on mono- or bi-dimensional silica gel eluted with different solvent systems as described later. The lipid bands are briefly exposed to iodine vapour, marked on the plate, and the lipids eluted from the silica gel in a glass-sintered Buchner funnel with chloroform-methanol (1:1 v/v) or single-phase Bligh–Dyer followed by a two-phase system (Ferrante et al., 1986). Polar lipid patterns can usually be obtained on both mono- and bi-dimensional thin-layer silica gel plates eluted with different solvent systems. The plates can be washed before use by development in chloroform-methanol (2:1, v/v) and reactivated at 110 °C for 1 h.

For rapid analysis of polar lipids, a simplified method is TLC on silica gel eluted in one dimension using chloroform-methanol-water (65:25:4, v/v) as the solvent system (De Rosa et al., 1988) or chloroform methanol-acetic acid-water (85:22.5:10:4, v/v) with single or double development (Ferrante et al., 1986; Torreblanca et al., 1986).

Collins et al. (1980) separated polar lipids using a bi-dimensional TLC silica gel procedure. Dried preparations of polar lipids were dissolved in small amounts of (chloroform-methanol (2:1, v/v) and the preparation spotted onto a corner of a thin-layer silical-gel plate which was developed in two dimensions; first using chloroform methanol-water (65:25:4, v/v) and secondly chloroform-methanol-glacial acetic acid-water (80:12:15:4, v/v). The plates were dried thoroughly after each development and the lipids detected using differential stains.

Kramer et al. (1987) separated polar lipids by two-dimensional TLC with chloroform-methanol-concentrated ammonia (6:3:1, v/v) followed by chloroform-acetone-methanol-acetic acid-water (10:4:2:1:1, v/v). Nishihara et al. (1989) used the solvent system chloroform-methanol-7 M aqueous ammonia (60:35:8, v/v) in a vertical direction and chloroform-methanol-acetic acid-water (85:30:15:5, v/v) in a horizontal direction. Kates and Deroo (1973) carried out analyses on TLC silica gel plates developed in jars lined with Whatman 3MM paper in the solvent system chloroform-90% acetic acid-methanol (30:30:4, v/v). Polar lipids can also be chromatographed on silicic acid-impregnated Whatman 3MM paper using an ascending solvent system, namely diisobutylketone-acetic acid-water (40:25:5, v/v; Kates & Deroo, 1973).

Total lipids are detected by charring at 150 °C for 10–15 min using plates sprayed with 50% (v/v) methanolic-sulphuric acid, 1% (w/v) cerium sulphate-1 M sulphuric acid or a 10% (w/v) solution of dodecano-molybdophosphoric acid in absolute ethanol; with these procedures all lipids appear as black spots on a clear background. Total lipids can also be detected with iodine vapour or by spraying plates to transparency with water.

Phospholipids are detected with Zinzadze reagent (Dittmer & Lester, 1964), where phosphate-containing lipids appear as light-blue spots on a white background within 5 min at room temperature. Similarly, amino-lipids are detected by spraying plates with a 0.2% (w/v) solution of nin-hydrin in butanol saturated with water followed by heating at 105 °C for 10 min, when lipids containing free amino groups appear as purple-mauve spots on a white background.

Glycolipids can be detected by spraying plates with l-naphthol reagent (Jacin & Mishkin, 1965) followed by heating at 120 °C for 10 min; they appear as purple spots on an orange-brown background. Lipids containing vicinal glycols are detected by spraying plates with a 1% (w/v) aqueous solution of sodium periodate until saturated. The sprayed plates are left at room temperature for 5 min, when the oxidation reaction is stopped by exposure to sulphur dioxide gas until the plates are decolorized. The plates are then sprayed lightly with aqueous Schiff reagent (Baddiley et al., 1956) and finally exposed again to sulphur dioxide. With this procedure, glycols appear immediately as purple-blue spots whereas glycolipids require at least 12 h at room temperature (Shaw, 1968). The paper chromatograms are developed by staining with rhodamine 6G and viewing under ultraviolet light (388 nm; Kates & Deroo, 1973).

7.2.3 Overall composition of archaeal lipids

The most striking feature of archaeal lipids is that mild alkaline hydrolysis (Dawson, 1954) releases neither fatty acids nor water-soluble phosphate esters. Even after more drastic hydrolysis, such as heating in boiling 0.6 N methanolic HCl for 4 h then in boiling 0.1 N NaOH for 1 h, fatty acids are absent from hydrolysis mixtures. The absence of fatty esters is confirmed by lack of ester absorption (1730–1750 cm^{-1}) in the infrared (IR) spectrum of the total lipids; the latter show strong absorption bands indicative of long-chain (2920, 2850 and 1460 cm^{-1}), OH (3300 cm^{-1}, broad signal), phosphate ester (P=O, 1230–1260 cm^{-1}), P—O—C, (1060–1090 cm^{-1}); P—O (1100 cm^{-1}) and isopropyl groups (doublet, 1365–1385 cm^{-1}).

^{1}H and ^{13}C-NMR spectra of total lipid extracts from archaea show signals of CH_2-O and $CH-O$ ($-C\underline{H}_2-O$ and $=C\underline{H}-O$ in the region 3.0–4.5 ppm; $-\underline{C}H_2-O$ and $>\underline{C}H-O$ in the region 62–78 ppm) and signals of long isopranoid chains ($-C\underline{H}_2$ centred at 1.25 ppm; $>C\underline{H}-$ centred at 1.6 ppm; $-C\underline{H}_3$ in the region 0.8–1.0 ppm; $\underline{C}H_2-$ and $\underline{C}H-$ in the region 23–45 ppm, and $-\underline{C}H_3$ in the region 17–22 ppm).

Therefore, a wealth of evidence exists which indicates that archaeal lipids lack fatty acids and are based on isopranoid alcohols, ether linked to polyols.

7.2.4 Degradative procedures for archaeal lipids

Acid methanolysis of archaeal lipids is carried out in anhydrous 1 N methanolic hydrochloride at 100 °C for 3 h in screw-cap tubes to obtain isopranoid ether core lipids. After cooling, equal volumes of chloroform and water are added, the contents mixed well, and the phases separated by centrifugation. Chloroform- and methanol-water-soluble products are analysed after drying either under a stream of nitrogen or under vacuum. The isopranoid core lipids, diethers or tetraethers, are recovered in the chloroform phase while sugars, polyols and phosphate esters are present in the methanol-water phase. Total hydrolysis of phosphate ester linkages, such as inositol phosphate esters, is obtained by treatment of archaeal lipids with 6 N HCl at 100 °C for 72 h.

Partial degradation of complex lipids by mild acid methanolysis is performed using 0.05 N methanolic hydrochloride at 60 °C for 15 h. Strong alkaline hydrolysis for the selective cleavage of phosphate ester linkages is done in 1 N NaOH at 100 °C for 3 h followed by partition with equal volumes of chloroform and methanol. The methanol-water phase is neutralised with Dowex 50 (H^+) resin before examination of the hydrolysis products. Mild alkaline hydrolysis of complex archaeal lipids is carried out at 17 °C for 15 min using the method of Dittmer and Wells (1969), or with sodium methoxide at room temperature for 1 h (Rouser et al., 1970).

Degradation of archaeal isopranoid ether lipids is performed with either hydroiodic acid (HI) or boron trichloride. The hydrolysis of isopranoid ether core lipids with HI at 110 °C for 5 h releases alkyl iodides, which are extracted from the reaction mixture with n-hexane. The n-hexane fraction is then washed with 10% NaCl, saturated Na_2CO_3 and 50% $Na_2S_2O_3$ to remove contaminants. The alkyl iodides are converted to the corresponding hydrocarbons by reduction with $LiAlH_4$ (Kates, 1972; De Rosa et al., 1976a, 1983b). Isopranoid hydrocarbons with deuterium on the terminal carbons, useful in assigning certain ions in the mass spectra, are prepared using $LiAl^2H_4$ (De Rosa et al., 1977a).

An alternative method for the conversion of alkyl iodides into hydro-

7. Archaeal Lipids

carbons uses zinc in acetic acid under reflux (Panganamala *et al.*, 1971). The alkyl iodides are converted to the corresponding acetates by refluxing with silver acetate in acetic acid for 24 h (Kates *et al.*, 1965). In turn, the acetates are converted to isopranoid alcohols by hydrolysis in 0.2 N methanolic sodium hydroxide at 100 °C for 2 h (De Rosa *et al.*, 1977a).

Degradation of isopranoid ether core lipids with liquefied boron trichloride in chloroform (1:1, v/v) releases alkyl chlorides and polyols (De Rosa

Table 7.2 Procedures used for structural characterization of archaeal lipids.

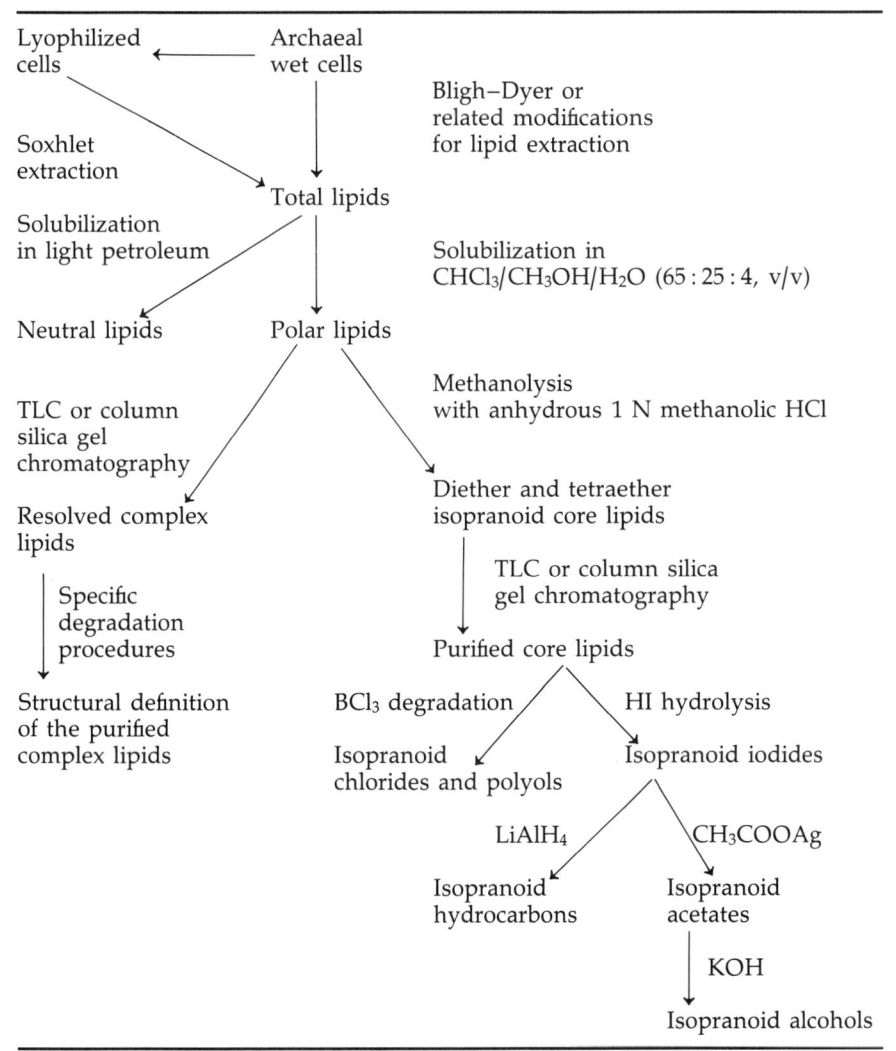

```
Lyophilized     ←——     Archaeal
cells                   wet cells
                                        Bligh–Dyer or
                                        related modifications
Soxhlet                                 for lipid extraction
extraction
                        ↓
                        Total lipids
Solubilization
in light petroleum                      Solubilization in
                                        CHCl₃/CH₃OH/H₂O (65:25:4, v/v)
                        ↓
Neutral lipids          Polar lipids

                                        Methanolysis
TLC or column                           with anhydrous 1 N methanolic HCl
silica gel
chromatography
                                        Diether and tetraether
Resolved complex                        isopranoid core lipids
lipids
                                            TLC or column silica
    Specific                                gel chromatography
    degradation
    procedures         Purified core lipids

Structural definition   BCl₃ degradation       HI hydrolysis
of the purified
complex lipids          Isopranoid             Isopranoid iodides
                        chlorides and polyols

                            LiAlH₄              CH₃COOAg

                        Isopranoid             Isopranoid
                        hydrocarbons           acetates

                                                    KOH

                                                Isopranoid alcohols
```

et al., 1977a). In a standard procedure, 10 ml of boron trichloride is added to samples in 10 ml chloroform and liquidized at $-70\,^{\circ}\mathrm{C}$, the tubes being capped tightly and the contents mixed well. After 24 h at room temperature, the samples are evaporated under nitrogen, 10 ml of methanol is added and the preparations again evaporated to dryness. The residues are partitioned between *n*-hexane and water and the two phases examined. If complex lipids are directly treated with liquidized boron trichloride, ether, ester and glycosidic bonds are broken while free polyols and carbohydrates remain unaltered. A general scheme for the extraction and chromatographic purification of archaeal lipids and the degradation procedures used for structural definition is shown in Table 7.2.

7.3 CORE LIPIDS

Isopranyl glycerol ether lipids represent the hydrophobic moiety of archaeal complex lipids whereas mainly fatty ester-linked glycerol lipids are found in bacteria and eukarya. Archaeal lipids are based on the ether linkages between glycerol, with 2,3-*sn* configuration, or other polyol(s) and isopranoid alcohols with 20, 25 or 40 carbon atoms (Woese & Wolfe, 1985; De Rosa *et al.*, 1986a; Langworthy & Pond, 1986; De Rosa & Gambacorta, 1988).

The complex lipids of archaea are mainly based on two classes of isopranoid ether core lipids classified as diethers and tetraethers. These compounds can be easily obtained by acid or alkaline hydrolysis of complex archaeal lipids. The following method was described by Ross *et al.* (1985) for rapid screening of core lipids.

Method 7.5 Acid methanolysis of core lipids for screening (from Ross *et al.*, 1985).

---

1. Freeze-dried biomass (100–200 mg) is mixed with 3.5 ml of toluene, 3.5 ml of methanol and 0.1 ml of concentrated sulphuric acid in Pyrex reaction tubes fitted with Teflon-lined caps.
2. The tubes are heated at $50\,^{\circ}\mathrm{C}$ overnight (16 h) and, after cooling, the preparation is extracted with 1.5 ml of *n*-hexane.
3. The *n*-hexane extract is methanolysed with a small amount of ammonium carbonate, transferred to a clean vial and left to stand at $4\,^{\circ}\mathrm{C}$.
4. Products of acid methanolysis are analysed by TLC on silica-gel plates as described later.

7.3.1 Diethers

7.3.1.a Structural types

The isopranyl diethers found in archaeal lipids are shown in Figure 7.1. 2,3-di-O-phytanyl-sn-glycerol (Figure 7.1a) originates by condensation, via an ether linkage, between two C_{20} isopranoid alcohols and a glycerol molecule. The three chiral centres of the 3,7,11,15-tetramethylhexadecyl chain have 3R, 7R and 11R configurations. 2,3-di-O-phytanyl-sn-glycerol is dextrorotatory and has a configuration opposite to that of diacylglycerols (Kates, 1972; Woese & Wolfe, 1985; De Rosa et al., 1986a; Langworthy & Pond, 1986; De Rosa & Gambacorta, 1988).

Diether b (Figure 7.1b) is structurally related to diether a (Figure 7.1a) but a hydroxyl group on C3 of the phytanyl chain at the sn-3 position of glycerol is present. The full chemical designation for this novel diether, isolated from the aceticlastic methanogen *Methanothrix concilii* GP6 (Ferrante et al., 1988a), is 2-O-[3,7,11,15-tetramethyl]hexadecyl-3-O- [3'-hydroxy-3',7',11',15'-tetramethyl]hexadecyl-sn-glycerol. Diethers c and d (Figure 7.1) have a structure similar to that of 2,3-di-O-phytanyl-sn-glycerol (Figure 7.1a), but differ in the nature of the isopranyl residues; one or two sesterterpanyl chains respectively substitute(s) the phytanyl residue(s) (De Rosa et al., 1986a; De Rosa & Gambacorta, 1986; 1988).

Diether e (Figure 7.1) has a macrocyclic structure with a 36-member ring, which originates from the condensation of a glycerol in the 2,3-sn configuration with 3,7,11,15,18,22,26,30-octamethyldotriacontane-1, 32-diol (Comita et al., 1984). Diether f (Figure 7.1) is a tetritol-diphytanyl-diether, which so far represents the only example among archaea of a core lipid component in which glycerol is absent. In this diether, the α and β groups of the tetritol form ether linkages with two C_{20} isopranoid alcohols having 3R, 7R and 11R configurations (De Rosa et al., 1986b). Mancuso et al. (1985), who screened the glycerol diethers of 25 monocultures of methanogenic bacteria by HI degradation with conversion of the alkyl iodides to acetate and alcohols, reported the presence of C_{20} phytanol and, as minor components, C_{15} and C_{25} isopranologues. Data on the fine structure of glycerol diethers based on these isopranoids is not yet available.

7.3.1.b Degradative procedures

The diethers shown in Figure 7.1 are obtained by acid hydrolysis of complex archaeal lipids. In particular, the preparation of diether b (Figure 7.1) from both purified complex lipids and crude lipid extracts requires some attention. Strong acid hydrolysis in 2.5% methanolic HCl results in the

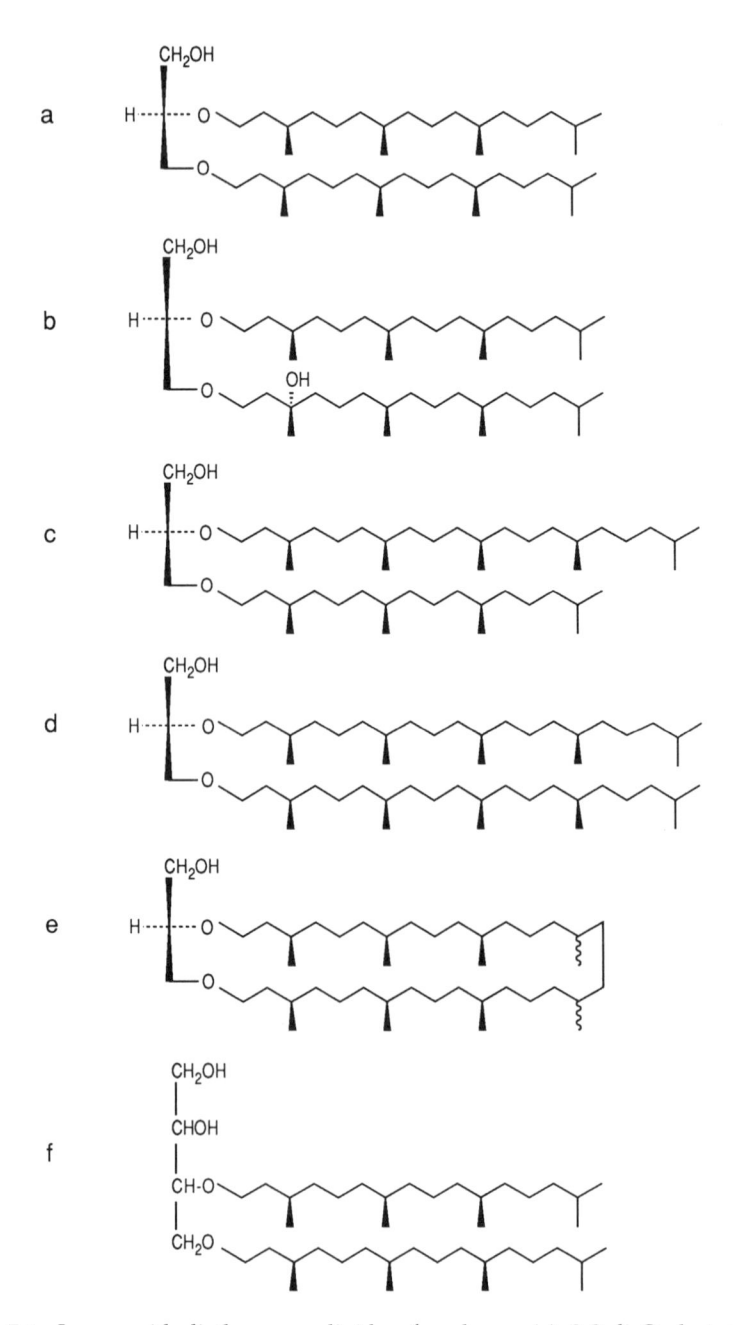

Figure 7.1 Isopranoid diether core lipids of archaea. (a) 2,3-di-O-phytanyl-sn-glycerol; (b) 2-O-[3,7,11,15-tetramethyl]hexadecyl-3-O-[3'-hydroxy-3',7',11',15'-tetramethyl]hexadecyl-sn-glycerol; (c) 2-O-sesterterpanyl-3-O-phytanyl-sn-glycerol; (d) 2,3-di-O-sesterterpanyl-sn-glycerol; (e) macrocyclic diether; (f) tetritol-diphytanyl-diether.

formation of several lipid artifacts while mild degradation in methanolic HCl (0.18%) yields only diether b (Figure 7.1.). The presence of a tertiary hydroxyl group in close proximity to the ether linkage probably makes the hydroxyl group more reactive and the ether linkage more susceptible to acid hydrolysis (Ferrante et al., 1988a). The petroleum ether-soluble unsaponifiable material obtained by acid hydrolysis of archaeal lipids is a viscous colourless oil having an IR spectrum that shows absorption bands due to OH (3400 cm^{-1} and 3528 cm^{-1} for the tertiary OH group of diether b; Figure 7.1), long-chain (2930, 2850, 1465, 750 cm^{-1}) with terminal isopropyl groups (1365 to 1385 cm^{-1}, doublet; absent in the diether e; Figure 7.1), ether C—O—C (1100 cm^{-1}), and primary alcoholic C—O groups (1050 cm^{-1}).

Hydriodic acid-LiAlH$_4$ degradation of the diethers shown in Figure 7.1 yields C$_{20}$ (diethers a, b, c, d and f), C$_{25}$ (diethers c and d) and C$_{40}$ (diether e) isopranoid alkanes. Furthermore, degradation of diethers with BCl$_3$ yields glycerol (diethers a–e) or tetritol (diether f) and C$_{20}$, C$_{25}$ alkyl chlorides (diethers a, b, c, d and f) or C$_{40}$ alkyl dichlorides (diether e). Hydriodic acid and BCl$_3$ degradation of diether b, due to the presence of a tertiary hydroxyl group on the C3 of the isopranoid chain, produces 1 mole of alkyl halide and 1 mole of an unsaturated phytanyl derivative with the respective loss of hydroxyl and halide groups. Milder hydrolytic conditions, such as 1 h with HI, results in partial hydrolysis of diether b, giving rise to 1 mole of an unsaturated phytanyl chain and 1 mole of β-mono-O-phytanylglycerol (Ferrante et al., 1988a).

7.3.1.c Chromatographic procedures

The chromatographic behaviour of diethers from archaeal lipids on silica gel plates using light petroleum-diethyl ether (85 : 15, v/v; Ross et al., 1981) is shown in Figure 7.2a. A better resolution of the C$_{20}$-C$_{20}$, C$_{20}$-C$_{25}$ and C$_{25}$-C$_{25}$ glycerol diethers is obtained using reverse phase RP$_{18}$ plates (Macherley Nagel, art. n. 811062) developed in methanol-acetone (Figure 7.2b; 1 : 1, v/v; Tindall, 1985) and TLC on silica gel plates using n-hexane-acetone (95 : 5, v/v) followed by redevelopment of the plates in the same direction, after drying, with toluene-acetone (97 : 3, v/v; Ross et al., 1985).

Preparative silica gel plates are developed with light petroleum-diethyl ether (85 : 15, v/v; Ross et al., 1981) until the solvent front has reached 18–19 cm; the plates are then allowed to air-dry for several hours in a fume cupboard or until the solvent has evaporated. Diethers are located by placing plates in a tank containing iodine crystals; after marking the resultant bands, the plates are placed in a hood until the iodine vapour has evaporated (Langworthy, 1982). The bands are then scraped off and the diethers eluted from the silica gel with 15 to 30 ml chloroform-methanol (2 : 1, v/v)

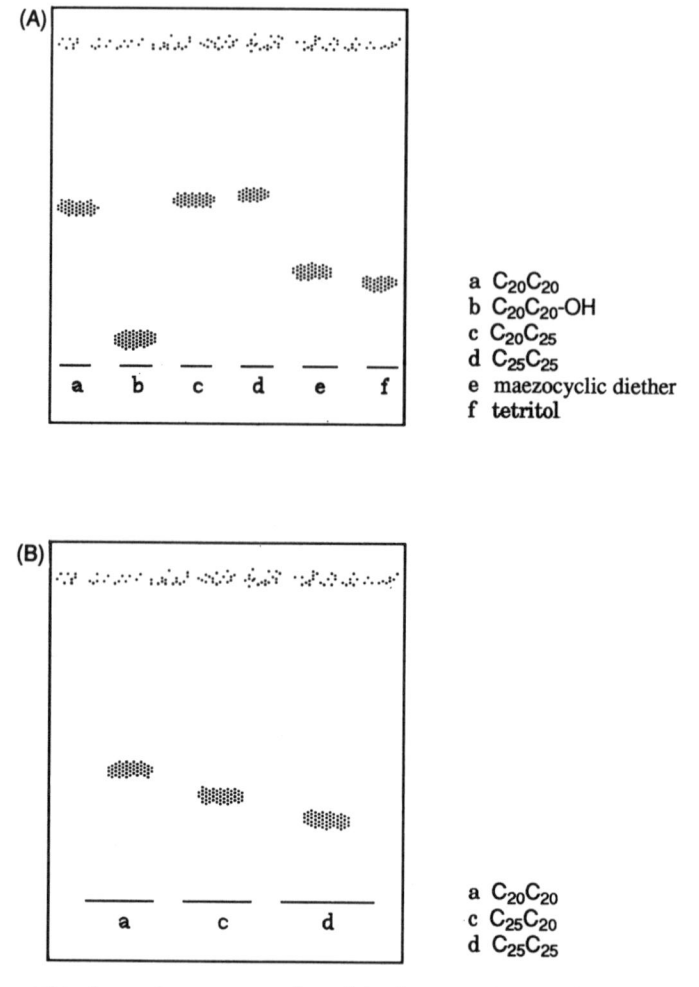

Figure 7.2 Thin layer chromatography of diether core lipids of archaea. (A) Silica gel chromatography developed in light petroleum-diethyl ether (85:15, v/v). (B) Reverse phase RP$_{18}$ plates developed in methanol-acetone (1:1, v/v). See Figure 7.1 for structure of compounds.

a C$_{20}$C$_{20}$
b C$_{20}$C$_{20}$-OH
c C$_{20}$C$_{25}$
d C$_{25}$C$_{25}$
e maezocyclic diether
f tetritol

a C$_{20}$C$_{20}$
c C$_{25}$C$_{20}$
d C$_{25}$C$_{25}$

using a sintered glass filter. When it is not necessary to recover the diethers from the thin-layer plates, components may be detected by spraying with 50% methanolic-H$_2$SO$_4$ followed by charring at 150 °C. Diethers are also resolved on silica gel columns eluted with chloroform-diethyl ether (98:2, v/v; De Rosa *et al.*, 1983b) using at least a 50:1 weight ratio of silicic acid to the material applied.

Gas-liquid chromatography (GLC) of long-chain acetates, alcohols, halides and hydrocarbons can be carried out using the following systems:

(a) a 1.5 m glass column (3 mm i.d.) packed with 1% OV-1 on Gas-Chrom (100–120 mesh), with the temperature programmed at 6 °C/min over 220 to 310 °C, and helium as carrier gas at 60 ml/min (De Rosa et al., 1977a);
(b) a 1.2 m glass column (4 mm i.d.) of 10% Apiezon L on Gas-Chrom Q at 197 °C (Kates et al., 1965);
(c) a 1.2 m glass column (4 mm i.d.) of butanediolsuccinate polyester on Gas-Chrom Q at 197 °C (Kates et al., 1965).

Some investigators use high-resolution glass capillary gas chromatography for the analysis of isopranoid hydrocarbons, their derivatives and isopranoid diether acetates. One method involves the use of a deactivated, 2.5 m (0.32 mm i.d.) cross-linked SE-52 coated glass capillary and a 30 m (0.32 mm i.d.) Duraband DB-5 coated fused-silica capillary column; a typical temperature programme involves injection on the column at 70 °C and maintenance of the column temperature at 70 °C for 3 min, 70 to 210 °C at 5 °C/min, 210 to 270 at 4 °C/min, 270 to 305 at 3 °C/min and 305 °C for 15 min (Comita et al., 1984). A second system uses a 20 m (4 mm, i.d.) 100/120 Supelcoport held isothermally at 250 °C (Ferrante et al., 1988b). A third method uses a 25 m (0.32 mm, i.d.) capillary column 007 series bonded phase, fused silica OV-17 with a starting temperature at 180 °C but set to increase to 240 °C at 2 °C/min.

In general, alkanes are the most convenient derivatives for the identification of isopranoid moieties of diethers. The corresponding monochloro- and monoiodo-derivatives give both broad and multiple peaks, a result of the loss of halides atoms at the high temperatures employed and for the formation of unstable double bonds. Gas-liquid chromatography of methylated sugar derivatives is carried out on a 1.2 m glass column (4 mm, i.d.) of 3% ECNSS-M on Gas-Chrom Q at 170 °C (Kates & Deroo, 1973). Quantitative determination of the underivatized diethers by GLC is carried out on a 2.0 m glass column (4 mm i.d.) of 5% SE-30 on Chromosorb WHP (80/100 mesh) at 320 °C (Tindall, 1985).

High-performance liquid chromatographic analyses are performed using a Microporasil column (30 cm × 3.9 mm i.d., flow rate of 1 ml/min for analytical work; 30 cm × 7.8 mm i.d., flow rate 5 ml/min for preparative work) and a differential refractometer as the detector system. Solvents include n-hexane/ethyl acetate (9 : 1, v/v) for glycerol diether a, c and d (Figure 7.1); n-hexane/ethyl acetate (87 : 3, v/v) for tetritol diether f (Figure 7.1; De Rosa et al., 1986b).

7.3.1.d NMR spectroscopy

^{13}C-NMR spectroscopy is, to a much greater extent than ^{1}H-NMR, an important tool for determining the structural identity of isopranoid diethers. The chemical shifts, multiplicities and assignments of the naturally abundant ^{13}C-NMR spectra of isopranoid alkanes obtained by

Table 7.3 ^{13}C chemical shifts (in δ from tetramethylsilane) and multiplicities of the C$_{20}$, C$_{25}$ and C$_{40}$ isopranoid hydrocarbons obtained by HI-LiAlH$_4$ degradation of ether core lipids of archaea and of the oxygen-bearing carbons in the intact diethers.

Carbon[a] no.	1	2	3
1,1'	11.32(q)	11.32(q)	11.32(q)
2,2'	29.71(t)	29.71(t)	29.71(t)
3,3'	34.73(d)	34.73(d)	34.73(d)
4,4'	37.02(t)	37.02(t)	37.02(t)
5,5'	24.35(t)	24.35(t)	24.35(t)
6,6'	37.46(t)	37.46(t)	37.46(t)
7,7'	32.85(d)	32.85(d)	32.85(d)
8,8'	37.46(t)	37.46(t)	37.46(t)
9,9'	24.53(t)	24.53(t)	24.53(t)
10,10'	37.46(t)	37.46(t)	37.46(t)
11,11'	32.85(d)	32.85(d)	32.85(d)
12,12'	37.46(t)	37.46(t)	37.46(t)
13,13'	24.79(t)	24.53(t)	24.53(t)
14,14'	39.34(t)	37.46(t)	37.60(t)
15,15'	28.02(d)	32.85(d)	33.12(d)
16,16'	22.68(q)	37.46(t)	34.35(t)
17,17'	19.41(q)	24.79(t)	19.41(q)
18,18'	19.80(q)	39.34(t)	19.80(q)
19,19'	19.80(q)	28.02(d)	19.80(q)
20,20'	22.68(q)	22.68(q)	19.80(q)
21	–	19.41(q)	–
22	–	19.80(q)	–
23	–	19.80(q)	–
24	–	19.80(q)	–
25	–	22.68(q)	–

[a] For structures and numbering of the hydrocarbons see Figure 7.3.

```
                    (70.14)                         (68.90)   H₃C   OH
         (71.19)  CH₂-O-CH₂~               CH₂-O-CH₂-CH₂-CH-CH₂-CH₂~
                  |                         |                (72.37)
         (78.48)  CH₂-O-CH₂~               CH₂-O-CH₂~
                  |       (68.64)           |
         (63.12)  CH₂-OH                   CH₂-OH

    (in diethers a, c, d, e: Figure 7.1)    (in diether b: Figure 7.1)
```

cleavage of diethers a to f (Figure 7.1) using the HI-LiAlH$_4$ degradation procedure are given in Table 7.3. The assignments (De Rosa et al., 1977b, 1980a) are based on chemical shift rules (Levy & Nelson, 1971; Stothers, 1972), comparison with appropriate model compounds (Christl et al., 1971) and observed multiplicities combined with additional spectral data obtained for the corresponding alcohols.

The ^{13}C-NMR spectra of the C$_{20}$ and C$_{25}$ isopranoid alkanes obtained from diethers a, b, c, d and f (Figure 7.1) show only 15 resolved signals for the strict or effective equivalence of many carbon atoms. In both cases the assignments are fully consistent with a regular head-to-tail isopranoidic structure. The ^{13}C-NMR spectrum of the C$_{40}$ isopranoid obtained from diether e (Figure 7.1) shows only 14 signals for the symmetry of its structure. The assignments are consistent with the 16,16'-biphytanyl structure and confirm the unusual head-to-head C$_{20}$—C$_{20}$ linkage. Thus, the chemical shifts of the pairs 16,16' and 14,14' differ from those of the biogenetically related pairs 6,6'; 8,8'; 10,10' and 12,12' by -3.11 ($-\gamma_2 + \delta_2$ effects) and $+0.14$ ($+\varepsilon_2$ effect) respectively; the chemical shift of pair 15,15' differs from the pairs 7,7' and 11,11' by 0.72 ($+\delta_3 + \varepsilon_3$ effects; Levy & Nelson, 1971; Stothers, 1972). The existence of two free tails at each end of the C$_{40}$ hydrocarbon is confirmed by the methyl signals at 11.32 and 19.41 ppm (1,1' and 17,17' respectively). The remaining methyls, 18,18'; 19,19' and 20,20', of the isopranoidic skeleton are all equivalent and give a ^{13}C signal at 19.80 ppm. The ^{13}C-NMR chemical shift assignments for the oxygen-bearing carbons in glycerol diethers a, c, d, e (Table 7.3) fully confirm the involvement of the α and β OH of glycerol in the ether linkage with long-chain isopranoidic alcohol(s).

7.3.1.e Mass spectrometry

Figure 7.3 shows the electron impact mass spectrometry (EI-MS) fragmentation pattern of C$_{20}$, C$_{25}$ and C$_{40}$ hydrocarbons obtained by degradation of diethers a to f (Figure 7.1) with HI to give C$_{20}$ and C$_{25}$ mono-iodides and C$_{40}$ di-iodides; the latter, in turn, were converted to the corresponding C$_{20}$, C$_{25}$ and C$_{40}$ hydrocarbons by reduction with LiAlH$_4$. Reduction with LiAl^{2}H$_4$ (De Rosa et al., 1977a) to obtain a deuterium on the terminal carbons is useful for assigning ions in the mass spectra of isopranoid hydrocarbons.

The EI-MS of the C$_{20}$ and C$_{25}$ hydrocarbons (M$^+$ m/z 282 and 352 respectively, ^{2}H$_1$ analogues 283 and 353) can be rationalized in terms of two series of cleavages, α to >CHMe groups, both with and without H transfer, removing successively 3, 3, 2, 3, 2 and 5 C atoms for the C$_{20}$ chain and 3, 3, 2, 3, 2, 3, 2 and 5 C atoms for the C$_{25}$. The fragmentation patterns indicate the presence of a regular head-to-tail structure in both isopranoids.

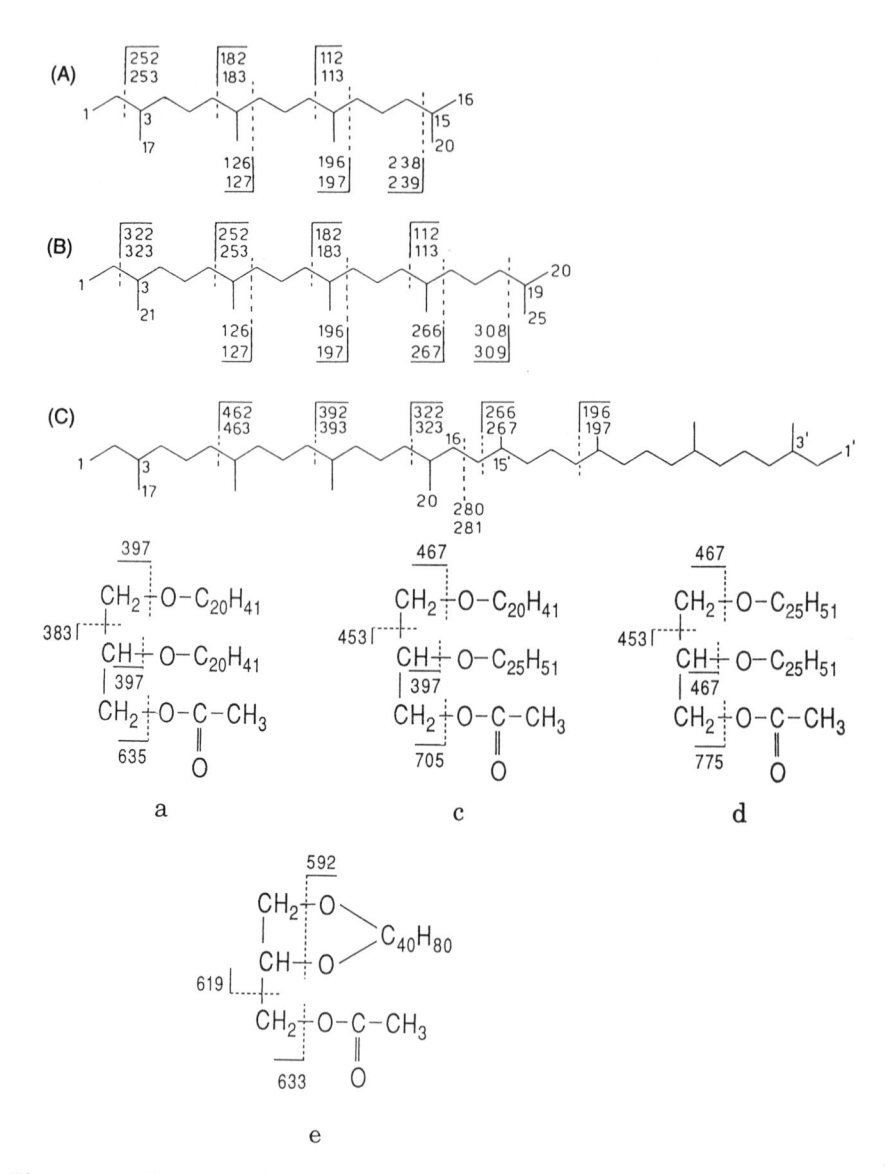

Figure 7.3 Electron impact mass spectrometry fragmentation patterns of: (A) C_{20}; (B) C_{25}; (C) C_{40} hydrocarbons obtained by HI-LiAlH$_4$ degradation of the diether core lipids of archaea and of acetate derivatives of the glycerol diethers based on C_{20} and C_{25} isopranoid chains (a, c, d and e).

The EI-MS of the $C_{40}H_{82}$ hydrocarbon (M^+ m/z 562, $^2H_2$ analogue 564; De Rosa et al., 1977a) can be rationalized in terms of two series of cleavages α to CHMe groups both with and without H transfer, removing successively 7, 5, 5, 4, and 5 C atoms, finally generating the base peaks at m/e 196/197. Diagnostic for the $C_{20}C_{20}$ head-to-head linkage is the median cleavage, again with or without H transfer, β to the two central CHMe groups, giving fragments at m/z 280/281. All the fragmentation peaks are shifted by 1 mass unit in the spectrum of $^2H_2$-hydrocarbon.

Figure 7.3 also shows the fragmentation pattern of acetate derivatives of the diethers a, c, d and e (Figure 7.1), which lack the molecular ion, and show peaks respectively at m/z 635, 705, 775 and 633 for the acetates of diethers a, c, d, and e, respectively (Figure 7.1) corresponding to the loss of the CH_3CO_2 group.

At m/z values higher than 250, there are a series of peaks originating from the loss of aliphatic chains. In particular, in diether c (Figure 7.1), the peaks at m/z 453 and 467 are associated with the loss of $CH_2-O-C_{20}H_{41}$ and $O-C_{20}H_{41}$, respectively; the former unequivocally shows that the C_{20} residue is located on the α-glycerol carbon, while the peak at m/z 397, associated with the loss of $O-C_{25}H_{51}$, places this residue on the β-carbon. The absence of a significant peak at m/z 383 in diether c confirms that these locations are unique.

The fragment at m/z 633 in the mass spectrum of the acetate of diether e, in comparison with the parent peak at m/z 635 in the acetate of the diether a (Figure 7.1), confirms the cyclic nature of the first molecule, while the fragments at m/z 619 and 592 clearly locate the two ether linkages on the α and β carbons of diether e.

Comita et al. (1984) used field desorption (FD-MS) and fast atom bombardment (FAB-MS) for structural elucidation of diether e (Figure 7.1). Samples for FD-MS were dissolved in CH_2,Cl_2, and applied to the emitter by dipping; the emitters used were benzonitrile-activated 10 μM tungsten wires. For FAB-MS, samples were dissolved in CH_2Cl_2/2-nitrophenyloctyl ether (1:1, v/v). The absence of a molecular ion in the EI-MS spectra of glyceroldiethers, the limited solubility of these molecules in matrices suitable for FAB, and the absence of fragmentation in their FD spectra rendered FD the preferred method of ionization for molecular weight determinations of isopranoid glycerol ethers.

The FD-MS of diethers a and e differ by 2 mass units with a m/z of 653 for the protonated molecular ion of diether a (Figure 7.1) and a m/z of 651 for the protonated ion of diether e (Figure 7.1), thereby confirming the cyclic structure of the latter molecule. The FAB-MS spectra of diethers a and d (Figure 7.1) show protonated ions at m/z 653 and 651, respectively.

Ferrante et al. (1988a) reported more significant data for the fully methylated diether b (Figure 7.1) using gas-liquid chromatography-chemical ionization mass spectrometry. The fully methylated diether b gave a molecular ion peak (M + 1) at m/z 697, and other significant ion fragments at m/z 665 (M + 1 −32; loss of methanol); m/z 387 (M+ 1 −310; loss of $C_{21}H_{43}O$); m/z 356, 341 and 297. In the ammonium desorption chemical ionization MS-spectrum, the compound f (Figure 7.1) showed a molecular ion at m/z 682 and a fragment at m/z 651 corresponding to the loss of $-CH_2OH$ (De Rosa et al., 1986b).

7.3.1.f Stereochemistry of diethers

All glycerol diethers have been shown to be dextrorotatory (α_D values ranging from +9 to +5) whereas synthetic sn-1,2-dialkyl glycerol ethers are laevorotatory. Natural glycerol diethers are therefore considered to have the sn-2,3-structure and configuration (Kates, 1978; De Rosa et al., 1983b; Ferrante et al., 1988a).

The absolute stereochemical configurations of the four chiral centres in the 3,7,11,15-tetramethylhexadecyl groups of diether a (Figure 7.1) were defined by Kates et al. (1967) as 3R, 7R and 11R after: (a) degradation of the diether to give a C_{20} hydrocarbon, C_{20} alcohol, C_{19} hydrocarbon and C_{19} acid—the latter was converted to the corresponding C_{18} ketone by the Barbier–Wieland degradation procedure—and (b) comparison of the α_D values of these products with authentic samples of phytol, phytane, pristane, pristanic acid and its C_{18} ketone derivative.

7.3.2 Tetraethers

7.3.2.a Structural types

The isopranyl tetraethers found in archaeal lipids are shown in Figure 7.4. Two series of tetraethers have been characterized as core lipids in archaea. In the first series, named glycerol-dialkyl-glycerol tetraethers (GDGTs), two sn-2,3-glycerol moieties are bridged, through ether linkages, by two isopranoid C_{40} diols formally derived from head-to-head linkage of two O-phytanyl residues. The more common structural type is glycerol tetraether a (Figure 7.4), which is characterized by a 72-member ring with 18 stereocentres; other members of this class contain up to four cyclopentane rings (Figure 7.4b–i) on the two (3R, 7R, 11R, 15S, 18S, 22R, 26R, 30R)-2,7,11,15,18,22,26,30-octamethyldotriacontane moieties formed by the formation of C—C bonds between methyls in 3,7,26,30 and methylenes 6,10,23 and 27, respectively (Woese & Wolfe, 1985: De Rosa & Gambacorta, 1986, 1988; Langworthy & Pond, 1986).

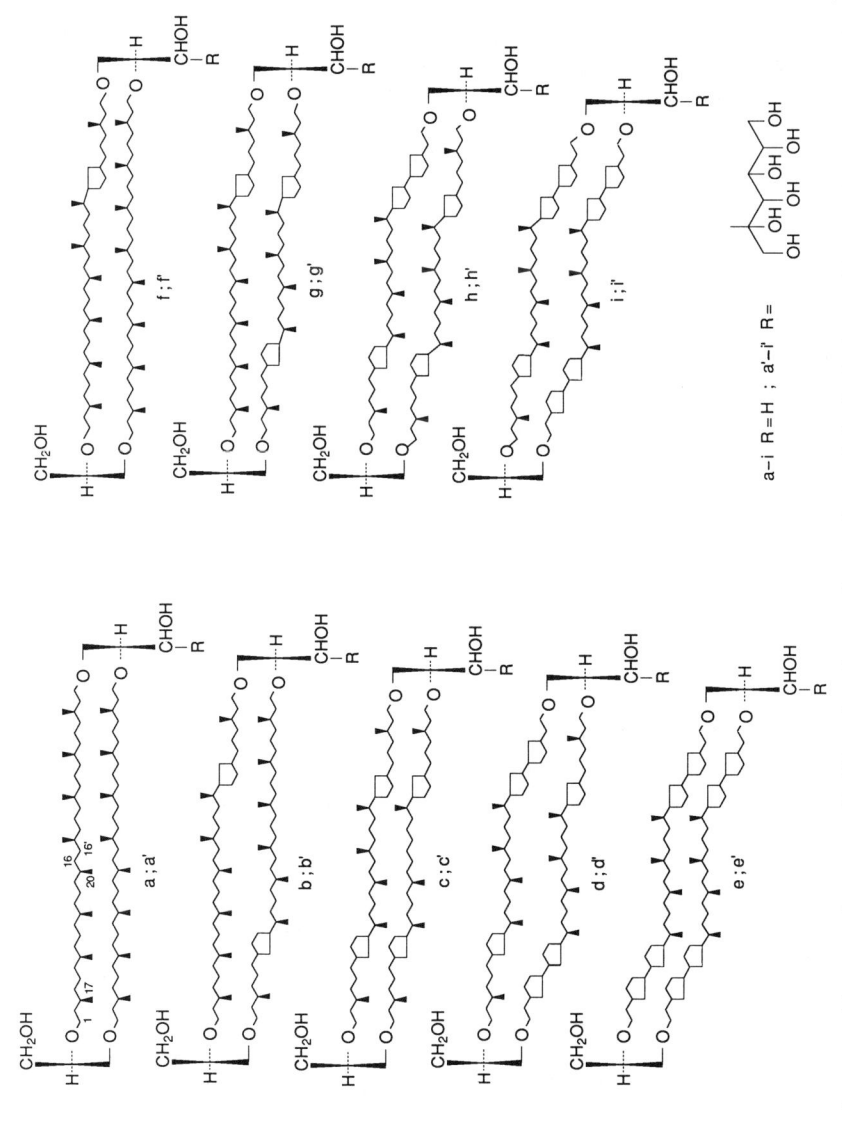

Figure 7.4 Isopranoid tetraether core lipids of archaea: (a–i) glycerol-dialkyl-glycerol tetraethers; (a′–i′) glycerol-dialkyl-nonitol tetraethers.

It is interesting to note the high degree of structural specificity of GDGTs. Indeed, from the five known differently cyclized C_{40} isopranoids, two of which are not symmetrical end-to-end, there are 27 possible pairs of which only nine have been observed. In particular, tetraethers a to c (Figure 7.4) have the two identical C_{40} isopranoids antiparallel, while in the series f to i (Figure 7.4) with two differently cyclized C_{40} chains, the more cyclized end of the C_{40} isopranoid is specifically linked to the primary carbinol of the glycerol (De Rosa et al., 1980a, 1983b). Recently, Thurl and Schafer (1988) reported the occurrence in *Thermoproteus tenax* of a new GDGT, a minor component of the tetraethers, having in the macrocyle an acyclic (3, Figure 7.3) and a bicyclic (b, Figure 7.7) C_{40} isopranoid chain. The orientation of the glycerol and C_{40} chains of this tetraether is currently unknown.

In the second class of tetraethers (Figure 7.4a'–i'), named glyceroldialkyl-nonitol tetraethers (GDNTs), a more complex branched polyol with nine carbon atoms replaces one of the glycerols. It is worth noting that the presence of two different polar heads increases the number of structural possibilities to 45, only nine of which have been detected in archaea (De Rosa et al., 1983b).

7.3.2.b Degradative procedures

The tetraethers shown in Figure 7.4 can be obtained by acid hydrolysis of complex archaeal lipids. The petroleum ether-soluble unsaponifiable material obtained is a viscous colourless oil with an IR spectrum similar to that previously reported for archaeal diethers. Hydroiodic acid-LiAlH$_4$ degradation of the tetraethers shown in Figure 7.4 yields five C_{40} isopranoid alkanes with up to four cyclopentane rings. Degradation of tetraethers with BCl$_3$ yields glycerol and C_{40} dichlorides in a molar ratio of 1:1 for tetraethers a to i (Figure 7.4) and glycerol, nonitol and C_{40} dichlorides in a molar ratio of 1:1:2 for tetraethers a' to i' (Figure 7.4).

7.3.2.c Chromatographic procedures

The TLC of the GDGT fractions obtained by acidic hydrolysis of *Pyrobaculum organotrophum*, *P. islandicum* and *Sulfolobus solfataricus* lipids is shown in Figure 7.5. In this chromatographic system the GDGTs are well resolved with respect to their degree of cyclization (Trincone et al., 1988). This behaviour is not observed on TLC of GDNT fractions given the higher polarity of such tetraethers. Preparative silica gel plates are prepared as previously reported for diethers. The two tetraether fractions, GDGTs and GDNTs, are resolved on silica gel columns using at least 50:1 weight ratio of silicic acid to material applied. Chloroform-diethyl ether (9:1, v/v) elutes the GDGT fraction and chloroform-methanol (95:5, v/v) the GDNT

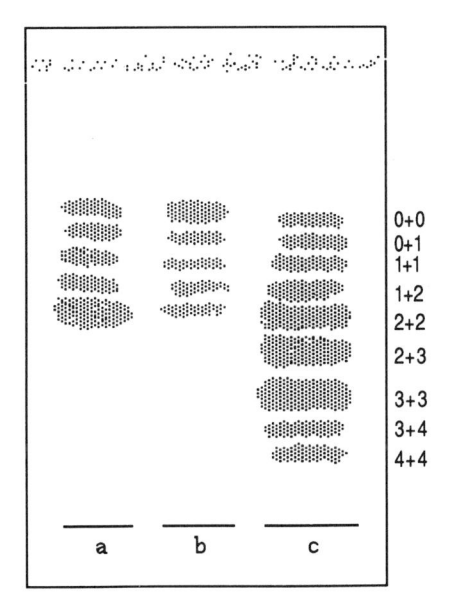

Figure 7.5 Silica gel thin-layer chromatography of tetraether core lipids (GDGTs) developed in n-hexane-ethyl acetate (7:3, v/v). (a) GDGTs from *Pyrobaculum organotrophum*; (b) GDGTs from *P. islandicum*; (c) GDGTs from *Sulfolobus solfataricus*. The numbers indicate the cyclopentane rings per C_{40} chain. See Figure 7.4a–i for the structures of the compounds.

fraction. Under these chromatographic conditions the GDGTs are poorly resolved with respect to the number of cyclopentane rings on the C_{40} chains.

Qualitative and quantitative analyses of tetraethers are obtained by HPLC. Thurl and Schafer (1988) reported the HPLC separation of the GDGT fraction as bis-p-nitrobenzoyl esters using a 10 μm Porasil column (3.9 × 300 mm) eluted isocratically with methylene chloride-acetone (997:3, v/v) at a flow rate of 0.7 ml/min. De Rosa *et al.* (1986b) described a simpler HPLC procedure to resolve the differently cyclized tetraethers of the underivatized GDGT fraction. These authors used n-hexane-ethyl acetate (6:4, v/v) on a Microporasil column (30 cm × 7.3 mm i.d., flow rate 5 ml/min for preparative work; 30 cm × 3.9 mm i.d., flow rate 1 ml/min for analytical work).

Given their high polarity, GDNTs are not easily resolved by HPLC. De Rosa *et al.* (1983b) described the HPLC resolution of fully acetylated derivatives of the GDNT fraction on a Microporasil column (30 cm × 7.8 mm i.d., flow 5 ml/min for preparative work; 30 cm × 3.9 mm i.d., flow rate 0.5 ml/min for analytical work), using n-hexane-ethyl acetate (8:2, v/v) as solvent. Eluants from the column were detected using a differential refractometer.

The HPLC profiles of GDGTs and GDNTs, as fully acetylated derivatives obtained from lipids of *Sulfolobus solfataricus*, are shown in Figure 7.6. Gas-liquid chromatography of C_{40} alkanes, alkyl di-halides and di-alcohols as acetates or trimethylsilyl derivatives can be performed at 320 °C on a 183 cm (6 mm i.d.) stainless-steel column packed with 5.5% SE-30 or on a 3.0% OV-101 column with 80–100 mesh GasChrom Q. Programmed temperature chromatography is performed from an initial temperature of 50 °C, for 12 min, followed by a programmed increase of 5 °C/min up to 340 °C.

Gas-liquid chromatography of isopranoid hydrocarbons can be carried out on a glass capillary column (25 m × 3.0 mm i.d.) of SE-30 with a temperature programmed at 100 to 300 °C, with a temperature increase of 4 °C/min; GDGTs, as trimethylsilyl derivatives, are chromatographed as described above using a 30 cm column (De Rosa *et al.*, 1976a; 1983b; Langworthy, 1977a). Trimethylsilyl derivatives of compounds containing free hydroxyl groups are prepared by reaction for 30 min with a mixture of pyridine/hexamethyldisilazane/trimethylchlorosilane/N, O-bis-(trimethylsilyl)trifluoroacetamide (2:2:1:1, v/v). The acetates of compounds containing free hydroxyl groups are prepared by incubating samples in pyridine-acetic anhydride (1:5, v/v) at 100 °C for 1 h.

Individual C_{40} alkanes can be identified by GLC from estimated chain length values (De Rosa *et al.*, 1976a). The equivalent chain length values are obtained by reference to logarithmic plots of the retention time versus chain length for a C_{20}–C_{40} *n*-paraffin mixture. In the series of C_{40} hydrocarbons that are differently cyclized, the retention time increases as the number of cyclopentane rings increases. This type of chromatographic behaviour is similar for SE-30 and OV-101 columns. The relative percentages of the alkanes may be computed by integration of peak areas. C_{40} hydrocarbons are the best derivatives for GLC analysis of the isopranoid moiety of tetraethers as the corresponding di-chlorides and di-iodides result in the formation of broad as well as multiple peaks both for the loss of halide atoms at the high temperatures employed and for unstable double-bond formation.

7.3.2.d NMR spectroscopy

^{13}C-NMR is the most important NMR technique used to define the structural identity of archaeal isopranoid tetraether core lipids. The chemical shifts, multiplicities and assignments of naturally abundant ^{13}C-NMR spectra of C_{40} isopranoid alkanes, obtained by HI-LiAlH$_4$ degradation of tetraethers a to i and a' to i' (Figure 7.4), are shown in Table 7.4; the data

7. Archaeal Lipids   221

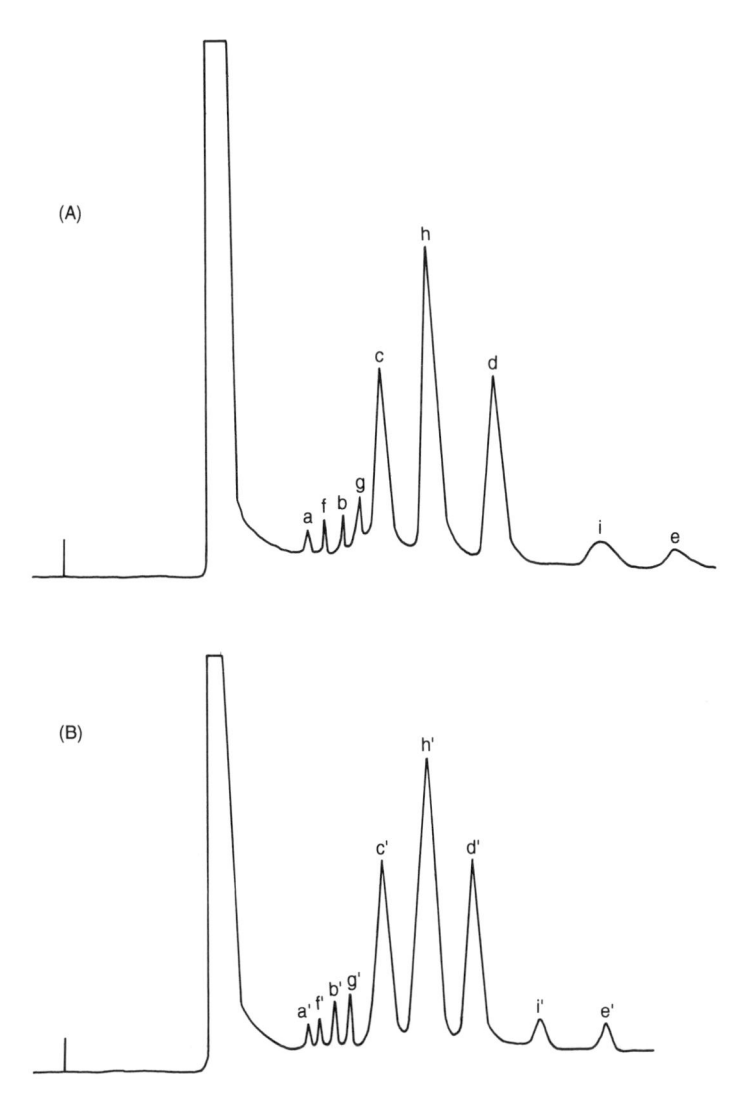

Figure 7.6 High pressure liquid chromatographic profiles on a Microporasil column (30 cm × 3.9 mm, i.d.) of: (A) glycerol-dialkyl-glycerol tetraethers (GDGTs) eluted by using n-hexane-ethyl acetate (6:4, v/v) as solvent at a flow rate of 1 ml/min; (B) glycerol-dialkyl-nonitol tetraethers (GDNTs) as fully acetylated derivatives by using n-hexane-ethyl acetate (8:2, v/v) as solvent at a flow rate of 0.5 ml/min. The GDGTs and GDNTs were obtained by acid hydrolysis of the lipids of *Sulfolobus solfataricus*. See Figure 7.4 for the structures of the compounds.

Table 7.4 Chemical shifts (in δ from tetramethylsilane) and multiplicities of the cyclized C$_{40}$ isopranoid hydrocarbons obtained by HI-LiAlH$_4$ degradation of ether core lipids of archaea.

Carbon[a] no.	a	b	c	d
1,1'	11.32(q)	11.32(q)	12.40(q); 11.32(q)	12.40(q)
2,2'	29.71(t)	29.71(t)	29.34(t); 29.71(t)	29.34(t)
3,3'	34.73(d)	34.73(d)	41.73(d); 34.73(d)	41.73(d)
4,4'	36.91(t); 37.02(t)	36.91(t)	31.46(t); 36.91(t)	31.46(t)
5,5'	25.87(t); 24.35(t)	25.87(t)	30.47(t); 25.87(t)	30.47(t)
6,6'	37.19(t); 37.46(t)	37.19(t)	46.43(d); 37.19(t)	46.43(d)
7,7'	39.13(d); 32.85(d)	39.13(d)	45.60(d); 39.13(d)	45.60(d)
8,8'	33.37(t); 37.46(t)	33.37(t)	32.52(t); 33.37(t)	32.52(t)
9,9'	31.28(t); 24.53(t)	31.28(t)	31.45(t); 31.28(t)	31.45(t)
10,10'	44.89(d); 37.46(t)	44.89(d)	45.15(d); 44.89(d)	45.15(d)
11,11'	38.29(d); 32.85(d)	38.29(d)	38.29(d)	38.29(d)
12,12'	35.77(t); 37.46(t)	35.77(t)	35.77(t)	35.77(t)
13,13'	24.53(t)	24.53(t)	24.53(t)	24.53(t)
14,14'	37.60(t)	37.60(t)	37.60(t)	37.60(t)
15,15'	33.12(d)	33.12(d)	33.12(d)	33.12(d)
16,16'	34.35(t)	34.35(t)	34.35(t)	34.35(t)
17,17'	19.41(q)	19.41(q)	39.20(t); 19.41(q)	39.20(t)
18,18'	36.00(t); 19.80(q)	36.00(t)	34.74(t); 36.00(t)	34.74(t)
19,19'	17.75(q); 19.80(q)	17.75(q)	17.75(q)	17.75(q)
20,20'	19.80(q)	19.80(q)	19.80(q)	19.80(q)

[a] For structures and numbering of C$_{40}$ hydrocarbons see Figure 7.7.
Chemical shifts and multiplicities of the acyclic hydrocarbon are reported in Table 7.3.

on the acyclic C$_{40}$ isopranoid are given in Table 7.3. The assignments (De Rosa et al., 1977b, 1980b) are based on chemical shift rules (Levy & Nelson, 1971; Stothers, 1972), comparison with appropriate model compounds (Christl et al., 1971) and observed multiplicities combined with additional spectral data obtained for the corresponding alcohols.

The spectra of the C$_{40}$ hydrocarbons labelled with ^{13}C from both 1-^{13}C and 2-^{13}C acetate (De Rosa et al., 1980a) are useful in ^{13}C assignments on a biogenetic basis. The ^{13}C-NMR spectrum of the acyclic C$_{40}$ hydrocarbon has been described in the diether section (Table 7.3). The ^{13}C-NMR spectrum of bicyclic hydrocarbon b (Table 7.4) shows 20 signals in agreement with the symmetric structure of the molecule. Of the four methyl signals, three are also present in the spectrum of acyclic hydrocarbon 3 (Table 7.3) and are assigned to carbons 1,1; 17,17' and 20,20', while the fourth at 17.75 ppm is assigned to the carbons 19,19', which are β to the cyclopentane rings. The carbons distant from the cyclopentane ring have the chemical shift values observed in the acyclic hydrocarbon 3 (Table 7.3).

7. Archaeal Lipids

Assignments for carbons near to or within the rings have been calculated from data for model compounds for both 1,3-cis and 1,3-trans stereochemistry (Christl et al., 1971). The data confirm the symmetric location of the two cyclopentane rings and indicate the *trans* stereochemistry in each, while the mutual stereochemistry of the two rings remains unknown. The ^{13}C spectrum of tetracyclic hydrocarbon d (Table 7.4) shows 20 signals in keeping with the symmetric structure of the molecule. Of the three methyl signals, two are also present in the spectrum of bicyclic hydrocarbon b (Table 7.4) and are assigned to carbons 19,19' and 20,20', while the third at 12.40 ppm is ascribed to the carbons 1,1' β to the cyclopentane rings. The ^{13}C chemical shift values of carbons 11 to 16 and 11' to 16' are the same as for the bicyclic C$_{40}$ hydrocarbon. In the ^{13}C-NMR spectra of the monocyclic and tricyclic asymmetric hydrocarbons a and c (Table 7.4) sets of signals given respectively by the acyclic, bicyclic and tetracyclic symmetric C$_{40}$ hydrocarbons (Table 7.4) are observed.

The ^{13}C-NMR chemical shift assignments for the oxygen-bearing carbons in the tetraethers a–i and a'–i' (Figure 7.4) are given in Table 7.5. These data give information on the orientation of the asymmetric monocyclic and tricyclic C$_{40}$ isopranoids in the tetraether macrocycle. In partial structures A to C (Table 7.5), the chemical shift of C-I of the isopranic chain

Table 7.5 ^{13}C Chemical shift values (in δ from tetramethylsilane) of oxygen-bearing carbons in intact tetraethers: basic components of complex lipids of archaea.

Partial structure	Tetraethers[a]	Carbons				
		1	2	3	I	II
A	a–c,f–h	71.19	78.48	63.12	70.14	68.64
B	d,h,i	71.26	78.48	63.12	71.26	68.64
C	e,i	71.26	78.61	63.12	71.26	69.90
D	a'–c',f',g'	/[b]	77.52	/	70.14	68.79
E	d',h'	/[b]	77.52	/	71.26	68.79
F	e',i'	/[b]	77.76	/	71.26	70.10

[a] For structures see Figure 7.4.
[b] Overlapping signals.

is higher by 1.4–1.5 units if it is linked to the primary carbinol of glycerol, and is higher by 1.1–1.2 units when it is in a chain with a ring on the third carbon; the chemical shifts of C(3) and C(2) of the glycerol are higher by 0.07 and 0.13 units respectively, if they are linked to the C_{40} isopranoid with a cyclopentane ring on the third carbon of the aliphatic chain.

The data reported for partial structures B and C (Tablet 7.5) indicate that the 'bicyclic end' of the tricyclic C_{40} isopranoid in GDGTs d, h and i (Figure 7.4) is linked to the primary carbinol of the glycerol. By analogy the orientation of the monocyclic C_{40} chain in GDGT b, f and g (Figure 7.4) is hypothesized under the same structural constraints. With similar considerations, GDNT i' (Figure 7.4), with one tricyclic and one tetracyclic C_{40} isopranoid, shows only signals given in partial structures B and F (Table 7.5), has the 'bicyclic end' of the tricyclic C_{40} isopranoid linked by an ether bond at the first carbon of the nonitol. This structural condition is also present in GDNT h' (Figure 7.4), which is based on a bicyclic and tricyclic C_{40} chain. In this tetraether, the only signals observed are those whose chemical shifts are given in partial structures A and E (Table 7.5). The structures assigned to the remaining GDNTs follow by simple analogy (De Rosa et al., 1977b; 1980b).

7.3.2.e Mass spectrometry

The EI-MS fragmentation pattern of C_{40} isopranoid hydrocarbons obtained by HI-LiAlH$_4$ degradation of tetraethers a–i and a'–i' (Figure 7.4) (De Rosa et al., 1977a; 1980b) are shown in Figure 7.7; the data relative to the C_{40} acyclic isopranoid were mentioned earlier (C, Figure 7.3). Reduction of the di-iodides with LiAl^{2}H$_4$ (De Rosa et al., 1977a) gives C_{40} hydrocarbons with deuterium atoms on the terminal carbons, which is useful in assigning ions in the mass spectrum. The EI-MS of the acyclic hydrocarbon $C_{40}H_{82}$ was described in Section 7.3.1.e.

The EI-MS of $C_{40}H_{80}$ monocyclic isopranoid a (Figure 7.7; m/z 560 and m/z 562 in the ^{2}H$_2$-isopranoid) shows significant differences in comparison with the fragmentation pattern of the parent acyclic C_{40} hydrocarbon (C, Figure 7.3). While the molecular ion and the first major fragment at m/z 460–461 show a formal unsaturation, the fragments at m/z 392–393 do not. Cleavage at the same bond also generates strong peaks at m/z 165–166–167 (base peak 166). Corresponding cleavages, seemingly at an adjacent bond and differing by 28 mass units, are also observed. The central cleavage of the hydrocarbon molecule only gives fragments at m/z 280–281 originated by the uncyclized moiety; the expected peaks at m/z 278–279 for the other cyclized part are absent presumably because this moiety generated similar fragments at m/z 166 and 194. As observed for the acyclic hydrocarbon (C,

7. Archaeal Lipids

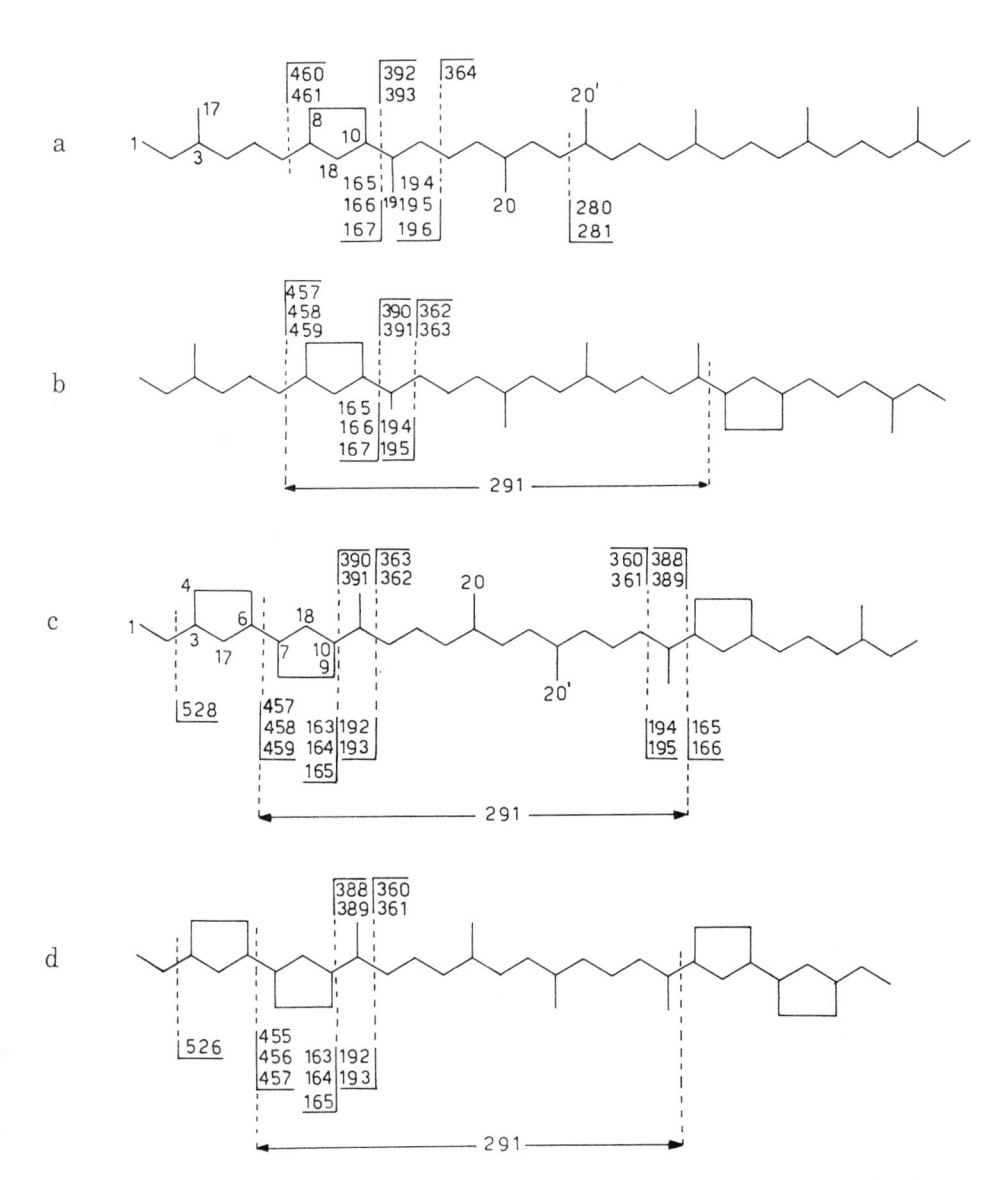

Figure 7.7 Electron impact-mass spectrometry fragmentation pattern of cyclized C_{40} hydrocarbons obtained by HI-LiAlH$_4$ degradation of tetraether core lipids of archaea. See Figure 7.3 for the fragmentation pattern of acyclic C_{40} hydrocarbons.

Figure 7.3), all the fragmentation peaks are shifted by one mass unit in the spectrum of the deuterated monocyclic isopranoid.

The EI-MS of $C_{40}H_{78}$ bicyclic isopranoid b (Figure 7.7; m/z 558, 560 in the $^2H_2$-hydrocarbon) and the fragments m/z 457–459, 390–391 and 362–363 (all shifted by one mass unit in the spectrum of the $^2H_2$-hydrocarbon) show a further deficit of two mass units in comparison with the fragmentation pattern of the parent monocyclic C_{40} hydrocarbon while the important counterparts at m/z 165–166 and 194–195 remain unmodified. The next major fragment in the spectrum has m/z 291; this corresponds to a C_{21} fragment with two formal unsaturations. Because this peak is not shifted in the spectrum of the $^2H_2$ hydrocarbon, it must originate by cleavages at both ends of the molecules retaining only one of the two rings originally present (De Rosa et al., 1977a, 1980b).

The EI-MS of $C_{40}H_{76}$ tricyclic isopranoid c (Figure 7.7; m/z 556, 558 in the $^2H_2$-hydrocarbon; De Rosa et al., 1980b) has the main features of the bicyclic hydrocarbon, but fragments at m/z 388–389, 360–361, 192–193 and 163–164, showing a further deficiency of two mass units when compared with data on the bicyclic hydrocarbon. This fragmentation pattern and the fragment at m/z 528 localize the third cyclopentane ring on the isopranoid skeleton. The structure of $C_{40}H_{74}$ tetracyclic isopranoid d (Figure 7.7; m/z 554, 556 in $^2H_2$-hydrocarbon; De Rosa et al., 1980b) is supported by similar data with significant differences arising from the symmetry of its structure.

7.3.2.f Stereochemistry of tetraethers

The GDGTs have been found to be dextrorotatory (α_D value = +7.5; De Rosa et al., 1976a) thereby indicating that both glycerols in the tetraethers have the sn-2,3-glycerol configuration according to the data of Kates (1978). Consequently, the two glycerol moieties are placed in a *trans* configuration as reported in Figure 7.4. In fact a *cis* configuration of the two glycerols, requiring an sn-1,2-glycerol and an sn-2,3-glycerol moiety, would theoretically result in zero molecular rotation. A similar stereochemistry occurs in GDNTs (De Rosa et al., 1983b). In fact, degradation of the nonitol moiety of GDNTs with $NaIO_4$ followed by reduction with $NaBH_4$ to obtain the corresponding synthetic GDGTs gives tetraethers with the same α_D values as the natural GDGTs. The absolute stereochemical configuration of the eight chiral centres of the 3,7,11,15,18,22,26,30-octamethyldotriacontane are defined as 3R, 7R, 11R, 15S, 18S, 22R, 26R and 30R by Heathcock et al. (1985).

7.3.3 Minor glycerol isoprenoid ether lipids

In addition to diethers and tetraethers (Figures 7.1 and 7.4), four minor isoprenoid ethers are found in archaeal lipids. The structures of monoether 3-O-phytanyl-sn-glycerol (Figure 7.8a; De Rosa et al., 1986b) and of a glycerol-trialkyl-glycerol tetraether (Figure 7.8b; De Rosa et al., 1983b) based upon two glycerols linked by four ether bonds with two molecules of (3R,7R,11R)-3,7,11,15-tetramethylesadecane-1-ol and one of (3R,7R,11R, 15S,18S,22R,26R,30R)-3,7,11,15,18,22,26,30-octamethyldotriacontane-1,32-diol are shown in Figure 7.8. These compounds are recovered after

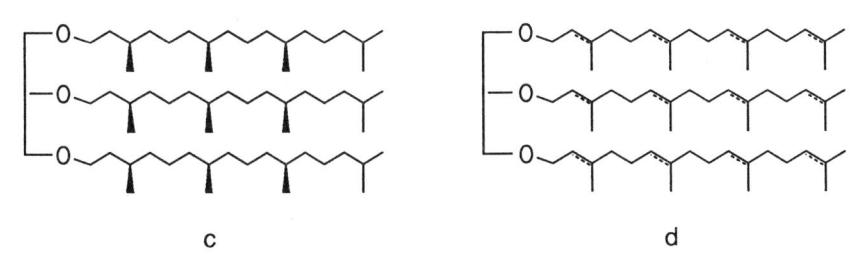

Figure 7.8 Isoprenoid ethers, minor components of archaeal lipids. (a) 3-O-phytanyl-sn-glycerol; (b) glycerol-trialkyl-glycerol tetraether; (c) tri-O-phytanylglycerol; (d) unsaturated tri-O-phytanylglycerols.

acid hydrolysis of the complex lipids of *Sulfolobus solfataricus* and are probably components of minor complex lipids of this microorganism. Monoether a (Figure 7.8) is also found in hydrolysates of the complex lipids of *Methanosarcina barkeri* (De Rosa et al., 1986b, 1989).

Triethers c and d (Figure 7.8), which are found in the neutral fraction of lipids of *S. solfataricus*, can be considered as counterparts of conventional triglycerides (De Rosa et al., 1976b). It is possible to establish the average value of the extent and location of unsaturations on the C_{20} chains, particularly using ^{1}H-NMR data. These data indicate that in triethers d (Figure 7.8) there are approximately six double bonds per molecule, one in the α, three in the β or γ positions, and two in the δ position of the three C_{20} residues.

7.3.4 Occurrence of diethers and tetraethers in archaeal lipids: taxonomic implications

The presence of isopranoid ether lipids and their structural features are useful markers for the classification and identification of archaea. The distribution of different ether core lipids in archaea is shown in Table 7.6. 2,3-di-O-phytanyl-*sn*-glycerol has been found, albeit in different amounts, in all archaea examined and can be considered to be the universal core lipid of these organisms. Indeed, this diether is the only type of core lipid present in the majority of neutral halophiles; it also occurs in some methanogenic and thermophilic coccoid forms. In extremely alkaliphilic halophiles living at pH 10 and in a few strains of neutrophilic halophiles diethers c and d (Figure 7.1) are also present in addition to 2,3-di-O-phytanyl-*sn*-glycerol.

The lipids of most methanogens, with the exception of members of the genera *Methanococcus* and *Methanothrix*, have only diethers based on diether a (Figure 7.1) and tetraether a (Figure 7.4). In contrast, the lipids of thermophilic archaea are mainly based on tetraethers (Figure 7.4) with the exception of those of *Pyrococcus woesei*, *Thermococcus celer* and isolate AN1; the latter have lipids based only on diether a (Figure 7.1).

Members of the order Sulfolobales display the widest spectrum of core lipids. In fact, the lipids of the genera *Acidianus*, *Desulfurolobus*, *Metallosphaera*, *Stygiolobus* and *Sulfolobus*, all of which are classified in the order Sulfolobales, are essentially based on GDGTs and GDNTs (Figure 7.4). This last group of tetraethers only occurs in the lipids of members of the order Sulfolobales and can be considered as a chemical marker for these microorganisms. The ratio of diethers/tetraethers, the ratio of GDGTs/GDNTs and the degree of cyclization on C_{40} isopranoid chains in the Sulfolobaceae vary according to the species and depend upon growth parameters such as aerobiosis, anaerobiosis, autotrophy, heterotrophy and

7. Archaeal Lipids

Table 7.6 Distribution of isopranyl ether lipids among archaea.

Microorganism	Compound[a]	Reference
Halobacterium cutirubrum	1a	Ross and Grant, (1985)
H. halobium	1a;1c	Woese and Wolfe (1985)
H. marismortui	1a	
H. saccharovorum	1a	Ross and Grant (1985)
H. salinarium	1a;1c	Woese and Wolfe (1985)
H. simoncinii	1a	
H. sodomense	1a	
H. trapanicum	1a;1c	
Halococcus morrhuae	1a;1c	
Haloancula californiae	1a	Torreblanca *et al.* (1986)
H. hispanica	1a	
H. sinaisensis	1a	
H. vallismortis	1a	Ross and Grant (1985)
Haloferax gibbonsii	1a	Torreblanca *et al.* (1986)
H. volcanii	1a	Ross and Grant (1985)
H. mediterranei	1a	
54R	1a	Lanzotti *et al.* (1989a)
Natronobacterium magadii	1a;1c	De Rosa *et al.* (1982)
N. pharaonis	1a;1c;1d	Tindall (1985)
N. gregoryi	1a;1c;1d	Ross and Grant (1985)
Natronococcus occultus	1a;1c	
sp.8	1a;1c;1d	De Rosa *et al.* (1983a)
Methanobacterium bryantii	1a;4a	Grant *et al.* (1985)
M. formicicum	1a;4a	
M. thermoautotrophicum	1a;4a	Woese and Wolfe (1985)
Methanobacterium strain M.O.H	1a;4a	
Methanobrevibacter arboriphilus	1a;4a	Grant *et al.* (1985)
M. ruminantium	1a;4a	
M. smithii	1a;4a	
Methanospirillum hungatei	1a;4a	
Methanospirillum strain AZ	1a;4a	Woese and Wolfe (1985)
Methanomicrobium mobile	1a;4a; unidentified	Grant *et al.* (1985)
Methanoplanus limicola	1a;4a	
Methanogenium cariaci	1a;4a	
M. marisnigri	1a;4a	
M. thermophilicum	1a;4a	
Methanolobus tindarius	1a;1c;4a[g]	Grant *et al.* (1985)
Methanothermus fervidus	1a;4a; unidentified	
M. sociabilis	1a;4;4b; unidentified	Laurer *et al.* (1986)
Methanothrix soehngenii	1a	Woese and Wolfe (1985)
M. concilii	1a;1b	Ferrante *et al.* (1988a)
Methanococcus strain PS	1a	Woese and Wolfe (1985)
M. jannaschii	1a;1e;4a	Comita *et al.* (1984)
M. thermolithotrophicus	1a	Grant *et al.* (1985)
M. vannielii	1a;4a[g]	

(continued)

Table 7.6 (Continued).

Microorganism	Compound[a]	Reference
M. voltae	1a	
Methanococcoides methylutens	1a; unidentified	Woese and Wolfe (1985)
Methanosarcina barkeri	1a;1c;1f;4c;4d;4h;8a	De Rosa et al. (1986b)
M. mazei	1a;1c;1f;4a;8a	Grant et al. (1985)
Archaeoglobus	1a;4a	Trincone et al. (1992)
Thermoplasma acidophilum	1a;GDGT[b]	Woese and Wolfe (1985)
Sulfolobus acidocaldarius	1a[g]; GDGT[c];GDNT[c]	Woese and Wolfe (1985)
S. solfataricus	1a;4a–i;4a'–i';8a;b	De Rosa et al. (1989)
Metallosphaera	1a;GDGT[c];GDNT[c]	Huber et al. (1989)
Stygiolobus azoricus	GDGT[f];GDNT[f]	Segerer et al. (1991)
Acidianus brierleyi:		
aerobically grown	1a[g];GDGT[c];GDNT[c]	Woese and Wolfe (1985)
anaerobically grown	1a; tetraethers[e]	Langworthy and Pond (1986)
A. infernus	isopranyl ether lipids[d]	Segerer et al. (1986)
Desulfurolobus ambivalens:		
aerobically grown	1a[g];4a–c;f–h;4c',g'	Trincone et al. (1989)
anaerobically grown	1a[g];4c,g;4c'	
Thermoproteus tenax	1a;4a–c,f–h	Thurl and Schafer (1988)
T. neutrophilus	1a; tetraethers[e]	Langworthy and Pond (1986)
Pyrobaculum	1a[g];4a–c;f,g	Trincone et al. (1992)
Desulfurococcus mobilis	1a;4a	Lanzotti et al. (1987)
D. mucosus	1a; tetraethers[e]	Woese and Wolfe (1985)
Pyrodictium occultum	1a; tetraethers[e]; unidentified	
Thermophilum pendens	1a; tetraethers[e]	Zillig et al. (1983)
T. librum	1a[g]; tetraethers[e]	Langworthy and Pond (1986)
Staphylothermus marinus F1	1a;GDGT[f]; unidentified	Fiala et al. (1986)
Pyrococcus furiosus Vc1	isopranyl ether lipids[d]	Fiala and Stetter (1986)
P. woesei	1a	Lanzotti et al. (1989c)
Thermococcus celer	1a	De Rosa et al. (1987)
Isolate ANI	1a	Lanzotti et al. (1989c)

[a] Designations refer to Figures 7.1, 7.4 and 7.8.
[b] Based on acyclic, monocyclic and bicyclic C_{40} chains.
[c] Based on acyclic to tetracyclic C_{40} chains.
[d] No data on the structures are available.
[e] No data on C_{40} cyclization and GDGT, GDNT presence.
[f] No data on C_{40} cyclization.
[g] In a trace amount.

temperature (De Rosa *et al.*, 1980c; De Rosa & Gambacorta, 1986, 1988; Langworthy & Pond, 1986; Scolastico *et al.*, 1986; Segerer *et al.*, 1991).

7.4 POLAR LIPIDS

7.4.1 Structure and distribution of polar lipid components in archaea: taxonomic implications

Complex polar lipids based on 2,3-di-O-phytanyl-*sn*-glycerol (Figure 7.1a) are found in all archaeal phenotypes, though differences occur in the nature of the attached polar heads.

7.4.1.a Halophiles

Halophilic archaea contain four basic complex lipid structures: isopranoid analogues of phosphatidylglycerol (PG); phosphatidylglycerophosphate (PGP); phosphatidylglycerosulphate (PGS; Figure 7.9); and a family of glycolipids and their sulphate derivatives which show differences in the number and type of the constituent sugars (Figure 7.9; De Rosa *et al.*, 1989). Using the FAB technique, Tsujimoto *et al.* (1989) and Fredrickson *et al.* (1989) revised the structure of PGP of *Halobacterium cutirubrum* and *Halobacterium halobium*, showing that in both microorganisms the major component of the polar lipids is a methyl ester of phosphatidylglycerophosphate, not the corresponding free acid reported earlier (PGP-Me; Figure 7.9).

It is surprising that aminolipids are not found in halophilic archaea since moderate halophilic and nonhalophilic bacteria usually contain either phosphatidylethanolamine or lipoamino acids (Kamekura & Kates, 1988). The major phospholipid present in the membrane lipids of halophilic archaea is PGP; PG and PGS usually occur in smaller amounts (Kamekura & Kates, 1988). It is worth noting that in these phospholipids both glycerols have a 2,3-*sn* stereochemistry (Figure 7.9; Kamekura & Kates, 1988; De Rosa *et al.*, 1989).

The glycolipids of halophilic archaea appear to be derived from the basic structural diglycosyl diether, mannosyl-glucosyl-diphytanylglycerol (DGD), by the substitution of a sugar or sulphate group at the 6 position of the mannose residue; this gives rise to triglycosyl diethers (TGD-1 and TGD-2; Figure 7.9) and sulphate diglycosyl diether (S-DGD-1; Figure 7.9) respectively. Another sulphated diglycosyl diether, different from S-DGD-1, has been detected in an extreme halophilic strain isolated from Japan. This compound is designated S-DGD-2, but its structure has still to be determined (Kamekura & Kates, 1988). Variation occurs amongst

Acidic lipids

	R_1	R_2
PG	3'-sn-glycerol	C_{20} or C_{25}
PGP	3'-sn-glycerol-1'-P	C_{20} or C_{25}
PGS	3'-sn-glycerol-1'-sulfate	C_{20}
PA	H	C_{20}
PL2	3'-sn-glycerol-1',2'-cyclicphosphate	C_{20}
PGP-Me	3'-sn-glycerol-1'-P-Me	C_{20}

$R_3 = C_{20}$ for all

Glycolipids

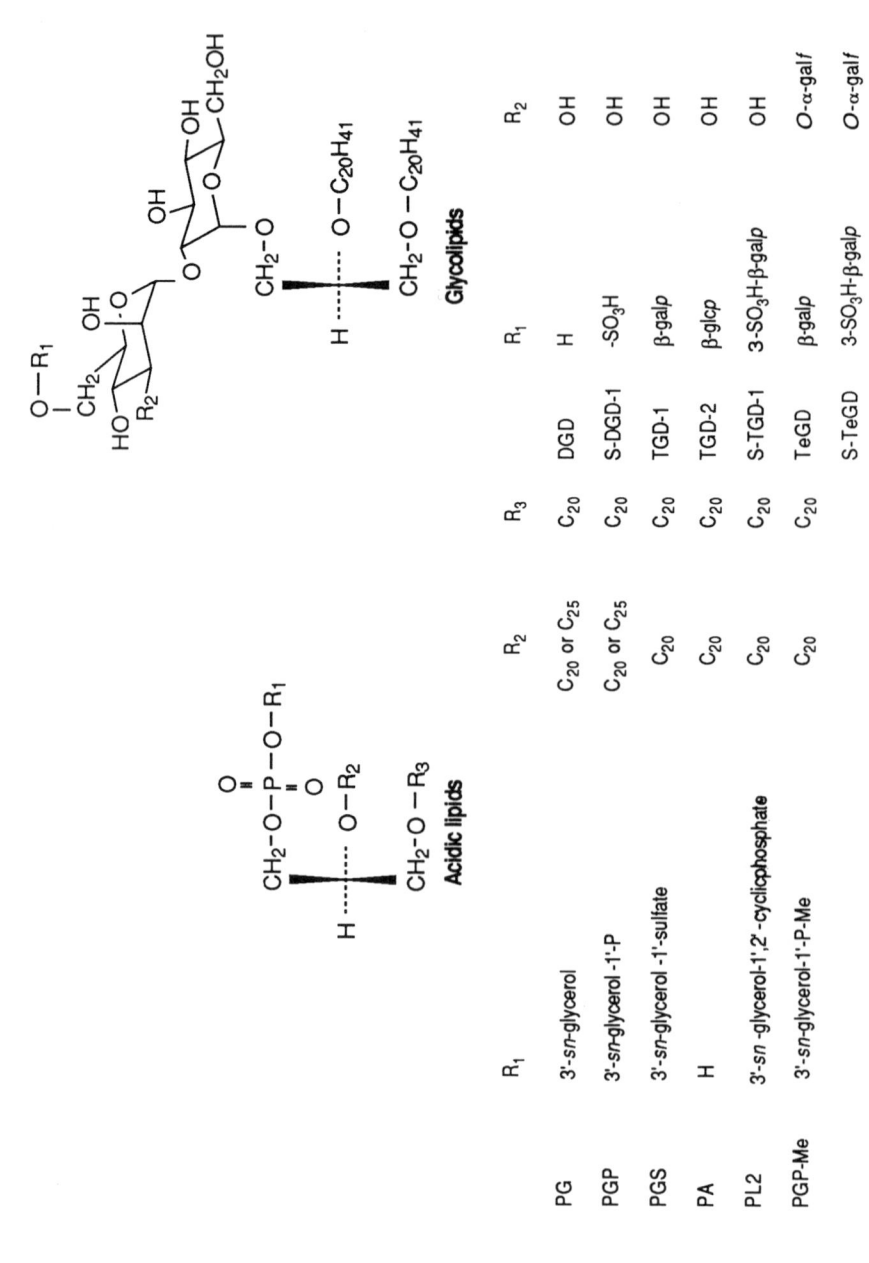

	R_1	R_2
DGD	H	OH
S-DGD-1	$-SO_3H$	OH
TGD-1	β-galp	OH
TGD-2	β-glcp	OH
S-TGD-1	3-SO_3H-β-galp	OH
TeGD	β-galp	O-α-galf
S-TeGD	3-SO_3H-β-galp	O-α-galf

Figure 7.9 Structures of complex lipids found in the halophilic phenotype. P = phosphate group.

triglycosyl glycolipids as the external sugar residue can be either galactose (TGD-1) or glucose (TGD-2). TGD-1 is derived from a sulphated triglycosyl derivative when the sulphate group is linked to the C3 position of the third sugar residue (S-TGD-1; Figure 7.9). Minor branched tetraglycosyl glycolipid (TeGD) and its sulphated derivatives (S-TeGD) are also found. These glycolipids, which have four sugars, show essentially the structure of the sulphated triglycosyl glycolipid with a galactose residue (TGD-1), but with the addition of a β-galactose residue to the 3 position of the mannose (Figure 7.9; Smallbone & Kates, 1981).

The polar lipid composition of nonalkaliphilic halobacteria has been found to be particularly useful in the classification and identification of these microorganisms. All strains examined to date contain PGP and PG as major phospholipids, but this is not the case with PGS. Furthermore, variation occurs in the type and amount of glycolipid components; the presence or absence of these and other minor unidentified lipids can be used to rapidly assign unknown isolates to particular groups.

Members of most *Halobacterium* species are characterized by the presence of PGS, S-TGD-1 and S-TeGD. *Halobacterium denitrificans, H. lacusprofundi, H. saccharovorum, H. sodomense* and *H. trapanicum*, however, contain neither triglycosyl nor tetraglycosyl lipids and hence have a polar lipid composition distinct from any of the currently accepted species of *Halobacterium*. The lipid data together with other biochemical and genetic features support the view that these species, currently considered *incertae sedis*, are in need of reclassification (Grant & Larsen, 1989; De Rosa & Gambacorta, unpublished data).

Haloarcula strains contain PGS, TGD-2 as the major glycolipid, and a minor unidentified glycolipid component, but lack sulphated glycolipids. *Haloferax* species are characterized by the presence of the sulphated glycolipids, S-DGD-1 and the absence of PGS. The predominant lipids of *Halococcus* species are PG, PGP, TGD-2 and a sulphated diglycosyl glycolipid, probably S-DGD-1 (Figure 7.9; Torreblanca *et al.*, 1986; Kamekura & Kates, 1986; Grant & Larsen, 1989).

Polar lipid extract from a neutrophilic archaea isolated from a salt mine not only contained PG, PGP, PGS and S-DGD-1 but also an analogue of phosphatidic acid (PA), 2,3-di-*O*-phytanyl-*sn*-glycero-1-phosphate that accounted for 16% of the total lipid composition (Figure 7.9; Lanzotti *et al.*, 1989a). In comparison with neutrophilic halophiles, haloalkaliphilic archaea belonging to the genera *Natronobacterium* and *Natronococcus* have a relatively simple polar lipid composition. In these organisms the major species of polar lipids are PGP and PG, but glycolipids are completely absent. In all species of these genera PGP elutes from silica columns in association with glycine betaine (De Rosa *et al.*, 1988). The latter forms an ionic complex with phospholipids when two charges are available for complex formation. Thus, phosphatidylglycerol (PG) does not form a complex nor does the newly cyclic phosphate derivative of PGP (PL2; Figure 7.9). Nicolaus et al. (1989) found that glycine betaine was associated with the membrane fraction and that its level increased along with a rise in the ratio of PG to PGP in the cell membrane when *Natronococcus occultus* was grown at different salt concentrations. Several minor unidentified phospholipids are also present in members of *Natronobacterium* and *Natronococcus*. In particular, *Natronococcus occultus* contains a significant amount of a PGP derivative with a 1',2'-cyclic phosphate (PL2; Figure 7.9; De Rosa et al., 1989; Lanzotti et al., 1989b).

Complex lipids based on $C_{20}C_{25}$ diether (Figure 7.1c) are currently found only as PG, PGP, and PL2; they occur essentially in haloalkaliphilic species (De Rosa & Gambacorta, 1988; Kamekura & Kates, 1988).

7.4.1.b Methanogens

Methanogenic archaea have complex lipids based on both diether and tetraether molecules. In contrast, halophilic and sulphur-dependent thermophilic species contain lipids based on diethers and tetraethers, respectively. Little is known about the structures of the polar lipids found in methanogenic archaea; structural analyses have been restricted to a consideration of the major polar lipids of a single species (Kushwaha et al., 1981; Ferrante et al., 1986; Nishihara et al., 1989). The complex lipids found in methanogenic archaea are generally classified into five groups—aminophospholipids, aminophosphoglycolipids, glycolipids, phosphoglycolipids and phospholipids—by means of specific staining tests (Nishihara & Koga, 1987).

In contrast with halophilic and thermophilic archaea, aminophospholipids are widely distributed in methanogens. The diphytanyl ether analogue of phosphatidylserine (PNL2b; Figure 7.10), for example, occurs as a major constituent in the lipids of members of the order Methanobacteriales (Morii et al., 1986; Koga et al., 1987). A phosphoethanolamine derivative of diphytanyl glycerol diether PNL1b has also been found in genera of this order (2, Figure 7.10; Kramer et al., 1987; Nishihara et al., 1989).

A novel 2,3-di-O-phytanyl-1-(phosphoryl-2-acetoamido-2-deoxy-β-D-glucopyranosyl)-sn-glycerol (3, Figure 7.10) (Ferrante et al. 1986) has been found in species belonging to the genus *Methanococcus*. Kushwaha et al. (1981) reported the presence of 2,3-di-O-phytanyl-1-(phosphoryl-1'sn-glycerol)-sn-glycerol, the diastereoisomer of PG found in extreme halophilic archaea, in *Methanospirillum hungatei*. In contrast, Ferrante et al. (1987) were unable to confirm this result but did find two new aminophospholipids, 2,3-di-O-phytanyl−1-[phosphoryl-2'-(1'-N, N-dimethylamino)-

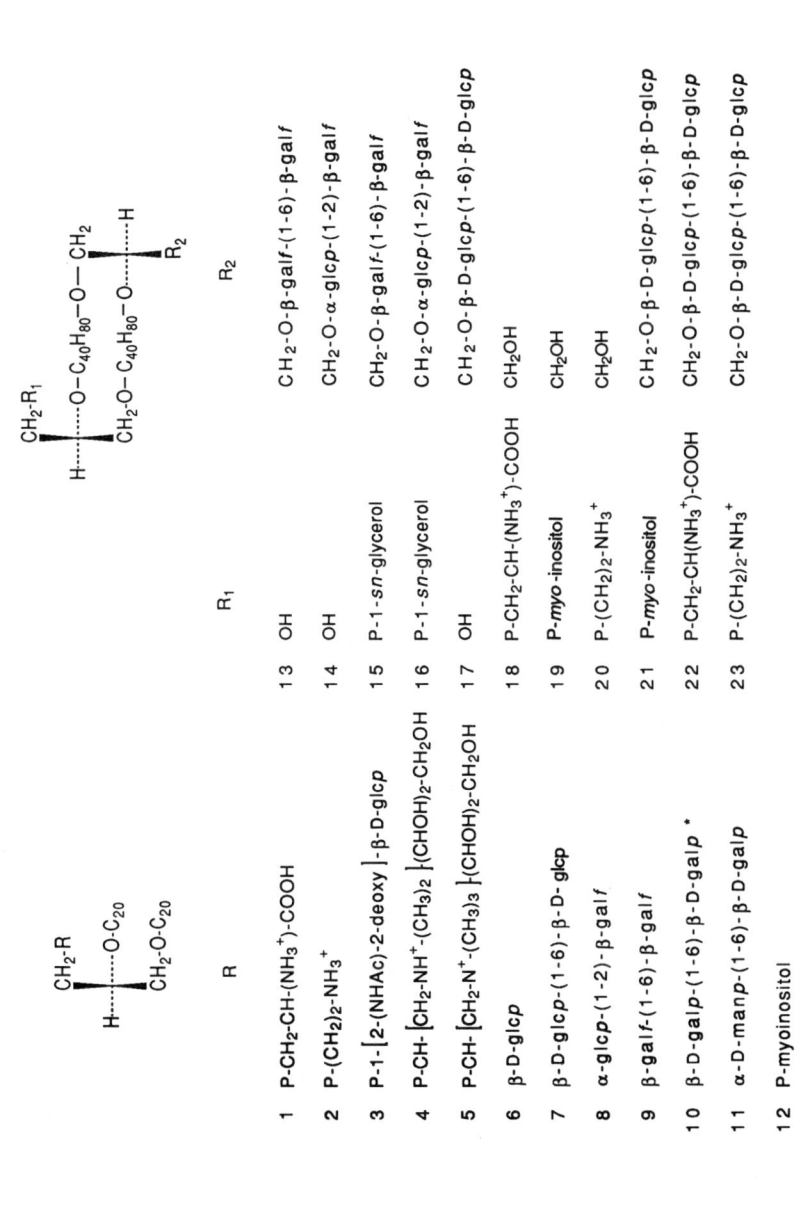

```
        CH₂-R                              CH₂-R₁
H┈┈┈┈┈┤                          H┈┈┈┈┈┤
        │─O-C₂₀                           │┈┈┈O-C₄₀H₈₀-O─CH₂
        CH₂-O-C₂₀                         CH₂-O─C₄₀H₈₀-O─┤┈┈┈H
                                                          R₂
```

	R		R₁	R₂
1	P-CH₂-CH-(NH₃⁺)-COOH	13	OH	CH₂-O-β-galf-(1-6)-β-galf
2	P-(CH₂)₂-NH₃⁺	14	OH	CH₂-O-α-glcp-(1-2)-β-galf
3	P-1-[2-(NHAc)-2-deoxy]-β-D-glcp	15	P-1-sn-glycerol	CH₂-O-β-galf-(1-6)-β-galf
4	P-CH-[CH₂-NH⁺-(CH₃)₂](CHOH)₂-CH₂OH	16	P-1-sn-glycerol	CH₂-O-α-glcp-(1-2)-β-galf
5	P-CH-[CH₂-N⁺-(CH₃)₃](CHOH)₂-CH₂OH	17	OH	CH₂-O-β-D-glcp-(1-6)-β-D-glcp
6	β-D-glcp	18	P-CH₂-CH-(NH₃⁺)-COOH	CH₂OH
7	β-D-glcp-(1-6)-β-D-glcp	19	P-myo-inositol	CH₂OH
8	α-glcp-(1-2)-β-galf	20	P-(CH₂)₂-NH₃⁺	CH₂OH
9	β-galf-(1-6)-β-galf	21	P-myo-inositol	CH₂-O-β-D-glcp-(1-6)-β-D-glcp
10	β-D-galp-(1-6)-β-D-galp *	22	P-CH₂-CH(NH₃⁺)-COOH	CH₂-O-β-D-glcp-(1-6)-β-D-glcp
11	α-D-manp-(1-6)-β-D-galp	23	P-(CH₂)₂-NH₃⁺	CH₂-O-β-D-glcp-(1-6)-β-D-glcp
12	P-myoinositol			

Figure 7.10 Structures of complex lipids found in the methanogenic phenotype. P = phosphate group.

235

2',3',4',5'-pentanetetrol-sn-glycerol 2(PPAD; 4, Figure 7.10) and 2,3-di-O-phytanyl-1-[phosphoryl-2'-(1'-N,N,N-trimethylamino)-2', 3', 4', 5'-pentanetetrol]-sn-glycerol (PPTAD; 5, Figure 7.10) in *M. hungatei*.

Methanogenic archaea contain glycolipids based on 2,3-di-O-phytanyl-sn-glycerol (Figure 7.1a); these compounds are monoglycosyl or diglycosyl derivatives of the diether. The glycosidic residues are glucose for the monoglycoside (MGD; 6, Figure 7.10) and glucose and/or galactose for the diglycosides (DGD), which are also named GL1b, DGD-I, DGD-II, (7-9, Figure 7.10, Kushwaha *et al.*, 1981; Ferrante *et al.*, 1986; Koga *et al.*, 1987; Nishihara *et al.*, 1989). Ferrante *et al.* (1988b) reported the structure of two new glycolipids in the aceticlastic methanogen, *Methanothrix concilii*, namely GL-I (10, Figure 7.10), which was identified as 2-O-phytanyl-3-O-3'[-hydroxy-3', 7', 11', 15'-tetramethyl]-hexadecyl-1-O[β-D-galactopyranosyl-1(1 → 6)-β-D-galactopyranosyl-sn-glycerol and GL-2, a galactose-mannose derivative of 2,3-di-O-phytanyl-sn-glycerol (11, Figure 7.10).

A phosphatidylinositol derivative of 2,3-di-O-phytanyl-sn-glycerol (Figure 7.1a; PL2b or PI; 12, Figure 7.10) (Koga *et al.*, 1987; Ferrante *et al.*, 1988b; Nishihara *et al.*, 1989) has also been detected in methanogenic archaea. With the exception of the genus *Methanococcus* the basic components of the complex lipids of methanogenic archaea are tetraether a (Figure 7.4) and 2,3-di-O-phytanyl-sn-glycerol (Figure 7.1a). The lipids of members of the genus *Methanococcus* are essentially based on diethers a and e (Figure 7.1).

Kushwaha *et al.* (1981) elucidated the structure of the complex lipids of *Methanospirillum hungatei* where the polar lipids based on tetraethers were limited to two glycolipids and two phospholipids. The glycolipids, DGT-I and DGT-II (13,14, Figure 7.10) have one of the free hydroxyl groups of GDGTa (Figure 7.4) linked glycosidically to a disaccharide residue. The phosphoglycolipids PGLI and PGLII (15,16, Figure 7.10), in which a 1-sn-glycerophosphate residue is attached to the opposite side on the tetraether molecule, are derived from the two glycolipids.

In addition to the lipids based on diether described above (1,2,7,12, Figure 7.10), *Methanobacterium thermoautotrophicum* contains major lipids based on tetraether core lipid a (Figure 7.4). One of these, GL1a (17, Figure 7.10), is a glycolipid in which one of the free hydroxyl groups of the GDGT a (Figure 7.4) is linked glycosidically to a glucose disaccharide residue. Three others are phospholipids derived from tetraether a (Figure 7.4), namely phosphoserine (PNL2a; 18, Figure 7.10), phospho-*myo*-inositol (PL2a; 19, Figure 7.10) and phosphoethanolamine (PNL1a; 20, Figure 7.10). The remaining lipids that have been characterized are phosphoglycolipid derivatives of GL1a, with two polar head groups of glucosylglucose and phospho-*myo*-inositol (PGL1; 21, Figure 7.10), or phosphoserine (PGL2; 22, Figure 7.10) or phosphoethanolamine (PNGL1; 23, Figure 7.10) attached separately on glycerol moieties (Nishihara et al., 1989).

The presence or absence of some of the complex lipids identified in methanogens may be of taxonomic value (Koga et al., 1987) but further comparative studies are needed to establish this. To this end it will be necessary to elucidate the structural and biosynthetic relationship of the complex lipids of archaea in general and amongst methanogens in particular with respect to polar lipids based on diethers and tetraethers. This will necessitate an analysis of the structure of a whole set of major polar lipids extracted from representative methanogenic strains.

7.4.1.c Thermophiles

The archaeal thermophilic phenotype comprises sulphur-dependent microorganisms, which form an independent kingdom in the domain archaea, and some intermediate genera classified in the kingdom accommodating halophilic and methanogenic strains. The former are classified in the orders Sulfolobales and Thermoproteales, the latter in the orders Archaeoglobales, Thermococcales and Thermoplasmatales (Woese, 1987; König, 1989; Trincone et al., 1992).

The complex lipids of thermophilic archaea are mainly based on tetraethers, with the exception of those found in the genera *Pyrococcus* and *Thermococcus* and in isolate AN1; the latter has polar lipids based on 2,3-di-O-phytanyl-sn-glycerol (Figure 7.1a). Members of the genera *Pyrococcus* and *Thermococcus* contain a phospholipid (which forms about 85% of the total complex lipid composition) in which the phospho-*myo*-inositol in the polar head is attached to the free hydroxyl of the glycerol moiety (12, Figure 7.10; De Rosa et al., 1987; Lanzotti et al., 1989c). In isolate AN1 this type of phospholipid accounts for 40% of the total complex lipids. This organism, however, also contains a novel phosphoglycolipid, which has been identified as 2,3-di-O-phytanyl-1-(3-phosphoryl-α-D-glucopyranosyl)-sn-glycerol, that represents the first example in the archaea of a glycolipid containing a phosphorylated sugar. These data underpin the relationships found between isolate AN1 and members of the genera *Pyrococcus* and *Thermococcus* highlighted by molecular systematic data (Woese, 1987). The chemical and molecular data justify the introduction of a separate order, the Thermococcales, for all of these microorganisms; this order may represent a third major division of the archaea. The remaining thermophilic species have complex lipids based on 2,3-di-O-phytanyl-sn-glycerol (Figure 7.1a), albeit in minor amounts. These compounds have still to be identified.

The complex lipids of *Thermoplasma acidophilum*, a moderately thermophilic microorganism phylogenetically related to the methanogenic archaea (Woese, 1987), are based on differently cyclized GDGTs (Figure 7.4). At least six different glycolipids and seven phosphorus-containing lipids have been detected but none of them have been fully characterized. The major component is a phosphoglycolipid derivative in which a 3-*sn*-glycerolphosphate and an unidentified monosaccharide are attached to the free hydroxyl groups of the GDGTs (1, Figure 7.11; Langworthy, 1985). Additional phosphoglycolipids occur only in minor amounts and appear to have up to three carbohydrate residues along with free amino groups, possibly as amino sugars (Langworthy, 1979). In addition, *T. acidophilum* has an unusual lipoglycan on its cell surface; this compound resembles the lipopolysaccharide of Gram-negative bacteria. The compound, (Man$p$-(α1 → 2)-Man$p$-(α1 → 4)-Man$p$-(α1 → 3)$_8$-Glc$p$-(α1 → 1)-*O*-(diglyceryl tetraether)-OH, has 24 mannose and one glucose residue(s) with the polysaccharide chain being attached to only one side of the tetraether molecule (Langworthy, 1985).

The complex lipids of thermophiles belonging to the second branch of the archaea, the 'sulphur-dependent' group, have been defined in the genera *Desulfurolobus*, *Metallosphaera* and *Sulfolobus* (König, 1988; Huber et al., 1989) of the order Sulfolobales, and in the genera *Desulfurococcus* and *Thermoproteus* of the order Thermoproteales.

Thermoproteus strains contain different glycolipids that are based both on 2,3-di-*O*-phytanyl-*sn*-glycerol (Figure 7.1a) and tetraethers of the GDGT types (Figure 7.4a,b,c,f,g) as core lipids and glucose residue(s) as polar head(s). The minor glycolipids, based on diether a (Figure 7.1), can probably be built up from at least four glucose residues (Thurl & Schafer, 1988). The glycolipids, based on GDGTs (Figure 7.4), have mono- (GL-1; 2, Figure 7.11), di- (GL-3; 3, Figure 7.11) and tri- (GL-4; 4, Figure 7.11) glucosyl residues. The glucose residues are attached chain-like with a 1 → 6 linkage to one of the two hydroxyl groups of the tetraethers.

Glycolipids 1 and 3 (2 and 3 in Figure 7.11) are a mixture of two compounds as they have tetraethers based on both acyclic chains (Figure 7.4a) or have one chain acyclic and one with a cyclopentane (Figure 7.4f). Glycolipid (4, Figure 7.11) is a mixture of lipids containing tetraethers with different degrees of cyclization (Figure 7.4a,b,c,f,g). The phospholipid fraction forms six bands on TLC but two main components, named PL-3 (5, Figure 7.11) and PL-5 (6, Figure 7.11) have been fully identified.

Phospholipids 3 and 5 are phosphoglycolipids with phosphoinositol and β-D-glucopyranosyl residue(s) attached at each side of the tetraether moiety, being phosphatidylinositol derivatives of GL-1 (2, Figure 7.11) and GL-4 (4, Figure 7.11), respectively (5, 6, Figure 7.11). In addition to these two phosphoglycolipids, small amounts of phospholipids without sugar

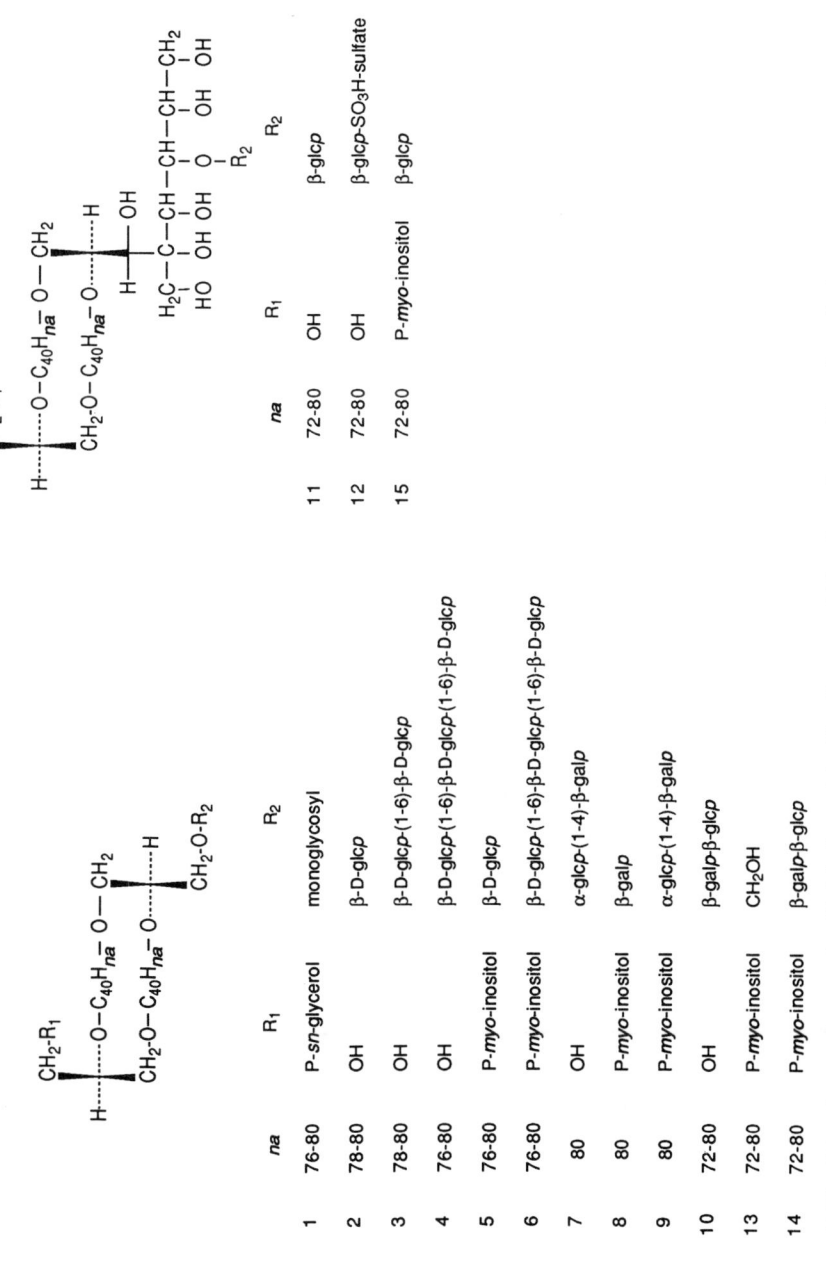

	na	R₁	R₂
1	76-80	P-sn-glycerol	monoglycosyl
2	78-80	OH	β-D-glcp
3	78-80	OH	β-D-glcp-(1-6)-β-D-glcp
4	76-80	OH	β-D-glcp-(1-6)-β-D-glcp-(1-6)-β-D-glcp
5	76-80	P-myo-inositol	β-D-glcp
6	76-80	P-myo-inositol	β-D-glcp-(1-6)-β-D-glcp-(1-6)-β-D-glcp
7	80	OH	α-glcp-(1-4)-β-galp
8	80	P-myo-inositol	β-galp
9	80	P-myo-inositol	α-glcp-(1-4)-β-galp
10	72-80	OH	β-galp-β-glcp
13	72-80	P-myo-inositol	CH₂OH
14	72-80	P-myo-inositol	β-galp-β-glcp

	na	R₁	R₂
11	72-80	OH	β-glcp
12	72-80	OH	β-glcp-SO₃H-sulfate
15	72-80	P-myo-inositol	β-glcp

Figure 7.11 Structures of complex lipids found in the thermophilic phenotype. P = phosphate group.

groups are present (Thurl & Schafer, 1988). Members of the genus *Thermoproteus* also have a phospholipid based on 2,3-di-*O*-phytanyl-*sn*-glycerol (Figure 7.1a). The complex lipids of members of the genus *Pyrobaculum* of the order Thermoproteales are quite similar to those found in the genus *Thermoproteus*, the sole difference being in the sequence (1–2) of the carbohydrate residues (Trincone *et al.*, 1992).

Members of the genus *Desulfurococcus* of the order Thermoproteales contain three identified complex lipids based on GDGT (Figure 7.4a). The first, a glycolipid, has one of the free hydroxyl groups of GDGT linked glycosidically to a glucose-galactose disaccharide (7, Figure 7.11). The second is a phosphoglycolipid where one of the free hydroxyls of GDGT is linked glycosidically to galactose and the other is esterified with the phospho-*myo*-inositol residue (8, Figure 7.11). The remaining compound is a phosphoglycolipid derived from glycolipid 7 (Figure 7.11), where the free hydroxyl of the GDGT moiety is esterified with the phospho-*myo*-inositol residue (9, Figure 7.11; Lanzotti *et al.*, 1987).

Archaea belonging to the genus *Sulfolobus* of the order Sulfolobales have complex lipids based on differently cyclized GDGTs and GDNTs (Figure 7.4). The glycolipids are both disaccharide and monosaccharide derivatives of tetraethers. The glycolipid based on GDGTs has a glucose-galactose disaccharide glycosidically linked to one of the free hydroxyl groups of the glycerol moiety (10, Figure 7.11). Similarly, the GDGT glycolipid has a glucose residue attached to the C6 of the nonitol moiety (11, Figure 7.11). Sulphoglycolipid 12 (Figure 7.10), which is derived from this glycolipid, has a sulphate group esterified to the glucose moiety.

The three phospholipids identified in *Sulfolobus* strains all contain a *myo*-inositol phosphate residue. The first is derived by GDGT in which one of the free hydroxyls of the glycerol moiety is esterified by phospho-*myo*-inositol (13, Figure 7.11). The remaining ones (14, 15, Figure 7.11) are derivatives of the glycolipids (10, 11, Figure 7.11) where the phospho-*myo*-inositol residue is esterified to the free hydroxyl group of the glycerol moiety (Langworthy, 1985; De Rosa *et al.*, 1989). Two unidentified polar lipids have been found in *Sulfolobus* strains grown autotrophically (Langworthy, 1977b).

Desulfurolobus strains, irrespective of whether they are grown aerobically or anaerobically, have complex lipids like those of members of the genus *Sulfolobus* (10–15, Figure 7.11) apart from the degree of cyclization in the aliphatic chains (De Rosa & Gambacorta, 1988; Trincone *et al.*, 1989). Members of the genus *Metallosphaera* of the order Sulfolobales contain similar core and complex lipids to those found in *Sulfolobus solfataricus* (Huber *et al.*, 1989) although the relative proportions of glycolipids (10, 11, Figure 7.11) and minor complex lipids are different.

The largest percentage of complex lipids of the sulphur-dependent archaea assigned to the genera *Desulfolobus*, *Desulfurococcus*, *Metallosphaera*, *Sulfolobus* and *Thermoproteus* occur as phosphoglycolipids which have a sugar residue and a phosphate group linked to opposite sides of the tetraether molecule. In addition, polar head groups are restricted to phospho-*myo*-inositol or to galactose or glucose or both. The different structures originate from the stereochemistry of the glycosidic bond and the location of the interglycosidic linkage.

7.4.1.d Conclusions

Some general conclusions can be drawn from work on the complex lipids of archaea. The diether polar lipids of these organisms are structurally analogous to their glycosyl or phosphatidyl diacylglycerol counterparts in bacteria and eukarya, with the exception that they contain ether linkages and are stereochemically mirror images. In contrast, the tetraether polar lipids are unique as the architecture of these bipolar amphipathic molecules does not have any counterpart amongst bacterial or eukaryotic lipids. Although relatively few complex lipid structures based on these unusual core lipids have been fully established, a series of structural constraints seem to operate. Tetraether polar lipids may have polar groups attached to one or both polyol moieties but usually the largest percentages occur as phosphoglycolipids where sugar residues and phosphate radicals are linked to opposite sides of the tetraether molecule. There are, therefore, no examples of symmetrical tetraether polar lipids. Further, with few exceptions, tetraether polar lipid structures contain carbohydrate residues.

The relationship amongst halophilic archaea between both core and polar composition and taxonomy has been described above. The structures of polar lipids found in the halophile phenotype are unique to archaea. Indeed, phosphoglycerol, as the polar residue of phospholipids, and mannose, as one of the sugar residues of glycolipids, are encountered very rarely amongst other archaea; the sole example is glycolipid 11 (Figure 7.10) found in the aceticlastic methanogen *Methanothrix concilii* (Ferrante et al., 1988b), which has mannose as the sugar residue. A related structure (12, Figure 7.11) has been found in *Sulfolobus* species—thermophilic sulphur-dependent archaea—but in these organisms the sulphate group esterifies the glucose moiety of the tetraether of GDNT types (Figure 7.4a',b'). Finally, in general, all of the polar lipids found in halophilic archaea, sulphated glycolipids and phospholipids, are acidic and hence impart a strong acidic character to the surface of these organisms. The function of sulphate lipids in halophiles is not clear; some workers have speculated that they may serve as proton donors for the functioning of the purple membrane possessing bacteriorhodopsin (Langworthy, 1985; Kamekura & Kates, 1988).

Aminolipids of methanogenic archaea are unique among archaea. These compounds are essentially phosphoderivatives of diether (Figure 7.1a) and tetraether (Figure 7.4a). The polar head group involved in the glycolipids, derivatives of both diether a (Figure 7.1) and tetraether a (Figure 7.4), are limited to glucose and/or galactose. Some tetraether glycolipids found in methanogens also occur in certain thermophilic sulphur-dependent archaeal species. The phospholipids and phosphoglycolipids, derivatives of both diether a (Figure 7.1) and tetraether a (Figure 7.4), are easily classified into three groups on the basis of their phosphoric ester polar head groups; namely phosphoserine, phosphoethanolamine and phosphoinositol. This last phosphate polar head group is also found in all phospholipids of the thermophilic phenotype; it occurs in members of the orders Sulfolobales, Thermococcales and in Thermoproteales.

Structural analysis of a whole set of polar lipids of some methanogenic archaea shows that the kind of polar head group found in diether lipids is also present in tetraether lipids and vice versa; one polar head group, found in tetraether lipids has the same stereochemical structure as that of the corresponding diether lipid. These observations suggest not only a structural regularity but also a biosynthetic relationship between diether and tetraether lipids. It is possible to speculate that the biosynthesis of tetraether polar lipids occurs by head-to-head condensation of diether polar lipids. Experiments performed by Nishihara *et al.* (1989) with labelled phosphorus support this biogenetic hypothesis.

Glycolipids are present in smaller amounts than acidic lipids in thermophilic sulphur-dependent archaea. It is likely that they may not only be biosynthetic precursors of phosphoglycolipids, but also may have specific functions in the cell membrane as, for instance, non-ionic lipids. Polar lipids both of the GDGT and GDNT types are mainly based on tetraethers. In particular, nonitol-glycerol tetraethers are useful markers for assessing the relationship of new organisms belonging to the order Sulfolobales. In fact, to date only members of this order have proved to possess complex lipids based on GDNTs (Langworthy & Pond, 1986; De Rosa & Gambacorta, 1988; Segerer *et al.*, 1991).

7.4.2 Methodologies used to define polar lipid structures

Despite the value of polar lipid patterns in archaeal systematics, much of the structural identification has been based merely on differential staining procedures and R_F values. Recently, detailed chemical work has been done on archaeal polar lipids by applying chemical degradation, synthesis and spectroscopic procedures (NMR, MS) to identified compounds (Kates, 1972; Kushwaha *et al.*, 1981; Ferrante, *et al.*, 1986, 1987, 1988b; Morii *et al.*,

1986; De Rosa et al., 1987, 1988; Kramer et al., 1987; Lanzotti et al., 1987, 1989a,c; Thurl & Schafer, 1988; Fredrickson et al., 1989; Tsujimoto et al., 1989). The general methodologies used to extract, purify and degrade polar lipids have been described in Section 7.2. This section is devoted to some aspects of NMR and MS spectroscopy of archaeal polar lipids.

7.4.2.a NMR spectroscopy

The chemical characterization of archaeal polar lipids based on ^{13}C-NMR and ^{1}H assignments has been reported recently. These studies herald the use of basic NMR techniques, such as selective decoupling experiments and distortionless enhancement by polarization transfer (DEPT) for analysis of multiplicities, two-dimensional ^{13}C-^{1}H shift correlation performed by polarization transfer ($^1J_{C-H}$), and X-H shift correlation (XHCORR).

De Rosa et al. (1988) and Ferrante et al. (1986) have determined the ^{13}C-NMR spectrum of PG (Figure 7.9). The low field part of the ^{13}C-NMR spectrum of this lipid reveals signals in the region 68.80–70.7 ppm due to the C1 of the phytanyl chain in the α and β positions of glycerol, respectively, while the signals at 65.9, 78.4 and 71.50 ppm are typical of C1, C2 and C3 of the glycerol. The remaining three signals, with chemical shifts at 63.4, 71.9 and 67.2 ppm, were assigned to the C1, C2 and C3 of the glycerol head group. The signals at 65.9 and 67.2 ppm in the carbon-decoupled ^{13}C-NMR spectrum are doublets with $J_{P-O-C} = 1$ Hz, indicating coupling with phosphorus. The ^{13}C-NMR spectrum of PGP (Figure 7.9; De Rosa et al., 1988) shows the signals reported for PG with the exception of the chemical shifts of C1 and C3 of the glycerol head group at 66.6 and 66.7 ppm, respectively; the latter are due to the attachment of the second phosphorus to the C1 of the glycerol head group. Both signals in the proton noise-decoupled ^{13}C-NMR spectrum are doublets with $J_{C-O-P} = 1$ HZ to the coupling with phosphorus.

The ^{13}C-NMR spectrum of PL2 (Figure 7.9; Lanzotti et al., 1989b), in which a cyclic phosphate occurs, shows identical resonances to PGP with the exception that the C2 signal of the glycerol head group has shifted by +4.3 ppm (74.6 ppm) and appeared as a doublet with $J_{C-O-P} = 1$ Hz in the proton noise-decoupled ^{13}C-NMR spectrum. ^{13}C-NMR assignments of PA (Figure 7.9) are very like those reported for PG but lack the signals of the glycerol polar group (Lanzotti et al., 1989a).

Kramer et al. (1987) described the NMR spectra of aminolipids 2 and 20 (Figure 7.10). The nature of the phosphorus linkage in these compounds is established by ^{31}P-NMR in which a signal at 2.3 ppm observed relative to H$_3$PO$_4$ is characteristic of a phosphate ester linkage. The chemical shifts of C1, C2 and C3 of glycerol are quite similar to those described above for PG, the $>$CH$_2$ near nitrogen having its chemical shift at 41.25 ppm, while the next nearest oxygen shifts at 62.20 ppm.

^{13}C-NMR studies have proved useful in establishing the structure of PPDAD and PPTAD (4, 5, Figure 7.10; Ferrante et al., 1987). At low field, the ^{13}C-NMR spectrum of PPDAD shows five additional signals with respect to PG (Figure 7.9), namely those at 73.18, 72.90, 72.12, 64.39 and 60.00 ppm, respectively. The signal with the chemical shift at 64.39 ppm, typical of a —CH$_2$OH group, is assigned to the C5 of the polyol, while signals at 73.18 and 72.9 ppm, typical of carbon atoms adjacent to hydroxyl groups ($>$CH—OH), are assigned to the C3 and C4 of the polyol, respectively. Signals of the glycerol backbone at 66.20 ppm (C1) and at 78.84 ppm (C2), as well as signals of the polyol at 72.12 and 73.18 ppm (C3), show splittings which can be explained by coupling with phosphorus. Proton-phosphorus correlation spectra clearly show that phosphate couples to the H2 of the polyol thereby unequivocally establishing the linkage of the phosphate to the C2 position of the polyol. The nitrogen in PPDAD (4, Figure 7.10) is linked at the C1 (60.00 ppm) and the two signals with chemical shifts at 42.25 ppm and 44.48 ppm are typical of chemically unequivalent methyl groups of the nitrogen, lying in nonidentical chemical environments.

The ^{13}C-NMR spectrum of PPTAD (5, Figure 7.10) suggests that this lipid is very similar to PPDAD. The only differences are in the chemical shifts of C1 of the polyol, which in the case of PPTAD is shifted to a lower field at 67.64 ppm, and in the methyls, which are now coalesced into a singlet at 54.14 ppm. The downfield shift of C1 is of the same magnitude as the shift of the methyls on nitrogen; this is in agreement with the presence of a $^+$N(CH$_3$)$_3$ group on the C1 of the polyol (Ferrante et al., 1987).

In the low field region of the spectrum, ^{13}C-NMR of PI (12, Figure 7.10) shows signals typical of glycerol and CH$_2$ or phytanyl chains. In addition, six signals are present, all originating from methane groups with chemical shift values falling within a narrow range of 70 to 77 ppm, suggesting that inositol may form the head group (De Rosa et al., 1987; Ferrante et al., 1988b; Lanzotti et al., 1989c). This was confirmed by ^{1}H-NMR, which showed coupling constant values typical of *myo*-inositol (Ferrante et al., 1988b).

Nuclear magnetic resonance studies are also useful in defining the nature and configuration of sugars attached to both diether and tetraether core lipids. Ferrante et al. (1986) found that the ^{13}C-NMR spectrum of glycolipid 6 (Figure 7.10) showed low field signals with chemical shifts typical of C1 substituted glucose with a one-bond coupling constant ^1J-(C1,H1) equal to 159 Hz; these findings are consistent with a β-configuration. This was further confirmed by ^{1}H-NMR, which showed the presence of a doublet at 4.38 ppm with a coupling constant of 7.8 Hz, typical of an anomeric proton with a β-configuration.

The ^{13}C-NMR spectrum of glycolipid 7 (Figure 7.10) is consistent with the presence of two glucose residues. The one-bond carbon proton coupling constants for the anomeric carbons of 160 and 159 Hz are indicative of two β-linked sugar residues; this was confirmed by ^{1}H-NMR studies, which showed constants for the anomeric protons ^1J(H1,H2) of 7.8 and 7.6 Hz (Ferrante et al., 1986).

The ^{13}C-NMR spectrum of glycolipid 11 (Figure 7.10) revealed signals with chemical shifts at 104.77 and 97.71 ppm, indicating the presence of a β-linked galactose and an α-linked mannose residue. The C3 signal of the galactose is shifted downfield in accordance with a 1 → 3 glycosidic linkage (Ferrante et al., 1988b).

The ^{13}C-NMR spectrum of glycolipid GL1 (Figure 7.10) shows signals at 104.73 and 104.56 ppm, respectively. These signals were attributed to two β-linked anomeric carbons, a result in agreement with the presence of two galactose residues. The signal of the C6 of galactose is shifted downfield by more than 6 ppm indicating a 1 → 6 glycosidic linkage. The ^{1}H-NMR spectrum shows two signals of equal intensity attributed to the anomeric protons at 4.37 and 4.39 ppm with coupling constants of $J_{1,2}$ = 7.6 Hz and $J_{1,2}$ = 7.6 Hz; these results are typical of two β-galactose moieties (Ferrante et al., 1988b).

The configuration of the glycosidic linkages in polar lipids 8, 13, 14, 15 and 16 (Figure 7.10) was determined by ^{13}C-NMR (Kushwaha et al., 1981). The ^{13}C-NMR of compounds 8, 14 and 16 (Figure 7.10) showed two resonances in the region of the anomeric carbons, i.e. at 99.1 (14 and 16), 98.5 (8) and 106.7 (14 and 16), 106.1 (8) ppm. The former is consistent with the resonance expected for C1 of an α-glucopyranosyl residue linked 1 → 2 to another sugar. Similarly, resonance at 106.7 ppm was assigned to the anomeric carbon of β-galactofuranose linked at its 2-position to another sugar. These conclusions are also supported by ^{1}H-NMR resonance data which show signals at 4.88 ppm and 4.92 ppm for a β-anomeric proton, and at 5.02 ppm for an α-anomeric proton. The ^{13}C-NMR spectrum of compounds 13 and 15 (Figure 7.10) showed anomeric carbon resonances at 109.0 and 108.4 ppm consistent with the presence of two β-galactofuranose groups attached to different residues at their anomeric carbons.

Ferrante et al. (1986) reported that the ^{13}C-NMR spectrum of compound 3 (Figure 7.10) showed low field signals with chemical shifts consistent with a structure having glucosamine as a polar head group. The ^{13}C signal of the carbonyl group from —NHCOCH$_3$ was not seen due to the weakness of the signal, while the methyl signal, appearing around 23 ppm, was overlapped by the phytanyl carbon signals. The chemical shifts at

94.54 and 54.61 ppm coupled with phosphorus are due to C1 and C2 of the glucose residue, respectively. The other chemical shifts at 71.69, 74.19 and 42.33 ppm are due to the couple C3 and CH, C3, C4, C5 and C6 of the polar head group. The ^{1}H-NMR spectrum showed the presence of a doublet at 4.19 ppm with coupling ^3J(H1,H2) of 10.5 Hz, typical of a β-glucosamine linkage.

The ^{13}C-NMR of compound 2 (Figure 7.11) gave a signal at 103.6 ppm indicating a β-glucosidic linkage (Thurl & Schafer, 1988). Similarly, the ^{13}C-NMR of compound 5 (Figure 7.11) revealed a signal at 61.6 ppm, assigned to the C6 atom of glucose, and a resonance at 103.6 ppm, indicating a β-glucosidic linkage.

Lanzotti et al. (1989c) described NMR studies on 2,3-di-O-phytanyl-sn-glycero-1-(α-D-glucopyranosyl-3-phosphate). The configuration at the anomeric centre of the glucose was inferred to be α on the basis of the chemical shifts of the anomeric proton and carbon (^{1}H-NMR: 4.88 ppm; ^{13}C-NMR: 98.3 ppm) and the value of the $^3J_{H1-H2}$ (1 Hz). Moreover, the downfield shift of 3 ppm of the C3 signal indicated that the phosphate residue was localized on the C3 position of the glucose. Chemical shifts at 71.5, 70.8, 72.6 and 61.8 ppm were assigned to C2, C4, C5 and C6 for the glucose moiety.

The relative configuration of the anomeric centres of compound 7 (Figure 7.11) was inferred to be β for galactose and α for glucose on the basis of the values of $^3J_{H1-H2}$ (7 Hz for the galactose and 1 HZ for the glucose). These data are consistent with the value of the chemical shift of the anomeric carbons in the ^{13}C-NMR spectrum (104.3 for β-galactose and 98.6 for α-glucose). ^{1}H-NMR showed two anomeric signals at 4.80 and 4.20 ppm for glucose and galactose, respectively. The downfield shift observed in the ^{13}C-NMR spectrum of the C4 of galactose (77.4 ppm) indicated that the α-D-glucopyranose moiety was linked to the 4-position of the β-D-galactopyranose. The terminal position of α-D-glucopyranose was confirmed both by the close similarity of its ^{13}C signals with those of methyl α-D-glucopyranoside and by ^{13}C spin-lattice relaxation time (T_1) measurements. In fact, the glucose carbon atoms displayed the largest T_1 values due to their high mobility at the end of the glycosidic chain. The chemical shifts at 70.4, 70.9, 74.0 and 61.3 ppm were due to C2, C3, C5 and C6 for the galactose moiety, while the signals at 71.8, 73.4, 70.1, 71.8 and 61.7 ppm can be ascribed to C2, C3, C4, C5 and C6 of the glucose residue (Lanzotti et al., 1987).

^{1}H and ^{13}C-NMR signals were in accordance with the structure of compound 8 (Figure 7.11) where two polar heads, β-galactopyranose and phospho-*myo*-inositol, are respectively linked to the opposite side of tetraether a (Figure 7.4). In comparison with the ^{13}C-NMR spectrum of compound 7 (Figure 7.11) the signals of the galactose moiety are the same with the exception of C4 (68.2 ppm). Moreover, in the ^{13}C-NMR spectrum of compound 8 (Figure 7.11) the C1 of the glycerol linked to phospho-*myo*-inositol appeared as a doublet with $J_{P-O-C} = 1$ Hz; it was shifted by 3 ppm with respect to the free C1 of glycerol as found in glycolipid 7 (Figure 7.11). Consistent with the presence of a *myo*-inositol head group were signals at 69.8, 70.2, 71.4, 73.4 and 72.6 ppm, the first being a doublet with $J_{C-O-P} = 1$ Hz in the proton noise-decoupled ^{13}C-NMR spectrum.

The ^{1}H and ^{13}C-NMR signals of compound 9 (Figure 7.11), in comparison with those of compounds 7 and 8 (Figure 7.11), clearly indicated that it was a phospho-*myo*-inositol derivative of compound 7 (Figure 7.11). In fact, all the ^{13}C resonances in compound 7 are present in compound 9 with the exception of the CH$_2$OH of the second glycerol, which shows a chemical shift and multiplicity identical to that for the corresponding signal in compound 8 (Figure 7.11; Lanzotti *et al.*, 1987).

7.4.2.b Mass spectrometry

Positive ion field desorption mass spectrometry (FD-MS) and both positive and negative fast atom bombardment (FAB-MS), of intact complex lipids produce ions which correspond to protonated molecules thereby revealing their molecular weights. These techniques are also sufficiently energetic to cleave phosphodiester and glycosidic linkages and to provide information on the components of polar lipids.

De Rosa *et al.* (1988) and Ferrante *et al.* (1987) described negative and positive FAB-MS for PG (Figure 7.9). In negative FAB they found an ion peak at m/z 805 (M—H)$^+$ for C$_{20}$, C$_{20}$ forms, and at m/z 875 (M—H)$^+$ for C$_{25}$, C$_{20}$ forms, and a very diagnostic fragment at m/z 731 (M—H—C$_3$H$_6$O$_2$) corresponding to the loss of glycerol. The positive FAB mass spectra showed M$^+$ at m/z 828 (M + Na$^+$) for the C$_{20}$, C$_{20}$ forms and at 898 (M + Na$^+$) for the C$_{25}$, C$_{20}$ forms. Also present in the spectra were fragmentation peaks at m/z 846 and m/z 776 (M—CH$_2$OH) for the C$_{25}$, C$_{20}$ and C$_{20}$, C$_{20}$ forms and at m/z 815 and m/z 745 for the loss of CHOH—CH$_2$OH from M$^+$ for the C$_{25}$, C$_{20}$ and C$_{20}$, C$_{20}$ forms. The mass spectra of PGP (Figure 7.9) showed the same peak as PG due to the loss of the additional phosphate group. All spectra show a regular series of peaks relating to the sequential cleavage of saturated isopranoid units.

Positive FAB-MS spectra are obtained by dissolving samples in a glycerol matrix, adding 0.1% (w/v) sodium acetate in methanol to the probe prior to bombardment with argon atoms with a kinetic energy equivalent to 2–6 keV. Negative FAB-MS spectra are obtained as for positive FAB-MS without the addition of sodium acetate (De Rosa *et al.*, 1988). According to Ferrante *et al.* (1986, 1987), negative FAB-MS spectra are obtained operating at 6 kV accelerating potential with a fast atom beam of xenon using a field ion source operating with a tube current of 1.5 mA, 9 keV, using glycerol as the matrix.

Recently, Fredrickson et al. (1989) and Tsujimoto et al. (1989) revised the structure of PGP (Figure 7.9) on the basis of data obtained using the tandem mass spectrometry (MS/MS) method. This approach has the advantage of discrete selection of ions with the first mass spectrometer and analysis of fragment ions derived from a single mass. FAB-MS/MS analyses can provide new information on the structure of intact polar ether lipids even when these compounds are present in mixtures at low concentrations. Fredrickson et al. (1989) used either glycerol as the matrix or a mixture of 1,1,3,3-tetramethylurea (TMU) and triethanolamine (TEA); the FAB gun (xenon) operated at a voltage of 6 kV. Positive-ion mass spectrometric analysis of PG (Figure 7.9), using glycerol as matrix, showed ions at m/z 829.9 and 851.8 corresponding to the protonate monosodium salt of PG and the monosodium salt cationized by Na, respectively.

The negative-ion FAB-MS spectrum of PG (Figure 7.9), with TMU-TEA as matrix, is simpler than the positive FAB spectrum, and shows ions at m/z 805.8 corresponding to the deprotonated PG molecule. The ion at m/z 827.7, with low abundance, can be explained as the deprotonated monosodium salt of PG. Both the positive and negative FAB-MS spectra of PGP (Figure 7.9) indicate that the molecular structure of this lipid is O-methylated (PGP—Me; Figure 7.9). Peaks at m/z 900 and 922 correspond to the deprotonated molecules $(M-1)^-$ of PGP-Me and of its sodium salt, respectively. The ion at m/z 111 clearly shows that the terminal phosphate group is methylated (Fredrickson et al., 1989; Tsujimoto et al., 1989).

The positive-ion mass spectrum of PGS (Figure 7.9) contains an ion at m/z 909.7 (low abundance) indicating the presence of the protonated monosodium salt of PGS. The ions at m/z 931.7 and m/z 953.6 may be due to Na cationization of both the monosodium and disodium salts of PGS respectively. The ions at m/z 829.9 and m/z 851.7 correspond to masses of fragments formed by the loss of the terminal SO_3 group from the protonated monosodium salt $(908.6 + H^+ - 80)^+$ and disodium salt $(930.7 + H^+ - 80)^+$ of PGS (Fredrickson et al., 1989).

The relative concentrations of polar lipids, which are judged from the relative abundance of the ions in FAB-MS spectra, are in general agreement with data from the literature. Lanzotti et al. (1989a,b,) found that negative FAB-MS spectra of PL2 and PA (Figure 7.9) had the same peaks as PG and PGP for the first compound while the second compound showed a molecular ion at m/z 731 $(M-H^+)$ consistent with the proposed structure.

Morii et al. (1986) obtained a molecular ion peak of PNL2b (1, Figure 7.10) using negative-ion FAB-MS with a crown-ether, 15-crown-5, added to glycerol, as a matrix. The spectrum showed peaks at m/z 818 (M−H), 731 (M−CH$_2$CH(NH$_3$)COOH + H), 433 (M-OC$_{20}$H$_{41}$-CH$_2$CH(NH$_3$)COOH) and 224 (M-20C$_{20}$H$_{41}$-H). Other ionization methods, such as electron impact (EI), field desorption (FD), do not give molecular ion peaks. The FD-MS spectrum of the acid form of PNL2b showed peaks at m/z 733 (diphytanylglycerophosphate + H)$^+$, 653 (diphytanylglycerol + H)$^+$ and at 634 (diphytanylglycerol - H$_2$O)$^+$. The masses of the peaks can be interpreted by pyrolysis of PNL2b at both sides of the phosphodiester bond rather than by fragmentation. The peaks are consistent with the presence of a phosphate group being considered in combination with the negative FAB-MS spectrum.

The FAB-MS spectra of PPDAD (4, Figure 7.10) identified in *Methanospirillum hungatei* by Ferrante et al. (1987) showed molecular ion [M$^-$] at m/z 893 much higher than that expected for a PG molecule, a result in contrast to the findings of Kushwaha et al. (1981; see above). The molecular weight of PPDAD was further confirmed by positive FAB-MS, which revealed a molecular ion peak at m/z 894 [M + H]$^+$.

Negative FAB-MS of PPTAD (5, Figure 7.10) gives an ion peak at m/z 892 (M-H-15) and, together with its fragment ions, it is identical to PPDAD (4, Figure 7.10). The molecular ion of PPTAD at m/z 909 [M + H]$^+$ was revealed by positive FAB-MS indicating that in negative FAB there is loss of a proton and a methyl group required to generate a negative ion (Ferrante et al., 1987).

Kramer et al. (1987) reported that the size of the side-chain of diether ethanolamine and tetraether ethanolamine (2, 18, Figure 7.10) can be established by FAB-MS. Each compound gives a strong M + 1 ion at m/z 776.7 and m/z 1425.1, respectively. The difference between these M + 1 ions and the corresponding glycerol diether (m/z 776.7 − 653.7 = 123.0) and glycerol tetraether (m/z 1425.1 − 1301.6 = 123.5) is basically the same, thus a 2-aminoethyl phosphate side-chain having a mass of m/z 123.01 can be attributed to both compounds.

Ferrante et al. (1988b) found that in the FAB-MS spectrum of Pl (12, Figure 7.10) an ion peak (M−H) was present at m/z 893. The fragment ion at m/z 731 was attributed to the diether analogue of phosphatidic acid with the loss of an ion fragment with a molecular weight of 162 similar to an inositol fragment.

Lanzotti et al. (1987) reported that for the glycolipids and phosphoglycolipids (7, 8, 9, Figure 7.11) found in *Desulfurococcus mobilis* the positive FAB-MS spectra gave ion peaks at m/z 1646, 1726 and 1888, (M + NA$^+$), respectively, consistent with the proposed structures.

7.5 ISOPRENOID QUINONES

7.5.1 Archaeal quinones: structural types

The physiological importance of isoprenoid quinones in electron transport (Brodie & Watanabe, 1966; Redfearn, 1966; Dunphy & Brodie, 1971; Yamada et al., 1977) has been known for many years but only relatively recently has the value of these compounds as taxonomic markers been appreciated (see Chapter 8; Watanuky & Aida, 1972; Shah & Collins, 1980; Collins & Jones, 1981; Collins et al., 1981; De Rosa & Gambacorta, 1988). The structure and distribution of the various types of isoprenoid quinones found in archaea is outlined below.

Isoprenoid quinones are recovered from the neutral fraction of total lipid extracts of archaea. The isoprenoid quinones of these organisms are based on naphthoquinone or benzothiophenquinone chromophores as

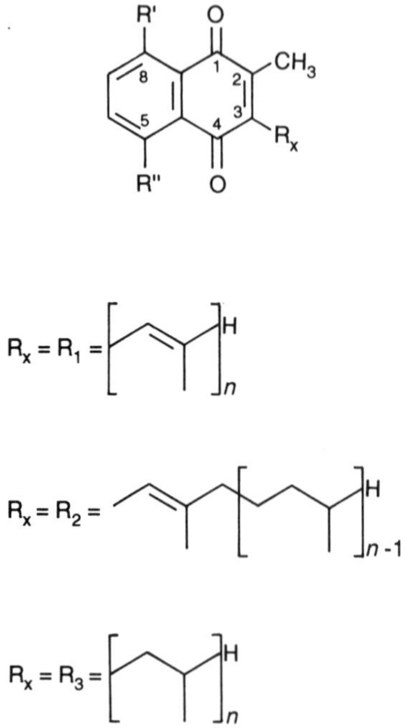

	n	R_x
MK-8	8	R_1
MK-7	7	R_1
MK-6	6	R_1
MK-5	5	R_1
MK-4	4	R_1
MK-8 (H_2)	8	R_1^*
MK-7 (H_2)	7	R_1^*
MK-7 ($7H_2$)	7	R_3
MK-6 ($6H_2$)	6	R_3
MK-5 ($5H_2$)	5	R_3
MK-4 ($4H_2$)	4	R_3
MK-6 ($5H_2$)	6	R_2
MK-5 ($4H_2$)	5	R_2
MK-4 ($3H_2$)	4	R_2
TK-7	7	R_1
TK-6 ($6H_2$)	6	R_3
TK-6 ($5H_2$)	6	R_2

MK-type R' and R" = H
TK-type R' and R" = H and CH_3 (not necessarily respectively)
$R_1^* = R_1$ with one double bond hydrogenated.

Figure 7.12 Archaeal quinones based on naphthoquinone or benzothiophenquinone chromophores.

7. Archaeal Lipids

shown in Figure 7.12. The naphthoquinones present in archaea are of MK (2-methyl-3-polyprenyl-1,4-naphthoquinones; Figure 7.12) and TK (2,5- or 8-dimethyl-3-polyprenyl-1,4-naphthoquinones; Figure 7.12) types; these compounds differ in the length and degree of saturation of the C3 polyprenyl side-chain.

Tindall et al. (1989, 1991) described two fully saturated menaquinones from archaea. They found an MK-7 (7H$_2$)—type quinone in the thermophile *Archaeoglobus fulgidus* and, most interestingly, a new MK-6(H$_{12}$) with a partially cyclized side-chain in *Pyrobaculum organotrophum*. The new compound was shown to correspond to 2-(14-(3-(1,5-dimethyl-hexyl)-cyclopentyl)-3,7,11-trimethyltetradecyl-3-methyl-1,4-naphthoquinone.

Benzothiophenquinones are present in all species belonging to the order Sulfolobales (De Rosa et al. 1975, 1977c; Collins & Langworthy, 1983; Lanzotti et al., 1986; Trincone et al., 1986). They are of three main types: namely, CQ (6-polyprenyl-5-methylthiobenzo[b]thiophene-47-quinone;

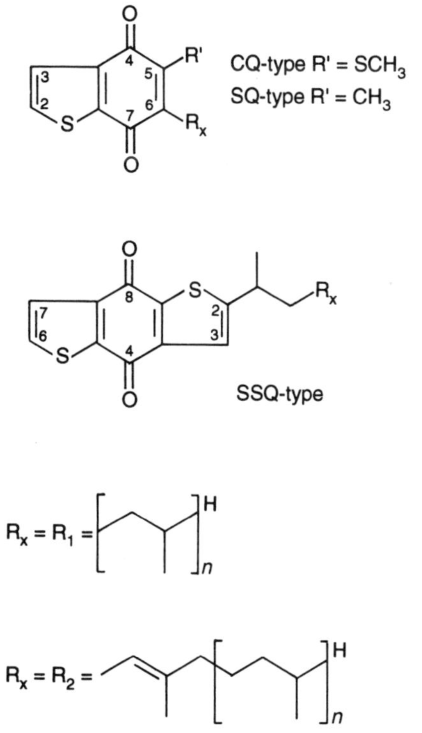

	n	R$_x$
CQ-6 (6H$_2$)	6	R$_1$
CQ-6 (5H$_2$)	6	R$_2$
CQ-6 (4H$_2$)	6	R$_2^*$
CQ-6 (3H$_2$)	6	R$_2^*$
CQ-5 (5H$_2$)	5	R$_1$
CQ-4 (4H$_2$)	4	R$_1$
SQ-6 (6H$_2$)	6	R$_1$
SQ-5 (5H$_2$)	5	R$_1$
SQ-4 (4H$_2$)	4	R$_1$
SQ-3 (3H$_2$)	3	R$_1$
SQ-6 (5H$_2$)	5	R$_2$
SSQ-5 (5H$_2$)	5	R$_1$

R$_2^*$ = R$_2$ with one or two additiional isoprenoidic double bonds

Figure 7.12 (*Continued*)

Table 7.7 ^{1}H and ^{13}C chemical shifts and multiplicities of the benzothiophenquinones of the order Sulfolobales (values are ppm).

C	CQ-6(6H$_2$) δ$_C$	CQ-6(6H$_2$) δ$_H$	SQ-6(6H$_2$) δ$_C$	SQ-6(6H$_2$) δ$_H$	SSQ-5(5H$_2$) δ$_C$	SSQ-5(5H$_2$) δ$_H$
2'	132.4D	7.45	132.6	7.50	133.2	7.58
3'	126.3D	7.35	126.1	7.43	126.7	7.54
4'	177.4S	–	182.0	–	182.6	–
5'	145.4S	–	141.2	–	143.1[b]	–
6'	149.8S	–	147.1	–	148.3[b]	–
7'	176.8S	–	181.0	–	181.9	–

Position	δC	δH	δC	δH	δC	δH
8'	143.2S	—	143.3	—	144.7[c]	—
9'	141.5S	—	141.4	—	143.0[c]	—
1	26.7T	2.75	29.6	2.49	122.3D	7.24
2	35.6T	—	35.7	—	133.4S	—
3	33.3D	—	33.3	—	29.8	2.98
4	36.9T	0.95 or 1.25	37.0	0.98 or 1.27	37.0	0.96 or 1.26
5,9,13,17	24.4T	1.15; 1.34	24.5	1.14; 1.30	24.6	1.10; 1.28
6,8,10,12,14,16,18,20	37.4T	0.95; 1.25	37.4	0.98; 1.20	37.6	0.98; 1.20
7,11,15,19	32.8D	1.32	32.8	1.28	32.8	1.26
21	24.8T	—	24.8	—	24.9	—
22	39.3T	1.05	39.4	1.04	39.5	1.03
23	27.9D	[a]	28.0	—	28.1	—
24	19.5Q	0.90	19.5	0.90	19.8	1.30
25,26,27,28	19.8Q	0.75	19.8	0.78	19.9	0.76
29,30	22.6Q	0.75	22.6	0.78	22.7	0.76
S-CH$_3$	22.7		22.7		22.8	
CH$_3$	17.9Q	2.52	14.1	2.05		

[a] No correlation observed.
[b,c] Assignment interchangeable.

253

Figure 7.12), SQ (6-polyprenyl-5-methylbenzo[b]thiophene-4,7-quinone; Figure 7.12) and SSQ (2-polyprenyl-benzo[1,2-b;4,5-b']dithiophene-4,8-quinone; Figure 7.12).

7.5.2 Chromatographic procedures

Neutral lipids (the acetone-soluble fraction of the total lipid extract) are treated with light petroleum and the soluble material chromatographed on a silica gel column eluted with light petroleum and a linear gradient of diethyl ether in light petroleum from 0 to 10%, v/v. Archaeal isoprenoid quinones are well resolved on thin-layer chromatographic plates developed with light petroleum/diethyl ether (95:5, v/v). Products, if colourless, are revealed using a UV lamp or by iodine vapour.

(High-performance liquid chromatographic analysis is done using a Microporasil column (30 cm × 3.9 mm i.d.) for analytical work and a 30 cm × 7.8 mm i.d. column for preparative work with a flow rate of 1.5 ml/min; the solvents used are n-hexane/ethyl acetate (99:1, v/v). Tindall et al. (1989) used an ODS-2 column (250 × 4.6 mm) with methanol as the mobile phase; eluants from the column were detected using a UV detector at 260 nm.

7.5.3 NMR spectroscopy

A detailed NMR study on the benzothiophenquinones of members of the genus *Sulfolobus* was given by Lanzotti et al. (1986). This study (Table 7.7) was based on multipulse one-dimensional and two-dimensional techniques; the signal multiplicities of CQ-6 (6H$_2$) (Figure 7.12) were determined from DEPT experiments; one-bond carbon-proton connectivities were based on two-dimensional carbon-proton shift correlation via $^1J_{CH}$ and chemical shifts of aromatic and quinoid carbons were fully assigned from two- and three-bond carbon-proton couplings, as available from a gated decoupled ^{13}C-NMR spectrum. ^{13}C-NMR signals of SQ-6(6H$_2$) and SSQ-5(5H$_2$) (Figure 7.12) were determined by comparison with those of CQ-6(6H$_2$). ^{1}H-NMR chemical shifts of archaeal naphthoquinones of MK and TK types are given in Table 7.8. Tindall et al. (1989) give a detailed ^{13}C-NMR of MK-7 (7H$_2$) from *Archaeoglobus fulgidus*.

7.5.4 Mass spectrometry

Mass spectrometry is important for determining the structure of archaeal isoprenoid quinones as it provides information on the molecular weight, the nature of the ring system, and the length and degree of saturation of the isoprenyl side-chain. The most diagnostic fragments found in EI-MS

Table 7.8 ^{1}H-NMR data of archaeal naphthoquinones of MK and TK types.

Compound[a]	MK-6(6H$_2$)	MK-6(5H$_2$)	TK-6(6H$_2$)	MK-7(7H$_2$)
Proton signal		Chemical shifts (ppm)		
CH$_2$ next to quinone	2.62	3.37	2.59	8.04–8.10
CH$_2$-C= (α isoprene unit)	–	5.10	–	7.65–7.70
CH = CH(CH$_3$)-CH$_2$ (α isoprene unit)	–	2.00	–	–
CH$_3$- (α isoprene unit)	0.87	1.77	0.87	2.17
CH$_2$ and CH-saturated	1.1–1.4	1.1–1.4	1.1–1.4	2.61
CH$_3$ saturated	0.85	0.85	0.85	–
Aromatic protons (C5 and/or C8)	8.07	8.07	8.00	–
Aromatic protons (C6 and/or C7)	7.68	7.68	7.52–7.45	0.97
Benzenoid CH$_3$	–	–	2.73	1.1–1.4
Quinonoid CH$_3$	2.19	2.19	2.16	0.83

[a] For structures see Figure 7.12.

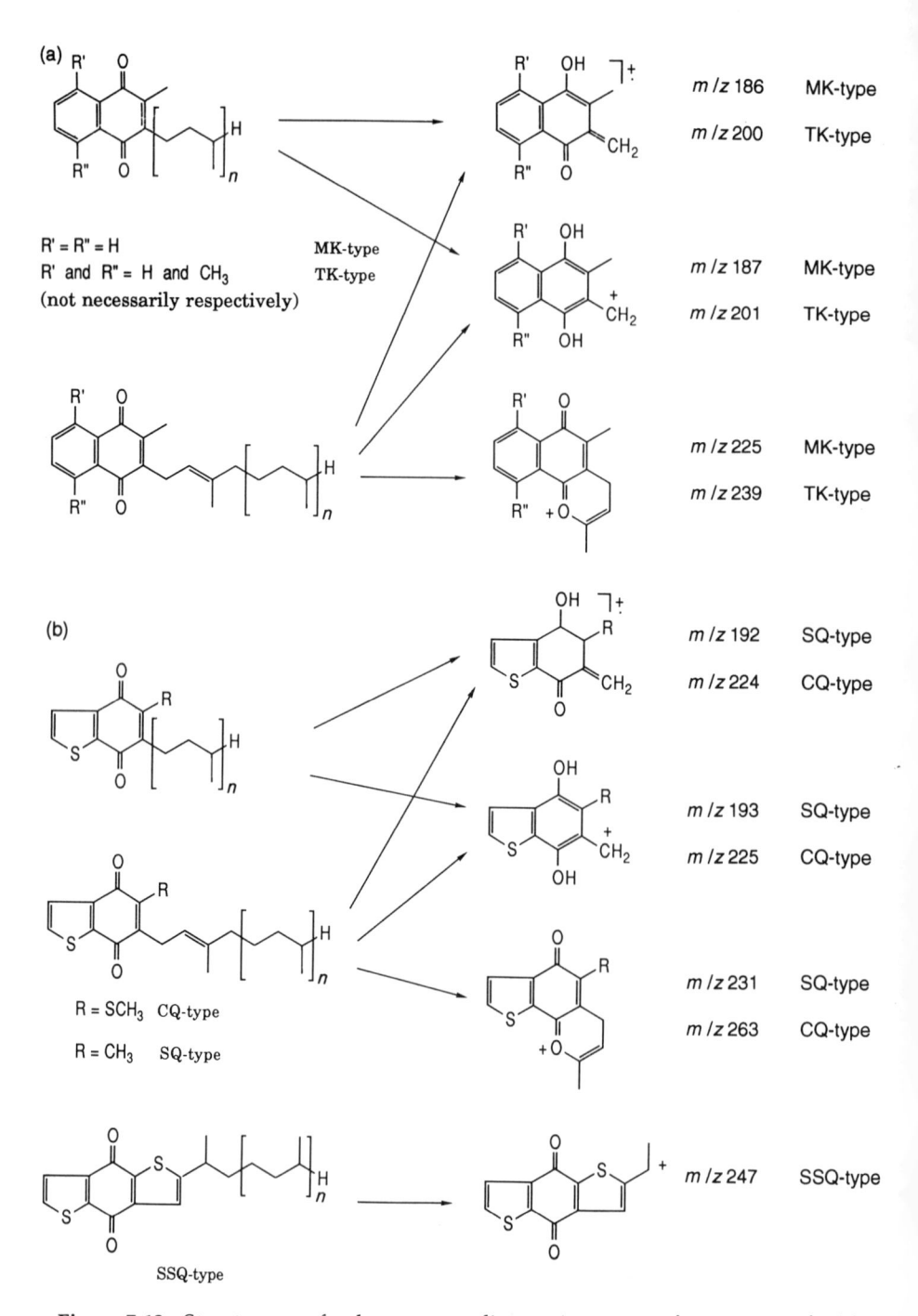

Figure 7.13 Structures of the more diagnostic mass fragments of (a) naphthoquinones of MK and TK types and (b) benzoquinones of SQ, CQ and SSQ types that occur in archaea. See Figure 7.12 for structures.

spectra of archaeal naphthoquinones and benzothiophenquinones (De Rosa et al., 1977c; Thurl et al., 1985; Trincone et al., 1986), are shown in Figure 7.13. The MS spectral analysis of MK-7(7H$_2$) from *Archaeoglobus fulgidus* reveals intense peaks at m/z 662 (M$^+$) and at m/z 187; a less intense peak at m/z 648 can be attributed to the loss of a CH$_3$ group from the naphthoquinone nucleus. The lack of fragmentation in the region m/z 187 to 648 is indicative of the absence of unsaturation in the isoprenoid chain (Thurl et al., 1986; Tindall et al., 1989).

7.5.5 Ultraviolet and visible spectroscopy

Ultraviolet and visible spectral data of archaeal naphthoquinones and benzothiophenquinones in different organic solvents are shown in Table 7.9. The coloured benzothiophenquinones of the CQ, SQ and SSQ types (Figure 7.12) range in colour from yellow for the SQ and SSQ types to red-orange for the CQ type. The UV-visible spectra of these unusual isoprenoid quinones are quite different from those of other described bacterial isoprenoid quinones. The UV-visible spectra of MK and TK naphthoquinones are quite similar with five absorption maxima (λ_{max}) at 242, 248, 260, 269 and 326 mm.

7.5.6 Quinone content in archaea: taxonomic considerations

Isoprenoid quinone studies on archaea have focused on halophilic and thermophilic strains; detailed studies on the quinone composition of methanogenic archaea have yet to be performed. The thermophilic archaea can be divided into two groups (Table 7.10) with respect to their quinone composition. Members of the genera *Acidianus*, *Desulfurolobus*, *Metallosphaera*, *Stygiolobus* and *Sulfolobus* have with few exceptions been based on the benzothiophene nucleus, whereas the genera *Archaeoglobus*,

Table 7.9 Ultraviolet and visible spectral data of archaeal naphthoquinones and benzothiophenquinones.

Compound[a]	Solvent	λ_{max} (nm)	References
CQ-6 (6H$_2$)	Methanol	241,282,333,471	De Rosa et al. (1975, 1977c)
CQ-6 (6H$_2$)	Hexane	239,273,280,320,456	Thurl et al. (1985)
CQ-6 (6H$_2$)	Iso-octane	237,272,278,322	Collins (1985)
SQ-6 (6H$_2$)	Hexane	225,270,279,321	Thurl et al. (1986)
SSQ-5 (5H$_2$)	Chloroform	250,297,346	Trincone et al. (1986)
MK type	Iso-octane	242,248,260,269,326	Collins and Jones (1981)
TK-7	Iso-octane	242,248,259,269,325	Collins (1985)

[a] See Figure 7.12 for structures.

Table 7.10 Distribution of isoprenoid quinones in archaea.

Microorganism	Major isoprenologue(s)	Minor components	Reference
Amoebacter morrhuae	MK-8		Collins et al. (1981)
Halobacterium cutirubrum	MK-8	MK-8(H_2), MK-7(H_2), MK-7	
H. halobium NCMB 736, NCMB 764, NCMB 777, and NCMB 2080	MK-8, MK-8(H_2)	MK-7(H_2), MK-7	
H. saccharovorum	MK-8, MK-8(H_2)		
H. salinarium	MK-8, MK-8(H_2)	MK-7(H_2), MK-7	
H. simoncinii ssp. neapolitanum	MK-8, MK-8(H_2)	MK-7(H_2), MK-7	
H. trapanicum	MK-8, MK-8(H_2)	MK-7(H_2), MK-7	
Halobacterium spp.	MK-8, MK-8(H_2)	MK-7(H_2), MK-7	
Halococcus morrhuae	MK-8, MK-8(H_2)	MK-7(H_2), MK-7	
Haloferax volcanii	MK-8, MK-8(H_2)	MK-7(H_2), MK-7	
Natronobacterium gregoryi	MK-8, MK-8(H_2)	MK-7(H_2), MK-7	
N. magadii	MK-8, MK-8(H_2)	MK-7(H_2), MK-7	
N. pharaonis	MK-8, MK-8(H_2)	MK-7(H_2), MK-7	
Thermoplasma acidophilum	MK-7, TK-7	MK-6, MK-5	Collins (1985)
Archaeoglobus fulgidus	MK-7($7H_2$)		Tindall et al. (1989)
Thermoproteus tenax	MK-6($6H_2$), MK-4($4H_2$)	MK-5($5H_2$), MK-6($5H_2$)	Thurl et al. (1985)
Pyrobaculum organotrophum	Δ14-cyclopentyl MK-6(H_{10})	MK-6($6H_2$)	Tindall et al. (1991)
Sulfolobus solfataricus	CQ-6($6H_2$)	SQ-6($6H_2$), SSQ-5($5H_2$)	De Rosa et al. (1977c); Tricone et al. (1986)
S. acidocaldarius	CQ-6($6H_2$)	CQ-6($5H_2$), CQ-5($5H_2$), CQ-4(H_2)	Collins and Langworthy (1983)
Metallosphaera	CQ-6($6H_2$)		Huber et al. (1989)
Acidianus brierleyi	CQ-6($6H_2$)	CQ-6($5H_2$), CQ-5($5H_2$), CQ-4($4H_2$), CQ-6($5H_2$), SQ-6($5H_2$)	Collins and Langworthy (1983); Segerer et al. (1986)
Stygiolobus azoricus	SQ-6($6H_2$)		Segerer et al. (1991)
Desulfurolobus ambivalens: anaerobically grown	SQ-6($6H_2$)		Tricone et al. (1989)
aerobically grown	CQ-6($6H_2$), SQ-6($6H_2$)	SSQ-5($5H_2$)	

Thermoplasma and Thermoproteus have only unsaturated or fully saturated isoprenoid naphthoquinones of the MK and TK types (Figure 7.12). In contrast, the isoprenoid quinone composition of halophilic archaea is more homogeneous; MK-8 and MK-8(H$_2$) are the major quinones found in halophilic archaea analysed to date; minor compounds of the quinone pool are generally MK-7 and MK-7(H$_2$).

ACKNOWLEDGEMENT

The authors are indebted to Mr E. Turco for drawing the figures.

REFERENCES

Baddiley, J., Buchanan, J.G., Hanschumaeher, R.E. and Prescott, J.F. (1956). Chemical studies on the biosynthesis of purine nucleotides. 1. The purification of β-glycyl-glucosylamine. *J. Chem. Soc.* 2818–923.
Bligh, E.G. and Dyer, W.J. (1959) A rapid method of total lipid extraction and purification. *Can. J. Biochem. Physiol.* **37**; 911–7.
Brodie, A.F. and Watanabe, T. (1966) Mode of action of vitamin K in microorganisms. *Vitam. Horm* **24**: 467–63.
Christl, N.M., Reich, M.I. and Roberts, J.D. (1971) NMR spectroscopy, carbon-13 chemical shifts of methylcyclopentanes, cyclopentanols and cyclopentyl acetates. *J. Am. Chem. Soc.* **93**: 3466–70.
Collins, M.D. (1985) Structure of thermoplasmaquinone from *Thermoplasma acidophilum*. *FEMS Microbiol. Lett.* **28**: 21–3.
Collins, M.D. and Jones, D. (1981) Distribution of isoprenoid quinone structural types in bacteria and their taxonomic implications. *Microbiol. Rev.* **45**: 316–54.
Collins, M.D. and Langworthy, T.A. (1983) Respiratory quinone composition of some acidophilic bacteria. *System. Appl. Microbiol.* **4**: 295–304.
Collins, M.D., Goodfellow, M. and Minnikin, D.E. (1980) Fatty acid, isoprenoid quinone and polar lipid composition in the classification of *Curtobacterium* and related taxa. *J. Gen. Microbiol.* **118**: 29–37.
Collins, M.D., Ross, H.N.M., Tindall, B.J. and Grant, W.D. (1981) Distribution of isoprenoid quinones in halophilic bacteria. *J. Appl. Bacteriol.* **50**: 559–65.
Comita, P.B., Gagosian, R.B., Pang, H. and Costello, C.E. (1984) Structural elucidation of a unique macrocyclic membrane lipid from a new, extremely thermophilic, deep-sea hydrothermal vent archaebacterium, *Methanococcus jannaschii*. *J. Biol. Chem.* **259**: 15234–41.
Dawson, R.M.C. (1954) The measurement of ^{32}P labelling of individual kephalins and lecithin in a small sample of tissue. *Biochim. Biophys. Acta* **14**: 374–80.
De Rosa, M. and Gambacorta, A. (1986) Lipid biogenesis in archaebacteria. In *Archaebacteria '85*, (Kandler, O. and Zillig, W., eds), pp. 278–85. Gustav Fischer Verlag: Stuttgart.
De Rosa, M. and Gambacorta, A. (1988) The lipids of archaebacteria. *Prog. Lipid Res.* **27**: 153–75.
De Rosa, M., Gambacorta, A. and Minale, L. (1975) A terpenoid 4,7-thianaphthenquinone from an extremely thermophilic and acidophilic microorganism. *Chem. Commun.* 392–3.
De Rosa, M., Gambacorta, A. and Bu'Lock, J.D. (1976a) The *Caldariella* group of extreme thermoacidophilic bacteria: Direct comparison of lipids in *Sulfolobus*, *Thermoplasma* and the MT strains. *Phytochemistry* **15**: 143–5.
De Rosa, M., De Rosa, S., Gambacorta, A. and Bu'Lock, J.D. (1976b) Isoprenoid triether lipids from *Caldariella*. *Phytochemistry* **15**: 1995–6.

De Rosa, M., De Rosa, S., Gambacorta, A., Minale, L. and Bu'Lock, J.D. (1977a) Chemical structure of the ether lipids of thermophilic acidophilic bacteria of the *Caldariella* group. *Phytochemistry* **16**: 1961–5.
De Rosa, M., De Rosa, S. and Gambacorta, A. (1977b) ^{13}C-NMR assignment and biosynthetic data for the ether lipids of *Caldariella*. *Phytochemistry* **16**: 1909–12.
De Rosa, M., De Rosa, S., Gambacorta, A., Minale, L., Thomson, R.H. and Worthington, R.D. (1977c) Caldariellaquinone, a unique benzo[b]thiophen-4,7-quinone from *Caldariella acidophila*, an extremely thermophilic and acidophilic bacterium. *J. Chem. Soc. Perkin* I: 653–657.
De Rosa, M., Gambacorta, A. and Nicolaus, B. (1980a) Regularity of isoprenoid biosynthesis in the ether lipids of archaebacteria. *Phytochemistry* **19**: 791–3.
De Rosa, M., De Rosa, S., Gambacorta, A. and Bu'Lock, J.D. (1980b) Structure of calditol, a new branched-chain nonitol, and of the derived tetraether lipids in thermoacidophilic archaebacteria of the *Caldariella* group. *Phytochemistry* **19**: 249–54.
De Rosa, M., Esposito, E., Gambacorta, A., Nicolaus, B. and Bu'Lock, J.D. (1980c) Effects of temperature on ether lipid composition of *Caldariella acidophila*. *Phytochemistry* **19**: 827–31.
De Rosa, M., Gambacorta, A., Nicolaus, B., Ross, H.N.M., Grant, W.D. and Bu'Lock, J.D. (1982) An asymmetric archaebacterial diether lipid from alkaliphilic halophiles. *J. Gen. Microbiol.* **128**: 343–8.
De Rosa, M., Gambacorta, A., Nicolaus, B. and Grant, W.D. (1983a) A C$_{25}$, C$_{25}$ diether core lipid from archaebacterial haloalkaliphiles. *J. Gen. Microbiol.* **129**: 2333–7.
De Rosa, M., Gambacorta, A., Nicolaus, B., Chappe, B. and Albrecht, P. (1983b) Isoprenoid ethers: Backbone of complex lipids of the archaebacterium *Sulfolobus solfataricus*. *Biochim. Biophys. Acta* **753**: 248–56.
De Rosa, M., Gambacorta, A. and Nicolaus, B. (1983c) A new type of cell membrane in thermophilic archaebacteria, based on bipolar ether lipids. *J. Membrane Sci.* **16**: 287–94.
De Rosa, M., Gambacorta, A. and Gliozzi, A. (1986a) Structure, biosynthesis and physicochemical properties of archaebacterial lipids. *Microbiol. Rev.* **50**: 70–80.
De Rosa, M., Gambacorta, A., Lanzotti, V., Trincone, A., Harris, J.E. and Grant, W.D. (1986b) A range of ether core lipids from the methanogenic archaebacterium *Methanosarcina barkeri*. *Biochim. Biophys. Acta* **875**: 487–92.
De Rosa, M., Gambacorta, A., Trincone, A., Basso, A., Zillig, W. and Holz, I. (1987) Lipids of *Thermococcus celer*, a sulfur-reducing archaebacterium: Structure and biosynthesis. *System. Appl. Microbiol.* **9**: 1–5.
De Rosa, M., Gambacorta, A., Grant, W.D., Lanzotti, V. and Nicolaus, B. (1988) Polar lipids and glycine betaine from haloalkaliphilic archaebacteria. *J. Gen. Microbiol.* **134**: 205–11.
De Rosa, M., Lanzotti, V., Nicolaus, B., Trincone, A. and Gambacorta, A. (1989) Lipids of archaebacteria: Structural and biosynthetic aspects. In *Microbiology of Extreme Environments and Its Potential for Biotechnology* (da Costa, M.S., Duarte, J.C. and Williams, R.A.D., eds), pp. 131–51. Elsevier Applied Science: London.
Dittmer, J.C. and Lester, R.L. (1964) A simple, specific spray for the detection of phospholipids on thin-layer chromatograms. *J. Lipid. Res.* **5**: 126–7.
Dittmer, J.C. and Wells, M.A. (1969) Quantitative and qualitative analysis of lipids and lipid components. In *Methods in Enzymology*, Volume 14 (Lowenstein, J.M;. ed), pp. 432–530. Academic Press: New York.
Dunphy, P.J. and Brodie, A.F. (1971) *Methods in Enzymology*, Volume 18. Academic Press: New York.
Ferrante, G., Ekiel, I. and Sprott, D.G. (1986) Structural characterization of the lipids of *Methanococcus voltae*, including a novel N-acetylglucosamine 1-phosphate diether. *J. Biol. Chem.* **36**: 17062–6.
Ferrante, G., Ekiel, I. and Sprott, D.G. (1987) Structure of diether lipids of *Methanospirillum hungatei* containing novel head groups N,N-dimethylamino and N,N,N-trimethylaminopentanetetrol. *Biochim. Biophys. Acta* **921**: 281–91.
Ferrante, G., Ekiel, I., Girishchandra, B.P. and Sprott, G.D. (1988a) A novel core lipid isolated from the aceticlastic methanogen, *Methanothrix concilii* GP6. *Biochim. Biophys. Acta* **963**: 173–82.

Ferrante, G., Ekiel, I., Girishchandra, B.P. and Sprott, G.D. (1988b) Structure of the major polar lipids isolated from the aceticlastic methanogen, *Methanothrix concilii* GP6. *Biochim. Biophys. Acta* **963**: 162–72.
Fiala, G. and Stetter, K.O. (1986) *Pyrococcus furiosus* sp. nov. represents a novel genus of marine heterotrophic archaebacteria growing optimally at 100 °C. *Arch. Microbiol.* **145**: 56–61.
Fiala, G., Stetter, K.O., Jannasch, H.W., Langworthy, T.A. and Madon, J. (1986) *Staphylothermus marinus* sp. nov. represents a novel genus of extremely thermophilic submarine heterotrophic archaebacteria growing up to 98 °C. *System. Appl. Microbiol.* **8**: 106–13.
Fredrickson, H.L., Leeuw, J.W., Tas, A.C., van der Greef, J., Lavos, G.F. and Boon, J.J. (1989) Fast atom bombardment (tandem) mass spectrometric analysis of intact polar ether lipids extractable from the extremely halophilic archaebacterium *Halobacterium cutirubrum*. *Biomed. Mass Spectrom.* **18**: 96–105.
Grant, W.D. and Larsen, H. (1989) Extremely halophilic archaeobacteria. In *Bergey's Manual of Systematic Bacteriology*, vol. 3 (Staley, J.T., Bryant, M.P., Pfennig, N. and Holt, J.G., eds), pp. 2216–30. Williams & Wilkins: Baltimore.
Grant, W.D., Pinch, G., Harris, J.E., De Rosa, M. and Gambacorta, A. (1985) Polar lipids in methanogen taxonomy. *J. Gen. Microbiol.* **131**: 3277–86.
Heathcock, C.H., Finkelstein, B.L., Aoki, T. and Poulter, C.D. (1985) Stereostructure of the archaebacterial C_{40} diol. *Science* **229**: 862–4.
Huber, G., Spinnler, C., Gambacorta, A. and Stetter, K.O. (1989) *Metallosphaera sedula* sp. nov. represents a new genus of aerobic, metal-mobilizing, thermophilic archaebacteria. *System. Appl. Microbiol.* **12**: 38–47.
Jacin, M. and Mishkin, A.R. (1965) Separation of carbohydrates on borate impregnated silica gel plates. *J. Chromat.* **18**: 170–6.
Jones, W.J., Nagle, D.P. and Whitman, W.B. (1987) Methanogens and the diversity of archaebacteria. *Microbiol. Rev.* **51**: 135–77.
Juez, G. (1988) Taxonomy of extremely halophilic archaebacteria. In *Halophilic Bacteria* (Rodriguez-Valera, F. ed), pp. 3–24. CRC Press: Boca Raton.
Kamekura, M. and Kates, M. (1988) Lipids of halophilic archaebacteria. In *Halophilic Bacteria* (Rodriguez-Valera, F. ed), pp. 25–54. CRC Press: Boca Raton.
Kates, M. (1972) *Ether Lipids, Chemistry and Biology*. Academic Press: New York.
Kates, M. (1978). The phytanyl ether-linked polar lipids and isoprenoid neutral lipids of extremely halophilic bacteria. *Prog. Chem. Fats Ether Lipids* **15**: 301–42.
Kates, M. (1986) Techniques of Lipidology. Elsevier: New York.
Kates, M., Joo, C.N., Palameta, B. and Shiez, T. (1967) Absolute stereochemical configuration of phytonyl (dihydrophytyl) groups in lipids of *Halobacterium cutirubrum*. *Biochemistry* **6**, 3329–38.
Kates, M. and Deroo, P.W. (1973). Structure determination of the glycolipid sulfate from the extreme halophilic *Halobacterium cutirubrum*. *J. Lipid Res* **14**: 438–45.
Kates, M., Yengoyan, L.S. and Sastry, P.S. (1965) A diether analogy of phosphatidyl glycerophosphate in *Halobacter cutirubrum*. *Biochim. Biophvs. Acta* **98**: 252–8.
Koga, Y., Ohga, M., Nishihara, M. and Morii, H. (1987) Distribution of a diphytanyl ether analog of phosphatidylserine and an ethanolamine-containing tetraether lipid in methanogenic bacteria. *System. Appl. Microbiol.* **9**: 176–82.
König, H. (1988) Archaeobacteria. In *Biotechnology* (Rhem, H.J., ed), pp. 697–728. Academic Press: New York.
Kramer, J.K.G., Sauer, F.D. and Blackwell, B.A. (1987) Structure of two new aminophospholipids from *Methanobacterium thermoautotrophicum*. *Biochem. J.* **245**: 139–43.
Kushwaha, S.C., Kates, M., Sprott, G.D. and Smith, I.C.P. (1981) Novel polar lipids from the methanogen *Methanospirillum hungatei* GP1. *Biochim. Biophys. Acta* **664**: 156–73.
Langworthy, T.A. (1977a) Long chain diglycerol tetraethers from *Thermoplasma acidophilum*. *Biochim. Biophys. Acta* **487**: 37–50.
Langworthy, T.A. (1977b) Comparative lipid composition of heterotrophically and autotrophically grown *Sulfolobus acidocaldarius*. *J. Bacteriol.* **130**: 1326–32.
Langworthy, T.A. (1979) Special features of *Thermoplasmas*. In *The Mycoplasma* (Barile, M. F. and Razin, R., eds), pp. 495–513. Academic Press: New York.

Langworth, T.A. (1982) Lipids of *Thermoplasma*. In *Methods in Enzymology*, Volume 88 (Packer, L., ed), pp. 396–406. Academic Press Inc.: New York.
Langworthy, T.A. (1985). Lipids of archaebacteria. In *The Bacteria* (Woese, C.R. and Wolfe, R.S., eds), pp. 459–91. Academic Press: New York.
Langworthy, T.A. and Pond, J.L. (1986) Archaebacterial ether lipids and chemotaxonomy. In *Archaebacteria '85*, (Kandler, O. and Zillig, W., eds), pp. 253–7. Gustav Fischer Verlag: Stuttgart.
Langworthy, T.A., Smith, P.F. and Mayberry, W.A. (1972) Lipids of *Thermoplasma acidophilum*. *J. Bacteriol.* **112**: 1193–200.
Langworthy, T.A., Mayberry, W.R. and Smith, P.F. (1974) Long chain glycerol diether and polyol dialkyl glycerol triether lipids of *Sulfolobus acidocaldarius*. *J. Bacteriol.* **119**: 106–16.
Lanzotti, V., Trincone, A., Gambacorta, A., De Rosa, M. and Breitmaier, E. (1986) ^{1}H and ^{13}C NMR assignment of benzothiophenquinones from the sulfur-oxidizing archaebacterium, *Sulfolobus solfataricus*. *Eur. J. Biochem.* **160**: 37–40.
Lanzotti, V., De Rosa, M., Trincone, A., Basso, A. and Zillig, W. (1987) Complex lipids from *Desulfurococcus mobilis*, a sulfur-reducing archaebacterium. *Biochim. Biophys. Acta* **922**: 95–102.
Lanzotti, V., Nicolaus, B., Trincone, A., De Rosa, M., Grant, W.D. and Gambacorta, A. (1989a) An isopranoid ether analogue of phosphatidic acid from a halophilic archaebacterium. *Biochim. Biophys. Acta* **1002**: 399–400.
Lanzotti, V., Nicolaus, B., Trincone, A., De Rosa, M., Grant, W.D. and Gambacorta, A. (1989b) A complex lipid with a cyclic phosphate from the archaebacterium *Natronococcus occultus*. *Biochim. Biophys. Acta* **1001**: 31–4.
Lanzotti, V., Trincone, A., Nicolaus, B., Zillig, W., De Rosa, M. and Gambacorta, A. (1989c) Complex lipids of *Pyrococcus* and AN1, thermophilic members of archaebacteria belonging to *Thermococcales*. *Biochim. Biophys. Acta* **1004**: 44–8.
Laurer, G., Kristjansson, J.K., Langworthy, T.A., König, H. and Stetter, K.O. (1986) *Methanothermus sociabilis* sp. nov., a second species within the *Methanothermaceae* growing at 97 °C. *System. Appl. Microbiol.* **8**: 100–5.
Levy, G.C. and Nelson, G.L. (1971). *Carbon-13 Nuclear Magnetic Resonance of Organic Chemists*. Wiley-Interscience: New York.
Mancuso, C.A., Odham, G., Westerdahl, G., Reeve, J.N. and White, D.C. (1985). C_{15}, C_{20}, and C_{25} isoprenoid homologues in glycerol diether phospholipids of methanogenic archaebacteria. *J. Lipid Res.* **26**: 1120–5.
Morii, H., Nishihara, M., Ohga, M. and Koga, Y. (1986) A diphytanyl ether analog of phosphatidylserine from a methanogenic bacterium, *Methanobrevibacter arboriphilus*. *J. Lipid Res.* **27**: 724–30.
Nicolaus, B., Lanzotti, V., Trincone, A., De Rosa, M., Grant, W.D. and Gambacorta, A. (1989) Polar lipid composition in halophilic archaebacteria in response to growth in different salt concentrations. *FEMS Microbiol. Lett.* **59**: 157–60.
Nishihara, M. and Koga, Y. (1987) Extraction and composition of polar lipids from the archaebacterium *Methanobacterium thermoautotrophicum*: Effective extraction of tetraether lipids by an acidified solvent. *J. Biochem.* **101**: 997–1005.
Nishihara, M., Morii, H. and Koga, Y. (1989) Heptads of polar lipids of an archaebacterium, *Methanobacterium thermoautotrophicum*: Structure and biosynthetic relationship. *Biochemistry* **28**: 95–102.
Panganamala, R.K., Sievert, C.F. and Cornwell, D.G. (1971) Quantitative estimation and identification of O-alkyl glycerol as alkyl iodides and their hydrocarbon derivatives. *Chem. Phys. Lipids* **7**: 336–44.
Redfearn, E.R. (1966) Mode of action of ubiquinones (coenzyme Q) in electron transport systems. *Vitam. Horm.* (N.Y.) **24**: 463–88.
Ross, H.N.M. and Grant, W.D. (1985) Nucleic acid studies on halophilic archaebacteria. *J. Gen. Microbiol.* **13**: 165–73.
Ross, H.N.M., Collins, M.D., Tindall, B.J. and Grant, W.D. (1981) A rapid procedure for the detection of archaebacterial lipids in halophilic bacteria. *J. Gen. Microbiol.* **123**: 75–80.
Ross, H.N.M., Grant, W.D. and Harris, J.E. (1985) Lipids in archaebacterial taxonomy. In *Chemical Methods in Bacterial Systematics* (Goodfellow, M. and Minnikin, D.E. eds), pp. 289–300. Academic Press: London.

Rouser, G., Kritchevsky, G., Siakotos, A.N. and Yamamoto, A. (1970) *Lipid Composition of the Brain and Its Subcellular Structures. An Introduction to Neuropathology*: Methods and Diagnosis, (pp. 691–753). Little, Brown: New York.
Scolastico, C., Sydimov, A., Potenza, D., De Rosa, M., Gambacorta, A. and Trincone, A. (1986) Cyclopentane ring formation in isoprenoid ether lipids of archaebacteria. In *Archaebacteria '85* (Kandler, O. and Zillig, W., eds), pp. 417–8. Gustav Fischer Verlag: New York.
Segerer, A., Neumer, A., Kristjansson, J.K. and Stetter, K.O. (1986) *Acidianus infernus* gen. nov., and *Acidianus brierley* comb. nov.: Facultatively aerobic, extremely acidophilic thermophilic sulfur-metabolizing archaebacteria. *Int. J. System. Bacteriol.* **36**: 559–64.
Segerer, A.H., Trincone, A., Gahrtz, M. and Stetter, K.O. (1991) *Stygiolobus azoricus* gen. nov., spec. nov. represents a novel genus of anaerobic extremely thermoacidophilic archaebacteria of the order *Sulfolobales*. *Int. J. System. Bacteriol.* **41**: 495–501.
Shah, H.N. and Collins, M.D. (1980) Fatty acid and isoprenoid quinone composition in the classification of *Bacteroides melaninogenicus* and related taxa. *J. Appl. Bacteriol.* **48**: 75–87.
Shaw, N. (1968) The detection of lipids on chromatograms with the Periodate—Schiff reagents. *Biochim. Biophys. Acta* **164**: 435–6.
Smallbone, B.W. and Kates, M. (1981) Structural identification of minor glycolipids in *Halobacterium cutirubrum*. *Biochim. Biophys. Acta* **665**: 551–8.
Stothers, J. B. (1972). *Carbon-13 NMR Spectroscopy*. Academic Press: New York.
Taber, H. (1980) *Vitamin K Metabolism and Vitamin K Dependent Proteins*. University Park Press: Baltimore, MDK.
Thurl, S. and Schafer, W. (1988) Lipids from the sulphur-dependent archaebacterium *Thermoproteus tenax*. *Biochim. Biophys. Acta* **961**: 253–61.
Thurl, S.,Buhrow, I. and Schafer, W. (1985) New types of menaquinones from the thermophilic archaebacterium *Thermoproteus tenax*. *Biol. Chem. Hoppe Seyler* **366**: 1079–83.
Thurl, S., Witke, W., Buhrow, I. and Schafer, W. (1986) Quinones from archaebacteria. II. Different types of quinones from sulphur-dependent archaebacteria. *Biol. Chem. Hoppe Seyler* **367**: 191–8.
Tindall, B.J. (1985) Qualitative and quantitative distribution of diether lipids in haloalkaliphilic archaebacteria. *System. Appl. Microbiol.* **6**: 243–6.
Tindall, B.J., Stetter, K.O. and Collins, M.D. (1989) A novel, fully saturated menaquinone from the thermophilic sulfate reducing archaebacterium *Archeoglobus fulgidus*. *J. Gen. Microbiol.* **135**: 693–6.
Tindall, B.J., Wray, V., Huber, H. and Collins, M.D. (1991) A novel, fully saturated cyclic menaquinone in the archaebacterium *Pyrobaculum organotrophicum*. *System. Appl. Microbiol.* **14**: 218–21.
Torreblanca, M., Rodriguez-Valera, F., Juez, G., Ventosa, A., Kamekura, M. and Kates, M. (1986) Classification of non-alkaliphilic halobacteria based on numerical taxonomy and polar lipid composition, and description of *Haloarcula* gen. nov. and *Haloferax* gen. nov. *System. Appl. Microbiol.* **8**: 89–99.
Trincone, A., Gambacorta, A., Lanzotti, V. and De Rosa, M. (1986) A new benzo-[1,2-b;4,5-b']dithiophene-4,8-quinone from the archaebacterium *Sulfolobus solfataricus*. *Chem. Commun.* 733.
Trincone, A., De Rosa, M., Gambacorta, A., Lanzotti, V., Nicolaus, B., Harris, J.E. and Grant, W.D. (1988) A simple chromatographic procedure for the detection of cyclized archaebacterial glycerol-bisdiphytanyl-glycerol tetraether core lipids. *J. Gen. Microbiol.* **134**: 3159–63.
Trincone, A., Lanzotti, V., Nicolaus, B., Zillig, W., De Rosa, M. and Gambacorta, A. (1989) Comparative lipid composition of aerobically and anaerobically grown *Desulfurolobus ambivalens*, an autotrophic thermophilic archaebacterium. *J. Gen. Microbiol.* **135**: 2751–7.
Trincone, A., Nicolaus, B., Palmieri, G., De Rosa, M., Huber, H., Stetter, K.O. and Gambacorta, A. (1992) Distribution of complex and core lipids within new hyperthermophilic members of the Archaea domain. *System. Appl. Microbiol.*, **15**: 11–17.
Tsujimoto, K., Yorimitsu, S., Takahashi, T. and Ohashi, M. (1989) Revised structure of a phospholipid obtained from *Halobacterium halobium*. *Chem. Commun.* 668–70.
Watanuky, M. and Aida, K. (1972) Significance of quinones in the classification of bacteria. *J. Gen. Microbiol.* **18**: 469–72.

Woese, C R. (1987) Bacterial evolution. *Microbiol. Rev.* **51**: 221–71.
Woese, C.R. and Wolfe, R.S. (1985) *The Bacteria: a Treatise on Structure and Function*. Academic Press: New York.
Yamada, Y., Takinami, H., Tahara, Y. and Kondo, K. (1977) The menaquinone system in the classification of radiation-resistance in micrococci. *J. Gen. Appl. Microbiol.* **23**: 105–8.
Zillig, W., Gierl, A., Schreiber, G., Wunderl, S., Janekovic, D., Stetter, K.O. and Klenk, H.P. (1983) The archaebacterium *Thermophilum pendens* represents a novel genus of the thermophilic, anaerobic sulfur respiring *Thermoproteales*. *System. Appl. Microbiol.* **4**: 79–87.

8
Isoprenoid Quinones

Matthew D. Collins
AFRC Institute of Food Research, Reading, UK

8.1 INTRODUCTION

The discovery by Henrik Dam (1929) of a blood clotting factor opened up the chapter on quinone biochemistry. Dam made this discovery while investigating the biochemistry of sterols. Chicks raised on an artificial diet with a very low sterol and lipid content tended to bleed easily and their blood showed delayed coagulation. Dam later proved that this syndrome had resulted from a deficiency in a new fat-soluble vitamin, which he named vitamin K. The structure of this compound was determined in 1939 and shown to be that of a lipid-soluble quinone (phylloquinone; Dam *et al.*, 1939; MacCorquodale *et al.*, 1939a,b). By the late 1950s it became clear that vitamin K was only one of a family of biologically active quinone compounds which were ubiquitous throughout the animal, plant and bacterial realms.

It is now well established that isoprenoid quinones occur in the cytoplasmic membranes of most prokaryotes. During the early 1960s it became apparent, primarily due to studies by Bishop, Crane, Page and their colleagues, that different bacteria not only synthesized different quinone classes (e.g. menaquinones, ubiquinones) but that the number of isoprene units in the multiprenyl side-chain often varied amongst taxa and that this structural variation might be of value in prokaryote systematics (Lester & Crane 1959; Page *et al.*, 1960; Bishop *et al.*, 1962).

Jeffries and his colleagues were the first to specifically demonstrate the taxonomic value of isoprenoid quinones during an examination of the menaquinone composition of some aerobic Gram-positive cocci (Jeffries *et al.*, 1967a,b,c). Since the early studies mentioned above, a large number of systematic investigations, most notably those of Collins and colleagues in the UK and Yamada and associates in Japan, have led to a resurgence of interest in microbial respiratory quinones. These studies have not only established isoprenoid quinones as some of the most useful markers in modern microbial systematics but have greatly expanded the known range of quinone structures, including a plethora of new complex hydrogenated prenologues, partially saturated positional isomers, and modified sidechains including cyclization and methylation, and types (e.g. methyl- and dimethyl-menaquinones).

Two major groups of bacterial isoprenoid quinones can be recognized on the basis of structural considerations: the benzoquinones and naphthoquinones. A third group of sulphur-containing quinones with structures based on benzo-thiopen-4,7-quinone can also be distinguished. These lipoquinones, however, have a much more restricted distribution, having been found only in certain acidophilic thermophilic archaea (see Chapter 7).

8.2 QUINONE STRUCTURES

8.2.1 Phylloquinones

Vitamin K_1 or phylloquinone was first isolated in 1939 from alfalfa by Dam and colleagues (Dam *et al.*, 1939) and Doisy and associates (MacCorquodale *et al.*, 1939a). The structure of vitamin K_1 was shown by degradation and synthetic studies to be 2-methyl-2-phytyl-1,4-naphthoquinone (Figure 8.1a) (MacCorquodale *et al.*, 1939b). Phylloquinone is normally associated

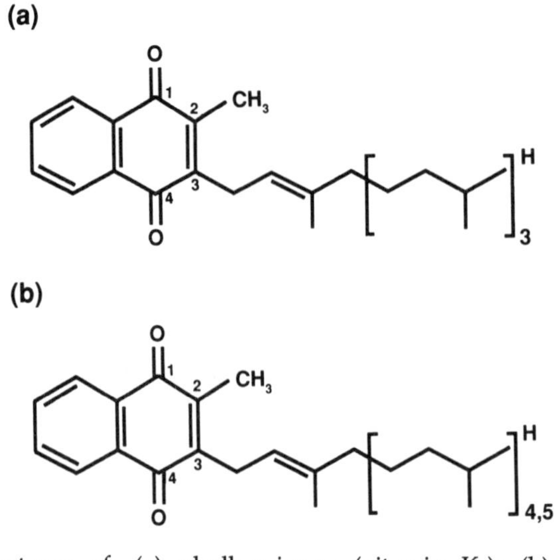

Figure 8.1 Structures of: (a) phylloquinone (vitamin K_1); (b) higher phylloquinone analogues from *Thermoproteus tenax*.

with the green tissues of plants and is not commonly found in bacteria. Until recently its presence within prokaryotes was thought to be limited to the cyanobacteria. Recently, however, phylloquinone (MK-4 (H$_6$)) and two higher phylloquinone analogues containing five or six units (i.e. MK-5 (H$_8$), MK-6 (H$_{10}$)) in which unit 1 adjacent to the ring system is unsaturated (Figure 8.1b) have been found as minor components in the thermophilic archaebacterium *Thermoproteus tenax* (Thurl et al., 1985). 5'-Monohydroxyphylloquinone has also been reported in some cyanobacteria (Law et al., 1973).

8.2.2 Menaquinones

Vitamin K$_2$, or menaquinone, was first isolated from bacterially putrefied fishmeal as a crystalline substance (MacCorquodale et al., 1939a; McKee et al., 1939). The structure was published in 1940 and thought to be 2-methyl-3-farnesyl-farnesyl-1,4-naphthoquinone (Binkley et al., 1940). However, it was not until 1958 that the plurality of menaquinones was established; Isler et al. (1958 a,b) found that the compound from fishmeal had a farnesyl-geranyl-geranyl side-chain, i.e. containing seven isoprene units (MK-7), with the related MK-6 present in only minor amounts. Naturally occurring menaquinones are now known to form a large class of molecules in which the length of the C3 isoprenyl side-chain varies from one up to 15 isoprene units, i.e. five to 75 carbon atoms (Figure 8.2a). In addition to variations in the length of the side-chain, varying degrees of saturation or hydrogenation of the C3 multiprenyl side-chain of menaquinones have been established (Collins & Jones, 1981a). For example, di- and tetra-hydrogenated menaquinones are quite widespread amongst actinomycetes. Even more highly saturated (e.g. hexa-, octa-hydrogenated) side-chains are common in sporoactinomycetes. Menaquinones with completely saturated side-chains have been found in certain archaea, e.g. *Thermoproteus tenax* (Thurl et al., 1985), *Sulfolobus ambivalens* (Thurl et al., 1986) and *Archaeoglobus fulgidus* (Tindall et al., 1989) (Figure 8.2b).

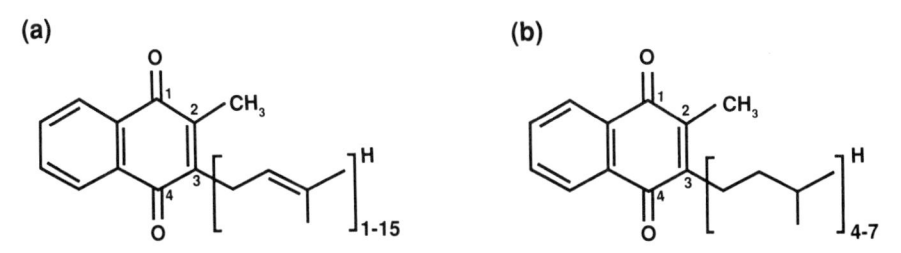

Figure 8.2 Structures of: (a) unsaturated menaquinones; (b) fully saturated menaquinones from archaea.

Recent studies indicate that the position or point of hydrogenation in multiprenyl side-chains can be very specific and consequently of taxonomic value. Although data are still somewhat fragmentary, it has been shown that some Gram-negative sulphate-reducing bacteria, which synthesize dihydrogenated menaquinones, preferentially hydrogenate the terminal (farthest from the ring system) isoprenoid unit; e.g. MK-5(V-H$_2$) in *Desulfobulbus propionicus*, MK-6(VI-H$_2$) in *Desulfovibrio africanus* and *Desulfovibrio salexigens* and MK-7(VII-H$_2$) in *Desulfovibrio baarsii* (Collins & Widdel, 1984, 1986) (Figure 8.3a). The dihydrogenated menaquinones of halophilic archaea also show end of chain saturation, e.g. MK-8(VIII-H$_2$) in *Halococcus morrhae* (Tindall & Collins, 1986). In contrast, the position of saturation of the dihydrogenated menaquinones of members of Gram-positive taxa invariably occurs in the second unit from the ring system, e.g. MK-8(II-H$_2$) in *Brevibacterium linens* and *Corynebacterium diphtheriae*, MK-9(II-H$_2$) from *Mycobacterium phlei* (Figure 8.3a,b).

Although the points of saturation of only a few tetrahydrogenated menaquinones have been unequivocally determined, all of those from Gram-positive bacteria have internally saturated units. For example, in MK-8(H$_4$) from *Nocardioides simplex* and MK-9(H$_4$) from *Oerskovia turbata* saturation involves the second and third isoprene units whereas in MK-9(H$_4$) from *Microtetraspora* (*Micropolyspora*) *angiospora* it is the third and eighth units which are saturated (Figures 8.4a and b respectively). Only a

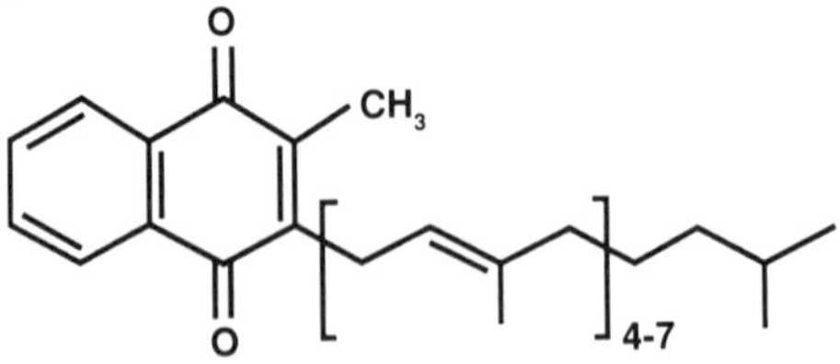

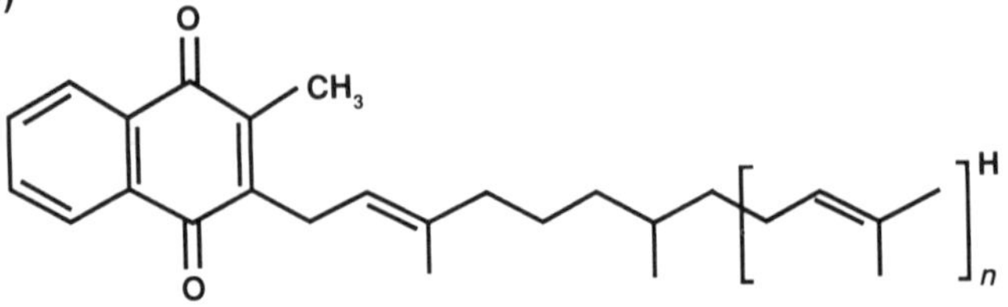

Figure 8.3 Structures of: (a) terminally saturated dihydrogenated menaquinones from Gram-negative bacteria and archaea; (b) II-dihydrogenated menaquinones from Gram-positive bacteria.

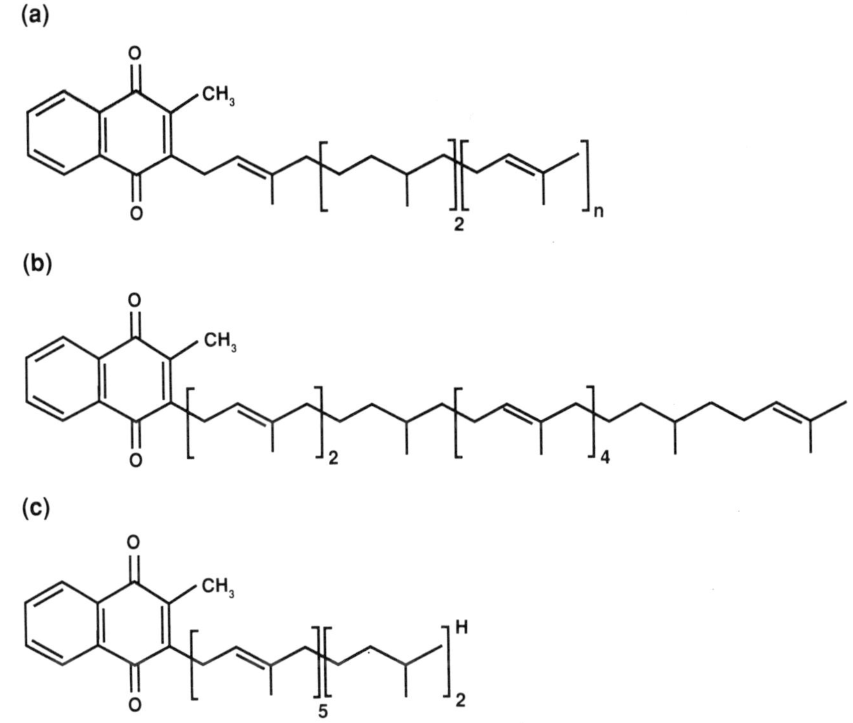

Figure 8.4 Structures of: (a) II, III-tetrahydrogenated menaquinones; (b) III, VIII-tetrahydrogenated MK-9 from *Microtetraspora* (*Micropolyspora*) *angiospora*; (c) VI, VII-tetrahydrogenated MK-7 from *Thermoleophilum album*.

single tetrahydrogenated menaquinone has been isolated from Gram-negative bacteria (Collins et al., 1986b): the aerobic, hydrocarbon-utilizing thermophile, *Thermoleophilum album*, possesses MK-7(H$_4$) in which the end of chain and adjacent (sixth) units are saturated, i.e. MK-7(VI,VII-H$_4$) (Figure 8.4c).

Hexahydrogenated menaquinones have been found only in Gram-positive bacteria. Three different isomers have been characterized: MK-9(II,III,VIII-H$_6$) from *Actinomadura madurae*, MK-9(II,III,IX-H$_6$) from *Streptomyces* spp. and MK-9(III,VIII,IX-H$_6$) from *Microtetraspora* (*Micropolyspora*) *angiospora* (Figure 8.5a,b,c). Menaquinones with partially saturated side-chains of known structure are given in Table 8.1.

Collins and associates described a new type of menaquinone from *Nocardia brasiliensis* in which the end two units of the side-chain were cyclized. Structural studies demonstrated the major cyclic quinone was a modified MK-8(H$_4$) and corresponded to 2-[3,7,11,15,19,23-hexamethyl-

Table 8.1 Menaquinones with partially saturated side-chains of known structure.

Taxon	Menaquinone	Reference
Domain: Bacteria		
GRAM-POSITIVE BACTERIA		
Corynebacterium diphtheriae	MK-8(II-H$_2$)	Azerad & Cyrot-Pelletier (1973)
Brevibacterium lipolyticum	MK-8(II,III-H$_4$)	Yamada *et al.* (1977)
Nocardioides albus	MK-8(II,III-H$_4$)	Collins, M.D., unpublished
Nocardia species[a]	MK-8(II,III,ω-cycl-H$_4$)[b]	Howarth *et al.* (1986)
	MK-8(II,ω-cycl-H$_2$)[c]	Collins *et al.* (1987)
Mycobacterium species	MK-9(II-H$_2$)	Azerad & Cyrot-Pelletier (1973)
Propionibacterium shermanii	MK-9(II,III-H$_4$)	Schwartz (1973)
Oerskovia turbata	MK-9(II,III-H$_4$)	Yamada *et al.* (1977)
Actinomadura madurae	MK-9(II,III,VIII-H$_6$)	Yamada *et al.* (1982a)
Microtetraspora angiospora	MK-9(III,VIII-H$_4$)	Collins *et al.* (1988)
	MK-9(III,VIII,IX-H$_6$)	Collins, M.D., unpublished
Streptomyces albus	MK-9(II,III,VIII,IX-H$_8$)	Yamada *et al.* (1982b)
Streptomyces olivaceus	MK-9(II,III,IX-H$_6$)	Batrakov *et al.* (1976)
	MK-9(II,III,VIII,IX-H$_8$)	
Glycomyces rutgersensis	MK-10(II-H$_2$)	Collins & Kroppenstedt (1987)
	MK-10(II,III-H$_4$)	
	MK-11(II-H$_2$)	
	MK-11(II,III-H$_4$)	
GRAM-NEGATIVE BACTERIA		
Desulfobulbus propionicus	MK-5(V-H$_2$)	Collins & Widdel (1984, 1986)
Desulfobulbus elongatus	MK-5(V-H$_2$)	Collins & Widdel (1986)
Desulfovibrio salexigens	MK-6(VI-H$_2$)	Collins & Widdel (1986)
Desulfovibrio africanus	MK-6(VI-H$_2$)	Collins & Widdel (1986)
Desulfobacter species	MK-7(VII-H$_2$)	Collins & Widdel (1986)
Thermoleohilum album	MK-7(VI,VII-H$_4$)	Collins *et al.* (1986)
Domain: Archaea		
Holococcus morrhuae	MK-8(VIII-H$_2$)	Tindall & Collins (1986)
Natronobacterium gregori	MK-8(VIII-H$_2$)	Collins & Tindall (1987)
	MMK-8(VIII-H$_2$)	
	DMMK-8(VIII-H$_2$)	

[a] Present in all authentic *Nocardia* species examined.
[b] II,III-tetrahydro-ω-(2,6,6-trimethylcyclohex-2-enylmethyl) menaquinone-6.
[c] II-dihydro-ω-(2,6,6-trimethylcyclohex-2-enylmethyl) menaquinone-6.

8. Isoprenoid Quinones

(a)

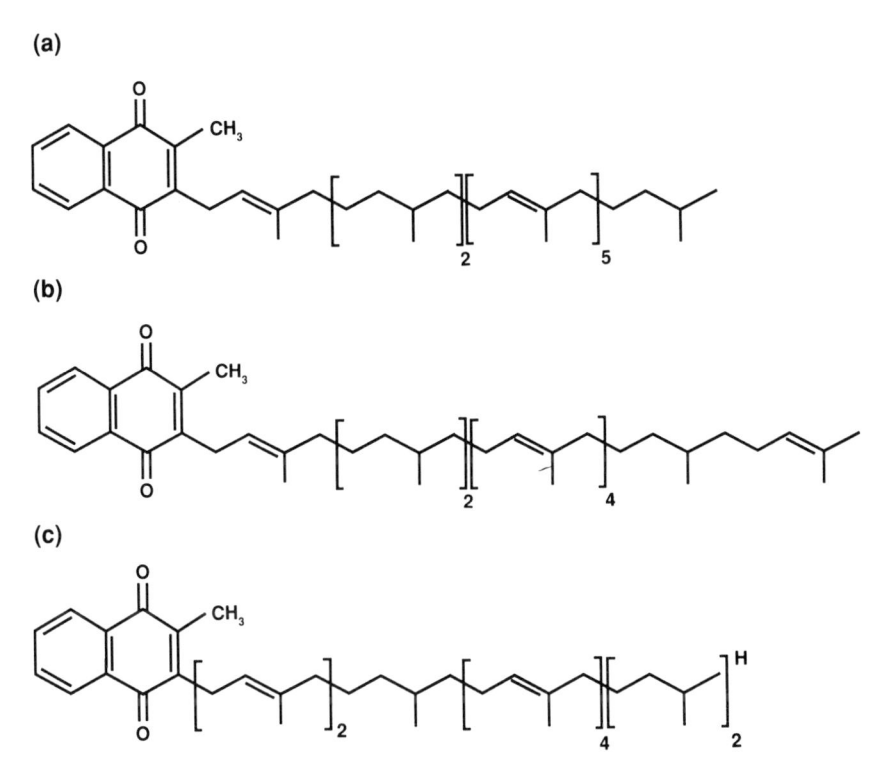

(b)

(c)

Figure 8.5 Structures of: (a) MK-9 (II, III, IX-H$_6$) from *Streptomyces* spp.; (b) MK-9 (II, III, VIII-H$_6$) from *Actinomadura madurae*; (c) MK-9 (III, VIII, IX-H$_6$) from *Microtetraspora* (*Micropolyspora*) *angiospora*.

25-(2,6,6-trimethylcyclohex-2-enyl)pentacose-2,14,18,22-tetraenyl]3-methyl-1,4-naphthoquinone (Figure 8.6a) (Howarth et al., 1986). A dihydrogenated isomer was subsequently found as a minor component (Figure 8.6b; Collins et al., 1987). These cyclic menaquinones have so far been found in all authentic *Nocardia* species examined and as such may prove a valuable marker for the genus. A novel, fully saturated cyclic menaquinone has also recently been characterized in the archaebacterium *Pyrobaculum organotrophum* (Tindall et al., 1991). The cyclic quinone was shown to correspond to 2-(14-[3-(1,5-dimethylhexyl)cyclopentyl]-3,7,11-trimethyltetradecyl)-3-methyl-1,4-naphthoquinone.

An unusual menaquinone, designated chlorobiumquinone, is produced by the green photosynthetic bacterium *Chlorobium thiosulphatophilum* (Frydman & Rapaport, 1963; Redfearn & Powls, 1967). This compound was initially thought to be a modified MK-7 in which the first methylene of the side-chain was absent (i.e. a C$_{34}$ side-chain with the first double bond of the chain conjugated to the ring), but subsequent work by Powls et al.

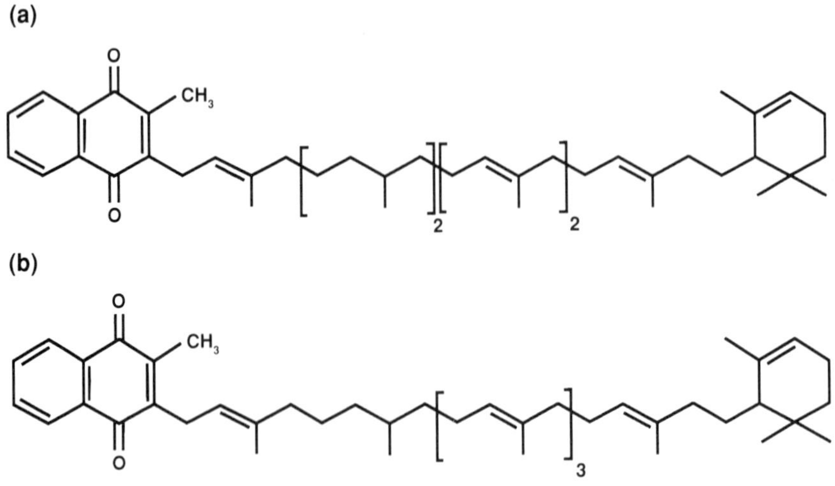

(a)

(b)

Figure 8.6 Structures of cyclic menaquinones from *Nocardia asteroides* and *Nocardia brasiliensis*.

(1968) has shown that chlorobiumquinone is in fact 1'-oxomenaquinone with seven isoprene units (1'-oxo-MK-7; Figure 8.7b). Chlorobiumquinone is the only example of a bacterial isoprenoid quinone containing a side-chain carbonyl group.

The only other oxygen-containing bacterial menaquinones described to date are two 2,3-epoxy menaquinones (Figure 8.7a) which are present as minor constituents in *Nocardia brasiliensis* (Collins et al., 1987). The origin of these epoxy menaquinones, however, is uncertain. It is possible that they derive from carboxylation reactions (i.e. similar to the formation of vitamin K_1 2,3-epoxide during the synthesis of γ-carboxyglutamates) or they may be simply artefacts of extraction/isolation.

8.2.3 Demethylmenaquinones

Baum and Dolin (1963) were the first to demonstrate the natural occurrence of demethylmenaquinones (Figure 8.8). The quinone (termed SFQ) isolated from *Enterococcus faecalis* was shown to be a monosubstituted 1,4-naphthoquinone in which the ring methyl was replaced by a hydrogen atom. The major component was subsequently found to contain a solanesyl side-chain (abbreviated DMK-9) with DMK-7 and DMK-8 present in minor amounts (Baum & Dolin, 1965). A series of demethylmenaquinones (DMK-5, DMK-6, DMK-7) was later found in *Haemophilus*

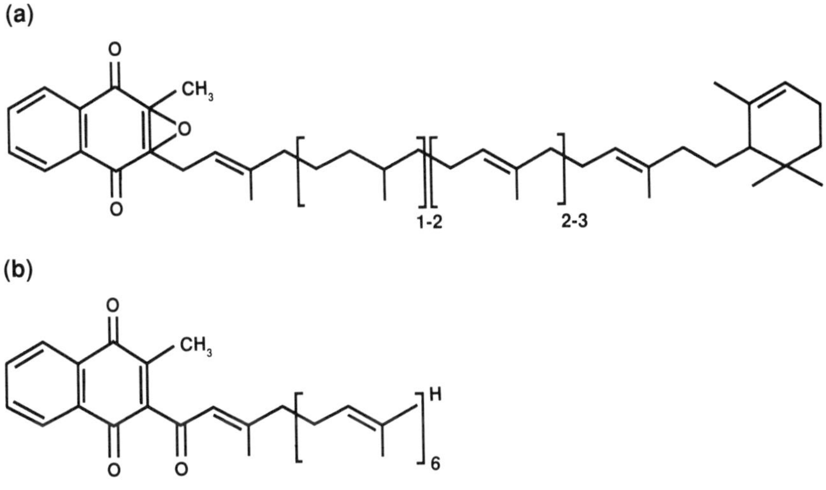

8. Isoprenoid Quinones

(a)

(b)

Figure 8.7 Structures of: (a) 2,3-epoxy-menaquinones from *Nocardia* species; (b) chlorobiumquinone (1'-oxo-MK-7).

parainfluenzae with the farnesyl-farnesyl naphthoquinone the predominant component (Lester *et al.*, 1964).

Demethylmenaquinones are now known to occur in members of many Gram-negative facultatively anaerobic taxa (e.g. *Escherichia*, *Pasteurella*, *Proteus*) although they are not nearly as widely distributed as the menaquinone series (Collins & Jones, 1981a). Demethylmenaquinones are rarely found in Gram-positive taxa and have not been reported in archaea. The first demethylmenaquinone containing a partially saturated side-chain, DMK-10(H$_4$), has recently been characterized as a minor component in the Gram-positive anaerobe *Actinomyces pyogenes* (Collins, M.D., unpublished).

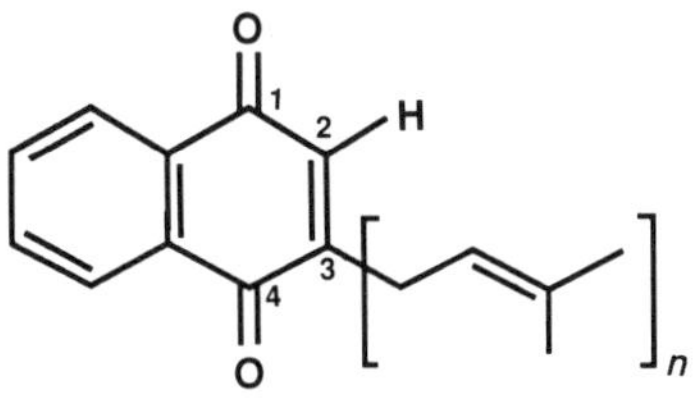

Figure 8.8 Structure of demethylmenaquinone.

8.2.4 Methylmenaquinones

Collins and Langworthy (1983) isolated a new naphthoquinone, designated thermoplasmaquinone, which possessed ultraviolet characteristics similar to that of menaquinones except the absorption at 325 nm shifted ~12 nm to higher wavelengths (λ_{max} 242, 248, 259, 269, 337 nm). Mass spectral studies showed that the quinone corresponded to MK-7 containing additional CH_2 (molecular formula $C_{47}H_{66}O_2$ compared with $C_{46}H_{64}O_2$ for MK-7). Although the detailed structure of the quinone was not established the presence of a base peak in the mass spectrum at m/z 239 was consistent with the extra CH_2 located in the ring system.

Carlone and Anet (1983) subsequently isolated a novel methyl-substituted menaquinone (MMK) from *Campylobacter jejuni*. This compound had similar UV and mass spectral characteristics to thermoplasmaquinone except that it contained six isoprene units. On the basis of proton nuclear magnetic resonance (^{1}H-NMR) analyses, Carlone and Anet (1983) established that the methyl substituent of the modified MK-6 from the *Campylobacter* was in the 5 or 8 position of the ring, i.e. (5 or 8)-dimethyl-3-farnesyl-farnesyl-1,4-naphthoquinone. Subsequent ^{1}H-NMR studies have shown that the methyl group of thermoplasmaquinone occurs in the 5 or 8 position of the ring and corresponds to 2,(5 or 8)-dimethyl-3-farnesyl-geranyl-geranyl-1,4-naphthoquinone (Collins, 1985b).

The naphthoquinones isolated from *Campylobacter* and *Thermoplasma* represent a new series of methylmenaquinones (Figure 8.9). Itoh *et al.* (1985) have recently shown by degradative studies that the methyl-substituent of MMK-7 from *Alteromonas putrefaciens* is located at position C8 of the ring system.

Methylmenaquinones have now been found in a variety of Gram-negative and Gram-positive bacteria; members with partially and fully saturated side-chains have also been found in certain archaea (Thurl *et al.*, 1985; Collins & Tindall, 1987). Although data on the distribution of methylmenaquinones are scant, there is every indication that these compounds are quite widely distributed amongst prokaryotes. Members of taxa reported to contain methylmenaquinones are listed in Table 8.2.

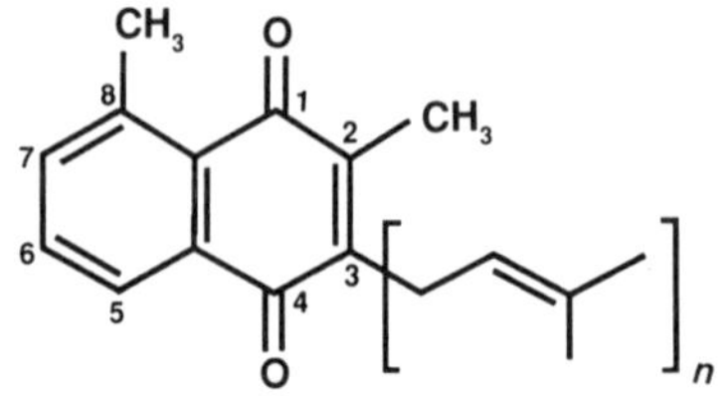

Figure 8.9 Structure of methylmenaquinone.

Table 8.2 Taxa containing methylated menaquinones.

Taxon	Menaquinone	Reference
Domain: Archaea		
Natronobacterium gregori	MMK-8, MMK-8(VIII-H$_2$) DMMK-8, DMMK-8(VIII-H$_2$)	Collins & Tindall (1987)
Thermoplasma acidophilum	MMK-7	Collins & Langworthy (1983); Collins (1985b)
Thermoproteus tenax	MMK-6(H$_{12}$), MMK-5(H$_{10}$)	Thurl et al. (1985)
Domain: Bacteria		
GRAM-NEGATIVE BACTERIA		
Alteromonas putrefaciens	MMK-7	Itoh et al. (1985)
Bacteroides gracilis[a]	MMK-6	Collins & Fernandez (1985)
Campylobacter coli	MMK-6	Collins et al. (1984); Moss et al. (1984)
Campylobacter concisus	MMK-6	Collins et al. (1984)
Campylobacter fecalis	MMK-6	Collins et al. (1984); Moss et al. (1984)
Campylobacter fetus ssp. *fetus*	MMK-6	Carlone & Anet (1983); Collins et al. (1984); Moss et al. (1984)
Campylobacter fetus ssp. *venerealis*	MMK-6	Collins et al. (1984); Moss et al. (1984)
Campylobacter hyointestinalis	MMK-6	Collins et al. (1984)
Campylobacter jejuni	MMK-6	Carlone & Anet (1983); Collins et al. (1984); Moss et al. (1984)
Campylobacter laridis	MMK-6	Collins et al. (1984); Moss et al. (1984)
Campylobacter sputorum ssp. *bubulus*	MMK-6	Collins et al. (1984); Moss et al. (1984)
Campylobacter sputorum ssp. *mucosalis*	MMK-6	Collins et al. (1984); Moss et al. (1984)
Campylobacter sputorum ssp. *sputorum*	MMK-6	Collins et al. (1984); Moss et al. (1984)
Spirillum sp[b]	MMK-6	Collins & Widdel (1986)
Wolinella succinogenes	MMK-6	Collins & Fernandez (1984)
Wolinella recta	MMK-6	Fernandez & Collins (1987)
GRAM-POSITIVE BACTERIA		
Eubacterium lentum	MMK-6, DMMK-6	Collins et al. (1985)

[a] *Bacteroides gracilis* is possibly more closely related to members of the genera *Wolinella* and *Campylobacter* than to the genus *Bacteroides*.
[b] Sulphate-reducing spirillum (5175) possibly related to *Campylobacter*.

8.2.5 Dimethylmenaquinones

Collins and associates described a new type of vitamin K in the Gram-positive anaerobe *Eubacterium lentum* in which two additional methyl groups were located in the ring system; the new compound was designated dimethylmenaquinone (Collins et al., 1985). ^{1}H-NMR analysis showed that the two methyls were located on adjacent carbon atoms and that the quinone corresponded to either 2,5 and 6- or 2,7 and 8-trimethyl-3-farnesyl-farnesyl-1,4-naphthoquinone (abbreviated DMMK-6; Figure 8.10). Two other dimethylmenaquinones, DMMK-8 and DMMK-8(H$_2$), have been found as minor components in the alkaliphilic halophilic archaebacterium *Natronobacterium gregori* (Collins & Tindall, 1987). Mass spectral and ^{1}H-NMR studies have shown that the point of saturation of the dihydrogenated DMMK is the terminal (8th) isoprenoid unit. Dimethylmenaquinones have not been reported in Gram-negative bacteria (Table 8.2).

8.2.6 Ubiquinones

The discovery of ubiquinones (coenzyme Q) was the result of independent studies by Morton and associates in England and by Crane and colleagues in the United States. In 1955 Morton's group isolated a nonsaponifiable lipid from tissues (vitamin A-deficient rats) which had an ultraviolet absorption maximum at 272 nm (Festenstein et al., 1955). The lipid was originally designated SA but later, due to its occurrence in many vegetable and animal materials, the term ubiquinone was used (Morton et al., 1957). Crane et al. (1957) isolated a lipid from mitochondria which from its oxidation-reduction behaviour and infrared spectrum was indicative of a quinone. This compound was designated Q$_{275}$, known later as coenzyme Q, due to its absorption maximum at 275 nm. It was soon realized that ubiquinone and coenzyme Q were identical, and the structure 2,3-dimethoxy-5-methyl-6-decaprenyl-1,4-benzoquinone was established by Folker and co-workers (Shunk et al., 1958; Wolf et al., 1958) and by Isler

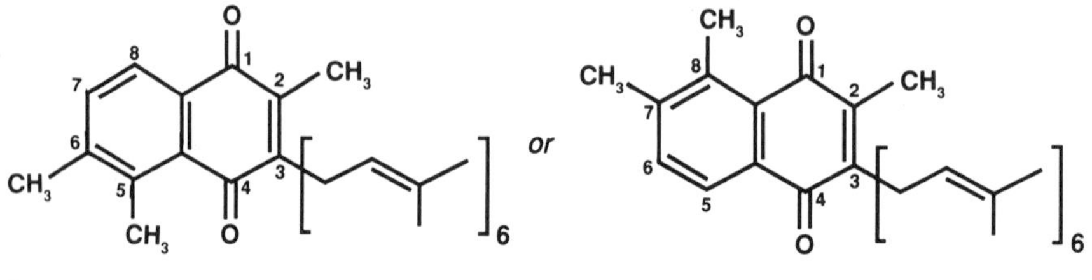

Figure 8.10 Structure of dimethylmenaquinone (DMMK-6) from *Eubacterium lentum*.

8. Isoprenoid Quinones

and co-workers (Morton et al., 1958). Ubiquinones (Figure 8.11a) form a large group of compounds that are widely distributed amongst animals, plants and microorganisms.

Ubiquinones have a more restricted distribution amongst prokaryotes than menaquinones (Collins & Jones, 1981a). They are present in many Gram-negative bacteria, but have not been found in either archaea or Gram-positive bacteria (although there have been a number of reports of ubiquinones within certain Gram-positive taxa, the structure of the quinone or taxonomic position of the species has subsequently been shown to be incorrect) (Collins & Jones, 1981a).

Ubiquinones with side-chains containing six to ten isoprene units have been found as major prenologues in bacteria. Those with smaller (one to five units) or larger (11 to 12 units) side-chains have also been found but generally as minor or trace components. However, ubiquinones with 11 to 14 isoprene units have been isolated as major components from some members of the genus *Legionella* (Collins & Gilbart, 1983; Gilbart & Collins, 1985).

Ubiquinones do not exhibit the same degree of structural variation as components of the menaquinone series. Unlike menaquinones, ubiquinones with partially saturated side-chains do not occur in bacteria though ubiquinones with terminally saturated side-chains do occur in some eukarya (Gale et al., 1963a,b: Lavate & Bentley, 1964; Lavate et al., 1965). Ubiquinones with epoxy groups in the side-chain have been reported in the phototroph *Rhodospirillum rubrum* (Figure 8.12; Friis et al., 1967) but as these compounds only constituted about 2% of the total ubiquinone content it is possible that they could be artifacts of the isolation procedure.

Unusual modifications of ubiquinone side-chains have been discovered in some obligate methane-oxidizing bacteria. Tamaoka et al. (1985) and Collins and Green (1985) independently reported the presence of a novel ubiquinone-8, designated 18-methylene ubiquinone-8, in which a vinylic CH_2 group was located at position C18 in the side-chain (Figure 8.13a). This quinone occurred in about half of the obligate methane-oxidizing

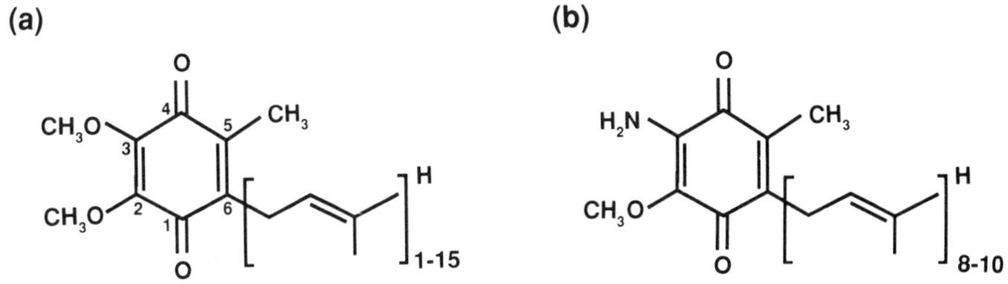

Figure 8.11 Structures of: (a) ubiquinone; (b) rhodoquinone.

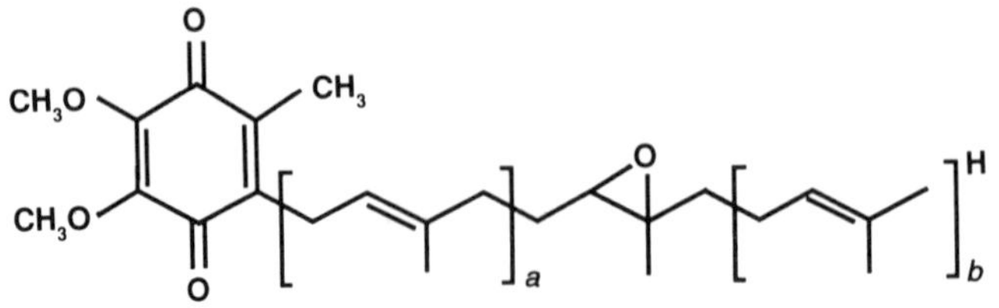

Figure 8.12 Structure of epoxyubiquinones ($a + b = 9$).

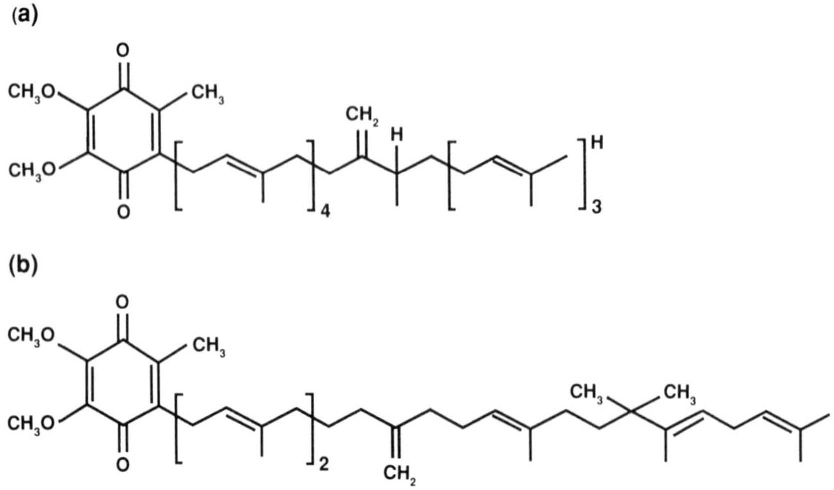

Figure 8.13 Structures of: (a) 18'-methylene-ubiquinone-8; (b) 11'-methylene- 18'-dimethyl-ubiquinone-6.

species examined (Collins & Green, 1985). Collins and associates also discovered a new ubiquinone-6-like molecule in *Methylomonas rubra* (Collins *et al.*, 1986), which they designated 11-methylene-18-dimethyl-ubiquinone-6. The hexaprenyl side-chain of this quinone contained a vinylic CH_2 group at position C'11 (instead of the normal methyl group) and two additional aliphatic methyl groups at position C'18 (Figure 8.13b).

Some phototrophic bacteria contain an unusual purple quinone designated rhodoquinone (Figure 8.11b). This compound was originally isolated from *Rhodospirillum rubrum* (Glover & Threlfall, 1962) and thought to be a hydroxy derivative in which the methoxyl of ubiquinone was replaced by a hydroxy group. However, Moore and Folkers (1966a,b) showed that the compound was an amino quinone related to ubiquinone-10, in which the methoxy group *para* to the side-chain was replaced by an amino group (Figure 8.11b); the compound was abbreviated RQ-10. Other prenologues

8. Isoprenoid Quinones

of rhodoquinone have been found in certain phototrophic bacteria, e.g. RQ-8 in *Rhodospirillum photometricum* and RQ-9 (plus RQ-10) in *Rhodophila globiformis* (Hiraishi & Hashino, 1984). Rhodoquinone is not confined to photosynthetic bacteria; it has been reported in eukarya, e.g. *Euglena gracilis* (Powls & Hemming, 1966a,b), *Astasia longa* (Begin-Heik & Blum, 1967) and *Ascaris lumbricoides* (Ozawa *et al.*, 1969: Sato & Ozawa, 1969).

A benzoquinone, designated mavioquinone, was found in small amounts in *Mycobacterium avium* by Beau *et al.* (1966). Mavioquinone, which has been shown to be 5-methoxy-2-methyl-3-(9,11,13,15-tetramethylheptadecyl)-1,4-benzoquinone (Scherrer *et al.*, 1977), differs chemically from other bacterial respiratory quinones as its side-chain is not isoprenoid in nature. This unusual compound has not been reported in any other bacterium.

8.2.7 Plastoquinones

A second major group of benzoquinones are the plastoquinones. Plastoquinone was originally isolated from alfalfa by Kofler (1946) but was not identified. It was only after the discovery of ubiquinone that plastoquinones were again recognized (Crane, 1959). The structure of plastoquinone was eventually elucidated by Kofler and his colleagues (Kofler *et al.*, 1959a,b) and Folkers and associates (Erickson *et al.*, 1959) and shown to be a 2,3-dimethyl-1,4-benzoquinone with a side-chain containing nine isoprene units (abbreviated PQ-9; Figure 8.14). Amongst prokaryotes, plastoquinones are apparently restricted to cyanobacteria (Collins & Jones, 1981a) but they have been found in the photosynthetic tissues of higher plants and brown, green and red algae. Plastoquinone is not found in phototrophic bacteria.

8.2.8 Sulphur-containing quinones

De Rosa *et al.* (1977) isolated a novel sulphur-containing quinone, designated caldariellaquinone, from the thermophilic, acidophilic

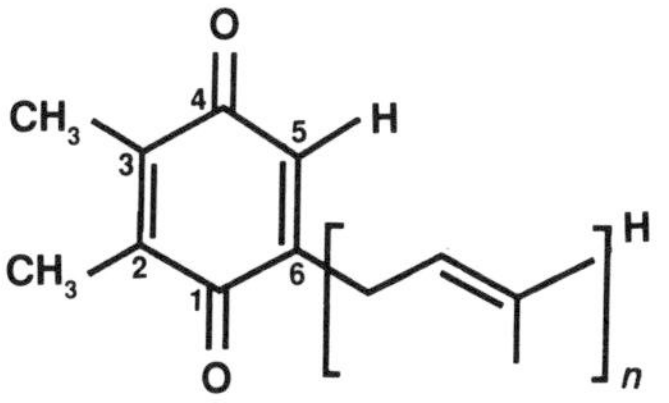

Figure 8.14 Structure of plastoquinone.

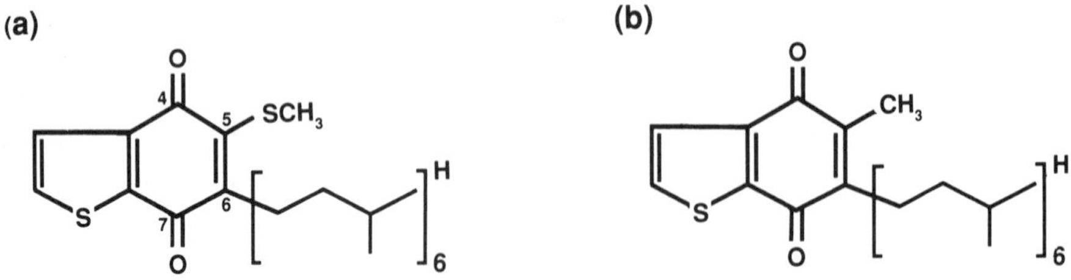

(a)

(b)

Figure 8.15 Structures of: (a) caldariellaquinone (CQ-6(H_{12})); (b) sulfolobusquinone (SQ-6(H_{12})).

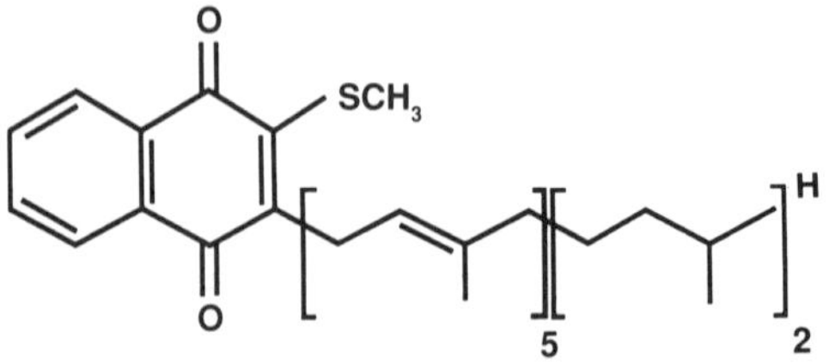

Figure 8.16 Structure of methionaquinone.

archaebacterium '*Caldariella acidophilum*' (*Sulfolobus acidocaldarius*). Caldariellaquinone was shown to be 6-(3,7,11,15,19,23-hexamethyl-tetra-cosyl)-5-methylthio-benzo- [b] -thiopen-4,7-quinone (abbreviated CQ-6(H_{12}); Figure 8.15a). Thurl et al. (1986) isolated a similar quinone, designated sulfolobusquinone, from *Sulfolobus ambivalens* (abbreviated SQ-6(H_{12}); Figure 8.15b). Caldariellaquinone and sulfolobusquinones containing monounsaturated side-chains (first unit unsaturated) have also been found as minor components (CQ-6(H_{10}), SQ-6(H_{10}); Thurl et al., 1986). A new benzo[1,2-b; 4,5-b']dithiophene-4,8-quinone has been reported as a trace component in *Sulfolobus solfataricus* (Trincone et al., 1986).

Ishii et al. (1983) described a novel sulphur-containing naphthoquinone, designated methionaquinone, from a thermophilic hydrogen bacterium. This compound was shown to be 2-methylthio-3-VI,VII-tetrahydroheptaprenyl-1,4-naphthoquinone (abbreviated MTK-7(VI,VII-H_4); Figure 8.16).

8.3 EXTRACTION AND PURIFICATION

Isoprenoid quinones are free lipids that can be readily extracted from bacterial cells using lipid solvents such as acetone, chloroform and hexane. Extraction of isoprenoid quinones is normally achieved with any one of these solvents or with a mixture of any two of them. Alkaline saponification is unnecessary and should be avoided as quinones are susceptible to strong acid or alkaline conditions. They are also somewhat susceptible to photo-oxidation in the presence of oxygen and strong light but it is not necessary to work in a nitrogen atmosphere or dim light. It is, however, good practice to conduct extraction and subsequent purification procedures fairly rapidly.

A large number of different systems have been used to extract quinones. The various procedures normally yield a complex mixture of lipids (the complexity of which depends on the method adopted) containing small amounts of non-lipid material from which the isoprenoid quinones can be readily purified by simple chromatographic procedures. The procedure employed by the author and probably the most widely used method is direct extraction of dry bacterial cells with chloroform-methanol (Collins et al., 1977; Collins, 1985a).

Method 8.1 Direct extraction of isoprenoid quinones with chloroform-methanol.

1. Approximately 50 to 100 mg of lyophilized cells are extracted with a small volume ($\sim$ 25–50 ml) of chloroform-methanol (2 : 1 v/v) for approximately 1 h using a magnetic stirrer.
2. The cell/solvent mixture is passed through filter paper (using a filter funnel) to remove cell debris.
3. The eluate is collected in a flask and evaporated to dryness under reduced pressure at $\sim 40\,°C$ on a rotary evaporator.

Another good extraction method is that of Krivánková and Dadák (1976) who used methanol-light petroleum (pentane). However, complete extraction of menaquinones using this procedure necessitates the use of perchloric acid though the latter is not necessary for ubiquinones (Krivánková & Dadák, 1976, 1980). A similar method has been described for the sequential extraction of isoprenoid quinones and polar lipids (Minnikin et al., 1984). An additional procedure which may be employed involves the direct extraction of cells (wet or lyophilized) with acetone (Lester et al., 1964; Redfearn & Burgos, 1966). In the case of Gram-positive organisms destruction of the cell wall by ultrasonic disintegration ($\sim$ 1 min in cold acetone) makes the cell more amenable to organic solvents.

Isoprenoid quinones can be readily isolated and purified from extracts by chromatographic procedures. Column chromatography, extensively used in the past, and still useful for large-scale preparations, has been largely superseded by thin-layer chromatographic techniques. The most widely employed procedure is thin-layer chromatography (TLC) using silica gel with hexane-diethylether (85 : 15, v/v) as developing solvent (Collins et al., 1977; Collins, 1985a, (see Method 8.2). Isoprenoid quinones are revealed (non-destructively) by quenching of UV light at 254 nm.

Method 8.2 Purification of isoprenoid quinones by thin-layer chromatography.

---

1. Resuspend lipid extract in a small volume of chloroform/methanol (2 : 1, v/v).
2. Apply extract (with Pasteur pipette or syringe) as a uniform streak (~5–8 cm long) to a silica-gel F_{254} sheet (plastic-backed cut to 10 × 10 cm).
3. Develop plate in hexane-diethyl ether (85 : 15 or 80 : 20, v/v); developing time ~20 min.
4. Allow plate to dry in air (~5 min), view isoprenoid quinones by brief irradiation with ultraviolet light at 254 nm. The isoprenoid quinones appear as dark-brown/purple bands on a green fluorescent background: menaquinones, R_F ~0.7; methylmenaquinones, R_F ~0.8 and ubiquinones, R_F ~0.3.
5. Scrape gel from plate with spatula and elute through a sintered glass filter with chloroform or other solvent.
6. Evaporate to dryness under a stream of nitrogen gas. Store at $-20\,^\circ$C.

Using this system menaquinones, methylmenaquinones and ubiquinones have R_F values of 0.7, 0.8 and 0.3, respectively. Other solvent systems may be used to facilitate particular separations. For example, the resolution of methyl- and dimethylmenaquinones is best achieved by triple development in a single-dimension using hexane-diethylether (20 : 1, v/v; Collins et al., 1985). Alternatively, isoprenoid quinones may be partially purified prior to analysis using silica cartridges (see Method 8.3).

Method 8.3 Purification of isoprenoid quinones by silica cartridge.

---

1. Pre-wash a silica cartridge (e.g. Sep-Pak, Waters Associates) with 10 ml hexane.
2. Dissolve lipid extract in hexane (1 ml) and load extract to cartridge head by either pushing the hexane solution with a syringe or applying vacuum.
3. Elute the cartridge with 10 ml of hexane and discard eluate.
4. Elute the cartridge with 5 ml of 10% (by vol.) diethyl ether in hexane and collect isoprenoid quinone in vial.
5. Evaporate to dryness either under a stream of nitrogen gas or on a rotary evaporator (40–45 $^\circ$C) under reduced pressure. Store at $-20\,^\circ$C.

This latter system is particularly useful when only small quantities of isoprenoid quinone are required, as with high-performance liquid chromatography (HPLC) linked to electrochemical detection (interfering pigments may be a problem for some organisms when this system is used for HPLC ultraviolet detection).

8.4 CHROMATOGRAPHIC ANALYSIS

The composition of natural mixtures of bacterial isoprenoid quinones can be investigated by a variety of chromatographic techniques. The methods used can be divided into two main types: partition chromatography (reverse-phase mode) and argentation chromatography.

8.4.1 Reverse-phase partition chromatography

Reverse-phase partition chromatography facilitates the separation of ubiquinone and menaquinone components according to their overall physical properties. The latter depend mainly on the length of the multiprenyl side-chains, and in the case of menaquinones, on the degree of saturation of the side-chain. In reverse-phase chromatography separations are achieved using non-polar (hydrophobic) stationary phases and polar developing solvents. In the past, reverse-phase partition paper chromatography was used extensively for the separation of isoprenoid quinone mixtures (Sommer & Kofler, 1966) although this method has now been superseded by reverse-phase partition thin-layer techniques (RPTLC). Until recently TLC plates were impregnated with non-polar stationary phase (usually a non-volatile hydrocarbon such as paraffin or hexadecane) by dipping the

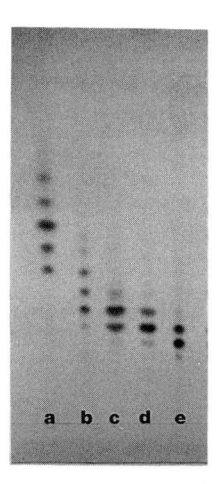

Figure 8.17 Reverse-phase partition TLC of ubiquinones: (a) Q-6 to Q-10; (b) *Legionella morrisei*; (c) *Legionella pneumophila*; (d) *Legionella micdadei*; (e) *Legionella sp.* (F268). Merck RP18 plate (10 × 10 cm) using acetone-acetonitrile (80:20, v/v).

plate in a solution of the hydrocarbon (in a volatile solvent) followed by evaporation of the solvent (e.g. Dunphy & Brodie, 1971) or by ascending chromatography (Hammond & White, 1969).

The methods outlined above were time consuming and further hampered by lack of reproducibility and loading problems. However, ready-made high-performance reverse-phase plates utilizing chemically bonded phases (C_8, C_{18}) are now available and have been employed extensively for lipoquinone analyses. The resolution, reproducibility and developing times of ready-made RPTLC plates are superior to those of conventional plates. The high loading capacity of these plates together with chemically bonded phase also facilitates preparative work (the gel is simply scraped from the plate and quinones eluted with solvent). The RPTLC methods employed by the author for menaquinone and ubiquinone analyses are outlined in Methods 8.4 and 8.5. An example of the separation of ubiquinones by RPTLC is shown in Figure 8.17.

Method 8.4 Reverse-phase thin-layer partition chromatography of ubiquinones.

---

1. Suspend ubiquinone sample in a small volume of acetone or chloroform-methanol (2:1, v/v).
2. Spot sample onto a Merck HPTLC RP18F_{254} (Art. 13724) plate (10 × 10 cm) with a fine pipette or glass capillary.
3. Develop plate in acetone-acetonitrile (4:1, v/v). Allow plate to dry and observe isoprenoid quinones under ultraviolet light at 254 nm. Isoprenoid quinones appear as dark spots on a blue fluorescent background. Alternatively, they can be revealed by spraying with molybdophosphoric acid in ethanol (10%) and heating at 140 °C for 1 to 3 min.

This system (Collins & Jones, 1981b) allows clear separation of Q-1 through to Q-15; identification of components can be achieved by comparing R_F values with known ubiquinones (Sigma Chemical Company Ltd). The system also gives good separation of rhodoquinone prenologues (acetone-water solvent systems may be employed for rhodoquinones, see Hiraishi & Hoshino, 1984).

Method 8.5 Reverse-phase thin layer partition chromatography of menaquinones.

---

1. Suspend menaquinone sample in a small volume of chloroform-methanol (2:1 v/v).
2. Spot sample onto a Merck HPTLC RP18F_{254} (Art. 13724) plate 10 × 10 cm with a fine pipette or glass capillary.
3. Develop plate in acetone.
4. Dry and observe isoprenoid quinones under ultraviolet light at 254 nm.

This chromatographic system facilitates the separation of menaquinone components with the same number of isoprene units but differing degrees of saturation. Saturation of one double bond causes a negative shift in R_F value, approximately

8. Isoprenoid Quinones

the effect of adding one isoprene unit (e.g. MK-9(H$_2$) behaves like MK-10). The presence of an additional methyl group in the ring system (methylmenaquinones) also causes a negative shift in R_F approximately equivalent to adding one isoprene unit (e.g. MMK-7 has a similar R_F to MK-8).

Thin-layer chromatographic techniques generate only qualitative or semi-quantitative data (relying upon the intensity of spots on chromatograms). In contrast, high-performance liquid chromatography can be used to generate quantitative data. The resolving power and sensitivity of HPLC is also superior to that of thin-layer techniques. The separation of isoprenoid quinone mixtures by HPLC is normally by the reverse-phase partition mode (RPHPLC). As with RPTLC, separation is determined by the lipophilic character of the isoprenoid quinones, which depends mainly on the length and degree of saturation of the multiprenyl side-chain.

Ubiquinone and menaquinone series may be separated isocratically using octadecylsilane (ODS) as stationary phase (e.g. Spherisorb ODS, 5 μm, columns 250 × 4.6 mm i.d.). Both types of compound are normally monitored with an ultraviolet detector (270–275 nm for Q, 269 nm for MK) and quantified using an on-line computer integrator. A large number of different mobile phases may be employed (see Tamaoka et al., 1983; Collins, 1985a; Kroppenstedt, 1985); the author prefers methanol/1-chlorobutane mixtures (e.g. 100:10 or 100:20, v/v) as they give good separation of members of menaquinone and ubiquinone series (the level of non-polar solvent can be altered to facilitate particular separations; Collins, 1985a; Gilbart & Collins, 1985). As with RPTLC, identification of components is normally achieved by comparing retention times with isoprenoid quinones of known structure. With ubiquinones this is relatively easy due to their simple structures and the commercial availability of commonly occurring prenologues. (If a particular ubiquinone is not available, because the relationship between common logarithm of retention time and the number of isoprene units is linear, the length of the side-chain of any unknown prenologue can be determined graphically; see Figure 8.18.)

The unequivocal assignment of menaquinone components is more difficult given their greater structural variability (also relatively few reference compounds are available commercially and menaquinone preparations from bacterial strains of established structure have to be used as standards). Saturated prenologues are more hydrophobic than their unsaturated counterparts and consequently are retained longer on the HPLC column. Cyclization can also result in longer elution times (Howarth et al., 1986).

As with ubiquinones, a linear relationship between log retention time and the number of units in the side-chain of equally hydrogenated menaquinones is observed under isocratic conditions (i.e. separate lines are obtained for unsaturated and each homologous series of di-, tetra-, hexa-,

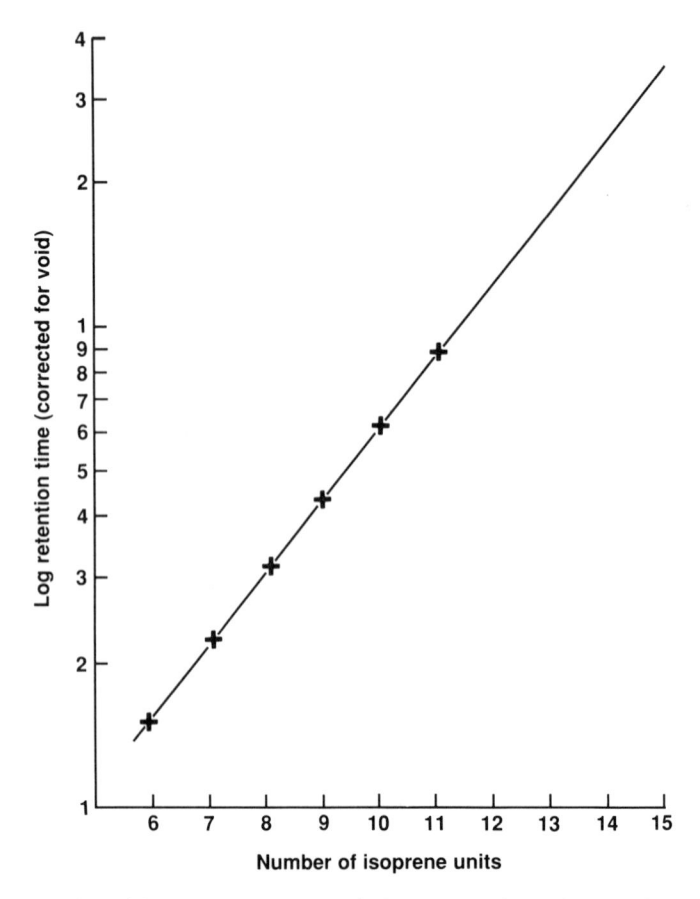

Figure 8.18 Plot of log retention time of ubiquinone homologies (Q-6 to Q-11).

octa-hydro-menaquinones (Kroppenstedt, 1985). This can result in near coincidence of elution times for some different menaquinone isomers. In the case of menaquinones further structural complications include ring demethylation and methylation (demethylmenaquinones elute before and methylmenaquinones after their normal menaquinone counterparts). However, the problem of ring demethylation/methylation can be easily overcome using ultraviolet spectroscopy. Examples of the separation of ubiquinones and menaquinones by RPHPLC are shown in Figures 8.19 and 8.20.

8.4.2 Argentation chromatography

Reverse-phase partition chromatography by itself is insufficient for the identification of partially saturated menaquinones. The method described

8. Isoprenoid Quinones

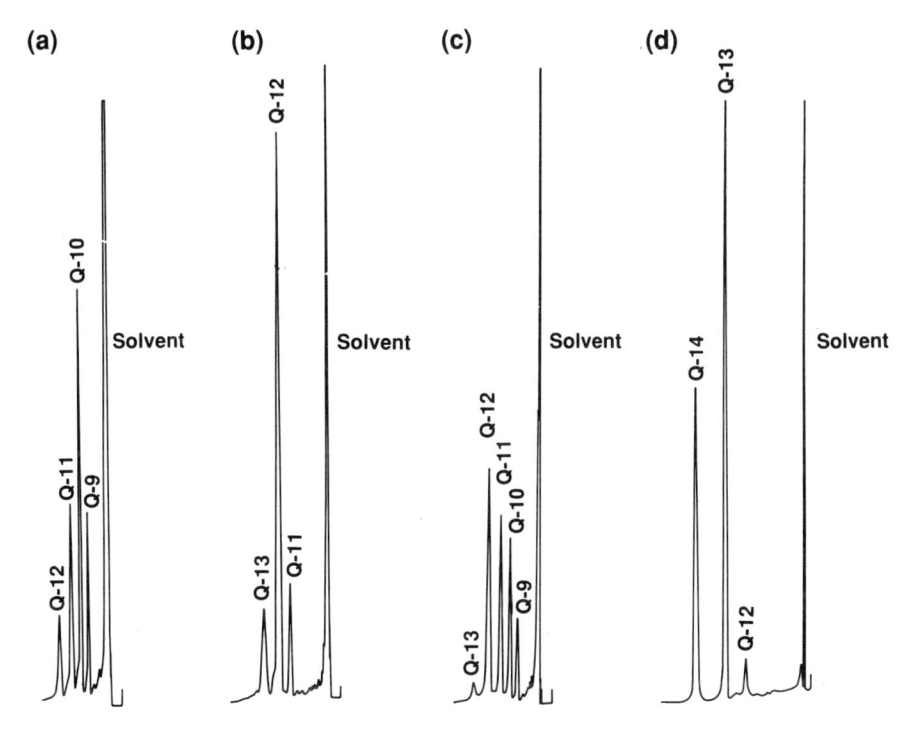

Figure 8.19 Reverse-phase high-performance liquid chromatography (RPHPLC) of ubiquinones from: (a) *Legionella oakridgensis*; (b) *L. rubrilucent*; (c) *L. sclinthelensi*; (d) *L. feeleii*. Spherisorb ODS (25 cm) column using methanol/1-chlorobutane (100:20, v/v) as mobile phase (1.5 ml/min).

above should therefore be used in conjunction with either physicochemical methods, such as mass spectrometry, or chromatographic procedures where the menaquinones display different separation behaviour. The capacity of silver ions to form complexes with olefinic bonds provides a means for the separation of menaquinone components according to the length and degree of saturation of their multiprenyl side-chains.

Argentation chromatographic analysis of menaquinones is normally based on thin-layer methods (Dunphy & Brodie, 1971). Superior separation/resolution, however, can now be achieved by liquid chromatography using silver-loaded ion-exchange columns (Kroppenstedt, 1985). Menaquinones (injection solvent ethanol) can be separated on a silver-loaded Nucleosil 5SA column (12.5 cm) using hot methanol as mobile phase. The column should be maintained at a constant temperature (50 to 65 °C, depending upon prenologues and separation required) using a precision thermostat (water jacket or oven). Menaquinones are monitored at 270 nm.

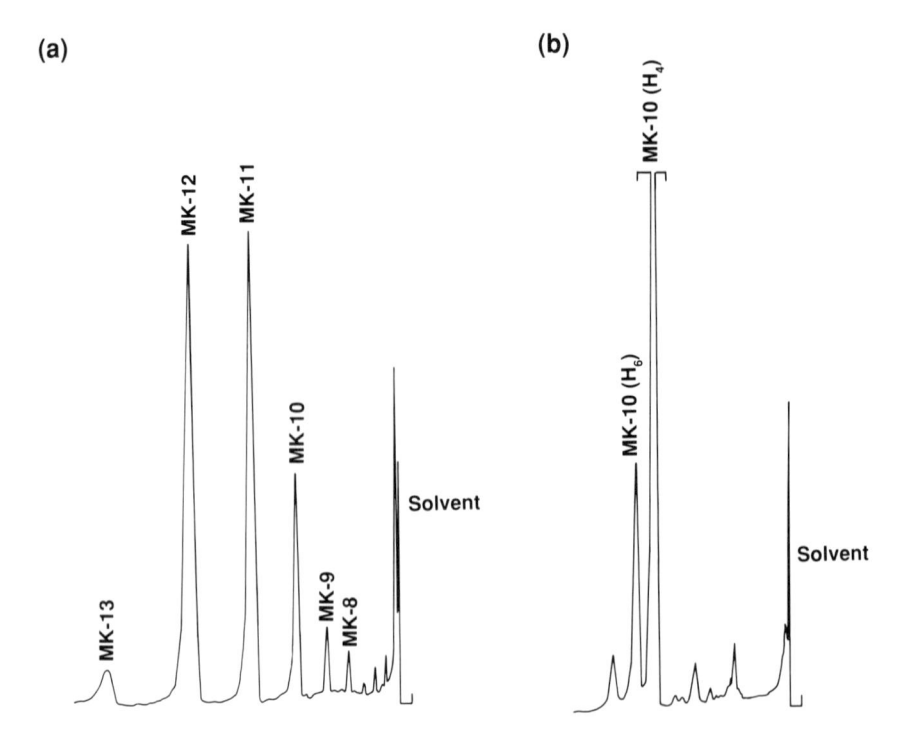

Figure 8.20 RPHPLC of menaquinones from: (a) *Aureobacterium liquefaciens*; (b) *Micromonospora purpurea*. Spherisorb ODS (25 cm) column using methanol/ 1-chlorobutane (100:10, v/v) as mobile phase (1.5 ml/min).

Separation by Ag^+ HPLC is determined primarily by the degree of unsaturation of the multiprenyl side-chain. Menaquinones with fewer double bonds are eluted before those with a larger number. Similarly, menaquinones with the same number of double bonds but longer isoprenoid chain length elute earlier. Ag^+ HPLC also facilitates the separation of stereoisomers of hydrogenated menaquinones, i.e. partially hydrogenated menaquinones which differ in the 'position' of saturation of the C3 multiprenyl side-chain (Kroppenstedt, 1982, 1985; Collins et al., 1988). The separation of menaquinone 'positional' isomers by Ag^+ HPLC is illustrated in Figures 8.21 and 8.22. By comparing the behaviour (retention time and direction of change) on both Ag^+ HPLC and RPHPLC it is possible to elucidate the structure of unknown hydrogenated menaquinones. Ag^+ HPLC has also been used for the analysis of bacterial ubiquinones (Collins, 1985a; Collins & Green, 1985; Collins et al., 1986a); an example is shown in Figure 8.23.

8. Isoprenoid Quinones

A chromatographic technique with considerable potential in respiratory isoprenoid quinone analysis is supercritical fluid chromatography (SFC). This technique has not been used extensively for lipoquinone analysis but the use of open-tubular capillary columns potentially offers high-resolution separations. An example of SFC separation of the ubiquinone series is shown in Figure 8.24.

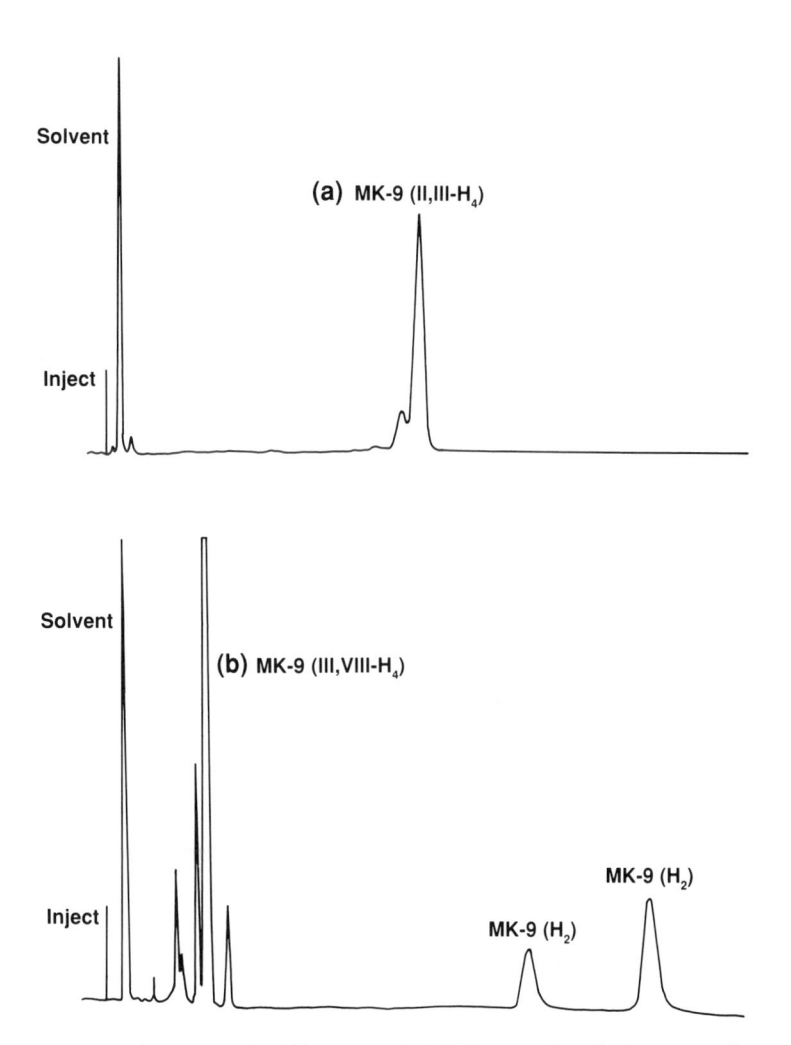

Figure 8.21 Ag$^+$HPLC of different MK-9(H$_4$) positional isomers from: (a) *Saccharopolyspora* (*Faenia*) *rectivirgula*; (b) *Microtetraspora* (*Micropolyspora*) *angiospora*. Silver-loaded Nucleosil 5SA (12.5 cm) column using methanol as mobile phase.

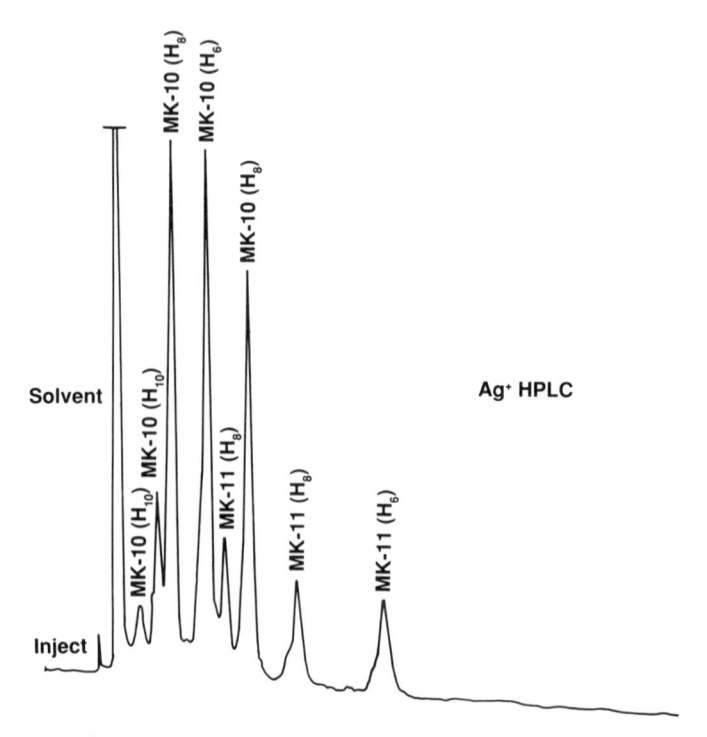

Figure 8.22 Ag⁺HPLC of menaquinone positional isomers from *Thermomonospora fusca*. Silver-loaded Nucleosil 5SA (12.5 cm) column using methanol as mobile phase.

8.5 PHYSICO-CHEMICAL ANALYSIS

8.5.1 Ultraviolet spectroscopy

Ultraviolet spectroscopy provides a simple method for establishing the class or category to which an unknown lipoquinone belongs. The various respiratory isoprenoid quinone types exhibit very characteristic ultraviolet absorption characteristics, details of which are summarized in Table 8.3.

The ultraviolet characteristics of menaquinones and related compounds are markedly affected by the nature of naphthoquinone ring substitution. Menaquinones and phylloquinones exhibit qualitatively identical spectra with five absorption maxima (λ_{max}) at 242, 248, 260, 269 and 326 nm and one point of inflection at 238 nm (Figure 8.25a). Removal of the methyl group at C2 of the naphthoquinone nucleus (as in demethylmenaquinones) causes a shift in the quinone absorption contribution to lower

8. Isoprenoid Quinones

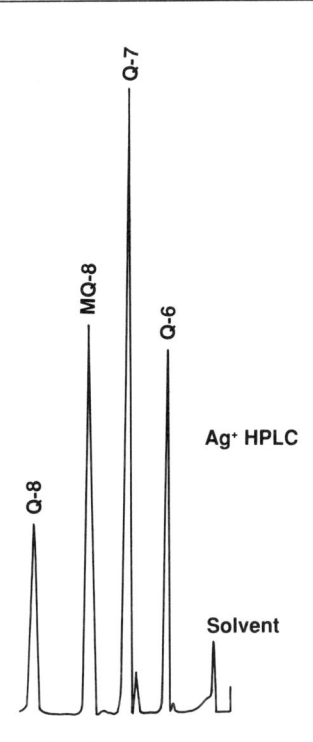

Figure 8.23 Ag⁺HPLC of ubiquinones (Q-6 to Q-8 and 18′-methylene-Q-8). Silver-loaded Nucleosil 5SA (12.5 cm) column using methanol as mobile phase.

wavelength of about 6 nm (λ_{max} 242, 248, 254, 263, 326, and 238 (inf) nm). The UV spectra of methylmenaquinones differ from those of the menaquinone series in that absorption at 326 nm shifts approximately 12 nm to higher wavelength (Figure 8.26a,b). The replacement of the methyl group at C2 of menaquinones with a methylthio group, as in methionaquinone, causes a marked alteration in absorption with λ_{max} at 238, 262 and 310 nm.

Ubiquinones produce a very simple ultraviolet spectrum with a single λ_{max} at 270 to 275 nm (Figure 8.25b). The replacement of the methoxyl group at C3 of ubiquinone by an amino group (as in rhodoquinone), results in a marked change in the spectrum with three absorption maxima at 253, 283 and 320 nm. Plastoquinone is readily distinguished from other benzoquinones in producing two absorption maxima at 254 and 262 nm (Figure 8.25c).

The unusual sulphur-containing lipoquinones, caldariellaquinone and sulfolobusquinone, exhibit absorption characteristics quite distinct from other respiratory quinones. Caldariellaquinone (Figure 8.25d) produces absorption maxima at 237, 272, 278 and 322 nm, whereas λ_{max} of sulfolobusquinone occur at 225, 270, 279 and 321 nm.

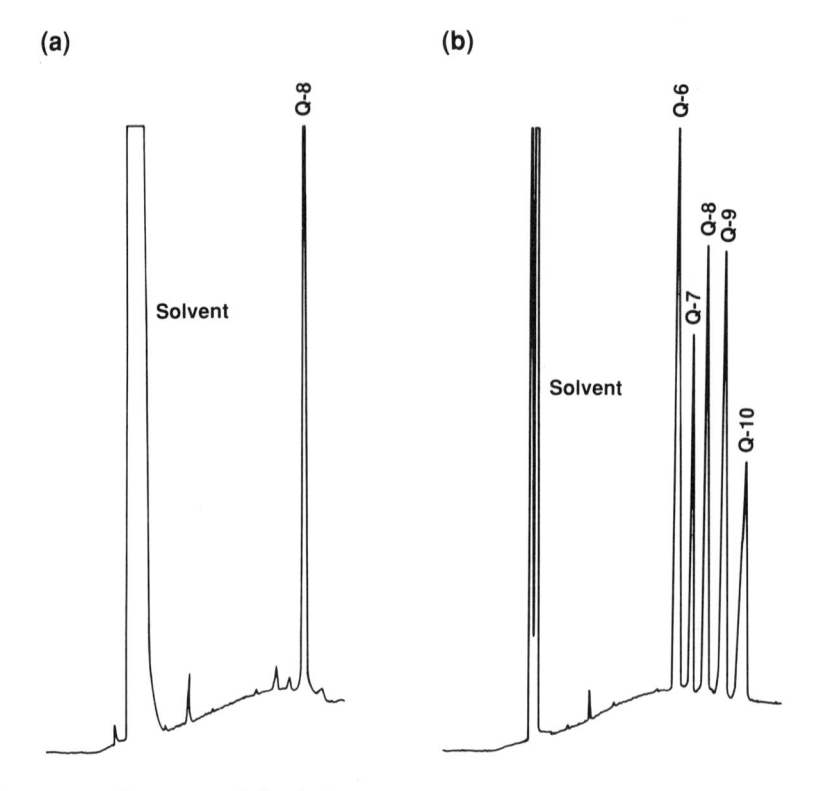

Figure 8.24 Supercritical fluid chromatogram of ubiquinones. Column 10 m SB-methyl 100 (50 μm i.d.). Mobile phase CO_2, initial density 0.2 g/ml, hold time 3 min, rising asymptotically to 0.65 g/ml with half-rise = 7 min.

Table 8.3 Ultraviolet absorption characteristics of respiratory quinone classes.

Quinone	Solvent	λ_{max}(nm)[a]
1,4-Naphthoquinones		
Menaquinone	Iso-octane	242, 248, 260, 269, 326, 238 (inf)
Phylloquinone	Iso-octane	242, 248, 260, 269, 326, 238 (inf)
Demethylmenaquinone	Iso-octane	242, 248, 254, 263, 326, 238 (inf)
Methylmenaquinone	Iso-octane	242, 248, 259, 269, 338
Dimethylmenaquinone	Iso-octane	244, 252, 261, 271, 342
Methionaquinone	Ethanol	238, 262, 310
Chlorobiumquinone	Ethanol	254, 265 (inf)
1,4-Benzoquinones		
Rhodoquinone	Ethanol	253, 283, 320, 500 (inf)
Ubiquinone	Ethanol	275
Plastoquinone	Iso-octane	254, 262
Benzo[b]thiophen-4,7-quinones		
Caldariellaquinone	Iso-octane	237, 272, 278, 322
Sulfolobusquinone	Hexane	225, 270, 279, 321

[a] inf = inflection point.

8. *Isoprenoid Quinones* 293

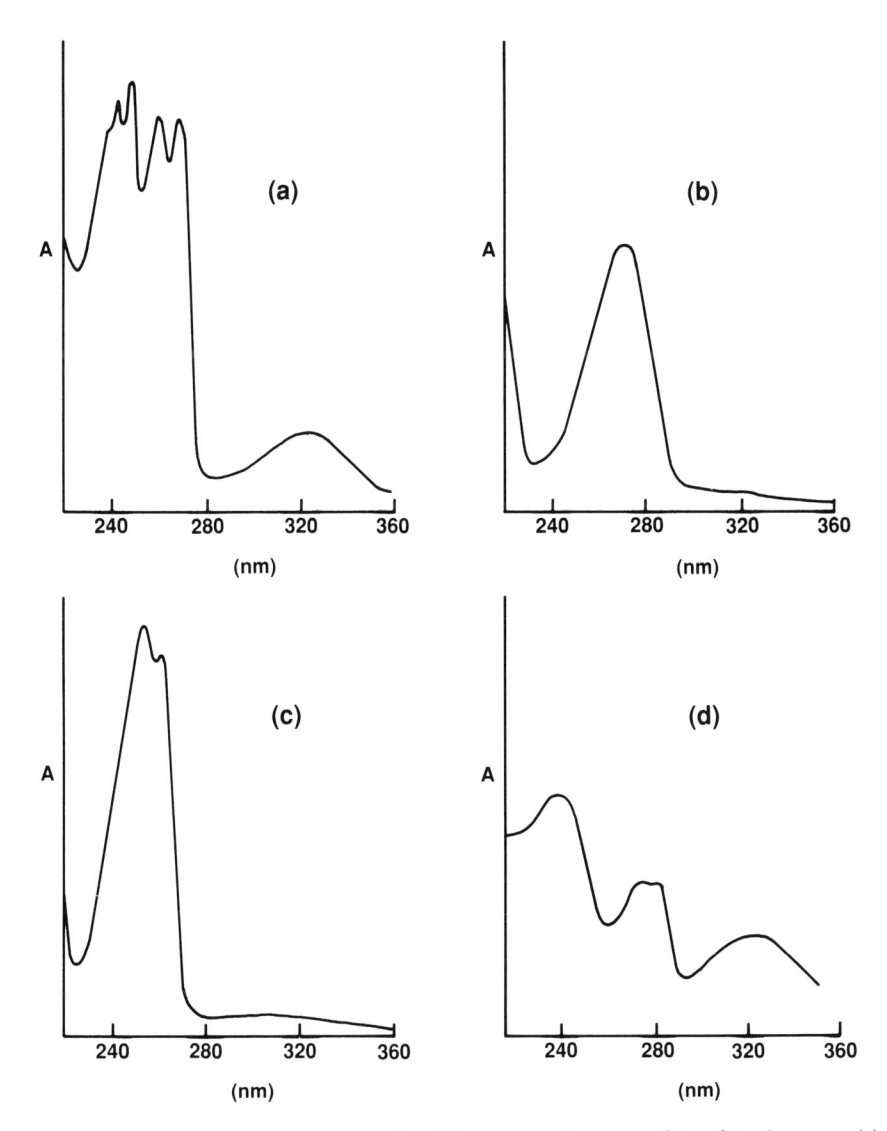

Figure 8.25 Ultraviolet spectra of: (a) menaquinone; (b) ubiquinone; (c) plastoquinone; (d) caldariellaquinone.

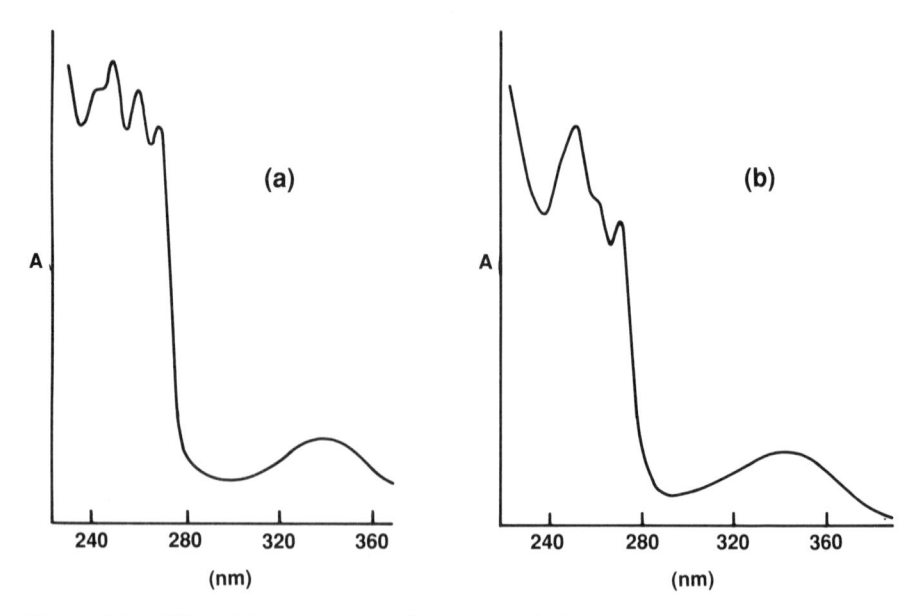

Figure 8.26 Ultraviolet spectra of: (a) methylmenaquinone; (b) dimethylmenaquinone.

8.5.2 Mass spectrometry

Direct probe mass spectrometry is an indispensable tool for the structural determination of isoprenoid quinones. Upon mass spectral analysis the various isoprenoid quinone types produce similar, relatively simple, fragmentation patterns (Collins, 1985a).

The base peak of most menaquinones occurs at m/z 225, with a second intense ion at m/z 187. The peak at m/z 225 corresponds to that of ion a (Figure 8.27) which is stabilized by extensive charge delocalization. The peak at m/z 187 can be assigned to that of ion b (Figure 8.27) and is generated from the corresponding hydroquinone. The base peak in the mass spectrum of demethylmenaquinones occurs at m/z 211, with a second intense peak at m/z 172 (ions c and d; Figure 8.27). Methylmenaquinones and dimethylmenaquinones have corresponding nuclear fragments at m/z 239 and 201, and at m/z 253 and 215 respectively (ions e to h; Figure 8.27) (Collins, 1985b; Collins et al., 1985). It should be noted that the base peak in the spectra of menaquinones and methylated analogues in which the first isoprene unit is saturated corresponds to the hydroquinone ion (m/z 187 for MK, 201 for MMK and 215 for DMMK). The sulphur-containing naphthoquinone, methionaquinone, produces an intense peak at m/z 257 (analogous to ion d of menaquinones) with a second peak at m/z 211 due

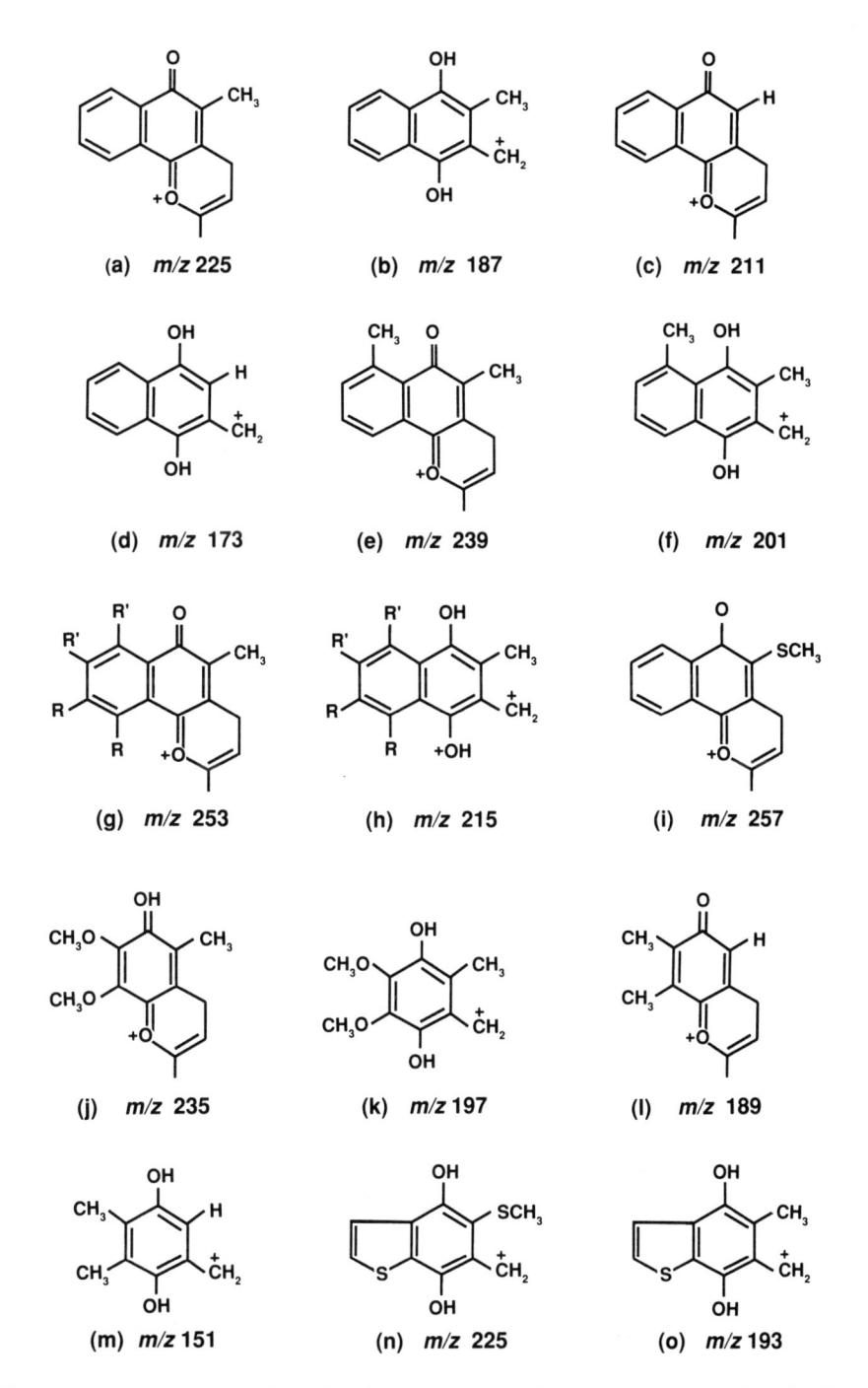

(a) m/z 225
(b) m/z 187
(c) m/z 211
(d) m/z 173
(e) m/z 239
(f) m/z 201
(g) m/z 253
(h) m/z 215
(i) m/z 257
(j) m/z 235
(k) m/z 197
(l) m/z 189
(m) m/z 151
(n) m/z 225
(o) m/z 193

Figure 8.27 Mass spectral nuclear fragments. R and R' correspond to H and CH$_3$, not necessarily respectively.

to the elimination of S=CH$_2$ (produced by hydrogen transfer) from ion i (Figure 8.27; Ishii et al., 1983). Ubiquinones produce strong ions at m/z 235 (base peak) and 197 and are similarly derived from the 2,3-dimethoxy-1,4-benzoquinone nucleus (ions j and k; Figure 8.27). The corresponding fragments for the plastoquinone series occur at m/z 189 and 151 (ions 1 and m; Figure 8.27).

In addition to the very intense fragments in the low mass region (the m/z values of which facilitate the assignment of the various isoprenoid quinone classes) all isoprenoid quinones give rise to a strong molecular ion (M$^+$). The m/z values of M$^+$ readily facilitate the determination of the length and degree of saturation of the multiprenyl side-chain. It is worth noting that the presence of intense M$^+$ also enables high resolution mass measurements to be performed and the establishment of molecular formulae. In addition to strong peaks corresponding to molecular ions, peaks of lower intensity due to the loss of methyl from M$^+$ are observed. A series of low intensity ions due to the fragmentation of the side-chain of unsaturated isoprenoid quinones are also formed; these weak side-chain fragments are generally less evident in ubiquinones than menaquinones.

Many prokaryotes (e.g. actinomycetes, some Gram-negative sulphate-reducing bacteria, halophilic archaea) contain menaquinones with partially saturated side-chains. Saturation of olefinic bonds within the side-chain causes a marked interruption of fragmentation. Although this alteration in fragmentation can be used to determine the point of saturation of dihydrogenated menaquinones and sometimes tetrahydrogenated side-chains, simple mass spectrometry cannot be used to determine the precise position of saturation of more complex hydrogenated forms. Such menaquinones can, however, be converted to chromenyl acetates and subjected to ozonolysis. Mass spectral analysis of the corresponding aldehydes, formed by reduction of ozonides, can provide information on the points of saturation of the side-chain (e.g. Azerad & Cyrot-Pelletier, 1973; Yamada et al., 1977, 1982 a,b).

The most precise approach for positional elucidation is MS/MS analysis. This technique gives rise to clear side-chain fragmentations from which it is relatively easy to deduce the exact positions of saturation of even highly hydrogenated forms (e.g. hexa- and octahydrogenation) (Collins et al., 1988; Ramsey et al., 1988). An example of a collision-induced decomposition (CID) spectrum of the molecular ion (m/z 788) of MK-9(H$_4$) is shown in Figure 8.28. It is apparent from the latter that A-type fragment ions arise by direct cleavage of single bonds that are allylic to two double bonds, whereas the B-series are rearranged ions in which a proton(s) is(are) transferred to the neutral fragment. Comparison of the two series of cleavages readily allows the points of saturation from the ring system to be located.

8. Isoprenoid Quinones

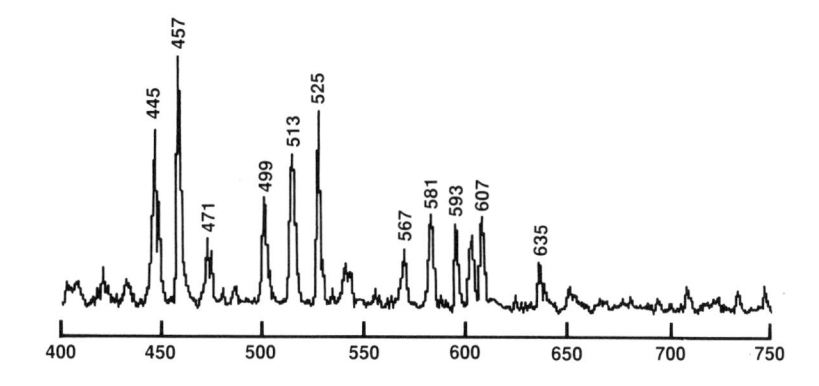

Figure 8.28 MS/MS of MK-9(III, VIII-H$_4$).

The sulphur-containing quinones, caldariellaquinone (CQ-6(H$_{12}$)) and sulfolobusquinone (SQ-6(H$_{12}$)), produce similar mass spectra to those of other lipoquinones except that the side-chain fragments are not observed due to the presence of fully saturated side-chains. The base peak of CQ-6(H$_{12}$) occurs at m/z 225/224 (Figure 8.27n). Intense fragments corresponding to M$^+$ occur together with low-intensity fragments at [M-15]$^+$ and [M-47]$^+$ characteristic of a methylthio group. Loss of the complete side-chain combined with H-transfer produces an intense peak at m/z 212. In contrast SQ-6(H$_{12}$) produces intense peaks at m/z 193/192 (Figure 8.27n) analogous to m/z 225/224 in CQ-6(H$_{12}$) due to the presence of only one sulphur atom in the quinone system. The spectrum of SQ-6(H$_{12}$) shows no fragment ion at [M-47]$^+$ and the loss of the side-chain by a fragmentation induced by the methylthio group as in CQ-6(H$_{12}$) is absent (Thurl et al., 1985).

It should be emphasized that although mass spectrometry is the most precise method for the structural analysis of isoprenoid quinones the technique does not always provide reliable quantitative information; the problem becomes particularly acute with complex mixtures of large prenologues and hydrogenated forms. Quantification should preferably be by high-performance liquid chromatographic techniques.

8.5.3 Nuclear magnetic resonance spectroscopy

One of the most useful techniques for the structural determination of respiratory quinones is proton nuclear magnetic resonance spectroscopy. Menaquinones and related 2,3-disubstituted naphthoquinones display highly characteristic ^{1}H-NMR spectra with complex absorption at δ 7.5 to 8.1 due to the presence of four adjacent ring protons (Figure 8.29). In the case of menaquinones with unsaturated side-chains complex overlapping absorption at δ 5.0 to 5.2, due to olefinic protons, and a doublet at δ 3.3, due to methylene adjacent to the ring (the latter is characteristic of

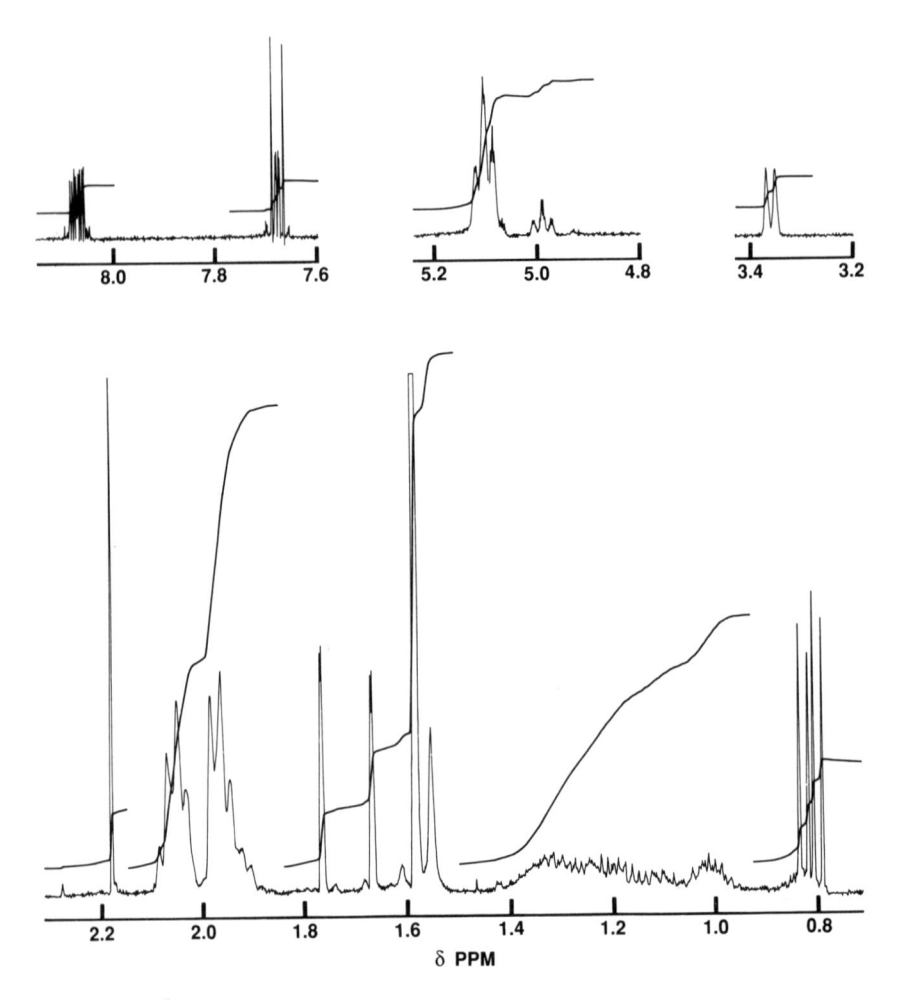

Figure 8.29 ^{1}H-NMR spectrum of MK-9(H$_4$).

quinones containing a double bond in the first isoprene unit), are observed. The methyl group at C2 on the ring produces a singlet at δ ~2.1. Strong resonances in the δ 1.9 to 2.1 and 1.5 to 1.8 regions are due to methylenes and methyl groups respectively adjacent to double bonds (Table 8.4).

Demethylmenaquinones produce similar spectra to menaquinones except that a triplet, due to long-range splitting, corresponding to quinoid hydrogen occurs at δ 6.7 whilst the singlet due to ring C2 methyl is absent

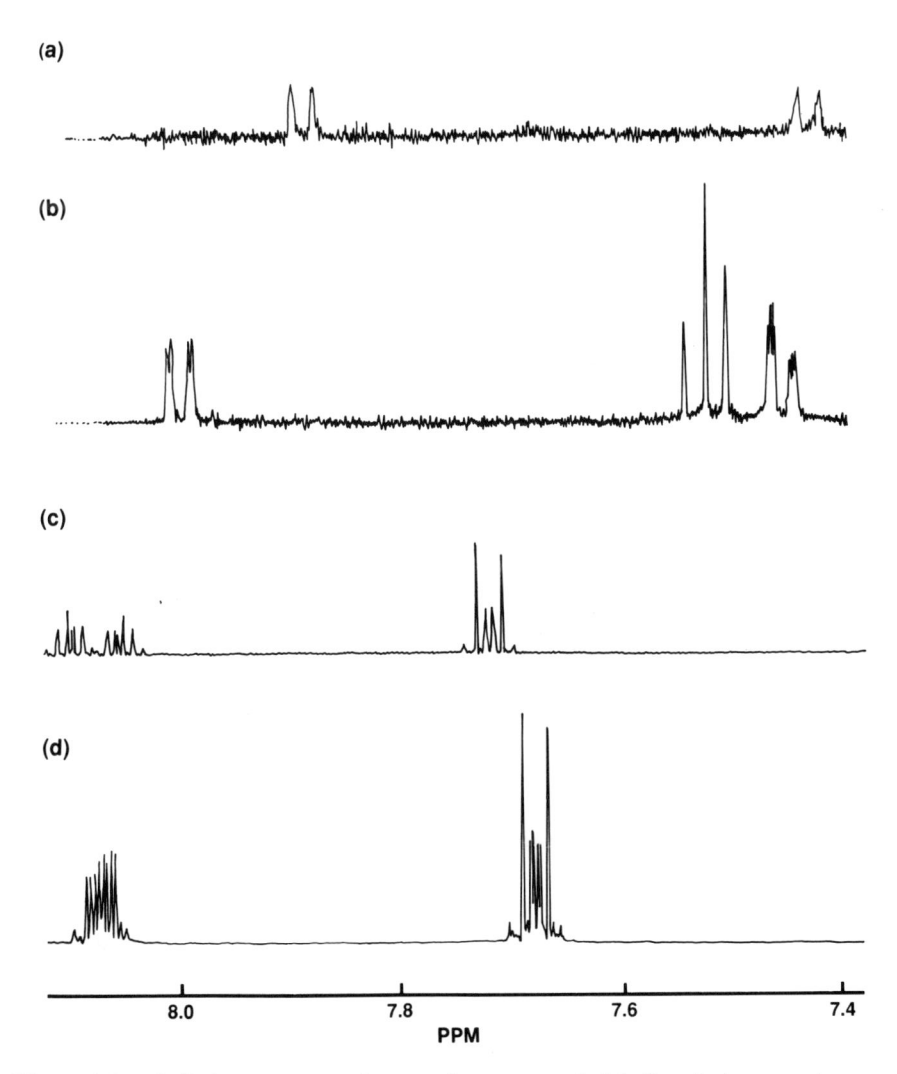

Figure 8.30 Splitting patterns of aromatic protons of: (a) dimethylmenaquinone; (b) methylmenaquinone; (c) demethylmenaquinone; (d) menaquinone.

Table 8.4 Chemical shifts (δ) and splitting patterns for ^{1}H-NMR spectra of menaquinones and related 1,4-naphthoquinones.[a]

Assignment	Phylloquinone	Unsaturated menaquinone	Partially saturated menaquinone	Demethylmenaquinone	Chlorobiumquinone[b]	Methylmenaquinone	Dimethylmenaquinone	Methionaquinone[c]
Aromatic hydrogens	7.6–8.1(m)	7.6–8.2(m)	7.6–8.2(m)	7.6–8.1(m)	7.7–8.1(m)	7.45(d) 7.52(t) 8.00(d)	7.9(d) 7.43(d)	7.79–8.08(m)
Quinonoid hydrogen (C-2)	—	—	—	6.7(t)[d]	—	—	—	—
Olefinic hydrogen adjacent to carbonyl	—	—	—	—	6.15(s)	—	—	—
Olefinic hydrogens	5.0–5.2(t)[e]	5.0–5.2(m)	5.0–5.2(m)	5.0–5.2(m)	4.9(m)	5.01–5.12(m)	4.9–5.2(m)	5.06(m)
Allylic methylene at C-1 (next to ring)	3.3–3.4(d)	3.3–3.4(d)	3.3–3.4(d)	3.3(d)	—	3.33(d)[f]	3.31(d)	3.6(d)
SCH$_3$ (ring)	—	—	—	—	—	—	—	2.64(s)
Ring methyl								
(C-5 or C-8)	—	—	—	—	—	2.73(s)	2.66(s)	—
(C-2)	2.1–2.2(s)	2.1–2.2(s)	2.1–2.2(s)	—	2.28(s)	2.15(s)	2.14(s)	—
(C6 or C7)	—	—	—	—	—	—	2.4(s)	—

Allylic methylenes	1.9–2.0(m)	1.9–2.1(m)	1.9–2.1(m)	1.9–2.1(m)	1.9–2.1(m)	1.9–2.05	1.9–2.1(m)	1.97(m)
trans-methyl 1st isoprene unit	1.8	1.8	1.8	1.8	2.2	1.77(bs)	1.77(bs)	1.81
cis end of chain methyl	–	1.65–1.7	1.65–1.7[g]	1.65–1.7	1.66	1.67[g]	1.67[g]	–
trans internal methyl groups	–	1.52–1.59(bs)	1.52–1.59(bs)	1.52–1.59(bs)	1.58(s)	1.52–1.59(bs)	1.5–1.6(bs)	1.57
Aliphatic methylenes and methine groups	1.1–1.3	–	1.1–1.3	–[h]	–	–[h]	–[h]	1.28
Aliphatic methyl groups (i.e. methyl groups on saturated carbon atoms)	0.8–0.9	–	0.8–0.9	–[h]	–	–[h]	–[h]	0.87

[a] Data from Powls et al. (1968);
[b] Data from Ishii et al. (1983);
[c] s, singlet; d, double; t, triplet; m, multiplet; b, broad;
[d] Due to long-range splitting;
[e] Due to single olefinic proton;
[f] J = 7 Hz;
[g] Signal absent if terminal isoprene unit is saturated (see Collins & Widdel, 1984);
[h] Signals present in partially saturated prenologues.

(Table. 8.4). The splitting patterns of the aromatic proton region of methylmenaquinones and dimethylmenaquinones are quite distinct from each other and from the menaquinones series (Figure 8.30). In the case of methylmenaquinones a doublet due to a peri proton (i.e. at C5 or C8) occurs at δ 8.0 ($J = 7.5$ Hz). A second doublet at δ 7.45 ($J = 7.5$ Hz) and a triplet at δ 7.52 (J = 7.5 Hz), corresponding to protons at C6 and C7 (not necessarily respectively), are also observed. In contrast, dimethylmenaquinones produce two aromatic proton doublets at δ 7.89 (C5 or C8) and δ 7.43 (C6 or C7).

Methylmenaquinones also differ from menaquinones in producing a second ring methyl signal at δ 2.7 whereas dimethylmenaquinones exhibit resonances at δ 2.4 and 2.6 due to the presence of two additional ring methyl groups. The partial ^{1}H-NMR spectrum of a dimethylmenaquinone is shown in Figure 8.31.

In the case of menaquinones (and methylated analogues) with partially or completely saturated side-chains, resonances occur at δ ~1.1 to 1.3 for

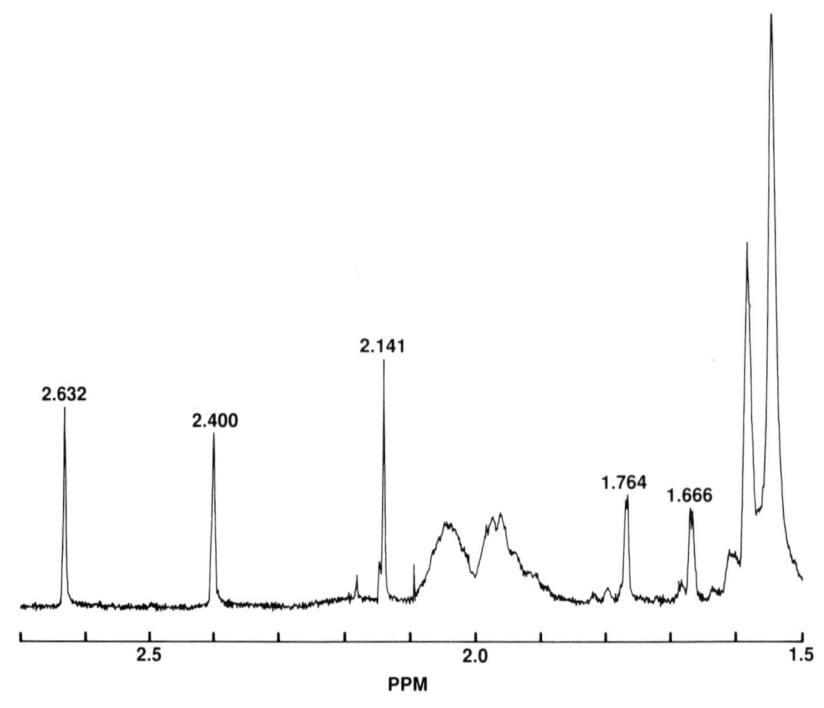

Figure 8.31 Partial ^{1}H-NMR spectrum of DMMK-8.

aliphatic ethylenes and methines and at δ 0.8 to 1.0 for aliphatic methyls (Figure 8.29). The integrals of the aliphatic methylene plus methine and aliphatic methyl signals provide information as to whether or not the terminal unit of the side-chain is saturated. The presence of end of chain saturation is also revealed by the absence of absorption at δ 1.67 due to cis-end of chain methyl. Menaquinones in which the first isoprene unit is saturated produce an aliphatic methyl doublet at significantly lower field (δ 0.98) whereas the doublet centred at δ 3.3 due to ring-methylene is replaced by a multiplet at δ 2.6 (see Table 8.4).

The ^{1}H spectra of ubiquinones differ from those of menaquinones in possessing intense signal at δ 3.9 to 4.0 due to the presence of two methoxy groups; absorption in the δ 7.5 to 8.1 region due to aromatic protons is absent. The remaining signals in the spectra of ubiquinones are generally similar to those of unsaturated menaquinones and are due to the presence of multiprenyl side-chains.

The sulphur-containing quinone, caldariellaquinone, produces a very characteristic ^{1}H spectrum. Two doublets centred at δ 7.58 and 7.51, derived from ring protons, and a singlet at δ 2.62, due to the presence of SCH$_3$, are observed. A methylene group adjacent to the ring system produces a multiplet at δ 2.8 whereas aliphatic methylenes and methines give rise to broad absorption centred at δ 1.2. Signals from protons of methyl groups on saturated carbon atoms occur in the region δ 0.8 to 1.0. The methyl group of the first isoprene unit produces a doublet at δ 0.97 whereas the remaining methyl groups give rise to overlapping doublets at ~0.83 to 0.87 (Collins, 1985a). The spectrum of sulfolobusquinone (SQ-6(H$_{12}$)) is essentially similar to that of caldariellaquinone except that the quinonoid methyl signal occurs at δ 2.12 and the allylic methylene multiplet is shifted to δ 2.6.

For the structural determination of most isoprenoid quinones simple ^{1}H-NMR is sufficient. However, a number of other more sophisticated techniques may be employed. Two-dimensional homonuclear shift correlation spectroscopy (^{1}H-COSY) is a relatively simple and reliable means of identifying all mutually coupled ^{1}H nuclei where the couplings are resolved. This technique has been found to be particularly useful in elucidating the structures of a number of problematic isoprenoid quinones (e.g. Collins et al., 1986; Howarth et al., 1986). An example of an ^{1}H-2D-COSY spectrum is shown in Figure 8.32. The ordinary spectrum lies along the diagonal (viewed from above). Off-diagonal or cross peaks occur when resonances are connected by J-coupling. ^{1}H-^{13}C heteronuclear shift correlation has also been employed for isoprenoid quinone analysis but this technique requires large amounts of sample (e.g. Howarth et al., 1986).

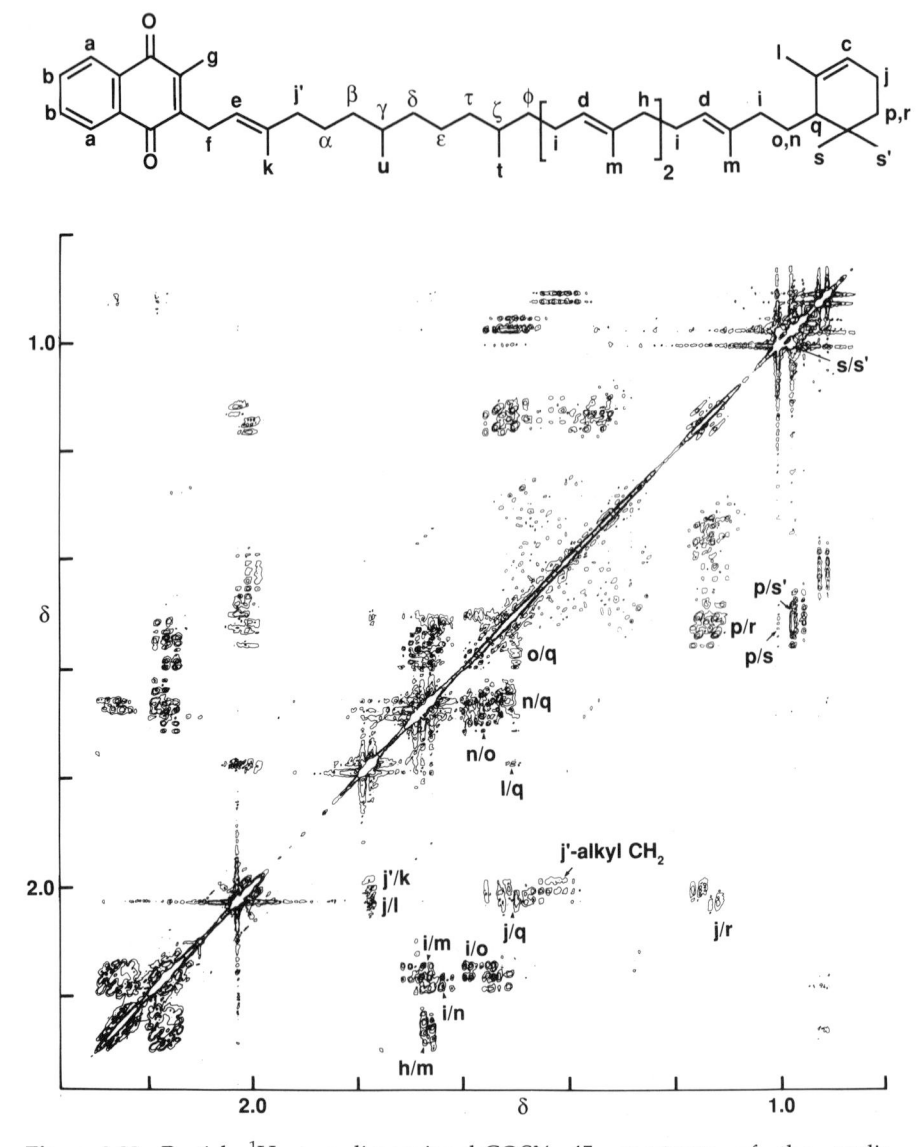

Figure 8.32 Partial ^{1}H two-dimensional-COSY 45 spectrum of the cyclic menaquinone from *Nocardia*.

8.6 CONCLUSIONS

Chemical methods now play a central role in microbial systematics. Knowledge of the distribution of isoprenoid quinone structural types in prokaryotes has increased dramatically over the past 10 to 15 years. Respiratory isoprenoid quinone analyses have made significant contributions to the systematics of many groups of prokaryotes and are now recognized to be exceedingly valuable biomarkers. Isoprenoid quinones are being increasingly incorporated into taxonomic schemes, including the description of new taxa, and there is every indication that advances in analytical methodology (e.g. HPLC, SFC, MS/MS) will provide systematists with a more rapid means of obtaining precise structural and quantitative information on isoprenoid quinones.

REFERENCES

Azerad, R. and Cyrot-Pelletier, M.O. (1973) Structure and configuration of the polyisoprenoid side-chain of dihydromenaquinones from myco- and corynebacteria. *Biochemie* **55**: 591–603.
Batrakov, S.G., Panosyan, A.G., Rosynov, B.V., Konova, I.V. and Bergelson, L.D. (1976) Menaquinones of *Actinomyces olivaceus*: On the structures of MK-9(H$_6$), MK-9(H$_8$), MK-8(H$_6$) and MK-8(H$_8$). *Bioorg. Khim.* **2**: 1538–46.
Baum, R.H. and Dolin, M.I. (1963) Isolation of a new naphthoquinone from *Streptococcus faecalis* 10Cl. *J. Biol. Chem.* **238**: 4109-4111.
Baum, R.H. and Dolin, M.I. (1965) Isolation of 2-solanesyl–1,4-naphthoquinone from *Streptococcus faecalis*. *J. Biol. Chem.* **240**: 3425–33.
Beau, P.S., Azerad, R. and Lederer, E. (1966) Isolement et characterisation des dihydromenaquinones des myco- et corynebacteries. *Bull. Soc. Chim. Biol.* **48**: 569–81.
Begin-Heik, N. and Blum, J.J. (1967) Oxygen toxicity in *Astasia*. *Biochem. J.* **105**: 813–9.
Brinkley, S.B., McKee, R.W., Thayer, S.A. and Doisy, E.A. (1940) The constitution of vitamin K$_2$. *J. Biol. Chem.* **133**: 721–9.
Bishop, D.H.L., Pandya, K.P. and King, H.K. (1962) Ubiquinone and vitamin K in bacteria. *Biochem. J.* **83**: 606–14.
Carlone, G.M. and Anet, F.A.L. (1983) Detection of menaquinone-6 and a novel methyl substituted menaquinone-6 in *Campylobacter jejuni* and *Campylobacter fetus* subsp. *fetus*. *J. Gen. Microbiol.* **129**: 3385–93.
Collins, M.D. (1985a) Analysis of isoprenoid quinones. *Methods Microbiol.* **18**: 329–66.
Collins, M.D. (1985b) Structure of thermoplasmaquinone from *Thermoplasma acidophilum*. *FEMS Microbiol. Lett.* **28**: 21–3.
Collins, M.D. and Fernandez, F. (1984) Menaquinone-6 and thermoplasmaquinone-6 in *Wolinella succinogenes*. *FEMS Microbiol. Lett.* **22**: 273–6.
Collins, M.D. and Fernandez, F. (1985) Co-occurrence of menaquione-6 and thermoplasmaquinone-6 in *Bacterioides gracilis*. *FEMS Microbiol. Lett.* **26**. 181–4.
Collins, M.D. and Gilbart, J. (1983) New members of the coenzyme Q series from Legionellaceae. *FEMS Microbiol. Lett.* **16**: 251–5.
Collins, M.D. and Green, P.N. (1985) Isolation and characterization of a novel coenzyme Q from some methane-oxidizing bacteria. *Biochem. Biophys. Res. Commun.* **133**: 1125–31.
Collins, M.D. and Jones, D. (1981a) Distribution of isoprenoid quinone structural types in bacteria and their taxonomic implications. *Microbiol. Rev.* **45**: 316–54.
Collins, M.D. and Jones, D. (1981b) A note on the separation of natural mixtures of bacterial ubiquinones using reverse-phase partition thin-layer chromatography and high performance liquid chromatography. *J. Appl. Bacteriol.* **51**: 316–34.

Collins, M.D. and Kroppenstedt, R.M. (1987) Structures of the partially saturated menaquinones of *Glycomyces rutgersensis*. *FEMS Microbiol. Lett.* **44**: 215–9.
Collins, M.D. and Langworthy, T.A. (1983) Respiratory quinone composition of some acidophilic bacteria. *System. Appl. Microbiol.* **4**: 295–304.
Collins, M.D. and Tindall, B.J. (1987) Occurrence of menaquinone and some novel methylated menaquinones in the alkalophilic, extremely halophilic archaebacterium *Natronobacterium gregori*. *FEMS Microbiol. Lett.* **4**: 307–12.
Collins, M.D. and Widdel, F. (1984) A new respiratory 2-methyl-3-V-dihydropentaprenyl-1,4-naphthoquinone, isolated from *Desulfobulbus propionicus*. *System. Appl. Microbiol.* **5**: 281–86.
Collins, M.D. and Widdel, F. (1986) Respiratory quinones of sulphate-reducing and sulphur-reducing bacteria: a systematic investigation. *System. Appl. Microbiol.* **8**: 8–18.
Collins, M.D., Pirouz, T., Goodfellow, M. and Minnikin, D.E., (1977) Distribution of menaquinones in actinomycetes and corynebacteria. *J. Gen. Microbiol.* **100**: 221–30.
Collins, M.D., Costas, M. and Owen, R.J. (1984) Isoprenoid quinone composition of representatives of the genus *Campylobacter*. *Arch Microbiol.* **137**: 168–70.
Collins, M.D., Fernandez, F. and Howarth, O.W. (1985) Isolation and characterization of a novel vitamin K. *Biochem. Biophys. Res. Commun.* **133**: 322–8.
Collins, M.D., Howarth, O.W. and Green, P.N. (1986a) Isolation and structural determination of a novel coenzyme from a methane-oxidizing bacterium. *Arch. Microbiol.* **146**: 263–6.
Collins, M.D., Howarth, O.W. and Perry, J.J. (1986b) A new respiratory quinone, 2-methyl-3-VI,VII-tetrahydroheptaprenyl-1,4-naphthoquinone isolated from *Thermoleophilum album*. *FEMS Microbiol. Lett.* **34**: 167–71.
Collins, M.D., Howarth, O.W., Grund, E. and Kroppenstedt, R.M. (1987) Isolation and structural determination of new members of the vitamin K_2 series from *Nocardia brasiliensis*. *FEMS Microbiol. Lett.* **41**: 35–9.
Collins, M.D., Kroppenstedt, R.M., Tamaoka, J., Komagata, K. and Kinoshita, T. (1988) Structures of the tetrahydrogenated menaquinones from *Actinomadura angiospora, Faenia rectivirgula* and *Saccharothrix australiensis*. *Curr. Microbiol.* **17**: 275–9.
Crane, F.L. (1959) Isolation of two quinones with coenzyme Q activity from alfalfa. *Plant Physiol.* **34**: 546–51.
Crane, F.L., Hatefi, Y., Lester, R.L. and Windmer, C. (1957) Isolation of a quinone from beef heart mitochondria. *Biochem. Biophys. Acta* **25**: 220–1.
Dam, H. (1929) Cholesterinstoffwechsel in Hühnereiern Hühnchen. *Biochem. Z.* **215**: 475–92.
Dam, H., Geiger, A., Glavind, J., Karrer, P., Karrer, W., Rothschild, E. and Salomon, H. (1939) Isolierung des Vitamins K in hochgereinigter Form. *Helv. Chim. Acta* **22**: 310–3.
De Rosa, M., De Rosa, S., Gambacorta, A., Minale, L., Thomson, R.H. and Worthington, R.D. (1977) Caldariellaquinone, a unique benzo-b-thiopen-4,7-quinone from *Caldariella acidophila*, an extremely thermophilic and acidophilic bacterium. *J. Chem. Soc. Perkin Trans.* **1**: 653–7.
Dunphy, P.J. and Brodie, A.F. (1971) The structure and function of quinones in respiratory metabolism. *Methods Enzymol.* **18C**: 407–61.
Erickson, R.E., Shunk, C.H., Trenner, N.R., Arison, B.H. and Folkers, K. (1959) Coenzyme Q. XI. The structure of solanesol. *J. Am. Chem. Soc.* **81**: 4999–5000.
Fernandez, F. and Collins, M.D. (1987) Vitamin K composition of anaerobic intestinal bacteria. *FEMS Microbiol. Lett.* **41**: 175–80.
Festenstein, G.N., Heaton, F.W., Lowe, J.S. and Morton, R.A. (1955) A constituent of the unsaponifiable portion of animal tissue lipids. *Biochem. J.* **59**: 558–66.
Fries, P., Davis, G.D. and Folkers, K. (1967) New epoxyubiquinones. *Biochemistry* **6**: 3618–24.
Frydman, B. and Rapaport, H. (1963) Non-chlorophyllous pigments of *Chlorella thiosulfatophilum* in chlorobium-quinone. *J. Am. Chem. Soc.* **85**: 823–5.
Gale, P.H., Arison, B.H., Trenner, N.R., Page, A.C. and Folkers, K. (1963a) Coenzyme Q. XXXVI. Isolation and characterization of coenzyme Q_{10}(H-10). *Biochemistry* **2**: 196–200.
Gale, P.H., Trenner, N.R., Arison, B.H., Page, A.C. and Folkers, K. (1963b) Coenzyme Q_{10}. XLIX. Characterization of coenzyme Q_{10}(H-10) from *Penicillium stipitatum*. *Biochem. Biophys. Res. Commun.* **12**: 414–17.
Gilbart, J. and Collins, M.D. (1985) High performance liquid chromatographic analysis of ubiquinones from some new *Legionella* species. *FEMS Microbiol. Lett.* **26**: 77–82.

Glover, J. and Threlfall, D.R. (1962) A new quinone (rhodoquinone) related to ubiquinone in the photosynthetic bacterium *Rhodospirillum rubrum*. *Biochem. J.* **85**: 14P–15P.
Hammond, R.K. and White, D.C. (1969) Separation of vitamin K$_2$ isoprenologues by reverse phase partition thin layer chromatography. *J. Chromatogr.* **45**: 446–452.
Hiraishi, A. and Hashino, Y. (1984) Distribution of rhodoquinone in *Rhodospirillaceae* and its taxonomic implications. *J. Gen. Appl. Microbiol.* **30**: 435–48.
Howarth, O.W., Grund, E., Kroppenstedt, R.M. and Collins, M.D. (1986) Structural determination of a new naturally occurring cyclic vitamin K. *Biochem. Biophys. Res. Commun.* **140**: 916–23.
Ishii, M., Kawasumi, T., Igarashi, Y., Kodama, T. and Minoda, Y. (1983) 2-Methylthio-1,4-naphthoquinone, a new quinone from an extremely thermophilic hydrogen bacterium. *Agric. Biol. Chem.* **47**: 167–9.
Isler, O., Ruegg, R., Chopard-dit-Jean, L.H., Winterstein, A. and Wiss, O. (1958a) Synthese von Vitamin K und isoprenologer Verbindungen. *Chimia* **12**: 69.
Isler, O., Ruegg, R., Chopard-dit-Jean, L.H., Winterstein, A. and Wiss, O. (1958b) Synthese und Isolierung von Vitamin K und isoprenologen Verbindungen. *Helv. Chim. Acta* **41**: 786–807.
Itoh, T., Funabashi, H., Katayama-Fujimura, Y., Iwasaki, S. and Kuraishi, H. (1985) Structure of methylmenaquinone-7 isolated from *Alteromonas putrefaciens* IAM 12079. *Biochem. Biophys. Acta* **840**: 51–5.
Jeffries, L., Cawthorne, M.A., Harris, M., Diplock, A.T., Green, J. and Price, S.A. (1967a) Menaquinone patterns in *Micrococcaceae*; their possible taxonomic significance. *Spisy Prirodored. Fak. Univ. J.E Purkňe Brno K40*, 230.
Jeffries, L., Cawthorne, M.A., Harris, M., Diplock, A.T., Green, J. and Price, S.A. (1967b) Distribution of menaquinones in aerobic *Micrococcaceae*. *Nature* (London) **215**: 257–9.
Jeffries, L., Harris, M. and Price, S.A. (1967c) Atypical menaquinone pattern in a strain of *Staphylococcus aureus*. *Nature* (London) **216**: 808–9.
Kofler, M. (1946) Über ein pflanzliches Chinon. In *Festschrift für Emil C. Barell*, pp. 199–212. F. Hoffmann La Roche: Basel.
Kofler, M., Langemann, A., Rüegg, R., Chopard-dit-Jean, L.H., Rayroud, A. and Isler, O. (1959a) Die Strukter eines pflanzlichen Chinons mit isoprenoider Seitenkette. *Helv. Chim. Acta* **42**: 1283–92.
Kofler, M., Langemann, A., Ruegg, R., Gloor, U., Schwieter, U., Würsch, J., Wiss, O. and Isler, O. (1959b) Struktur und partial synthese des pflanzlichen Chinons mit isoprenoider Seitenkette. *Helv. Chim. Acta* **42**: 2252–4.
Krivánková, L. and Dadák, V. (1976) An improved extraction method of MK from cells of *Staphylococcus epidermidis*. *Scripta Fac. Sci. Nat. Ujep Brunensis, Chemia* **6**: 79–88.
Krivánková, L. and Dadák, V. (1980) Semimicro extraction of ubiquinone and menaquinone from bacteria. *Methods Enzymol.* **67**: 111–14.
Kroppenstedt, R.M. (1982) Separation of bacterial menaquinones by HPLC using reverse phase (RP18) and a silver loaded ion exchanger as stationary phases. *J. Liquid Chromatogr.* **5**: 2359–67.
Kroppenstedt, R.M. (1985) Fatty acid and menaquinone analysis of actinomycetes and related organisms. In *Chemical Methods in Bacterial Systematics* (Goodfellow, M. and Minnikin, D.E., eds), pp. 173–99. Academic Press: London.
Lavate, W.V. and Bentley, R. (1964) Distribution of normal isoprenologs of coenzyme Q and dihydrocoenzyme Q$_{10}$ in various moulds. *Arch. Biochem Biophys.* **108**: 287.
Lavate, W.V., Dyer, J.R., Springer, C.M. and Bentley, R. (1965) Studies on coenzyme Q. The isolation, characterization and general properties of a partially reduced coenzyme Q$_{10}$ from *Penicillium stipitatum*. *J. Biol. Chem.* **240**: 524–31.
Law, A., Thomas, G. and Threlfall, D.R. (1973) 5-Monohydroxyphylloquinone from *Anacystis* and *Euglena*. *Phytochemistry* **12**: 1999–2004.
Lester, R.L. and Crane, F.L. (1959) The natural occurrence of coenzyme Q and related compounds. *J. Biol. Chem.* **234**: 2169–75.
Lester, R.L., White, D.C. and Smith, S.L. (1964) The 2-demethyl vitamin K$_2$'s. A new group of naphthoquinones isolated from *Haemophilus parainfluenzae*. *Biochemistry* **3**: 949–54.

MacCorquodale, D.W., Binkley, S.B., McKee, R.W., Thayer, S.A. and Doisy, E.A. (1939a) Inactivation of vitamin K by light. *Proc. Soc. Exp. Biol. Med.* **40**: 482–3.
MacCorquodale, D.W., McKee, R.W., Binkley, S.B., Cheney, L.C., Holcomb, W.F., Thayer, S.A. and Doisy, E.A. (1939b) Identification of vitamin K$_1$ (alfalfa). *J. Biol. Chem.* **130**: 433.
McKee, R.W., Binkley, S.B., Thayer, S.A., MacCorquodale, D.W. and Doisy, E.A. (1939) The isolation of vitamin K$_2$. *J Biol. Chem.* **131**: 327–344.
Minnikin, D.E., O'Donnell, A.G., Goodfellow, M., Alderson, G., Athalye, M., Schaal, A. and Parlett, J.H. (1984) An integrated procedure for extracting bacterial isoprenoid quinones and polar lipids. *J. Microbiol. Methods* **2**: 233–41.
Moore, H.W. and Folkers, K. (1966a) New method for structural assignments of hydroxy analogs of coenzyme Q. *J. Am. Chem. Soc.* **88**: 564–7.
Moore, H.W. and Folkers, K. (1966b) Structure of rhodoquinone. *J. Am. Chem. Soc.* **88**: 567–70.
Morton, R.A., Wilson, G.M., Lowe, J.S. and Leat, W.M.F. (1957) Ubiquinone. *Chem. Ind.* (London): 1649–50.
Morton, R.A., Gloor, U., Schinler, O., Wilson, G.M., Chopard-dit-Jean, L.H., Hemming, L.H., Isler, O., Leat, W.M.F., Pennock, J.F., Rüegg, R., Schweiter, U. and Wiss, O. (1958) Die Struktur des Ubichinons aus Schweineherzen. *Helv. Chim. Acta* **41**: 2343.
Moss, C.W., Kai, A., Lambert, M.A. and Patton, C. (1984) Isoprenoid quinone content and cellular fatty acid composition of *Campylobacter* species. *J. Clin. Microbiol.* **19**: 772–6.
Ozawa, H., Sato, M., Natori, S. and Ogawa, H. (1969) Occurrence of rhodoquinone-9 in the muscle of *Ascaris lumbricoides* var. *suis*. *Experimentia* **25**: 484–5.
Page, A.C., Gale, P., Wallick, H., Walton, R.B., McDaniel, L.E., Woodruff, H.B. and Folkers, K. (1960) Coenzyme Q. XVII. Isolation of coenzyme Q$_{10}$ from bacterial fermentations. *Arch. Biochem. Biophys.* **89**: 318–21.
Powls, R. and Hemming, F.W. (1966a) The chemistry of rhodoquinones. *Phytochemistry* **5**: 1235.
Powls, R. and Hemming, F.W. (1966b) Biosynthesis of quinones. *Phytochemistry* **5**: 1249.
Powls, R., Redfearn, E.R. and Trippett, S. (1968) The structure of chlorobiumquinone. *Biochem. Biophys. Res. Commun.* **33**: 408–11.
Ramsey, E.D., Collins, M.D. and Mellon, F.A. (1988) Tandem mass spectrometric studies of menaquinones. Potential for rapid location of points of saturation. *Rapid Comm. Mass. Spec.* **2**: 273–4.
Redfearn, E.R. and Burgos, J. (1966) Ubiquinone (Coenzyme Q) and the respiratory chain. *Nature* (London) **209**: 711–13.
Redfearn, E.R. and Powls, R. (1967) The quinones of green photosynthetic bacteria. *Biochem. J.* **106**: 50.
Sato, M. and Ozawa, H. (1969) Occurrence of ubiquinone and rhodoquinone in parasitic nematodes. *Metastrongylus elongatus* and *Ascaris lumbricoides* var. *suis*. *J. Biochem.* **65**: 861–7.
Scherrer, F., Anderson, H.A. and Azead, R. (1977) Mavioquinone—a new quinone from *Mycobacterium avium*. *J. Chem. Soc.* **D**: 127–30.
Schwartz, A.C. (1973) Terpenoid quinones of the anaerobic *Propionibacterium shermanii* I.(II,III)-Tetrahydromenaquinone-9. *Arch. Microbiol.* **91**: 273–9.
Sunk, C.H., Linn, B.O., Wong, E.L., Wittreich, P.E., Robinson, F.M. and Folkers, K. (1958) Coenzyme Q. II. Synthesis of 6-farnesyl- and 6-phytyl-derivatives of 2,3-dimethoxy-5-methylbenzoquinone and related analogs. *J. Am. Chem. Soc.* **80**: 4753.
Sommer, P. and Kofler, M. (1966) Physicochemical properties and methods of analysis of phylloquinones, menaquinones, ubiquinones, plastoquinones, menadione and related compounds. *Vitam. Horm.* (N.Y.) **24**: 349–99.
Tamaoka, J., Katayama-Fujimura, Y. and Kuraishi, H. (1983) Analysis of bacterial menaquinone mixtures by high performance liquid chromatography. *J. Appl. Bacteriol.* **54**: 31–6.
Tamaoka, J., Komagata, K., Kinoshita, T., Shen, G., Kodama, T., Imai, T. and Minoda, Y. (1985) Methylubiquinones, a new isoprenoid quinone in methane-oxidizing bacterium strain H-2. *FEMS Microbiol. Lett.* **29**: 151–4.
Thurl, S., Buhrow, I. and Schäfer, W. (1985) New types of menaquinones from the thermophilic archaebacterium *Thermoproteus tenax*. *Biol. Chem. Hoppe-Seyler* **366**: 1079–83.

Thurl, S., Witke, W., Buhrow, I. and Schäfer, W. (1986) Different types of quinones from sulphur-dependent archaebacteria. *Biol. Chem. Hoppe-Seyler* **367**: 191–7.
Tindall, B.J. and Collins, M.D. (1986) Structure of 2-methyl-3-VIII-dihydroocaprenyl-1,4-naphthoquinone from *Halococcus morrhuae*. *FEMS Microbiol. Lett.* **37**: 117–9.
Tindall, B.J., Stetter, K.O. and Collins, M.D. (1989) A novel, fully saturated menaquinone from the thermophilic, suphate-reducing archaeobacterium *Archaeoglobus fulgidus*. *J. Gen Microbiol.* **135**: 693–6.
Tindall, B.J., Wray, V., Huber, R. and Collins, M.D. (1991) A novel, fully saturated cyclic menaquinone in the archaebacterium *Pyrobaculum organotrophum*. *Syst. Appl. Microbiol.* **14**: 218–21.
Trincone, A., Gambacorta, A., Lanzotti, V. and De Rosa, M. (1986) A new benzo[1,2-b; 4,5-b']dithiophen-4,8-quinone from the archaebacterium *Sulfolobus solfataricus*. *J.Chem. Soc. Commun.* **10**: 733.
Wolf, D.E., Hoffman, C.H., Trenner, N.R., Arison, B.H., Shunk, C.H., Linn, B.O., McPherson, J.F. and Folkers, K. (1958) Coenzyme Q.I. Structure studies on the coenzyme Q group. *J. Am. Chem. Soc.* **80**: 4752.
Yamada, Y., Inouye, G., Tahara, Y. and Kondo, K. (1977) The structure of the menaquinones with a tetrahydrogenated isoprenoid side-chain. *Biochem. Biophys. Acta* **488**: 280–4.
Yamada, Y., Aoki, K. and Tahara, Y. (1982a) The structure of the hexahydrogenated isoprenoid side-chain menaquinone with nine isoprene units isolated from *Actinomadura madurae*. *J. Gen. Appl. Microbiol.* **28**: 321–9.
Yamada, Y., Hou, C.F., Sasaki, J., Tahara, Y. and Yoshioka, H. (1982b) The structure of the octahydrogenated isoprenoid side-chain menaquinone with nine isoprene units isolated from *Streptomyces albus*. *J. Gen. Appl. Microbiol.* **28**: 519–29.

9
Analysis of Cytochromes

Robert K. Poole
Division of Life Sciences, King's College London, UK

9.1 INTRODUCTION

A cytochrome is a haemoprotein whose characteristic mode of action involves electron transport. This is accomplished by a reversible redox change in the haem prosthetic group, generally between the Fe(II) and Fe(III) states.

Cytochromes are widely, but not universally, distributed in microorganisms. The mitochondrial cytochromes of eukaryotic microorganisms exhibit remarkable conservatism in microbial groups. With few exceptions, which are described by Lloyd (1974) and Lloyd and Edwards (1978), the respiratory chains in the mitochondria of these organisms comprise aa_3-type cytochrome oxidase and cytochromes of the *b* and *c* types. Although distinctive patterns are exhibited by some eukaryotic microbes, notably the protozoa and fungi, they are probably of little value in any taxonomic study. However, the cytochromes of prokaryotic microbes are diverse and the combinations of cytochromes observed have enabled some correlations to be made between cytochrome patterns and taxonomic status. It should be recognized from the outset that cytochrome analysis *by itself* is of little value in bacterial taxonomy; however, cytochrome patterns should be part of the minimal description of any bacterium. In this restricted sense they are important in systematics. Examples of the usefulness of this approach and of some pitfalls are presented later in this chapter.

David Keilin is credited with the rediscovery of cytochromes and the appreciation of their function in cellular respiration in various animal, yeast and bacterial cells. His classical spectroscopic examinations of the absorption bands of cytochromes showed that they underwent reversible changes of oxidation and reduction. Warburg at first disputed this proposal, but subsequently measured the relative ability of light of different wavelengths to relieve the inhibition of respiration by carbon

Chemical Methods in Prokaryotic Systematics. Edited by M. Goodfellow and A.G. O'Donnell
© 1994 John Wiley and Sons Ltd.

monoxide. In doing so, he obtained the photochemical action spectrum and paved the way for more sophisticated spectroscopic studies by Chance and his collaborators in the 1950s. In all of these early studies, examination of microbial cytochromes played a seminal role (for an account, see Poole, 1988). Although cytochromes were at first thought to be absent in anaerobic bacteria, they were later shown to function in anaerobic respiration (i.e. electron transfer to terminal acceptors other than oxygen) and in photosynthesis.

Optical spectroscopy of various redox states and of the effects of respiratory inhibitors, such as carbon monoxide, cyanide and azide, remains the most powerful means, and the most accessible, of studying cytochrome presence and function in microbial cells. These methods have the considerable advantage that they can be applied to modest quantities of intact cells with relatively little preparation. Although non-spectroscopic methods can be of value and are sometimes based on the catalytic function of cytochromes, this chapter is restricted largely to the use of visible spectrophotometry and attempts to provide guidelines for the appropriate use of such techniques in the identification and quantification of bacterial cytochromes. Much of the information presented here was collected in an earlier volume on components of bacterial cells that are of taxonomic interest (Jones & Poole, 1985). During the time since the preparation of that article, there have been several important developments in this field which are included here, namely the reassignment of certain visible absorption bands to cytochrome types, an explosion of interest in molecular genetic studies of bacterial cytochromes, and the availability of a laboratory manual devoted to practical aspects of spectrophotometry (Harris & Bashford, 1987). Description of the roles of cytochromes in bacterial energetics is outside the scope of this chapter and the reader is referred to Anthony (1988).

9.2 STRUCTURE AND NOMENCLATURE

With the possible exception of iron-sulphur proteins, haemoproteins are the most widespread and best understood of electron transfer components. Haem is composed of four pyrrole rings joined by $-CH=$ bridges to form porphyrin, a complex planar structure. A single iron atom within the porphyrin cycles between the reduced (FeII) and oxidized (FeIII) states and transfers one electron. The iron centre is generally six-coordinate with four equatorial coordination sites occupied by the nitrogens of the pyrrole ring and the fifth and sixth sites filled with donor atoms from groups in amino acid residues of the protein. The important donors in these axial positions are the imidazole and $-SH$ groups of histidine or cysteine residues respectively. The iron is generally low spin, reflecting the presence of these strong ligands, and in this configuration is appropriate for the facilitation of electron transfer. Other positions around the edge of the haem are available for substitution and it is the nature of these substituents that is primarily responsible for the characteristic spectroscopic and functional properties of the haemoprotein. Those cytochromes that bind a ligand, such as the oxygen-binding and -reducing terminal oxidases, generally have a five-co-ordinate, high-spin haem.

Hemoproteins may be conveniently classified by either their function or the type of haem present and the spin state of that haem. Thus, according to the first classification, it is possible to distinguish cytochromes, which by definition are involved in electron transfer processes, from oxygen carriers such as myoglobin and haemoglobin, oxygenases which are responsible for the incorporation of oxygen atoms into substrates, and the hydroperoxidases, individually referred to as peroxidases or catalases.

This chapter is devoted primarily to the cytochromes but, where appropriate, reference will be made to other haemoproteins that may be mistaken for cytochromes. Studies of a number of bacterial systems in the past few years have shed new light on the interpretation of haemoprotein spectra and have led to the view that several high-spin haemoproteins have previously and erroneously been referred to as cytochromes. Five classes of cytochromes are recognized:

(a) Cytochromes a. These include the well-characterized cytochromes a and a_3 of the oxidase complex in mitochondria and several bacteria. The prosthetic group is haem A.
(b) Cytochromes b. These include various non-autoxidizable cytochromes b and cytochrome P450. In these proteins, the haem prosthetic group (haem B or protohaem) is the same as that in haemoglobin and myoglobin.
(c) Cytochromes c. These include a variety of cytochrome types in which the haem prosthetic group (haem C or mesohaem) is covalently attached to the protein via thioether bridges.
(d) Cytochromes d. These include cytochrome oxidase d of several bacteria (previously called cytochrome a_2) and a form of nitrite reductase (cytochrome cd_1) in which the haem prosthetic group (haem d or d_1) is a dihydroporphyrin, or chlorin (Timkovich et al., 1985).
(e) Cytochromes o. Until recently, the ligand-binding haem of cytochrome oxidase(s) of the o-type was believed to be haem B. Indeed, the spectrum of cytochrome o is very similar to that of a high-spin cytochrome b. Recent studies reveal, however, a novel type of haem (O) related more to haem A (Puustinen & Wikstrom, 1991).

The present confused situation with respect to cytochrome nomenclature dates back to the early observations of Keilin and his contemporaries and studies of the distinctive four-banded spectra of various cell types. Keilin named the four components he observed using his spectroscope a, b, c and d (working from high to low wavelengths). These bands were subsequently found to be due to the composite absorbances of *three* haem types that are now referred to as cytochromes a, b, and c.

Because the reader of this chapter is more likely to be concerned with the taxonomy and nomenclature of bacteria rather than of their cytochromes, a description of the current recommendations for cytochrome nomenclature will be brief. The interested reader is referred to the 1989 Recommendations on Nomenclature of Electron-Transfer Proteins of the IUB Nomenclature Committee (Palmer & Reedijk, 1992). Use of the small, unprimed, italicized letter should imply that the haem prosthetic group is in a haemochrome linkage. In other words, the haem should be a low-spin compound in which the fifth and sixth coordination places are occupied by strong field ligands, regardless of the oxidation state of the iron. To indicate that in both the oxidized and reduced forms the haem group is *not* in a haemochrome linkage, a primed, lower case italicized letter should be used, e.g. c'. Where a cytochrome has two or more different haem groups attached to a specific protein, each different haem may be indicated, e.g. as in *Escherichia coli* terminal oxidase bd or *Pseudomonas* oxidase cd_1. Where a cytochrome has two or more of the same haem groups attached to a specific protein, but in distinct environments, both types of linkage should be indicated by using both the unprimed and primed lower case italicized letter. Thus, cytochrome c oxidase (EC 1.9.3.1) is considered to contain both a haem A in haemochrome-type linkage (called cytochrome a) and a haem A in a non-haemochrome-type linkage (called cytochrome a_3). The latter is the ligand-reactive component of the complex. Using the convention suggested above, this oxidase complex should be called cytochrome aa'. Although this nomenclature is unlikely to be universally accepted in the near future, the recommended notation should be noted. Bacterial cytochromes of the o-type have until recently been regarded as b-type cytochromes, i.e. having protohaem or haem B, and the nomenclature b' has been proposed to signify the ligand-binding character of these cytochromes (Palmer & Reedijk, 1992). However, recent results (Puustinen & Wikstrom, 1991) show that the haem is not of the B-type, but a unique haem O. Revised nomenclature is likely to follow. Cytochrome bo' or oo', according to whether the second, ligand-unreactive haem is B or O, respectively, (Puustineen *et al.*, 1992) would be appropriate names. In general, the name of a cytochrome need not indicate the number of identical haem centres per molecule of protein. For example, there is still some uncertainty as to whether the cytochrome bd complex of *Escherichia coli* contains one or more moles of haem d. However, not evident from its name is the fact that this complex certainly contains two different kinds of b-type cytochrome.

The main practical tests to be adopted in determining which group a cytochrome belongs to should be: (a) the position of the alpha band of the pyridine Fe(II) haemochrome and, (b) the ether solubility of the product after treatment of the cytochrome with acidified acetone or acidified methylethyl ketone. Thus, cytochromes of the a, b and d type give a soluble product after treatment with acetone HCl in ether, whereas cytochromes of type c and d_1 give an insoluble product. The pyridine ferrohaemochromes in alkaline solution give α-absorption bands as follows: cytochromes a, 580–590 nm; cytochromes b, 556–558 nm; cytochromes c, 549–551 nm (two thioether links) or 553 nm (single thioether link); cytochromes d, 600–620 nm.

It is unlikely that the worker who wishes to use cytochromes for taxonomic or descriptive purposes will purify a cytochrome to allow it to be properly identified and named. Nevertheless, it is important to be aware of the dangers of assigning names to cytochromes on the basis of spectral data obtained with whole cells or other complex systems. The requirements are emphasized in the procedure for naming a cytochrome described in the 1989 Recommendations for the Nomenclature of Electron-Transfer Proteins. The first step is to determine the nature of the haem group and thus the class into which the cytochrome falls. The name is then expanded to indicate the source (in parentheses) as in cytochrome b_1 (*Bacterium X*). Other adjectives or descriptions may be added such as 'periplasmic' and 'high potential'. Any cytochromes not adequately described by the current names of well-established cytochromes should be given a name based on the α-band wavelength (nm) and written as, for example, cytochrome c-554. Note that the α-band cited should be determined at *room* temperature, not 77 K, and should be obtained from *absolute* absorption spectra under carefully controlled and described conditions. It is important to calibrate or check the wavelength drive of the spectrophotometer using holmium or didymium 'filters' or potassium chromates in 0.05 M KOH (see Poole & Bashford, 1987). The spectrum of purified horse heart cytochrome c in phosphate buffer, pH 7.0, is also a convenient standard taking the λ_{max} in the reduced state to be at 550.25 nm. Purified cytochromes sometimes display asymmetrical or split α or β peaks which are described, for example, as cytochrome c-555 (550). This indicates a major peak at 555 nm and a minor peak or shoulder at 550 nm.

Additional characterization of cytochromes may be valuable, especially when in complex preparations such as membranes or multi-haem purified complexes. The spectrophotometric measurement of midpoint potentials ($E_{m,7}$) at room temperature is a relatively common and useful approach and the potential may be quoted to expand the description (e.g. 'high potential' component). However, redox potentials determined at low temperatures by electron spin resonance spectroscopy should not be applied to spectral features recorded at room temperature.

9.3 ANALYTICAL METHODS

9.3.1 Spectrophotometer design considerations

For the analysis of cytochromes, only the visible region of the electromagnetic radiation spectrum, between about 400 and 700 nm, is generally of interest, although occasionally investigation of non-haem signals, especially of copper, at longer wavelengths is warranted. Furthermore, for most routine purposes, workers are concerned only with scanning spectrophotometers, in which the absorption of light by a sample is monitored as the wavelength of the incident light is continuously varied by a monochromator. The resulting absorbance spectrum is conventionally presented on a chart recorder (or computer) as the light absorbed (A) on the ordinate versus the wavelength (λ, in nm). Clearly there are numerous recording spectrophotometers on the market, not only of vastly different sophistication and price but also of markedly different suitability for the recording of cytochrome spectra in suspensions of bacteria. It is frequently a great surprise to tyros in this field that a microprocessor-driven, touch-button, 'slick' spectrophotometer which, of course, yields an admirable spectrum of purified cytochrome c, is enormously disappointing when challenged with opaque suspensions of intact microorganisms. Good performance under such optically unfavourable conditions is generally attributable to optical layout and in particular to the prime requirement that a turbid sample must be placed as close to the detector as is feasible, so that as much scattered light as possible is received at the detector.

9.3.1.a Components of a scanning spectrophotometer

A conventional scanning spectrophotometer (Figure 9.1) comprises:

(a) One or more light sources emitting a continuous range of frequencies in the uv-visible range.
(b) A monochromator, driven to scan the desired wavelength range.
(c) A sample or cuvette holder.
(d) A detector (a photomultiplier or, increasingly commonly, a silicone diode or diode array).

9. *Analysis of Cytochromes* 317

(e) The hardware and perhaps software required to coordinate wavelength scanning with absorbance measurements and to plot the resulting spectrum.

Automatic scanning instruments must also compensate for wavelength-dependent lamp intensity; this is achieved either by altering the slit width (the physical aperture through which light reaches the sample) or, in most modern instruments, by automatic gain control of the signal amplifiers.

9.3.1.b The requirement for difference spectra

Instruments designed for wavelength scanning generally employ some means for scanning a difference spectrum. This is the absorption spectrum

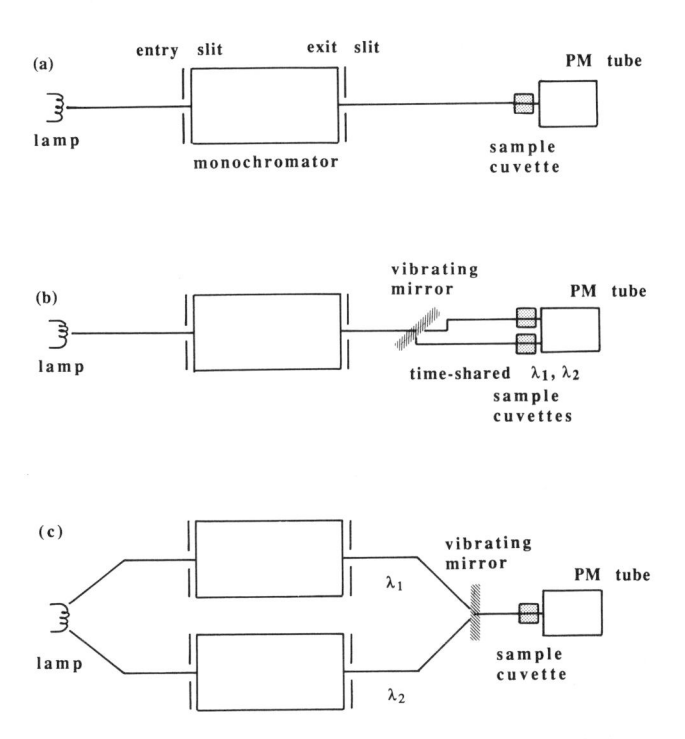

Figure 9.1 Block diagrams of optical configurations of (a) single beam, (b) split-beam (dual-beam) and (c) dual-wavelength spectrophotometers. Static mirrors, lenses and filters are not shown. PM = photomultiplier. Note positions of cuvettes close to PM tube in all configurations, suitable for use with turbid samples. In (b) λ_1 and λ_2 are time-shared between two cuvettes. In (c) the scanning mode requires λ_1 (reference) to be fixed, while λ_2 scans the desired wavelength range; λ_1 and λ_2 are time-shared by the single cuvette.

of a sample, from which another spectrum has been subtracted. The latter reference spectrum may be of:

(a) a chemically different state (e.g. reduced, oxidized) of a similar sample, or
(b) a material that has no distinctive optical properties in the region of interest but which roughly matches the light scattering properties of the actual sample. Diluted milk or a fine suspension of scattering particles are frequently used, for example, to mimic the light scattering properties of microbial cell suspensions. Confusingly, perhaps, a difference spectrum of the latter type is generally called an absolute spectrum.

The requirements for recording a difference spectrum when dealing with suspensions of microbial cells or particulate fractions such as membranes can be appreciated by considering the example presented in Figure 9 of Poole and Bashford (1987). Recording the 'absolute' spectrum, with buffer as reference, of a reduced yeast suspension of modest turbidity (apparent absorbance at 390 nm of, say, 2) reveals an almost featureless spectrum in the α/β regions of interest. This situation cannot be rectified merely by increasing the gain or sensitivity of the instrument, since this approach would amplify the signal (α and β cytochrome absorption bands) as well as the considerable light scattering on which these signals are superimposed. Instead, the turbidity must be cancelled out, revealing the bands of interest, which constitute only about 2% of the 'apparent absorbance' (for justification of this term, see Fewson et al., 1984).

Several optical solutions are available (Figure 9.1) often within a single instrument. The main options are described below:

(a) Split-beam or double-beam spectrophotometry. Here, a monochromatic beam of light, which generally scans a predetermined wavelength range with time, is split or time-shared between two cuvettes, in a sample (or test) position and in a reference position. 'Double-beam' also describes this mode of operation, but must be distinguished from double- or dual-wavelength (see below). Split-beam instruments are usually distinguishable by the presence in the sample compartment of two cuvette holders side by side. The single detector receives light passing through each sample and the difference spectrum (test *minus* reference) is plotted.
(b) Dual-wavelength spectrophotometry. A single sample is alternately illuminated with two 'chopped' wavelengths (λ_1, λ_2). The reference wavelength (λ_2) will preferably be at an isosbestic point (see below) or in some other relatively featureless region of the anticipated spectrum,

9. Analysis of Cytochromes

so that absorbance at that reference wavelength will remain constant, despite changes in the sample elicited by, say, redox changes, and thus serve as a measure of sample turbidity. If λ_1 is scanned and λ_2 fixed, we have the basis of a dual-wavelength scanning spectrophotometer. Since only one sample is used, difference spectra can be obtained only by recording spectra in succession of two or more samples and computing appropriate differences.

(c) Single beam. With the decline in the cost of microprocessors and microcomputers, there has been a resurgence of interest in single-beam instruments (i.e. using one cuvette and one unchopped light beam) for wavelength scanning. A reference spectrum is 'stored' in a computer, part of or interfaced with the spectrophotometer, and is subtracted from subsequent scans to give the desired difference spectrum.

9.3.1.c Stray light

This may be defined as 'white' light received at the detector that arises from the monochromator itself or from light leaks in the sample and detector region. In commercial spectrophotometers, both sources should have been minimized by sound construction and good design and there is little that the user can do to improve matters. Nevertheless, it is prudent to determine the range in which the sample concentration is proportional to the absorbance measured, i.e. the range over which the Beer–Lambert law applies. An additional check is to use neutral density filters, the absorbance of which in the light path should be strictly additive.

9.3.1.d Low-temperature accessories

A great boon in the application of visible wavelength-scanning spectrophotometry for microbial cytochromes is a facility for recording difference spectra at very low temperatures, generally that of liquid nitrogen (77 K). The main effects are sharpening and enhancement of the spectral features (Wilson, 1967). In addition to band sharpening and narrowing resulting directly from the low temperature, enhancement is generally also obtained by the increase in the effective light path of the cuvette contents that results from freezing and multiple internal reflections (Vincent et al., 1982).

Such attachments are available for some (generally top-of-the-range) spectrophotometers but not all, but it is surprisingly easy and effective to construct devices for most split-beam, dual-wavelength or single-beam instruments. The basic requirements are: (a) a clear-sided, miniature Dewar vessel for liquid nitrogen (which can be constructed to specification by some laboratory glass firms); (b) a cuvette holder, generally metal, for maintaining the position(s) of the cuvette(s) in the light beam(s) and in contact with the liquid nitrogen; and (c) cuvette(s) constructed so as to allow repeated freezing and thawing. Refinements might include a means for blowing dry air or nitrogen gas over the Dewar surfaces to minimize frosting, and replaceable or removable cuvettes to allow the recording of various difference spectra using a minimum number of sample cuvettes. Practical details are given by Jones and Poole (1985) and Poole and Bashford (1987); the former review also cites references to alternative earlier designs and configurations. These sources also give details of an apparatus for maintaining frozen samples at other subzero temperatures and obtaining low-temperature photodissociation spectra. Although the last approach is a valuable adjunct to conventional carbon monoxide difference spectra, it is probably beyond the requirements of those wishing to use cytochrome analyses for descriptive or taxonomic purposes.

9.3.2 Application of spectrophotometric techniques

9.3.2.a Difference spectra of redox states

The most valuable single analysis is undoubtedly the recording of a reduced *minus* oxidized difference spectrum. The most distinctive absorption bands of a cytochrome are generally elicited by the reduced [Fe(II)] state (although ligand-bound states and 'oxygenated' forms are often diagnostic), whilst the oxidized [Fe(III)] form is relatively featureless, particularly in the visible region between 500 and 700 nm. Consequently, the reduced *minus* oxidized difference spectrum comprises absorption peaks that can, with proper care (see Sections 9.4.3 and 9.4.4), be used to identify and quantify the cytochromes present in intact cells or derived sub cellular fractions.

Reduction of a sample is most easily achieved by addition of a powerful reducing agent such as sodium dithionite. Although addition of a few grains of *fresh* solid to a 1–3 ml sample is satisfactory for the vast majority of instances, problems may be encountered with (a) reduction and bleaching of non-cytochrome components with very broad absorption bands, so that a marked skewing of the spectrum results or (b) failure of dithionite to penetrate the cells. Sodium borohydride is also a useful reductant although the hydrogen bubbles generated on the cuvette walls can be troublesome. Alternatively, the sample can be allowed to become anoxic by respiration of suitable oxidizable and permeant substrates such as glucose, succinate or glycerol.

Oxidation is generally achieved by a chemical oxidant or vigorous aeration of the sample just before scanning the spectrum. Sometimes, problems will be experienced with samples of intact cells that are inadequately washed or harvested from exponential cultures, and which contain significant levels of endogenous substrates. These may be oxidized rapidly, partially re-reducing cytochrome components of the respiratory chain. Incomplete oxidation will underestimate cytochrome concentration measured from the difference spectrum or, worse in the present application, fail to reveal a cytochrome component at all. Hydrogen peroxide addition (with a trace of catalase if the cell suspension is catalase-negative) can serve to generate oxygen in solution and help alleviate such problems but oxygen bubbles or, in extreme cases, foaming, may be troublesome. Chemical oxidants such as potassium ferricyanide or hexachloroiridate may be used but both are highly coloured and will probably preclude measurements in the Soret (γ) region below about 500 nm. The author favours ammonium persulphate ($S_2O_8^{2-}/2SO_4^{2-}$, +2.0 V), which is colourless and may be added as the proverbial 'few grains' directly to a cuvette.

In recent years, there has been a growing awareness of the formation of stable oxygenated and 'peroxy' forms of microbial haemoproteins that may result from the use of oxygen (air) or peroxide as oxidants. These problems appear particularly acute in the case of cytochrome d-type oxidases (see Section 9.4.1) and of haemoglobin-like pigments that can be mistaken for cytochromes (see Section 9.4.2).

9.3.2.b Spectra of ligand-bound forms

The binding of carbon monoxide provides a convenient means for identifying haem proteins in which the iron atom has an absent or replaceable sixth ligand. Carbon monoxide binding to the Fe(II) state causes a pronounced change in the absorption spectrum of the haem. Cyanide binds preferentially to Fe(III) haem and can also give useful spectroscopic signatures.

The classical application of CO is in the identification of terminal oxidases, i.e. the O_2-trapping and O_2-reducing enzymes of respiratory chains, of which carbon monoxide is a competitive inhibitor. The carbon monoxide-Fe(II) compounds, as well as having distinctive spectral properties, are photodissociable and this property forms the basis of important techniques, especially the recording of photochemical action spectra (Lloyd & Scott, 1987) and the initiation of the reactions of oxidases with O_2 at subzero temperatures (Chance, 1978; Jones & Poole, 1985).

However, the most common practical application of CO binding to oxidases and other cytochromes is the recording of the reduced-plus-CO *minus* reduced difference spectrum (or just 'CO difference spectrum'). In the split-beam mode, two samples of the cell suspension or subcellular fraction are reduced with dithionite or a suitable physiological substrate and the difference between the cuvettes recorded to give a reduced *minus* reduced baseline. The 'front' or sample cuvette is then bubbled with a fine stream of CO gas, generally for 1–5 min. The top of the cuvette or the needle or tube delivering the CO to it may be very lightly smeared with silicone antifoam to avoid foaming and loss of the sample. The difference spectrum is now scanned, i.e. reduced-plus-CO *minus* reduced, and may reveal a complex but characteristic spectrum comprising both peaks and troughs. The peaks represent the formation of Fe(II)-CO adducts, whilst the troughs are due to the loss of absorbance (with respect to the reduced form) of those components that have reacted with CO. Haemoproteins that have not reacted with CO are 'invisible' in this kind of difference spectrum. An alternative to the split-beam mode of operation is to use a dual-wavelength scanning mode. Here the baseline is obtained by computing the difference between two successive scans of the same reduced sample; the CO difference spectrum is obtained by subsequently treating the same sample with CO and plotting the difference between the CO-bubbled and pre-bubbled states.

The recording of photodissociation spectra is beyond the scope of the present chapter but low-temperature attachments can often be modified relatively easily for this application (see Poole & Bashford, 1987).

9.3.2.c How to record difference spectra of microbial samples—a practical guide

It is clearly impossible to provide a detailed protocol that would be applicable to the analysis of all microbial suspensions. In all cases, however, decisions have to be made about the operating conditions of the spectrophotometer and the preparation of the sample.

The wavelength range will generally be selected on the basis of published optical properties of similar samples. The most useful absorption bands of reduced cytochromes, called γ, β and α (from low to high wavelengths—see later), lie between 400 and 700 nm. This wavelength range also encompasses the absorption bands of ligand-bound forms such as those of carbon monoxide and cyanide. In this visible region, a tungsten-halogen light source is appropriate; filter settings should be those recommended by the manufacturer of the spectrophotometer. The visible range allows use of most window materials for the cuvettes; glass or plastic will be found most useful and practical. Choice of instrument response time, scanning speed and spectral band width are closely interrelated. Starting or typical values for these variables might be a scan rate of 1–4 nm/s, a spectral band width of 2 nm and a response time of 0.1 s. Brief justification of these choices is given in the paragraph below, but for further details the reader is referred to the article by Poole and Bashford (1987).

The spectral band width used for recording the spectrum is frequently

9. Analysis of Cytochromes

ignored, not known or not quoted when the spectrum is published. The ability of the spectrophotometer to distinguish between two absorption bands close together will depend on (a) the widths of the bands under study and (b) on the width of the spectral region passed by the monochromator. The former is generally large and is the 'natural bandwidth'. It is defined as the width in nanometres at half the height of the sample absorption peak, and is an intrinsic sample characteristic. It is not changed by instrument settings other than low temperature. In contrast, the spectral bandwidth is a property of the instrument and will either be given or can be calculated. It is defined as the band of wavelengths containing the central half of the entire band of wavelengths passed by the exit slit of the monochromator. It is this slit that can be adjusted by the user and the adjustment will be graduated in spectral bandwidth (nm), slit width (mm) or both. A typical monochromator might have a reciprocal dispersion of 2 or 4 nm/mm. This means that an exit slit 1 mm wide would pass light of 2 or 4 nm spectral bandwidth, respectively. If the spectral bandwidth is known or measured and the natural bandwidth is known or assumed, the error in peak height measured by the spectrophotometer can be calculated. Figure 9.2 shows the increasing error in measured peak height as a function of the ratio of spectral and natural bandwidths. If, for example, a cytochrome at room temperature has a natural bandwidth of 10 nm and if the spectral bandwidth of the instrument is set to 2 nm (as suggested above) then the observed peak height/true peak height will be greater

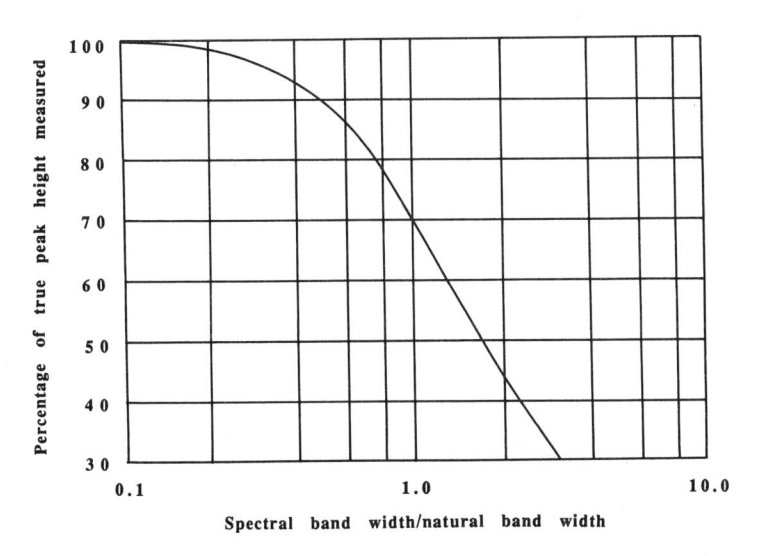

Figure 9.2 Estimation of the error in measured spectral peak height for a given monochromator slit width and natural band width.

than 0.95, i.e. the error will be less than 5%. It is clear that spectral bandwidth as selected by the operator, and natural bandwidth, as determined by the chromophore under investigation or the temperature in use, can markedly affect the precision with which chromophore quantification can be made.

It might appear, therefore, that it would always be preferable to set the spectral bandwidth as narrow as possible, but the undesirable consequence of this is a decrease in light intensity at the sample and at the detector and consequently an increase in instrument noise. Instrument noise is also influenced by the response time of the spectrophotometer so that some compromise is needed in setting these conditions. Scanning speed is also intimately related to instrument response time. Selection of too fast a scan and/or too slow a response time will cause spectral distortion. Poole and Bashford (1987) present spectra that dramatically illustrate this artefact. When a fast scan speed is desirable for some reason, such as lability of a sample (perhaps an oxygenated form of a haemoprotein) or when there is a danger of cells settling or clumping, a fast scan speed can be counteracted with a fast response time, again provided that the noise inevitably introduced is not troublesome.

It is imperative to be aware of those factors that can affect the positions as well as the intensities of absorption peaks in cytochrome spectra. Clearly, miscalibration of the instrument can result in a wavelength error in either direction and suitable calibration checks have already been outlined. If the scan speed is too fast, there will be wavelength shift in the direction of the scanning, i.e. in scanning from low to high wavelengths the shift will be to longer wavelengths. The desirability of recording a baseline before scanning a difference spectrum has already been stressed. In extreme cases, baseline irregularities may be mistaken for spectral peaks and troughs and a steeply sloping baseline can cause a wavelength shift of a component in the direction of the slope. In other words, when a baseline is sloped down at high wavelengths the apparent absorption maximum will be shifted to high wavelengths. Baseline flatness can be improved (a) electronically, (b) by extra care given to sample preparation, particularly with respect to matching the two samples, and (c) occasionally, by altering the configuration of the detector or photomultiplier. In the Aminco DW2, for example, rotating the detector can reduce the effect of non-uniformity in the photocathode sensitivity. A quartz diffuser plate in front of the detector may achieve a similar result.

9.3.3 Techniques other than wavelength scanning and their applications

The presence of iron in haem allows the application of a whole arsenal of instrumental techniques, such as electronic and electron spin resonance spectroscopy. These provide information about the geometry of a metal site and the nature of the ligands binding the metal. Much useful information may be obtained from Raman and infrared spectroscopy on metal-ligand stretching frequencies and properties of small molecules coordinated to the haem. The resonance Raman spectra of haem groups are dominated by the vibrational modes of the porphyrin ring. Most applications of infrared spectroscopy have centred on oxygenated haem intermediates. Both techniques have been used to study the vibrational spectrum of the carbon monoxide adduct of haemoproteins. A brief overview of the application of such physical techniques to metalloproteins can be found in Hughes and Poole (1989), with references to more detailed accounts. More specialist spectrophotometric techniques such as stopped flow methods for the study of fast reactions and the recording of photochemical action spectra for the identification of terminal oxidases are covered by Eccleston (1987) and Lloyd and Scott (1987), respectively. Here, only simple laboratory staining procedures and selective media that are of some value in bacterial systematics and which exploit properties of cytochrome components of the respiratory chain are described.

9.3.3.a The oxidase test

This test is a simple but useful chemotaxonomic tool. It is based on the ability of terminal oxidase systems to catalyse the aerobic synthesis of indophenol blue from α-naphthol and DMPD (N,N-dimethyl-p-phenylenediamine). An agar plate of the culture is flooded with a mixture of the substrates for 15 s and then drained. The appearance of a deep-blue coloration by indophenol blue is taken as a positive result and generally interpreted as signifying the presence of a membrane-bound high redox potential cytochrome c linked to an active cytochrome oxidase, usually cytochrome aa_3. The test can also be applied to bacterial suspensions and used semi-quantitatively as a measure of the activity of the terminal oxidase system. The test can also be performed with TMPD (N,N,N',N'-tetramethyl-p-phenylenediamine) instead of DMPD and has been used to select a cytochrome oxidase d-deficient mutant of *Escherichia coli* (Green & Gennis, 1983). The demonstrated effectiveness of the method in isolating such a mutant in an organism that contains neither cytochrome c nor cytochrome c oxidase casts some doubt on the basis of the assay. It is possible that the usual classification of *Escherichia coli* as an oxidase-negative organism may reflect the inability of the substrate or substrates to penetrate the cell envelope. The procedure described by Green and Gennis (1983) relies on permeabilization of the envelope.

Despite this uncertainty, the test has been used with some success to screen for cytochrome c-deficient mutants of various bacteria including the aerobic chemoheterotrophs *Azotobacter vinelandii*, *Paracoccus denitrificans* and *Pseudomonas* AM1 and the facultative phototroph, *Rhodopseudomonas capsulata*. It has also been used to screen for oxidase-deficient mutants of *P. denitrificans* and *R. capsulata* (for references see Jones & Poole, 1985).

9.3.3.b The peroxidase test

Although a discrete class of haem-containing enzymes (the hydroperoxidases) are largely responsible *in vivo* for removing biologically deleterious alkyl and hydrogen peroxides, most partially denatured haemoproteins, including cytochromes and globins, will exhibit some peroxidase activity. This activity can be linked to the oxidation of an artificial substrate such as the carcinogenic benzidene, orthotolidine or 3,3',5,5'-TMBZ (3,3',5,5'-tetramethylbenzidine) to yield a greenish-blue product. This chromogenic reaction allows the detection and localisation of haemoprotein bands on SDS polyacrylamide gels. Briefly, SDS polyacrylamide gels of samples denatured in the absence of mercaptans are immersed in a solution of methanolic TMBZ and acetate buffer (pH 5) and incubated in the dark for 1–2 h. Hydrogen peroxide is then added and the haemoprotein bands become stained within 30 min. The stain can subsequently be removed by soaking the gels in sodium sulphite and repeated washings with 30% (v/v) isopropanol. The de-stained gels can then be re-stained for protein using more conventional procedures. References and applications are detailed by Jones and Poole (1985).

9.3.3.c Selective media

The differential sensitivity of terminal oxidases to various respiratory poisons such as cyanide (Poole, 1983) has prompted the development of solid media for discriminating between organisms whose respiratory chains are terminated by different oxidases. For example, cytochrome *d* is widely known to be more resistant to inhibition by classical respiratory poisons, such as cyanide and azide, and the presence of heavy metal ions than are cytochromes of the *o* or aa_3 type. In one application, azide and zinc ions were incorporated in a low-phosphate medium (to prevent precipitation of zinc phosphate) that proved useful in the isolation and screening of cytochrome *d* deficiency in *E. coli* mutants (Poole *et al.*, 1989).

9.3.3.d Fluorescence techniques

Cytochrome *c* can be detected on SDS polyacrylamide gels by its ability to fluoresce when excited at 360 or 366 nm. Usefully, the method is specific

9. Analysis of Cytochromes

for c-type cytochromes if electrophoresis is carried out in the presence of mercaptans, which remove the non-covalently bound haems of a, b and d-type cytochromes, but not the covalently bound mesohaem of c-type cytochromes. The procedure is described by Jones and Poole (1985), who also cite applications of this technique.

9.4 ASSIGNMENT OF VISIBLE ABSORPTION BANDS TO CYTOCHROMES AND OTHER HAEMOPROTEINS

9.4.1 Overview of cytochrome types

It is the ability of the haem iron to undergo oxidation and reduction that allows cytochromes to participate in cellular redox reactions. Thus, in both respiratory and photosynthetic electron transport, reducing equivalents are transferred from a donor to a higher redox potential acceptor with concomitant establishment of an electrochemical proton gradient or protonmotive force (Δp) across the H^+-impermeable membrane. Δp is utilized in the performance of 'useful work' (Harold, 1986) such as ATP synthesis, solute translocation via secondary transport systems, reversed electron transfer, transhydrogenase activity and motility.

All bacterial respiratory chains appear to contain one or, more frequently, several cytochromes b. As in mitochondria, these cytochromes frequently function close to quinone and serve to transfer electrons from the iron-sulphur centres of the primary dehydrogenases (e.g. NADH, lactate or succinate dehydrogenases) on the cytoplasmic side of the membrane to cytochrome c on the outside face. As in mitochondria, cytochromes b are generally integral membrane proteins and rather difficult to purify. Cytochromes b with more specialized functions occur in certain bacteria. These include the low potential cytochromes b of some chemolithotrophs (e.g. *Nitrobacter*, *Thiobacillus*) involved in reversed electron transfer to NAD^+ (Wood, 1988) and the cytochromes b of particularly low E_m values that function in the respiratory chains of anaerobes such as the sulphate-reducers *Desulfotomaculum* and *Desulfovibrio* and in fumarate reducers such as *Vibrio succinogenes* and some methanogens (Hamilton, 1988). The term 'cytochrome b_1' has sometimes been used to describe the broad absorption maximum at 560 nm in *E. coli*, but this band reflects contributions from several components with different optical and functional properties and use of the term should be discontinued.

A major role of c-type cytochromes in bacteria is, as in mitochondria, to transfer electrons from the quinone-cytochrome b system to the cytochrome oxidases or to other terminal reductases. Particularly high levels of c-type cytochromes are present in specialized respiratory chains such as those of methylotrophic and sulphur- or ammonia-oxidizing bacteria, where these cytochromes accept electrons from primary dehydrogenases or cytochrome c reductases located on the periplasmic side of the membrane (Wood, 1988). In contrast, many chemoheterotrophic bacteria lack significant concentrations of a high redox potential c-type cytochrome. Examples are *E. coli* and other enterobacteria. In general, the oxidases in these organisms are quinol oxidases rather than cytochrome c oxidases. A large number of sequences are available for cytochromes c, but many are from eukaryotic sources. Nevertheless, there are enough data to allow comparison of the purple photosynthetic bacteria. A dendrogram for the cytochrome c data is in good agreement with a 16S rRNA dendrogram (see Fox & Stackebrandt, 1987).

A number of bacterial c-type cytochromes have been given distinguishing subscripts, which are briefly explained below (for references see Jones & Poole, 1985; Pettigrew & Moore, 1987; Palmer & Reedjik, 1992:

(a) Cytochrome c_2 is a small soluble cytochrome with the same folding and binding of the haem as in mitochondrial cytochrome c, but with a higher redox potential. In photosynthetic bacteria such as *Rhodopseudomonas viridis* and *Chromatium vinosum*, it transports electrons from the bc complex to the membrane bound c-type cytochrome, which donates electrons to the photosystem.
(b) Cytochrome c_3 is a low-potential cytochrome, which can exist as a dimer. It exists in tri- and tetra-haem forms as a monomer and is found in the strictly anaerobic sulphate- and sulphur-reducing bacteria where it participates in sulphate respiration coupled to phosphorylation. The trihaem form was formerly called cytochrome c_7. It is possible that the cytochromes c_{552} from *Escherichia coli* may be related.
(c) Cytochrome c_4 is a dihaem high-potential protein found in *Azotobacter vinelandii* and some *Pseudomonas* species.
(d) *Azotobacter* also contains cytochrome c_5, a dimeric protein.

Terminal oxidases transfer electrons from either a high-potential cytochrome c (cytochrome c oxidases) or from quinols (quinol oxidases) to molecular oxygen, with the formation of water. A distinctive feature of most bacterial respiratory chains is that they are branched, each branch being terminated by a different terminal oxidase. The concentrations of the cytochrome components, particularly the terminal oxidases, of the various branches is frequently under the control of environmental conditions, but this is poorly understood. It is of great importance, however, in attempting to draw any taxonomic conclusions from cytochrome composition. The optical absorption spectrum will change dramatically according to the growth history of the culture being investigated. Bacterial cytochrome oxidases have been extensively reviewed by Poole (1988) and Yamanaka (1992).

Bacterial oxidases of the aa_3 type exhibit many similarities to their mitochondrial counterparts. They contain two haems, a and a_3, two or three copper atoms and perhaps other metals in sub-stoichiometric amounts. Strikingly, however, their subunit composition is much simpler, comprising at most three subunits in comparison with the seven to thirteen present in the mitochondrial enzyme. Haems a and a_3 are readily distinguished by the ability of haem a_3 to bind carbon monoxide, cyanide and other ligands. The optical properties of the coppers are less clear, but there is general agreement that only Cu_A (the copper associated with haem a) shows any significant optical signature at around 830 nm. Some bacteria contain two types of cytochrome aa_3 complex, an aa_3 and caa_3 (Garcia-Horsman et al., 1991). Although the purified forms of the two types are distinct, the caa_3-type exhibiting a peak in the reduced form at 551 nm due to cytochrome c, the two types are less easy to distinguish in intact cells. However, in *Bacillus subtilis*, the α-band of the aa_3-type is at 601 nm, whereas that of the caa_3-type is at 605 nm (for a review of these and other cytochromes of this organism, see von Wachenfeldt and Hederstedt, 1992).

Most cytochrome a_3-containing oxidases are cytochrome c oxidases, but within this class complexes that can be described as ba_3 or caa_3 have been distinguished (Saraste et al., 1991). *Sulfolobus acidocaldarius* appears unique in having the same complement of redox centres (a, a_3 and 2Cu) as in the 'classical' aa_3 type, but contained in a single subunit; it is a quinol oxidase (see Section 9.5).

Cytochrome a_1 remains the most enigmatic of the putative cytochrome oxidases in bacteria. The name cytochrome a_1 should be regarded as applying to a group of haemoproteins, probably quite unrelated, which share a distinctive α absorption maximum between 587 and 597 nm. A proposed classification of this group has been made by Poole et al. (1985). Cytochrome a_1 is frequently reported as occurring in *E. coli*. However, no haem A has been detected in this organism. The α absorption band previously attributable to cytochrome 'a_1' contains contributions from at least two haemoproteins, a soluble, high-spin b-type cytochrome (haemoprotein b-590), which is a hydroperoxidase, and a high-spin b-type cytochrome (cytochrome b-595), which is a component of the cytochrome bd oxidase complex. The best evidence for a true cytochrome a_1-type oxidase comes from *Acetobacter aceti* (Matsushita et al., 1990), in which haems a_1 and b plus a copper atom constitute a quinol oxidase.

Cytochrome 'o' (see comments in Section 9.2 on nomenclature) is probably the most widespread oxidase in bacteria (Poole, 1983; Kranz & Gennis, 1985). It contains as prosthetic group at least one mole of a novel haem O, in addition to a further mole of haem B or O (Puustinen et al., 1992). It has also been reported in some protozoa. Cytochromes of the o-type have been purified from several bacteria. The *E. coli* oxidase is the best characterized. It comprises four subunits, two moles of haem and one copper atom. Some of the polypeptide subunits show an unexpected and striking homology with certain subunits of aa_3-type oxidases (Chepuri *et al.*, 1990).

Cytochrome d was previously known as cytochrome a_2. It is present in many aerobic bacteria, especially when grown with a limited oxygen supply. It occurs, for example, in *Azotobacter vinelandii* and *E. coli*. It has been purified from *E. coli* and *Photobacterium phosphoreum* and comprises two subunits. It contains three types of haems: haem d (a chlorin; Timkovich *et al.*, 1985), a low-spin cytochrome b-558 and a high-spin cytochrome b-595. There is no copper. The optical absorption bands of cytochrome d are very distinctive (see later) and it forms several stable compounds with oxygen, peroxide and other ligands.

Cytochrome cd_1 is a dissimilatory nitrite reductase present in denitrifying bacteria such as *Paracoccus denitrificans* and *Pseudomonas aeruginosa*. Two forms have been identified in the marine, aerobic, photosynthetic bacterium *Erythrobacter* sp. Och 114 (Doi *et al.*, 1989), a finding that has been interpreted as evidence of an evolutionary relationship between pseudomonads and other denitrifying bacteria. The haem d_1 (Chang *et al.*, 1986) is non-covalently associated, while the haem c is covalently bound to the polypeptide. Again, the optical absorption bands are quite distinctive. In addition to its nitrite reductase function, it will reduce oxygen to water and has been used as a model system for the investigation of haem/haem interactions and intramolecular electron flow.

9.4.2 Non-cytochrome haemoproteins

Cytochromes are haemoproteins that transfer electrons. As used presently, the name includes all intracellular haemoproteins, with the exception of haemoglobin, myoglobin, the peroxidases, catalases, nitrite and sulphite reductases and tryptophan-2,3-dioxygenase. However, since the spectral properties of some non-cytochrome haemoproteins may be readily confused with cytochrome absorption bands, a brief overview of their properties is desirable. Most of the proteins in this group are CO-binding b-type haemoproteins. A provisional key to distinguishing them is provided by Wood (1984).

Tryptophan-2,3-dioxygenase catalyses the incorporation of both oxygen atoms from molecular oxygen into L-tryptophan, giving L-formyl kynurenine as product. The enzyme has been studied most extensively in *Pseudomonas* species. The ferrous form has an absorption maximum at 588 nm and other features characteristic of high-spin b-type haems (see later).

The putative oxygen buffers, i.e. those proteins in bacteria that may have a function somewhat analogous to haemoglobin or myoglobin, include the soluble *Vitreoscilla* haemoglobin (VHb; for a review, see Webster, 1987). Until 1986 this protein was referred to as *Vitreoscilla* cytochrome *o*. The reduced state shows a pure high-spin spectrum and it reduces oxygen only slowly. Oxygen is bound reversibly to give a spectrum like that of oxymyoglobin. The subunit molecular weight is small (13 000). The native protein is a dimer, with one haem per polypeptide chain. Recently a protein with extensive sequence homology to the *Vitreoscilla* protein has been identified in *E. coli* (Vasudevan *et al.*, 1991). The protein is much larger (M_r about 44 000) and, although the gene encoding the protein has been cloned and sequenced, the protein's physiological function is poorly understood. It binds oxygen to form an oxygenated complex that is only transiently stable in the presence of reductant (Ioannidis *et al.*, 1992). It binds CO but is unlikely to be detectable in intact cells of normal strains because its signals are masked by those of other CO-binding pigments. The protein also contains FAD, but the absorption properties of the flavin are overshadowed by those of the haem. It is similar to flavohaemoproteins in yeast (Zhu & Riggs, 1992) and *Alcaligenes eutrophus* (Probst *et al.*, 1979). A haemoprotein *b*-558 has been purified from *Acetobacter suboxydans*, which also appears to form a stable oxy-compound (Iwasaki, 1966).

Generally, catalases are not reducible by dithionite and will not contribute significantly to difference spectra. However, the catalase from *Comamonas compransoris* gives a CO difference spectrum with a peak at 424 nm and a trough at 444 nm. The protein is a dimer of M_r 150 000 (Nies & Schlegel, 1982). From *E. coli*, a soluble *b*-type haem has been described and named haemoprotein *b*-590. This protein has a spectrum similar to that of tryptophan–2,3-dioxygenase and has peroxidase activity (Poole *et al.*, 1986). It is probably the same as hydroperoxidase-1, purified previously from this organism.

Sulphite reductases, which have a characteristic sirohaem, have been reviewed by Hamilton (1988). Sirohaem is also a component of nitrite reductase (Lin & Kuritzkes, 1984). The haem is of the isobacteriochlorin-type and unusually hydrophilic.

9.4.3 Interpretation of spectra

This section will attempt to offer some guidance to those with little experience in the interpretation of the optical spectra of bacterial cytochromes and related haemoproteins. It will be assumed (a) that the user will have recourse to spectral scanning only in the visible wavelength range (400 to 700 nm); (b) that reduced minus oxidized and CO difference spectra will be run; and (c) that further guidance will be sought in cases of doubt. A visual guide to supplement this description was presented by Jones and Poole (1985).

9.4.3.a Reduced minus oxidized difference spectra

The Soret (or γ) region extends from about 400 to 450 nm (Figure 9.3). Here absorption bands are more prominent than in the β and α regions but the problem of spectral overlap is greatest. Cytochromes c in the low-spin state give strong γ-bands between about 415 and 420 nm. This band is generally seen fused with the γ-band of low-spin b-type cytochromes (approximately 430 nm), frequently generating a broad asymmetrical band centred between 420 and 430 nm. A distinct shoulder of lower intensity at higher wavelengths, say 440 nm (which might be clearly split at 77 K), is most frequently attributed to a-type cytochromes, such as aa_3. However, some high-spin b-type cytochromes can also give peaks here (Poole et al., 1986). High spin c-types absorb maximally near 425–433 nm, so that interpretation of the γ region is generally problematic, although CO difference spectra can be helpful in this region (see below).

The low-intensity absorbances between about 510 and 540 nm are attributable to the β-bands of low-spin cytochromes (Figure 9.3). In high-spin b-type proteins such as myoglobin or haemoprotein b-590 ($E.\ coli$), the β-band is at longer wavelengths (see later). Within the β region the bands of c-type cytochromes are generally at lower wavelengths than those of the b-type cytochromes. However, since even purified cytochromes (e.g. horse heart cytochrome c) can exhibit split β-bands, the fusion of peaks in this region makes it the least useful for cytochrome identification and virtually useless for cytochrome quantitation.

The α-bands are caused by the same electronic transition as the β-bands but in the case of most bacterial low-spin cytochromes are more intense and useful (Figure 9.3). The major contributions to the 540 to 570 nm region are from cytochromes c (about 548–555 nm) and cytochromes b (556–566 nm). A much broader band with a maximum at 550 to 560 nm may lie under such peaks and arise from the β-band of high-spin b-type haemoproteins. The α-band of such components is generally weak and observed only as a shoulder at higher wavelengths near 590 nm. However, in the case of haemoprotein b-590 ($E.\ coli$), cytochrome b-595 (a component of cytochrome oxidase complexes of the bd-type) and tryptophan 2,3-dioxygenase, the α-band is more pronounced and is seen in difference spectra at 588 to 595 nm.

The problems of nomenclature of the a_1-like pigments with an α-band near 590 nm have already been raised. Although a haem A-containing cytochrome that resembles 'a_1' has been purified from *Nitrobacter*, the cytochrome is not an oxidase. Recently, in other bacteria previously

9. Analysis of Cytochromes

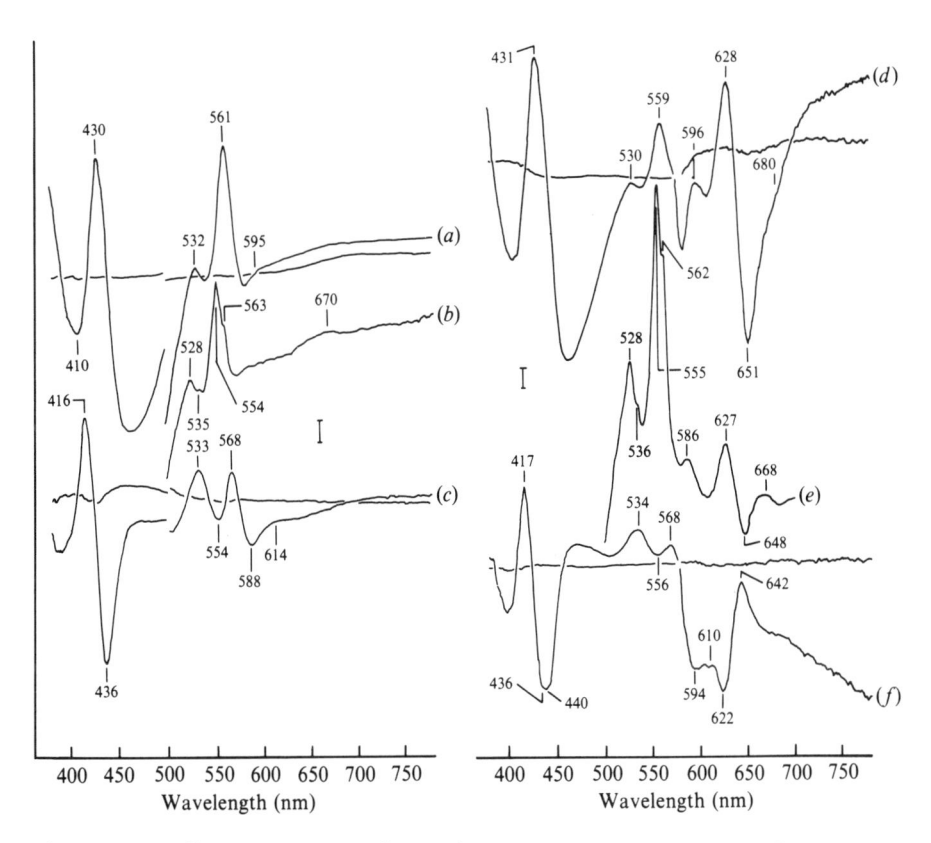

Figure 9.3 Difference spectra of *Acetobacter pasteurianus* grown under oxygen-sufficient (a–c) or oxygen-limited (d–f) conditions. Spectra (a) and (d) are room temperature reduced (dithionite) *minus* oxidized (persulphate) spectra; (b) and (e) are 77 K reduced *minus* oxidized spectra; (c) and (f) are room temperature CO-reduced *minus* reduced difference spectra. Unlabelled spectra are baselines, oxidized *minus* oxidized or reduced *minus* reduced, as appropriate. 77 K spectra were recorded at 4 nm/s and a spectral band width of 1 nm while room-temperature spectra were recorded at 5.7 nm/s, a spectral band width of 8 nm and a reference wavelength of 500 nm in the dual-wavelength mode. The vertical bar represents, from 380 to 500 or 570 nm, ΔA 0.04 (a, d), 0.02 (b, e) and from above these wavelengths 0.02, 0.019, 0.01, 0.004, 0.008 and 0.004 in (a) to (f) respectively. Cells were suspended to a protein concentration of 40 to 50 mg/ml. The pathlength was 2 mm at 77 K and 10 mm at room temperature.

The spectra illustrate (i) the marked influence of growth conditions on spectral form, (ii) the intensification and band sharpening at 77 K (see regions above 500 nm), (iii) the distinctive bands of cytochrome *d* (d to f), (iv) a CO spectrum due largely to *b*-type cytochrome(s), and (v) the absence of an aa_3-type oxidase. Reproduced from Williams & Poole (1987) by permission of the Society for General Microbiology.

thought to contain 'cytochrome a_1', the a-band near 590 nm has been attributed to hydroperoxidase. This is the case for *E. coli* (Poole *et al.*, 1986), *Halobacterium halobium* (Fukumori *et al.*, 1985) and *Rhizobium japonicum* bacteroids (Appleby & Poole, 1991).

The α-bands beyond about 590 nm are due to cytochromes a and/or a_3 (600–605 nm), d (630 nm) and d_1 (Poole, 1983).

Unusual contributions to the reduced minus oxidized spectrum include unattributed bands between 610 and 620 nm (Poole, 1983), the 503 nm pigment (for a review, see Ingledew & Poole, 1984), a highly distinctive 578 band in *Leptospirillum* (Hart *et al.*, 1991), believed to arise from a zinc-containing protein, and the bands attributable to cytochrome P-450 (for reviews, see Nozaki *et al.*, 1982).

A special warning about the nomenclature of signals due to cytochrome d is in order. Reduced minus 'oxidized' spectra of cells containing this oxidase generally show a distinct trough at about 650 nm and, frequently, a weaker trough at about 680 nm (Figure 9.3). Neither is due to the fully oxidized (i.e. Fe(III)) haem d. The 650 nm trough is now attributed to the 'oxy form' or oxygenated intermediate of cytochrome d (Poole *et al.*, 1983). This form is easily generated when a cell or membrane suspension is aerated or shaken and is quite stable under physiological conditions. Some suggested extinction coefficients for quantifying cytochrome d propose measurements of peak height (~ 630 nm) *minus* trough depth (~ 650 nm) but this cannot be recommended unless the oxy form is known to be fully formed. A broad trough at about 680 nm is sometimes observed as a shoulder on the 650 nm trough. This is believed to be due to a further intermediate and can be generated by treating the cytochrome d-containing preparation with hydrogen peroxide (Poole & Williams, 1988).

9.4.3.b CO-reduced minus reduced difference spectra

The chief usefulness of the CO difference spectrum is that most cytochromes are *not* CO-reactive so that the spectrum has contributions from fewer components. This is particularly valuable in attempting to define the cytochromes present in intact cell preparations. It cannot be emphasized too strongly that although all terminal oxidases are expected to bind CO, not all CO-binding haemoproteins are oxidases. Wood (1984) surveys a wide range of CO-binding b- and c-type haemoproteins. Other techniques such as photochemical action spectroscopy, measurement of redox potentials and oxygen affinities, fast kinetic measurements and observation of reaction intermediates at subzero temperatures are needed to demonstrate oxidase function. Since CO complexes are always low-spin, the CO complexes of such diverse proteins as myoglobin and horseradish peroxidase are rather similar (Wood, 1984).

Nevertheless, the CO difference spectra of low-spin and high-spin b- and c-type cytochromes are distinct. In the low-spin case, the 500 to 600 nm region is dominated by a deep trough that mirrors the α-band of the reduced state. In the high-spin case, the spectrum shows a series of peaks and troughs. A trough at 594 nm coincides with the (usually) hidden band of the uncomplexed protein. Peaks at about 579 nm and 538 nm are due to the α- and β-bands respectively of the CO complex. The trough at about 558 nm is mainly due to the broad β-band of the uncomplexed protein. It must be emphasized that in complex mixtures of high- and low-spin haemoproteins, such as might be expected in intact cells (Figure 9.3), these bands will be superimposed and confused, frustrating attempts to determine spinstate or haem concentration as described by Wood (1984).

The γ-bands in CO-difference spectra of high and low-spin cytochromes can also be useful (Figure 9.3), with the above caveats. For the low-spin state, the peak is near 416 nm and the trough near 427 nm; for the high-spin state the bands are shifted to about 422 and 438 nm, respectively. The high-spin trough may be as high as 440 nm or more and thus confused with the trough due to cytochromes a or a_3 at about 440 nm in such spectra. Cytochrome d has a weak Soret band (Poole, 1983).

The CO difference spectrum clearly reveals cytochromes a_3 and d (Figure 9.3) in the α-region (>600 nm). Cytochrome a_3 reacts with CO to give a CO complex absorbing at about 590 nm; the difference spectrum thus reveals a peak at this wavelength and a trough at 600–605 nm due to 'bleaching' of the reduced a_3 absorbance. Similarly, cytochrome d reacts to give a compound absorbing near 640 nm; the accompanying trough is typically at 622–624 nm.

9.4.4 Cytochrome quantification

A summary of the recommended wavelengths for identifying cytochromes in bacteria and proposed absorption coefficients for quantifying the cytochromes in difference spectra is presented in Table 9.1. A more comprehensive listing was given by Jones and Poole (1985). It should be noted that many of the extinction coefficients cited, particularly in the latter review, have not been obtained from microbial systems, or at best have been determined for one organism and the values extrapolated to cytochromes of the same type in other species.

It is customary to measure and record the protein concentration of the cell suspension used for cytochrome quantification. In the author's laboratory, the modifications of the Lowry method described by Markwell *et al.* (1978) have been found most useful. The method is applicable to intact cells without separate digestion with alkali, and the ability to measure protein in the presence of moderate concentrations of sucrose makes the method

Table 9.1 A summary of measuring wavelengths and proposed absorption coefficients (ε) for quantifying bacterial cytochromes in difference spectra.

Cytochrome	Spectral conditions	Wavelength pair (nm)	ε (mM^{-1}cm^{-1})	Source	Reference
aa_3	Reduced minus oxidized	605 *minus* 630	11.7	*Paracoccus denitrificans*	Ludwig and Schatz (1980)
a_3	CO difference	430 *minus* 444	91	Various, including yeast, *Bacillus subtilis* 'Consensus'	Chance (1957)
Low-spin *b*- and *c*-types	CO difference	α-peak *minus* α-trough, e.g. 556 *minus* 572	25		Wood (1984)
High-spin *b*- and *c*-types	CO difference	γ-peak[a] *minus* γ-trough, e.g. 420 *minus* 440	177 215	myoglobin ('consensus')	Wood (1984)
d	Reduced minus oxidized	628 *minus* 649	18.8	*E. coli*	Kita *et al.* (1986)
	CO difference	628 *minus* 605	7.5	*E. coli*	Miller and Gennis (1986)
	CO difference	622 *minus* 642	12.6	*E. coli*	Kita *et al.* (1986)
o	CO difference	γ-peak *minus* γ-trough (e.g. 415 *minus* 430)	145	*E. coli*	Kita *et al.* (1984)

[a] Danger of overlap from other CO-binding pigments.

particularly useful where cells have been suspended in sucrose, perhaps for measurements at 77 K.

9.5 TAXONOMIC CONSIDERATIONS

There is little doubt that no bacterium can be identified by its cytochrome composition alone. However, the description of the cytochrome pattern of an organism is an important part of its total description, and cytochrome analyses have been used increasingly as a taxonomic tool or aid. The most widely used approach is the description of the cytochrome pattern of an organism, principally from measurements of reduced *minus* oxidized, or reduced plus CO *minus* reduced difference spectra. Much of the work that has been reported in this area is unreliable. Even some recent reports must be assessed carefully. Unfortunately, the quality of the spectral data is often poor and authors rush to inappropriate conclusions about the identity of cytochromes that contribute to the complex spectra generally observed. Recommendations regarding cytochrome nomenclature summarized previously in this chapter are generally ignored and there are many spurious claims about the existence of cytochrome *o*, for example (which must be an oxidase), and cytochrome a_1, whose identity has been questioned increasingly in recent years.

It is clearly of interest to examine whether the distribution of cytochrome patterns correlates with current views on bacterial phylogeny (Stackebrandt, 1991). Unfortunately, too few genera have been examined with sufficient rigour to allow many useful correlations to be made. Nevertheless, a number of general conclusions can be, and have been, drawn regarding the cytochrome patterns exhibited by various groups of bacteria. This work has been reviewed previously by Jones (1977, 1980) and Jones and Poole (1985). Exceptions to these general patterns are occasionally observed, however, and it is perhaps these exceptions that are the most interesting and useful. Except where indicated, references are given in these earlier accounts.

The Gram-positive heterotrophs have been regarded by Jones (1980) as a rather homogeneous grouping with respect to cytochrome content. Cytochromes *b*, *c*, aa_3 and *o* form the predominant pattern, as in *Bacillus subtilis* (von Wachenfeldt & Hederstedt, 1992); i.e. their cytochrome systems superficially resemble those of eukaryotic mitochondria with the addition of a second oxidase, cytochrome *o*. Some species appear to contain only one terminal oxidase and a significant number appear to lack cytochrome *c*. Presumably, *a*-type cytochromes serve to reoxidize quinones (as demonstrated for *Bacillus subtilis*) or cytochromes *b* in a manner quite unlike that in the mitochondrial system. Similar claims have been made for the archaebacteria *Halobacterium halobium* and *Sulfolobus acidocaldarius* (see later). Other bacteria in this group appear to contain cytochrome *d*. Cytochrome *d* has been reported in all *Listeria* spp. reported to date (Feresu & Jones, 1988), in *Brochothrix thermosphacta*, which also contains cytochrome *aa*$_3$ (Gil et al., 1992), and the alkalophile *Bacillus firmus* (Hicks et al., 1991). Low levels of cytochrome *d* are also present in some micrococci and related actinomycetes.

A complication that must always be borne in mind is that the cytochrome composition of bacteria can vary dramatically according to growth conditions. Thus, in *Arthrobacter globiformis*, logarithmic phase cells contain cytochromes *b, c, aa*$_3$ and *o*, whereas under oxygen-limited conditions cytochrome *d* and an *a*$_1$-like cytochrome are additionally synthesized. *Bacillus popilliae* is also an exception to the general pattern reported for Gram-positive heterotrophs; this bacterium contains cytochromes *d, o* and *a*$_1$ in place of *aa*$_3$ and *o*.

The Gram-negative chemoheterotrophs exhibit much less consistent patterns of cytochrome content. Most of these organisms exhibit a cytochrome *b, o, a*$_1$*, d* pattern; cytochrome *c* is usually absent. Some species of *Pseudomonas*, however, synthesize cytochrome oxidase *aa*$_3$, as do most methylotrophs. Cytochrome *aa*$_3$ also occurs in *Rhizobium* (free-living cells) and in *Paracoccus denitrificans* (alpha subclass). In fact the latter organism contains the best characterized prokaryotic *aa*$_3$ type oxidase. Some species of *Pseudomonas* are unusual in exhibiting unbranched respiratory systems with cytochrome *o* as the apparent sole oxidase. Cytochrome oxidases of the *o*- and *d*-types are widespread in Gram-negative bacteria, but immunological experiments suggest that the spectrally similar cytochromes are not always closely related. Proteins closely related to the cytochrome *d* complex of *E. coli* appear to be widely distributed in, for example, *Serratia marcescens, Enterobacter aerogenes, Azotobacter vinelandii* and other bacteria related to *E. coli* in the gamma subclass of the Proteobacteria. Cross-reactive antigens were not detected in several alpha subclass bacteria (a *Rhodospirillum* sp., *Rhodopseudomonas spheroides* and *Paracoccus denitrificans*), but neither were antigens detected in a further gamma subclass member, *Pseudomonas fluorescens* (Kranz & Gennis, 1985). Similar experiments with antibodies raised against the cytochrome *o*-type oxidase of *E. coli* revealed the presence of related proteins only in *Klebsiella pneumoniae* and *Salmonella typhimurium*; even other members of the family Enterobacteriaceae gave negative results, as did *Vitreoscilla*, whose membrane-bound cytochrome *o* is functionally very similar to that of *E. coli* (Poole, 1988). Studies of this type are limited but should help to shed light on the possible value of cytochrome patterns in bacterial taxonomy.

Chemolithotrophic bacteria generally exhibit the *bcaa*$_3$*o* cytochrome pattern. In these organisms cytochrome *c* is involved in electron transfer from the oxidation of high redox potential substrates. *Thiobacillus ferrooxidans* remains one of the very few bacteria in which a claim for cytochrome a_1 continues to be taken seriously. This organism has an α absorption maximum below 600 nm and there is good biophysical evidence for an a-type cytochrome acting as functional oxidase. The so-called cytochrome a_1 from *Nitrobacter* is a nitrite dehydrogenase.

The phototrophs contain *b*- and *c*-type cytochromes which are involved in cyclic electron transfer. Facultative phototrophs such as *Rhodospirillum* and *Rhodopseudomonas* synthesize a conventional respiratory chain during aerobic growth in the dark with cytochromes b, c, aa_3 and o. Comparison of cytochrome c sequences are in agreement with a 16S rRNA dendrogram for relationships within the purple photosynthetic bacteria (Fox & Stackebrandt, 1987).

Following Postgate's pioneering studies, the existence of cytochromes in certain anaerobic bacteria has become well established. Cytochromes of the *c*- and *b*-type are found in *Desulfovibrio* and *b*-types also occur in *Desulfotomaculum* and bacteria that respire anaerobically with fumarate as electron acceptor. In other anaerobic bacteria, the functions of cytochromes are not understood (for references, see Kühn & Gottschalk, 1983). Anaerobic chemoheterotrophs generally lack cytochrome oxidases. A few obligate anaerobes (e.g. *Desulfotomaculum, Propionibacterium*) synthesize low levels of cytochromes a_1 and/or d. It has been suggested that the low levels of cytochrome oxidases may enable them to scavenge molecular oxygen and maintain an anaerobic environment. This is consistent with the hypothesis of respiratory protection invoked for nitrogenase protection in *Azotobacter vinelandii* (Kelly et al., 1990). Anaerobically, *Paracoccus denitrificans* and the denitrifying pseudomonads synthesize cytochrome cd_1 when grown on nitrate or nitrite. *Wolinella curva* and *Wolinella recta* (recently reclassified as *Campylobacter curvus* and *Campylobacter rectus*, respectively) as well as *Bacteroides ureolyticus* and *Bacteroides gracili* have been described as anaerobes but all are now known to respire microaerophilically and contain *b*- and *c*-type cytochromes (Han et al., 1992), resembling other *Campylobacter* species. Among methanogens, cytochromes appear to be present only in those bacteria that grow on substrates containing the methyl group, suggesting that cytochromes are involved in the catabolism of these substrates, possibly in oxidation of the methyl group to CO_2 (Kühn & Gottschalk, 1983).

Studies of cytochromes in the archaea are in their infancy. That the archaea are probably of early evolutionary origin and are distinct from both eukaryia and bacteria is well known. A striking characteristic of the group is the distinctive chemistry of their cell walls and membranes, and the presence of unusual electron carriers, such as F_{430}, but there has been little evidence until recently to suggest that this is mirrored in unique cytochromes. The methanogens were once thought to lack all cytochromes, but a membrane-bound, spectrally conventional, CO-binding cytochrome b_{559} was reported in *Methanosarcina barkeri* (Kühn et al., 1979). Subsequently, at least two membrane-bound b-type cytochromes and a cytochrome c were resolved in several *Methanosarcina* species (Kühn & Gottschalk, 1983). The cytochromes in several methanogens have been surveyed (Kühn et al., 1983).

Recent studies of the respiratory chain of the extreme halophile, *Halobacterium halobium*, have identified two b-type cytochromes (Gradin & Colmsjö, 1987), cytochrome o (Hartsel et al., 1988) and, in the exponential phase of growth, cytochrome aa_3 (Fujiwara et al., 1987). Early studies did not detect aa_3 but rather cytochromes 'o' and 'a_1' as terminal oxidases (Cheah, 1970), underlining the caution necessary in describing cytochrome patterns and defining and varying growth conditions.

The thermoacidophilic archaebacterium *Sulfolobus acidocaldarius* possesses a respiratory chain (Anemüller et al., 1985) and an aa_3-type cytochrome complex (Anemüller & Schäfer, 1989), spectrally similar to bacterial and eukaryotic cytochrome oxidases of this type. However, further studies (Anemüller & Schäfer, 1990) show this oxidase to be unique, thus far, to the archaea. It has a single subunit, exhibits unusual electron spin resonance spectra that suggest unusual haem-copper coupling, and oxidizes caldariella quinol, not reduced cytochrome c. Membranes of the thermoacidophile *Thermoplasma acidophilum* (Gartner, 1991) contain cytochromes of the b,d and perhaps a_1-type. Previously, exclusively fermentative pathways were proposed (for references, see Anemüller et al., 1985). Thus, its cytochrome composition appears to be more like that of certain Gram-negative bacteria than, say, *Sulfolobus acidocaldarius*.

9.6 SUMMARY

The diversity of cytochromes and other haemoproteins in bacteria and the combination of cytochromes to give different patterns that reflect in part established taxonomic distinctions, makes the analysis of cytochromes a useful adjunct to other methods. A complete description and identification of the haemoproteins in an isolate will require facilities and expertise beyond those available in many laboratories. Nevertheless, a commercial split-beam or dual-wavelength scanning spectrophotometer can be used, with due attention to sample preparation and operating conditions, to record reduced *minus* oxidized or CO plus reduced *minus* reduced difference spectra of highly turbid suspensions of whole cells. Such instruments can generally be used, or adapted, to record spectra at 77 K. These spectra should reveal the main classes of cytochrome types present (a-, b-, c- or d-types) and may indicate the presence of other haemoproteins such as hydroperoxidase. Carbon monoxide difference spectra can indicate *potential* terminal oxidases and allow a provisional assignment of spin-state for CO-binding pigments. Appropriate extinction coefficients will allow estimates and comparisons to be made of cytochrome concentration. It is probable that the increasing availability of cloned cytochrome and haemoprotein genes from various bacteria will prove useful in probing genomic digests of other species for cytochrome types.

ACKNOWLEDGEMENTS

RKP thanks the SERC and the Royal Society for grants, and Dr C.W. Jones for introducing him to the area of cytochromes as taxonomic tools. Parts of this chapter rely heavily on the material provided by Dr Jones in previous accounts.

REFERENCES

Anemüller, S. and Schäfer, G. (1989) Cytochrome aa_3 from the thermoacidophilic archaebacterium *Sulfolobus acidocaldarius*. *FEBS Lett.* **244**: 451–5.

Anemüller, S. and Schäfer, G. (1990) Cytochrome aa_3 from *Sulfolobus acidocaldarius*, a single-subunit, quinol-oxidizing archaebacterial terminal oxidase. *Eur. J. Biochem.* **191**: 297–305.

Anemüller, S. Lübben, M. and Schäfer, G. (1985) The respiratory system of *Sulfolobus acidocaldarius*, a thermoacidophilic archaebacterium. *FEBS Lett.* **193**: 83–7.

Anthony, C. (ed) (1988) *Bacterial Energy Transduction*. Academic Press: London.

Appleby, C.A. and Poole, R.K. (1991) Characterization of a soluble catalase-peroxidase haemoprotein b-590, previously identified as 'cytochrome a_1' from *Bradyrhizobium japonicum* bacteroids. *FEMS Microbiol. Lett.* **78**: 325–32.

Chance, B. (1957) Techniques for the assay of the respiratory enzymes. *Meth. Enzymol.* **4**: 273–329.

Chance, B. (1978) Cytochrome kinetics at low temperatures; trapping and ligand exchange. *Meth. Enzymol.* **54**: 102–11.

Chang, C.K., Timkovich, R. and Wu, W. (1986) Evidence that heme d_1 is a 1,3-porphyrindione. *Biochemistry* **25**: 8447–53.

Chea, K.S. (1970). The membrane-bound carbon monoxide-reactive hemoproteins in the extreme halophiles. *Biochim. Biophys. Acta* **197**: 84–6.

Chepuri, V., Lemieux, L., Au, D.C.T. and Gennis, R.B. (1990) The sequence of the *cyo* operon indicates substantial structural similarities between the cytochrome-*o* ubiquinol oxidase of *Escherichia coli* and the aa_3-type family of cytochrome-*c* oxidases. *J. Biol. Chem.* **265**: 11185–92.

Doi, M., Shioi, Y., Morita, M. and Takamiya, K. (1989) Two types of cytochrome cd_1 in the aerobic photosynthetic bacterium, *Erythrobacter* sp. Och-114. *Eur. J. Biochem.* **184**: 521–7.

Eccleston, J.F. (1987) Stopped-flow spectrophotometric techniques. In *Spectrophotometry and Spectrofluorimetry* (Harris, D.A. and Bashford, C.L., eds), pp. 137–64. IRL Press: Oxford.

Feresu, S.B. and Jones, D. (1988) Taxonomic studies on *Brochothrix*, *Erysipelothrix*, *Listeria* and atypical lactobacilli. *J. Gen. Microbiol.* **134**: 1165–83.

Fewson, C.A., Poole, R.K. and Thurston, C.F. (1984) Spectrophotometry in microbiology: Symbols and terminology. Scattered thoughts on opaque problems. *Soc. Gen. Microbiol. Quart.* **11**: 87–9.

Fox, G.E. and Stackebrandt, E. (1987) The application of 16S rRNA cataloguing and 5S rRNA sequencing in bacterial systematics. In *Methods in Microbiology*, vol. 19 (Colwell, R.R. and Grigorova, R. eds), pp. 405–58. Academic Press: London.

Fujiwara, T., Fukumori, Y. and Yamanaka, T. (1987) aa_3-Type cytochrome *c* oxidase occurs in *Halobacterium halobium* and its activity is inhibited by higher concentrations of salts. *Plant Cell Physiol.* **28**: 29–36.

Fukumori, Y., Fujiwara, T., Okada-Takahashi, Y., Mukohota, Y. and Yamanaka, T. (1985) Purification and properties of a hydroperoxidase from *Halobacterium halobium* L-33. *J. Biochem.* **98**: 1055–61.

Garcia-Horsman, J.A., Barquera, B. and Escamilla, J.E. (1991) Two different aa_3-type cytochromes can be purified from the bacterium *Bacillus cereus*. *Eur. J. Biochem.* **199**: 761–8.

Gartner, P. (1991) Characterization of a quinol-oxidase activity in crude extracts of *Thermoplasma acidophilum* and isolation of an 18-kDa cytochrome. *Eur. J. Biochem.* **200**: 215–22.

Gil, A., Kroll, R. and Poole, R.K. (1992) The cytochrome composition of the meat spoilage organism *Brochothrix thermosphacta*: Identification of cytochrome a_3- and *d*-type terminal oxidases under various conditions. *Archiv. Microbiol.* **158**: 226–33.

Gradin, C.H. and Colmsjö, A. (1987) Oxidation-reduction potentials and absorption spectra of two *b*-type cytochromes from the halophilic archaebacterium *Halobacterium halobium*. *Archiv. Biochem. Biophys.* **256**: 515–22.

Green, G.N. and Gennis, R.B. (1983) Isolation and characterization of an *Escherichia coli* mutant lacking cytochrome *d* terminal oxidase. *J. Bacteriol.* **154**: 1269–75.

Hamilton, W.A. (1988) Energy transduction in anaerobic bacteria. In *Bacterial Energy Transduction* (Anthony, C. ed), pp. 83–149. Academic Press: London.

Han, Y.-H., Smibert, R.M. and Krieg, N.R. (1992) Cytochrome composition and oxygen-dependent respiration-driven proton translocation in *Wolinella curva*, *Wolinella recta*, *Bacteroides ureolyticus*, and *Bacteroides gracilis*. *Can. J. Microbiol.* **38**: 104–10.

Harold, F.M. (1986) *The Vital Force: A Study of Bioenergetics*. W.H. Freeman: New York.

Harris, D.A. and Bashford, C.L. (eds) (1987) *Spectrophotometry and Spectrofluorimetry*. IRL Press: Oxford.

Hart, A., Murrell, J.C., Poole, R.K. and Norris, P.R. (1991) An acid-stable cytochrome in iron-oxidizing *Leptospirillum ferrooxidans*. *FEMS Microbiol. Lett.* **81**: 89–94.

Hartsel, S.C., Kolodziej, B.J. and Cassim, J.Y. (1988) Spectral evidence for cytochrome *o* in the brown membrane of *Halobacterium halobium*. *Archiv. Biochem. Biophys.* **264**: 74–81.

Hicks, D.B., Plass, R.J. and Quirk, P.G. (1991) Evidence for multiple terminal oxidases, including cytochrome *d*, in facultatively alkaliphilic *Bacillus firmus* OF4. *J. Bacteriol.* **173**: 5010–6.

Hughes, M.N. and Poole, R.K. (1989) *Metals and Micro-organisms*. Chapman and Hall: London.

Ingledew, W.J. and Poole, R.K. (1984) The respiratory chains of *Escherichia coli*. *Microbiol. Rev.* **48**: 222–71.

Ioannidis, I., Cooper, C. and Poole, R.K. (1992) Spectroscopic studies on an oxygen-binding hemoglobin-like flavohaemoprotein from *Escherichia coli*. *Biochem J.* **288**: 649–55.

Iwasaki, H. (1966) Lactate oxidation system in *Acetobacter suboxydans*, with special reference to carbon monoxide-binding pigment. *Plant Cell Physiol.* **7**: 199–216.

Jones, C.W. (1977) Aerobic respiratory systems in bacteria. In *Microbial Energetics* (Haddock, B.A. and Hamilton, W.A., eds), pp. 23–59. Cambridge University Press: Cambridge.

Jones, C.W. (1980) Cytochrome patterns in classification and identification including their relevance to the oxidase test. In *Microbiological Classification and Identification* (Goodfellow, M. and Board R.G., eds), pp. 127–38. Academic Press: New York.

Jones, C.W. and Poole, R.K. (1985) Analysis of cytochromes. In *Methods in Microbiology*, vol. 18, (Gottshalk, G., ed.), pp 285–328. Academic Press: London.

Kelly, M.J.S., Poole, R.K., Yates, M.G. and Kennedy, C. (1990) Cloning and mutagenesis of genes encoding the cytochrome *bd* complex in *Azotobacter vinelandii*. Mutants deficient in cytochrome *d* are unable to fix nitrogen in air. *J. Bacteriol.* **172**: 6010–9.

Kita, K., Konishi, K. and Anraku, Y. (1984) Terminal oxidases of *Escherichia coli* respiratory chain. I. Purification and properties of cytochrome b_{562}-*o* complex from cells in the early exponential phase of aerobic growth. *J. Biol. Chem.* **259**: 3368–74.

Kita, K., Konishi, K. and Anraku, Y. (1986) Purification and properties of two terminal oxidase complexes of *Escherichia coli* aerobic respiratory chain. *Meth. Enzymol.* **126**: 94–113.
Kranz, R.G. and Gennis, R.B. (1985) Immunological investigation of the distribution of cytochromes related to the two terminal oxidases of *Escherichia coli* in other Gram-negative bacteria. *J. Bacteriol.* **161**: 709–13.
Kühn, W. and Gottschalk, G. (1983) Characterization of the cytochromes occurring in *Methanosarcina* species. *Eur. J. Biochem.* **135**: 89–94.
Kühn, W., Fiebig, K., Walther, R. and Gottschalk, G. (1979) Presence of a cytochrome b_{559} in *Methanosarcina barkeri*. *FEBS Lett.* **105**: 271–4.
Kühn, W., Fiebig, K., Hippe, H., Mah, R.A., Huser, B.A. and Gottschalk, G. (1983). Distribution of cytochromes in methanogenic bacteria. *FEMS Microbiol. Letts.* **20**: 407–10.
Lin, E.C.C. and Kuritzkes, D.R. (1984) Pathways for anaerobic electron transport. In *Escherichia coli and Salmonella typhimurium. Cellular and Molecular Biology*, vol. 1 (Neidhardt, F.C., ed.-in-chief), pp. 201–21. American Society for Microbiology: Washington DC.
Lloyd, D. (1974) *The Mitochondria of Microorganisms*. Academic Press: London.
Lloyd, D. and Edwards, S.W. (1978) Electron transport pathways alternative to the main phosphorylating respiratory chain. In *Functions of Alternative Terminal Oxidases* (Degn, H., Lloyd, D., Hill, G.C., eds), pp. 1–10. Pergamon Press: Oxford.
Lloyd, D. and Scott, R.I. (1987) The determination of photochemical action spectra. In *Spectrophotometry and Spectrofluorimetry* (Harris, D.A. and Bashford, C.L., eds), pp. 165–72. IRL Press: Oxford.
Ludwig, B. and Schatz, G. (1980) A two-subunit cytochrome c oxidase (cytochrome aa_3) from *Paracoccus denitrificans*. *Proc. Natl. Acad. Sci. USA* **77**: 196–200.
Markwell, M.A.K., Haas, S.M., Bieber, L.L. and Tolbert, N.G. (1978) A modification of the Lowry procedure to simplify protein determination in membrane and lipoprotein samples. *Anal. Biochem.* **87**: 206–10.
Matsushita, K., Shinagawa, E., Adachi, O. and Ameyama, M. (1990) Cytochrome a_1 of *Acetobacter aceti* is a cytochrome ba functioning as ubiquinol oxidase. *Proc. Natl. Acad. Sci. USA* **87**: 9863–7.
Miller, M.J. and Gennis, R.B. (1986) Purification and reconstitution of the cytochrome d terminal oxidase complex from *Escherichia coli*. *Meth. Enzymol.* **126**: 87–94.
Nies, D. and Schlegel, H.G. (1982) Catalase from *Comamonas compransoris*. *J. Gen. Appl. Microbiol.* **28**: 311–19.
Nozaki, M., Yamamoto, S., Ishimura, Y., Coon, M.J., Ernster, L. and Estabrook, R.W. (eds) (1982) *Oxygenases and Oxygen Metabolism*. Academic Press: New York.
Palmer, G. and Reedijk, J. (1992) Nomenclature of electron-transfer proteins. Recommendations 1989. *J. Biol. Chem.* **267**: 665–77.
Pettigrew, G.W. and Moore, G.R. (1987) *Cytochromes c. Biological Aspects*. Springer-Verlag: Berlin.
Poole, R.K. (1983) Bacterial cytochrome oxidases: A structurally and functionally diverse group of electron transfer proteins. *Biochim. Biophys. Acta* **726**: 205–43.
Poole, R.K. (1988) Bacterial cytochrome oxidases. In *Bacterial Energy Transduction* (Anthony, C., ed), pp. 231–91. Academic Press: London.
Poole, R.K. and Bashford, C.L. (1987) Spectra. In *Spectrophotometry and Spectrofluorimetry* (Harris, D.A., and Bashford, C.L., eds), pp. 23–48. IRL Press: Oxford.
Poole, R.K. and Williams, H.D. (1988) Formation of the 680 nm-absorbing form of the cytochrome bd oxidase complex of *Escherichia coli* by reaction of hydrogen peroxide with the oxidized form. *FEBS Lett* **243**: 243–6.
Poole, R.K., Kumar, C., Salmon, I. and Chance, B. (1983). The 650 nm chromophore in *Escherichia coli* is an "oxy"- or oxygenated compound, not the oxidized form of cytochrome oxidase d: A hypothesis. *J. Gen. Microbiol.* **129**: 1335–44.
Poole, R.K., Baines, B.S. and Williams, H.D. (1985) Sensor sensationalism? Alternative views on the nature and role of "cytochrome a_1" in bacteria. *Microbiol. Sci.* **2**: 21–3.
Poole, R.K., Baines, B.S. and Appleby, C.A. (1986) Haemoprotein b-590 (*Escherichia coli*), a reducible catalase and peroxidase: Evidence for its close relationship to hydroperoxidase I and a "cytochrome a_1b" preparation. *J. Gen. Microbiol.* **132**: 1525–39.

Poole, R.K., Williams, H.D., Downie, J.A. and Gibson, F. (1989) Mutations affecting the cytochrome d-containing oxidase complex of Escherichia coli K12: Identification and mapping of a fourth locus, cydD. *J. Gen. Microbiol.* **135**: 1865–74.

Probst, I., Wolf, G. and Schlegel, H.G. (1979) An oxygen-binding flavohaemoprotein from Alcaligenes eutrophus. *Biochim. Biophys. Acta* **576**: 471–8.

Puustinen, A. and Wikstrom, M. (1991) The heme groups of cytochrome o from Escherichia coli. *Proc. Natl. Acad. Sci. USA* **88**: 6122–6.

Puustinen, A., Morgan, J.E., Verkhovsky, M., Thomas, J.W., Gennis, R.B. and Wikstrom, M. (1992) The low-spin heme site of cytochrome o from Escherichia coli is promiscuous with respect to heme type. *Biochemistry* **31**: 10363–9.

Saraste, M., Holm, L., Lemieux, L., Luben, M. and van der Oost, J. (1991) The happy family of cytochrome oxidases. *Biochem. Soc. Trans.* **19**: 608–12.

Stackebrandt, E. (1991) Unifying phylogeny and phenotypic diversity. In *The Prokaryotes* (Balows, A., Trüper, H.G., Dworkin, M., Harder, W. and Schleifer, K.H., eds), pp. 19–47. Springer-Verlag: New York.

Timkovich, R., Cork, M.S., Gennis, R.B. and Johnson, P.Y. (1985) Proposed structure of heme d, a prosthetic group of bacterial terminal oxidases. *J. Am. Chem. Soc.* **107**: 6069–75.

Vasudevan, S.G., Armarego, W.L.F., Shaw, D.C., Lilley, P.E., Dixon, N.E. and Poole, R.K. (1991) Isolation and nucleotide sequence of the hmp gene that encodes a haemoglobin-like protein in Escherichia coli. *Mol. Genet.* **226**: 49–58.

Vincent, J.-C., Kumar, C. and Chance, B. (1982) Quantitative visible spectroscopy at low temperatures: A systematic examination. *Anal. Biochem.* **126**: 86–93.

von Wachenfeldt, C. and Hederstedt, L. (1992) Molecular biology of Bacillus subtilis cytochromes. *FEMS Microbiol. Rev.* **100**: 91–100.

Webster, D.A. (1987) Structure and function of bacterial haemoglobin and related proteins. In *Advances in Inorganic Biochemistry*, vol. 7, *Heme Proteins* (Eichhorn, G.L. and Marzilli, C.G. eds), pp. 245–65. Elsevier: New York.

Williams, H.D. and Poole, R.K. (1987) The cytochromes of Acetobacter pasteurianus NCIB 6428. Evidence of a role for a cytochrome a_1-like haemoprotein in electron transfer to cytochrome oxidase d. *J. Gen. Microbiol.* **133**: 2461–72.

Wilson, D.F. (1967) Effect of temperature on the spectral properties of some ferrocytochromes. *Archiv. Biochem. Biophys.* **121**: 757–68

Wood, P.M. (1984) Bacterial proteins with CO-binding b- or c- type haem. Functions and absorption spectroscopy. *Biochim. Biophys. Acta* **768**: 293–317.

Wood, P.M. (1988) Chemolithotrophy. In *Bacterial Energy Transduction* (Anthony, C., ed), pp. 183–230. Academic Press: London.

Yamanaka, T. (1992) *The Biochemistry of Bacterial Cytochromes*. Japan Scientific Societies Press: Tokyo.

Zhu, H. and Riggs, A.F. (1992) Yeast flavohemoglobin is an ancient protein related to globins and a reductase family. *Proc. Natl. Acad. Sci. USA* **89**: 5015–9.

10
Analysis of Pigments: Bacteriochlorophylls

William R. Richards
Simon Fraser University, British Columbia, Canada

10.1 INTRODUCTION

There are two major classes of photosynthetic prokaryotes: those which carry out oxygenic (oxygen-evolving) photosynthesis and have two photosystems related to higher plants, and those which carry out anoxygenic photosynthesis and have single photosystems. Both groups contain carotenoids and tetrapyrrole pigments involved in solar energy conversion. Oxygenic photosynthetic prokaryotes include the cyanobacteria and the prochlorophytes; the tetrapyrrole pigments found in the former are chlorophyll *a* and the phycobilins, while chlorophylls *a* and *b* are found in the latter. Anoxygenic photosynthetic prokaryotes include the purple, green and brown photosynthetic bacteria; the magnesium tetrapyrrole pigments found in all species are referred to as bacteriochlorophylls (abbreviated Bchls). This chapter will be concerned primarily with the photosynthetic prokaryotes in the latter category. A variety of books and chapters have dealt with various aspects of the biology and biochemistry of the photosynthetic prokaryotes and their chlorophylls; among them are those by Clayton and Sistrom (1978), Carr and Whitton (1982), Ormerod (1983), Tindall and Grant (1986), Imhoff (1988), Olson *et al.* (1988), Harashima *et al.* (1989), Lewin and Cheng (1989), Drews and Dawes (1990), Amesz (1991), Drews and Imhoff (1991), Scheer (1991), Smith (1991) and Stolz (1991).

Figure 10.1 shows the structures of the known Bchls. These may be divided into three subgroups, as outlined below.

10.1.1 Bacteriochlorophyll *a*

Bacteriochlorophyll *a* is a magnesium tetrahydroporphyrin (also called a bacteriochlorin) structurally related to 2-acetyl-2-desvinylchlorophyll *a* by

---

Chemical Methods in Prokaryotic Systematics. Edited by M. Goodfellow and A.G. O'Donnell
© 1994 John Wiley and Sons Ltd.

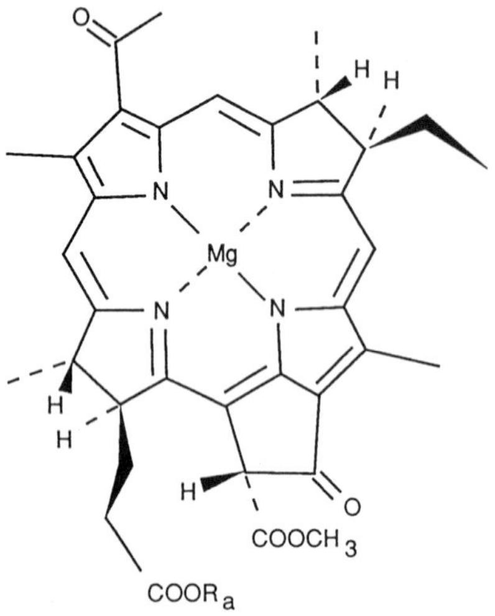

Bacteriochlorophyll *a*

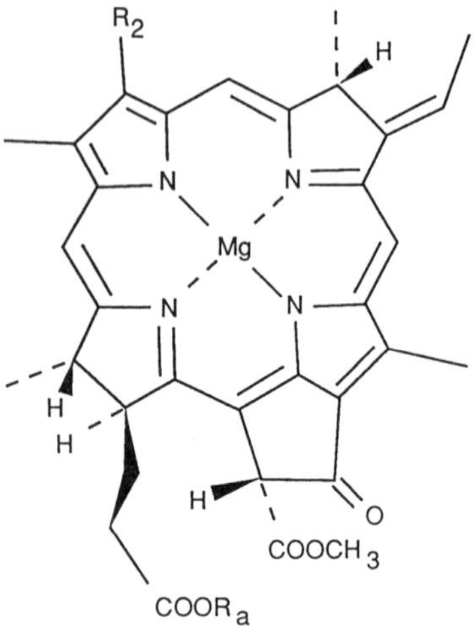

Bacteriochlorophyll *b* : $R_2 = -\overset{O}{\overset{\|}{C}}CH_3$

Bacteriochlorophyll *g* : $R_2 = -CH{=}CH_2$

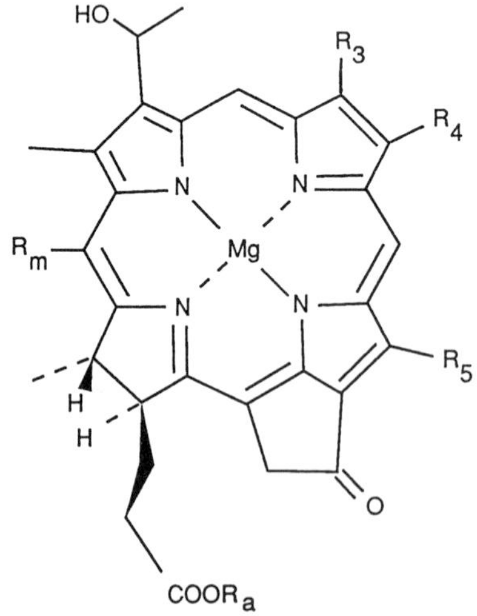

Bchl	$-R_3$	$-R_4$	$-R_5$	$-R_m$
c	$-CH_3$	$-CH_2CH_3$	$-CH_3$	$-CH_3$
	$-CH_3$	$-CH_2CH_3$, $-CH_2CH_2CH_3$, $-CH_2CH(CH_3)_2$	$-CH_2CH_3$	$-CH_3$
d	$-CH_3$	$-CH_2CH_3$, $-CH_2CH_2CH_3$, $-CH_2CH(CH_3)_2$, $-CH_2C(CH_3)_3$	$-CH_3$ or $-CH_2CH_3$	$-H$
e	$-CH{=}O$	$-CH_2CH_3$, $-CH_2CH_2CH_3$, $-CH_2CH(CH_3)_2$	$-CH_2CH_3$	$-CH_3$

Bacteriochlorophylls *c*, *d*, and *e*

Figure 10.1 Structures of the bacteriochlorophylls. The numbering system of Fischer & Stern (1940) is used throughout: pyrrole ring I is in the upper lefthand corner (containing positions 1 and 2), and the α-methine position is at the top, with the numbering and lettering progressing clockwise. R_a refers to the esterifying alcohol, which depends upon the nature of the bacteriochlorophyll and the species in which it is found.

a reduction of pyrrole ring II. It is the most widely distributed of all the Bchls, and is found in all but a handful of species of anoxygenic photosynthetic bacteria (those containing either Bchl *b* or *g*). In all species of purple bacteria in which it is found, it is the sole Bchl. In green and brown bacteria, however, it is found in conjunction with much larger amounts of Bchl *c*, *d* or *e*. Crystalline Bchl *a* is stable indefinitely in the dark; neutral or slightly acidic solutions are also fairly stable in the dark, as are deoxygenated solutions in the light (Lindsay Smith & Calvin, 1966), particularly when in concentrated solutions (Mauzerall, 1978). However, Bchl *a* undergoes a base-catalysed auto-oxidation in the dark sometimes referred to as 'allomerization'. This reaction has been well-characterized in the case of chlorophyll *a* (Seely, 1966), and an allomerization product of Bchl *a* has been identified (Brereton et al., 1980). Basic conditions can also lead to epimerization of C10 (containing the carbomethoxy substituent) via enolate formation with the 9-keto group (Seely, 1966; Katz et al., 1968). Actually, all five of the chiral carbons of Bchl *a* can epimerize to some extent. Scholz and Ballschmiter (1982) have observed at least seven diastereomers of Bchl *a* during its purification, and the extent of epimerization was greatly increased by heat. The reactivity of Bchl *a* towards oxygen is greatly increased in the light (photo-oxidation), where Bchl *a* undergoes oxidation to chlorins; however, 2-acetyl-2-desvinylchlorophyll *a* is only a minor product (Lindsay Smith & Calvin, 1966). Brereton and Sanders (1983b) have also demonstrated that Bchl *a* is much more unstable towards photo-oxidation when it is dissolved in solvents in which it is 5-coordinate (acetone, ether and tetrahydrofuran) than in solvents in which it is 6-coordinate (methanol and pyridine). Bacteriochlorophyll *a* is more unstable towards light than is its magnesium-free bacteriopheophytin (abbreviated Bphy) (Mauzerall, 1978). Bacteriochlorophyll *a* is rather more stable than chlorophyll *a* towards loss of magnesium (or 'pheophytination') by acid, the order of decreasing stability of the magnesium chelates of tetrapyrroles being: tetrahydroporphyrin > chlorin > porphyrin (Mauzerall, 1978).

10.1.2 Bacteriochlorophylls *b* and *g*

Bacteriochlorophylls *b* and *g* are isomeric with magnesium dihydroporphyrins derived by isomerization of the 3,4-double bond in pyrrole ring II to a 4-ethylidene group. Hence, Bchl *b* is isomeric with 2-acetyl-2-desvinylchlorophyll *a*, and Bchl *g* is isomeric with chlorophyll *a* (with the exception that they may possess different esterifying alcohols on the 7-propionic acid). Bacteriochlorophylls *b* and *g* are each found in a relatively small number of species of photosynthetic bacteria, where they exist as the sole Bchl. Bacteriochlorophyll *b* was first isolated from *Rhodopseudomonas viridis* by Eimhjellen et al. (1963), and Bchl *g* from *Heliobacterium chlorum* by

Gest and Flavinger (1983). Both are much more unstable than Bchl *a* towards photodecomposition to chlorins, due to the presence of the 4-ethylidene in both. After purification of Bchl *b* by paper chromatography and elution by acetone, Eimhjellen *et al.* (1963) found that it was rapidly converted to a compound spectroscopically identical to 2-acetyl-2-desvinylchlorophyll *a*. Steiner *et al.* (1983) have found, however, that the latter is formed by photo-isomerization only under strictly anaerobic conditions, whereas the major products in the presence of oxygen were the 4-acetyl and 4-α-methoxyethyl derivatives of 2-acetyl-2-desvinyl-4-desethylchlorophyll *a*. Exposure of Bchl *g* to light caused its complete transformation within 3 min to a compound spectroscopically identical to chlorophyll *a* (Michalski *et al.*, 1987). However, Brockmann and Lipinski (1983) had earlier identified the decomposition products of Bchl *g* as the 4-acetyl and 4-α-hydroxyethyl derivatives of 4-desethylchlorophyll *a*. Because of its extreme sensitivity, all operations during the isolation and purification of Bchl *g* must be conducted in the dark with rigorous exclusion of air. It is indeed remarkable how such photosensitive compounds as Bchls *b* and *g* are able to carry out their photosynthetic functions in these bacteria, which are somehow able to protect them by their particular *in vivo* protein environments. Removal of magnesium from either Bchl *b* or *g* greatly increases its stability towards light. Photo-isomerization of Bphy *g* to pheophytin *a* occurred with a half-time of 42 min, while Bphy *g* was stable in the dark for several days even in the presence of oxygen (Michalski *et al.*, 1987). Similarly, methyl bacteriopheophorbide *b* is approximately 10 times more stable in the light than Bchl *b* (Denning & Richards, unpublished).

10.1.3 Bacteriochlorophylls *c*, *d* and *e*

Bacteriochlorophylls *c*, *d* and *e* are magnesium dihydroporphyrins (chlorins) found only in the green and brown bacteria. Although they are the predominant Bchls in the organisms in which they are found, Bchls *c*, *d* and *e* have only light-harvesting (LH) functions. All of these bacteria also contain Bchl *a*, which has not only an LH function, but is also found in the reaction centre (RC) of these organisms. In addition to having different esterifying alcohols on the 7-propionic acid, Bchls *c*, *d* and *e* all differ from chlorophyll *a* by having an α-hydroxyethyl (rather than a vinyl) group in the 2-position, and having no carbomethoxy group on C10 (i.e. they are derivatives of pyrochlorophyll *a*). In fact, each is not a single chlorophyll at all but a series of homologues with different alkyl substituents in the 4- (either ethyl, *n*-propyl, isobutyl or neopentyl) and/or 5- (either methyl or ethyl) positions (cf. Figure 10.1). The Bchl *c* homologues all differ from the Bchl *d* homologues by containing a methyl substituent on the δ-methine position. The Bchl *e* homologues are all 3-formyl derivatives of the Bchl *c* homologues. In this respect, Bchl *e* bears the same relationship to Bchl *c* that chlorophyll *b* does to chlorophyll *a*. The designation 'Bchl *f*' is reserved for 3-formyl derivatives of Bchl *d*, should a species containing such a pigment ever be found (Gloe *et al.*, 1975). Bacteriochlorophylls *c*, *d* and *e* are much more stable towards oxidation than Bchls *a*, *b* and *g* because the former are chlorins, not tetrahydroporphyrins (as in Bchl *a*), and do not contain an ethylidene substituent (as in Bchls *b* and *g*). However, Bchl *c* can be photo-oxidized in the presence of oxygen to a linear bilitriene (Kenner *et al.*, 1976), as can its magnesium-free derivative (Brown *et al.*, 1980). Since they possess no 10-carbomethoxy group, they are not subject to base-catalysed allomerization (reaction with oxygen) at C10. This means that the esterifying alcohol on the 7-propionic acid of Bchls *c*, *d* and *e* can be removed by base-catalysed hydrolysis without loss of magnesium, yielding the corresponding bacteriochlorophyllide. If sodium methoxide in methanol is used, methyl bacteriochlorophyllide is formed. Since the central magnesium atom is more unstable in the presence of dilute acids than is that of Bchl *a*, much of the purification work has been done on magnesium-free derivatives, either the bacteriopheophorbide (abbreviated Bpho), or the methyl bacteriopheophorbide (abbreviated Bmph), formed by the action of concentrated aqueous acids or 5% sulphuric acid in methanol, respectively, on the Bchl.

10.2 ANALYTICAL METHODS

The isolation and purification of Bchls *a*, *b* and *g* is simplified by the fact that they are the sole chlorophylls present in the organisms that possess them, but is complicated by their instabilities. Hence, they must be separated not only from carotenoids, quinones and other colourless lipophilic substances, but also from their many decomposition products. On the other hand, while Bchls *c*, *d* and *e* are somewhat more stable, their purification is greatly complicated by the fact that each Bchl is actually a mixture of homologues with different alkyl substituents at the 4- and 5-positions, each of which may also exist as a pair of diastereomeric isomers with different chiralities of the 2-α-hydroxyethyl group, and/or have different esterifying alcohols on the 7-propionic acid. Their separation into purified fractions has, therefore, proved to be extremely difficult.

10.2.1 Isolation and estimation

Bacteriochlorophylls are easily extracted from bacteria by a variety of different organic solvents (Svec, 1991). The centrifuged (wet-packed) bacteria may be extracted immediately, or they may be frozen or freeze-dried and kept for future extraction. Some of the solvent systems which have been used for the extraction of all of the known Bchls are given in Table 10.1. When diethyl ether and dioxane are used, care must be taken to remove peroxides (e.g. by distillation from $FeSO_4$). Small samples can be extracted in hand-held glass homogenizers. Larger samples, however, may require extraction in explosion-proof, water-cooled blenders. Addition of an antioxidant is a useful precaution to employ, particularly with large-scale extractions. For the very labile Bchls *b* and *g*, exposure to heat, light and oxygen must be kept to a minimum. Brockmann and Lipinski (1983) allowed a frozen cell paste of *Heliobacterium chlorum* to warm up to 20 °C under pyridine in the dark for the extraction of Bchl *g*. For the Bchls that are more sensitive to acid, but less sensitive to bases (Bchls *c*, *d* and *e*), extracts can be treated with dilute solutions of a weak base (e.g. 0.01 M K_2HPO_4 or $MgCO_3$) to neutralize extracted organic acids (Mathewson *et al.*, 1963a).

An improved method for the preparation of Bchl *a* has been reported by Brereton and Sanders (1983b). This method used a procedure developed by Iriyama *et al.* (1974) in which the chlorophyll is precipitated as a coordination polymer with dioxane. Following chromatography, Brereton and Sanders (1983b) were able to obtain reproducibly high yields of indefinitely stable microcrystalline Bchl *a*. Dioxane precipitation has also been used

Table 10.1 Some solvent systems used to extract bacteriochlorophylls.

Solvent system	v/v	Bchl	Reference
Methanol	–	a	Smith and Benitez (1955)
Acetone/methanol	7:2	a	Cohen-Bazire *et al.* (1957)
Methanol saturated with H_2S	–	a	Kaplan and Silberman (1959)
Methanol/ether/petroleum ether	5:2:1	a,c	Strain and Svec (1966)
Isopropanol	–	a,b	Scholz and Ballschmiter (1982)
Methanol saturated with H_2S	–	b	Eimhjellen *et al.* (1963)
Methanol/ether/petroleum ether	5:2:1	b	Scheer *et al.* (1974)
Acetone/0.01 M K_2HPO_4	4:1	c	Mathewson *et al.* (1963a)
Methanol/tetrahydrofuran	9:1	d	Hughes and Holt (1962)
Methanol/water	4:1	c,d	Stanier and Smith (1960)
Acetone or acetone/CCl_4	27:5	c,d,e	Gloe *et al.* (1975)
Methanol	–	e	Brockmann (1976)
Pyridine	–	g	Brockmann and Lipinski (1983)
Methanol/water/pyridine	95:4:1	g	Michalski *et al.* (1987)
Ethanol/ether/hexane	5:2:1	g	Michalski *et al.* (1987)

by Scholz and Ballschmiter (1981a) during the preparation of single diastereomers of Bchl *a* (Scholz & Ballschmiter, 1981b, 1982).

Bacteriochlorophylls *a, c* and *d* may be estimated from their extinction coefficients in a variety of solvents (Table 10.2). Table 10.2 also summarizes the *in vivo* maxima commonly exhibited by all of the Bchls. Since extraction is seldom quantitative, and if estimation is all that is required, Bchl *a* may be estimated *in situ*. This procedure suffers from the fact that Bchl *a* exists in more than one spectral form (cf. Table 10.2), and the relative amounts of these forms vary with the species under examination as well as with the growth conditions employed (Drews, 1985). Nevertheless, several methods of *in situ* estimation have been developed for whole cells (Sojka *et al.*, 1970) and chromatophores (Sistrom, 1964; Clayton, 1966; Neufang *et al.*, 1982) of a variety of purple bacteria. The study by Neufang *et al.* (1982) used the *in vivo* Bchl *a* maximum at 375 nm rather than the far-red maxima used in the earlier studies.

10.2.2 Purification by conventional methods

Techniques for the separation of a variety of plant and bacterial chlorophylls by column, paper and thin-layer chromatography (TLC) have been reviewed by Strain and Svec (1975), Sestak (1980, 1982), Cavaleiro and Smith (1986), Brockmann and Risch (1991), and Shioi (1991). Paper and TLC techniques are more useful for analysing the purity of Bchl samples than for purification per se; however, preparative thick-layer chromatography may be carried out for large-scale purifications. Polar TLC absorbants such as silica gel and alumina often lead to loss of magnesium from Bchls, while absorbants such as sucrose, cellulose (which gives separations similar to paper chromatography), and reversed-phase silica or powdered polyethylene are less likely to cause decomposition. Reversed-phase silica TLC was used to separate Bchls *a* and *b* with different esterifying alcohols (Scholz & Ballschmiter, 1982). Silica gel TLC has been used, however, to distinguish Bphys (or Bmphs) *d* and *e* (Gloe *et al.*, 1975), and Brockmann *et al.* (1976) were able to resolve Bmph *e* into its homologues by silica gel TLC.

The same precautions observed with TLC with respect to the purification of the Bchls on polar absorbants must also be observed with column chromatography. Tswett was the first to use powdered sugar (sucrose) for the separation of plant chlorophylls and carotenoids in 1906. In fact, the word 'chromatography' was coined for the technique as a result of his work. Even though sucrose was the first chromatographic absorbant ever used, it is still one of the most effective because it leads to very little decomposition. Powdered sucrose columns have been especially effective with the more labile Bchls, e.g. Bchl *a* (Sauer *et al.*, 1966; Strain & Svec, 1966;

Table 10.2 *In vivo* and organic solvent absorption maxima for bacteriochlorophylls.

Bchl	*In vivo* absorption Maxima (nm)	Ref[a]	Solvent	λ_{max} (nm)	ε_{mM} ([b])	λ_{max} (nm)	ε_{mM} ([b])	Ref[a]
a	800, 820, 850, 870–890	1	CCl$_4$	780	(88)	361.5	(61.7)	6
			Ethanol	773	(62)	365.5	(58.5)	6
			Ether	770	(96)	357	(73.5)	6
			Acetone	770	(69)	358	(66)	6
			Dioxane	772	(115)	—	—	7
			Acetone/methanol (7:2)	767	(76)	—	—	7
b	1020	2	Ether	792	—	—	—	7
			Acetone	794	—	407, 368	—	8
			Isopropanol	800	—	400, 370	—	9
c	745–760	2	Ether	660	(91)	432	(141)	10
	736–749	3	Acetone	662.5	(74)	433	(115)	10
			Methanol	670	(69)	435	(79)	10
d	725–745	2	Ether	650	(90)	425	(115)	10
	725–736	4[c]	Acetone	654	(78)	427	(100)	10
			Methanol	659	(65)	427	(66)	10
e	715–725	2	Acetone	646–648	—	456–459	—	2
g	778, 793, 808	5	Ether	766	—	405, 365	—	11

[a] References: 1, Drews (1985); 2, Gloe et al. (1975); 3, Gloe and Risch (1978); 4, Smith and Goff (1985); 5, van Dorssen et al. (1985); 6, Sauer et al. (1966); 7, Clayton (1966); 8, Eimhjellen et al. (1963); 9, Scholz and Ballschmiter (1982); 10, Stanier and Smith (1960); 11, Michalski et al. (1987).
[b] Units are mM^{-1} cm^{-1}.
[c] In one newly isolated strain, the *in vivo* maximum for Bchl *d* was as low as 714 nm (cf. Smith & Goff, 1985).

Brereton & Sanders, 1983b), Bchl *b* (Scheer *et al.*, 1974), and Bchl *g* (Michalski *et al.*, 1987), although silica gel has also been used to separate Bphy *g* from its decomposition products (Brockmann & Lipinski, 1983). Other carbohydrate-based absorbants to have been used with Bchl *a* include DEAE-cellulose and Sephadex LH-20 (Sato & Murata, 1978), Sephasorb HP ultrafine (Scholz & Ballschmiter, 1981a, 1982), and Sepharose CL-6B (Omata & Murata, 1983).

Reversed-phase chromatography in powdered polyethylene columns was first used for chlorophylls *a* and *b* by Anderson and Calvin (1962). The polyethylene must first be washed with acetone or methanol before employing aqueous mixtures of either solvent for elution. Hence, the chromatography is probably a partition between a stationary organic (acetone or methanol) phase held by the polyethylene, and a moving aqueous acetone or methanol phase. This method is particularly effective when used in conjunction with a polar column such as powdered sucrose because the order of elution of chlorophylls and carotenoids is reversed. For example, Bchl *a* was first purified by chromatography on sucrose with 0–20% (v/v) ether in petroleum ether, followed by chromatography on powdered polyethylene with 80–90% (v/v) aqueous methanol (Mathewson *et al.*, 1963b; Richards, 1966). Powdered polyethylene columns have also proved very effective for the purification of polycarboxylated porphyrins (Richards & Rapoport, 1966) and monocarboxylated intermediates of Bchl *a* biosynthesis (Richards & Lascelles, 1969; Pudek & Richards, 1975).

Bacteriochlorophylls *c* and *d* were first separated from carotenoids and other lipophilic substances by chromatography on powdered sucrose (Goodwin, 1955; Stanier & Smith, 1960) or cellulose (Kaplan & Silberman, 1959). A similar removal of impurities from Bchl *c* can be obtained by a precipitation and washing procedure employing petroleum ether and *iso*-octane (Mathewson *et al.*, 1963a; Richards, 1966). Ordinary column chromatography on sucrose or alumina failed to separate Bchl *c* or Bpho *c* into their homologous components, while chromatography on powdered polyethylene gave only partial resolution (Richards, 1966). Risch *et al.* (1978) have, however, described a separation of Bmph *e* homologues on silica gel columns. Holt and co-workers used liquid-liquid partition between aqueous HCl and ether in powdered Celite columns to effect separation of Bphos *d* (Hughes & Holt, 1962) and *c* (Holt *et al.*, 1962, 1966) into six fractions each. This method of column chromatography is, however, quite cumbersome and inefficient (Chapman *et al.*, 1971); an alternative method of resolution is countercurrent distribution (CCD) of Bphos *d* (Hughes & Holt, 1962) and *c* (Richards & Rapoport, 1967). The latter authors used 2.2 M HCl-ether for separation of the two major components; faster running minor components could be resolved by a second CCD run in 3.0 M HCl-ether, and slower running minor components by a second CCD run in 1.8 M HCl-ether (Richards, 1966). Countercurrent distribution has also proved effective for separation of polycarboxylated porphyrins (Richards & Rapoport, 1966) and intermediates of Bchl *a* biosynthesis (Pudek & Richards, 1975).

10.2.3 Purification by HPLC

High-performance liquid chromatography (HPLC) has proven extremely effective for resolution of the multiple components of many Bchls (Shioi, 1991). HPLC with silica-based columns was first used in 1975 to separate porphyrin and chlorophyll derivatives (Evans *et al.*, 1975), and many research groups have contributed to the improvement of HPLC techniques over the past 15 years. Reversed-phase alkylsilica columns were used to separate pheophytin *a* with different esterifying alcohols (Schoch *et al.*, 1978), and chlorophylls *a* and *b* (together with several of their derivatives) (Burke & Aronoff, 1981). Octadecylsilica (ODS) columns used in the presence of ion-pairing reagents (Fuesler *et al.*, 1982) have allowed the separation of closely related unesterified chlorophyll precursors, including the monovinyl and divinyl derivatives of both protochlorophyllide and chlorophyllide (Hanamoto & Castelfranco, 1983). Shioi and co-workers have used ODS columns with 100% methanol (Shioi *et al.*, 1983) in the presence of ion-suppressing reagents (Shioi *et al.*, 1984) to achieve the separation of chlorophylls *a* and *b*, protochlorophyll *a*, and Bchl *a* species with a variety of esterifying alcohols. Reversed-phase ODS columns have also been used for the separation of phytol and geranylgeraniol esters of Bchl *a* into single diastereomers (Scholz & Ballschmiter, 1981b), for the resolution of Bchl *b* containing different esterifying alcohols (Steiner *et al.*, 1981, 1982), and for the purification of Bchl *g* (Michalski *et al.*, 1987).

Smith and co-workers have used C18 Bondapak ODS columns to resolve Bmph *c* into its components. Having already removed the naturally occurring esterifying alcohols, they were able to resolve the methyl esters into six fractions, including (during recycling of the HPLC eluate) separation of diastereomeric isomers differing in the stereochemistry of the 2-α-hydroxyethyl substituent (Smith *et al.*, 1980a,c,d, 1983a). For Bmph *d*, Smith and co-workers combined a preliminary separation by preparative thick-layer silica gel chromatography with reversed-phase HPLC to resolve the methyl esters into eight fractions (Smith *et al.*, 1982, 1983b; Smith & Goff, 1985). Reversed-phase HPLC has also been used for the separation of Bchl *c* homologues from *Chloroflexus aurantiacus* (Brune *et al.*, 1987). Finally, Smith and Simpson (1986) resolved Bmph *e* into three homologues by HPLC, but were unable to further resolve them into diastereomers without first chemically converting them to the corresponding Bmph

10. Analysis of Pigments: Bacteriochlorophylls

c homologues. Cavaleiro and Smith (1986) have reviewed the HPLC methods used with the Bchls of green and brown bacteria.

Strouse and co-workers were first to report the use of powdered polyethylene in giant (up to 3.5 in diameter × 9 ft length!) preparative HPLC columns for the separation of Bchl c components (Chow et al., 1978). Using these columns, they were able to resolve Bchl c into a total of 14 different components, consisting of four different Bchl c homologues, two of which were each esterified with six different alcohols (Caple et al., 1978). Shioi and Beale (1987) have recently adapted powdered polyethylene to much smaller columns to effect the separation of a whole range of esterified and unesterified chlorophylls and chlorophyll derivatives, including Bchl a esterified with four different alcohols. These workers used a simple isocratic acetone-water elution system with a range of acetone concentrations from only 60 to 82% (v/v) to effect separations of their diverse collection of tetrapyrroles. They found that the resolution was increased upon lowering the temperature to 15–20 °C. Using fluorometry as their detection system, Shioi and Beale (1987) were able to detect as little as 100 fmol or as much as 100 pmol on a single column. Different tetrapyrroles required different excitation and emission maxima; hence, variable excitation and emission settings on the fluorescence monitor were essential. Polyethylene was found to be much more chemically stable and inert than ODS and, since it did not irreversibly bind the added tetrapyrroles, resulted in their quantitative recovery (Shioi & Beale, 1987). Also, the preparation of the elution system was much less complex and time-consuming than for ODS columns using either ion-suppression or ion-pairing methods. Polyethylene columns are not commercially available, but powdered polyethylene can be hand-packed into empty columns. Since Chow et al. (1978) report that polyethylene is ~ 1000 times less expensive than ODS, its advantages are obvious if it is desired to scale up separations to preparative- or semipreparative-sized columns.

10.2.4 Characterization of bacteriochlorophylls

Determining the oxidation state and functional groups of tetrapyrroles is greatly facilitated by ^{1}H-NMR (Scheer & Katz, 1975) and visible spectroscopy (Smith, 1975; Weiss, 1978). Porphyrins, chlorins, tetrahydroporphyrins, and their metallo-derivatives have very characteristic visible spectra which are affected in predictable ways by the presence of conjugated functional groups such as vinyl, formyl and acetyl. A great deal of information can be learned from the spectra and chromatographic behaviour of unknown tetrapyrroles by comparing them with derivatives of known chlorophylls formed by carrying out reactions known to alter the side-chains or oxidation state of the porphyrin ring in a characteristic manner (Seely, 1966; Fuhrhop & Smith, 1975; Fuhrhop, 1978). An example of the utilization of this technique occurred quite early during the characterization of the structure of Bchl *a*. As more Bchls were subsequently discovered, their structures were also initially related to the structures of chlorophylls *a* and *b* and Bchl *a*. Although none of the newly discovered Bchls has proved to have a structure radically different from the previously characterized chlorophylls, a few challenges have been presented to the natural product chemists who studied their structures. These included:

(a) Separation of the homologues of Bchls *c*, *d* and *e*, and determination of the alkyl substituents in their 4- and 5-positions.
(b) Determination of the nature and location of the *meso*-alkyl substituent in Bchls *c* and *e*.
(c) Determination of the chirality of the 2a-carbon in the 2-α-hydroxyethyl substituent of Bchls *c*, *d* and *e*.
(d) Determination of the 4-ethylidene substituent in Bchls *b* and *g*.
(e) Characterization of the variety of esterifying alcohols on the 7-propionic acid of many of the Bchls.

10.2.4.a Bacteriochlorophyll a

When Bchl *a* was first studied, it was found that it was chemically oxidized to a green compound (initially called 'bacterioviridin') spectroscopically identical to 2-acetyl-2-desvinylchlorophyll *a*, which could be prepared from chlorophyll *a*. Hence, the basic structure of Bchl *a* was correctly assigned as early as 1940 by Hans Fischer and his co-workers (Fischer & Stern, 1940). The location of the two 'extra' hydrogens was later confirmed by Golden *et al*. (1958); however, the important consideration of the stereochemistry of the five chiral centres of Bchl *a* (excluding the esterifying alcohol, phytol) had yet to be resolved. The absolute configurations of the chiral centres of Bchl *a* were determined to be 3R,4R,7S,8S,10R nearly simultaneously by both Brockmann (1968a,b) and Fleming (1968). These determinations were based on the oxidation of pyrrole rings II and IV of Bchl *a* to substituted succinimides (which were subsequently converted to derivatives of succinic acid, the absolute configurations of which were known), and on the known *trans* arrangement of the 10-carbomethoxy group (Brockmann, 1968a,b; Fleming, 1968). [See Brockmann (1978) for a review of the stereochemical work.]

Subsequent to this initial structural work, which was primarily chemical in nature, a number of instrumental techniques have been applied to Bchl *a* structural and aggregation studies. Katz and co-workers have assigned the ^{1}H-NMR spectrum of Bchl *a* (Katz *et al*., 1966) and studied its state of aggregation by changes in ^{1}H-NMR chemical shifts upon the addition of polar solvents (Katz, 1973; Katz & Janson, 1973). The natural abundance ^{13}C-NMR spectrum of Bchl *a* was assigned by Brereton and Sanders (1983c) and used by them to study its coordination and aggregation state (Brereton & Sanders, 1983a). The ligation state of Bchl *a* has also been studied by absorption and resonance Raman spectroscopy (Callahan & Cotton, 1987). A high precision structure of Bmph *a* has also recently been determined by X-ray crystallography (Barkigia *et al.*, 1989).

The structures of many intermediates in the biosynthesis of Bchl *a* have been assigned by comparing them both spectroscopically and chromatographically with derivatives of chlorophyll *a* and Bchl *a* (cf. references in Richards & Lascelles, 1969; Pudek & Richards, 1975). In these studies, the assumption was always made that the substituents were in equivalent positions to the parent chlorophyll of known structure. Houghton *et al.* (1983) have, however, carried out an analysis by nuclear Overhauser enhancement ^{1}H-NMR on the magnesium-free derivative of an intermediate of Bchl *a* biosynthesis (2-α-hydroxyethyl-2-desvinylchlorophyllide *a*), and demonstrated the validity of this assumption.

10.2.4.b Bacteriochlorophylls b and g

The close relationship of Bchl *b* to Bchl *a* had been obvious since it was first isolated from *Rhodopseudomonas viridis* by Eimhjellen *et al.* (1963). This close similarity was proved when Brockmann and Kleber (1970) converted both to methyl 2-acetyl-2-desvinylpyropheophorbide *a*. Hence, Brockmann and Kleber (1970) suggested that Bchl *b* might be the 3,4-*cis* isomer of Bchl *a*. Scheer *et al.* (1974), however, demonstrated the presence of a 4-ethylidene group in pyrrole ring II by the formation of acetaldehyde from it by ozonolysis, hydration of it to a 4-α-hydroxyethyl group, ^{1}H-NMR, and mass spectrometry, in which Bchl *b* was shown to have two mass units less than Bchl *a*. Risch (1981) has analysed Bchl *b* by nuclear Overhauser effect difference spectroscopy and assigned the E-configuration to the 4-ethylidene double bond by analogy with the stereochemistry of phycocyanobilin.

The structure of Bchl *g* from *Heliobacterium chlorum* was first studied by Brockmann and Lipinski (1983). Because of its extreme instability towards oxygen (especially in the presence of light), two decomposition products were studied, after first removing magnesium. These products were shown to be the 4-acetyl- and 4-α-hydroxyethyl-derivatives of 4-desethylpheophorbide *a* by ^{1}H-NMR. Chiralities of 7S,8S were confirmed by ^{1}H-NMR (Katz *et al.*, 1968), and 10R by circular dichroism (CD) spectroscopy (Briat *et al.*, 1967); all were the same as those of chlorophyll *a*. Because of the similar reactivities of Bchl *g* and *b*, Brockmann and Lipinski (1983) proposed that Bchl *g* had an analogous 4-ethylidene group to that found in Bchl *b* (cf. Figure 10.1). Comparison of the ^{1}H-NMR of Bphys *b* and *g* confirmed the 4-E-ethylidine configuration (Risch, 1981), and comparison of the CD spectra of Bchls *a*, *b* and *g* confirmed the 3R chirality (Brockmann & Lipinski, 1983). Mass spectrometry by Michalski *et al.* (1987) has confirmed the isometric nature of the magnesium tetrapyrrole portions of Bchl *g* and chlorophyll *a*.

10.2.4.c Bacteriochlorophylls c, d, and e

Holt and co-workers were the first to carry out structural studies on Bchls *c* and *d*, each of which was isolated from a different strain of *Chlorobium*. It was known as early as 1961 that both Bchls *c* and *d* were related to 2-desvinyl-2-α-hydroxyethylpyrochlorophyll *a*, but were esterified with an alcohol different from phytol (Holt & Morley, 1960a,b; Holt & Hughes, 1961). Neither gave a positive test for a methoxy group or a Molisch phase test, and neither could be converted to derivatives related to chlorin e_6, hence, they both lacked a 10-carbomethoxy group. The presence of a conjugated 9-keto group was indicated by its reduction with $NaBH_4$, and by the oxidation of the Bphos to compounds spectroscopically similar to phylloerythrin derived from chlorophyll *a*. The presence of the α-hydroxyethyl group was indicated by its oxidation to a conjugated acetyl, dehydration to vinyl, and reduction to ethyl (Et). Also, there was an indication that both Bchl *c* (Holt & Morley, 1960b; Morley & Holt, 1961) and Bchl *d* (Holt & Hughes, 1961) contained a pyrrole ring substituted with a methyl (Me) and an *n*-propyl (*n*-Pr) group. The big surprise came when it was reported that both Bpho *d* (Hughes & Holt, 1962) and Bpho *c* (Holt *et al.*, 1962) could be separated into at least six different fractions by Celite liquid-liquid partition chromatography (see Section 10.2.2). The latter study also demonstrated that a *meso*-alkyl substituent must be present on one of the three methine carbons in Bchl *c*. Oxidation of C9 and C10 of the main fraction (5) of Bpho *c* to carboxylic acids followed by decarboxylation and oxidation to a porphyrin yielded a compound similar to phylloporphyrin (a γ-methylporphyrin derived from chlorophyll *a*). The ^{1}H-NMR of Bpho *c* indicated that both it and δ-chloromesopheophorbide *a* lacked a δ-methine proton when compared to pheophorbide *a*; it also had the same number of Me groups as pheophorbide *a*, but contained an extra Et group. No evidence had yet been obtained for the nature of the substituent in the 5-position; hence, Holt *et al.* (1962) proposed that Bchl *c* possessed δ-Et and 5-Me groups (or vice versa).

Mathewson *et al.* (1963a) presented ^{1}H-NMR evidence that was complementary to that of Holt *et al.* (1962), but concluded that the *meso*-alkyl substituent was more likely found on the α- or β-methine than on the δ-methine. This feeling was based on the observed exchangeability for deuterium of the δ-methine hydrogen of pheophorbide *a* in CH_3COOD

(Woodward & Skaric, 1961) and of chlorophyll a in CH₃OD/CCl₄ (Katz et al., 1962). Since both Bphos c and d exhibited one exchangeable hydrogen in CH₃COOD (Mathewson et al., 1963a), it seemed likely that each possessed a δ-methine hydrogen adjacent to the reduced pyrrole ring IV. This likelihood was supported by the demonstration that chlorophylls a and b and Bchl c exhibited one (while Bchl a exhibited three) exchangeable hydrogens in CH₃OD/CCl₄ (Mathewson et al., 1963b).

Conclusive proof for the presence of δ-methyl groups in Bchl c has, however, been provided by the following studies:

(a) Synthetic work by Archibald et al. (1963, 1966) which has confirmed the presence of δ-methyl groups in fractions 5 and 6 of Bchl c (cf. below),
(b) Reductive degradative experiments by Chapman et al. (1971) in which methine carbons (together with any *meso*-alkyl groups) are retained in the resulting alkylated monopyrroles.
(c) ¹³C-NMR analyses which indicated that the δ- and γ-methine carbons were clearly quaternary in Bchl c derivatives, and which related a trifluoroacetic acid-induced shift in the absorption of the α- and β-methine carbons in methyl pyropheophorbide a, to a similar shift of the two tertiary methine carbons in Bmph c (Smith & Unsworth, 1975; Kenner et al., 1976).
(d) Comparison of the ¹H- and ¹³C-NMR spectra of Bmphs c and d, which indicated a majority of atoms in the vicinity of the δ-methine to have undergone chemical shift changes, thereby indicating an alkyl substituent on the δ-methine (Brockmann, 1976).
(e) Synthesis of Bmph c (4-Et, 5-Me) from chlorin e_6 trimethyl ester, which took advantage of the known reactivity of the δ-methine position towards electrophilic substitution (Brockmann et al., 1979; Smith et al., 1980a,b,c; Jürgens & Brockmann, 1982). Kenner et al. (1976) have also provided evidence that it is the α-methine hydrogen which undergoes exchange with deuterium in Bchl c, perhaps due to the electron-donating influence of the δ-methyl substituent, while the β-methine hydrogen does not exchange, perhaps due to the influence of the electron-withdrawing 9-keto substituent.

Holt and co-workers proposed the structures shown in Table 10.3 for the six fractions each of Bchl d (Holt et al., 1963; Purdie & Holt, 1965) and Bchl c (Holt et al., 1966). Based on the structures of the substituted maleimides isolated after chromic acid oxidation of the separated Bpho fractions, they inferred that the latter differed by the nature of their alkyl substituents in the 4- and 5-positions only (Bchl d) or the 4-, 5-, and δ-positions (Bchl c) (see Figure 10.1). Pyrrole rings I and III were normally lost during chromic acid oxidation; however, the substituents on pyrrole rings IV and II were

Table 10.3 Structures proposed for the separated fractions of bacteriochlorophylls c and d.

A. Bacteriochlorophyll c

	According to Holt et al. (1966)				According to Smith et al. (1980e)				
	Alkyl substituents[a]				Alkyl substituents[a]				
Band	R_4	R_5	R_m	%	R_4	R_5	R_m	Chirality of C2a	%
1	i-Bu	Et	Et[b]	0.5	i-Bu	Et	Me	S	4.5
2	i-Bu	Et	Me	0.5	i-Bu	Et	Me	R	<0.1
3	n-Pr	Et	Et[b]	2.0	n-Pr	Et	Me	S	5.3
4	n-Pr	Et	Me	16.0	n-Pr	Et	Me	R	18.3
5	Et	Et	Me	71.0	Et	Et	Me	R	71.7
6	Et	Me	Me	10.0	Et	Me	Me	R	0.2

B. Bacteriochlorophyll d

	According to Purdie and Holt (1965)				According to Smith and Goff (1985)				
	Alkyl substituents[a]				Alkyl substituents[a]				
Band	R_4	R_5	R_m	%	R_4	R_5	R_m	Chirality of C2a	%[c]
—	—	—	—	—	neoPn	Et	H	S	1.5
—	—	—	—	—	neoPn	Me	H	S	0.2
1	i-Bu	Et	H	10.2	i-Bu	Et	H	S	19.0
2	n-Pr	Et	H	33.6	n-Pr	Me	H	R	35.7
3	i-Bu	Me	H	14.4	i-Bu	Et	H	S	12.7
4	Et	Et	H	23.1	Et	Et	H	R	19.8
5	n-Pr	Me	H	11.0	n-Pr	Me	H	R	7.7
6	Et	Me	H	7.6	Et	Me	H	R	3.4

[a] Refer to Figure 10.1.
[b] Incorrectly assigned based on the work of Smith et al. (1980e).
[c] Calculated from the data of Table 5 (sample 2) of Smith and Goff (1985), assuming a 3:1 ratio of 5-Et to 5-Me homologues.

determined as follows: ring IV yielded *trans*-dihydrohematinic acid from all 12 fractions. Its stereochemistry was later shown by Brockmann (1976) to be 2S,3S, thus demonstrating the stereochemical relatedness of Bchls *c* and *d* to chlorophyll *a* and Bchl *a*. Ring II yielded either methylethyl-, methyl-*n*-propyl-, or methylisobutylmaleimide from *pairs* of fractions from both Bpho *d* (4 and 6, 2 and 5, and 1 and 3, respectively) and Bpho *c* (5 and 6, 3 and 4, and 1 and 2, respectively). [This was the first indication of isobutyl (*i*-Bu) groups in these Bchls.] If the hydroxyethyl group was first reduced to Et before oxidation, methylethylmaleimide was isolated from fractions previously lacking it, thus confirming the substituents on ring I in these fractions.

The substituents on ring III were more difficult to determine, and required conversion of the Bpho *d* fractions to pyroporphyrins (and Bpho *c* fractions to phyllochlorins) by a reaction sequence analogous to that described above for the formation of phylloporphyrin from Bpho *c*. Ethylmaleimide was isolated from the oxidation of fractions 1, 2 and 4 of pyroporphyrin *d*, indicating the presence of a 5-Et group in these fractions. The presence of the 5-Me group in fraction 6 of Bpho *d* was confirmed by its conversion to pyropheophorbide *a* (Holt *et al.*, 1963), while the structures of fractions 1 to 5 were confirmed by comparison of corresponding derivatives with authentic samples of synthetic pyroporphyrins (fractions 1 to 4) or 4-desethyl-4-*n*-propyldesoxophylloerythrin (fraction 5) (Archibald *et al.*, 1963, 1966). Ethylmaleimide was also isolated from fractions 1 to 5 of the phyllochlorins *c*, while methylmaleimide (citraconimide) was isolated from fraction 6 (Holt *et al.*, 1966). The structures of fractions 5 and 6 of the phylloporphyrins *c* were confirmed by comparison with authentic samples of synthetic 5-desmethyl-5-ethyl-δ-phylloporphyrin IX and δ-phylloporphyrin IX, respectively (Archibald *et al.*, 1963, 1966). Holt *et al.* (1966) proposed δ-Me groups for fractions 2 and 4, based on mass spectrometric data of these fractions provided by Smith (1967), and δ-Et groups for fractions 1 and 3 (as the most likely way of distinguishing them from fractions 2 and 4, respectively), even though there was no evidence for the presence of *meso*-Et groups in any fraction. Subsequently, a δ-ethylphylloporphyrin prepared by Cox *et al.* (1971) failed to match a phylloporphyrin from any of the Bpho *c* fractions. The validity of the structures proposed for fractions 1 and 3 thus came seriously into question.

Strouse and co-workers separated Bchl *c* from *Chlorobium limicola* f. *thiosulfatophilum* by preparative-scale HPLC, and identified only four major homologues (corresponding to Holt's fractions 2, 4, 5 and 6 of Table 10.3) by mass spectrometry (Caple *et al.*, 1978). Smith *et al.* (1980d) also could not find mass spectrometric evidence for more than the same four homologues in Bmph *c* prepared from *Prosthecochloris aestuarii* (formerly '*Chloropseudomonas ethylicum*', but later shown to be a syntrophic mixture of *P. aestuarii* and *Desulfovibrio acetoxidans*; Pfennig & Biebl, 1976). Although one of Holt's 'δ-ethyl' fractions (3) should be isomeric with fraction 2 and, therefore, undetectable by mass spectrometry, no mass peak corresponding to Holt's fraction 1 was observed. Smith *et al.* (1980d) also reported the resolution of Bmph *c* into four fractions by reversed-phase ODS HPLC. This time, one band (corresponding to Holt's fraction 4) was partially resolved into components which appeared to be isomers by mass spectral analysis. The mystery was solved when it was realized that the HPLC system could resolve the R- and S-diastereomeric epimers of the 2-α-hydroxyethyl group (Smith *et al.*, 1980a,c). The resolution was enhanced by recycling through the HPLC until a complete separation was attained. The two 'fraction 4' isomers were shown by Smith *et al.* (1980e) to be the 2aS- and 2aR-isomers of the (4-*n*-Pr,5-Et) homologue (renamed fractions 3 and 4, respectively). The same study also demonstrated that the Bmphs *c* in the other three fractions could also be resolved into two epimers after chemical racemization of the 2-α-hydroxyethyl group. On the assumption that the 2aR-epimer was always the faster running of the two, the naturally occurring diastereomers were assigned as shown in Table 10.3. [This also necessitated renaming the major (2aS) epimer of the (4-*i*-Bu,5-Et) homologue as Fraction 1.] This assumption was based in part by work (Brockmann & Risch, 1974; Risch & Brockmann, 1976) which assigned the chirality of C2a as R in both Bchls *c* and *d* on the basis of Horeau analyses. This was later confirmed by oxidation of the 2a-*O*-benzoyl-derivatives of each to 2-*O*-benzoyl-R-lactic acid (Brockmann & Tacke-Karimdadian, 1979). This work was done on unseparated fractions; hence, small amounts (<10%) of the 2aS-epimer could have gone undetected (Smith *et al.*, 1980e). However, it indicated that the major fractions of Bchl *c* (4 and 5) must have had the 2aR-configuration. The percentages of each of the fractions was found to be strain-dependent, with trace amounts of what appeared to be Bmph *d* homologues also present (Smith *et al.*, 1983a). This latter study also reported the ^{1}H-NMR spectra for all of the separated Bmph *c* fractions.

Smith *et al.* (1982) have analysed the 2a-chiralities of the Bmph *d* fractions from *Chlorobium vibrioforme* f. *thiosulfatophilum* and found that, whereas all four 4-Et and 4-*n*-Pr homologues were R, the two 4-*i*-Bu homologues were both S (see Table 10.3). This was extended by the discovery of two new homologues bearing neopentyl (neoPn) substituents, both of which were also S (Smith *et al.*, 1983b). The absolute configurations of three of the homologues were determined by single crystal X-ray analysis (Smith *et al.*, 1982, 1983b), which confirmed the assignments based on HPLC mobilities. Finally, the ^{1}H-NMR spectra for all eight fractions (and the ^{13}C-NMR spectra for all but the 4-neoPn homologues) were assigned by Smith and Goff (1985). The ^{13}C-chemical shifts of the α- and δ-methine were correlated with the absolute configuration of the 2a carbons so that the 2a-chiralities for all fractions of the Bmphs *d* could be confirmed (Smith & Goff, 1985).

Gloe and Risch (1978) have studied Bchl *c* from the green nonsulphur bacterium, *Cloroflexus aurantiacus*, by mass spectrometry, and found that it contained almost exclusively *one* homologue (4-Et,5-Me). It is, therefore, much more closely related to chlorophyll *a* than are the Bchl *c* homologues from the green sulphur bacteria. Brune *et al.* (1987), however, have recently separated Bchl *c* from *Cloroflexus aurantiacus* and found that, whereas the (4-Et,5-Me) homologue was the major fraction, three other homologues were also present, but were not characterized. The main fraction consisted of almost equal amounts of the 2aR- and 2aS-epimers (Brune *et al.*, 1987); however, Bobe *et al.* (1990) have recently determined that the ratio of 2aR to 2aS could be as high as 70:30 under certain growth conditions. Minor amounts of other Bchl *c* fractions were probably due to esterifying alcohols other than stearyl alcohol (Risch *et al.*, 1979; Fages *et al.*, 1990). Finally, Bobe *et al.* (1990) have detected minor amounts of the (4-Et,5-Me) homologue of Bchl *d* in *Cloroflexus aurantiacus* as well.

Bacteriochlorophyll *e* was first isolated from *Chlorobium phaeovibrioides* by Gloe *et al.* (1975), who compared Bmph *e* with methyl pyropheophorbide *b* by ^{1}H- and ^{13}C-NMR, infrared and visible spectroscopy (Gloe *et al.*, 1975; Brockmann, 1976). The spectroscopic studies were consistent with Bchl *e* being a 3-formyl derivative of Bchl *c*, and mass spectrometry indicated that at least three different homologues were present. The chiralities of C7 and C8 (Brockmann, 1976) and C2a (Risch & Brockmann, 1976) were found to be the same as the major fractions of Bchls *c* and *d*. Brockmann *et al.* (1976) then proved the above structural assignments by chemical conversion of Bmph *e* into meso-Bmph *c* by reduction of the 3-formyl group (and, coincidentally, the 2-α-hydroxyethyl group), and comparison of the optical rotatory dispersion (ORD) spectrum of Bmph *e* with that of methyl pyropheophorbide *b*. Risch *et al.* (1978) resolved Bmph *e* into three fractions by silica gel column chromatography; each was characterized by ^{1}H-NMR and mass spectrometry. Only the 5-Et (but no 5-Me) homologues were present (see Figure 10.1). Smith and Simpson (1986) have resolved the Bmph *e* homologues from *Chlorobium phaeobacteroides* by reversed-phase HPLC after first converting them to Bmph *c* homologues by reduction of the 3-formyl group (*without* coincidental reduction of the 2-α-hydroxyethyl group). Both 2a-chiralities of all three homologues were present, the (4-Et,5-Et) homologue was 95% R, the (4-*n*-Pr,5-Et) homologue was 60% S, and the (4-*i*-Bu,5-Et) homologue was >98% S (Smith & Simpson, 1986; Simpson & Smith, 1988). Finally, Risch *et al.* (1988) have synthesized a potentially naturally occurring homologue of Bmph *f* from chlorophyll *b*, and described a method for the resolution of 2aR,2aS-diastereomers.

10.2.5 Characterization of the esterifying alcohols

The esterifying alcohol of chlorophylls *a* and *b* and Bchl *a* was identified as phytol by Fischer and co-workers (Fischer & Stern, 1940). Its absolute configuration was later determined to be 7R,11R-Δ^{2t}-phytaenol. Fischer worked with a strain of *Thiocystis*; however, the assumption seems to have been made that all species of purple bacteria esterified Bchl *a* with the same alcohol, until Katz *et al.* (1972) reported that whereas *Chromatium vinosum* and *Rhodobacter sphaeroides* did indeed employ phytol as the esterifying alcohol (thus forming Bchl a_P), *Rhodospirillum rubrum* employed all-*trans*-geranylgeraniol (forming Bchl a_{GG}). Analysis of the esterifying alcohols was carried out by mass spectrometry, and minor amounts of alcohols with 2, 4, 6 and 8 extra mass units indicated the presence in *R. rubrum* of C_{20} isoprenoid alcohols with from 0 to 3 double bonds in addition to all-*trans*-geranylgeraniol (Katz *et al.*, 1972). The presence of Bchl a_{GG} in *R. rubrum* was subsequently confirmed by Brockmann *et al.* (1973). However, Walter *et al.* (1979) have isolated the RC complex from *R. rubrum* G-9$^+$, and demonstrated that even though Bchl *a* was present in the RC in the form of its geranylgeraniol ester, Bphy *a* was esterified with phytol!

Künzler and Pfennig (1973) examined a total of 13 species of purple non-sulphur bacteria, finding Bchl a_P in all species except *Rhodospirillum rubrum* (where Bchl a_{GG} was present in all ten strains examined), and one of two strains of *Rhodospirillum photometricum* (which contained approximately equal amounts of Bchl a_{GG} and Bchl a_P). Gloe and Pfennig (1974) then examined a total of 15 species of purple sulphur and seven species of green sulphur bacteria, finding phytol to be the sole or major esterifying alcohol for Bchl *a* in all species examined, with only traces of Bchl a_{GG} found in seven out of 38 strains (in a total of six species) of the purple sulphur bacteria. Bacteriochlorophyll a_P has also been found in the *aerobic* photosynthetic bacteria, *Erythrobacter longus* and *Erythrobacter* sp. OCh114 and OCh175 (Harashima *et al.*, 1980; Shiba & Abe, 1987). Finally, trace amounts of Bchl *a* esterified with geranylgeraniol, dihydrogeranylgeraniol, and tetrahydrogeranylgeraniol were separated from Bchl a_P by reverse-phase HPLC from both *Rhodopseudomonas palustris* (Shioi *et al.*, 1983) and *Rhodobacter sphaeroides* (Shioi & Beale, 1987); however, these may have been present as biosynthetic precursors of Bchl a_P.

Brockmann and Kleber (1970) had originally established the esterifying alcohol of Bchl *b* from *Rhodopseudomonas viridis* to be phytol by a chromatographic analysis. This was later confirmed by a ^{1}H-NMR analysis by Scheer *et al.* (1974). Gloe and Pfennig (1974) also found phytol to be the major esterifying alcohol of Bchl *b* in both *Rhodopseudomonas viridis* and *Thiocapsa pfennigii*, but also found traces of another ester taken to be Bchl b_{GG}. Using GC-mass spectrometry followed by ozonolysis, the major esterifying alcohol of *Ectothiorhodospira halochloris*, however, was found to be $\Delta^{2,10}$-phytadienol by Steiner et al. (1981).

In contrast to the purple bacteria, which employ C_{20} isoprenoid alcohols exclusively for esterification of Bchls *a* and *b*, green sulphur bacteria employ predominantly C_{15} isoprenoid alcohols for esterification of Bchls *c*, *d* and *e* (but C_{20} alcohols for the Bchl *a* which they also form!). Holt and co-workers first noticed that the esterifying alcohols of both Bchls *c* and *d* were different from phytol (Holt & Morley, 1960a,b; Holt & Hughes, 1961). The esterifying alcohol of Bchl *c* from *Chlorobium limicola* f. *thiosulfatophilum* was identified as *trans,trans*-farnesol (Bchl c_F) by Rapoport and Hamlow (1961). This was later confirmed by Holt et al. (1963) for Bchls *c* and *d* from two different strains of *Chlorobium*, and Katz et al. (1972) for Bchl *c* from *Prosthecochloris aestuarii*. Strouse and co-workers have carried out large-scale HPLC separations of the Bchl *c* homologues from *C. limicola* f: *thiosulfatophilum* (Chow et al., 1978), and have determined that, while *trans,trans*-farnesol was present as the major esterifying alcohol of all four homologues (present to the extent of 91.3% of all alcohols), two of the homologues (4-Et,5-Et and 4-*i*-Pr,5-Et) were also esterified with the C_{20} alcohols, phytol (3.3%), $\Delta^{2,6}$-phytadienol (1.5%), and all-*trans*-geranylgeraniol (0.2%), a C_{16} fatty alcohol, Δ^{9c}-hexadecen-1-ol (3.1%), and an unusual C_{16} furan derivative, 4-undecyl-2-furanomethanol (0.6%) (Caple et al., 1978).

Do these minor esterifying alcohols somehow earmark their Bchls *c* for special functions within the cell? The work of Braumann et al. (1986) indicates that this may be so, in at least one respect. The reaction centre (RC) of *Prosthecochloris aestuarii* contains Bchl *a* and an unidentified pigment (Bchl 663), spectroscopically identical to Bchl *c*, which may function as an electron acceptor in the RC (Braumann et al., 1986). Although it was first tentatively identified as Bphy *c* (Vasmel et al., 1983), Bchl 663 seems to be a more lipophilic analogue of Bchl *c*. Preliminary evidence suggested that it was Bchl *c* esterified with phytol rather than farnesol, although the possibility remained that it might also have been an as yet unknown homologue of Bchl *c* (Braumann et al., 1986). Smith and Goff have unpublished evidence for the presence of multiple esters of Bchl *d* in *Chlorobium vibrioforme* f. *thiosulfatophilum* as well (reported in Smith et al., 1983a). Also, Gloe and Risch (1978) have determined that, while the major homologue (4-Et,5-Me) of Bchl *c* in the green nonsulphur bacterium, *Chloroflexus aurantiacus*, was esterified with the $C_{18:0}$ fatty alcohol, stearyl alcohol (Bchl c_S), minor amounts of the Bchl *c* esters of phytol, geranylgeraniol, cetyl alcohol ($C_{16:0}$), and oleyl alcohol ($C_{18:1}$) were also present in *C. aurantiacus* (Risch et al., 1979; Fages et al., 1990). Finally, Brockmann et al. (1976) have determined that the major esterifying alcohol was *trans,trans*-farnesol for all three of the Bchl *e* homologues (Bchl e_F) from *Chlorobium phaeovibrioides*.

Brockmann and Lipinski (1983) initially reported, based on a ¹H-NMR analysis, that the esterifying alcohol of Bchl *g* from *Heliobacterium chlorum* was geranylgeraniol. However, Michalski *et al.* (1987) have recently found by mass spectrometry that *H. chlorum* employs *trans,trans*-farnesol for esterification of Bchl *g* (Bchl g_F). Hence, the Heliobacteriaceae may prove to be similar to the green sulphur bacteria in this respect.

10.3 TAXONOMIC CONSIDERATIONS

Photosynthetic prokaryotes may carry out either oxygenic or anoxygenic photosynthesis. A proposal to place all photosynthetic prokaryotes in an interim set of two higher taxa based on this property (to be known as Oxyphotobacteria and Anoxyphotobacteria, respectively) was made by Murray in *Bergey's Manual of Systematic Bacteriology* in 1984 (Murray, 1984). Such a suggestion, which attempted to create a higher taxon to encompass *all* of the purple, green and brown bacteria, was frustrated by the extreme diversity of these organisms. For example, within the anoxygenic phototrophs are species which may be obligate anaerobic photolithotrophs, facultative photoheterotrophs or obligate aerobic heterotrophs that use photosynthesis to enhance aerobic growth. They may have marine, freshwater or soil habitats; may be non-motile or motile by means of flagella or gliding; may divide by binary fission or budding; may be capable of forming capsules or heat-resistant spores; and may exist in an unassociated state, associated in filaments or synotropic consortia, or attached to solid supports by stalks. They may be capable of sulphur oxidation, methylotrophy, denitrification or nitrogen fixation (either alone or as symbionts of the stem nodules of leguminous plants). Finally, phototrophic bacteria are known which can live in exceptionally adverse growth conditions, including extremes of temperature and concentrations of acid, base, sulphide, and/or salt. Whether this is due to modern adaptation or to the inheritance of such tolerances from ancestors that evolved when such conditions were common on earth is a moot point. However, thermophilic (or thermotolerant) and halophilic representatives are known for nearly all of the major families of the phototrophic eubacteria.

The ability to carry out photosynthesis is so widely spread throughout the prokaryotic world that the thought has been expressed (Woese *et al.*, 1985a) that *all* bacteria may have had a common photosynthetic ancestry. (To date, there are no known photosynthetic archaea except for a unique type of light-driven ATP synthesis carried out by some halophilic archaea containing bacteriorhodopsin.) Analysis of the sequences of the 16S ribosomal RNA (rRNA) of many different bacteria has revealed that many photosynthetic bacteria are more closely related to various species of *non-photosynthetic* bacteria than they are to other photosynthetic bacteria which were previously classified within the same genus (Gibson et al., 1979)! This discovery has meant either that the ability to carry out photosynthesis has been lost by the modern, non-photosynthetic descendants of a photosynthetic ancestor, or that the ability to carry out photosynthesis has been somehow passed from photosynthetic to distantly related non-photosynthetic bacteria by the lateral transfer of genetic information.

There are five major classes of photosynthetic prokaryotes which can be distinguished by the nature of their photosynthetically active tetrapyrrole pigments. These are:

(a) Purple bacteria, consisting of three families (Chromatiaceae, Ectothiorhodospiraceae and Rhodospirillaceae), which carry out anoxygenic photosynthesis and contain either Bchl *a* or (rarely) Bchl *b*.
(b) Green and brown bacteria, consisting of two families (Chlorobiaceae and Chloroflexaceae), which carry out anoxygenic photosynthesis and contain a small amount of Bchl *a* in addition to a majority of Bchl *c*, *d* or *e*.
(c) Heliobacteria, which carry out anoxygenic photosynthesis and contain Bchl *g*.
(d) Cyanobacteria, which carry out oxygenic photosynthesis and contain chlorophyll *a* and phycobilins, but no chlorophyll *b*.
(e) Prochlorophytes, which carry out oxygenic photosynthesis and contain both chlorophylls *a* and *b*, but no phycobilins.

Subclasses of prokaryotes within these five major classes cannot, however, be distinguished by the nature of the chlorophylls they contain. For example, most purple bacteria contain Bchl *a*, while Bchl *b* seems to be limited to a very small number of purple bacteria. However, these include species in all three of the purple bacterial families (e.g. *Rhodopseudomonas sulfoviridis* and *Rhodopseudomonas viridis* in the Rhodospirillaceae, *Ectothiorhodospira abdelmalekii* and *Ectothiorhodospira halochloris* in the Ectothiorhodospiraceae, and *Thiocapsa pfennigii* in the Chromatiaceae). In the case of green bacteria, Bchl *c* is found in species of both the Chlorobiaceae and Chloroflexaceae, while Bchl *c*, *d* or *e* may be found in species of nearly every genus of the Chlorobiaceae. The possession of a particular Bchl is, therefore, of no special significance to the taxonomy of either the purple or green bacteria.

A brief discussion of each of the five major classes follows:

10.3.1 Purple bacteria

Historically, the purple bacteria (called the suborder Rhodospirillinae by Pfennig & Trüper, 1971) were divided into two families, based on their photometabolism. Those purple bacteria which use reduced sulphur compounds (e.g. sulphide) as a source of electrons for obligate photolithotrophic CO_2 fixation under anaerobic or microaerophilic conditions were placed in the family Thiorhodaceae (literally 'sulphur purple' bacteria). Those purple bacteria which use reduced organic compounds for photoheterotrophic growth under microaerophilic or semi-aerobic conditions were placed in the family Athiorhodaceae (literally 'nonsulphur purple' bacteria). Most of these latter bacteria have alternative modes of metabolism and are thus facultative phototrophs. They are able to grow heterotrophically in the dark under fully aerobic conditions, in which case the formation of Bchl *a* and photosynthetic membranes is repressed. Some are also able to grow anaerobically in the dark in the presence of a fermentable substrate, while others can also grow anaerobically in the light using hydrogen gas (and/or reduced sulphur compounds) as a source of electrons.

In recent years, as the use of 16S rRNA sequences to establish bacterial phylogeny has become widespread (Fox *et al.*, 1977, 1980), it has become apparent that the traditional methods for the classification of the purple bacteria were insufficient to ascertain relatedness among the various members (Gibson *et al.*, 1979). Although the division between sulphur and nonsulphur bacteria has been confirmed by 16S rRNA sequence analyses, there is now a total of three subdivisions containing purple bacteria, referred to as the alpha, beta and gamma subdivisions (Woese *et al.*, 1984a,b, 1985b) (Table 10.4). The purple nonsulphur bacteria (which have been placed in the family Rhodospirillaceae) fall into two of these subdivisions (alpha and beta), with further subclassification into subgroups, for example the subgroups alpha-1 to alpha-4. The purple sulphur bacteria, which have been divided into two separate families, Chromatiaceae (which accumulate elemental sulphur within the cell) and Ectothiorhodospiraceae (which deposit sulphur outside the cell) (Fowler *et al.*, 1984; Imhoff, 1984), are all in the gamma subdivision. This reorganization of purple bacterial phylogeny has necessitated the reorganization and renaming of several purple bacterial genera (Imhoff *et al.*, 1984) causing many workers in the field to have to learn new names for their pet bacteria! It also became apparent that many *non-photosynthetic* bacteria were more closely related to a particular purple bacterial genus than the latter was to other purple bacterial genera previously thought to be closely related. Woese (1987) and Stackebrandt *et al.*, (1988a) have reviewed the phylogenetic relationships between the photosynthetic bacteria and many closely related species.

Woese (1987) has defined a total of ten bacterial phyla, fully half of which consist (either entirely or in part) of photosynthetic representatives. A new term 'Proteobacteria' has been proposed as a class name for the phylum containing 'the purple bacteria and their relatives' (Stackebrandt et al., 1988b).

Table 10.4 Phylogeny of photosynthetic bacteria based on 16S ribosomal RNA sequence homologies.[a]

Family, subdivision, and subgroup	Photosynthetic genera	Representative non-photosynthetic species
A. PROTEOBACTERIA: PURPLE PHOTOSYNTHETIC BACTERIA AND THEIR RELATIVES Rhodospirillaceae Alpha subdivision		
Alpha-1	*Rhodospirillum* *Rhodopila*	*Acidophilium rubrum* *Azospirillum brasilense* *Thiobacillus acidophilus*
Alpha-2	*Methylobacterium* *Rhizobium* BTai1 *Rhodomicrobium* *Rhodopseudomonas*	*Agrobacterium tumefacians* *Azorhizobium caulinodans* *Bradyrhizobium japonicum* *Pseudomonas diminuta* *Nitrobacter winogradskyi* *Rhizobium leguminosarum*
Alpha-3	*Rhodobacter* *Roseobacter*[b]	*Gemmobacter aquatilis* *Paracoccus denitrificans* *Thiobacillus versutus*
Alpha-4	*Erythrobacter longus*	None known
Rhodospirillaceae Beta subdivision[c]		
Beta-1	*Rhodocyclus gelatinosus*	*Aquaspirillum aquaticum* *Pseudomonas acidovorans* *Pseudomonas testosteroni* *Sphaerotilus natans* *Thiobacillus intermedius*
Beta-2	*Rhodocyclus tenuis* *Rhodocyclus purpureus*	*Alcaligenes eutrophus* *Aquaspirillum bengal* *Chromobacterium violaceum* *Pseudomonas cepacia* *Thiobacillus denitrificans*

(*continued*)

Table 10.4 (*Continued*)

Family, subdivision, and subgroup	Photosynthetic genera	Representative non-photosynthetic species

Chromatiaceae
  Gamma subdivision[d]
    Gamma-1
	Amoebobacter
	Chromatium
	Lamprocystis
	Thiocapsa
	Thiocystis
	Thiodictyon
	Thiospirillum
		Nitrosococcus oceanus

Ectothiorhodospiraceae
  Gamma subdivision
    Gamma-1
	Ectothiorhodospira	None known

Delta subdivision	None known	Myxococcus group Bdellovibrio group *Desulfovibrio desulfuricans* *Desulfuromonas* sp.
B. GREEN PHOTOSYNTHETIC BACTERIA, HELIOBACTERIA AND THEIR RELATIVES		
Chlorobiaceae	*Chlorobium* *Chloroherpeton* *Clathrochloris* *Pelodictyon* *Prosthecochloris*	*Flavobacterium aquatile* *Flexibacter flexilis*
Chloroflexaceae	*Chloroflexus*	*Herpetosiphon aurantiacus* *Thermomicrobium roseum*
Heliobacteriaceae	*Heliobacillus* *Heliobacterium*	*Thermoanaerobium brockii* *Acetogenium kivui*

[a] Selected from the 16S rRNA sequence analyses of Gibson *et al.* (1985), Woese (1987), Stackebrandt *et al.* (1988a), Woese *et al.* (1990), Young *et al.* (1991), and Lane *et al.* (1992).
[b] *Roseobacter denitrificans* (formerly *Erythrobacter* OCh114) is thought to be related to the alpha-3 subgroup, although no rRNA sequence analysis has yet been carried out (Shiba, 1991).
[c] Organisms in the beta subdivision, but separated from the beta-1 and beta-2 subgroups, include *Nitrosococcus mobilis*, *Nitrosolobus multiformis*, *Nitrosomonas europaea*, *Nitrosospira* sp., *Nitrosovibrio tenuis*, *Spirillum volutans*, *Thiobacillus ferrooxidans*, and *Thiobacillus thiooxidans*.
[d] The gamma subdivision also contains the legionella group (in gamma-2), and a variety of other non-photosynthetic bacteria, including acinetobacteria, enterics, the leucothrix group, oceanospirilla, fluorescent pseudomonads, vibrios and the xanthomonas group (all in gamma-3).

10. Analysis of Pigments: Bacteriochlorophylls

In the late 1970s and early 1980s, several species of aerobic bacteria (which were not previously thought to be capable of photosynthesis) were found to contain Bchl *a*. These included (a) two marine organisms, *Erythrobacter longus* and *Erythrobacter* sp. OCh114 (Shiba *et al.*, 1979; Harashima *et al.*, 1980; Shiba & Simidu, 1982), and (b) several facultative methylotrophic bacteria, *Protaminobacter ruber* (Sato, 1978), *Pseudomonas* AM-1 (Sato, 1978), a radiation-resistant bacterium *Pseudomonas radiora* (Nishimura *et al.*, 1981), and several strains of the genus *Protomonas* (Urakami & Komagata, 1984). The two marine bacteria contained between 10 and 50% of the Bchl found in purple bacteria, but were found to be able to form it only under *aerobic* conditions (Shiba, 1984, 1987; Harashima *et al.*, 1987). *Erythrobacter* sp. OCh114 is capable of forming intracytoplasmic membranes (ICM) (Iba *et al.*, 1988), and both have been found to possess light-harvesting (LH) and reaction centre (RC) complexes related to those of the recognized purple bacteria (Shimada *et al.*, 1985; Takamiya *et al.*, 1987). However, photophosphorylation and light-dependent CO_2 fixation, while present, were found to be very feeble (Shiba, 1989a). These bacteria seem to employ light only to enhance aerobic growth, and are unable to grow completely photoautotrophically. The methylotrophic bacteria contained only 1% or less of the Bchl content of purple bacteria. *Protaminobacter ruber* forms Bchl aerobically in the dark after having first been grown aerobically in the light (Sato & Shimizu, 1979; Sato *et al.*, 1985). Initially thought to be non-photosynthetic, membrane fractions containing a photochemically active RC have now also been isolated from both *Protaminobacter ruber* (Takamiya & Okamura, 1984) and *Pseudomonas radiora* (Nishimura *et al.*, 1989). A monograph summarizing the background work on the aerobic photosynthetic bacteria has been produced by Harashima *et al.*, (1989).

Erythrobacter longus is orange in colour, whereas most other aerobic bacteria containing Bchl are pink. These differences in colour are due primarily to differences in carotenoids (Harashima, 1989). More than 70% of the carotenoids of *E. longus* are unidentified polar carotenoids not bound to Bchl-protein complexes. Those that are bound are either unique entities or carotenoids (such as zeaxanthin) normally found only in cyanobacteria or higher plants. On the other hand, *Erythrobacter* sp. OCh114 contains oxo-derivatives of the spirilloxanthin series commonly found in the alpha-3 subgroup of the purple bacteria. This and a number of other biochemical differences (summarized by Shiba, 1989b) have indicated that the two marine bacteria do not belong in the same genus; hence, *Erythrobacter* sp. OCh114 has been renamed *Roseobacter denitrificans* (Shiba, 1991). Although no RNA analysis has yet been carried out, it and another bacterium in the same genus, *Roseobacter litoralis*, are thought to be related to the alpha-3 subgroup of proteobacteria. *Erythrobacter longus* contains a cytochrome

aa_3 which is completely different from the cytochrome aa_3 found in some members of the alpha-3 subgroup (e.g. *Rhodobacter sphaeroides* and *Paracoccus denitrificans*). Hence, *E. longus* has been placed in the alpha-4 subgroup as the only species of the sole genus represented (Woese, 1987). (However, another orange-coloured aerobic bacterium containing Bchl (*Erythrobacter* sp. OCh175), described by Shiba and Abe (1987), appears to be similar to *E. longus*.)

Finally, several additional strains of *freshwater* aerobic bacteria containing Bchl *a* have recently been characterized by Yurkov et al. (1991a). On the basis of DNA-DNA hybridization analysis and other microbiological characteristics, none of these freshwater strains appeared to be related to the previously described marine bacteria. Hence, it was proposed that all be placed into two new genera, the orange-coloured strains becoming five species of the genus *Erythromicrobium* (Yurkov et al., 1991a), and a pink-coloured strain becoming *Roseococcus thiosulfatophilus* (Yurkov & Gorlenko, 1991). No doubt many more aerobic photosynthetic bacteria await discovery. (It is interesting that at least two members of the Rhodospirillaceae, *Rhodobacter sulfidophilus* (Shioi & Doi, 1989) and *Rhodospirillum centenum* (Yildiz et al., 1991) have recently been described which are able to form Bchl and a functional photosynthetic apparatus under completely aerobic growth conditions. Furthermore, *Rhodobacter sulfidophilus* is able to carry out photosynthesis under *both* aerobic and anaerobic conditions, and may thus be a link between the microaerophilic and aerophilic photosynthetic bacteria.)

Methylotrophic bacteria are phylogenetically an extremely diverse group which are scattered among the proteobacteria in the alpha, beta and gamma subdivisions. Based on 5S rRNA sequence analysis, Wolfrum and Stolp (1987) could distinguish two clusters within the so-called 'group I' methylotrophic bacteria (which utilize the ribulose monophosphate pathway of formaldehyde assimilation): obligate methanol-oxidizing bacteria were related to the beta subdivision and obligate methane-oxidizing bacteria to the gamma subdivision. These relationships were confirmed by the 16S rRNA sequence analyses of Tsuji et al. (1990), who also extended the analysis to 'group II' methylotrophic bacteria (which utilize the serine pathway of formaldehyde assimilation). These organisms were found to form two clusters within the alpha-2 subgroup of proteobacteria: one cluster was composed of non-pigmented, mostly obligate methane-oxidizing bacteria, and the other of pink-pigmented facultative methanol-oxidizing bacteria. (At least one purple photosynthetic bacterium in the alpha-2 subgroup (*Rhodopseudomonas acidophila*) has been shown by Siefert and Pfennig (1979) to be capable of semiaerobic methylotrophic growth.) Some, but not all of the pink-pigmented methylotrophs have been shown to contain Bchl; however, due to their demonstrated similarities, it has been proposed that all be placed within the same genus, *Methylobacterium* (Hood *et al.*, 1987, Green *et al.*, 1988). A total of six species containing Bchl have been validated (summarized in Harashima *et al.*, 1989), including *Methylobacterium rhodesianum* (previously *Protaminobacter ruber*), *Methylobacterium radiotolerans* (previously *Pseudomonas radiora*), and *Methylobacterium extorquens* (previously *Protomonas extorquens* also known as *Pseudomonas* AM-1).

Recently, the first photosynthetic symbiotic nitrogen-fixing *Rhizobium* species was isolated from the stem nodules of *Aeschynomene indica* by Eaglesham *et al.* (1990). This strain (BTai1) was also found to form Bchl and an active RC *only* under aerobic conditions when cultivated *ex planta* under a light-dark cycle (Evans *et al.*, 1990). Based on 16S rRNA sequences, it has been placed in the alpha-2 subgroup of the proteobacteria, very closely related to *Rhodopseudomonas palustris* and the soybean symbiont, *Bradyrhizobium japonicum* (Young *et al.*, 1991). The latter authors, in fact, even suggested that all three species should belong to the same genus, which would result in the two symbionts being reclassified as species of *Rhodopseudomonas*! Barring such a radical change, however, strain BTai1 should more accurately be referred to as *Bradyrhizobium* sp. (*Aeschynomene*). The *Bradyrhizobium* group was also found to be more closely related to *Azorhizobium caulinodans*, a stem-nodulating symbiont of *Sesbania rostrata*, than to other *Rhizobium* species (Young *et al.*, 1991). Attempts to demonstrate Bchl in *A. caulinodans* and other carotenoid-containing *Rhizobium* species have, however, been unsuccessful (Evans *et al.*, 1990).

Classification of purple bacteria based on the morphology of their photosynthetic membranes corresponds only in part to the phylogenies determined by 16S rRNA sequence analyses, and has, therefore, only a limited value in the taxonomy of purple bacteria. For example, most organisms in the alpha-1 subgroup exhibit a grana-like stacked arrangement of their photosynthetic intracytoplasmic membranes (ICM), those in the alpha-2 subgroup exhibit a parallel orientation of ICM, while those in the alpha-3 subgroup have ICM of the vesicular type (Woese *et al.*, 1984a). However, the latter ICM type is also shared by several members of the alpha-1 subgroup (including *Rhodospirillum rubrum* and *Rhodopila globiformis*), and all members of the Chromatiaceae (in the gamma-1 subgroup), except for *Thiocapsa pfennigii* which has a tubular array of ICM. Finally, all members of the Ectothiorhodospiraceae (also in the gamma-1 subgroup) have grana-like membrane stacks also exhibited by members of the alpha-1 subgroup (Collins & Remsen, 1991; Hawthornthwaite & Cogdell, 1991).

The photosynthetic apparatus of *Roseobacter denitrificans* (*Erythrobacter* sp. OCh114) consists of large membranous vesicles formed only under aerobic-dark conditions (Iba *et al.*, 1988; Harashima & Takamiya, 1989). The vesicles are sometimes seen to be continuous with the cytoplasmic membrane (CM) and are similar in appearance to those found in the alpha-3 subgroup, but are approximately twice as large. Vesicular ICM were also observed in *Methylobacterium rhodesianum* (*Protaminobacter ruber*) (Sato & Shimizu, 1979) and *Rhizobium* sp. BTai1 (Evans *et al.*, 1990) even though these aerobic photosynthetic bacteria are more closely related to the alpha-2 subgroup. However, other group II methylotrophs are reported to contain parallel ICM around the cell periphery (Tsuji *et al.*, 1990) similar to members of the alpha-2 subgroup. Methylotrophs in both groups I and II possess ICM whether they form Bchl or not (Davies & Whittenbury, 1970). Well-formed ICM are also found in nitrifying bacteria which oxidize either ammonia (the *Nitroso*-group) or nitrite (the *Nitro*-group) (Watson & Mandel, 1971), species of the free-living nitrogen-fixing genus *Azotobacter* (Oppenheim & Marcus, 1970), and a hydrocarbon-oxidizing species of the genus *Acinetobacter* (Kennedy & Finnerty, 1975). Although none of these latter strains are known to contain Bchl, many have been placed within the proteobacteria (see Table 10.4) and may have inherited their ICM from their photosynthetic ancestors.

The situation is quite different in the case of *Erythrobacter longus*, in which no ICM has yet been visualized (Shiba, 1989b), and *Erythromicrobium sibiricus*, in which irregular vesicular or looplike invaginations of the CM are formed (Yurkov *et al.*, 1991b). These morphologies are most similar to those of species of the genus *Rhodocyclus*, which are classified in the beta subdivision (Woese *et al.*, 1984b), and whose photosynthetic apparatus is either found on the CM only, or in the form of a limited number of irregular invaginations of the CM (de Boer, 1969). Lee (1987) has found, however, that *Rhodocyclus gelatinosus* (beta-1 subgroup) could be induced to form significant membrane invaginations under conditions of 3000 lux illumination, whereas *Rhodocyclus tenuis* (beta-2 subgroup) could not. In fact, recent DNA-rRNA hybridization experiments by Willems *et al.* (1991) indicated that *Rhodocyclus gelatinosus* was only very distantly related to the only two other species of the genus, *Rhodocyclus purpureus* and *Rhodocyclus tenuis* (both in the beta-2 subgroup). These authors have, therefore, proposed that *Rhodocyclus gelatinosus* be placed in a new genus and renamed *Rubrivivax gelatinosus*. Finally, Hiraishi *et al.* (1991) have identified several strains of a '*Rhodocyclus gelatinosus*-like' photosynthetic bacterium, which they found to be unlike any of the previously described species of *Rhodocyclus* on the basis of DNA-DNA homology, and have proposed a new genus and species, *Rhodoferax fermentans*, for this organism.

Species differences can also be observed in the nature of the LH Bchl-protein complexes contained within the photosynthetic membranes. Three major types of LH complexes are found in the purple bacteria (Drews, 1985; Zuber & Brunisholz, 1991). These consist of a core LH complex, which absorbs light maximally at a single wavelength (either between 870 and 890 nm for Bchl *a* (B880) or between 1010 and 1020 nm for Bchl *b* (B1015), and two types of variable LH complexes which absorb maximally at two different wavelengths (either 800 and 850 nm (B800–850) or 800 and 820 to 830 nm (B800–820). Most of these LH complexes are composed of two different types of small molecular mass (5–8 kDa) protein subunits, referred to as the α- and β-subunits. Light-harvesting proteins from a large number of purple bacteria have now been sequenced (Zuber & Brunisholz, 1991), and all show a high degree of homology within these two subunit types. Different species of purple bacteria may contain either one, two or all three of these distinctly different LH complexes, and may, therefore, be classified into three groups on this basis. Some representative species in each of these three groups are as follows:

(a) Group I: *Rhodospirillum rubrum* (alpha-1) and *Rhodopseudomonas marina* and *Rhodopseudomonas viridis* (alpha-2) form only the core LH complex (B880 or Bl015).
(b) Group II: *Rhodopseudomonas palustris* (alpha-2), *Rhodobacter capsulatus* and *Rhodobacter sphaeroides* (alpha-3), and *Rhodocyclus gelatinosus* (beta-1) all form the core (B880) and one variable LH complex (B800–850).
(c) Group III; *Rhodopseudomonas acidophila* (alpha-2) and *Chromatium vinosum* (gamma-1) both form the core (B880) and two different variable LH complexes (B800–850 and B800–820).

As can be seen, the possession of a particular set of LH complexes does not correlate well with the phylogenetic relationships determined by 16S rRNA sequence analysis; hence, the classification above is of little taxonomic value. In addition, some of the complexes exist in more than one form (e.g. forms I and Il for B800–850 in *Rhodopseudomonas palustris*, *Rhodopseudomonas acidophila* and *Chromatium vinosum*) which are variably expressed under different conditions of light and/or temperature (Hawthornthwaite & Cogdell, 1991). Again, however, these differences cut across the subdivisional lines. In the case of the aerobic photosynthetic bacteria, *Erythrobacter longus* contains only a core LH complex (B865), while *Erythrobacter* sp. OCh175 and *Roseobacter denitrificans* (*Erythrobacter* sp. OCh114) have both the core complex and a second LH complex which absorbs at both 802 and 844 nm in OCh175 (Shiba & Abe, 1987), but *only* at 806 nm in OCh114 (Shimada *et al.*, 1985). Finally, new types of LH complexes continue to be discovered: a new purple nonsulphur bacterium, *Chromatium tepidum*, has recently been shown to have an *in vivo* maximum of 917 nm for one of its Bchl a-containing LH complexes, presumably allowing this organism to absorb light in a spectral region not used by other organisms containing either Bchl *a* or Bchl *b* (Nozawa *et al.*, 1986).

In contrast to the LH complexes, RC complexes are very similar in most, if not all, purple bacteria, and, except for differences between Bchl *a*- and *b*-containing organisms, have very similar spectra. Those RCs examined in detail have been found to contain four Bchl (*a* or *b*), two Bphy (*a* or *b*), two quinones (ubiquinone and/or menaquinone), and one iron atom bound by two intrinsic membrane proteins referred to as the L- and M-subunits (Drews, 1985; Feher *et al.*, 1989). An H-subunit is also present which binds no pigment, and the RC of some species is also isolated as a complex with a tetrahaeme *c*-type cytochrome. The primary electron donor of the RC (P870) has been shown to be a dimer of Bchl, with one of the two Bphy molecules and both of the quinones (Q_A and Q_B) being electron acceptors. The L- and M-subunits of *Rhodobacter sphaeroides*, *Rhodobacter capsulatus* and *Rhodopseudomonas viridis* have been found to have significant homology with two analogous proteins (D1 and D2) found in the RC of PS-II of higher plants (Komiya *et al.*, 1988; Michel & Deisenhofer, 1988). The genes for the L- and M-subunits of the RC of the aerobic photosynthetic bacterium *Roseobacter denitrificans* (*Erythrobacter* sp. OCh114) have also recently been sequenced by Liebetanz *et al.* (1991) and found to be homologous to those of the purple bacteria.

Again, it might seem that analysis of the RCs would be of little taxonomic value. However, Dijkman *et al.* (1988) measured parameters during absorbance-detected magnetic resonance (ADMR)-monitored triplet difference spectroscopy of the RCs in membranes from several different photosynthetic bacteria. They were able to distinguish between species of Chromatiaceae and Rhodopirillaceae, as well as between species of Chlorobiaceae and Chloroflexaceae. They suggested that after more species had been examined, finer distinctions may be able to be made, and this technique may have some use as a future taxonomic tool (Dijkman *et al.*, 1988).

10.3.2 Green and brown bacteria

Historically, the green and brown bacteria (called the suborder Chlorobiinae by Pfennig & Trüper, 1971) were classified as strictly anaerobic obligate photolithotrophs which used reduced sulphur compounds (e.g. sulphide, elemental sulphur, and thiosulphate) as sources of electrons for CO_2 fixation. With the discovery by Pierson and Castenholz (1974) of *Chloroflexus aurantiacus* (which was able to grow heterotrophically as well as photoheterotrophically under microaerophilic conditions), the Chlorobiinae have been divided into two families: the green sulphur bacteria (Chlorobiaceae) and the green nonsulphur bacteria (Chloroflexaceae). The majority of the LH Bchls (*c*, *d* or *e*) of green and brown bacteria in both families are found in unusual sac-like vesicles called 'chlorosomes', which are usually found in close proximity to the CM (see Amesz, 1991 for a review).

As was the case with the purple bacteria, the recent use of 16S rRNA sequences has confirmed the validity of the division between the green sulphur and the green nonsulphur bacteria (Gibson et al., 1985; Witt et al., 1989). Gibson et al. (1985) also found that two non-photosynthetic species *Herpetosiphon aurantiacus* and *Thermomicrobium roseum*, are very closely related to the green nonsulphur bacterium, *Chloroflexus aurantiacus*, on the basis of 16S rRNA sequence analyses. Both *Chloroflexus aurantiacus* and *Thermomicrobium roseum* are thermophilic, and all three share the ability to form long, gliding filaments. The evidence indicates that the Chloroflexaceae and their relatives represent an extremely ancient branch away from the Chlorobiaceae, and both should be considered as separate bacterial phyla (Gibson et al., 1985; Oyaizu et al., 1987, Woese, 1987). The evidence also points to a thermophilic ancestor of all of the bacteria (Oyaizu et al., 1987). The close relationship of *Chloroflexus aurantiacus* with *Thermomicrobium roseum* was confirmed by van den Eynde et al. (1990) by sequence analysis of 5S rRNA; however, their analysis indicated that *Herpetosiphon aurantiacus* was not closely related to the other two.

Sequence analyses of 16S rRNA have also suggested a relationship between the Chlorobiaceae and the non-photosynthetic flavobacteria and their relatives, although both are still considered to be separate bacterial phyla (Woese, 1987). This relatedness has recently been confirmed by analysis of 23S rRNA sequences (Woese et al., 1990). However, analysis of 5S rRNA sequences by van den Eynde et al. (1990) demonstrated a relationship between the green sulphur bacteria, several species of the genus *Bacteroides*, and *Cytophaga aquatalis*, but failed to demonstrate any relatedness between a number of other members of the '*Bacteroides-Cytophaga-Flavobacterium* and relatives' phylum.

Chloroflexus aurantiacus is the only member of the family Chloroflexaceae which has been well characterized, and thus it has been used for most of the biochemical studies carried out on this family. However, many other '*Chloroflexus*-like' photosynthetic organisms have been described in the literature (but most, unfortunately, not maintained in pure culture), including '*Chloroflexus*-like' mesophilic, halophilic and sulphide-oxidizing, obligately photoautotrophic strains (Giovannoni et al., 1987; Gorlenko, 1988; Mack & Pierson, 1988). Several other green bacteria, which have been assigned to other genera, share the ability to form long, gliding filaments. One such bacterium, *Chloroherpeton thalassium*, which contains both Bchl a and c and forms chlorosomes, has, however, been placed with the Chlorobiaceae based on 16S rRNA analysis (Gibson et al., 1985). A related species, *Chloroherpeton limnophilum*, has also recently been reported (Eichler & Pfennig, 1988).

Several species of the genera *Oscillochloris* and *Chloronema* have been considered to be candidates for membership in the Chloroflexaceae family (Gorlenko, 1988). In both *Oscillochloris* and *Chloronema*, chlorosomes are found attached not only to the CM, but to numerous internal membranes and cell septa as well (Sprague & Fuller, 1991). However, unpublished studies by Keppen (quoted in Gorlenko, 1988) revealed a total lack of DNA homology between *Oscillochloris trichoides* and both *Chloroflexus aurantiacus* and *Chlorobium limicola* f. *thiosulfatophilum*. It should be noted that 16S rRNA sequence analyses of these strains have yet to be carried out (Stackebrandt *et al.*, 1988a). Another thermophilic gliding filamentous bacterium, *Heliothrix oregonensis*, contains only a Bchl *a*-like pigment and is unable to form chlorosomes (Pierson *et al.*, 1985; Castenholz, 1988). Sequence analysis of 5S rRNA indicated that it was related more closely to *Chloroflexus aurantiacus* than to any other type of photosynthetic bacterium examined, but that it should be placed in a new genus (Pierson *et al.*, 1985). However, again, sequence analysis of 16S rRNA has yet to be carried out. If a close phylogenetic relationship to *Chloroflexus aurantiacus* is confirmed, *Heliothrix oregonensis* may thus either be a Chloroflexaceae species that has lost the ability to form Bchl *c* and chlorosomes, or one which has not yet acquired such membrane differentiation (Hawthornthwaite & Cogdell, 1991).

The chlorosomes of both Chloroflexaceae (Staehelin *et al.*, 1980) and Chlorobiaceae (Feick & Fuller, 1984) are quite similar in their overall morphology, which, rather surprisingly, gives no hint of the deep phylogenetic difference between the two families. However, there is a lack of cross-reactivity of antisera prepared against three major chlorosome proteins from these two families, which does provide evidence for their evolutionary divergence (Stoltz *et al.*, 1990). The chlorosomes of both are bounded by a monolayer formed from phospho- and glycolipids and several different protein subunits, rather than a bilayer membrane. The antenna Bchls *c*, *d* or *e* are aggregated into rod-like elements which run lengthwise through the oblong chlorosome vesicles. The organization of the antenna Bchls *c*, *d* and *e* in the chlorosomes of green bacteria has been the subject of a good deal of speculation, and a controversy over the exact function of several small molecular mass (5.7–7.5 kDa) chlorosome polypeptides has yet to be resolved (see Holzwarth *et al.*, 1990; Wullink *et al.*, 1991; Niedermeier *et al.*, 1992). Chlorobiaceae also contain a water-soluble LH complex containing Bchl *a* (B808) which forms a paracrystalline 'baseplate' between the chlorosome and the CM, where the RC is located. This complex is not found in Chloroflexaceae; however, both families seem to possess another LH complex containing Bchl *a* (B792 or B794), which may also be located in the area of connection between the CM and the chlorosome, and may act as an intermediary in the transfer of light energy from the antenna Bchl *c* in the chlorosome to the RC complexes in the CM (Betti *et al.*, 1982; Gerola & Olson, 1986).

The distinction between Bchls *c* and *d* (then known as chlorobium chlorophylls 660 and 650, respectively, after their absorption maxima in ether) was initially made by Stanier and Smith (1960). Using two closely related strains (named *Chlorobium thiosulfatophilum*-660 and -650, respectively), they found that Bchl *c* was produced by a rod-shaped strain (Tassajara 6230), now known as *Chlorobium limicola* f. *thiosulfatophilum*, while Bchl *d* was produced by a vibrio-shaped strain (Lascelles 8327), now known as *Chlorobium vibrioforme* f. *thiosulfatophilum*. Other species of green bacteria have since been described which contain one or the other of these two Bchls, including *Chloroflexus aurantiacus*, which contains Bchl *c* (Pierson & Castenholz, 1974). Bchl *e* was originally discovered in two species of sulphur bacteria (*Chlorobium phaeobacteroides* and *Chlorobium phaeovibrioides*) which appear brown due to the presence of unusual carotenoids (Gloe *et al.*, 1975). Although initially it seemed it would prove rather limited in its distribution, Bchl *e* has now been found in several species of at least two other genera (e.g. *Pelodictyon phaeum* and *Prosthecochloris phaeoasteroidea*), and new strains of brown sulphur bacteria containing Bchl *e* continue to be reported (e.g. *Pelodictyon phaeoclathratiforme*; Overmann & Pfennig, 1989).

Although containing predominantly Bchl *c*, small amounts of Bchl *d* were reported to be present in the 6320 strain of *Chlorobium* (Richards & Rapoport, 1966), in *Prosthecochloris aestuarii* (Smith *et al.*, 1983a), and in *Chloroflexus aurantiacus* (Brune *et al.*, 1987). Also, Broch-Due and Ormerod (1978) had reported that their 8327 strain of *Chlorobium* had shifted from the production of Bchl *d* to the production of Bchl *c* during several months of continuous culture under low-light conditions. This prompted Smith and Bobe (1987) to re-examine their own culture of the 8327 strain of *Chlorobium*, which had been in continuous culture in their laboratory for a period of years. To their surprise, they found that it had also shifted to the production of predominantly Bchl *c* with an absorbance maximum at 750 nm! In addition, they found that the ratio of homologues containing 5-Et groups to those containing 5-Me groups (once about 3:1) had changed to about 16:1. Fortunately, they had saved culture samples during the period of the change (1980–84), and the pigment content of these samples was re-examined (Bobe *et al.*, 1990). In the 5-Et series of homologues, an increase in the 4-*i*-Bu homologue of Bchl *d* occurred as early as 1981, with the 4-*n*-Pr homologue of Bchl *c* becoming predominant two years later. Huster and Smith (1990) also found an increase in the degree of methylation of the Bchl in the 8327 strain during the 4-month growth period of a single culture. Earlier, the pigment content of a newly isolated strain (B1–20) of *Chlorobium vibrioforme* f. *thiosulfatophilum* had also been observed to change over a period of 14 months from one in which the (4-Et,5-Et) homologue of Bchl *d* predominated (~80%) to one in which the (4-*n*-Pr,5-Et) and (4-*i*-Bu,5-Et) homologues of Bchl *d* were major components, and the *in vivo* absorbance maximum had shifted from 714 to 728 nm

(Smith & Goff, 1985). Hence, a shift from the synthesis of Bchl d to Bchl c (and/or an increase in the degree of methylation at the 4- and 5-positions) may be an adaptive response of a given strain of green bacteria to growth conditions, rather than a permanent property of the strain, and the possession of either Bchl c or d is, therefore, of no particular taxonomic significance. Simpson and Smith (1988) have also observed the presence of minor amounts of compounds presumed to be derived from Bchl f in the Bchl e-forming *Chlorobium phaeovibrioides*. Perhaps this bacterium has also undergone an adaptive shift from Bchl f to Bchl e!

Aggregates of the (4-i-Bu,5-Et) homologue of Bchl c have been found to exhibit a longer wavelength of absorption than the (4-n-Pr,5-Et) homologue (Causgrove *et al.*, 1990). This observation correlates well with those of Smith and co-workers (Smith & Goff, 1985; Smith & Bobe, 1987; Bobe *et al.*, 1990) that increased methylation in the 4- and 5-positions of Bchl c or d increased the wavelength of the far-red absorption maximum *in vivo*. Bobe *et al.* (1990) proposed a model whereby increased methylation leads to a stabilization of the Bchl c (or d) aggregates, which are also stabilized by the S-configuration (and destabilized by the R-configuration) of the 2a-carbon of the 2-α-hydroxyethyl group. In this context, it is of interest that Otte *et al.* (1991) have recently determined that rod elements containing aggregates with a higher proportion of longer-wavelength absorbing Bchls increase with decreasing distance from the CM, presumably in order to facilitate more efficient light energy transfer to the RC. Hence, regulation of the degree of methylation of the 4- and 5-positions may be an important method for the regulation of light harvesting in the green bacteria.

In spite of the similarities observed in the structure of the chlorosomes of green sulphur and green nonsulphur bacteria, the photosynthetic components in their CMs are strikingly different, and do reflect the deep division between the two families indicated by 16S rRNA sequence analyses. In the case of *Chloroflexus aurantiacus*, the RC complex (P865) located in the CM contains two polypeptides, three Bchl a, three Bphy a, and one to two menaquinone molecules. The two polypeptides have been sequenced (Ovchinnikov *et al.*, 1988a,b; Shiozawa *et al.*, 1989) and shown to have a fairly high degree of homology (40–42%) to the L- and M-subunits of the purple bacterial RC, as well as polypeptides D1 and D2 in the RC of PS-II of higher plants (Komiya *et al.*, 1988). No H-subunit or c-type cytochrome was isolated as part of the RC complex, however. An intrinsic LH complex of Bchl a (B806–866) is also present in the CM. It is composed of two small molecular mass protein subunits which are homologous to the core LH complexes (B880 and B1015) of purple bacteria (Zuber & Brunisholz, 1991), and no doubt associates with the RC in the membrane.

In the case of the green sulphur bacteria, the RC has not been as well characterized, but appears to be similar to the RC of PS-I of higher plants and cyanobacteria. Purified antenna-RC preparations have recently been obtained from the CM of *Chlorobium limicola* f. *thiosulfatophilum* (Nitschke et al., 1990a; Feiler et al., 1992) and found to contain about 40 Bchl *a* per phototrap (P840). It also contained three iron-sulphur centres (similar, but not identical, to the Fe-S$_A$, Fe-S$_B$ and Fe-S$_X$ centres of higher plants) of low enough potential to be able to reduce NAD$^+$ in the light. Evidence for two additional electron acceptors (A$_0$ and A$_1$) present in the RC complex, one of which (A$_1$) may be a quinone, was presented by Nitschke et al. (1987). There have been a number of proposals as to the identity of the other electron acceptor (A$_0$), including Bphy *c*, Bphy *a*, Bchl *c*, and even chlorophyll *a*. Braumann et al. (1986) analysed the RC of *Prosthecochloris aestuarii* and characterized one such component (which they termed Bchl 663) as the phytyl ester of Bchl *c*; moreover, they found no Bphy *a* nor Bphy *c* in the RC of this organism.

10.3.3 Heliobacteria

Heliobacterium chlorum was discovered only recently (Gest & Flavinger, 1983), but it was immediately apparent that it did not fit into any of the previously known groups of photosynthetic prokaryotes. Although brownish-green in colour and with a gliding motility similar to *Chloroflexus*, it was a strictly anaerobic photoheterotroph, did not have chlorosomes (or indeed any extensively developed ICM) (Miller et al., 1986; Sprague & Fuller, 1991), and contained the previously unknown Bchl *g* as its sole Bchl. As seen in freeze-fracture electron micrographs, the cell membranes were densely packed with uniformly-sized particles and contained nearly 70% protein (Fuller et al., 1985). On the basis of 16S rRNA sequence analysis, it was found to be most closely related to the Gram-positive bacteria within the *Bacillus-Lactobacillus-Streptococcus* branch, although *H. chlorum* itself is apparently not Gram-positive (Woese et al., 1985a). Thus, another major bacterial phylum has been found to have a photosynthetic member, and further strengthens the case for a common photosynthetic ancestry for all bacteria (Woese et al., 1984a). Although the structure of Bchl *g* is closely related to that of Bchl *b* (Figure 10.1), analysis of the *in vivo* spectrum of *H. chlorum* reveals LH Bchl *g*-protein complex(es) with the rather low absorption maxima of 778, 793 and 808 nm (van Dorssen et al., 1985), rather than the greatly far-red shifted absorption maximum of 1020 nm in the Bchl *b*-containing LH complexes (see Table 10.2). The molecular basis for this difference awaits further studies. Recently, three additional strains of Heliobacteriaceae have been isolated: *Heliobacillus mobilis* (Beer-Romero & Gest, 1987) and two species of *Heliobacterium*, *H. gestii* and *H. fasciculum* (Ormerod et al., 1989). The latter two strains both form heat-resistant endospores.

Antenna-RC complexes have been recently isolated from the CM of *Heliobacillus mobilis* and *Heliobacterium chlorum* (van de Meent *et al.*, 1990), and *Heliobacillus mobilis* and *Heliobacterium gestii* (Trost & Blankenship, 1989; Trost *et al.*, 1992). The purified complexes were found to be similar in all three species studied, and consisted of a single polypeptide which contained about 30–60 antenna Bchl *g* molecules per phototrap (P800). Most of the evidence indicates that there are no additional LH complexes present in the membranes. The primary electron donor is probably a Bchl *g* dimer (Prince *et al.*, 1985), although Kobayashi *et al.*, (1991a,b) have recently provided evidence that it is a dimer of Bchl *g'* (the 10S epimer of Bchl *g*). Also, van de Meent *et al.* (1991) have characterized a hydrated and oxidized derivative of Bchl *g* (4-desethyl-4-α-hydroxyethylchlorophyll a_F), present in the ratio of about two per P800, which may be an electron acceptor. In addition, approximately 1.4 menaquinones per P800 were present in the purified complex, plus iron and acid-labile sulphur (Trost & Blankenship, 1989). Sequence analysis of a peptide fragment from the RC protein of *Heliobacillus mobilis* (Trost *et al.*, 1992) indicated a significant homology with both RC proteins (A1 and A2) of green plants and cyanobacteria, including the two highly conserved cysteine residues thought to bind the Fe-S_x electron acceptor. Hence, the RC of Heliobacteriaceae may be fundamentally similar to the RCs of both the green sulphur bacteria and PS-I of green plants. The primary donor P800, and electron acceptors A_0, A_1 and Fe-S_x may, therefore, all be bound to a homodimer of the RC protein. This model remains to be established definitively, however, as no photo-induced reductions of these putative acceptors can be observed in purified RCs, and there is no evidence for dimerization of the RC protein. Photo-induced EPR signals, possibly due to a quinone radical (Fischer, 1990) and two Fe-S centres (Nitschke *et al.*, 1990b) have, however, been observed in intact membranes from *Heliobacterium chlorum*.

Based on recent experience, it is likely that numerous other interesting anoxygenic photosynthetic prokaryotes await discovery. Hence, a further increase in the number of classes of such organisms can be anticipated and, perhaps also an increase in the number of previously unknown Bchls.

10.3.4 Cyanobacteria

The biology of the cyanobacteria is treated in detail in a book edited by Carr and Whitton (1982). The taxonomic scheme of Rippka *et al.* (1979) divides the cyanobacteria into five sections: unicellular cyanobacteria reproducing either by binary fission or budding (section I) or by multiple fission (section II); and filamentous cyanobacteria either containing vegetative cells only (section III) or capable of heterocyst formation without (section IV) or with (section V) true branching. The cyanobacteria comprise a phylogenetically coherent class (or phylum) which contains no known non-photosynthetic representatives (Woese, 1987). Analysis of 16S rRNA, c-type cytochromes, and D-ribulose-1,5-bisphosphate carboxylase/oxygenase (rubisco) sequences favours a divergence of cyanobacteria from the purple bacterial line *after* the divergence of the latter from the green bacterial line (Doolittle, 1982). A more extensive sequence analysis of the 16S rRNA from 30 cyanobacteria (including the cyanelle of the phytoflagellate *Cyanophora paradoxa*) has recently been carried out by Giovannoni et al. (1988). They found less divergence among the various cyanobacteria examined than between the members of many other bacterial phyla. This indicated that, rather than representing a deep evolutionary branch, many modern cyanobacterial lineages arose in a more recent expansive evolutionary radiation (Giovannoni et al., 1988). Their results also confirmed that sections II, IV and V of the five sections of cyanobacteria defined by Rippka et al. (1979) were phylogenetically coherent, but that representatives of sections I and III were dispersed throughout the cyanobacterial tree, and were thus not natural groupings.

Giovannoni et al. (1988) have also compared the sequences they obtained for the 16S rRNA of cyanobacteria to those of the green chloroplasts of higher plants, green algae and *Euglena*. These comparisons indicated that the green chloroplasts and all of the cyanobacteria that they examined form a single coherent phylogenetic group (even though the cyanobacteria contain phycobilisomes and no chlorophyll *b*). The closest relationship, however, was between green chloroplasts and the cyanelle of *Cyanophora paradoxa*. Sequence data has also been used to demonstrate a closer relatedness between most cyanobacteria and the red and brown chloroplasts of algae lacking chlorophyll *b*, than the chloroplasts of higher plants and green algae (Doolittle & Bonen, 1981; Schwartz & Dayhoff, 1981; Doolittle, 1982). These data also demonstrated a much more distant relatedness between chloroplast and *cytoplasmic* rRNA, favoring the endosymbiotic theory for the origin of all chloroplasts. On the other hand, analysis of the sequences of the genes encoding the large and small subunits of rubisco by Morden and Golden (1991) has shown a close relationship between the red and brown chloroplasts and the beta-subclass of proteobacteria. On the basis of both 16S rRNA and rubisco sequence analyses, Markowicz and Loiseaux-de Goër (1991) have postulated that red and brown chloroplasts may, therefore, have had a composite polyphylogenetic origin from at least two different ancestors (a cyanobacterial and a beta-proteobacterial). Moreover, green chloroplasts (except for euglenophytes) may have had *only* a cyanobacterial ancestor (which was, however, different from that of the red and brown chloroplasts).

10.3.5 Prochlorophytes

Lewin first reported the discovery of a prokaryotic alga, now called *Prochloron didemni*, with a green plant-type photosynthetic apparatus containing chlorophyll *b* and no phycobilisomes (Lewin, 1975; Lewin & Withers, 1975). Attention immediately focused on this organism as a possible 'missing link' in the evolution of green plant chloroplasts (Lewin, 1981). Unfortunately, strains of *Prochloron* cannot be cultured in the laboratory, and it remains, for the most part, a 'microbial enigma' (Lewin & Cheng, 1989). In 1986, however, another prochlorophyte, *Prochlorothrix hollandica*, was isolated which *can* be easily cultured in the laboratory (Burger-Wiersma *et al.*, 1986). Florenzano *et al.* (1986) proposed that the newly described prochlorophytes be placed in a new order to be known as Prochlorales. More recently, it has been found that many previously unknown prochlorophytes can easily be strained from the open ocean (Chisholm *et al.*, 1988). Two strains of these bacteria have now been cultured and given the name *Prochlorococcus marinus* (Chisholm *et al.*, 1992). These strains have been shown by Goericke and Repeta (1992) to contain two 'divinylchlorophylls' (the 4-desethyl-4-vinyl-derivatives of both chlorophylls *a* and *b*), zeaxanthin, α-carotene (rather than β-carotene), and a protochlorophyll(ide) spectroscopically similar to magnesium 2,4-divinylpheoporphyrin a_5. Perhaps even more surprises can be expected as the search for additional members of this interesting group continues!

Similarities between the genes encoding both the D1 polypeptide of PS-II (Morden & Golden, 1989) and the chlorophyll *a*/*b*-binding proteins (Bullerjahn, 1989) of *Prochlorothrix hollandica* and green chloroplasts, indicated that both share a common evolutionary origin about 1 billion years ago (Bullerjahn, 1989). However, the first detailed analyses of 16S rRNA sequences of several species of *Prochloron* (Seewaldt & Stackebrandt, 1982; Stackebrandt *et al.*, 1982), *Prochlorothrix hollandica* (Turner *et al.*, 1989), and *Prochlorococcus marinus* (Urbach *et al.*, 1992) have indicated that the three known prochlorophyte genera are polyphyletic and are much more closely related to diverse species of cyanobacteria than they are to each other. Furthermore, they are only distantly related to the green chloroplasts of higher plants. Although yielding a less-than-certain branching pattern, sequence analysis of DNA-dependent RNA polymerases by Palenik and Haselkorn (1992) has confirmed the wide divergence in the phylogenetic relationship between the prochlorophyte genera. Hence, a phycobilisome$^-$/ chlorophyll b^+ phenotype seems to have arisen independently at least three (and possibly four) times during the evolution of the green chloroplast ancestor and the three prochlorophyte genera (Urbach *et al.*, 1992). These authors also feel that the order Prochlorales can no longer be justified, and that the prochlorophytes should, therefore, be classified within the cyanobacterial phylum.

10.3.6 The evolution of the chlorophylls of photosynthetic prokaryotes

The nature of the 'ancestral chlorophyll' of the first photosynthetic prokaryotes will probably never be known. The recapitulation hypothesis states that the modern-day biosynthetic pathway of a biomolecule recapitulates its own evolution. Larkum (1991) has summarized the application of this hypothesis to chlorophylls. If one assumes that the evolution of the early tetrapyrrole intermediates of chlorophyll synthesis paralleled that of haem up to the production of protoporphyrin IX, then magnesium protoporphyrin IX might have been the first photoactive pigment specific to the chlorophyll pathway to have evolved. Acquiring the ability to elaborate the exocyclic cyclopentenone ring would have afforded magnesium 2,4-divinylpheoporphyrin a_5 (also called 'divinylprotochlorophyllide'), which absorbs with a higher extinction coefficient in the visible (red) region of the spectrum. Its formation would have been facilitated by acquisition of the ability to form the 6-β-ketopropionate derivative of magnesium protoporphyrin IX monomethyl ester, while subsequent reduction of the 4-vinyl group would have yielded 'monovinyl' protochlorophyllide (Richards, 1993). Noting the structural similarity between these two protochlorophyllides and chlorophylls c_1 and c_2, Larkum (1991) has suggested that the latter may also have preceded chlorophyll a as ancestral chlorophylls. Evolution of the ability to form chlorophyllide a by reduction of ring IV of the protochlorophyllide macrocycle would have further increased the extinction coefficient of the red band of the photoactive pigment (as well as shifting its absorption maximum to higher wavelengths). Finally, evolution of the ability to carry out the esterification of the 7-propionic acid residue of either protochlorophyllide or chlorophyllide a would have enhanced the hydrophobicity (and hence the membrane solubility) of the resulting protochlorophyll (Olson, 1981) or chlorophyll a (Larkum, 1991), further increasing their usefulness as photoactive pigments. It may be that ancient photosynthetic prokaryotes experimented with all of the pigments mentioned above in prototype photoactive RCs.

Because of the extremely ancient branch between the green sulphur and nonsulphur bacteria (predicted on the basis of 16S rRNA sequence analyses), Oyaizu *et al.* (1987) and Woese (1987) have suggested that *Chloroflexus*-like organisms may have been the first truly successful photosynthetic prokaryotes. In fact, *Chloroflexus*-like organisms may have been responsible for the 3.5 billion-year-old stromatolites, rather than cyanobacteria which appear to be of a more recent origin (Oyaizu et al., 1987). The cyanobacteria are thought to have branched from the line leading to the Bchl g-containing heliobacteria (and all other Gram-positive bacteria) after the latter had previously branched from the purple bacteria line (Woese, 1987). This scheme of prokaryotic evolution would suggest three possible alternatives:

(a) Bchl a had evolved from chlorophyll a by the time of the separation of the green sulphur and nonsulphur bacteria (either as the sole RC chlorophyll or as one of a mixture of photoactive chlorophylls, including those currently acting as modern-day RC chlorophylls, chlorophyll a, Bchl a and Bchl g.
(b) Bchl a evolved from chlorophyll a at least *three* different times (at the branch points of the green nonsulphur, green sulphur and purple bacteria).
(c) Bchl a evolved later and the genetic information required to form it from chlorophyll a was passed between the above bacterial groups by a process of lateral transfer.

It is currently not possible to decide between these different alternatives.

As chlorophyll a is such a successful pigment for modern-day cyanobacterial photosynthesis, why should the pathway have evolved to the level of Bchl a at all? The modern-day biosynthesis of Bchl a is nearly identical to that of chlorophyll a up to the production of chlorophyllide a (Richards & Lascelles, 1969; Pudek & Richards, 1975). Bacteriochlorophyllide a is formed from chlorophyllide a by an additional three biochemical steps, followed by subsequent esterification to yield Bchl a. Bchl a has a much longer wavelength for its red absorption maximum than chlorophyll a (770 vs. 660 nm), and the *in vivo* maximum in early organisms could have been at a much longer wavelength. (In modern-day purple bacteria, it is shifted way out to 900 nm in the near-infrared.) Also, Bchl a has a much *lower* wavelength of absorption in the Soret region of the spectrum than does chlorophyll a (360 vs. 430 nm). It would, therefore, be much more effective in the absorption of near ultraviolet light, which was much more abundant at times before the advent of oxygen in the atmosphere.

Similar considerations would also apply to Bchl g as well as to Bchl a. The evolution of the ability to form Bchl g may have occurred by alteration of an intermediate in the biosynthetic pathway of Bchl a (for example, by reduction of divinylchlorophyllide a followed by esterification of the resulting bacteriochlorophyllide g), or by isomerization of chlorophyll a. If a variety of RC chlorophylls were available to early photosynthetic prokaryotes, such an arrangement would have allowed for a much greater efficiency of energy absorption before other pigments currently acting as LH pigments came into existence. These LH pigments (chlorophylls b, c_1 and c_2, Bchl c, d and e, phycobilins, and the photoactive carotenoids) may only have evolved at much later dates (Larkum, 1991). Hence, the major branching events may then have corresponded to a particular line of photosynthetic prokaryotes selecting one or the other of the major RC chlorophylls from the available pool, in order to give it a particular selective advantage.

In modern-day photosynthetic organisms, RCs of two different types (PS-I and PS-II) are found in the cyanobacteria, prochlorophytes and the chloroplasts of algae and higher plants, while an RC of only the PS-I type is found in green sulphur bacteria and heliobacteria, and an RC of only the PS-II type is found in green nonsulphur bacteria and purple bacteria. Their fundamentally different nature makes it unlikely that one could have evolved from the other, and the two may have had completely separate origins. (Margulies (1991) has, however, detected some sequence similarity between helix X of the A2 polypeptide of the PS-I RC and helix IV of the D2 polypeptide of the PS-II RC, implying a possible evolutionary relatedness.) Doolittle (1982) presented a scheme for the molecular evolution of the photosynthetic prokaryotes which represented the 'ancestral photosynthetic bacterium' as having a PS-I type of photosystem. If *Chloroflexus*-like green nonsulphur bacteria are taken to be the most ancient photosynthetic prokaryotes, however, the first RC may have been of the PS-II type. Nevertheless, it is possible that the photosynthetic progenitor contained RCs of *both* the PS-I and PS-II types by the time of the separation of the green sulphur and nonsulphur bacteria. It is even possible that such a 'dual RC progenitor' later developed the ability to evolve oxygen (Mathis, 1990), although the composition of its RC chlorophylls cannot be guessed. Buick (1992) has expressed the opinion that the 2.7 billion-year-old stromatolites he studied were the result of metabolism by oxygenic photosynthesis.

The following is a possible order for branching events which corresponds to the scheme of bacterial evolution presented by Woese (1987):

(a) The first three branches from the 'dual RC progenitor' all settled upon Bchl a (by whatever mechanism) as their sole RC chlorophyll:

  (i) The first branch retained PS-II only and initiated a line leading to the modern-day green nonsulphur bacteria.
  (ii) The second branch retained PS-I only and initiated a line leading to the modern-day green sulphur bacteria. Bacteriochlorophylls c, d and e may have evolved subsequent to this branching event in order to increase the efficiency of light-harvesting in these organisms. A chlorosome LH system containing Bchl c may have been

transferred to the Chloroflexaceae at a later date by lateral transfer of genetic information.

(iii) The third branch retained PS-II only and initiated a line leading to the purple bacteria and their relatives, with subsequent branches leading to the alpha, beta and gamma subdivisions of the proteobacteria (with the development of Bchl b in some species).

(b) The next branch from the 'dual RC progenitor' settled upon Bchl g as its sole RC chlorophyll. This branch retained PS-I only, and initiated a line leading to modern-day heliobacteria.

(c) The 'dual RC progenitor' eventually settled upon chlorophyll a as its sole RC chlorophyll and retained both PS-I and PS-II. Following the addition of phycobiliproteins as LH pigments, numerous subsequent branching events (of uncertain order) would have generated the many different cyanobacterial lines, the line leading to the modern-day cyanelle of *Cyanophora paradoxa*, and other lines branching from some of the above, subsequent to the following events:

(i) The evolution of the ability to form chlorophylls c_1 and c_2 from protochlorophyllide and divinylprotochlorophyllide, respectively. Branches with members retaining chlorophyll a and either chlorophyll c_1 and c_2 (chromophytes), phycobiliproteins (rhodophytes), or both (cryptophytes) would have initiated lines leading to the prokaryotic ancestors of modern-day brown and red chloroplasts.

(ii) The evolution of the ability to form chlorophyll b from chlorophyll a (possibly at four or five independent points). Branches with members retaining both chlorophylls a and b (but not phycobiliproteins) would have initiated lines leading to modern-day prochlorophytes and the prokaryotic ancestors of modern-day green chloroplasts.

REFERENCES

Amesz, J. (1991) Green photosynthetic bacteria and heliobacteria. In *Variations in Autotrophic Life* (Shively, J.M. and Barton, L.L., eds), pp. 99–119. Academic Press: London.

Anderson, A.F.H. and Calvin, M. (1962) An improved method for separation and purification of chlorophyll a. *Nature* (London) **194**: 285–6.

Archibald, J.L., MacDonald, S.F. and Shaw, K.B. (1963) Synthetic porphyrins related to *Chlorobium* chlorophylls. *J. Am. Chem. Soc.* **85**: 644.

Archibald, J.L., Walker, D.M., Shaw, K.B., Markovac, A. and MacDonald, S.F. (1966) The synthesis of porphyrins derived from *Chlorobium* chlorophylls. *Can. J. Chem.* **41**: 345–62.

Barkigia, K.M., Gottfried, D.S., Boxer, S.G. and Fajer, J. (1989) A high precision structure of a bacteriochlorophyll derivative, methyl bacteriopheophorbide a. *J. Am. Chem. Soc.* **111**: 6444–6446.

Beer-Romero, P. and Gest, H. (1987) *Heliobacillus mobilis*: A peritrichously flagellated anoxyphototroph containing bacteriochlorophyll *g*. *FEMS Microbiol. Lett.* **41**: 109–14.
Betti, J.A., Blankenship, R.E., Natarajan, L.V., Dickinson, L.C. and Fuller, R.C. (1982) Antenna organization and evidence for the function of a new antenna pigment species in the green bacterium *Chloroflexus aurantiacus*. *Biochim. Biophys. Acta* **680**: 194–201.
Bobe, F.W., Pfennig, N., Swanson, K.L. and Smith, K.M. (1990) Red shift of absorption maxima in *Chlorobiineae* through enzymic methylation of their antenna bacteriochlorophylls. *Biochemistry* **29**: 4340–8.
Braumann, T., Vasmel, H., Grimme, L.H and Amesz, J. (1986) Pigment composition of the photosynthetic membrane and reaction center of the green bacterium *Prosthecochloris aestuarii*. *Biochim. Biophys. Acta* **848**: 83–91.
Brereton, R.G. and Sanders, J.K.M. (1983a) Co-ordination and aggregation of bacteriochlorophyll *a*: An N.M.R. and electronic absorption study. *J. Chem. Soc. Perkin Trans.* I: 423–30.
Brereton, R. G. and Sanders, J.K.M. (1983b) Bacteriochlorophyll *a*: Influence of axial co-ordination on reactivity and stability. Design of an improved extraction procedure. *J. Chem. Soc. Perkin Trans.* I: 431–4.
Brereton, R.G and Sanders, J.K.M. (1983c) Bacteriochlorophyll *a*: Assignment of the natural abundance ^{13}C N.M.R. spectrum. Use of power spectra. *J. Chem. Soc. Perkin Trans.* I: 435–8.
Brereton, R.G., Rajananda, V., Blake, T.J., Sanders, J.K.M. and Williams, D.H. (1980) In beam electron impact mass spectrometry: The structure of a bacteriochlorophyll allomer. *Tetrahedron Lett.* **21**: 1671–4.
Briat, B., Schooley, D.A., Records, R., Bunnenberg, E. and Djerassi, C. (1967) Magnetic circular dichroism studies. III. Investigation of some optically active chlorines. *J. Am. Chem. Soc.* **89**: 6170–7.
Broch-Due, M. and Ormerod, J.G. (1978) Isolation of a bacteriochlorophyll *c* mutant from *Chlorobium* with bacteriochlorophyll *d* by cultivation at low light intensity. *FEMS Microbiol. Lett.* **3**: 305–8.
Brockmann, H., Jr. (1968a) The absolute configuration of chlorophyll. *Angew. Chem. Int. Edit.* **7**: 222–2.
Brockmann, H., Jr. (1968b) The absolute configuration of bacteriochlorophyll. *Angew. Chem. Int. Edit.* **7**: 222.
Brockmann, H., Jr. (1976) Bacteriochlorophyll *e*: Structure and stereochemistry of a new type of chlorophyll from *Chlorobiaceae Phil. Trans. R. Soc. London, Ser.* B **273**: 277–85.
Brockmann, H., Jr. (1978) Stereochemistry and absolute configuration of chlorophylls and linear tetrapyrroles. In *The Porphyrins*, vol. II (Dolphin, D., ed), Ch. 10, Academic Press: New York.
Brockmann, H., Jr. and Kleber, I. (1970) Bacteriochlorophyll *b*. *Tetrahedron Lett.* **25**: 2195–8.
Brockmann, H., Jr. and Lipinski, A. (1983) Bacteriochlorophyll *g*: A new bacteriochlorophyll from *Heliobacterium chlorum*. *Arch. Microbiol.* **136**: 17–19.
Brockmann, H., Jr. and Risch, N. (1974) Modified Horeau analysis for determination of the chirality of amines, alcohols, and carboxylic acids. *Angew. Chem. Int. Edit.* **13**: 664–5.
Brockmann, H. and Risch, N. (1991) Preparative chromatography. In *Chlorophylls* (Scheer, H., ed), pp. 103–14. CRC Press: Boca Raton, FL.
Brockmann, H., Jr. and Tacke-Karimdadian, R. (1979) Oxidativer Abbau von Bacteriochlorophyll *d*, Bestätigung der Konstitution und Bestimmung der Absoluten Konfiguration. *Liebigs Ann. Chem.* 419–430.
Brockmann, H., Jr., Knobloch, G., Schweer, I. and Trowitzsch, W. (1973) Die Alkoholkomponente des Bacteriochlorophyll *a* aus *Rhodospirillum rubrum*. *Archiv. Microbiol.* **90**: 161–4.
Brockmann, H., Jr., Gloe, A., Risch, N. and Trowitzsch, W. (1976) Bacteriochlorophyll *e*, ein Neues Chlorophyll aus Braunen Arten von *Chlorobiaceae*. *Liebigs Ann. Chem.* 566–77.
Brockmann, H., Jr., Jürgens, U. and Thomas, U. (1979) Partialsynthese eines Bacteriophäophorbide-*c*-methylesters. *Tetrahedron Lett.* 2133–6.
Brown, S.B., Smith, K.M., Bisset, G.M.F. and Troxler, R.F. (1980) Mechanism of photooxidation of bacteriochlorophyll *c* derivatives. A possible model for natural chlorophyll breakdown. *J. Biol. Chem.* **255**: 8063–8.

Brune, D.C., Nozawa, T. and Blankenship, R.E. (1987) Antenna organization in green photosynthetic bacteria. 1. Oligomeric bacteriochlorophyll c as a model for the 740 nm absorbing bacteriochlorophyll c in *Chloroflexus aurantiacus*. *Biochemistry* **26**: 8644–52.

Buick, R. (1992) The antiquity of oxygenic photosynthesis. Evidence from stromatolites in sulphate-deficient archaean lakes. *Science* **255**: 74–7.

Bullerjahn, G.S. (1989) Prochlorophytes and the evolution of chloroplasts. In *The Spectrum*, vol. 2 (no. 3), (Green, P., ed), pp. 10–12. Center for Photochemical Sciences: Bowling Green, OH.

Burger-Wiersma, T., Veenhuis, M., Korthals, H.J., Van de Wiel, C.C.M. and Mur, L.R. (1986) A new prokaryote containing chlorophylls *a* and *b*. *Nature* (London) **320**: 262–4.

Burke, S. and Aronoff, S. (1981) High-performance liquid chromatography of the plasmid pigments and some of their immediate derivatives. *Anal. Biochem.* **114**: 367–70.

Callahan, P.M. and Cotton, T.M. (1987) Assignment of bacteriochlorophyll *a* ligation state from absorption and resonance raman spectra. *J. Am. Chem. Soc.* **109**: 7001–7.

Caple, M.B., Chow, H.-C. and Strouse, C.E. (1978) Photosynthetic pigments of green sulfur bacteria. The esterifying alcohols of bacteriochlorophylls c from *Chlorobium limicola*. *J. Biol. Chem.* **253**: 6730–7.

Carr, N.G, and Whitton, B.A. (eds) (1982) *The Biology of Cyanobacteria*. Blackwell Scientific Publications: Oxford.

Castenholz, R.W. (1988) The green sulfur and nonsulfur bacteria of hot springs. In *Green Photosynthetic Bacteria* (Olson, J.M., Ormerod, J.G., Amesz, J., Stackebrandt, E. and Trüper, H.G., eds), pp. 243–56. Plenum Press: New York.

Causgrove T.P., Brune, D.C., Blankenship, R.E. and Olson, J.M. (1990) Fluorescence lifetimes of dimers and higher oligomers of bacteriochlorophyll c from *Chlorobium limicola*. *Photosynth Res.* **25**: 1–10.

Cavaleiro, J.A.S. and Smith, K.M. (1986) Chromatography of chlorophylls and bacteriochlorophylls. *Talanta* **33**: 963–71.

Chapman, R.A., Roomi, M.W., Morton, T.C., Krajcarski, D.T. and MacDonald, S.F. (1971) An analytical reduction of porphyrins to pyrroles. *Can. J. Chem* **49**: 3544–64.

Chisholm, S.W., Olson, R.J., Zettler, E.R., Goericke, R., Waterbury, J.B. and Welschmeyer, N.A. (1988) A novel free-living prochlorophyte abundant in the oceanic euphotic zone. *Nature* (London) **334**: 340–3.

Chisholm, S.W., Frankel, S.L., Goericke, R., Olson, R.J., Palenik, B., Waterbury, J.B., West-Johnson, L. and Zettler, E.R. (1992) *Prochlorococcus marinus* nov. gen. nov. sp.: An oxyphototrophic marine prokaryote containing divinyl chlorophyll *a* and *b*. *Arch. Microbiol.* **157**: 297–300.

Chow, H.-C., Caple, M.B and Strouse, C.E. (1978) Polyethylene powder as a stationary phase for preparative-scale reversed-phase high-performance liquid chromatography. *J. Chromatogr.* **151**: 357–62.

Clayton, R.K. (1966) Spectroscopic analysis of bacteriochlorophylls *in vitro* and *in vivo*. *Photochem. Photobiol.* **5**: 669–77.

Clayton, R.K. and Sistrom, W.R., (eds) (1978) *The Photosynthetic Bacteria*. Plenum Press: New York.

Cohen-Bazire, G., Sistrom, W.R. and Stanier, R.Y. (1957) Kinetic studies of pigment synthesis by non-sulfur purple bacteria. *J. Cell. Comp. Physiol.* **49**: 25–68.

Collins, M.L.P. and Remsen, C.C. (1991) The purple phototrophic bacteria. In *Structure of Phototrophic Prokaryotes* (Stolz, J. F., ed), pp. 49–77. CRC Press: Boca Raton, FL.

Cox, M.T., Jackson, A.H. and Kenner, G.W. (1971) Pyrroles and related compounds. Part XIX. Synthesis of phylloporphyrins related to *Chlorobium* chlorophyll (660). *J. Chem Soc.* **C**: 1974–81.

Davies, S.L. and Whittenbury, R. (1970). Fine structure of methane and other hydrocarbon-utilizing bacteria. *J. Gen. Microbiol.* **61**: 227–32.

de Boer, W.E. (1969) On ultrastructures in *Rhodopseudomonas gelatinosa* and *Rhodospirillum tenue*. *Antonie van Leeuwenhoek. J. Microbiol. Serol.* **35**: 241–2.

Dijkman, J.A., den Blanken, H.J. and Hoff, A.J. (1988). Towards a new taxonomy of photosynthetic bacteria: ADMR-monitored triplet difference spectroscopy of reaction center pigment-protein complexes. *Isr. J. Chem.* **28**: 141–48.

Doolittle, W.F. (1982) Molecular evolution. In *The Biology of Cyanobacteria* (Carr, N.G. and Whitton, B.A., eds), pp. 307–31. Blackwell Scientific Publications: Oxford.
Doolittle, W.F. and Bonen, L. (1981) Molecular sequence data indicating an endosymbiotic origin for plastids. *Ann. N.Y. Acad. Sci.* **361**: 248–59.
Drews, G. (1985) Structure and functional organization of light-harvesting complexes and photochemical reaction centers in membranes of phototrophic bacteria. *Microbiol. Rev.* **49**: 59–70.
Drews, G. and Dawes, E.A. (eds) (1990) *Molecular Biology of Membrane-Bound Complexes in Phototrophic Bacteria.* Plenum Press: New York.
Drews, G. and Imhoff, J.F. (1991) Phototrophic purple bacteria. In *Variations in Autotrophic Life* (Shively, J.M. and Barton, L.L., eds), pp. 51–97. Academic Press: London.
Eaglesham, A.R.J., Ellis, J.M., Evans, W.R., Fleischman, D.E., Hungria, M. and Hardy, R.W.F. (1990) The first photosynthetic N$_2$-fixing *Rhizobium*: Characteristics. In *Nitrogen Fixation: Achievements and Objectives* (Gresshoff, P.M., Roth, L.E., Stacey, G. and Newton, W.E., eds), pp. 805–12. Chapman and Hall: New York.
Eichler, B. and Pfennig, N. (1988) A new green sulfur bacterium from a freshwater pond. In *Green Photosynthetic Bacteria* (Olson, J.M., Ormerod, J.G., Amesz, J., Stackebrandt, E. and Trüper, H.G., eds), pp. 233–6. Plenum Press: New York.
Eimhjellen, K.E., Aasmundrud, O. and Jensen, A. (1963) A new bacterial chlorophyll. *Biochem. Biophys. Res. Commun.* **10**: 232–6.
Evans, N., Games, D.E., Jackson, A.H. and Matlin, S.A. (1975) Applications of HPLC and field desorption mass spectrometry in studies of natural porphyrins and chlorophyll derivatives. *J Chromatogr.* **115**: 325–33.
Evans, W.R., Fleischman, D.E., Calvert, H.E., Pyati, P.V., Alter, G.M. and Rao, N.S.S. (1990) Bacteriochlorophyll and photosynthetic reaction centers in *Rhizobium* strain BTai 1. *Appl. Environ. Microbiol.* **56**: 3445–9.
Fages, F., Griebenow, N., Griebenow, K., Holzwarth, A R. and Schaffner, K. (1990) Characterization of light-harvesting pigments of *Chloroflexus aurantiacus*. Two new chlorophylls, oleyl (octadec-9-enyl) and cetyl (hexadecanyl) bacteriochlorophyllides c. *J. Chem. Soc. Perkin Trans.* **I**: 2791–8.
Feher, G., Allen, J.P., Okamura, M.Y. and Rees, D.C. (1989) Structure and function of bacterial photosynthetic reaction centers; *Nature* (London) **339**: 111–6.
Feick, R. and Fuller, R.C. (1984) Topography of the photosynthetic apparatus in *Chloroflexus aurantiacus*. *Biochemistry* **23**: 3693–700.
Feiler, U., Nitschke, W. and Michel, H. (1992) Characterization of an improved reaction center preparation from the photosynthetic green sulfur bacterium *Chlorobium* containing the iron sulfur centers F$_A$ and F$_B$ and a bound cytochrome subunit. *Biochemistry* **31**: 2608–14.
Fischer, H. and Stern, A. (1940) Bacteriochlorophyll und seine Derivate. In *Die Chemie des Pyrrols*, vol. II, 2nd part (Fischer, H. and Orth, H., eds), pp. 305–21. Akademische Verlag: Leipzig.
Fischer, M.R. (1990) Photosynthetic electron transfer in *Heliobacterium chlorum* studied by EPR spectroscopy. *Biochim. Biophys. Acta* **1015**: 471–81.
Fleming, I. (1968) The absolute configuration and the structure of chlorophyll and bacteriochlorophyll. *J. Chem Soc.* **C**: 2765–70.
Florenzano, G., Balloni, W. and Materassi, R. (1986) Nomenclature of *Prochloron didemni* (Lewin 1977) sp. nov., nom. rev., *Prochloron* (Lewin 1976) gen. nov., nom. rev., *Prochloraceae* fam. nov., *Prochlorales* ord. nov., nom. rev. in the Class *Photobacteria* (Gibbons and Murray 1978). *Int. J. Syst. Bacteriol.* **36**: 498–502.
Fowler, V.J., Pfennig, N., Schubert, W. and Stackebrandt, E. (1984) Towards a phylogeny of phototrophic purple sulfur bacteria: 16S ribosomal RNA oligonucleotide cataloging of 11 species of *Chromatiaceae*. *Arch. Microbiol.* **139**: 382–7.
Fox, G.E., Pechman, K.J. and Woese, C.R. (1977) Comparative cataloguing of 16S ribosomal ribonucleic acid: Molecular approach to procaryotic systematics. *Int. J. Syst. Bacteriol.* **27**: 44–57.
Fox, G.E., Stackebrandt, E., Hespell, R.B., Gibson, J., Maniloff, J., Dyer, T.A., Wolfe, R.S., Balch, W.E., Tanner, R., Magrum, L., Zablen, L.B., Blakemore, R., Gupta, R., Bonen, L.,

Lewis, B.J., Stahl, D.A, Luehrsen, K.R., Chen, K.N. and Woese, C.R. (1980) The phylogeny of prokaryotes. *Science* **209**: 457–63.

Fuesler, T.P., Hanamoto, C.M. and Castelfranco, P.A. (1982) Separation of Mg-protoporphyrin IX and Mg-protoporphyrin IX monomethyl ester synthesized *de novo* by developing cucumber etioplasts. *Plant Physiol.* **69**: 421–23.

Fuhrhop, J.-H. (1978) Irreversible reactions on the porphyrin periphery (excluding oxidations, reductions, and photochemical reactions). In *The Porphyrins*, vol. II (Dolphin, D., ed), ch. 5. Academic Press: New York.

Fuhrhop, J.-H. and Smith, K.M. (1975) Laboratory methods. In *Porphyrins and Metalloporphyrins* (Smith, K.M., ed), ch. 19. Elsevier: Amsterdam.

Fuller, R.C., Sprague, S.G., Gest, H. and Blankenship, R.E. (1985) A unique photosynthetic reaction center from *Heliobacterium chlorum*. *FEBS Lett.* **182**: 345–9.

Gerola, P.D. and Olson, J.M. (1986) A new bacteriochlorophyll *a*-protein complex associated with chlorosomes of green sulfur bacteria. *Biochim. Biophys. Acta* **848**: 69–76.

Gest, H. and Flavinger, J.L. (1983) *Heliobacterium chlorum* gen. nov., sp. nov.: An anoxygenic brownish-green photosynthetic bacterium containing a "new" form of bacteriochlorophyll. *Arch. Microbiol.* **136**: 11–16.

Gibson, J., Stackebrandt, E., Zablen, L.B., Gupta, R. and Woese, C.R. (1979) A phylogenetic analysis of the purple photosynthetic bacteria. *Curr. Microbiol.* **3**: 59–64.

Gibson, J., Ludwig, W., Stackebrandt, E. and Woese, C.R. (1985) The phylogeny of the green photosynthetic bacteria. Absence of a close relationship between *Chlorobium* and *Chloroflexus*. *Syst. Appl. Microbiol.* **6**: 152–6.

Giovannoni, S.J., Ward, D.M., Revsbech, N.P. and Castenholz, R.W. (1987) Obligately phototrophic *Chloroflexus*: Primary production in anaerobic, hot spring microbial mats. *Arch. Microbiol.* **147**: 80–7.

Giovannoni, S.J., Turner, S., Olsen, G.J., Barnes, S., Lane, D.J. and Pace, N.R. (1988) Evolutionary relationships among cyanobacteria and green chloroplasts. *J. Bacteriol.* **170**: 3584–92.

Gloe, A. and Pfennig, N. (1974) Das Vorkommen von Phytol und Geranylgeraniol in den Bacteriochlorophyllen Roter und Grüner Schwefelbakterien. *Arch. Microbiol.* **96**: 93–101.

Gloe, A. and Risch, N. (1978) Bacteriochlorophyll c_S: A new bacteriochlorophyll from *Chloroflexus aurantiacus*. *Arch. Microbiol.* **118**: 153–6.

Gloe, A., Pfennig, N., Brockmann, H., Jr. and Trowitzsch, W. (1975) A new bacteriochlorophyll from brown-colored *Chlorobiaceae*. *Arch. Microbiol.* **102**: 103–10.

Goericke, R. and Repeta, D. (1992) The pigments of *Prochlorococcus marinus*: The presence of divinyl chlorophyll *a* and *b* in a marine procaryote. *Limnol. Oceanogr.* **34**: 425–33.

Golden, J.H., Linstead, R.P. and Whitham, G.H (1958) Chlorophyll and related compounds. Part VII. The structure of bacteriochlorophyll. *J. Chem. Soc.* 1725–32.

Goodwin, T.W. (1955) Bacteriochlorophyll and *Chlorobium* chlorophyll. *Biochim. Biophys. Acta* **18**: 309–10.

Gorlenko, V.M. (1988) Ecological niches of green sulfur and gliding bacteria. In *Green Photosynthetic Bacteria* (Olson, J.M., Ormerod, J.G., Amesz, J., Stackebrandt, E. and Trüper, H.G., eds), pp. 257–68. Plenum Press: New York.

Green, P.N., Bousfield, I.J. and Hood, D. (1988) Three new *Methylobacterium* species: *M. rhodesianum* sp. nov., *M. zatmanii* sp. nov., and *M. fujisawaense* sp. nov. *Int. J. Syst. Bacteriol.* **38**: 124–7.

Hanamoto, C.M. and Castelfranco, P.A. (1983) Separation of monovinyl and divinyl protochlorophyllides and chlorophyllides from etiolated and phototransformed cucumber cotyledons. *Plant Physiol.* **73**: 79–81.

Harashima, K. (1989) Carotenoids, quinones and other lipids. In *Aerobic Photosynthetic Bacteria* (Harashima, K., Shiba, T. and Murata, N., eds), pp. 125–48. Japan Scientific Societies Press: Tokyo.

Harashima, K. and Takamiya, K. (1989) Photosynthesis and photosynthetic apparatus. In *Aerobic Photosynthetic Bacteria* (Harashima, K., Shiba, T. and Murata, N., eds), pp. 39–72. Japan Scientific Societies Press: Tokyo.

Harashima, K., Hayasaki, J., Takaomi, I. and Shiba, T. (1980) O_2-stimulated synthesis of bacteriochlorophyll and carotenoids in marine bacteria. *Plant Cell Physiol.* **21**: 1283–94.

Harashima, K., Kawazoe, K., Yoshida, I. and Kamata, H. (1987) Light-stimulated aerobic growth of *Erythrobacter* species OCh114. *Plant Cell Physiol.* **28**: 365–74.
Harashima, K., Shiba, T. and Murata, N. (eds) (1989) *Aerobic Photosynthetic Bacteria*. Japan Scientific Societies Press: Tokyo.
Hawthornthwaite, A.M. and Cogdell, R.J. (1991) Bacteriochlorophyll-binding proteins. In *Chlorophylls* (Scheer, H., ed), pp. 493–528. CRC Press: Boca Raton, FL.
Hiraishi, A., Hoshino, Y. and Satoh, T. (1991) *Rhodoferax fermentans* gen. nov., sp. nov., a phototrophic purple nonsulfur bacterium previously referred to as the *Rhodocyclus gelatinosus*-like group. *Arch. Microbiol.* **155**: 330–6.
Holt, A.S. and Hughes, D.W. (1961) Studies of *Chlorobium* chlorophylls. III. *Chlorobium* chlorophyll (650). *J. Am. Chem. Soc.* **83**: 499–500.
Holt, A.S. and Morley, H.V. (1960a) *Chlorobium* chlorophyll. *J. Am. Chem. Soc.* **82**: 500–501.
Holt, A.S. and Morley, H.V. (1960b) Recent studies of chlorophyll chemistry. In *Kaiser Foundation Symposium on Comparative Biology*, vol. I (Allen, M.B., ed), pp. 169–79. Academic Press: New York.
Holt, A.S., Hughes, D.W., Kende, H.J. and Purdie, J.W. (1962) Studies of *Chlorobium* chlorophylls. V. *Chlorobium* chlorophyll (660). *J. Am. Chem. Soc.* **84**: 2835–6.
Holt, A.S., Hughes, D.W., Kende, H.J. and Purdie, J.W. (1963) Chlorophylls of green photosynthetic bacteria. *Plant Cell Physiol.* **4**: 49–55.
Holt, A.S., Purdie, J.W. and Wasley, J.W.F. (1966) Structures of *Chlorobium* chlorophylls (660). *Can. J. Chem.* **44**: 88–93.
Holzwarth, A.R., Griebenow, K. and Schaffner, K. (1990) A photosynthetic antenna system which contains a protein-free chromophore aggregate. *Z. Naturforsch. Sect. C Biosci.* **45**: 203–6.
Hood, D.W., Dow, C.S. and Green, P.N. (1987) DNA:DNA hybridization studies on the pink-pigmented facultative methylotrophs. *J. Gen. Microbiol.* **133**: 709–20.
Houghton, J.D., Jones, O.T.G., Quirke, J.M.E., Murray, M. and Honeybourne, C.L. (1983) 2-(1-hydroxyethyl)-2-devinyl chlorophyllide *a*: Characterisation by nuclear Overhauser enhancement proton magnetic resonance of a novel pigment obtained from mutants of *Rhodopseudomonas sphaeroides*. *Tetrahedron Lett.* **24**: 5703–6.
Hughes, D.W. and Holt, A.S. (1962) Studies of *Chlorobium* chlorophylls. IV. Preparative liquid-liquid partition chromatography of porphyrins and chlorophyll derivatives and its use to resolve *Chlorobium* pheophorbide (650) into six components. *Can. J. Chem.* **40**: 171–6.
Huster, M.S. and Smith, K.M. (1990) Biosynthetic studies of substituent homologation in bacteriochlorophylls *c* and *d*. *Biochemistry* **29**: 4348–55.
Iba, K., Takamiya, K.-I., Toh, Y. and Nishimura, M. (1988) Roles of bacteriochlorophyll and carotenoid synthesis in formation of intracytoplasmic membrane systems and pigment-protein complexes in an aerobic photosynthetic bacterium *Erythrobacter* sp. strain OCh114. *J. Bacteriol.* **170**: 1843–7.
Imhoff, J.F. (1984) Reassignment of the genus *Ectothiorhodospira* Pelsh 1936 to a new family, *Ectothiorhodospiraceae* fam. nov., and emended description of the *Chromatiaceae* Bavendamm 1924. *Int. J. Syst. Bacteriol.* **34**: 338–9.
Imhoff, J.F. (1988) Halophilic phototrophic bacteria. In *Halophilic Bacteria*, vol. I (Rodriguez-Valera F., ed), ch. 4. CRC Press: Boca Raton, FL.
Imhoff, J.F., Trüper, H.G. and Pfennig, N. (1984) Rearrangement of the species and genera of the phototrophic "purple nonsulfur bacteria". *Int. J. Syst. Bacteriol.* **34**: 340–3.
Iriyama, K., Ogura, N. and Takamiya, A. (1974) A simple method for extraction and partial purification of chlorophyll from plant material, using dioxane. *J. Biochem.* (Tokyo) **76**: 901–4.
Jürgens, U. and Brockmann, H., Jr. (1982) Partialsynthese eines [3-Ethyl]-bacteriophäophorbid-c-methylesters. *Liebigs Ann. Chem.* 472–9.
Kaplan, I.R. and Silberman, H. (1959) Spectroscopy of bacterial chlorophylls separated by paper and cellulose column chromatography. *Arch. Biochem. Biophys.* **80**: 114–24.
Katz, J.J. (1973) Chlorophyll interactions and light conversion in photosynthesis. *Naturwissenschaften* **60**: 32–9.
Katz, J.J. and Janson, T.R. (1973) Chlorophyll-chlorophyll interactions from proton and carbon-13 NMR spectroscopy. *Ann. N.Y. Acad. Sci.* **206**: 579–603.

Katz, J.J., Thomas, M.R. and Strain, H.H. (1962) Site of exchangeable hydrogen in chlorophyll *a* from proton magnetic resonance measurements of deuteriochlorophyll *a*. *J. Am. Chem. Soc.* **84**: 3587.
Katz, J.J., Dougherty, R.C. and Boucher, L.J. (1966) Infrared and nuclear magnetic resonance spectroscopy of chlorophyll. In *The Chlorophylls* (Vernon, L.P. and Seely, G.R., eds), ch. 7. Academic Press: New York.
Katz, J.J., Norman, G.D., Svec, W.A. and Strain, H.H. (1968) Chlorophyll diastereoisomers. The nature of chlorophylls *a'* and *b'* and evidence for bacteriochlorophyll epimers from proton magnetic resonance studies. *J. Am. Chem. Soc.* **90**: 6841–5.
Katz, J.J., Strain, H.H., Harkness, A.L., Studier, M.H., Svec, W.A., Janson, T.R. and Cope, B.T. (1972) Esterifying alcohols in the chlorophylls of purple photosynthetic bacteria. A new chlorophyll, bacteriochlorophyll (*gg*), *all-trans*-geranylgeranyl bacteriochlorophyllide *a*. *J. Am. Chem. Soc.* **94**: 7938–9.
Kennedy, R.S. and Finnerty, W.R. (1975) Microbial assimilation of hydrocarbons. II. Intracytoplasmic membrane induction in *Acinetobacter sp*. *Arch. Microbiol.* **102**: 85–90.
Kenner, G.W., Rimmer, J., Smith, K.M. and Unsworth, J.K. (1976) Studies on the biosynthesis of the *Chlorobium* chlorophylls. *Phil. Trans. R. Soc. London, Ser. B* **273**: 255–76.
Kobayashi, M., van de Meent, E.J., Erkelens, C., Amesz, J., Ikegami, I. and Watanabe, T. (1991a) Bacteriochlorophyll *g* epimer as a possible reaction center component of heliobacteria. *Biochim. Biophys. Acta* **1057**: 89–96.
Kobayashi, M., Watanabe, T., Ikegami, I., van de Meent, E.J. and Amesz, J. (1991b) Enrichment of bacteriochlorophyll *g'* in membranes of *Heliobacterium chlorum* by ether extraction. Unequivocal evidence for its existence *in vivo*. *FEBS Lett.* **284**: 129–31.
Komiya, H., Yeates, T.O., Rees, D.C., Allen, J.P. and Feher, G. (1988) Structure of the reaction center from *Rhodobacter sphaeroides* R-26 and 2.4.1: Symmetry relations and sequence comparisons between different species. *Proc. Natl. Acad. Sci. USA* **85**: 9012–16.
Künzler, A. and Pfennig, N. (1973) Das Vorkommen von Bacteriochlorophyll a_P und a_{GG} in Stämmen aller Arten der *Rhodospirillaceae*. *Arch. Microbiol.* **91**: 83–6.
Lane, D.J., Harrison, A.P., Jr., Stahl, D., Pace, B., Giovannoni, S.J., Olsen, G.J. and Pace, N.R. (1992) Evolutionary relationships among sulfur- and iron-oxidizing eubacteria. *J. Bacteriol.* **174**: 269–78.
Larkum, A.W.D. (1991) Evolution of chlorophylls. In *Chlorophylls* (Scheer, H., ed), pp. 367–83. CRC Press: Boca Raton, FL.
Lee, H.-S. (1987) Photosynthetic membranes of *Rhodocyclus gelatinosus* and *Rhodocyclus tenuis*. *Korean J. Microbiol.* **25**: 144–7.
Lewin, R.A. (1975) A marine *Synechocystis* (*Cyanophyta, Chroococcales*) epizoic on ascidians. *Phycologia* **14**: 153–60.
Lewin, R.A. (1981) *Prochloron* and the theory of symbiogenesis. *Ann. N.Y. Acad. Sci.* **361**: 325–9.
Lewin, R.A. and Cheng, L. (eds) (1989) *Prochloron: A Microbial Enigma*. Chapman and Hall: New York.
Lewin, R.A. and Withers, N.W (1975) Extraordinary pigment composition of a prokaryotic alga. *Nature* (London) **256**: 735–37.
Liebetanz, R., Hornberger, U. and Drews, G. (1991) Organization of the genes coding for the reaction center L and M subunits and B870 antenna polypeptides alpha and beta from the aerobic photosynthetic bacterium *Erythrobacter* sp. OCh114. *Mol. Microbiol.* **5**: 1459–68.
Lindsay Smith, J.R. and Calvin, M. (1966) Studies on the chemical and photochemical oxidation of bacteriochlorophyll. *J. Am. Chem. Soc.* **88**: 4500–6.
Mack, E.E. and Pierson, B.K. (1988) Preliminary characterization of a temperate marine member of the *Chloroflexaceae*. In *Green Photosynthetic Bacteria* (Olson, J.M., Ormerod, J.G., Amesz, J., Stackebrandt, E. and Trüper, H.G., eds), pp. 237–41. Plenum Press: New York.
Margulies, M.M. (1991) Sequence similarity between photosystems I and II. Identification of a photosystem I reaction center transmembrane helix that is similar to transmembrane helix IV of the D2 subunit of photosystem II and the M subunit of the nonsulfur purple and flexible green bacteria. *Photosynth Res.* **29**: 133–48.
Markowicz, Y. and Loiseaux-de Goër, S. (1991) Plastid genomes of the *Rhodophyta* and *Chromophyta* constitute a distinct lineage which differs from that of the *Chlorophyta* and have a

composite phylogenetic origin, perhaps like that of the *Euglenophyta*. *Curr. Genet.* **20**: 427–30.
Mathewson, J.H., Richards, W.R. and Rapoport, H. (1963a) *Chlorobium* chlorophylls. Nuclear magnetic resonance studies on a *Chlorobium* pheophorbide-660 and -650. *J. Am. Chem. Soc.* **85**: 364–5.
Mathewson, J.H., Richards, W.R. and Rapoport, H. (1963b) A nuclear magnetic resonance study of hydrogen exchange at methine positions in chlorophyll *a*, chlorophyll *b*, *Chlorobium* chlorophyll-660, and bacteriochlorophyll. *Biochem. Biophys. Res. Commun.* **13**: 1–5.
Mathis, P. (1990) Compared structure of plant and bacterial photosynthetic reaction centers. Evolutionary implications. *Biochim. Biophys. Acta* **1018**: 163–7.
Mauzerall, D. (1978) Bacteriochlorophyll and photosynthetic evolution. In *The Photosynthetic Bacteria* (Clayton, R.K. and Sistrom, W.R., eds), ch. 11. Plenum Press: New York.
Michalski, T.J., Hunt J.E., Bowman, M.K., Smith, U., Bardeen, K., Gest, H., Norris, J.R. and Katz, J.J. (1987) Bacteriopheophytin *g*: Properties and some speculations on a possible primary role for bacteriochlorophylls *b* and *g* in the biogenesis of chlorophylls. *Proc. Natl. Acad. Sci. USA* **84**: 2570–4.
Michel, H. and Deisenhofer, J. (1988) Relevance of the photosynthetic reaction center from purple bacteria to the structure of photosystem II. *Biochemistry* **27**: 1–7.
Miller, K.R., Jacob, J.S., Smith, U., Kolaczkowski, S. and Bowman M.K. (1986) *Heliobacterium chlorum*. Cell organization and structure. *Arch. Microbiol.* **146**: 111–4.
Morden, C.W. and Golden, S.S. (1989) *psbA* Genes indicate common ancestry of prochlorophytes and chloroplasts. *Nature* (London) **337**: 382–5.
Morden, C.W. and Golden, S.S. (1991) Sequence analysis and phylogenetic reconstruction of the genes encoding the large and small subunits of ribulose-1,5-bisphosphate carboxylase/oxygenase from the chlorophyll *b*-containing prokaryote *Prochlorothrix hollandica*. *J. Mol. Evol.* **32**: 379–95.
Morley, H.V. and Holt, A.S. (1961) Studies on *Chlorobium* chlorophylls. II. The resolution of oxidation products of *Chlorobium* pheophorbide (660) by gas-liquid partition chromatography. *Can. J. Chem.* **39**: 755–60.
Murray, R.G.E. (1984). The higher taxa, or a place for everything? In *Bergey's Manual of Systematic Bacteriology*, vol. 1 (Kreig, N.R. and Holt, J.G., eds), p. 33. Williams and Wilkins: Baltimore.
Neufang, H., Müller, H. and Knobloch, K. (1982) The quantitation of bacteriochlorophyll in chromatophore suspensions from purple bacteria. *Biochim. Biophys. Acta* **681**: 327–9.
Niedermeier, G., Scheer, H. and Feick, R.G. (1992) The functional role of protein in the organization of bacteriochlorophyll *c* in chlorosomes of *Chloroflexus aurantiacus*. *Eur. J. Biochem.* **204**: 685–92.
Nishimura, Y., Shimadzu, M. and Iizuka, H. (1981) Bacteriochlorophyll formation in radiation resistant *Pseudomonas radiora*. *J. Gen. Appl. Microbiol.* **27**: 427–30.
Nishimura, Y., Mukasa, S., Iizuka, H. and Shimada, K. (1989) Isolation and characterization of bacteriochlorophyll-protein complexes from an aerobic bacterium *Pseudomonas radiora*. *Arch. Microbiol.* **152**: 1–5.
Nitschke, W., Feiler, U., Lockau, W. and Hauska, G. (1987) The photosystem of the green sulfur bacterium *Chlorobium limicola* contains two early electron acceptors similar to photosystem I. *FEBS Lett.* **218**: 283–6.
Nitschke, W., Feiler, U. and Rutherford, W.A. (1990a) Photosynthetic reaction center of green sulfur bacteria studied by EPR. *Biochemistry* **29**: 3834–42.
Nitschke, W., Setif, P., Liebl, U., Feiler, U. and Rutherford, W.A. (1990b) Reaction center photochemistry of *Heliobacterium chlorum*. *Biochemistry* **29**: 11079–88.
Nozawa, T., Fukada, T., Hatano, M. and Madigan, M.T. (1986) Organization of intracytoplasmic membranes in a novel thermophilic photosynthetic bacterium as revealed by absorption circular dichroism and emission spectra. *Biochim. Biophys. Acta* **852**: 191–7.
Olson, J.M. (1981) Evolution of photosynthetic and respiratory prokaryotes and organelles. *Ann. N.Y. Acad. Sci.* **361**: 8–19.
Olson, J.M., Ormerod, J.G., Amesz, J., Stackebrandt, E. and Trüper, H.G. (eds) (1988) *Green Photosynthetic Bacteria*. Plenum Press: New York.

Omata, T. and Murata, N. (1983) Preparation of chlorophyll *a*, chlorophyll *b* and bacteriochlorophyll *a* by column chromatography with DEAE-Sepharose CL-6B and Sepharose CL-6B. *Plant Cell Physiol.* **24**: 1093–100.
Oppenheim, J. and Marcus, L. (1970) Correlation of ultrastructure in *Azotobacter vinelandii* with nitrogen source for growth. *J. Bacteriol.* **101**: 286–91.
Ormerod, J.G. (ed) (1983) *The Phototrophic Bacteria: Anaerobic Life in the Light.* Blackwell Scientific Publications: Oxford.
Ormerod, J., Nesbakken, T. and Torgersen, Y. (1989) Phototrophic bacteria that form heat resistant endospores. *Physiol. Plant.* **76**: A137.
Otte, S.C.M., van der Heiden, J.C., Pfennig, N. and Amesz, J. (1991) A comparative study of the optical characteristics of intact cells of photosynthetic green sulfur bacteria containing bacteriochlorophyll *c*, *d*, or *e*. *Photosynth. Res.* **2**: 77–88.
Ovchinnikov, Yu.A., Abdulaev, N.G., Shmuckler, B.E., Zargarov, A.A., Kutuzov M.A., Telezhinskaya, I.N., Levina, N.B. and Zolotarev, A. S. (1988a) Photosynthetic reaction center of *Chloroflexus aurantiacus*. Primary structure of M-subunit. *FEBS Lett.* **232**: 364–8.
Ovchinnikov, Yu.A., Abdulaev, N.G., Zolotarev, A.S., Shmuckler, B.E., Zargarov, A.A., Kutuzov, M.A., Telezhinskaya, I.N. and Levina, N.B. (1988b) Photosynthetic reaction center of *Chloroflexus aurantiacus*. I. Primary structure of L-subunit. *FEBS Lett.* **231**: 237–42.
Overmann, J. and Pfennig, N. (1989) *Pelodictyon phaeoclathratiforme* sp. nov.: A new brown-colored member of the *Chlorobiaceae* forming net-like colonies. *Arch. Microbiol.* **152**: 401–6.
Oyaizu, H., Debrunner-Vossbrinck, B., Mandelco, L., Studier, J.A. and Woese, C.R. (1987) The green non-sulfur bacteria: A deep branching lineage in the eubacterial line of descent. *Syst. Appl. Microbiol.* **9**: 47–53.
Palenik, B. and Haselkorn, R. (1992) Multiple evolutionary origins of prochlorophytes, the chlorophyll *b*-containing prokaryotes. *Nature* (London) **355**: 265–7.
Pfennig, N. and Biebl, H. (1976) *Desulfuromonas acetoxidans* gen. nov. and sp. nov., a new anaerobic, sulfur-reducing, acetate-oxidizing bacterium. *Arch. Microbiol.* **110**: 3–12.
Pfennig, N. and Trüper, H.G. (1971) Higher taxa of the phototrophic bacteria. *Int. J. Syst. Bacteriol.* **21**: 17–8.
Pierson, B.K. and Castenholz, R.W. (1974) Studies of pigment and growth in *Chloroflexus aurantiacus*, a phototrophic filamentous bacterium. *Arch. Microbiol.* **100**: 276–82.
Pierson, B.K., Giovannoni, S.J., Stahl, D.A. and Castenholz, R.W. (1985) *Heliothrix oregonensis* gen. nov., sp. nov. A phototrophic filamentous gliding bacterium containing bacteriochlorophyll *a*. *Arch Microbiol.* **142**: 164–7.
Prince, R.C., Gest, H. and Blankenship, R.E. (1985) Thermodynamic properties of the photochemical reaction center of *Heliobacterium chlorum*. *Biochim. Biophys. Acta* **810**: 377–84.
Pudek, M.R. and Richards, W.R. (1975) A possible alternate pathway of bacteriochlorophyll biosynthesis in a mutant of *Rhodopseudomonas sphaeroides*. *Biochemistry* **14**: 3132–7.
Purdie, J.W. and Holt, A.S. (1965) Structures of *Chlorobium* chlorophylls (650). *Can. J. Chem.* **43**: 3347–53.
Rapoport, H. and Hamlow, H.P. (1961) *Chlorobium* chlorophyll-660. The esterifying alcohol. *Biochem. Biophys. Res. Commun.* **6**: 134–7.
Richards, W.R. (1966) Studies on *Chlorobium* chlorophyll-660. Thesis. University of California, Berkeley, CA.
Richards, W.R. (1993) Biosynthesis of the chlorophyll chromophore of pigmented thylakoid proteins. In *Pigment-Protein Complexes in Plastids: Synthesis and Assembly* (Sundqvist, C. and Ryberg, M., eds). Academic Press: New York pp. 91–178.
Richards, W.R. and Lascelles, J. (1969) The biosynthesis of bacteriochlorophyll. The characterization of latter stage intermediates from mutants of *Rhodopseudomonas spheroides*. *Biochemistry* **8**: 3473–82.
Richards, W.R. and Rapoport, H. (1966) The biosynthesis of *Chlorobium* chlorophylls-660. The isolation and purification of porphyrins from *Chlorobium thiosulfatophilum*-660. *Biochemistry* **5**: 1079–89.
Richards, W.R. and Rapoport, H. (1967) The biosynthesis of *Chlorobium* chlorophylls-660. The production of magnesium protoporphyrin monomethyl ester, bacteriochlorophyll, and *Chlorobium* pheoporphyrins by *Chlorobium thiosulfatophilum*-660. *Biochemistry* **6**: 3830–40.

Rippka R., Deruelles, J., Waterbury, J.B., Herdman, M. and Stanier, R.Y. (1979). Genetic assignments, strain histories and properties of pure cultures of cyanobacteria. *J. Gen. Microbiol.* **111**: 1–61.
Risch, N. (1981) Bacteriochlorophyll b. Determination of its configuration by nuclear Overhauser effect difference spectroscopy. *J. Chem. Res* **(S)**: 116–7.
Risch, N. and Brockmann, H., Jr. (1976) Die absolute Konfiguration der Bacteriochlorophylle c, d, und e an C-2'. *Liebigs Ann. Chem.* 578–83.
Risch, N., Kemmer, T. and Brockmann, H., Jr. (1978) Chromatographische Trennung und Charakterisierung der Bacteriomethylphäophorbide e. *Liebigs Ann. Chem.* 585–94.
Risch, N., Brockmann, H., Jr. and Gloe, A. (1979) Strukturaufklärung von Neuartigen Bacteriochlorophyllen aus *Chloroflexus aurantiacus*. *Liebigs Ann. Chem.* 408–18.
Risch, N., Köster, B., Schormann, A., Siemens, T. and Brockmann, H., Jr. (1988) Bacteriochlorophyll f. Partialsynthese und Eigenschaften einiger Derivate. *Liebigs Ann Chem.* 343–8.
Sato, K. (1978) Bacteriochlorophyll formation by facultative methylotrophs. *Protaminobacter ruber* and *Pseudomonas* AM-1. *FEBS Lett.* **85**: 207–10.
Sato, K. and Shimizu, S. (1979) The conditions for bacteriochlorophyll formation and the ultrastructure of a methanol-utilizing bacterium, *Protaminobacer ruber*, classified as non-photosynthetic bacteria. *Agric. Biol. Chem.* **43**: 1669–76.
Sato, K., Hagiwara, K. and Shimizu, S. (1985) Effect of cultural conditions on tetrapyrrole formation (especially bacteriochlorophyll formation) in a facultative methylotroph *Protaminobacter ruber*. *Agric. Biol. Chem.* **49**: 1–6.
Sato, N. and Murata. N. (1978) Preparation of chlorophyll *a*, chlorophyll *b*, and bacteriochlorophyll *a* by means of column chromatography with DEAE cellulose. *Biochim. Biophys. Acta* **501**: 103–11.
Sauer, K., Lindsay-Smith, J.R. and Schultz, A.J. (1966) The dimerization of chlorophyll *a*, chlorophyll *b*, and bacteriochlorophyll in solution. *J. Am. Chem. Soc.* **88**: 2681–8.
Scheer, H. (ed) (1991) *Chlorophylls*. CRC Press: Boca Raton, FL.
Scheer, H. and Katz, J.J. (1975) Nuclear magnetic resonance spectroscopy of porphyrins and metalloporphyrins. In *Porphyrins and Metalloporphyrins* (Smith, K.M., ed), ch. 10. Elsevier: Amsterdam.
Scheer, H., Svec, W.A., Cope, B.T., Studier, M.H., Scott, R.G. and Katz, J.J. (1974) Structure of bacteriochlorophyll *b*. *J. Am. Chem. Soc.* **96**: 3714–16.
Schoch, S., Lempert, U., Wieschoff, H. and Scheer, H. (1978) High-performance liquid chromatography of tetrapyrrole pigments. Pheophytins esterified with different diterpene alcohols, isomeric biliverdins and synthetic bilins. *J. Chromatogr.* **125**: 357–64.
Scholz, B. and Ballschmiter, K. (1981a) Preparation and reversed-phase high-performance liquid chromatography of chlorophylls. *J. Chromatogr.* **208**: 148–55.
Scholz, B. and Ballschmiter, K. (1981b) Do all eight diastereomeric bacteriochlorophylls exist in nature? *Angew. Chem. Int. Edit.* **20**: 956–8.
Scholz, B. and Ballschmiter, K. (1982) Chromatographic separation and analytical characterization of bacteriochlorophylls a_P, a_{GG} and *b*. *J. Chromatogr.* **252**: 269–82.
Schwartz, R.M. and Dayhoff, M.O. (1981) Chloroplast origins: Inferences from protein and nucleic acid sequences. *Ann. N.Y. Acad. Sci.* **361**: 261–72.
Seely, G.R. (1966) The structure and chemistry of functional groups. In *The Chlorophylls* (Vernon, L.P., and Seely, G.R., eds), ch. 3. Academic Press: New York.
Seewaldt, E. and Stackebrandt, E. (1982) Partial sequence of 16S ribosomal RNA and the phylogeny of *Prochloron*. *Nature* (London) **295**: 618–20.
Sestak, Z. (1980) Paper chromatography of chloroplast pigments (chlorophylls and carotenoids)—Part 3. *Photosynthetica* **14**: 239–70.
Sestak, Z. (1982) Thin layer chromatography of chlorophylls—2. *Photosynthetica* **16**: 568–617.
Shiba, T. (1984) Utilization of light energy by the strictly aerobic bacterium *Erythrobacter* sp. OCh114. *J. Gen. Appl. Microbiol.* **30**: 239–44.
Shiba, T. (1987) Oxygen regulation of bacteriochlorophyll synthesis in the aerobic bacterium *Erythrobacter*. *Plant Cell Physiol.* **28**: 1313–20.
Shiba, T. (1989a) Overview of the aerobic photosynthetic bacteria. In *Aerobic Photosynthetic Bacteria* (Harashima, K., Shiba, T. and Murata, N., eds), pp. 1–8. Japan Scientific Societies Press: Tokyo.

Shiba, T. (1989b), Taxonomy and ecology of marine bacteria. In *Aerobic Photosynthetic Bacteria* (Harashima, K., Shiba, T. and Murata, N., eds), pp. 9–23. Japan Scientific Societies Press: Tokyo.

Shiba, T. (1991) *Roseobacter litoralis* gen. nov., sp. nov, and *Roseobacter denitrificans* sp. nov., aerobic pink-pigmented bacteria which contain bacteriochlorophyll *a*. *Syst. Appl. Microbiol.* **14**: 140–5.

Shiba, T. and Abe, K. (1987) An aerobic bacterium containing bacteriochlorophyll-proteins showing absorption maxima of 802, 844, and 862 nm in the near infrared region. *Agric. Biol. Chem.* **51**: 945–6.

Shiba, T. and Simidu, U. (1982) *Erythrobacter longus* gen. nov., sp. nov., an aerobic bacterium which contains bacteriochlorophyll *a*. *Int. J. Syst. Bacteriol.* **32**: 211–17.

Shiba, T., Simidu, U, and Taga, N. (1979) Distribution of aerobic bacteria which contain bacteriochlorophyll *a*. *Appl. Environ. Microbiol.* **38**: 43–5.

Shimada, K., Hayashi, H. and Tasumi, M. (1985) Bacteriochlorophyll-protein complexes of aerobic bacteria *Erythrobacter longus* and *Erythrobacter* sp. OCh114. *Arch. Microbiol.* **143**: 244–7.

Shioi, Y. (1991) Analytical chromatography of chlorophylls. In *Chlorophylls* (Scheer, H., ed), pp. 59–88. CRC Press: Boca Raton, FL.

Shioi, Y. and Beale, S.I. (1987) Polyethylene-based high-performance liquid chromatography of chloroplast pigments: Resolution of mono- and divinyl chlorophyllides and other pigment mixtures. *Anal. Biochem.* **162**: 493–9.

Shioi, Y. and Doi, M. (1989) Aerobic and anaerobic photosynthesis and bacteriochlorophyll formation in *Rhodobacter sulfidophilus*. *Physiol. Plant.* **76**: A137.

Shioi, Y., Fukae, R. and Sasa, T. (1983) Chlorophyll analysis by high-performance liquid chromatography. *Biochim. Biophys. Acta* **722**: 72–9.

Shioi, Y., Doi, M. and Sasa, T. (1984) Separation of non-esterified chlorophylls by ion-suppression high-performance liquid chromatography. *J. Chromatogr.* **298**: 141–9.

Shiozawa, J.A., Lottspeich, F., Oesterhelt, D. and Feick, R. (1989) The primary structure of the *Chloroflexus aurantiacus* reaction-center polypeptides. *Eur. J. Biochem.* **180**: 75–84.

Siefert, E. and Pfennig, N. (1979) Chemoautotrophic growth of *Rhodopseudomonas* species with hydrogen and chemotrophic utilization of methanol and formate. *Arch Microbiol.* **122**: 177–82.

Simpson, D.J. and Smith, K.M. (1988) Structures and transformations of the bacteriochlorophylls *e* and their bacteriopheophorbides. *J. Am. Chem. Soc.* **110**: 1753–8.

Sistrom, W.R. (1964) Calculations of the absorption coefficients of the individual components of the spectra of bacteriochlorophyll in chromatophore preparations of *Rhodopseudomonas spheroides*. *Biochim. Biophys. Acta* **79**: 419–21.

Smith, J.H.C. and Benitez, A. (1955). Chlorophylls: Analysis in plant materials. In *Modern Methods of Plant Analysis*, vol. IV (Paech, K. and Tracy, M.V., eds), pp. 142–96. Julius Springer: Berlin.

Smith, K.M. (1967) Synthetic and spectroscopic studies on porphyrins and related tetrapyrroles. Thesis. University of Liverpool, U.K.

Smith, K.M. (1975) General features of the structure and chemistry of porphyrin compounds. In *Porphyrins and Metalloporphyrins* (Smith, K.M., ed), ch. 1. Elsevier: Amsterdam.

Smith, K.M. (1991) The structure and biosynthesis of bacteriochlorophylls. In *New Comprehensive Biochemisry. Biosynthesis of Tetrapyrroles*, vol. 19. (Jordan, P.M., ed), pp. 237–55. Elsevier: Amsterdam.

Smith, K.M. and Bobe, F.W. (1987) Light adaptation of bacteriochlorophyll-*d* producing bacteria by enzymic methylation of their antenna pigments. *J. Chem. Soc. Chem. Commun.* 276–7.

Smith, K.M. and Goff, D.A. (1985) Bacteriochlorophylls-*d* from *Chlorobium vibrioforme*: Chromatographic separations and structural assignments of the methyl bacteriopheophorbides. *J. Chem. Soc. Perkin Trans.* **I**: 1099–114.

Smith, K.M. and Simpson, D.J. (1986) Stereochemistry of the bacteriochlorophyll-*e* homologues. *J. Chem. Soc. Chem. Commun.* 1682–4.

Smith, K.M. and Unsworth, J.F. (1975) The nuclear magnetic resonance spectra of porphyrins. IX. Carbon-13 NMR spectra of some chlorins and other chlorophyll degradation products. *Tetrahedron* **31**: 367–76.

Smith, K.M., Bisset, G.M.F. and Bushell, M.J. (1980a) Partial synthesis of optically pure methyl bacteriopheophorbides c and d from methyl pheophorbide a. *J. Org. Chem.* **45**: 2218–24.
Smith, K.M., Bisset, G.M.F. and Bushell, M.J. (1980b) *meso*-Methylporphyrins and chlorins. *Bioorg Chem.* **9**: 1–26.
Smith, K.M., Bisset, G.M.F. and Bushell, M.J. (1980c) Recent studies on the *Chlorobium* chlorophylls (bacteriochlorophylls-c). *Int. J. Biochem.* **12**: 695–700.
Smith, K.M., Bushell, M.J., Rimmer, J. and Unsworth, J.F. (1980d) Bacteriochlorophylls c from *Chloropseudomonas ethylicum*. Composition and NMR studies of the pheophorbides and derivatives. *J. Am. Chem. Soc.* **102**: 2437–48.
Smith, K.M., Kehres, L.A. and Tabba H.D. (1980e) Structures of the bacteriochlorophyll c homologues: Solution to a longstanding problem. *J. Am. Chem. Soc.* **102**: 7149–51.
Smith, K.M., Goff, D.A., Fajer, J. and Barkigia, K.M. (1982) Chirality and structures of bacteriochlorophylls d. *J. Am. Chem. Soc.* **104**: 3747–9.
Smith, K.M., Craig, G.W., Kehres, L.A. and Pfennig, N. (1983a) Reversed-phase high-pressure liquid chromatography and structural assignments of the bacteriochlorophylls-c. *J. Chromatogr.* **281**: 209–24.
Smith, K.M., Goff, D.A., Fajer, J. and Barkigia, K.M. (1983b) Isolation and characterization of two new bacteriochlorophylls d bearing neopentyl substituents. *J. Am. Chem. Soc.* **105**: 1674–6.
Sojka, G.A., Freeze, H.H. and Gest, H. (1970) Quantitative estimation of bacteriochlorophyll *in situ*. *Arch. Biochem. Biophys.* **136**: 578–80.
Sprague, S.G. and Fuller, R.C. (1991) The green phototrophic bacteria and heliobacteria. In: *Structure of Phototrophic Prokaryotes* (Stolz, J.F., ed), pp. 79–103. CRC Press: Boca Raton, FL.
Stackebrandt, E., Seewaldt, E., Fowler, V.J. and Schleifer, K.-H. (1982) The relatedness of *Prochloron* sp. isolated from different didemnid ascidian hosts. *Arch. Microbiol.* **132**: 216–17.
Stackebrandt, E., Embley, T.M. and Weckkesser, J. (1988a) Phylogenetic, evolutionary, and taxonomic aspects of phototrophic eubacteria. In *Green Photosynthetic Bacteria* (Olson, J.M., Ormerod, J.G., Amesz, J., Stackebrandt, E. and Trüper, H.G., eds), pp. 201–15. Plenum Press: New York.
Stackebrandt, E., Murray, R.G.E. and Trüper, H.G. (1988b) *Proteobacteria* classis nov., a name for the phylogenetic taxon that includes the "Purple Bacteria and Their Relatives". *Int. J. Syst. Bacteriol.* **38**: 321–5.
Staehelin, L.A., Golecki, J.R. and Drews, G. (1980) Supramolecular organization of chlorosomes (*Chlorobium* vesicles) and of their membrane attachment sites in *Chlorobium limicola*. *Biochim. Biophys. Acta* **589**: 30–45.
Stanier, R.Y. and Smith, J.H.C. (1960) The chlorophylls of green bacteria. *Biochim. Biophys. Acta* **41**: 478–84.
Steiner, R., Schäfer, W., Blos, I., Wieschhoff, H. and Scheer, H. (1981) $\Delta^{2,10}$-Phytadienol as esterifying alcohol of bacteriochlorophyll b from *Ectothiorhodospira halochloris*. *Z. Naturforsch. Sect. C Biosci.* **36**: 417–20.
Steiner, R., Wieschhoff, H. and Scheer, H. (1982) High-performance liquid chromatography of bacteriochlorophyll b and its derivatives as an aid for structure analysis. *J. Chromatogr.* **242**: 127–34.
Steiner, R., Cmiel, E. and Scheer, H. (1983) Chemistry of bacteriochlorophyll b: Identification of some (photo)oxidation products. *Z. Naturforsch. Sect. C Biosci.* **38**: 748–52.
Stolz, J.F. (ed) (1991) *Structure of Phototrophic Prokaryotes*. CRC Press: Boca Raton, FL.
Stolz, J.F., Fuller, R.C. and Redlinger, T.E. (1990) Pigment-protein diversity in chlorosomes of green phototrophic bacteria. *Arch. Microbiol.* **154**: 422–7.
Strain, H.H. and Svec, W.A. (1966) Extraction, separation, estimation, and isolation of the chlorophylls. In *The Chlorophylls* (Vernon, L.P. and Seely, G.R., eds), ch. 2. Academic Press: New York.
Strain, H.H. and Svec, W.A. (1975) Chromatography of chlorophylls and related porphyrins. In *Chromatography*, 3rd edn (Heftmann, E., ed), p. 744. Van Nostrand Reinhold: New York.
Svec, W.A. (1991) The distribution and extraction of the chlorophylls. In: *Chlorophylls* (Scheer, H., ed), pp. 89–102. CRC Press: Boca Raton, FL.

Takamiya, K.-I. and Okamura, K. (1984) Photochemical activities and photosynthetic ATP formation in membrane preparation from a facultative methylotroph *Protaminobacter ruber* strain NR-1. *Arch Microbiol.* **140**: 21–6.

Takamiya, K.-I., Iba, K. and Okamura, K. (1987) Reaction center complex from an aerobic photosynthetic bacterium *Erythrobacter* sp. OCh114. *Biochim. Biophys. Acta* **890**: 127–33.

Tindall, B.J. and Grant, W.D. (1986) The anoxygenic phototrophic bacteria. In *Anaerobic Bacteria in Habitats other Than Man* (Barnes, E.M. and Mead, G.C., eds), pp. 115–55. Blackwell Scientific Publications: Oxford.

Trost, J.T. and Blankenship, R.E. (1989) Isolation of a photoactive photosynthetic reaction center-core antenna complex from *Heliobacillus mobilis. Biochemistry* **28**: 9898–904.

Trost, J.T., Brune, D.C. and Blankenship, R.E. (1992) Protein sequences and redox titrations indicate that the electron acceptors in reaction centers from heliobacteria are similar to photosystem I. *Photosynth. Res.* **32**: 11–22.

Tsuji, K., Tsien, H.C., Hanson, R.S., Depalma, S.R., Scholtz, R. and Laroche, R. (1990) 16S Ribosomal sequence analysis for determination of phylogenetic relationship among methylotrophs. *J. Gen. Microbiol.* **136**: 1–10.

Turner, S., Burger-Wiersma, T., Giovannoni, S.J., Mur, L.R. and Pace, N.R. (1989) The relationship of a prochlorophyte *Prochlorothrix hollandica* to green chloroplasts. *Nature* (London) **337**: 380–2.

Urakami, T. and Komagata, K. (1984) *Protomonas*, a new genus of facultatively methylotrophic bacteria. *Int. J. Syst. Bacteriol.* **34**: 188–201.

Urbach, E., Robertson, D.L. and Chisholm, S.W. (1992) Multiple evolutionary origins of prochlorophytes within the cyanobacterial radiation. *Nature* (London) **355**: 267–70.

van de Meent, E.J., Kleinherenbrink, F.A.M. and Amesz, J. (1990) Purification and properties of an antenna-reaction center complex from heliobacteria. *Biochim. Biophys. Acta* **1015**: 223–30.

van de Meent, E.J., Kobayashi, M., Erkelens, C., van Veelen, P.A., Amesz, J. and Watanabe, T. (1991) Identification of 81-hydroxychlorophyll *a* as a functional reaction center pigment in heliobacteria. *Biochim. Biophys. Acta* **1058**: 356–62.

van den Eynde, H., van de Peer, Y., Perry, J. and de Wachter, R. (1990) 5S Ribosomal sequences of representatives of the genera *Chlorobium, Prosthecochloris, Thermomicrobium, Cytophaga, Flavobacterium, Flexibacter,* and *Saprospira* and a discussion of the evolution of eubacteria in general. *J. Gen. Microbiol.* **136**: 11–8.

van Dorssen, R.J., Vasmel, H. and Amesz, J. (1985) Antenna organization and energy transfer in membranes of *Heliobacterium chlorum. Biochim. Biophys. Acta* **809**: 199–203.

Vasmel, H., Swarthoff, T., Kramer, H.J.M. and Amesz, J. (1983) Isolation and properties of a pigment-protein complex associated with the reaction center of the green photosynthetic sulfur bacterium *Prosthecochloris aestuarii. Biochim. Biophys. Acta* **725**: 361–7.

Walter, E., Schreiber, S., Zass, E. and Eschenmoser, A. (1979) Bakteriochlorophyll a_{GG} und Bakteriophäophytin, a_P in den photosynthetischen Reactionzentren von *Rhodospirillum rubrum* G-9$^+$. *Helv. Chim. Acta* **62**: 899–920.

Watson, S.W. and Mandel, M. (1971) Comparison of the morphology and deoxyribonucleic acid composition of 27 strains of nitrifying bacteria. *J. Bacteriol.* **107**: 563–9.

Weiss, C. (1978) Optical spectra of chlorophylls. In *The Porphyrins*, vol. III (Dolphin, D., ed), ch. 3. Academic Press: New York.

Willems, A., Gillis, M. and de Ley, J. (1991) Transfer of *Rhodocyclus gelatinosus* to *Rubrivivax gelatinosus* gen. nov., comb. nov., and phylogenetic relationships with *Leptothrix, Sphaerotilus natans, Pseudomonas saccharophila,* and *Alcaligenes latus. Int. J. Syst. Bacteriol.* **41**: 65–73.

Witt, D., Bergstein-ben Dan, T. and Stackebrandt, E. (1989) Nucleotide sequence of 16S ribosomal RNA and phylogenetic position of the green sulfur bacterium *Clathrochloris sulfurica. Arch. Microbiol.* **152**: 206–8.

Woese, C.R. (1987) Bacterial evolution. *Microbiol. Rev.* **51**: 221–71.

Woese, C.R., Stackebrandt, E., Weisburg, W.G., Paster, B.J., Madigan, M.T., Fowler, V.J., Hahn, C.M., Blanz, P., Gupta, R., Nealson, K.H. and Fox, G.E. (1984a) The phylogeny of purple bacteria: The alpha subdivision. *Syst. Appl. Microbiol.* **5**: 315–26.

Woese, C.R., Weisburg, W.G., Paster, B.J., Hahn, C.M., Tanner, R.S., Krieg, N.R.,

Koops, H.-P., Harms, H. and Stackebrandt, E. (1984b) The phylogeny of purple bacteria: The beta subdivision. *Syst. Appl. Microbiol.* **5**: 327–36.

Woese, C.R., Debrunner-Vossbrinck, B.A., Oyaizu, H., Stackebrandt, E. and Ludwig, W. (1985a) Gram-positive bacteria: Possible photosynthetic ancestry. *Science* **229**: 762–5.

Woese, C.R., Weisburg, W.G., Hahn, C.M., Paster, B.J., Zablen, L.B., Lewis, B.J. Macke, T.J., Ludwig, W. and Stackebrandt, E. (1985b) The phylogeny of purple bacteria: The gamma subdivision. *Syst. Appl. Microbiol.* **6**: 25–33.

Woese, C.R., Mandelco, L., Yang, D., Gherna, R. and Madigan, M.T. (1990) The case for relationship of the flavobacteria and their relatives to the green sulfur bacteria. *Syst. Appl. Microbiol.* **13**: 258–62.

Wolfrum, T. and Stolp, H. (1987) Comparative studies on 5S RNA sequences of ribulose monophosphate-type methylotrophic bacteria. *Syst. Appl. Microbiol.* **9**: 273–6.

Woodward, R.B. and Skaric, V. (1961) A new aspect of the chemistry of chlorins. *J. Am. Chem. Soc.* **83**: 4676–8.

Wullink, W., Knudsen, J., Olson, J.M., Redlinger, T.E. and van Bruggen, E.F.J. (1991) Localization of polypeptides in isolated chlorosomes from green phototrophic bacteria by immuno-gold labeling electron microscopy. *Biochim. Biophys. Acta* **1060**: 97–105.

Yildiz, F.H., Gest, H. and Bauer, C.E. (1991). Attenuated effect of oxygen on photopigment synthesis in *Rhodospirillum centenum*. *J. Bacteriol.* **173**: 5502–6.

Young, J.P.W., Downer, H.L. and Eardly, B.D. (1991) Phylogeny of the phototrophic *Rhizobium* strain BTai1 by polymerase chain reaction-based sequencing of a 16S ribosomal RNA gene segment. *J. Bacteriol.* **173**: 2271–7.

Yurkov V.V. and Gorlenko, V.M. (1991) A new genus of freshwater aerobic bacteriochlorophyll *a*-containing bacteria. *Microbiology* (Moscow) **60**: 628–32.

Yurkov, V.V., Lysenko, A.M. and Gorlenko, V.M. (1991a) Hybridization analysis of the classification of bacteriochlorophyll *a*-containing freshwater aerobic bacteria. *Microbiology* (Moscow) **60**: 362–6.

Yurkov, V.V., Mityushina, L.L. and Gorlenko, V.M. (1991b) Ultrastructure of the aerobic bacterium *Erythrobacter sibiricus* which contains bacteriochlorophyll *a*. *Microbiology* (Moscow) **60**: 234–8.

Zuber, H. and Brunisholz, R.A. (1991) Structure and function of antenna polypeptides and chlorophyll-protein complexes: Principles and variability. In *Chlorophylls* (Scheer, H., ed), pp. 627–703. CRC Press: Boca Raton, FL.

11

Analysis of Pigments: Carotenoids and Related Polyenes

Karin Schmidt[1], Ann Connor[2] and George Britton[2]
[1]Georg-August-Universität, Göttingen, Federal Republic of Germany
[2]University of Liverpool, Liverpool, UK

11.1 INTRODUCTION

11.1.1 Carotenoids

Carotenoids are yellow-orange pigments which are widely distributed in plants and animals as well as in prokaryotes. All phototrophic prokaryotes contain carotenoids in their photosynthetic apparatus, and any non-photosynthetic bacteria that are coloured yellow-orange are likely to be pigmented by carotenoids or, rarely, by other polyenes (see Section 11.13). The nature of these pigments and their quantitative composition provide extremely valuable criteria for classification.

11.1.2 Carotenoid structure and nomenclature

Carotenoids are almost all tetraterpenes that have a symmetrical C_{40} skeleton, which may be either acyclic or possess a six-membered ring (occasionally five-membered) at one or both ends of the molecule. The hydrocarbons are generally known as *carotenes*; derivatives containing one or more oxygen functions (e.g. aldehyde, carboxylic acid, epoxy, hydroxy, keto, methoxy groups) are known as *xanthophylls*.

About 200 of the 500 or more naturally occurring carotenoids examined have been found in prokaryotes. Traditionally, these compounds have been given trivial names, frequently derived from the name of the biological source from which they were first isolated. A semi-systematic nomenclature which conveys structural information has now been devised (IUPAC-IUB, 1971). According to this scheme, the carotenoid molecule is considered in two halves, and each half is designated by a Greek letter

---

Chemical Methods in Prokaryotic Systematics. Edited by M. Goodfellow and A.G. O'Donnell
© 1994 John Wiley and Sons Ltd.

prefix. Each individual carotenoid is then named as a derivative of the parent carotene, specified by the Greek letters describing its two end-groups, with conventional prefixes and suffixes being used to indicate changes in hydrogenation level and the presence of substituent groups. Chirality is designated conventionally by the (R, S) system. The seven end-groups that have been recognized are illustrated in Figure 11.1. The full structures and numbering of acyclic lycopene (ψ,ψ-carotene) and dicyclic β-carotene (β,β-carotene) are illustrated in Figure 11.2. The C_{30} carotenoids found in a few non-photosynthetic bacteria are considered to be 4,4'-diapocarotenoids. The C_{45} and C_{50} carotenoids are named as C_5-substituted derivatives of the normal C_{40} skeleton.

Because of the extensive double bond system in the molecule, any carotenoid can, theoretically, exist in a large number of geometrical isometric forms (*cis-trans* or Z-E isomers). Most carotenoids occur naturally in the

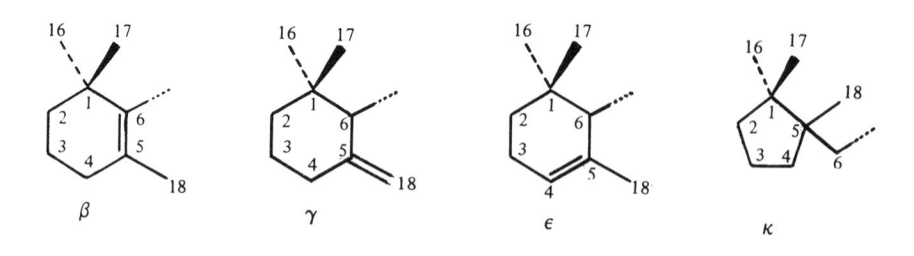

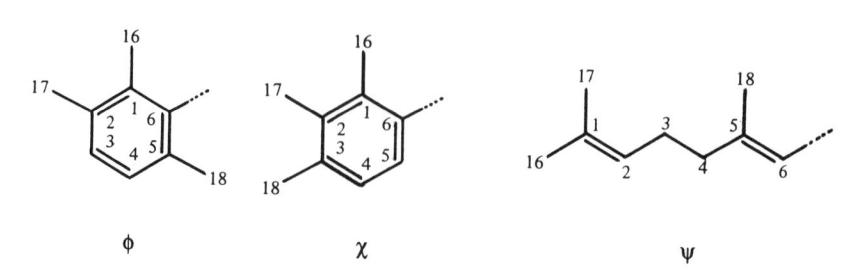

Figure 11.1 General carotenoid structure: the seven end-groups.

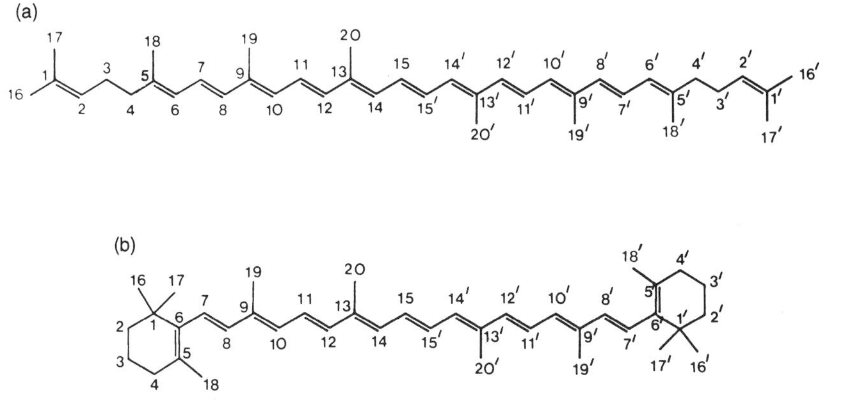

Figure 11.2 Structures and numbering scheme for (a) lycopene and (b) β-carotene.

all-*trans* (all-*E*) form, but the presence of *cis* (*Z*) isomers, usually in small amounts, must always be considered.

The structures and lists of the semi-systematic names of the most important carotenoids referred to in this chapter are given in Appendix 11.1. The reader should consult Isler (1971) or Weedon and Moss (1994) for details of the application of the new nomenclature. A list of all known natural carotenoids (up to 1986), with their structures and extensive references to their occurrence and properties, has been compiled (Pfander, 1987).

For a survey of the carotenoid field in general, the monograph of Isler (1971) is still a monumental and useful source of information. Complementary to this is a volume by Goodwin (1980), which includes comprehensive tabulated data on the occurrence and distribution of carotenoids including chapters on those found in prokaryotes. A book entitled *Carotenoids in Photosynthesis* (Young & Britton, 1993) is also useful in this respect as well as providing details of the functioning of carotenoids in photosynthesis. The proceedings of the 'International Symposium on Carotenoids' series report progress in all aspects of carotenoid science (IUPAC, 1979, 1985, 1991; Britton & Goodwin, 1982; Krinsky *et al.*, 1990). In a new series of books devoted to carotenoids a practical approach is adopted; the first volume gives full details of all methods used in the isolation and analysis of carotenoids (Britton *et al.*, 1994).

11.1.3 Carotenoid biosynthesis

Carotenoids, as tetraterpenes, are synthesized by the normal isoprenoid pathway from two molecules of the C_{20} intermediate geranylgeranyl diphosphate via the C_{40} prephytoene diphosphate and the first C_{40}

hydrocarbon, phytoene. Phytoene then undergoes a series of desaturation reactions. In each of these, two hydrogen atoms are removed and the conjugated chromophore is extended by two double bonds, to form, sequentially, phytofluene, ζ-carotene or its unsymmetrical isomer 7,8,11,12-tetrahydrolycopene, neurosporene and lycopene. Cyclization, the alternative C1,2 hydration and the introduction of other oxygen functions or C_5 substituents occur late in the biosynthetic sequence. A general outline of carotenoid biosynthesis is given in Figure 11.3.

The C_{30} 4,4'-diapocarotenoids are, biosynthetically, triterpenes formed from two molecules of the C_{15} farnesyl diphosphate via the phytoene analogue didehydrosqualene. The C_{45} and C_{50} carotenoids are synthesized by the normal C_{40} pathway, with the additional one or two C_5 units being added late in the biosynthesis, e.g. during cyclization. Details of carotenoid biosynthesis can be found in many review articles (e.g. Goodwin, 1971; Britton, 1976, 1988, 1990, 1991; Schmidt, 1978; Davies, 1980).

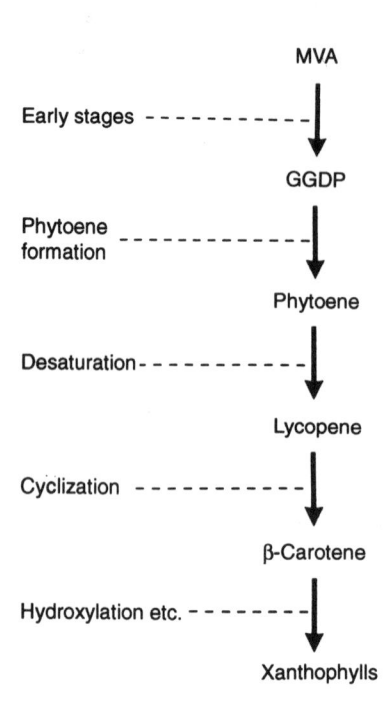

```
                              MVA
      Early stages  - - - - - - - - - ↓
                              GGDP
        Phytoene    _ _ _ _ _ _ _ _ _ ↓
        formation
                            Phytoene
      Desaturation - - - - - - - - - ↓
                            Lycopene
       Cyclization  - - - - - - - - - ↓
                           β-Carotene
   Hydroxylation etc. - - - - - - - - ↓
                          Xanthophylls
```

Figure 11.3 Summary of the carotenoid biosynthetic pathway.

11.2 DISTRIBUTION AND TAXONOMY

Prokaryotes can be divided into three main physiological groups:

(a) oxygenic phototrophs (cyanobacteria);
(b) anoxygenic phototrophs (Chlorobiaceae, Chloroflexaceae, Chromatiaceae and Rhodospirillaceae);
(c) non-photosynthetic prokaryotes.

11.2.1 Cyanobacteria

Cyanobacteria contain chlorophyll a and a collection of carotenoids in their photosynthetic pigment-protein complexes. Carotenoid is also frequently present elsewhere in the cell, for example in the cell wall and in chlorophyll-less carotenoid-proteins.

The carotenoids most characteristic of cyanobacteria are β-carotene and its oxygenated (oxo- and hydroxy-) derivatives, especially caloxanthin, echinenone, nostoxanthin and zeaxanthin, with other, structurally-related carotenoids usually present in small amounts. Ring types other than β-rings have not been found. Also characteristic are carotenoid glycosides, notably monocyclic myxoxanthophyll and acyclic oscillaxanthin; different sugar residues may be present in these compounds, which are probably located outside the photosynthetic membranes. The main differences between species are quantitative variations in overall carotenoid composition.

11.2.2 Anoxygenic phototrophs

Approximately 80 different carotenoids have been found in phototrophic bacteria; their distribution has been discussed in detail elsewhere (Schmidt, 1978). These carotenoids, especially those found in members of the family Rhodospirillaceae, are mostly acyclic; the main compounds are substituted at the C1 (and C1') position by tertiary methoxy or hydroxy groups, the latter sometimes being glycosylated. These carotenoids are considered to be formed by two pathways, the 'normal' spirilloxanthin pathway via lycopene and the 'alternative' spirilloxanthin pathway via neurosporene and spheroidene. Under aerobic conditions, some species introduce conjugated oxo groups at C2 (e.g. spheroidenone).

Although several species may utilize the same biosynthetic pathway, quantitative carotenoid composition may differ considerably, because of different relative activities of the enzymes involved in elaboration of the end-groups. Compounds that are 'intermediates' in some species may be the main carotenoids in others, e.g. lycopene and rhodopin in *Rhodospirillum rubrum* and *Rhodomicrobium vannielii*, respectively. Cyclic carotenoids containing β-rings do occur, albeit rarely. β-Carotene and canthaxanthin have been found, but carotenoids with aryl end-groups occur more frequently and are characteristic of many sulphur bacteria. These compounds may be hydrocarbons (chlorobactene) or oxygenated compounds such as okenone. Carotenoids are mostly located, together with bacteriochlorophylls, in the pigment-protein complexes of photosynthetic membranes. The different complexes have different pigment compositions, which vary with the age of the cultures and with environmental conditions. Carotenoids have also been found in the cell wall of some species. For details of the location and functioning of carotenoids in the pigment-protein complexes, see Young (1993). The main features of the distribution of carotenoids in the phototrophic bacteria are summarized in Table 11.1.

Recently isolated strains of marine organotrophic photosynthetic bacteria which will not grow anaerobically even in the light were originally classified in a new genus, *Erythrobacter*, but are now designated *Roseobacter*. These organisms contain bacteriochlorophyll and carotenoids of the spirilloxanthin-spheroidene series, but some also contain β-ring carotenoids typical of cyanobacteria.

11.2.3 Non-photosynthetic prokaryotes

Although many non-photosynthetic bacteria do not produce carotenoids, there are many that do. These carotenoids are localized mainly in

Table 11.1 Main features of carotenoid distribution in photosynthetic bacteria.

(a) 'Normal' spirilloxanthin series
(Main carotenoids: lycopene, rhodopin, anhydrorhodovibrin, rhodovibrin, spirilloxanthin)

Rhodospirillaceae:	Chromatiaceae:
Rhodomicrobium vannielii[a]	*Amoebobacter pendens*
Rhodospirillum rubrum	*Amoebobacter roseus*
Rhodospirillum fulvum[b]	*Chromatium minutissimum*
Rhodopseudomonas palustris	*Chromatium vinosum*
Rhodopseudomonas photometricum	*Ectothiorhodospira halochloris*[b]
	Ectothiorhodospira halophila
	Ectothiorhodospira mobilis
	Ectothiorhodospira shaposhnikovii
	Thiospirillum jenense

11. Analysis of Pigments: Carotenoids and Related Polyenes

Table 11.1 (Continued).

(b) *Spirilloxanthin series + rhodopinal series*
(Main carotenoids as (a), plus lycopen-20-al, lycopen-20-ol, rhodopin-20-al, rhodopin-20-ol)

Rhodospirillaceae:
   Rhodocyclus purpureus
   Rhodocyclus tenuis
   Rhodopseudomonas acidophila[a,b]

Chromatiaceae:
   Chromatium violascens
   Chromatium warmingii
   Lamprocystis roseopersicina
   Thiocapsa pfennigii
   Thiocystis violacea
   Thiodictyon bacillosum
   Thiodictyon elegans

(c) *Spheroidene-spirilloxanthin series*
(Main carotenoids: neurosporene, spheroidene, spheroidenone, hydroxyspheroidene, hydroxyspheroidenone; rhodovibrin and spirilloxanthin often present)

Rhodospirillaceae:
   Rhodobacter capsulatus
   Rhodobacter sphaeroides
   Rhodocyclus gelatinosus
   Rhodopseudomonas blastica
   Rhodopseudomonas sulphidophilus

(d) *1,2-Dihydrocarotenoid series*
(Main carotenoids: 1,2-dihydroneurosporene, neurosporene, 1,2-dihydrolycopene, lycopene)

Rhodospirillaceae:
   Rhodopseudomonas viridis

(e) *Okenone and related 4-oxo-carotenoid series*
(Main carotenoids: okenone, thiothece-HO-474)

Rhodospirillaceae:
   Rhodopila globiformis

Chromatiaceae:
   Chromatium minus
   Chromatium okenii
   Chromatium purpuratum
   Chromatium weissii
   Thiocapsa strain 5811
   Thiocystis gelatinosus

(f) *Aryl and alicyclic carotenes series*
(Main carotenoids: chlorobactene, isorenieratene, β-carotene, γ-carotene)

Chlorobiaceae:
   Chlorobium limcola f. *thiosulfatophilum*
   Chlorobium phaeobacteriodes
   Chlorobium phaeovibriodes
   Chlorobium vibreoforme

Chloroflexaceae:
   Chloroflexus aurantiacus[b]

---

[a] β-Carotene also present.
[b] Glycosides present.

membranes but also, sometimes, in cell walls. Carotenoid biosynthesis is often photo-induced, but is also affected by many other factors, notably carbon and nitrogen sources, minerals, temperature and growth factors, particularly thiamine. Culture conditions must therefore be carefully regulated and specified when carotenoid content and composition are reported.

Detailed studies of the carotenoids of non-photosynthetic bacteria are sparse and taxonomic trends difficult to recognize. The variety of carotenoid structures encountered is large. Monocyclic and dicyclic β-ring carotenoids are most common. ε-Rings are rare; reports of their occurrence should be reinvestigated by modern methods. Arylcarotenoids with ϕ- or χ-end-groups are not uncommon. It is characteristic of many bacteria that they produce C_{30}, or C_{45} and C_{50} carotenoids rather than the conventional

Table 11.2 Main carotenoids in some representative non-photosynthetic bacteria.

Group	Representative species	Main carotenoid
Gliding bacteria	*Flexibacter* sp.	Flexixanthin (337)[a]
	Flavobacterium spp.	Zeaxanthin (119)
	Flavobacterium breve	Flexirubin[b]
	Myxococcus fulvus	Myxobactone (309)
Gram-negative, aerobic, rods and cocci	*Halobacterium halobium*	Bacterioruberin (456)
	Pseudomonas xanthochrus	Zeaxanthin (119)
	Rhizobium lupini	2,3,2',3'-Tetrahydroxy-β,β-caroten-4-one (361)
	Xanthomonas juglandis	Xanthomonadin[b]
Gram-positive cocci	*Micrococcus luteus*	Sarcinaxanthin (C_{50}) (441)
	Micrococcus roseus	Canthaxanthin (380)
	Staphylococcus aureus	4,4'-Diaponeurosporen-4-oic acid (C_{30}) (526)
	Streptococcus faecium	4,4'-Diaponeurosporene (C_{30}) (515)
Actinomycetes	*Arthrobacter glacialis*	'A.g.470' (C_{50}) (451)
	Brevibacterium linens	3,3'-Dihydroxy-isorenieratene (161)
	Cellulomonas dehydrogenans	Decaprenoxanthin (C_{50}) (447)
	Curtobacterium flaccumfaciens pv. poinsettiae (formerly *Corynebacterium poinsettiae*)	'C.p.450' (C_{50}) (439)
	Mycobacterium aureum	Isorenieratene (24)
	Mycobacterium phlei	Phleixanthophyll (147)
Mycoplasmas	*Mycoplasma laidlawii*	Neurosporene (34)

[a] Number of this carotenoid in the list of Pfander (1987).
[b] Non-carotenoid polyene (see Section 11.13).

11. Analysis of Pigments: Carotenoids and Related Polyenes 411

C_{40} structures. Conjugates such as glycosides and their acyl esters also appear to be common. The main features of the meagre knowledge of carotenoid distribution in non-photosynthetic bacteria are summarized in Table 11.2.

11.3 STRATEGY FOR DETERMINING CAROTENOID COMPOSITION

The overall strategy for carotenoid analysis is simple. Carotenoids are lipid substances and have to be extracted from cells with organic solvent. At this stage, total carotenoid content can be determined. The individual carotenoids are then separated, purified, identified and assayed. Separation and purification can be achieved by classical chromatographic procedures: column, thin-layer and paper. The individual carotenoids are recovered, estimated spectrophotometrically and identified first by their chromatographic properties, co-chromatography with authentic standards, UV-visible absorption spectra and simple chemical tests, and characterized fully by MS, NMR and other spectroscopic methods.

This strategy is still widely used and reliable, but suffers from the following disadvantages: it is time-consuming for routine screening; minor components of a mixture may be missed in small-scale work; and losses of carotenoids during chromatography can be substantial and variable, so that the accuracy of quantitative determinations may be suspect.

The recommended method for routine screening and quantitative analysis of carotenoid composition is now high-performance liquid chromatography (HPLC). Particularly with reversed-phase procedures carotenoid losses are negligible, very small samples can be analysed (nanogram levels) and a total extract can be analysed and a full quantitative carotenoid composition obtained in 30 to 60 min. Retention times provide a valuable first means of identification and, if the absorption spectrum of each compound can be determined on-line by use of a suitable detector, the procedure provides a very powerful and efficient means of analysing large numbers of samples in a relatively short time, and is therefore ideal for taxonomic studies. It must be stressed, however, that this does not constitute rigorous identification or characterization of carotenoids. For this, it is always necessary to purify larger samples and subject them to detailed MS and NMR analysis. Even with HPLC, minor components of a mixture can easily be missed.

Recommended procedures will be described and discussed, and some specific examples given. Further details are available in many other articles (e.g. Britton & Goodwin, 1971; Davies, 1976; Britton, 1985; Liaaen-Jensen

& Andrewes, 1985; Goodwin & Britton, 1988; Britton & Young, 1993; Britton et al., 1994).

11.4 GENERAL METHODS

11.4.1 Properties and handling of carotenoids

Carotenoids are like other groups of higher isoprenoids with respect to their solubility properties. Carotenes are typical non-polar hydrocarbons; xanthophylls are more polar, yet are still insoluble in water. The special and characteristic properties of carotenoids are a consequence of the conjugated double bond system, which constitutes the light-absorbing chromophore. This conjugated polyene system also renders carotenoids sensitive to oxygen, light, heat, acids and, in some cases, alkali, and especially to combinations of these factors, e.g. light and oxygen. Stringent precautions must be observed if losses of material or unwanted structural changes are to be avoided. Speed of manipulation is important; all procedures that inevitably introduce risks of oxidation, isomerization and the like should be carried out as rapidly as possible. An extensive account of the isolation of carotenoids, including the special handling methods required and the production of artefacts if these methods are not applied, is given by Schiedt and Liaaen-Jensen (1994).

11.4.2 Protection against oxygen

The need to prevent oxidation of carotenoids cannot be over-emphasized. Oxygen, especially in combination with light and/or heat, is the most destructive factor. The acyclic carotenoids, particularly phytofluene and ζ-carotene, seem to be the most labile under oxidizing conditions. The presence of even traces of oxygen in stored samples, or peroxides in solvents (especially diethyl ether) or of any oxidizing agents even in crude samples containing carotenoids can rapidly lead to bleaching or to the formation of artefacts such as apocarotenals or epoxides. Carotenoid samples or extracts must always be stored in the complete absence of oxygen, either *in vacuo* or in an inert atmosphere (Ar or N_2). If solutions must be stored they should be flushed with nitrogen for a few minutes.

11.4.3 Protection against light and heat

Exposure of carotenoids to light or heat should be avoided if possible, although direct photo-isomerization of most carotenoids during normal manipulations occurs comparatively slowly and usually not to any great

11. Analysis of Pigments: Carotenoids and Related Polyenes

extent. With extracts of phototrophic organisms which contain chlorophylls, photo-isomerization, sensitized by chlorophyll and proceeding via the carotenoid triplet state, can occur very rapidly so that appreciable amounts of carotenoid Z-isomers (*cis*-isomers) can be produced during even brief exposure of the extract to light. Even in the absence of such a sensitizer, direct sunlight or UV-light may cause some geometrical isomerization and must be avoided. Low-intensity diffuse daylight or subdued artificial light is acceptable for most carotenoid manipulations, but precautions should be taken to exclude light during chromatography. Thus, glass chromatography columns should be wrapped in aluminium foil or black paper, and developing tanks for thin-layer chromatography covered with black cloth or kept in a dark cupboard.

Carotenoids may also undergo isomerization and structural modification if heated, either as solids or in solution. They should never be subjected to excessive heat. Solvents with low boiling points should be used whenever possible, as these can subsequently be removed without the need for excessive heating. Solvents should be evaporated on a rotary evaporator at a temperature not exceeding 40 °C.

11.4.4 Avoidance of acid and alkali

There is a risk of oxidative decomposition, *cis-trans* isomerization or, in some cases, dehydration if carotenoids are subjected to acid conditions, especially in the presence of light and/or oxygen. Neutral chromatographic adsorbents should be used, and the use of chloroform, which usually contains traces of HCl, should be avoided. Acidic reagents and strong acids should not be used in rooms where carotenoids are being handled. Most carotenoids are stable to alkali but some, notably those containing a 3-hydroxy-4-oxo-β-ring, are altered by treatment with even weak alkali. Saponification must be avoided if it is suspected that any such compounds, or carotenoid acyl esters, may be present.

11.4.5 Purity of solvents, adsorbents and reagents

Pure solvents and reagents should be used for all manipulations. Solvents should be dried and redistilled, and peroxides must be removed from diethyl ether by distillation from, for example, reduced iron powder or calcium hydride. Solvents should be stored in airtight, dark glass bottles. Because of its toxicity, the use of benzene is not recommended; the less toxic toluene is usually a satisfactory substitute.

The sensitive spectroscopic methods now available require purification to be extremely rigorous. When small samples (e.g. a few micrograms) of a carotenoid are being prepared for analysis by mass spectrometry or NMR spectroscopy, impurities can be introduced during the purification procedure. Rigorous purification of solvents by double distillation and filtration through an activated material such as alumina are recommended. Thin-layer chromatography plates should be pre-washed with a solvent at least as strong in polarity as that which is to be used for elution of the carotenoids.

Small amounts of plasticizers, especially phthalates, are readily dissolved by organic solvents, and can present a major problem. All contact of samples, solvents and the like with plastic material must be avoided. Apparatus such as plastic or polythene wash bottles, pipettes and stoppers should never be used in carotenoid work. Solvents, even water, should not be stored in plastic bottles.

11.5 EXTRACTION AND SAPONIFICATION OF CAROTENOIDS

Carotenoids are extracted from harvested cells with a water-miscible organic solvent, usually acetone, ethanol or methanol, or a mixture of these. Extraction with chloroform-methanol mixtures, a method widely used for other lipids, is not recommended for carotenoids because of the likely presence of HCl in the chloroform.

In some cases, carotenoid can be extracted efficiently simply by soaking the cell pellet in solvent at room temperature under N_2 for a few hours. When very polar carotenoids, especially glycosides, are present, or when the carotenoids are more tenaciously held in the cell wall or membrane, more drastic treatment is required. In such cases cells may need to be disrupted by, for example, French press, ultrasonic disintegration, shaking or grinding with glass beads or other abrasive material, treatment with enzymes (especially lysozyme) followed by osmotic shock, or even direct saponification. Lyophilization of cell material may also facilitate extraction of pigments. In some cases, however, similarly efficient extraction may be achieved, without cell disruption, by the cautious use of hot solvent. The brief exposure to heat may be less harmful than the effects of prolonged and inefficient extraction with cold solvent, or of cell disruption.

Solvent extraction should be repeated. In some cases, a first extraction serves mainly to remove water, or to extract bacteriochlorophyll, but liberates little carotenoid. The bulk of the carotenoid is then removed by a second extraction.

Method 11.1 Extraction of carotenoids from cells.

1. Cells are harvested by centrifugation under nitrogen and the extracting solvent, acetone-methanol (7:2) is added.

11. Analysis of Pigments: Carotenoids and Related Polyenes

2. If necessary the solvent is rapidly heated to boiling, with shaking or stirring, and the mixture left to stand for 10 min in the dark under N_2.
3. Cell debris is removed by filtration under suction and is re-extracted with acetone-methanol until no more colour is removed.
4. The combined extracts are transferred to a separating funnel, an approximately equal volume of ether is added, followed by water or 3% sodium chloride solution until the two phases separate.
5. The ether layer, containing the carotenoids and other lipids, is washed two to three times with water to remove the acetone/methanol and then evaporated on a rotary evaporator.
6. The residue is dissolved in ether or dichloromethane, transferred to a small vial and evaporated to dryness under a stream of N_2, a little ethanol being added, if necessary, to aid removal of any residual water. The extract may be stored at $-20\,°C$ under N_2.
7. If the extract is to be analysed by HPLC, it is redissolved in dichloromethane and filtered through a cotton wool plug in a Pasteur pipette to remove particulate matter. The sample is again evaporated to dryness under a stream of N_2 and redissolved in the required solvent for HPLC.

Saponification may, in some cases, be an advantageous way of removing unwanted acyl lipids or chlorophylls, but should be avoided if carotenoid esters or alkali-labile carotenoids may be present. If acetone has been used for the extraction, all traces of it must be removed before the extract is saponified. Under saponification conditions, acetone polymerizes to give oils which can become serious impurities. Also methyl ketone artefacts can be produced from carotenoid aldehydes by aldol condensation with traces of acetone.

Method 11.2 Saponification of carotenoid extracts.

1. The carotenoid-containing extract is dissolved in ethanol (or methanol) and sufficient aqueous KOH solution (60% w/v) added to bring the final overall KOH concentration to 5–6%.
2. The mixture is then allowed to stand under N_2 at room temperature or below for 10 to 12 h or overnight, in the dark.
3. The saponified pigments are transferred to ether and the ethereal solution is washed exhaustively with water or NaCl solution to remove alkali, and then evaporated to dryness.

11.6 ISOLATION AND PURIFICATION: CHROMATOGRAPHY

Column, thin-layer and paper chromatography and HPLC are widely used with carotenoids (Pfander et al., 1994). Gas-liquid chromatography, however, is not suitable because of the instability of the compounds. In order to devise the best separation and purification procedure for a particular extract, a rapid preliminary examination by TLC is recommended. This gives an indication of the number and variety of carotenoids present. Column chromatography is then used, if necessary, to separate the extract into fractions containing compounds of similar polarity. Thin-layer chromatography is used to isolate and purify the individual carotenoids in each fraction or in the total extract. Different adsorbents should be used; a general strategy that will usually yield a pure carotenoid is to use successive TLC on silica gel, MgO-kieselguhr G, and silica again, but with a different solvent. Samples for MS and NMR analysis need further treatment to bring them to the rigorous state of purity required.

11.6.1 Choice of adsorbent

The most useful adsorbents for column and thin-layer chromatography of carotenoids are alumina, silica and basic inorganic materials such as MgO and $Ca(OH)_2$. Separation on silica and alumina depends upon polarity; carotenoids containing polar substituents are most strongly adsorbed. In contrast, separation on, for example, MgO or $Ca(OH)_2$, is determined by the number and arrangement of double bonds in the molecule. Carotenoids with the most extensive conjugated double bond system are most strongly adsorbed on MgO, e.g. lycopene > neurosporene > ζ-carotene. Spirilloxanthin and other acyclic carotenoids with a very long chromophore may crystallize on the adsorbent. They are also so strongly adsorbed that they are difficult to elute, and losses during the chromatography can be substantial. Acyclic carotenoids are much more strongly adsorbed than cyclic ones having the same number of double bonds (lycopene > γ-carotene > β-carotene), and β-ring compounds are more strongly held than the corresponding ε-ring isomers (e.g. β-carotene > α-carotene). The presence of polar groups, even hydroxy-groups, is less important; the acyclic hydrocarbon lycopene is much more strongly adsorbed than the dicyclic diol, zeaxanthin.

Other basic adsorbents, especially $Ca(OH)_2$ and $ZnCO_3$, are useful for separating geometrical isomers that may be difficult or impossible to resolve on other adsorbents.

The instability of some carotenoids, however, can cause problems even with commonly used adsorbents. Thus, on alumina, some carotenoids, especially those containing a 3-hydroxy-4-oxo-β-ring (e.g. astaxanthin) undergo irreversible oxidation to the corresponding 2,3-didehydro-(diosphenol) derivatives, which are virtually impossible to elute. Indiscriminate use of activated alumina (grade 0 or I) can cause geometrical isomerization. Both alumina and silica, when used for carotenoid chromatography, should be neutral, to avoid the risk of acid or base-catalysed structural modifications.

Basic adsorbents such as MgO should not be used without dilution with a filter aid (celite) or binder (kieselguhr G), if polymerization of acetone in the developing solvent is to be avoided. Microcell C has been considered a useful and inert adsorbent, but it is now known to catalyse structural modifications, e.g. β-carotene to isocryptoxanthin.

11.6.2 Preliminary analysis by thin-layer chromatography

An unfamiliar extract may be subjected to preliminary investigation by TLC. This can be achieved very rapidly, economically and on a small scale by using either small pieces (5 × 2 cm) cut from aluminium foil-backed silica gel plates or silica gel-coated microscope slides.

Method 11.3 TLC for preliminary analysis of extracts.

A single spot of extract is applied to the plate, which is then developed successively in a series of solvents of increasing polarity to reveal the presence of carotenoids of different polarity groups. (A suitable series of solvents is listed in Table 11.3, together with the group of carotenoids that each would reveal.)

A small jar or beaker is used as developing chamber, and development time is approximately 2 min, with 2–5 ml of each solvent. From these preliminary results a detailed purification strategy is devised.

11.6.3 Column chromatography

Alumina and silica (silicic acid) are the adsorbents usually used for column chromatography of carotenoid-containing extracts. Other adsorbents such

Table 11.3 List of solvents for chromatography of different classes of carotenoids. These solvents are suitable for elution from columns of alumina (Grade III) or for TLC on silica gel G.

Solvent[a]	Classes of carotenoids
P, 1% E/P or 10% T/P	Carotene hydrocarbons
10% E/P	Carotenol esters
20–30% E/P	Monomethoxy-, dimethoxycarotenoids and their mono-oxo derivatives
50% E/P	Monohydroxycarotenoids
60% E/P	Dioxocarotenoids, mono- and dimethoxy-dioxocarotenoids
E or 1% EtOH/E	Dihydroxy, trihydroxy- and tetrahydroxycarotenoids
5–20% EtOH/E	Carotenoid glycosides

[a] P, light petroleum or hexane; E, diethyl ether; T, toluene; EtOH, ethanol.

as MgO, Ca(OH)$_2$ may sometimes be useful for separating difficult mixtures or geometrical isomers on a preparative scale.

Method 11.4 Column chromatography on alumina.

1. The required amount of neutral activated alumina (e.g. 10 g for up to 200 mg lipid) is weighed into a beaker or flask, covered with dry, redistilled petrol and deactivated to Brockmann grade III by pipetting in 0.6 ml (6% w/w) of water and stirring gently for 3 to 4 min to achieve even distribution of the water.
2. The slurry of alumina in the petrol is then used to pack the column. The ratio of column length to diameter should be 10 to 20 : 1 for best separation. A 1–2 cm layer of sand may be added to the top of a large column to prevent blockage if any solid material should be present or if any carotenoid should precipitate or crystallize from the extract. The sand layer can be stirred to restore solvent flow without the actual column being disturbed.
3. The lipid extract is dissolved in the minimal volume of petrol, and the solution carefully applied to the column. Long-chromophore acyclic carotenoids do not readily dissolve in petrol, so extracts containing them may need to be dissolved first in a little toluene or dichloromethane, and petrol then added to the solution so that the overall toluene or dichloromethane concentration does not exceed 10%.
4. Elution is continued with this solvent until all the carotenes have been eluted and the eluate becomes colourless. The solvent is then changed and elution continued with ether-petrol mixtures of increasing ether content, to give fractions of increasing polarity. (Suitable solvent mixtures and the groups of carotenoids that each will elute are listed in Table 11.3.)
5. Each fraction is then evaporated and the individual carotenoids are separated and purified by TLC.

11.6.4 Thin-layer chromatography

Thin-layer chromatography is widely used to purify carotenoids for spectroscopic analysis, and is also used for preliminary identification by comparison with authentic standards. A particular advantage is that, apart from the biosynthetic intermediates, phytoene and phytofluene, carotenoids are coloured and can be seen on TLC plates with great sensitivity. Phytoene and phytofluene and their derivatives are located by examining chromatograms under UV light. Phytofluene fluoresces intensely greenish-white. The weak violet fluorescence of phytoene is not so easily seen, but phytoene can readily be detected by its ability to quench the fluorescence of silica gel GF254 plates or of any TLC plates that have been sprayed with a fluorescent organic dye such as Rhodamine 6G.

The lability of carotenoids is, however, a disadvantage and general TLC practice may need to be modified to avoid excessive losses. Carotenoids are most vulnerable to the destructive effects of light and oxygen while they are adsorbed on thin layers of active materials. Samples must therefore be

11. Analysis of Pigments: Carotenoids and Related Polyenes

applied to chromatograms as rapidly as possible and eluted immediately after plates are removed from the solvent tank. A gentle stream of nitrogen may be directed onto chromatograms as the samples are being applied, and the developing chamber may be flushed with nitrogen and sealed to provide an inert atmosphere during development, which is always carried out in the dark.

11.6.4.a TLC on silica gel

Diethyl ether-light petroleum mixtures are very suitable for most carotenoid separations, and the solvent composition that was used to elute a fraction from a preliminary alumina column usually gives effective separations on TLC (Table 11.3). Good resolution is also obtained with acetone-petrol mixtures (1% of acetone is used in place of each 5% of ether in the solvent mixtures listed in Table 11.3). Other mixtures, e.g. of methanol-toluene, propan-2-ol/petrol, or ethyl acetate-carbon tetrachloride, may be used for a second TLC on silica. For carotenoid glycosides ethyl acetate/propan-2-ol/water (60:40:2) is recommended.

After separation, the individual carotenoid bands are recovered from the developed chromatogram by elution with ether or acetone, a little ethanol being added, if necessary, for more polar carotenoids.

Method 11.5 Elution of carotenoid bands following separation.

1. The carotenoid zone is rapidly scraped from the plate and eluted with a solvent more polar than that used for development.
2. The adsorbent is removed by filtration or centrifugation. A very simple but effective method is to place the adsorbent in a small glass funnel with a small plug of absorbent (defatted) cotton wool in the stem. Ether or ethanol is then added dropwise until the surface of the adsorbent is colourless.
3. Elution is continued with ether until all the carotenoid has been recovered (usually 2–5 ml).
4. The solvent is removed under a gentle stream of nitrogen.

11.6.4.b TLC on magnesium oxide

Magnesium oxide does not adhere well, if at all, to glass plates. A 1:1 mixture of MgO and kieselguhr G gives stable layers which are extremely valuable for carotenoid TLC, and allow separation of mixtures which co-chromatograph on silica, e.g. α- and β-carotenes, lutein and zeaxanthin, chloroxanthin and rhodopin. Thin-layer chromatography on MgO does have some disadvantages, notably that carotenoids tend to run as broad, tailing bands especially when the chromatogram is overloaded (which happens very easily). Also carotenoids may bind irreversibly to MgO unless they are eluted as rapidly as possible, preferably before the solvent has fully evaporated.

The most effective solvents for TLC on MgO-kieselguhr G are mixtures of petrol with acetone and/or toluene. The adsorptive strength of MgO is somewhat variable, so the composition of the solvent needs to be optimized for each batch. As a rough guideline, the following solvent mixtures are usually approximately correct: 4% acetone in petrol for α- and β-carotene, 25% acetone in petrol for zeaxanthin, and acetone-toluene-petrol (1:1:4) or (1:1:1) for lycopene and spirilloxanthin, respectively. Elution from MgO is best achieved as described above for silica TLC, but with acetone, plus some toluene and ethanol for the strongly adsorbed carotenoids.

As well as separating mixtures of carotenoids, TLC on MgO also provides an efficient means of removing colourless, non-carotenoid impurities. A developing solvent is used such that the carotenoid leaves the origin but runs with a low R_F value. Colourless impurities usually run virtually with the solvent front and are thus easily removed.

11.6.5 Paper chromatography

Paper chromatography is useful for rapid preliminary analysis of carotenoid extracts, but it is now mainly used for analysing isomeric sets obtained by I_2-catalysed photo-isomerization of pure carotenoids and for following the course of chemical reactions that are specific to certain functional groups, e.g. acetylation of hydroxycarotenoids. This work normally uses specially prepared impregnated papers, the most effective ones being those filled with kieselguhr or alumina (Schleicher and Schull Nos. 287 and 288, respectively).

Method 11.6 Paper chromatography for carotenoids.

Circular papers are used (18 cm diameter) and chromatograms are developed horizontally between two Petri dishes, solvent being supplied via a small filter paper wick to the centre of the paper. Samples are applied as small spots and the chromatogram is developed for a short time with acetone, to concentrate the sample into a narrow line to give optimum resolution. The acetone is evaporated rapidly under a stream of N_2 and the chromatogram developed in the required solvent. The usual solvents employed are mixtures of acetone (2–25%) in petrol. Details are given by Jensen and Liaaen-Jensen (1959). Carotenoid bands are then cut out and eluted (usually with acetone) for spectroscopic analysis.

11.6.6 Purification of samples for mass spectrometry and nuclear magnetic resonance spectroscopy

The chromatographic procedures described above will usually give a carotenoid sample free from other contaminating carotenoids and suitable for UV-visible light absorption spectroscopic and spectrophotometric analysis. Substantial amounts of impurities originating from the chromatographic adsorbents can be introduced into the sample, however, and these must be excluded or removed before the sample is submitted for MS or NMR analysis.

Purification of solvents to be used for preparing these samples must be extremely rigorous, and TLC plates must be prewashed by running in solvent at least as polar as that which is to be used for chromatography. If possible, samples should be eluted from TLC adsorbents with ether; acetone or ethanol are more likely to extract contaminants, so if they are required only the minimum amounts should be used. Carotenoid samples from TLC or PC should be given an additional purification immediately before being subjected to MS or NMR analysis. A satisfactory procedure is to filter the sample through a small column of neutral alumina (3–4 cm in a Pasteur pipette is usually sufficient), the activity of which has been adjusted by addition of water so that the carotenoid will be eluted with ether or ether-petrol mixtures.

The purified sample is applied to the column in petrol, and elution is begun with petrol or a solvent mixture not sufficiently polar to remove the carotenoid. The eluate is discarded, and the carotenoid is then eluted with a more polar solvent mixture, evaporated and, if intended for NMR analysis, dried under vacuum to remove traces of solvent. A small column of silica may be used in a similar way for those carotenoids that are not stable on alumina.

11.6.7 Identification by co-chromatography

Co-chromatography with an authentic sample is a useful early step in identifying a carotenoid. Paper chromatography and TLC can be used, and co-chromatographic comparison should be sought in at least two systems, e.g. on silica and MgO.

11.7 HIGH-PERFORMANCE LIQUID CHROMATOGRAPHY

By virtue of its versatility, sensitivity and reproducibility, HPLC has become the method of choice for carotenoid analysis, both qualitative and quantitative, and is ideal for the systematic screening of bacterial pigment composition. Carotenoids are readily detected by their UV-visible light absorption. The procedure therefore becomes extremely powerful when a photodiode array detector is available. This permits simultaneous detection and monitoring at any chosen wavelength and also continuous determination and memorizing of absorption spectra during the progress of the chromatography thereby facilitating identification.

11.7.1 Reversed-phase HPLC

Reversed-phase partition chromatography is most suitable for routine analysis of carotenoid composition. Because the column materials are inert, there is little risk of structural modification or decomposition of the carotenoids or of chlorophylls if these are present. The stationary phases most commonly used are those with bonded C_{18} hydrocarbon chains (ODS; octadecylsilyl). Although isocratic procedures with simple solvent mixtures consisting of an appropriate concentration of water (usually 5 to 10%) in methanol or acetonitrile (more rarely acetone or propan-2-ol) can be used, more complex solvent gradient programmes are usually required to give optimal resolution of the mixtures of pigments that are present in many natural extracts.

A gradient procedure that is used routinely for analysing plant chloroplast pigments (Goodwin & Britton, 1988) is also very useful for screening pigment composition of cyanobacteria and non-photosynthetic bacteria. This procedure employs a linear gradient (0–100%) of ethyl

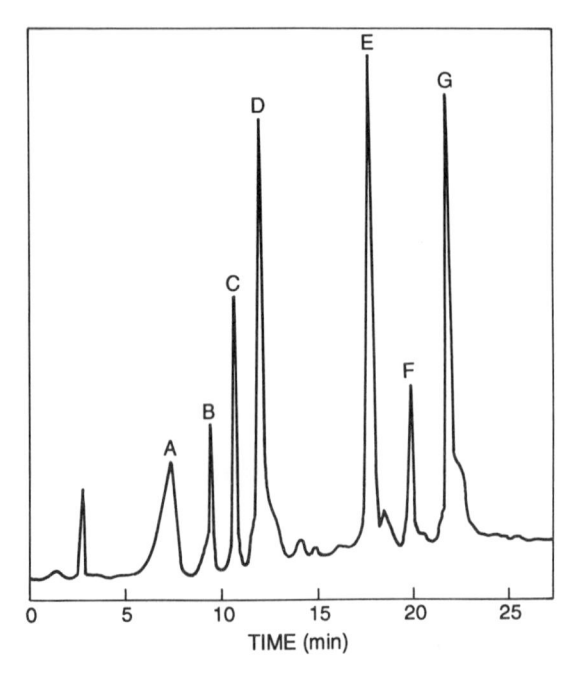

Figure 11.4 Reversed-phase HPLC chromatogram of the pigments of a cyanobacterium. A, myxoxanthophyll; B, nostoxanthin; C, caloxanthin; D, zeaxanthin; E, chlorophyll *a*; F, echinenone; G, β-carotene. Conditions: column, Spherisorb-ODS2, 25 × 0.46 cm; solvent, linear gradient, 0 to 100% ethyl acetate in acetonitrile-water (9 : 1, containing 0.5% triethylamine); flow rate 1 ml/min.

11. Analysis of Pigments: Carotenoids and Related Polyenes 423

acetate in acetonitrile-water (9:1) over 25 min, and gives good resolution of both xanthophylls and carotenes in the same run. The separation of carotenoids of a cyanobacterium by this method is illustrated in Figure 11.4. By appropriate modification of the solvent gradient programme this procedure is easily adapted to improve the resolution of difficult mixtures or to separate different collections of carotenoids. A procedure consisting of isocratic elution with 23% ethyl acetate: 77% acetonitrile-water (9:1) for 15 min followed by a linear gradient of 23 to 100% ethyl acetate over 30 min provides a very useful method for analysing pigment composition (both carotenoids and bacteriochlorophylls) in phototrophic bacteria, as illustrated in Figure 11.5.

In general, the least polar carotenoids have the longest retention times on reversed-phase HPLC, e.g. dihydroxycarotenoids are eluted before monohydroxycarotenoids, with carotenes eluted last of all. However, the

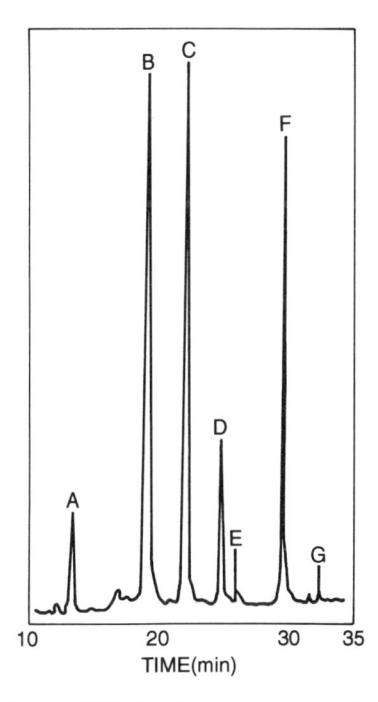

Figure 11.5 Reversed-phase HPLC chromatogram of the pigments of the phototrophic bacterium *Rhodocyclus gelatinosus*. A, hydroxyspheroidenone; B, hydroxyspheroidene; C, bacteriochlorophyll *a*; D, spheroidenone; E, spirilloxanthin; F, spheroidene; G, neurosporene. Conditions: column, Spherisorb-ODS2, 25 × 0.46 cm; solvent, 23% ethyl acetate/77% acetonitrile-water (9:1) for 15 min followed by linear gradient 23 to 100% ethyl acetate over 30 min; flow rate 1 ml/min.

factors affecting separation are more subtle than this. Particularly in the acyclic series, carotenoids bearing the same substituent groups but with different length chromophores are well resolved; e.g. spheroidene and anhydrorhodovibrin; the more saturated compounds have the longer retention times. Compounds with substituent groups in both end-groups are less strongly held than those with the same substituent groups located together in one end-group; the unsubstituted hydrocarbon end-group interacts strongly with the C_{18} stationary phase. Acyclic carotenoids are generally more strongly retained than corresponding cyclic ones with the same functional groups. Some typical retention times are listed in Table 11.4. These should be taken as indicating relative elution patterns and not as absolute values that can be repeated in all HPLC separations.

Reversed-phase HPLC is an excellent method for distinguishing between C_{40} and C_{50} carotenoids with the same chromophore and functional groups. Thus, decaprenoxanthin and 'C.p. 450' are eluted approximately 4 min later than the corresponding C_{40} carotenoids, lactucaxanthin and zeaxanthin.

11.7.2 Normal-phase (adsorption) HPLC

Adsorption HPLC on silica columns is also widely used for carotenoids. Separation is achieved on the same basis as on silica TLC, i.e. the presence of polar functional groups. The order of elution is therefore approximately the opposite of that with RP-HPLC. The mechanisms of interaction with the columns are different, however, so some combinations of carotenoids,

Table 11.4 Typical retention times for the main acyclic carotenoids in photosynthetic bacteria on reversed-phase HPLC.

Carotenoid	Retention time (min)
Hydroxyspheroidenone	13
Rhodovibrin	18.5
Hydroxyspheroidene	19.5
Rhodopin	24
Spirilloxanthin	25
Spheroidenone	26
Anhydrorhodovibrin	27.5
Spheroidene	29.5
Lycopene	30.5
Neurosporene	32
[Bacteriochlorophyll-phytyl]	21

Conditions: Spherisorb-ODS2 column (25×0.46 cm); solvents 23% ethyl acetate/77% acetonitrile-water (9:1), isocratic over 15 min, followed by linear gradient 23 to 100% ethyl acetate over 30 min; flow rate 1 ml/min.

11. Analysis of Pigments: Carotenoids and Related Polyenes

e.g. lutein and zeaxanthin, are eluted in the same order in both systems. Carotenes are so weakly held that their separation is difficult to achieve.

The use of silica columns is not strongly recommended for routine quantitative analysis because some degradation of carotenoids may occur, and chlorophylls are particularly liable to decomposition. Nevertheless a solvent programme consisting of a complex gradient of propan-2-ol in hexane is useful for resolving carotenoids with two or more hydroxy groups (Goodwin & Britton, 1988), and shallow gradients of ethyl acetate in hexane can be used to separate less polar carotenoids. The most useful feature of HPLC on silica is the ability to separate geometrical isomers, which are virtually unresolved by reversed-phase columns, by simple isocratic solvent systems. This is illustrated in Figure 11.6 by the resolution of Z-E isomers of okenone with a simple ethyl acetate-hexane mixture. All traces of water and alcohols, which adhere tenaciously to silica columns, must be removed by long equilibration with hexane, if efficient and reproducible separation of the isomers is to be achieved.

Columns containing bonded nitrile (CN) groups are also useful. They do not provide an easily reproducible process for the routine screening of

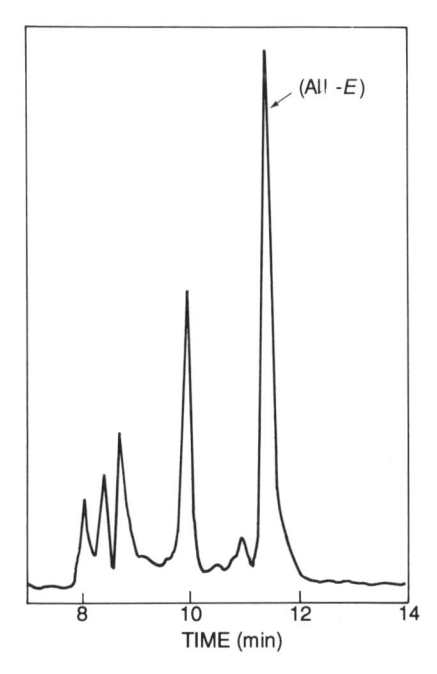

Figure 11.6 HPLC separation of geometrical isomers of okenone. Conditions: column, Spherisorb-S5 silica (25 × 0.46 cm); solvent, 7% ethyl acetate in hexane; flow rate 1 ml/min.

extracts containing carotenoids of widely different polarities, but will separate efficiently geometrical isomers and mixtures of carotenoids of similar polarity. The general use of nitrile columns is outlined by Rüedi (1985).

11.8 ULTRAVIOLET-VISIBLE LIGHT ABSORPTION SPECTROSCOPY

Because of their long conjugated double bond systems the carotenoids show strong absorption of ultraviolet and visible light. Absorption spectra are determined routinely; both the position of the absorption maxima (λ_{max}) and the shape or fine structure of the spectrum are characteristic of the chromophore of the molecule and provide valuable information for identifying carotenoids (Britton, 1994).

11.8.1 Position of the absorption maxima

The absorption spectra of most carotenoids exhibit three maxima. Spectra of apolar carotenoids are usually determined in petrol or hexane, those of the more polar xanthophylls in ethanol. Values of λ_{max} are markedly dependent on the solvent. The values recorded in petrol, hexane and ethanol are almost identical, but those recorded in acetone are greater by ~4 nm, those recorded in chloroform or benzene greater by 10–12 nm, and those in carbon disulphide greater by as much as 35–40 nm. When spectra are determined on-line during HPLC, it must be remembered that the values for λ_{max} in the eluting solvent frequently do not correspond to the published values recorded in a pure solvent. The λ_{max} values for all chromophores that have been found to date in the carotenoids of prokaryotes are given in Table 11.5. Comprehensive tables giving λ_{max} values for a wider range of carotenoids in a variety of solvents are available in other articles (Davies, 1976; De Ritter, 1981; Britton, 1994). In any given solvent, λ_{max} values increase as the length of the chromophore increases. This effect is clearly seen in the acyclic series as the chromophore increases from three conjugated double bonds in phytoene (λ_{max} 276, 286, 297 nm) to 13 in spirilloxanthin (λ_{max} 465, 493, 525 nm). Non-conjugated double bonds, e.g. the C-4,5 double bond of the ε-ring, do not contribute to the chromophore; ε,ε-carotene therefore is essentially a conjugated nonaene and has an absorption spectrum almost identical to that of the acyclic conjugated nonaene, neurosporene.

Extension of the conjugated double bond system into a ring (particularly the C5,6 double bond of the β-ring) does extend the chromophore but, because the ring double bond is not coplanar with the main polyene chain, the λ_{max} occur at shorter wavelengths than those of the acyclic carotenoid with the same number of conjugated double bonds. Thus, although they are all conjugated undecaenes, the acyclic, monocyclic and dicyclic lycopene, γ-carotene and β-carotene have λ_{max} at 444, 470, 502 nm, at 437, 462, 494 nm and at 425, 450, 478 nm, respectively.

Carbonyl groups, in conjugation with the polyene system, also extend the chromophore. A ring keto-group increases λ_{max} by 10 to 20 nm, so that echinenone and canthaxanthin, the 4-oxo and 4,4'-dioxo derivatives of β-carotene, have λ_{max} at 461 and 478 nm, respectively. The addition of a keto-group to an acyclic chromophore increases λ_{max} by a much greater amount, ~30 nm; spheroidene and spheroidenone have λ_{max} at 455 nm and 487 nm, respectively.

Other substituents, such as hydroxy- and methoxy-groups do not affect the chromophore. All such substituted carotenoids therefore have λ_{max} virtually identical to those of the parent hydrocarbon with the same chromophore. For example, β-carotene and its hydroxy-derivatives, β-cryptoxanthin, zeaxanthin, isocryptoxanthin, isozeaxanthin, caloxanthin and nostoxanthin, all have virtually identical spectra with λ_{max} at 425, 450 and 478 nm.

11.8.2 Spectral fine structure

The overall shape or fine structure of the spectrum is also diagnostic and generally reflects the degree of conformational homogeneity that the chromophore can achieve. Thus the spectra of acyclic compounds in which the conjugated double bond system is free to adopt an almost planar conformation are characterized by sharp maxima and minima (= fine structure). The degree of fine structure decreases a little when the chromophore exceeds nine double bonds. Cyclic carotenoids in which conjugation does not extend into the rings behave as linear polyenes and have similar well-defined spectral fine structure.

When conjugation extends into a β-ring, steric strain causes the ring to adopt a preferred conformation in which the ring double bond is not coplanar with the π-electron system of the polyene chain. This restriction of conformational flexibility is even more pronounced in carotenoids that have carbonyl groups in conjugation with the polyene chain. Spectral fine structure therefore decreases in the order lycopene > γ-carotene > β-carotene > canthaxanthin (Figure 11.7). The spectrum of canthaxanthin in ethanol has only a single, rounded, almost symmetrical absorption peak, but a slight degree of fine structure remains if the spectrum is determined in a non-polar solvent such as light petroleum.

Table 11.5 Wavelengths of maximal absorption (λ_{max}) for different carotenoid chromophores in a range of solvents. (Compounds are given numbers according to the list in Pfander, 1987. For structures, see Appendix 11.1.)

Chromophore type/no. of conj. double bonds	Example (number)[a]	λ_{max} (nm)/ (solvent)[b]	Carotenoids with similar spectra
Acyclic			
3	(44) (all-E)-Phytoene	276,286,297 (H,P,E) 280,291,303 (C,T)	45,107,108
5	(15Z)-Phytoene (42) Phytofluene	275,285,296 (H,P) 331,347,367 (H,P,E) 337,354,374 (C,T)	43,105,106,178
7	(38) ζ-Carotene	378,400,425 (H,P,E) 384,405,430 (C,T)	41,177
8 9	(37) 7,8,11,12-Tetrahydrolycopene (102) 11',12'-Dihydrospheroidene (34) Neurosporene	374,395,420 (H,P) 391,412,439 (P) 415,440,470 (H,P,E) 424,451,480 (C,T)	39,103,104,175,176 35,40,100,101,173,174
10	(97) Spheroidene	429,454,486 (H,P,E) 433,458,490 (A) 440,467,501 (B,C) 457,486,527 (CS$_2$)	96,98,171,172
11	(31) Lycopene	445,470,502 (H,P,E) 448,474,505 (A) 458,484,518 (C,T) 477,508,548 (CS$_2$)	33,36,93,94,95 111,164,169,170,454
12	(91) Anhydrorhodovibrin	454,483,514 (H,P,E) 459,488,522 (A) 471,499,533 (B,C) 493,521,559 (CS$_2$)	32,90,92,167,168,438
13	(166) Spirilloxanthin	465,493,525 (H,P,E) 468,499,533 (A) 480,510,548 (B) 500,532,570 (CS$_2$)	165,208,209,453,455,456,457,458

Monocyclic			
β+6	(14) 7′,8′,11′,12′-Tetrahydro-γ-carotene	358,378,397 (P)	
β+8	(13) β-Zeacarotene	407,427,454 (H,P,E) 414,439,465 (C)	73
β+10	(12) γ-Carotene	438,461,491 (H,P,E) 446,476,510 (B,C) 463,496,533 (CS$_2$)	72,79,80
β+11	(146) Plectaniaxanthin	447,473,504 (P) 454,478,509 (A)	75,76,77,78,140, 141,142,147,192,193, 194,195,207,440
β+12	(11) Torulene	457,486,519 (H,P,E) 469,502,539 (B,C) 491,525,563 (CS$_2$)	84
ε+6	(437) Nonaprenoxanthin	354,373,394 (H)	
ε+8	(22) α-Zeacarotene	398,421,449 (H)	435
ε+10	(21) δ-Carotene	430,456,489 (H,P) 440,470,503 (C) 457,490,526 (CS$_2$)	434
ε+11	(451) A.g. 470[c]	443,470,502 (H)	
φ+10	(27) Chlorobactene	435,461,491 (P) 450,476,508 (B) 470,496,530 (CS$_2$)	88,89
Dicyclic			
β+9+β	(3) β-Carotene	426,450,478 (H,P,E) 429,452,480 (A) 435,462,487 (B,C) 450,485,520 (CS$_2$)	55,57,112,119,124, 125,129,182,196, 433,439
β+9+ε	(7) α-Carotene	423,445,474 (H,P,E) 424,448,476 (A) 433,457,484 (B,C) 445,477,508 (CS$_2$)	60,133

(*continued*)

Table 11.5 (*Continued*)

Chromophore type/no. of conj. double bonds	Example (number)[a]	λ_{max} (nm)/ (solvent)[b]	Carotenoids with similar spectra
$\beta + 9 + \phi$	(9) β-Isorenieratene	440,464,492 (B) 456,487,508 (CS$_2$)	
$\gamma + 9 + \gamma$	(441) Sarcinaxanthin	415,440,470 (H,E) 423,451,480 (B,C) 442,469,494 (CS$_2$)	442,443
$\varepsilon + 9 + \varepsilon$	(447) Decaprenoxanthin	415,440,470 (H,E) 423,451,481 (B,C) 442,469,494 (CS$_2$)	20,149,445,448, 449
$\phi + 9 + \phi$	(24) Isorenieratene	426,449,479 (P,H) 428,460,495 (C) 452,484,520 (CS$_2$)	87,161
$\phi + 9 + \chi$	(26) Renieratene	457,476,507 (B) 463,496,532 (CS$_2$)	

Chromophore with conjugated carbonyl group(s)
Acyclic

7 + oxo	(322) R.g. keto V[c]	399,420,446 (H)	
9 + oxo	(321) R.g. keto VI[c]	(437),461,490 (H)	347
10 + oxo	(319) Spheroidenone	460,482,513 (H,P) 488 (E) 500 (532) (B,C) (495),520,553 (CS$_2$)	343,344
11 + oxo	(346) Thiothece-HO-484	458,484,515 (P)	
12 + oxo	(345) R.g. Keto-II[c]	(470),495,527 (H)	320
13 + oxo	(342) 2-Keto-HO-spirilloxanthin	488,516,550 (P)	
oxo + 11 + oxo	(417) R.g. Keto-III[c]	461,489,523 (H)	
oxo + 13 + oxo	(416) 2,2'-Diketospirilloxanthin	498,528,560 (P)	
11-(20-al)	(277) Rhodopin-20-al	495 (H) 505 (A)	273,278,279,280

Monocyclic			
$\beta + 10 + $ oxo	(314) Thiothece-HO-474	447,474,505 (P)	
$\beta + 11 + $ oxo	(312) 2'-Dehydroplectaniaxanthin	495(522) (H)	
4-oxo-β + 10	(287) 4-Keto-γ-carotene	465 (H,P,E)	307,310,336
		468 (A)	
4-oxo-β + 11	(337) Flexixanthin	483,510 (A)	308,309,338,339,355,358
4-oxo-β + 12	(286) 4-Ketotorulene	(465),489,519 (P)	
		(470),491,518 (E)	
χ + 10 + oxo	(317) Okenone	(460),484,516 (P)	
		(465),487,518 (A)	
Dicyclic			
4-oxo-β + 9 + β	(283) Echinenone	(432),459,(483) (P,H)	294,295,296,324,
		457–461 (E)	325,348,361
		460 (A)	
		470–473 (B,C)	
		488–494 (CS$_2$)	
4-oxo-β + 9 + 4-oxo-β	(380) Canthaxanthin	468 (H,P)	391,396,398,
		478 (E)	406,421
		480 (A)	
		484 (B,C)	
		502 (CS$_2$)	
3,3'-dioxoretro-β	(424) Rhodoxanthin	458,489,524 (H)	
(10 c.d.b.)		496,538 (E)	
		482,510,546 (C)	
		491,525,564 (CS$_2$)	

[a] Compounds are given numbers according to the list in Pfander (1987). For structures see Appendix 11.1.
[b] Abbreviations: A, acetone; B, benzene; C, chloroform; CS$_2$, carbon disulphide; E, ethanol; H, hexane; P, light petroleum.
[c] R.g. keto-II, R.g. keto-III, etc. (from *Rhodopseudomonas globiformis*) and A.g. 470 (from *Arthrobacter glacialis*): these were the names originally given to these compounds, and are the trivial designations still in use.
The C$_{30}$ 4,4'-diapocarotenoids have λ_{max} virtually identical to those of the C$_{40}$ carotenoids with the same chromophore.

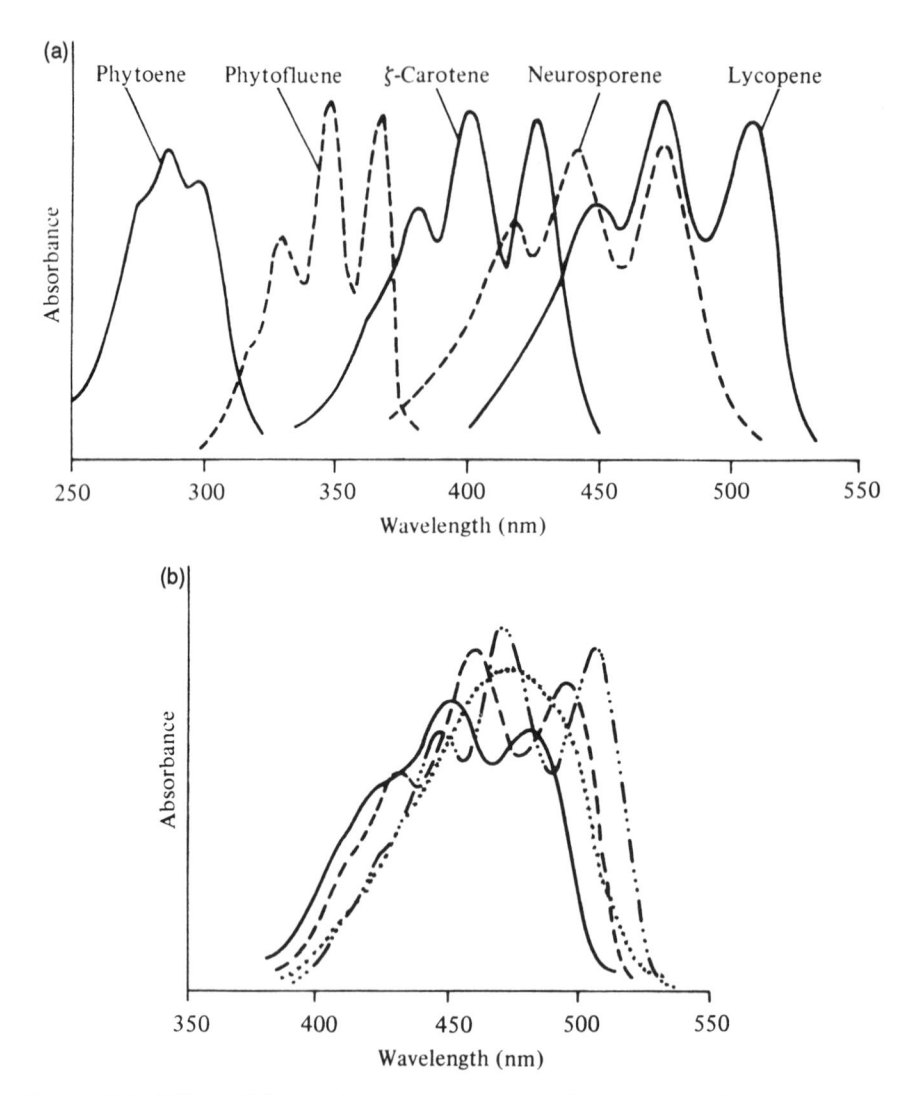

Figure 11.7 Effect of increasing conjugation, cyclization and the presence of conjugated keto-groups on λ_{max} and spectral fine structure of (a) phytoene, phytofluene, ζ-carotene, neurosporene and lycopene; and (b) lycopene —···—···, γ-carotene ———, β-carotene ——— and canthaxanthin ····.

11.8.3 Geometrical isomers

The absorption spectra of carotenoids containing one or more cis or Z double bonds in the chromophore show several characteristic differences from the spectrum of the all-*trans* compound (Vetter *et al.*, 1971; Zechmeister, 1962). For the cis isomers, the λ_{max} are generally 1 to 5 nm lower, the spectral fine structure is decreased and a new absorption peak, usually referred to as the 'cis-peak', appears at a characteristic wavelength in the UV region 142 ± 2 nm below the longest wavelength peak in the main visible absorption region. These effects, especially the intensity of the 'cis-peak' are greatest when the cis-double bond is located at or near the centre of the chromophore, and are illustrated for anhydrorhodovibrin in Figure 11.8. With the very long chromophore acyclic carotenoids, such as spirilloxanthin, a peak is usually present in the 'cis-peak' region of the spectrum of the all-*trans* isomer, though this peak is usually of lower intensity than in the spectra of the cis-isomers.

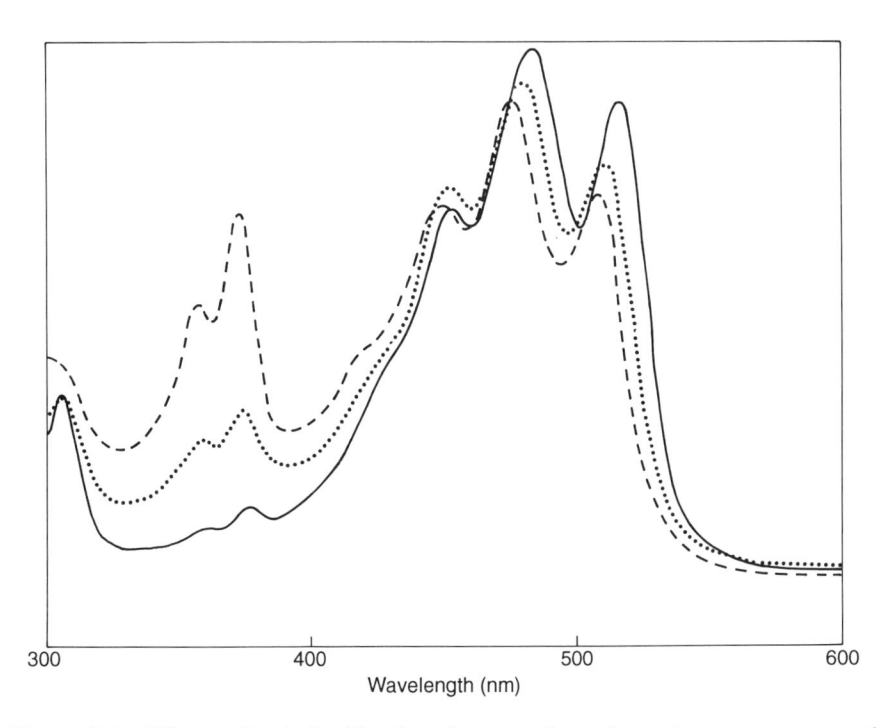

Figure 11.8 Effect of cis-double bonds on the absorption spectrum of anhydrorhodovibrin. ——— (all-*E*); · · · ·, ––– (*Z*)-isomers.

11.9 QUANTITATIVE DETERMINATION

11.9.1 Spectrophotometry

Spectrophotometric analysis is normally used for the quantitative determination of carotenoids. The purified carotenoid is dissolved in an accurately known volume of solvent, usually petrol or hexane for carotenes, ethanol for xanthophylls, though other solvents may need to be used to achieve complete solution of crystalline samples. The absorbance, A, (= extinction) of the solution in a 1 cm light path cuvette is then determined at an appropriate wavelength (usually λ_{max}) and the amount of carotenoid present is calculated from the equation:

$$x = Ay/(A^{1\%}_{1cm} \times 100)$$

where x is the mass of carotenoid (g), y the volume of solution (ml), A the measured absorbance, and $A^{1\%}_{1cm}$ is the specific absorption coefficient, i.e, the absorbance of a solution of 1 g of that carotenoid in 100 ml of solution. Tables of $A^{1\%}_{1cm}$ values have been published (Davies, 1976; De Ritter, 1981); Table 11.6 gives the values for some of the most common bacterial carotenoids. An arbitrary value of 2500 is often taken when no experimentally determined value has been reported, for an unknown compound, or to give an estimate of the total carotenoid content of an extract.

11.9.2 Quantitative determination by high-performance liquid chromatography

High-performance liquid chromatography provides the most sensitive, accurate and reproducible method for quantitative analysis of carotenoids, particularly when the instrumentation includes automatic integration facilities for measuring peak areas. The relative amounts of each component in the chromatogram can be determined, provided the peak area can be calculated for each component at its λ_{max} by use of a multi-wavelength detector. If monitoring is possible at only a single wavelength, corrections must be made for the difference between the absorbance at that wavelength and at λ_{max} for each pigment.

The procedure outlined above allows the relative proportions of each component in a mixture to be determined, but for the estimation of absolute amounts or concentrations, calibration is necessary. This can be achieved by injecting known amounts (determined spectrophotometrically) of pure carotenoid, determining peak areas and creating a calibration graph (usually linear over the range 20 to 2000 pmol), a procedure which allows the amount of each carotenoid to be estimated with great precision,

11. Analysis of Pigments: Carotenoids and Related Polyenes

at the nanogram level. An alternative strategy is to use an internal standard. The usual procedure is to add to the sample an accurately known amount of a pure standard compound. The amount of each component in the chromatogram is then estimated by comparing its peak area with that

Table 11.6 Values of $A_{1cm}^{1\%}$ for some important carotenoids.

Carotenoid	(No.)[a]	$A_{1cm}^{1\%}$ (λ max)	Solvent[b]
Anhydrorhodovibrin	(91)	2700 (483)	P
Bacterioruberin	(456)	2350 (494)	P
		2540 (498)	A
Canthaxanthin	(380)	2200 (466)	P
α-Carotene	(7)	2800 (445)	P
β-Carotene	(3)	2592 (450)	P
		2620 (450)	E
		2396 (465)	C
ζ-Carotene	(38)	2555 (400)	H
β-Cryptoxanthin	(55)	2386 (450)	P
2,2'-Diketospirilloxanthin	(416)	2000 (518)	P
Echinenone	(238)	2158 (458)	P
2-Hydroxyplectaniaxanthin	(192)	2445 (476)	A
Hydroxyspheroidenone	(343)	2070 (501)	B
Isorenieratene	(24)	2080 (448)	P
Isozeaxanthin	(129)	2400 (450)	P
Lutein	(133)	2550 (445)	E
Lycopene	(31)	3450 (470)	P
Myxoxanthophyll	(193)	2160 (478)	A
Neurosporene	(34)	2918 (440)	H
Okenone	(317)	2320 (484)	P
Phytoene	(44)	1250 (285)	P
Phytofluene	(42)	1350 (348)	P
Plectaniaxanthin	(146)	2505 (474)	A
Rhodopin	(93)	2989 (474)	Cy
Rhodoxanthin	(424)	2500 (490)	H
Rubixanthin	(72)	2750 (460)	P
Saproxanthin	(140)	2920 (479)	A
Spheroidene	(97)	2785 (468)	B
Spirilloxanthin	(166)	2470 (510)	B
7,8,11,12-Tetrahydrolycopene	(37)	2519 (395)	H
Torulene	(11)	3240 (480)	P
β-Zeacarotene	(13)	2520 (428)	P
		1940 (428)	H
Zeaxanthin	(119)	2348 (450)	P
		2540 (450)	E
		2340 (453)	A

[a] Number of carotenoid in the list of Pfander (1987).
[b] Solvents: A, acetone; B, benzene; C, chloroform; Cy, cyclohexane; E, ethanol and P, light petroleum.

of the standard, again correcting for the difference between absorbance at λ_{max} and the monitoring wavelength, if necessary.

The internal standard used may be a purely synthetic compound or a pure natural compound. The main criteria for choosing a standard are that it should not be present in the extract under investigation, it should clearly separate from all components in that extract, its stability and λ_{max} should be similar to those of the natural components, and its $A_{1cm}^{1\%}$ should be accurately known. An alternative strategy is to use as an internal standard the substance or one of the substances under investigation. The experimental sample is first analysed by HPLC and the peak areas are determined in the normal way. Then a known amount of the pure standard compound is added to the sample and a second HPLC analysis is performed. The increment in the peak area for the standard compound is then related to the amount that was added, to give a calibration that can be used to determine the amounts of all substances present in the extract. If automatic integration is not available, compounds separated by HPLC may be collected and the concentration of each pigment determined in a spectrophotometer. The virtually quantitative recovery of carotenoids from HPLC allows greater accuracy than is possible when other separation methods (such as TLC) are used.

11.10 DIAGNOSTIC CHEMICAL REACTIONS AND FORMATION OF DERIVATIVES

A good deal of structural information may be obtained by judicious use of simple chemical reactions, which can be performed on a very small scale and monitored spectroscopically or chromatographically.

11.10.1 Iodine-catalysed geometrical isomerization

Iodine-catalysed photo-isomerization of any carotenoid will produce a characteristic pseudo-equilibrium mixture of geometrical isomers, which can help to identify that carotenoid.

Method 11.7 Iodine-catalysed photo-isomerization for carotenoid analysis.

1. The carotenoid is dissolved in petrol (if solubility in petrol is low, the pigment may first be dissolved in a little acetone before petrol is added) and a solution of iodine in petrol is added, such that the final concentration of I_2 is approximately 10% that of the carotenoid.

11. Analysis of Pigments: Carotenoids and Related Polyenes

2. The reaction mixture is incubated under N_2 at room temperature in diffuse light. Samples are taken at 15 min intervals and the isomerization is followed spectroscopically until equilibrium is reached. The iodine may be removed by filtration through alumina or by washing with 0.1 M sodium thiosulphate.
3. The solution is then concentrated, the isomers are separated by HPLC or PC and the absorption spectrum of each is determined. Purity can be checked by subjecting each of the separated isomers to similar I_2-catalysed reactions, which should produce the same pseudo-equilibrium mixture.

11.10.2 Acetylation or silylation of hydroxy groups

Primary and secondary hydroxy groups are readily acetylated, and the number of such groups present in a carotenoid can be determined by following the time-course of acetylation. Tertiary hydroxy groups cannot be acetylated but their presence and number can be determined by silylation.

Method 11.8 Acetylation of hydroxy groups.

1. The purified carotenoid is dissolved in a few drops of dry pyridine and one drop of acetic anhydride is added.
2. The mixture is incubated in the dark at room temperature under N_2. The course of the reaction is followed by analysing (PC, TLC or HPLC) a sample every 15 min so that the number of intermediates and hence the number of primary or secondary hydroxy groups can be determined.
3. Acetylation is normally complete after 16 to 18 h and the peracetate product may be analysed by mass spectrometry (see Section 11.11.1 below). An increase of 42 mass units in the molecular mass is given for each acetate group introduced.

Method 11.9 Silylation of hydroxy groups.

The carotenoid is dissolved in dry pyridine (1 ml), and hexamethyldisilazane (0.4 ml) and trimethylchlorosilane (0.2 ml) are quickly added. The reaction proceeds very rapidly and intermediates can only be detected by performing the reaction at $-35\,°C$ and taking samples as quickly as possible (every 1 to 2 s). Products can be purified by PC, TLC or HPLC. The increase in molecular mass is 73 for each hydroxy group silylated.

11.10.3 Reduction of carbonyl groups by sodium borohydride

Carotenoids containing aldehyde or keto groups are readily reduced by $NaBH_4$ to give the corresponding alcohol. The reduction products have greater polarity and their absorption spectra show λ_{max} at lower wavelengths and have increased fine structure.

Method 11.10 Reduction of carbonyl groups by sodium borohydride.

1. The carotenoid is dissolved in ethanol and a few milligrams of NaBH$_4$ added.
2. The spectrum is redetermined after intervals of 30 s, 5 min and 30 min, or until no further spectral change is seen. Rapid reduction, complete within 30 s, usually indicates the presence of an aldehyde group; complete reduction of a ketocarotenoid normally takes 5 to 30 min.
3. The product can be isolated by transfer to ether, followed by chromatography. Chromatographic comparison with an authentic sample of the alcohol product can be used to aid identification.

11.10.4 Dehydration: elimination of allylic hydroxy groups

If allylic hydroxy groups are present in a natural carotenoid, or have been introduced by NaBH$_4$ reduction of a conjugated carbonyl group, the compound normally will undergo ready dehydration by treatment with HCl in chloroform. This process introduces a new double bond in conjugation and hence causes a spectral shift to longer wavelength. The product is also substantially less polar.

Method 11.11 Dehydration of allylic hydroxy groups.

1. The purified pigment is dissolved in chloroform (1.4 ml) and 0.4 ml of acidified chloroform (produced by bubbling HCl gas into CHCl$_3$ for 30 to 60 min) is added, to give a final HCl concentration of 0.003 to 0.03 M.
2. The reaction is allowed to proceed at room temperature for 5 to 15 min, during which time spectral changes may be monitored.
3. On completion of the reaction, the pigment is transferred to ether, residual acid removed by washing with NaHCO$_3$, and the product(s) are purified by chromatography and identified by their chromatographic and spectroscopic properties.

11.10.5 Carotenoid glycosides: liberation and identification of sugar residues

Carotenoids which exhibit a very high degree of polarity, which is greatly reduced by acetylation, are likely to be glycosides. The sugar can frequently be identified simply by hydrolysis of the glycoside, though this may destroy the carotenoid.

Method 11.12 Identification of sugars in glycosides.

1. Typically 2 to 4 mg of pigment is dissolved in methanol (5 ml) and treated for 20 h at room temperature with acidified methanol (prepared by bubbling HCl into methanol for 30 to 60 min) to give a final HCl concentration of 0.15 M.

2. The reaction mixture is then diluted with methanol and the solvent evaporated *in vacuo* over KOH.
3. The methyl glycoside product is hydrolysed with aqueous 0.04 M polystyrene sulphonic acid under reflux at 100 °C.
4. The liberated sugar is then identified (tentatively) by chromatographic comparison with authentic standards, especially by GLC of the derivatized product.

This approach is now rarely used, because full characterization of a carotenoid glycoside, i.e. identification of both the carotenoid and sugar moieties, can be achieved by NMR, with much smaller quantities of sample.

11.11 OTHER SPECTROSCOPIC METHODS

Chromatography, especially HPLC, and UV-visible absorption spectroscopy provide very valuable first means of identification of carotenoids. This, however, does not constitute rigorous characterization. For this, other sophisticated methods, particularly mass spectrometry (MS) and nuclear magnetic resonance spectroscopy (NMR) must be used, together with circular dichroism (CD) if the chirality of a carotenoid is to be determined. Infrared (IR) spectroscopy has also been widely used, but the information given about carotenoid structure is usually limited. An outline of the main features and applications of these techniques will be given here. The interpretation of the large amount of data which these methods generate requires a great deal of experience, and the characterization of novel or unusual carotenoids is best left to experts.

11.11.1 Mass spectrometry

Whenever possible, all identifications should be supported by MS data, now readily obtainable in most institutions. Although chemical ionization (CI) has been used, most MS work on carotenoids has employed ionization by electron impact (EI). Carotenoids have very low volatility, and samples are usually inserted by means of a direct probe, which is heated to 200 to 220 °C. Sample requirements are small; a good mass spectrum can be obtained from only 1 to 2 μg of pure carotenoid, though a realistic goal is to obtain 10 to 20 μg of the compound. Only an outline of some of the most important diagnostic features of the MS of carotenoids is given here. Further details of fragmentations and their mechanisms are given by Moss and Weedon (1976), by Vetter *et al.* (1971), and by Enzell and Back (1994).

11.11.1.a *Molecular mass and formula*

Almost all carotenoids give good molecular ions so that the molecular mass can be obtained, and molecular formulae can be determined with a high-resolution instrument. This is particularly valuable with the nonphotosynthetic bacteria because it will immediately distinguish C_{30}, C_{45} and C_{50} carotenoids from the conventional compounds.

11.11.1.b Fragmentations

Many fragmentations have been identified which are diagnostic of particular structural features. Some of these provide information primarily about the carbon skeleton of the molecule; others identify the presence of particular functional groups or end-groups.

11.11.1.c Polyene chain fragmentations

All carotenoids undergo reactions in which the polyene chain is folded and portions of the chain are then excised. The most intensively studied are the losses of toluene (92 mass units) and *m*-xylene (106 mass units) which occur almost universally with carotenoids. Similar losses of 79 and 158 mass units are also frequently seen. The abundance ratio of the (M-92) and (M-106) fragment ions can give a good indication of the carbon skeleton of a carotenoid (Table 11.7). Carotenoids with a higher degree of saturation, e.g. phytoene, phytofluene and ζ-carotene, or bearing substituted side-chains, show analogous losses of 94 or [92 + X] mass units (e.g. benzaldehyde and *m*-tolualdehyde for rhodopin-20-al), respectively.

11.11.1.d Functional groups and end-groups

As with other classes of compounds, the presence of functional groups in a carotenoid is indicated by characteristic fragmentations; e.g. hydroxy and methoxy groups, especially if in an allylic position, give rise to strong losses of water (18 m.u.) or methanol (32 m.u.), respectively. Acetylation

Table 11.7 Approximate abundance ratio of the [M-92] : [M-106] fragment ions in the EI-mass spectra for different carotenoid skeletons.

Carotenoid type	Ratio [M-92] : [M-106]
Dicyclic (β,β; β,ε)	2 : 1–10 : 1
Monocyclic (β,ψ)	0.6 : 1–1 : 1
Acyclic (11 c.d.b.)[a]	0.3 : 1
(12 c.d.b.)	0.06 : 1
(13 c.d.b.)	0.02 : 1
Monoaryl (ϕ,ψ)	1 : 1
(χ,ψ)	0.15 : 1

[a] c.d.b. = conjugated double bonds.

11. Analysis of Pigments: Carotenoids and Related Polyenes

and silylation increase the molecular mass by 42 and 73 mass units, respectively, for each hydroxy group reacting. Other fragmentation reactions characteristic of particular carotenoid end-groups are summarized in Table 11.8.

Especially useful are the so-called 'bis-allylic' fragmentations. Those single bonds in the carotenoid molecule which are in a position allylic to the main polyene chain and also to an isolated double bond in an end-group undergo particularly facile cleavage (e.g. C3,4 of lycopene, C7,8 of ζ-carotene and C11,12 of phytoene). This allows the length and position of the conjugated chromophore in any acyclic carotenoid to be

Table 11.8 Some characteristic end-group fragmentations that are of diagnostic value in mass spectra of carotenoids.

End-group	Fragment ions
	M-205
	M-137
	M-69
RO–	(R=H) 59, M-18, M-58 (R=CH₃), 73, M-32, M-73
CH₃O–	73, M-32, M-73, M-101
CH₃O–	M-32, M-101, M-129
	M-56, M-123
	133, M-133

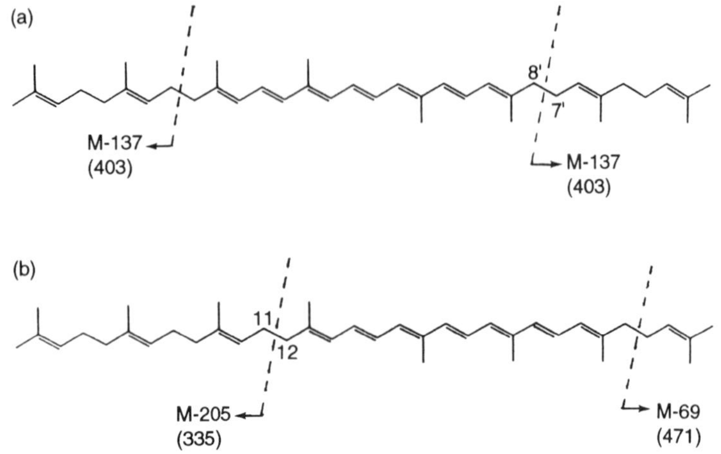

Figure 11.9 Use of 'bis-allylic' bond cleavage to distinguish between ζ-carotene and its 'asymmetric' isomer, 7,8,11,12-tetrahydrolycopene by mass spectrometry.

determined. The value of this is illustrated by the ease with which ζ-carotene and its 'unsymmetrical' isomer, 7,8,11,12-tetrahydrolycopene can be distinguished by MS (Figure 11.9). A typical carotenoid mass spectrum is illustrated in Figure 11.10.

11.11.2 Nuclear magnetic resonance spectroscopy

Nuclear magnetic resonance spectroscopy is undoubtedly the most powerful technique for investigating carotenoid structures, and its use is essential for full characterization and structure elucidation. Both proton (^{1}H)-NMR and ^{13}C-NMR are used routinely by chemists in such work. Moss and Weedon (1976) discussed the general features of NMR spectra of carotenoids, and presented a considerable amount of tabulated data. A survey of progress since then was given by Goodwin and Britton (1988) and further tabulated values for ^{1}H- and ^{13}C-NMR assignments of a range of carotenoid end-groups were presented. More detailed authoritative articles by Englert (1982, 1985, 1991, 1994) are also available. The interested reader is recommended to consult these, and to survey the primary literature for full details of the NMR investigations and assignments for individual carotenoids. Selected tabulated data are not presented in this chapter; the use of limited information by inexperienced workers can easily lead to erroneous conclusions. The interpretation of NMR spectra of carotenoids is such a sophisticated and specialized field that the help and advice of experts should always be sought.

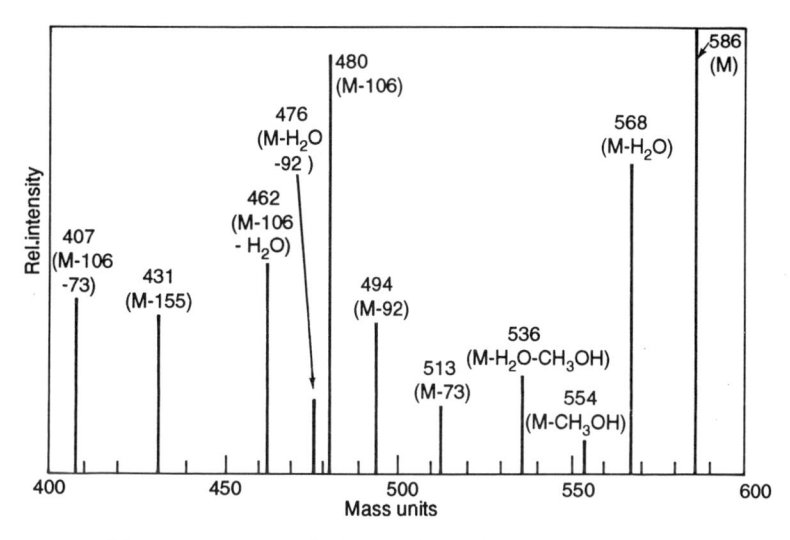

11. Analysis of Pigments: Carotenoids and Related Polyenes

Figure 11.10 Mass spectrum of hydroxyspheroidene.

With a modern, high-resolution PFT NMR instrument, good ^{1}H spectra can be obtained from about 100 to 200 µg of sample, whereas ^{13}C spectra require 0.5 to 5 mg, depending on the amount of information needed.

It is essential that carotenoid samples for NMR analysis are rigorously purified and that all traces of residual solvent are removed under vacuum.

11.11.3 Circular dichroism

Full characterization of a carotenoid includes determination of its chirality. This is usually achieved by a combination of NMR spectroscopy and circular dichroism (CD), perhaps together with chemical correlations. Again, CD study and interpretation should be left to experts. A brief general summary of the use of CD in the carotenoid field is given by Goodwin and Britton (1988) and expert reviews of the subject have been presented (e.g. Noack, 1982; Noack & Buchecker, 1994).

11.12 APPLICATION OF CAROTENOID ANALYSIS IN PRACTICE

The method used by Züllig (1985) for determining the abundance of carotenoids in lake sediments illustrates how the techniques described above may be applied to practical situations.

Method 11.13 Determination of carotenoids in lake sediments (Züllig, 1985).

1. Samples (5 to 10 g of sediment) are extracted by stirring with acetone-ethanol (1:1, 100 ml) for 5-6 h under N_2 in the dark at room temperature.
2. The sediment is removed by brief centrifugation or by filtration.
3. An aliquot (3-4 ml) of the crude extract is acidified with one drop of 10% HCl.
4. After 30 s the absorbance at 450 and 600 nm is determined, and the carotenoid content calculated from the formula:

$$\text{mg carotenoid} = \frac{[A_{450} - A_{600}] \times 10 \times \text{volume}}{A_{1cm}^{1\%}}$$

($A_{1cm}^{1\%}$ is taken as 2250, or as 1890 for samples containing 15-30% of oscillaxanthin.)

5. The crude extracts are taken to dryness and analysed by TLC on silica Kiesel gel 60, Merck: 341845 with 20% acetone in petrol (b.p. 110-140 °C) or other suitable mixtures. The solvent is allowed to reach a height of 15-16 cm (approx. 1 h) and the chromatogram is then dried under N_2 and redeveloped in the same solvent. Very polar pigments such as carotenoid glycosides and degraded chlorophylls are left at or near the origin, but clear and reproducible patterns of the less polar pigments are obtained (Figure 11.11).

From these chromatograms, the abundance of β-carotene and echinenone indicated the presence of cyanobacteria in the sediments, spheroidene and spheroidenone showed the presence of *Rhodobacter* species or related members of the family Rhodospirillaceae. The likely presence of *Chromatium* and *Chlorobium* species is indicated by the detection of okenone and arylcarotenes, respectively. Different pigment patterns indicated the variation in dominant microbial populations in sediments from different depths and ages.

11.13 NON-CAROTENOID POLYENES

A small number of bacteria produce yellow pigments which are very similar in properties to carotenoids but are now known to be aryl polyenes which are not isoprenoid in origin and which lack the lateral methyl substituents. Typical structures are illustrated in Figure 11.12.

11.13.1 Flexirubin-type pigments

Approximately 50 different flexirubin-type pigments have been found. These all consist of a polyenoic acid chromophore terminated by a hydroxyphenyl group and esterified with a second substituted phenol, of

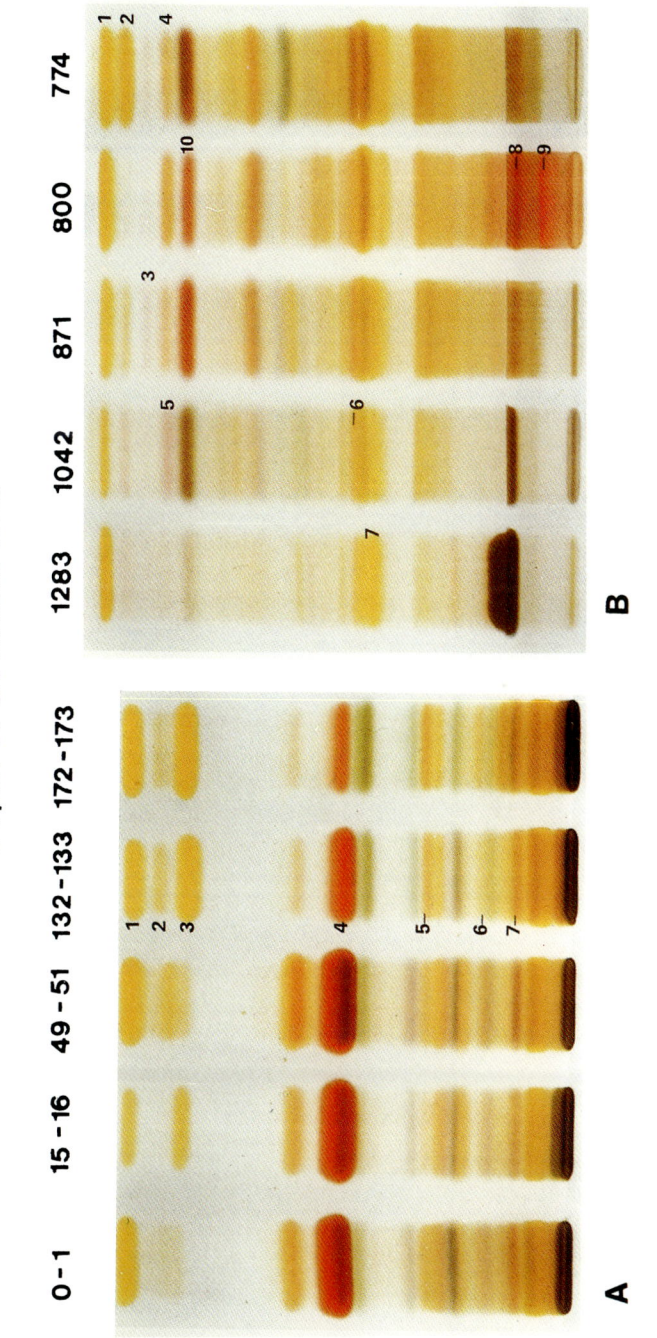

Figure 11.11 Thin-layer chromatograms (Silica, Merck No. 11798) of sediment extracts from two lakes (Switzerland). A: Lago Cadagno. Solvent, 20% acetone in petrol. 1 = β-carotene, 2 = isorenieratene, 3 = spheroidene, 4 = okenone, 5 = rhodopin, 6 = hydroxyspheroidenone, 7 = rhodovibrin. B: Lobsigensee. Solvent, hexane:2-propanol (:ethanol) = 80:20 (:5). 1 = β-carotene, 2 = isorenieratene, 3 = spheroidenone, 4 = echinenone, 5 = lycopenal, 6 = lutein, 7 = alloxanthin, 8 = myxoxanthophyll, 9 = oscillaxanthin, 10 = okenone. Reproduced with permission from Züllig (1988) *Berichte der St Gallischen Naturwissenschaftlichen Gesellschaft* **83**, 8–68.

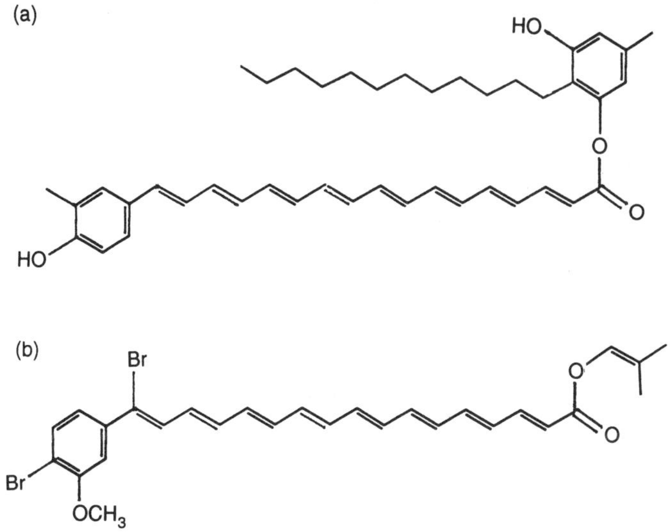

Figure 11.12 Structures of non-carotenoid polyenes (a) flexirubin and xanthomonadin.

the resorcinol type. The variety of structures is created by variations in the length of the polyene chain (6–8 double bonds) and in the substituents on the two aromatic rings. These pigments have been found in some *Cytophaga*, *Flavobacterium* and *Flexibacter* species, where they are located in the outer membrane of the Gram-negative cell wall. The structures and properties of the flexirubins have been reviewed (Achenbach *et al.*, 1978).

11.13.2 Xanthomonadins

The xanthomonadins are also arylpolyenoic acid esters which are characterized by the presence of bromide substituents on the aryl ring and also the polyene chain. The nature of the esterifying alcohol is not clear. The occurrence of these compounds has only been established in some *Xanthomonas* strains. The isolation and characterization of xanthomonadin I are outlined by Andrewes *et al.* (1976).

11.13.3 Properties and methods

Like the carotenoids, these polyene pigments are lipid-soluble and are extracted and purified in the same way as carotenoids. The general procedure is to extract with acetone and then separate and purify the pigments by normal column or thin-layer chromatographic methods. Most work with flexirubins has used TLC on silica gel H. Systematic HPLC work has not been reported, but the preparative separation of the dimethyl ethers of flexirubin pigments by a reversed-phase (C$_{18}$) HPLC procedure has been described.

The stability and properties of these pigments are very similar to those of the carotenoids. Hydrolysis of the ester link requires vigorous saponification conditions or reductive cleavage with LiAlH$_4$. The most useful distinguishing feature is the behaviour under alkaline conditions. The absorption spectra of these aryl polyenes are markedly dependent on pH, and their colour changes from yellow to blue in alkali, due to formation of the phenolate ions (Figure 11.13). (N.B. The phenolic carotenoid, 3,3'-dihydroxyisorenieratene shows similar behaviour.)

Mass spectrometric and ^{1}H-NMR data have been obtained for both the flexirubin and xanthomonadin pigments. A particularly useful diagnostic feature in the MS is the loss of benzene (78 m.u.) by fragmentation of the unmethylated polyene chain by a mechanism apparently analogous to the losses of toluene and *m*-xylene in carotenoid mass spectra. The original

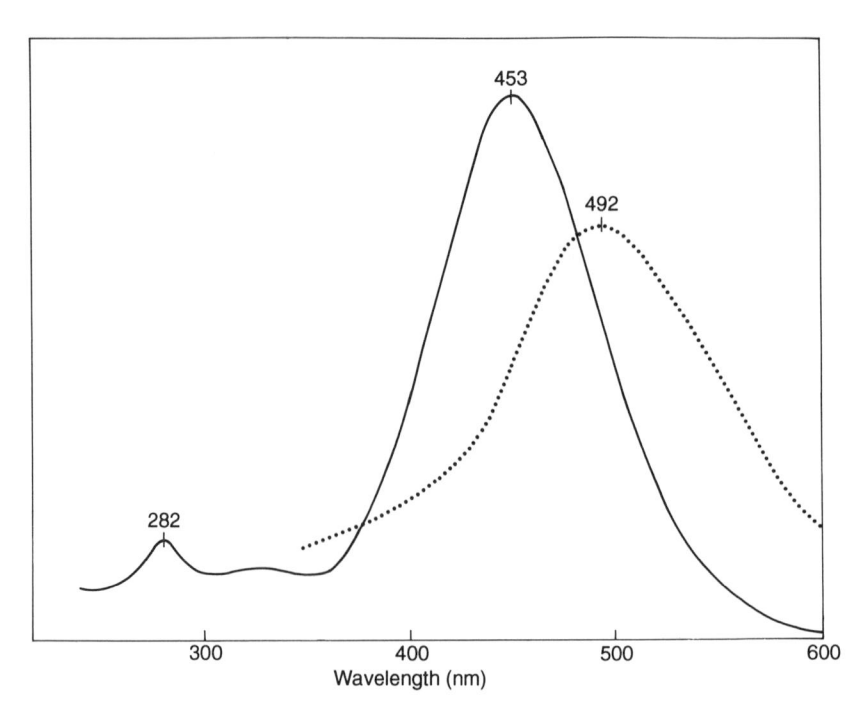

Figure 11.13 Light absorption spectrum of flexirubin in neutral (———) and alkaline (.....) solution.

literature should be consulted for details of the spectroscopic properties (Achenbach et al., 1978; Andrewes et al., 1976).

REFERENCES

Achenbach, H., Kohl, W. and Reichenbach, H. (1978) The flexirubin-type pigments—a novel class of natural pigments from gliding bacteria. *Rev. Latinoameric. Quim.* **9**: 111–24.

Andrewes, A.G., Jenkins, C.L., Starr, M.P., Shepherd, J. and Hope, H. (1976) Structure of xanthomonadin. I. A novel dibrominated aryl-polyene pigment produced by the bacterium *Xanthomonas juglandis*. *Tetrahedron Lett.* **45**: 4023–4.

Britton, G. (1976) Biosynthesis of carotenoids. In *Chemistry and Biochemistry of Plant Pigments*, 2nd edn (Goodwin, T.W., ed), pp. 262–327. Academic Press: London.

Britton, G. (1985) General carotenoid methods. *Meth. Enzymol.* **111**: 113–49.

Britton, G. (1988) Biosynthesis of carotenoids. In *Plant Pigments* (Goodwin, T. W., ed), pp. 133–82. Academic Press: London.

Britton, G. (1990) Carotenoid biosynthesis—an overview. In *Carotenoids: Chemistry and Biology* (Krinsky, N.I., Mathews-Roth, M.M. and Taylor, R.F., eds), pp. 167–84. Plenum Press: New York.

Britton, G. (1991) Carotenoids. In *Methods in Plant Biochemistry*, vol. 7 (Banthorpe, D.V. and Charlwood, B.V., eds), pp. 473–518. Academic Press: London.

Britton, G. (1994) UV-Visible spectroscopy. In *Carotenoids*, vol. 1B (Britton, G., Pfander, H. and Liaaen-Jensen, S., eds), in press. Birkhäuser: Basel.

Britton, G. and Goodwin, T.W. (1971) Biosynthesis of carotenoids. *Meth. Enzymol.* **18C**: 654–701.

Britton, G. and Goodwin, T.W. (eds) (1982) *Carotenoid Chemistry and Biochemistry*. Pergamon: Oxford.

Britton, G. and Young, A.J. (1993) Methods for the isolation and analysis of carotenoids. In *Carotenoids in Photosynthesis* (Young, A.J. and Britton, G., eds), pp. 409–88. Chapman and Hall: London.

Britton, G., Pfander, H. and Liaaen-Jensen, S. (eds) (1994) *Carotenoids*. Birkhäuser: Basel.

Davies, B.H. (1976) Carotenoids. In *Chemistry and Biochemistry of Plant Pigments*, 2nd edn (Goodwin, T.W., ed), pp. 38–165. Academic Press: London.

Davies, B.H. (1980) Carotenoid biosynthesis. In *Pigments in Plants*, 2nd edn (Czygan, F.-C., ed), pp. 31–56. Fischer Verlag: Stuttgart.

De Ritter, E. (1981) Absorption maxima and $E^{1\%}_{1cm}$ values for carotenoids. In *Carotenoids as Colorants and Vitamin A Precursors* (Bauernfeind, J.C., ed), pp. 883–923. Academic Press: New York.

Englert, G. (1982) NMR of carotenoids. In *Carotenoid Chemistry and Biochemistry* (Britton, G. and Goodwin, T.W., eds), pp. 107–34. Pergamon: Oxford.

Englert, G. (1985) NMR of carotenoids—New experimental techniques. *Pure Appl. Chem.* **57**: 801–21.

Englert, G. (1991) NMR of carotenoids: Novel experimental techniques. *Pure Appl. Chem.* **63**: 59–70.

Englert, G. (1993) NMR spectroscopy. In *Carotenoids*, vol. 1B (Britton, G., Pfander, H. and Liaaen-Jensen, S., eds), in press. Birkhäuser: Basel.

Enzell, C.R. and Bromann, S. (1993) Mass spectrometry. In *Carotenoids*, vol. 1B (Britton, G., Pfander, H. and Liaaen-Jensen, S., eds), in press. Birkhäuser: Basel.

Goodwin, T.W. (1971) Biosynthesis. In *Carotenoids* (Isler, O., ed), pp. 577–636. Birkhäuser: Basel.

Goodwin, T.W. (1980) *The Biochemistry of the Carotenoids, Vol. 1, Plants*, 2nd edn. Chapman and Hall: London.

Goodwin, T.W. and Britton, G. (1988) Distribution and analysis of carotenoids. In *Plant Pigments* (Goodwin, T.W., ed), pp. 61–132. Academic Press: London.

Isler, O. (ed) (1971) *Carotenoids*. Birkhäuser: Basel.
IUPAC (1979) Carotenoids—5 (Madison 1978). *Pure Appl. Chem.* **51**: 435–675; 857–86.
IUPAC (1985) Proceedings of the 7th International Symposium on Carotenoids, Munich 1984. *Pure Appl. Chem.* **57**: 639–821.
IUPAC (1991) Proceedings of the 9th International Symposium on Carotenoids, Kyoto 1990. *Pure Appl. Chem.* **63**: 1–176.
IUPAC Commission on the Nomenclature of Organic Chemistry and IUPAC-IUB Commission on Biochemical Nomenclature (1971) Rules for the nomenclature of carotenoids. In *Carotenoids* (Isler, O., ed), pp. 851–64. Birkhäuser: Basel.
Jensen, A. and Liaaen-Jensen, S. (1959) Quantitative paper chromatography of carotenoids. *Acta Chem. Scand.* **13**: 1863–68.
Krinsky, N.I., Mathews-Roth, M.M. and Taylor, R.F. (eds) (1990) *Carotenoids: Chemistry and Biology*. Plenum Press: New York.
Liaaen-Jensen, S. and Andrewes, A.G. (1985) Analysis of carotenoids and related polyene pigments. *Meth. Microbiol.* **18**: 235–55.
Moss, G.P. and Weedon, B.C.L. (1976) Chemistry of the carotenoids. In *Chemistry and Biochemistry of Plant Pigments*, 2nd edn (Goodwin, T.W., ed), pp. 149–224. Academic Press: London.
Noack, K. (1982) Circular dichroism of carotenoids and its use in investigations of their structures, configurations and conformations. In *Carotenoid Chemistry and Biochemistry* (Britton, G. and Goodwin, T.W., eds), pp. 135–53. Pergamon: Oxford.
Noack, K. and Buchecker, R. (1994) Circular dichroism. In *Carotenoids*, vol. 1B (Britton, G., Pfander, H. and Liaaen-Jensen, S., eds). Birkhäuser: Basel.
Pfander, H. (ed) (1987) *Key to Carotenoids (Straub)*, 2nd edn. Birkhäuser: Basel.
Pfander, H., Schiedt, K., Bernhard, K. and Riesen, R. (1993) Chromatography. In *Carotenoids*, vol. 1A (Britton, G., Pfander, H. and Liaaen-Jensen, S., eds), in press. Birkhäuser: Basel.
Rüedi, P. (1985) HPLC, a powerful tool in carotenoid research. *Pure Appl. Chem.* **57**: 793–800.
Schiedt, K. and Liaaen-Jensen, S. (1994). Isolation and analysis. In *Carotenoids*, vol. 1A (Britton, G., Pfander, H. and Liaaen-Jensen, S., eds), in press. Birkhäuser: Basel.
Schmidt, K. (1978) Biosynthesis of carotenoids. In *The Photosynthetic Bacteria* (Clayton, R.K. and Sistrom, W.R., eds), pp. 729–49. Plenum Press: New York.
Vetter, W., Englert, G., Rigassi, N. and Schwieter, U. (1971). Spectroscopic methods. In *Carotenoids* (Isler, O., ed), pp. 189–266. Birkhäuser: Basel.
Weedon, B.C.L. and Moss, G.P. (1994) Structure, stereochemistry and nomenclature. In *Carotenoids*, vol. 1A (Britton, G., Pfander, H. and Liaaen-Jensen, S., eds), in press. Birkhäuser: Basel.
Young, A.J. (1993) Carotenoids in pigment-protein complexes. In *Carotenoids in Photosynthesis* (Young, A.J. and Britton, G., eds), pp. 72–95. Chapman and Hall: London.
Young, A.J. and Britton, G. (eds) (1993) *Carotenoids in Photosynthesis*. Chapman and Hall: London.
Zechmeister, L. (1962) Cis-trans *Isomeric Carotenoids, Vitamins A and Arylpolyenes*. Springer: Vienna.
Züllig, H. (1985) Pigmente phototropher Bakterien in Seesedimenten und ihre Bedeutung für die Seeforschung. *Schweiz. Z. Hydrologie* **47**: 87–126.

APPENDIX 11.1 SEMI-SYSTEMATIC NAMES AND STRUCTURES OF CAROTENOIDS REFERRED TO IN THE CHAPTER*

3 β-Carotene
  β,β-Carotene

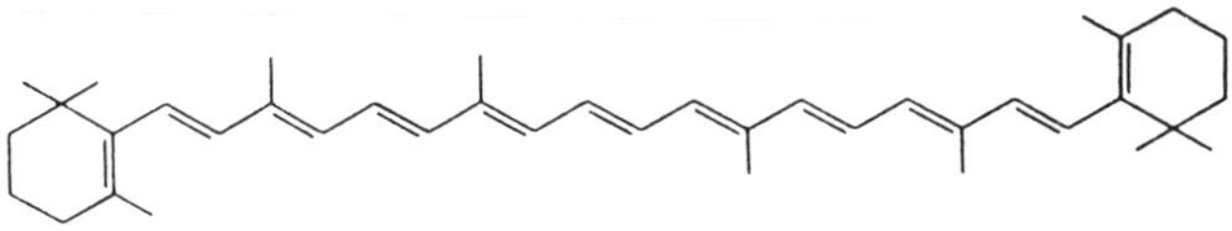

7 α-Carotene
  β,ε-Carotene

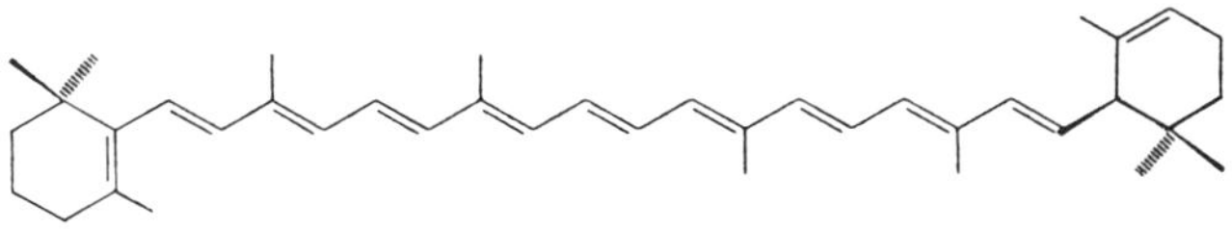

9 β-Isorenieratene
  β,φ-Carotene

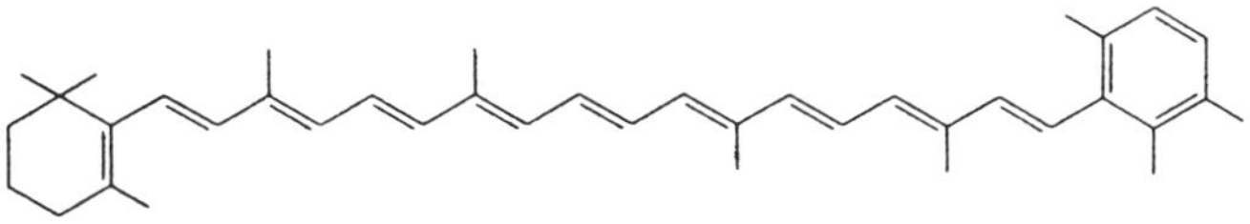

11 Torulene
  3′,4′-Didehydro-β,ψ-carotene

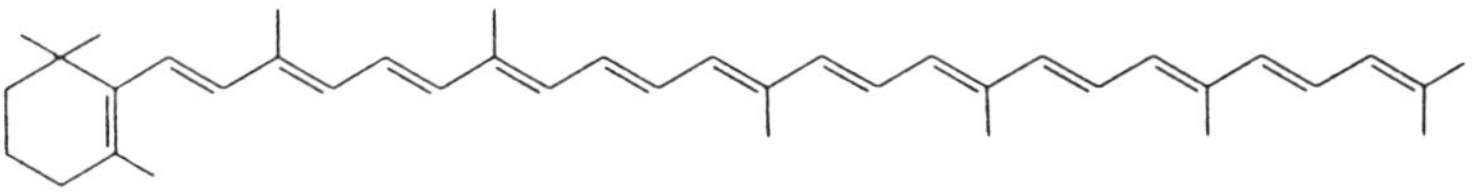

12 γ-Carotene
  β,ψ-Carotene

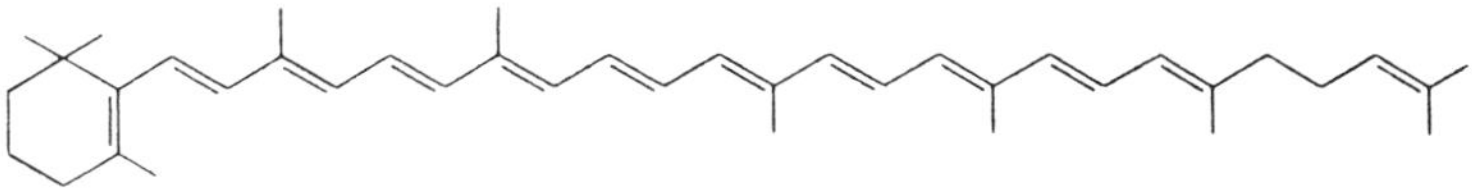

---

*The number given for each carotenoid is that in the list of Pfander (1987).

13 β-Zeacarotene
   7',8'-Dihydro-β,ψ-carotene

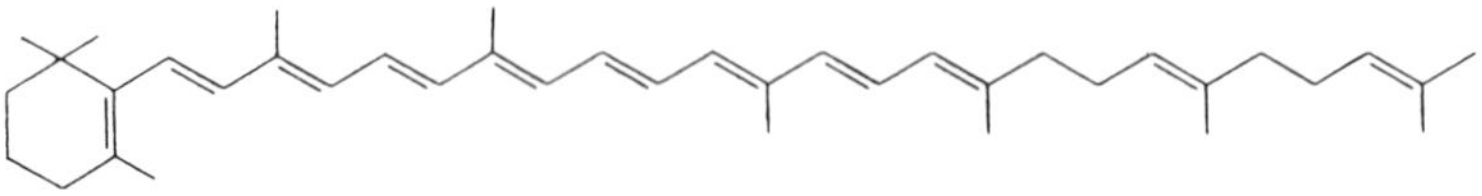

14 Tetrahydro-γ-carotene
   7',8',11',12'-Tetrahydro-β-ψ-carotene

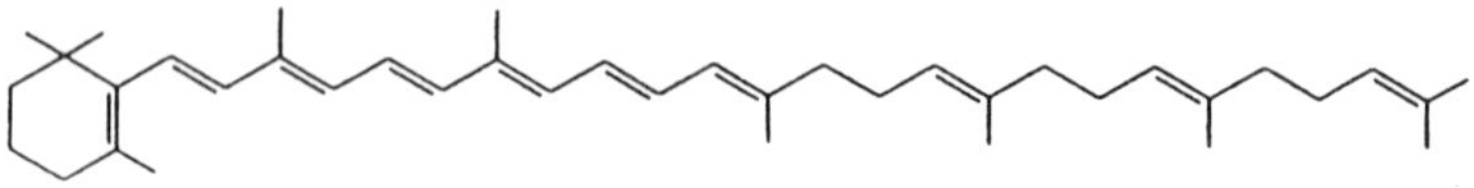

21 δ-Carotene
   ε,ψ-Carotene

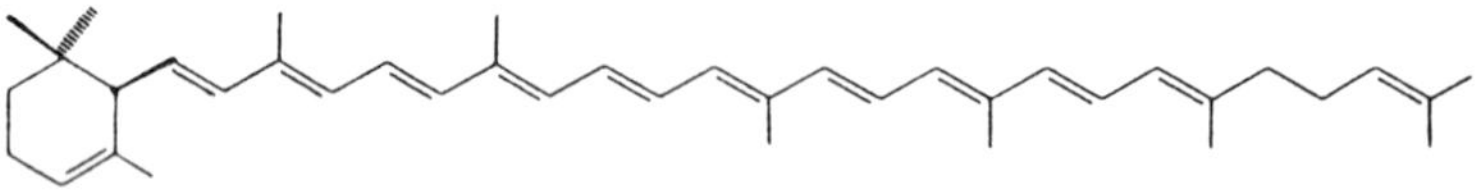

24 Isorenieratene
   φ,φ-Carotene

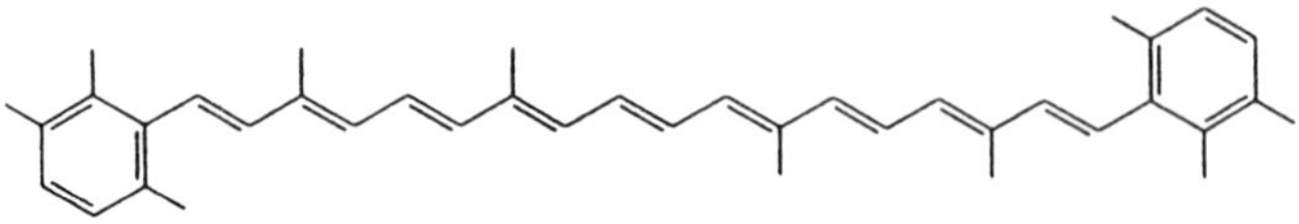

26 Renieratene
   φ,χ-Carotene

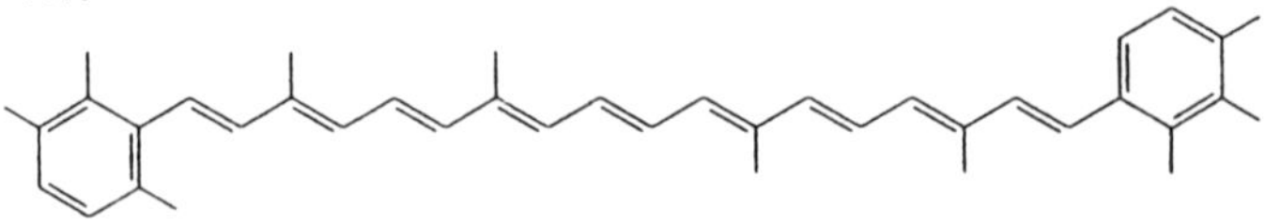

27 Chlorobactene
   φ,ψ-Carotene

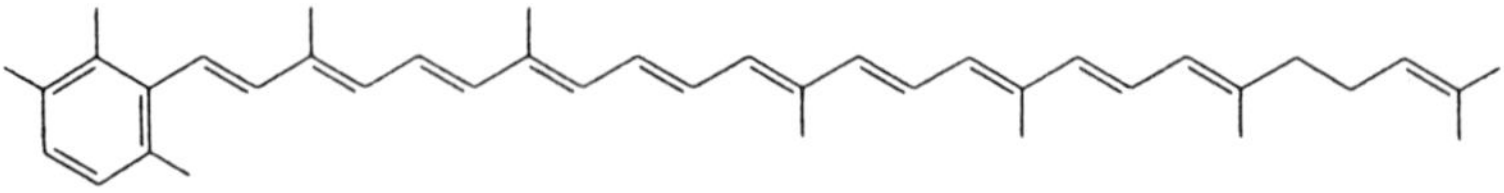

31 Lycopene
   ψ,ψ-Carotene

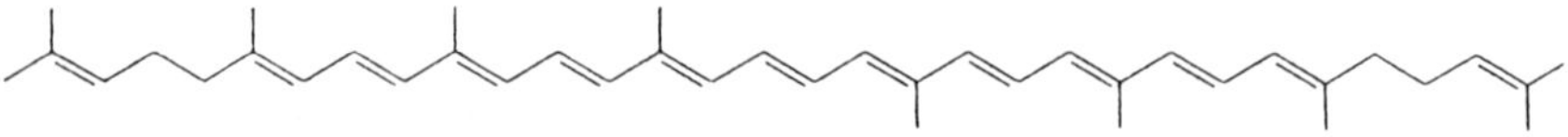

11. Analysis of Pigments: Carotenoids and Related Polyenes 451

34 Neurosporene
7,8-Dihydro-ψ,ψ-carotene

37 'Asymm. ζ-carotene'
7,8,11,12-Tetrahydro-ψ,ψ-carotene

38 ζ-Carotene
7,8,7',8'-Tetrahydro-ψ,ψ-carotene

42 Phytofluene
7,8,11,12,7',8'-Hexahydro-ψ,ψ-carotene

44 Phytoene
7,8,11,12,7',8',11',12'-Octahydro-ψ,ψ-carotene

55 β-Cryptoxanthin
β,β-Caroten-3-ol

57 Isocryptoxanthin
β,β-Caroten-4-ol

72 Rubixanthin
   β,ψ-Caroten-3-ol

91 Anhydrorhodovibrin
   1-Methoxy-3,4-didehydro-1,2-dihydro-ψ,ψ-carotene

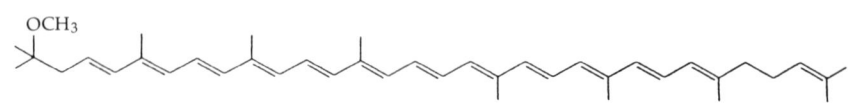

93 Rhodopin
   1,2-Dihydro-ψ,ψ-caroten-1-ol

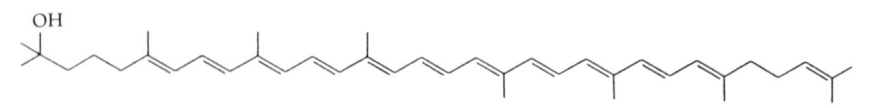

97 Spheroidene
   1-Methoxy-3,4-didehydro-1,2,7′,8′-tetrahydro-ψ,ψ-carotene

100 Chloroxanthin
   1,2,7′,8′-Tetrahydro-ψ,ψ-caroten-1-ol

102 11′,12′-Dihydrospheroidene
   1-Methoxy-3,4-didehydro-1,2,7′,8′,11′,12′-hexahydro-ψ,ψ-carotene

11. Analysis of Pigments: Carotenoids and Related Polyenes

119 Zeaxanthin
   β,β-Carotene-3,3'-diol

129 Isozeaxanthin
   β,β-Carotene-4,4'-diol

133 Lutein
   β,ε-Carotene-3,3'-diol

140 Saproxanthin
   3',4'-Didehydro-1',2'-dihydro-β,ψ-carotene-3,1'-diol

146 Plectaniaxanthin
   3',4'-Didehydro-1',2'-dihydro-β,ψ-carotene-1',2'-diol

147 Phleixanthophyll
1'-(β-D-Glucopyranosyloxy)-3',4'-didehydro-1',2'-dihydro-β,ψ-caroten-2'-ol

159 Lactucaxanthin
ε,ε-Carotene-3,3'-diol

161 3,3'-Dihydroxyisorenieratene
φ,φ-Carotene-3,3'-diol

166 Spirilloxanthin
1,1'-Dimethoxy-3,4,3',4'-tetradehydro-1,2,1',2'-tetrahydro-ψ,ψ-carotene

167 Rhodovibrin
1'-Methoxy-3',4'-didehydro-1,2,1',2'-tetrahydro-ψ,ψ-caroten-1-ol

182 Caloxanthin
β,β-Carotene-2,3,3'-triol

11. Analysis of Pigments: Carotenoids and Related Polyenes

192 2-Hydroxyplectaniaxanthin
3',4'-Didehydro-1',2'-dihydro-β,ψ-carotene-2,1',2'-triol

193 Myxoxanthophyll
2'-(β-L-Rhamnopyranosyloxy)-3',4'-didehydro-1',2'-dihydro-β,ψ-carotene-3,1'-diol

196 Nostoxanthin
β,β-Carotene-2,3,2',3'-tetrol

208 Oscillaxanthin
2,2'-Bis(β-L-rhamnopyranosyloxy)-3,4,3',4'-tetradehydro-1,2,1',2'-tetrahydro-ψ,ψ-carotene-1,1'-diol

277 Rhodopin-20-al
13-cis-1-Hydroxy-1,2-dihydro-ψ,ψ-caroten-20-al

283 Echinenone
   β,β-Caroten-4-one

286 4-Ketotorulene
   3',4'-Didehydro-β,ψ-caroten-4-one

287 4-Keto-γ-carotene
   β,ψ-Caroten-4-one

309 Myxobactone
   1'-Glucosyloxy-3',4'-didehydro-1',2'-dihydro-β,ψ-caroten-4-one

   $C_6H_{11}O_5$

312 2'-Dehydroplectaniaxanthin
   1'-Hydroxy-3',4'-didehydro-1',2'-dihydro-β,ψ-caroten-2'-one

314 Thiothece-474
   1'-Methoxy-1',2'-dihydro-β,ψ-caroten-4'-one

11. Analysis of Pigments: Carotenoids and Related Polyenes

317 Okenone
1'-Methoxy-1',2'-dihydro-χ,ψ-caroten-4'-one

319 Spheroidenone
1-Methoxy-3,4-didehydro-1,2,7',8'-tetrahydro-ψ,ψ-caroten-2-one

321 R.g. Keto-VI
1-Methoxy-1,2,7',8'-tetrahydro-ψ,ψ-caroten-4-one

322 R.g. Keto-V
1-Methoxy-1,2,7',8',11',12'-hexahydro-ψ,ψ-caroten-4-one

337 Flexixanthin
3,1'-Dihydroxy-3',4'-didehydro-1',2'-dihydro-β,ψ-caroten-4-one

342 2-Keto-HO-spirilloxanthin
1'-Hydroxy-1-methoxy-3,4,3',4'-tetradehydro-1,2,1',2'-tetrahydro-ψ,ψ-caroten-2-one

343 Hydroxyspheroidenone
1'-Hydroxy-1-methoxy-3,4-didehydro-1,2,1',2',7',8'-hexahydro-
ψ,ψ-caroten-2-one

345 R.g. Keto-II
1,1'-Dimethoxy-3',4'-didehydro-1,2,1',2'-tetrahydro-ψ,ψ-caroten-4-one

346 Thiothece-OH-484
1'-Hydroxy-1-methoxy-1,2,1',2'-tetrahydro-ψ,ψ-caroten-4-one

361
2,3,2',3'-Tetrahydroxy-β,β-caroten-4-one

380 Canthaxanthin
β,β-Carotene-4,4'-dione

11. Analysis of Pigments: Carotenoids and Related Polyenes

406 Astaxanthin
  3,3′-Dihydroxy-β,β-carotene-4,4′-dione

416 2,2′-Diketospirilloxanthin
  1,1′-Dimethoxy-3,4,3′,4′-tetradehydro-1,2,1′,2′-tetrahydro-ψ,ψ-carotene-2,2′-dione

417 R.g. Keto-III
  1,1′-Dimethoxy-1,2,1′,2′-tetrahydro-ψ,ψ-carotene-4,4′-dione

424 Rhodoxanthin
  4′,5′-Didehydro-4,5′-*retro*-β,β-carotene-3,3′-dione

437 Nonaprenoxanthin
  2-(4-Hydroxy-3-methylbut-2-enyl)-7′,8′,11′,12′-tetrahydro-ε,ψ-carotene

439 C.p. 450
2,2'-Bis(4-hydroxy-3-methylbut-2-enyl)-β,β-carotene

441 Sarcinaxanthin
2,2'-Bis(4-hydroxy-3-methylbut-2-enyl)-γ,γ-carotene

447 Decaprenoxanthin
2,2'-Bis(4-hydroxy-3-methylbut-2-enyl)-ε,ε-carotene

451 A.g. 470
2-(4-Hydroxy-3-methylbut-2-enyl)-2'-(3-methylbut-2-enyl)-3',4'-didehydro-1',2'-dihydro-ε,ψ-caroten-1'-ol

456 Bacterioruberin
2,2'-Bis(3-hydroxy-3-methylbutyl)-3,4,3',4'-tetradehydro-1,2,1',2'-tetrahydro-ψ,ψ-carotene-1,1'-diol

11. Analysis of Pigments: Carotenoids and Related Polyenes

515 4,4'-Diaponeurosporene
    7,8-Dihydro-4,4'-diapocarotene

526 4,4'-Diaponeurosporen-4-oic acid
    7',8'-Dihydro-4,4'-diapocaroten-4-oic acid

12

Determination of DNA Base Composition

Jin Tamaoka
The RIKEN Institute, Hirosawa, Japan

12.1 INTRODUCTION

Marmur and Doty (1962) reported a simplified method to determine DNA base composition. The latter has subsequently become an important criterion in microbial taxonomy. Indeed, the guanine (G) plus cytosine (C) ratio of DNA is one of the few characters which can be used in the description of any cultivable microorganism. DNA base composition has proved useful for both classification and identification of prokaryotes.

The base composition of bacterial chromosomal DNA ranges from about 25% to 80% G + C. In general, the DNA base composition of strains of a species shows a narrow range of about 1 to 3 mole % G + C (Mandel, 1966; Tamaoka, 1984). Once the DNA base ratio of a strain has been determined, the number of species to be compared for identification of that strain decreases as microorganisms with base composition differences will have different chromosomal DNA and hence belong to separate species. The DNA base ratios of almost all prokaryotic species are available in texts such as *Bergey's Manual of Systematic Bacteriology* (Krieg & Holt, 1984; Sneath et al., 1986; Staley et al., 1989; Williams et al., 1989). Measuring DNA base composition has been made much easier with the application of new methods and the development of new instrumentation as outlined below.

12.2 METHODOLOGY

12.2.1 Isolation of DNA

Precise determinations of DNA base compositions require DNA free of protein, polysaccharide and RNA. Two well-known methods are available

---

Chemical Methods in Prokaryotic Systematics. Edited by M. Goodfellow and A.G. O'Donnell
© 1994 John Wiley and Sons Ltd.

Method 12.1 Isolation of high molecular weight and double-stranded DNA.

1. Harvest cultured cells at middle or late phase of logarithmic growth.
2. Wash harvested cells twice with saline-EDTA[a].
3. Suspend 1 g of wet cells in 10 ml of lytic solution[b] in a glass centrifuge tube (COREX No. 8446) and keep the suspension at 37 °C for 30 min.
4. Add 1 ml of Tris-SDS solution[c], mix well and heat the suspension to 60 °C for 5 min.
5. Add 2 ml of phenol and shake gently for 1 min.
6. Cool the solution in ice.
7. Add 2 ml of chloroform and shake gently for 1 min.
8. Centrifuge at $10\,000 \times g$ for 5 min.
9. Transfer the upper layer to another tube.
10. Add 5 ml of standard saline citrate[d] (SSC) solution to the precipitated lysate, mix well, heat at 50 °C, cool and centrifuge at $10\,000 \times g$ for 5 min. Transfer the upper layer to the collected solution.
11. Add 5 ml of chloroform to the collected solution, shake gently for 1 min.
12. Centrifuge at $10\,000 \times g$ for 5 min and transfer the upper layer to another tube using a pipette.
13. Add 5 ml of chloroform to the solution and shake gently for 1 min.
14. Centrifuge at $12\,000 \times g$ for 10 min and transfer the upper layer to a small beaker.
15. While stirring the solution with a glass rod, add 2 vols of cold ethanol and spool the DNA with a glass rod.
16. Rinse the DNA with 70% then 99% (v/v) ethanol.
17. Dry the DNA then dissolve it in 5 ml of $0.1 \times$ SSC.
18. Add 0.3 ml of RNase solution[e] and keep at 37 °C for 20 min.
19. Add 0.3 ml of proteinase K solution[f] and keep at 37 °C for 20 min.
20. Add 0.5 ml of $10 \times$ SSC, 2 ml of phenol and 2 ml of chloroform then shake gently for 1 min.
21. Centrifuge at $12\,000 \times g$ for 10 min and transfer the upper layer to another tube.
22. Add 5 ml of chloroform to the solution and shake gently for 1 min.
23. Centrifuge at $12\,000 \times g$ for 10 min then transfer the upper layer to a small beaker.
24. While stirring the solution with a glass rod, add 10 ml of cold ethanol and spool the DNA with a glass rod.
25. Rinse the DNA with 70% then 99% (v/v) ethanol.
26. Dry the DNA, then store at -20 °C in ethanol.

[a] Saline-EDTA: 0.15 NaCl + 0.1 EDTA, pH 8.0.
[b] Lytic solution: (Achromopeptidase 0.5 mg + lysozyme 0.75 mg)/ml of 10 mM Tris-HCl buffer, pH 8.0. Use freshly prepared solution.
[c] Tris-SDS buffer: 1 M Tris + 10% (w/v) SDS.
[d] Standard saline citrate: 0.15 M NaCl + 15 mM trisodium citrate, pH 7.0.
[e] RNase solution (RNase A 1 mg + RNase T1 400 units)/ml of 50 mM Tris-HCl, pH 7.5. Keep at -20 °C.
[f] Proteinase solution: Proteinase K (Sigma) 4 mg/ml of 50 mM Tris-HCl, pH 7.5. Use freshly prepared solution.

for the isolation of DNA (Marmur, 1961; Daito & Miura, 1963) though other methods have been described (Johnson, 1991). In the two commonly applied methods either chloroform or phenol are used to denature proteins. Phenol is more effective than chloroform, but the latter is better for removing polysaccharides from samples. Current isolation methods, therefore, use phenol and chloroform. A procedure for isolating DNA is given in Method 12.1.

Achromopeptidase is added to lytic solutions for efficient cell lysis (step 3). Similarly, ribonuclease (RNase) T_1 is added to RNase solution for the effective hydrolysis of RNA (step 18). Hydrolysis of RNA with RNase A and T1 is more effective than treatment two or three times with RNase A only. Proteinase K is used for the hydrolysis of residual proteins and peptides (step 19).

The general method outlined in Method 12.1 can be used to isolate up to 2 mg of high molecular weight DNA from 1 g of wet cells. However, the isolation of DNA is not always successful with this method given the ineffectiveness of cell lysis. If the cell lysate at step 4 is not viscous then cell lysis can be considered to have been ineffective. The following modifications of the procedure can then be used to achieve effective lysis: (a) use cells in the middle or late phase of logarithmic growth as lytic enzymes (lysozyme and achromopeptidase) are more active against such cells than ones at the stationary phase of growth; (b) freezing and thawing of cells sometimes makes them more sensitive to lytic enzymes; (c) the addition of glycine to the medium (up to 2 g/l) weakens the walls of some coryneform actinomycetes (Yamada & Komagata, 1970); (d) if cells are not lysed by one of these techniques, physical disruption becomes the last resort.

Physical disruption techniques include ultrasonication, grinding cells with alumina and the use of a French press. The grinding method is a somewhat mild method which allows the isolation of relatively high molecular weight DNA: after washing cells with saline-EDTA, grind cells (2 g) and alumina (3 g) in a pestle and mortar for several minutes until the mixture shows viscosity, i.e. when cell lysis occurs. After lysis, add 8 ml of SSC buffer, mix well, and continue from step 4.

High molecular weight, double-stranded DNA is needed for the analysis of base composition using the thermal melting point (T_m) method. Low molecular weight, single-stranded DNA is required for the high-performance liquid chromatography (HPLC) method. Isolation of DNA for the HPLC method can be simplified as shown in Method 12.2; about 10 μg of DNA is enough for such analyses.

Method 12.2 Simplified method for the isolation of DNA and HPLC analysis.

1. Harvest about 50 mg of cultured cells at the middle or late phase of logarithmic growth in a polypropylene tube.
2. Wash the harvested cells twice with 0.5 ml of saline-EDTA.
3. Suspend cells in 0.5 ml of lytic solution and keep the suspension at 37 °C for 30 min.
4. Add 50 μl of Tris-SDS solution, mix well, and heat the suspension to 60 °C for 5 min.
5. Add 0.2 ml of phenol and shake gently for 1 min.
6. Cool the solution in ice.
7. Add 0.22 ml of chloroform and shake gently for 1 min.
8. Centrifuge at 10 000 $\times g$ for 5 min.
9. Transfer the upper layer to another tube.
10. Add 0.5 ml of chloroform to the solution and shake gently for 1 min.
11. Centrifuge at 10 000 $\times g$ for 5 min then transfer the upper layer to another tube.
12. Add 0.5 ml of chloroform to the solution and shake gently for 1 min.
13. Centrifuge at 12 000 $\times g$ for 10 min and transfer the upper layer to another tube.
14. Add 50 μl of RNase solution and keep at 37 °C for 20 min.
15. Add 50 μl of proteinase K solution and keep at 37 °C for 20 min.
16. Add 0.2 ml of phenol and 0.2 ml of chloroform and shake gently for 1 min.
17. Centrifuge at 12 000 $\times g$ for 10 min then transfer the upper layer to another tube.
18. Add 0.5 ml of chloroform to the solution and shake gently for 1 min.
19. Centrifuge at 12 000 $\times g$ for 10 min then transfer about 0.3 ml of the upper layer to another tube.
20. Add 0.7 ml of cold ethanol to the solution and mix gently.
21. Using a toothpick mop up precipitated DNA and transfer it to another tube.
22. Rinse the DNA with 70% then 99% (v/v) ethanol.
23. Dry the DNA using a vacuum desiccator or Speed Vac.
24. Add 50 μl of distilled water and dissolve the DNA by heating at 60 °C for 1 h.

With the simplified isolation procedure the spooling of DNA (steps 15 and 24 in Method 12.1) is omitted and all treatments are carried out in a polypropylene tube. This procedure takes about 4 h and yields about 10 μg of isolated and purified DNA.

12.2.2 DNA base composition

DNA base composition can be determined directly or indirectly. A direct method involves the determination of the four deoxyribonucleosides by HPLC following enzymatic hydrolysis of DNA (the HPLC method; Kaneko, 1977; Katayama-Fujimura et al., 1985; Tamaoka & Komagata, 1984). In indirect methods, base composition can be calculated from the

12. Determination of DNA Base Composition

hyperchromic shift accompanying thermal denaturation of DNA (the T_m method; Marmur & Doty, 1962), or from buoyant density measurements in caesium chloride (the B_d method; Schildkraut et al., 1962). The T_m and B_d methods were developed in 1962 and have been used widely in microbial taxonomy.

The HPLC method is popular because it allows the rapid and precise determination of base compositions with ease (Tamaoka & Komagata, 1984; Mesbah et al., 1989). DNA is hydrolysed into nucleosides using nuclease P1 (EC 3.1.3.30) and alkaline phosphatase (EC 3.1.3.1) (Method 12.3) and the four deoxyribonucleosides measured by HPLC (Table 12.1).

Method 12.3 Hydrolysis of DNA.

1. Prepare sample of DNA solution; about 0.5–1.0 g of DNA/litre of distilled water (OD_{260} = 10–20). Several micrograms of DNA are sufficient for one determination.
2. Heat the DNA solution in boiling water for 5 min and cool in ice.
3. Add 10 μl of the DNA solution to a polypropylene tube.
4. Add 10 μl of nuclease P1 solution,[a] and keep at 50 °C for 1 h.
5. Add 10 μl of alkaline phosphatase solution[b] and keep at 37 °C for 1 h.

[a] Nuclease P1 solution: nuclease P1 0.1 mg or 40 units/ml of 40 mM CH_3COONa + 12 mM $ZnSO_4$, pH 5.3. Store at 4 °C.
[b] Alkaline phosphatase solution: alkaline phosphatase 2.4 units/ml of 0.1 M Tris-HCl, pH 8.1.

With this procedure, the amounts of nuclease P_1 and alkaline phosphatase used are five or ten times greater than those needed for complete hydrolysis of 1 μg DNA. After DNA hydrolysis, the sample should be analysed within a day or stored at 4 °C. The hydrolysed sample should not be frozen as the solubilities of deoxyribonucleosides are low and re-dissolving nucleosides is not easy.

The selection of a column is the most significant factor for the HPLC method. The separation of the four deoxyribonucleosides and the four ribonucleosides is complete under the conditions shown in Figure 12.1 but the use of other columns may give different chromatograms. If good

Table 12.1 Conditions for high-performance liquid chromatography.

Detector wave length	270 nm
Column	Nakarai Cosmosil packed column $5C_{18}$ (150 × 4.6 mm)
Column temperature	Room temperature
Eluent	0.2 M $NH_4H_2PO_4$-acetonitrile (20 : 1, v/v)
Flow rate	1 ml/min
Sample	5–10 μl

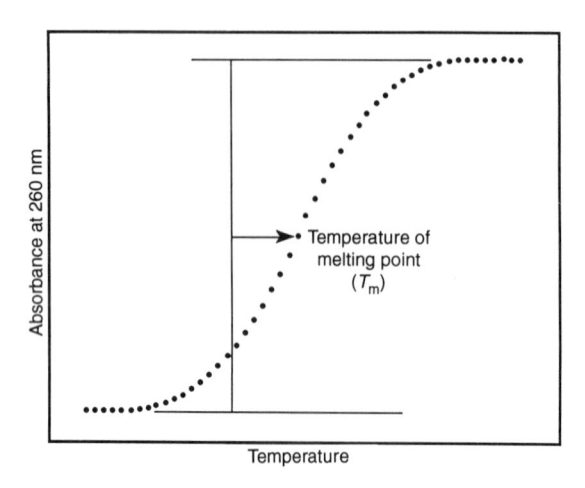

Figure 12.1 A melting curve of DNA.

separation is not achieved the column should be changed or the effluent composition altered:

(a) Decreasing the acetonitrile concentration improves the separation of cytidine, deoxycytidine and uridine making the elution time of deoxyadenosine longer.
(b) Increasing the $NH_4H_2PO_4$ concentration facilitates the separation of deoxyguanosine and deoxythymidine.

12.3 TAXONOMIC IMPLICATIONS

The determination of mean DNA base composition was the first unique feature of DNA that was seen to have taxonomic importance. The DNA base composition of a strain is constant and, given the application of the HPLC method described above, is highly reproducible (standard deviation <0.5 mole % G+C). Initially, DNA base composition studies provided an easy and useful way of distinguishing between phenetically similar but genetically different strains (Colwell & Mandel, 1964; Silvestri & Hill, 1965). It is, however, self-evident given the extensive nature of prokaryotic diversity that the mole % G+C contents of archaea and bacteria are not the preserve of single species.

The importance of the mole % G+C content of DNA in the systematics of prokaryotes is that it can be weighted as an excluding characteristic. Thus, if two organisms have DNAs with markedly different base composition values they must belong to different taxa. It is, however, important to realize that two organisms with similar base compositions are not necessarily closely related as mole % G + C determinations do not take into account the linear arrangement of nucleotides in DNA.

DNA base composition studies are particularly important in highlighting taxa that are in need of taxonomic revision. This point is exemplified by the genus *Bacillus*, which currently encompasses organisms with DNA base compositions that range from 32 to 69 mole % G + C (Slepecky & Hemphill, 1991). Indeed, despite its limitations as a taxonomic criterion, DNA base composition is correctly seen as one of the required characteristics for the minimum descriptions of genera and species (Lévy-Frébault & Portaels, 1992).

REFERENCES

Colwell, R.R. and Mandel, M. (1964) Adansonian analysis and deoxyribonucleic acid base composition of some Gram-negative bacteria. *J. Bacteriol.* **87**: 1412–22.
Johnson, J.L. (1991) Isolation and purification of nucleic acids. In *Nucleic Acid Techniques in Bacterial Systematics* (Stackebrandt, E. and Goodfellow, M., eds), pp. 1–19. John Wiley: Chichester.
Kaneko, T. (1977) Estimation of base composition of a deoxyribonucleic acid from ultraviolet spectrum of its hydrolyzate. *Agric. Biol. Chem.* **4**: 2277–83.
Katayama-Fujimura, Y., Komatsu, Y., Kuraishi, H. and Kaneko, T. (1984) Estimation of DNA base composition by high performance liquid chromatography of its nuclease P1 hydrolysate. *Agric. Biol. Chem.* **48**: 3169–72.
Krieg, N.R. and Holt, J.G. (eds) (1984) *Bergey's Manual of Systematic Bacteriology*, vol. 1. Williams & Wilkins: Baltimore.
Lévy-Frébault, V.V. and Portaels, F. (1992) Proposals for recommended minimal standards for the genus *Mycobacterium* and for newly described slowly growing *Mycobacterium* species. *Int. J. Syst. Bacteriol.* **42**: 315–23.
Mandel, M. (1966) Deoxyribonucleic acid base composition in the genus *Pseudomonas*. *J. Gen. Microbiol.* **43**: 273–92.
Marmur, J. (1961) A procedure for the isolation of deoxyribonucleic acid from microorganisms. *J. Mol. Biol.* **3**: 208–18.
Marmur, J. and Doty, P. (1962) Determination of the base composition of deoxyribonucleic acid from its thermal denaturation temperature. *J. Mol. Biol.* **5**: 109–18.
Mesbah, M., Premachandran, U. and Whitman, W.B. (1989) Precise measurement of the G + C content of deoxyribonucleic acid by high-performance liquid chromatography. *Int. J. Syst. Bacteriol.* **39**: 159–67.
Saito, H. and Miura, K. (1963) Preparation of transforming deoxyribonucleic acid by phenol treatment. *Biochim. Biophys. Acta* **72**: 619–29.
Schildkraut, C.L., Marmur, J. and Doty, P. (1962) Determination of the base composition of deoxyribonucleic acid from its buoyant density in CsCl. *J. Mol. Biol.* **4**: 430–443.
Silvestri, L. and Hill, L.R. (1965) Agreement between deoxyribonucleic acid base composition and taxonomic classification of Gram-positive cocci. *J. Bacteriol.* **90**: 136–40.
Slepecky, R.A. and Hemphill, H.E. (1991) The genus *Bacillus*—Nonmedical. In *The Prokaryotes*, 2nd edn. (Balows, A., Trüper, H.G., Dworkin, M., Harder, W. and Schleifer, K.H., eds), pp. 1663–96. Springer Verlag: Stuttgart.
Sneath, P.H.A., Mair, N.S., Sharpe, M.E. and Holt, J.G. (eds) (1986) *Bergey's Manual of Systematic Bacteriology*, vol. 2. Williams & Wilkins: Baltimore.
Staley, J.T., Bryant, M.P., Pfennig, N. and Holt, J.G. (eds) (1989) *Bergey's Manual of Systematic Bacteriology*, vol. 3. Williams & Wilkins: Baltimore.

Tamaoka, J. and Komagata, K. (1984) Determination of DNA base composition by reverse-phase high performance liquid chromatography. *FEMS Microbiol. Lett.* **25**: 125–8.

Williams, S.T., Sharpe, M.E. and Holt, J.G. (1989) *Bergey's Manual of Systematic Bacteriology*, vol. 4. Williams & Wilkins: Baltimore.

Yamada, K. and Komagata, K. (1970) Taxonomic studies on coryneform bacteria. III. DNA base composition of coryneform bacteria. *J. Gen. Appl. Microbiol.* **16**: 215–24.

13
Enzymes in Taxonomy and Diagnostic Bacteriology

Arthur L. James
University of Northumbria, Newcastle upon Tyne, UK

13.1 INTRODUCTION

Since the earliest recorded demonstration of enzymatic activities in microorganisms such as the yeast, *Saccharomyces*, and the proposal of the name 'enzyme' a vast number of microorganisms have been examined and their metabolic activities demonstrated. Furthermore, a large number of enzymes have been isolated and characterized, their kinetic and other parameters evaluated, and, frequently, their subunit structure and three-dimensional arrangement elucidated. Whilst such studies have often been of primarily academic interest, they have yielded a tremendous body of knowledge on enzyme distribution in microorganisms and have led, in an ever increasing degree, to the use of isolated microbial enzymes for commercial, industrial and diagnostic purposes.

The differential distribution of enzymes amongst microbial species provides a convenient method for the characterization and identification of individual taxa. For over half a century, diagnostic tests have been employed which are dependent on the possession of particular enzymes, for example the esculin test (Wasilauskas, 1971; Facklam, 1972), indole production (Kovacs, 1928; Miller & Wright, 1982; James *et al.*, 1986b), urea hydrolysis and tests for the detection of lysine and ornithine decarboxylases (Ewing *et al.*, 1960; James *et al.*, 1986a). Initially, such test procedures were developed and used empirically. The underlying biochemical basis and enzyme involvement only became apparent at a later date. The relative rapidity and ease of applying such tests has led to their wide usage, particularly in kit form, for identification purposes.

Early examples of such diagnostic packages were the API 20E (API-BioMerieux, La Balme les Grottes, France), widely used even today for the

---

Chemical Methods in Prokaryotic Systematics. Edited by M. Goodfellow and A.G. O'Donnell
© 1994 John Wiley and Sons Ltd.

differentiation and identification of enterobacteria, as well as other products in this range of API identification kits including API-Zyme (Humble et al., 1977) which has been extensively applied by taxonomists and others to the study of enzyme distribution patterns (Maddocks & Greenan, 1975: Bascomb, 1976, 1980; D'Amato et al., 1978; Kilian, 1978; Williams & Shah, 1980).

Most of the methods mentioned above are based on the detection of preformed (constitutive) enzymes: certain limitations apply when dealing with those enzymes which are inducible. In certain cases these problems can be overcome by a judicious choice of inducer. Thus, in the ONPG (2-nitrophenyl-B-D-galactoside hydrolysis) test for β-galactosidase-containing organisms, addition of a gratuitous inducer such as methyl or isopropyl-β-D-thiogalactoside is recommended.

In certain tests, the end result is not merely the product of a single enzymatic process but of a set of sequential ones. This is true of the arginine dihydrolase (ADH) test and specific fermentation tests. In this latter case, a very large number of possible carbohydrates and related substances can be employed. For a positive reaction, evidenced by acid production and consequent change in indicator colour, the carbohydrate used must be enzymatically convertible into glucose, fructose or more usually at the phosphorylated level for subsequent channelling into the glycolytic sequence. It is not surprising, therefore, that such fermentation tests, whilst being of considerable value, are frequently slow to give a visible end result and exhibit a greater or lesser degree of variability.

From the taxonomic perspective, a very wide range of criteria have traditionally been employed for classification and identification. These have included such aspects as micromorphology, staining characteristics, spore formation, shape and colour, colonial morphology and pigmentation, substrate utilization studies, growth inhibition and antibiosis. These gross phenomena may be attributed to the interplay of a variety of enzymes, in most cases involving a highly complex pattern.

More overtly biochemical and enzymatic tests used by taxonomists have included lecithinase production, pectin hydrolysis (pectin esterase), hydrolysis of chitin, nitrate reduction, hydrogen sulphide formation and hippurate hydrolysis (MacFaddin, 1980). However, such tests are but a small fraction of the total number of accessible enzymatic procedures available to the taxonomist. Each year, the number and type of substrates available for detection of specific enzymatic activities are increasing thereby providing a wider range of readily applied procedures for establishing taxonomic relationships.

The use of enzymes as taxonomic indicators has certain potential advantages over several other methods. These include the following:

(a) Ease of performance.

(b) Flexibility for use in a variety of situations, e.g. liquid media, agar-based plates, microtitre wells.
(c) Applicability to diverse organisms, e.g. aerobic/anaerobic, fast/slow growing, mycelial or discrete.
(d) Lack of necessity for complex instrumentation or for expensive and purified antisera.

It would be unwise, however, to minimize the problems inherent in basing a taxonomic system on enzyme distribution. As with other bacterial proteins, the type of enzyme elaborated by markers of particular taxa is an expression of the genome. This is revealed by inspection of a linkage map for organisms such as *Escherichia coli* or *Salmonella typhimurium*.

Whilst most enzymes are composed of more than one polypeptide chain, the synthesis of each is regulated via transcription and translation processes. Subsequent association of polypeptide chains and post-translational modification yields the active enzyme.

It would be tempting to conclude that the enzyme complement of a given microorganism is an accurate reflection of the genetic potential of that organism. However, it is well known that microorganisms can frequently regulate both the amount of enzyme produced and also the activity of enzymes by a variety of mechanisms. The implications of this with respect to the use of enzyme tests for classification and identification are considerable (Ison et al., 1982). One consequence relates to choice of media for growth. Enzyme patterns from a given microorganism grown under different cultural conditions may exhibit differences due to induction of catabolic enzymes or repression of particular biosynthetic pathways (Kersters & De Ley, 1971).

13.2 CLASSES OF ENZYMES OF KNOWN TAXONOMIC OR IDENTIFICATION VALUE

To date, most diagnostic enzymes have been those catalysing hydrolytic processes. There are several reasons for this, in particular the ready accessibility of substrates and the diversity of representative enzymes of this class.

Six major classes of enzyme are recognized. The principal biochemical properties of each and particular features of value for taxonomy are outlined below.

13.2.1 Oxidoreductases

Oxidoreductases catalyse reactions of the type:

$$AH_2 + B \rightleftharpoons A + BH_2$$

From the above equation it appears that such reactions involve oxidation-reduction pairs and are reversible. It may be difficult in practice, however, to recognize the redox systems involved and reversibility may be difficult to demonstrate due to the magnitude of the equilibrium constant if the equilibrium lies very much over to one side.

This class of enzymes includes the NADP-dependent dehydrogenases such as lactate and sorbitol dehydrogenase, as well as the flavoprotein enzymes such as D-amino acid oxidase and succinate dehydrogenase. The former enzyme, like other oxidases, can transfer electrons to molecular oxygen via the flavoprotein prosthetic group. Other oxidoreductases used in systematics include catalase, cytochrome oxidase, nitrate reductase and peroxidase. The former enzyme (catalase) which catalyses the disproportionation of hydrogen peroxide into water and oxygen is of considerable value in the differentiation of many microbial taxa.

13.2.2 Transferases

These enzymes permit the transfer of a functional group (X) between molecules in an exchange reaction, i.e.

$$AX + B \rightleftharpoons A + BX$$

Such reactions are frequently highly reversible although the equilibrium may be disturbed for detection purposes or *in vivo* by removal of one of the products by subsequent further reaction.

Examples of such enzymes include the aminotransferases (transaminases) of, for instance, *Escherichia coli*, and various transacetylases such as chloramphenicol acetyltransferase, which has therapeutic importance. Such enzymes do not appear to be widely used diagnostically or taxonomically for similar reasons to the dehydrogenases. Detection in cell extracts is a more facile proposition.

13.2.3 Hydrolases

This group contains the greatest range of enzymes of taxonomic value. Hydrolases catalyse the reaction:

$$\text{A-X} + \text{HOH} \rightarrow \text{A-OH} + \text{HX}$$

Such reactions are usually unidirectional or at least proceed a long way towards completion. Thermodynamically they proceed with a relatively large decline in free energy associated with a large value for the equilibrium constant, K (10^2–10^5). Furthermore, the aqueous environment means that the high molar concentration of water ($\simeq 50$ M) ensures that the reaction will be driven to effective completion. Even more important for practical purposes is the broad specificity which this class of enzymes exhibits. Thus phosphatases may be detected by use of a number of suitable substrates all of which are phosphate esters but with substantially different core molecules chosen to have advantageous characteristics for detection: e.g. intense coloration when released; strong fluorescence; chemiluminescence or other physico-chemical characteristics. This question of specificity of enzyme action has many implications in the design of suitable substrates, particularly those used for demonstration of peptidolytic activity where steric factors play a predominant role.

13.2.4 Lyases

Functional groups are removed by these enzymes non-hydrolytically. Typical examples include the carboxy-lyases (decarboxylases) and deaminases.

$$AB \to A + B$$

Both the above enzyme types have been employed for characterization of particular microorganisms, particularly among members of the family Enterobacteriaceae, and to some extent for taxonomic purposes (Schofield & Schaal, 1980). The lysine and ornithine decarboxylase tests have considerable importance in the detection of *Salmonella* spp., whilst the tryptophan deaminase test has value in delineating members of the *Proteus-Providencia* group.

13.2.5 Isomerases

As the name suggests, these catalyse the interconversion of two related molecular species typified by:

$$AB \rightleftharpoons BA$$

Examples include triose phosphate isomerase and phosphoglucomutase, which are widely distributed enzymes involved in carbohydrate utilization. Likewise, the enzymes catalysing interconversion of chiral amino acids, e.g. L-alanine and D-alanine, are termed racemases since they generate an equilibrium mixture of the two forms (racemic modification). They are currently of little value in either classification or identification.

13.2.6 Ligases

These enzymes catalyse the linkage of two compounds to create a more complex molecule. Such reactions are invariably endergonic (proceed with an increase in free energy) and hence require the intermediacy of a nucleotide complex usually derived from ATP.

$A + B \rightleftharpoons A - B$

$ATP \rightleftharpoons ADP + P_i$

This group of enzymes has so far found little value in systematics.

13.3 CHOICE OF SUBSTRATE, SPECIFICITY OF ENZYME ACTION, NATURAL AND SYNTHETIC SUBSTRATES

In the natural state, the microbial cell or mycelial mass possesses a certain number of enzymes. These may operate on the cellular components or intermediary metabolites, in which case the enzymes are most likely to be internal/cytoplasmic. Alternatively the enzymes may be concerned with the obtaining of nutrients by dissimilation of large molecules in which case they are likely to be extracytoplasmic or even, as in the case of some macromolecular hydrolases, extracellular. In the vast majority of cases the substrate transformed will yield products which are not suited to rapid and ready detection.

The value of the hydrolase class of enzymes in diagnostic microbiology has already been mentioned. This class comprises some 30 to 40% of known enzymes and may be divided into various subclasses such as the esterases, glycosidases, phosphatases, proteases/peptidases and sulphatases. All are hydrolytic so that, apart from the substrate and enzyme, the only other component needed is water. Additionally, in the microbial cell, the enzyme is usually accessible, being extracytoplasmic, or extracellular or non-localized in the cytoplasm, i.e. not bound to intracellular structures.

Considerable attention has been given to the design of substrates for the detection of such enzymes for a variety of purposes, many of which are outside the field of microbiology. Thus, phosphatases which cleave organic phosphate esters have broad specificity (group specificity) and have been detected or quantified by a large number of substrates including the phosphate esters of phenol, naphthol, naphthol ASBI, *p*-nitrophenol, phenolphthalein, thymolphthalein and 4-methylumbelliferone. In general, the microbial mass or biological fluid containing the suspected enzyme is incubated for a given time at a suitable temperature with a dilute solution of the substrate in appropriately buffered medium and the released product detected either colorimetrically or fluorimetrically. In the above example, the first three products are simple phenols and are colourless. Consequently, a second stage is needed in the development of colour. This usually consists of adding a stable diazonium salt such as Fast-Blue RR salt, Fast Red or Fast Black to couple with the released phenolic compound and hence give an intensely coloured azo-dye. Such procedures form the basis of many detection systems, particularly the API-Zyme and AN-IDENT kits. *p*-Nitrophenol and the two phthalein dyes are highly coloured *per se* (provided that the pH is neutral to alkaline) and this permits ready detection. 4-Methylumbelliferone is colourless but extremely fluorescent, emitting a sky-blue fluorescence on excitation with a low-pressure mercury arc lamp.

The glycosidases represent a large group of hydrolytic enzymes which have great microbial importance. Their role in the natural state probably involves the degradation of external carbohydrate sources such as hyaluronic acid, mannans, starches and xylans and the turnover of microbial polysaccharides. Two subgroups of the glycosidases exist, defined by the method of linkage of the 'sugar unit' to the aglycone—the non-carbohydrate moiety. Thus, β-glycosidases act on β-glycosides in which the aglycone is linked to C1 of the sugar molecule in the manner shown in Figure 13.1, while in α-glycosidases the corresponding α-glycosides have a different configuration (Figure 13.1).

A number of naturally occurring glycosides have been used for classification and in diagnostic microbiology for many years. These include amygdalin, arbutin, esculin and salicin. These are all β-linked glycosides with the sugar being β-D-glucose. The aglycone released may be detected in a variety of ways. Thus, esculin is readily detected by the dark-brown/black colour produced in the presence of added ferric salts. Arbutin degradation may be similarly detected but usually requires a longer incubation period. Amygdalin and salicin utilization is frequently detected by acid production from the released glucose that is metabolized.

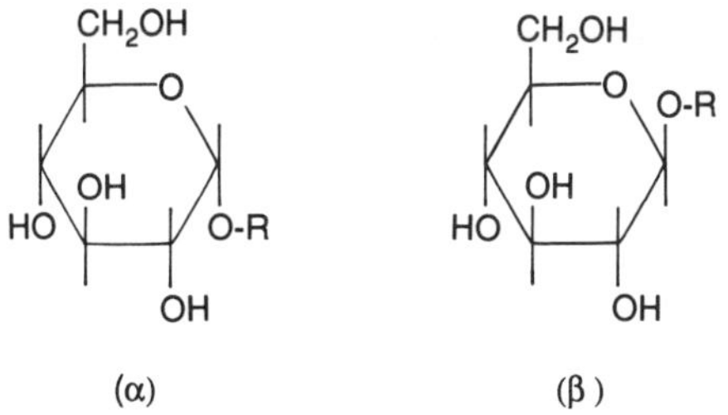

Figure 13.1 α and β glucosides. R denotes the non-carbohydrate or aglycone portion.

It is apparent that such procedures, despite their value in the detection of β-glucosidase, lack general applicability. A number of synthetic aglycones have consequently been employed linked to the appropriate sugar moiety for more general glycosidase detection.

The enzyme β-galactosidase has received more attention than possibly any other bacterial enzyme. The problem of 'late lactose fermenters' in diagnostic microbiology led to the development of the now widely used ONPG test. The substrate (2-nitrophenyl-β-D galactoside) has many advantages over simply detecting lactose fermentation by pH indicator means. It is readily taken into the cell regardless of the permease and is hydrolyzed by the β-galactosidase at a rapid rate and with a K_m 30 times lower than lactose; i.e. the substrate is hydrolysed at half maximal velocity at a concentration of 1/30th that of lactose. This, coupled with the obvious yellow colour of the *ortho*-nitrophenol released, ensures ready detection (Le Minor & Ben Hamida, 1962).

Other substrates for detection of the enzyme have included β-galactosides of the naphthols, phthalein dyes and the indolic derivatives, notably indoxyl-β-galactoside and X-gal (5-bromo-4-chloro-indolyl-β-D galactoside). Indolic galactosides such as X-gal have advantages particularly for detection of organisms possessing the enzyme on agar plates since the aglycone released is oxidized rapidly by air to produce a blue/purple indigoid dye on the colony mass. It has proved its value on many occasions, particularly with respect to examining *E. coli* colonies plated out from gene fusion experiments where successful fusion colonies are distinguished by their lack of colour.

β-Galactosidase has also been detected fluorimetrically by use of fluorogenic galactosides, in particular those of fluorescein, resorufin and 4-methylumbelliferone. In most cases detection of the enzyme is assisted by the addition of a gratuitous inducer such as isopropyl thiogalactoside (IPTG). However, in the case of particularly sensitive fluorogenic substrates this may not be necessary, the endogenous levels of enzyme being enough to permit detection.

β-Glucuronidase is another enzyme on which much interest has centred over the last decade. This interest has arisen largely from Killian's observation of the highly restricted distribution of the enzyme to *Escherichia coli* and *Shigella sonnei* within the family Enterobacteriaceae and its value in the differentiation of streptococci (Kilian & Bulow, 1976). It is now possible to buy a range of both α- and β-glycosidase substrates which are capable of being used in a variety of modes for microbial classification and identification. Similar considerations apply to substrates for other hydrolytic enzymes such as esterases, lipases, peptidases, phosphatases and sulphatases.

The case of peptidases and other proteolytic enzymes requires special discussion. Whereas the aforementioned hydrolytic enzymes operate on derivatives of phenolic or other hydroxyl-based aglycones, this is not the case with peptidases. The nature of the peptide bond ($-CO-NH-$) necessitates use of substrates in which an amino acid or small peptide is linked via the carboxyl group to a chromogenic or fluorogenic amine. Various amines have been used for this purpose including β-naphthylamine and its methoxy derivative, p-nitroaniline, as well as the highly fluorescent 7-amino-4-methylcoumarin. The latter is particularly useful in permitting detection of very low levels of specific peptidases. The distribution of aminopeptidases and other proteolytic enzymes has been studied on a number of occasions (Muftic, 1967: McIntyre et al., 1975; Watson, 1976).

13.4 DETECTION METHODS

In the detection of specific enzymes, visualization of the substrate-product transformation is central. While selectivity of the procedure resides essentially in correct choice of substrate and conditions of the test, the sensitivity of the method is largely governed by the nature of the substrate and the method of visualization used. Broadly speaking, two major methods are employed. Visual observation and colorimetric analysis, on the one hand, are convenient procedures for routine characterization and identification. On the other hand, fluorescence observation and fluorimetric measurement are useful procedures, being both rapid and sensitive, thereby permitting the use of smaller inocula and allowing speciation or identification of relatively slow-growing organisms such as mycobacteria (Grange, 1977) and streptomycetes (Goodfellow et al., 1987) in a reasonable time scale rather than waiting the 20 days frequently required by conventional taxonomic procedures.

13.4.1 Visual observation

Substrates for selected enzymatic activities based on such core molecules as nitrophenols, p-nitroanilides and indoxyls can be directly loaded in suitably buffered solution into a series of tubes, inoculated and incubated for a set period of time at optimal temperatures. The appearance of colour is automatic in these cases so that time can be chosen to give the best differentiation of positive and negative cultures. Where subsequent reagent addition is needed, trial experiments should be performed to assess a convenient time for incubation followed by reagent addition. Thus, naphthol- or naphthylamine-based substrates may be detected by addition of azo-coupling reagents such as Fast Blue RR salt or alternatively by addition of

4-dimethylaminocinnamaldehyde, which yields a purple-red Schiff's base with the released amine.

Since the cost of many of the substrates employed is quite high, microtubes or even microtitre plates may be used to accommodate the tests. Substrates may be dried into the wells and subsequently reconstituted with buffer prior to incubation. Where a quantitative assessment is required or a numerical value generated, the use of microtitre plates read by an ELISA reader has potential applications. The use of strip tests based on selected chromogenic substrates simplifies the procedure considerably. API market a number of strips for diagnostic uses and also many for more investigational purposes. In particular, the API-ZYM method has been widely applied (Tharagonnet et al., 1977: Williams & Shah, 1980; Slots, 1981; Laughon et al., 1982).

13.4.2 Fluorescence techniques

Fluorescence is caused by absorption of electromagnetic radiation—ultraviolet, visible, infrared—leading to promotion of electrons from the ground state to an excited state. The latter has a finite lifetime, measured usually in microseconds to nanoseconds, during which time some loss of vibrational energy occurs. The residual energy is either lost by collisional deactivation or by re-emission of radiant energy (fluorescence). The loss of vibrational energy means that fluorescence energy is always less than the energy of absorption and the wavelength maximum, λ_F, is correspondingly longer than the absorption maximum, λ_A. Since absorption of energy leads, in these cases, to electronic excitation, the wavelength of absorption is often designated λ_x.

An important feature of fluorescence emission is the sensitivity of detection or measurement. With highly fluorescent molecules such as fluorescein, 4-methylumbelliferone or 7-amino-4-methylcoumarin the quantum yield Φ_F is high and approaches unity under optimal conditions. In addition, the molecules have a high molar absorption coefficient ε. The product $\Phi_F \varepsilon$ primarily determines analytical sensitivity and hence detection limits and is the major reason for the use of derivatives of these molecules as fluorogenic substrates for bacterial classification and identification and clinical/biochemical utilization.

A wide range of fluorogenic substrates are now available from leading manufacturers and distributors (Bachem, CRB, Novobiochem, Sigma). These substrates are based primarily on two 'core molecules' namely 4-methylumbelliferone (4-MU) and 7-amino-4-methylcoumarin (7-AMC) although a more restricted range of substrates based on 6-aminoquinoline, fluorescein, indoxyl and resorufin are also available. 4-Methylumbelliferone in particular has been widely employed (Grange &

13. Enzymes in Taxonomy and Diagnostic Bacteriology

Clark, 1977; Grange & McIntyre, 1979; Godsey et al., 1981; Feng & Hartman, 1982).

The coumarinic compounds 4-MU and 7-AMC have, like other fluorophores, structural features which predispose towards fluorescence. These are planarity, molecular rigidity, electron delocalization via an efficient conjugated system and the presence of at least one electron-releasing group. These points are illustrated in Figure 13.2. Derivatization of the OH-group in 4-MU or of the NH$_2$-group in 7-AMC prevents electron release from these groups and hence greatly diminishes fluorescence. It has been estimated that L-leucyl-7AMC (Figure 13.3) has only 0.3% of the fluorescence of 7-AMC measured at 430 nm.

The mode of usage of fluorogenic substrates can be varied to suit individual requirements. A tube technique may be used as for chromogenic substrates but may create problems due to the natural fluorescence of many glass materials under ultraviolet illumination. Furthermore, the

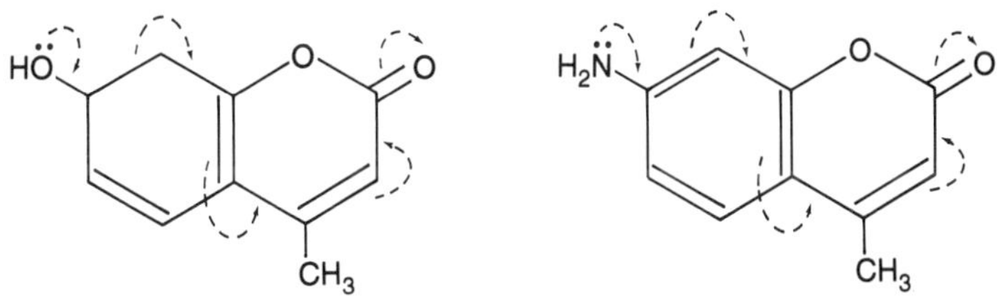

Figure 13.2 Structural formulae for (i) 4-methylumbelliferone and (ii) 7-amino-4-methylcourmarin showing electron delocalization producing intense fluorescence.

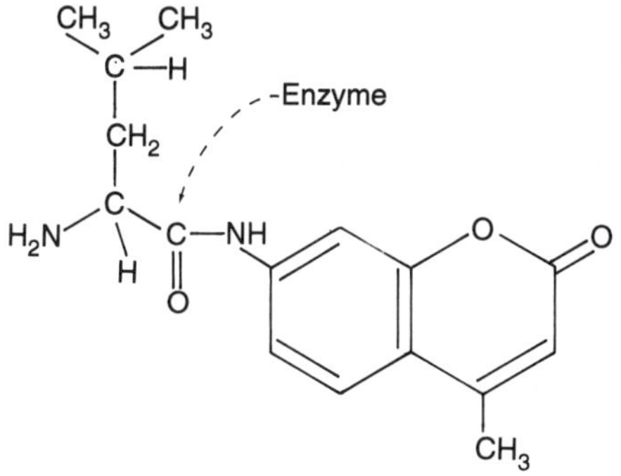

Figure 13.3 Structural formula for substrate L-leucyl-7AMC showing position of enzymatic hydrolysis.

expense of many glycosidase and peptidase substrates precludes the use of large volumes of substrate even though the concentrations needed are low. The use of microtitre plates offers a number of advantages but here again the plates must lack inherent fluorescence. Those plates designed for fluorescence immunoassay work should be satisfactory.

A useful and relatively simple technique has been used successfully both in routine classification and identification investigations. This is a modification of the procedure of Slifkin and Gil (1983) and consists of rubbing a colony or precipitate of the organism onto positions towards the periphery of a 9 cm Whatman No. 3 filter paper contained in a Petri dish cover. Several zones of the organism can thus be created in a clock formation. One drop of 12 different substrates in buffered medium is added to each zone and the whole assembly is incubated at a selected temperature for 15 to 30 min using the base of the Petri dish to protect and to maintain humidity. The paper is subsequently inspected using an ultraviolet lamp or cabinet employing long-wavelength (366 nm) ultraviolet radiation from a mercury arc. Hydrolysis of substrates is detected by observation of zones of bright-blue fluorescence against a dark background. The test can obviously be modified to examine each of several organisms against a given substrate. In this case it is convenient to place a spot of the substrate centrally on the paper to check for inherent fluorescence and for non-enzymatic hydrolysis.

In the case of substrates based on 4-MU, fluorescence may be enhanced by addition of a drop of saturated sodium bicarbonate solution post-incubation. This generates the ionized form of methylumbelliferone, which has a more intense fluorescence than the undissociated molecule. However, in a few cases, alkaline hydrolysis of the substrate is possible so that a check on the substrate blank should also be made.

A more elegant application of fluorogenic substrates is available from and produced by Sensititre (Sensititre Ltd., East Grinstead, UK). For identification purposes a number of substrates having discriminatory value are dried into the wells of a 96-position plate. With 32 positions occupied by different substrates this allows three organisms to be completely identified down to species level after a 4-h incubation period. Fluorescence is generated by production of 4-MU or 7-AMC or related fluorogens having excitation maxima in the near UV and emission maxima around 420–440 nm. In some cases 4-MU acts as a fluorescent indicator due to the pH dependence of fluorescence. This is useful for decarboxylase, urease and fermentation tests. The fluorescence reader is essentially a filter fluorimeter: the fluorescence is measured by a photoelectric transducer and the signal is converted to a digital output. The cut-off value can be set in order to distinguish positive from negative or weak positive organisms. A multipoint inoculator is available to dose the wells with sample, thus speeding up the process. The

13. Enzymes in Taxonomy and Diagnostic Bacteriology 483

fluorescence reader is microcomputerized and the fluorescence intensity signals generated can be displayed on a dot matrix/thermal printer.

The procedure outlined above has great potential for both classification and identification of microorganisms. Great flexibility can be achieved by correct choice of fluorogenic substrates for a given application. This has already been demonstrated by work on the numerical taxonomy of staphylococci based largely on fluorescence techniques and the use of the Sensititre instrument (Figure 13.4) (Bovill, 1991).

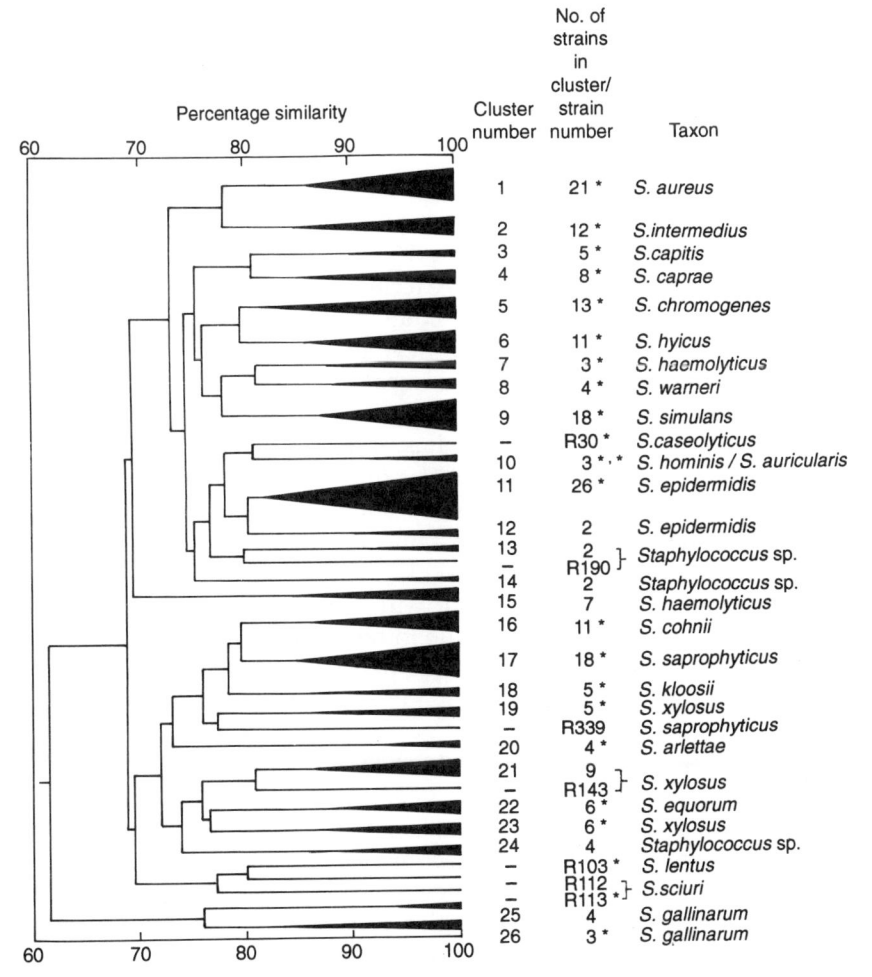

Cluster number	No. of strains in cluster/ strain number	Taxon
1	21*	S. aureus
2	12*	S. intermedius
3	5*	S. capitis
4	8*	S. caprae
5	13*	S. chromogenes
6	11*	S. hyicus
7	3*	S. haemolyticus
8	4*	S. warneri
9	18*	S. simulans
–	R30*	S. caseolyticus
10	3*,*	S. hominis / S. auricularis
11	26*	S. epidermidis
12	2	S. epidermidis
13	2	Staphylococcus sp.
–	R190	Staphylococcus sp.
14	2	Staphylococcus sp.
15	7	S. haemolyticus
16	11*	S. cohnii
17	18*	S. saprophyticus
18	5*	S. kloosii
19	5*	S. xylosus
–	R339	S. saprophyticus
20	4*	S. arlettae
21	9	S. xylosus
–	R143	S. xylosus
22	6*	S. equorum
23	6*	S. xylosus
24	4	Staphylococcus sp.
–	R103*	S. lentus
–	R112, R113 *	S. sciuri
25	4	S. gallinarum
26	3*	S. gallinarum

Figure 13.4 Abbreviated dendrogram showing relationships between representative staphylococci obtained using the simple matching coefficient and unweighted pair-group method with arithmetic averages algorithm. *Clusters containing type strains.

For taxonomic purposes, however, fluorescence techniques are not without drawbacks. These relate mainly to three areas:

(a) Fluorogenic substrates are invariably expensive, some very expensive. Those based on 7-AMC are particularly costly, especially where a di- or tripeptidyl sequence is involved. Greater use of these compounds in research and clinical diagnosis may well reduce their cost in relative terms.
(b) Certain fluorogenic compounds are relatively unstable. The short-chain C_1 to C_4 esters of 4-MU are too unstable to be used in this context and more stable derivatives are needed.
(c) Some compounds are relatively water insoluble and consequently require initial dissolution in a small volume of a high dielectric constant solvent such as dimethylformamide (DMF) or dimethyl sulphoxide (DMSO). Dilution with buffer may then yield a homogeneous solution. This is a particular problem with long-chain esters of 4-MU and with some endopeptidase substrates due to the lipophilic nature of the N-terminal blocking group (see later).
(d) Whilst the fluorescence reader is excellent for use with substrates based on 4-MU, 7-AMC and related coumarinic compounds, it cannot at present deal with fluorophores having markedly different excitation and fluorescence maxima. This is not at present a great drawback since most fluorogenic substrates have coumarinic nuclei. Use of other fluorophores would necessitate dual or even multiple wavelength facilities and may necessitate a move to a dual monochromator system with a significant increase in complexity and cost.
(e) Provided that a reasonably constant inoculum can be used, reproducible results can be obtained. With most non-filamentous organisms this can be readily ensured, but with, for instance, fungi and many actinomycetes problems arise due to non-homogeneity, clumping and related manifestations. How great a problem this is appears uncertain. Short-term sonication appears to improve the situation but may create other problems.

13.4.3 Multipoint techniques

The use of multipoint inoculation techniques on agar-based media is well known to most diagnostic bacteriologists. Several workers have attempted to utilize the technique for identification purposes. Much of this work has been in the area of enterobacterial identification where biochemical diagnostic tests are well documented and data readily available. At least one commercial system is available for identification and for susceptibility testing. The MAST system (MAST Ltd., Bootle, UK) utilizes a number of

13. Enzymes in Taxonomy and Diagnostic Bacteriology

multipoint pins (up to 36) to inoculate that number of samples onto nutrient agar plates incorporating other constituents for the demonstration of a particular biochemical characteristic, e.g. carbohydrate fermentation, esculin hydrolysis, urease or H_2S production. Coloured zones are produced at or around the positive organism, these contrasting with corresponding negative organisms and with the agar background. The number and position of such organisms are ascertained by using a colour camera linked to a video display and a digitizing computer.

The system outlined above is limited by the contrast available and it is unfortunate that this is not always satisfactory. Thus, while the esculin and the deaminase plates are eminently satisfactory, producing dark zones around positive colonies and light-coloured negatives, the urease and decarboxylase tests suffer from excessive zone spreading and indeterminate colour changes in the indicators used. The system nevertheless has potential application in microbial systematics if appropriate substrates are used. To be of value in numerical taxonomy, a 36-pin inoculation system would appear to be essential. This necessitates use of substrates which give high-contrast zones of colour around positive organisms with little or no diffusion so that 'nearest neighbour' interference is not observed. The esculin plate comes near to meeting these criteria.

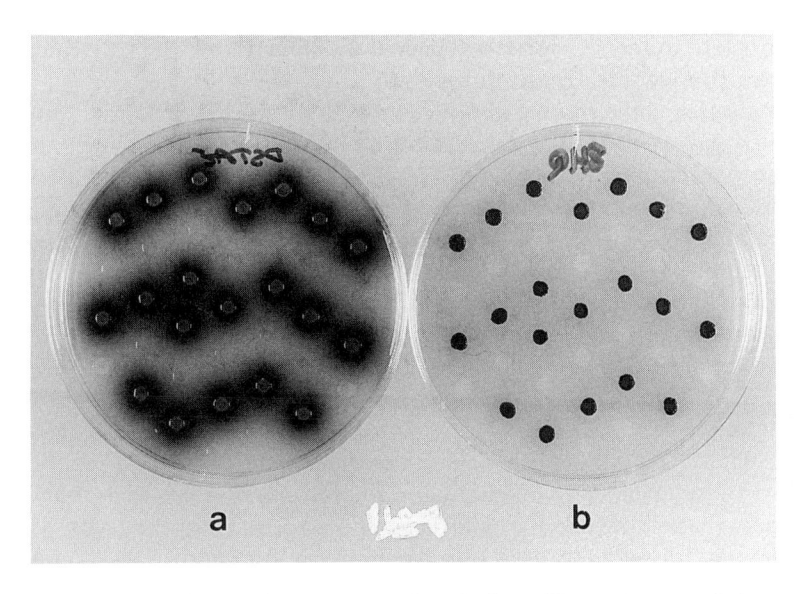

Figure 13.5 Equivalent multipoint inoculated plates illustrating much improved contrast obtained by use of 8-hydroxyquinoline β-D-glucoside (b) compared with esculin (a), auxill reagent, ferric ammonium citrate. (Photography courtesy of P. Yeoman.)

It has recently been demonstrated that very high-contrast plates can be produced by incorporation of derivatives of 8-hydroxyquinoline (8-HQ) in the presence of iron salts. Thus, β-glucosidase, β-galactosidase, β-glucuronidase and aryl sulphatase activities have been demonstrated using the appropriate 8-HQ derivative (James & Yeoman, 1987, 1988). The contrast between the jet-black positive colonies and pale negatives on a straw-coloured background is readily observed (Figure 13.5) or interpreted by an Image Analyser such as that of Perceptive Instruments (Little Yeldham, Halstead, UK) who market a 'Domino System' for microbiological applications.

13.5 AREAS FOR DEVELOPMENT

13.5.1 Hydrolytic enzymes

Although a wide range of 4-MU substrates are available from various sources, there is still room for others which are capable of demonstrating different enzymatic activities. Thus the range of glycosides could be greatly extended to cover glycuronides other than the β-D-glucuronide at present available and including galacturonides, iduronides and mannuronides. Similarly derivatives of deoxy and dideoxy sugars which have been shown to be present in markers of many bacterial taxa have yet to be synthesized although the technical procedures exist.

In a similar manner long-chain fatty acids found in bacteria could be derivatized with 4-MU or other fluorogens. In this fashion *iso*, *anteiso*, cyclopropanoid and hydroxy fatty acyl derivatives could be created and their pattern of enzymatic cleavage used to provide additional taxonomic information. The pattern of distribution of enzymes cleaving long-chain esters among staphylococci has been shown to be substantially diverse across the genus, e.g. the C_{18} acids elaidic, oleic and stearic. Polyunsaturated derivatives were not available for comparison (Bovill, 1991).

The range of peptidase substrates that can be obtained is even greater, with 7-AMC derivatives available for demonstration of most of the known aminopeptidases and many endopeptidases. Most of the literature on the use of chromogenic and fluorogenic substrates for endopeptidases refers to enzymes of mammalian origin although it is clear that such enzymes have microbial counterparts, at least as regards determining peptide sequence if not biochemical function. The nomenclature and structural complexity of these substrates is often confusing to the microbial systematist. It is, therefore, appropriate to attempt clarification.

The peptide chain of a normal protein may contain a hundred or more amino acid residues with one N-terminal and one C-terminal group; i.e. there is a free amino group at one end and a free carboxyl group at the other. Enzymes operating on the N-terminus are the aminopeptidases while those operating on the C-terminus are the carboxypeptidases. The latter are less amenable to study taxonomically due to substrate limitations.

Removal of the N-terminal amino acids, for instance by L-leucine aminopeptidase (LAP) or L-alanine aminopeptidase, occurs by a hydrolytic cleavage of the terminal peptide group (Lazdunski et al., 1975). Other enzymes exist which cleave at a point further down the chain. Thus, the dipeptidyl aminopeptidases remove a two-amino acid fragment intact. The substrates L-leucyl-7-AMC and Gly-L-Pro-7-AMC are substrates for leucine aminopeptidase and dipeptidyl aminopeptidase IV respectively.

Yet other enzymes cleave at a point in the interior of the polypeptide chain. These are the endopeptidases typified by the mammalian enzymes chymotrypsin, collagenase, elastase, trypsin and the intracellular cathepsins. The point of hydrolytic cleavage is signified by the adjacent amino acid sequence. Sometimes a single amino acid will act as an enzyme recognition site. Alternatively, a sequence of two, three or more amino acids may be necessary or be more effective. Thus, for Cathepsin B a number of substrates have been reported, namely:

Benzoyl-Arg-4-MβNA

Z-Arg-Arg-AMC

Z-Phe-Arg-AMC

Of these, the first, 4-methoxy-β-naphthylamide (4-MβNA), has an alternative fluorescent moiety. Note the importance of the terminal arginine in all cases. Both dipeptide and single amino acids serve as recognition sites. With endopeptidase substrates the amino group of the auxiliary amino acid needs to be blocked (Figure 13.6) to prevent the substrate from being cleaved by aminopeptidase action. A number of blocking groups are used. These are often of the type illustrated above and employed in the synthetic procedure used to produce the substrate. Thus, benzoyl, benzyloxycarbonyl (Z) or tertiary butoxycarbonyl (tBOC) groups frequently appear in this context.

Alternatively such groups may be removed and replaced by other more water-solubilizing groups. Paramount among these are the glutaryl and succinyl groups, which possess free ionizable carboxyls. Thus, chymotrypsin-like activity has been demonstrated by use of the

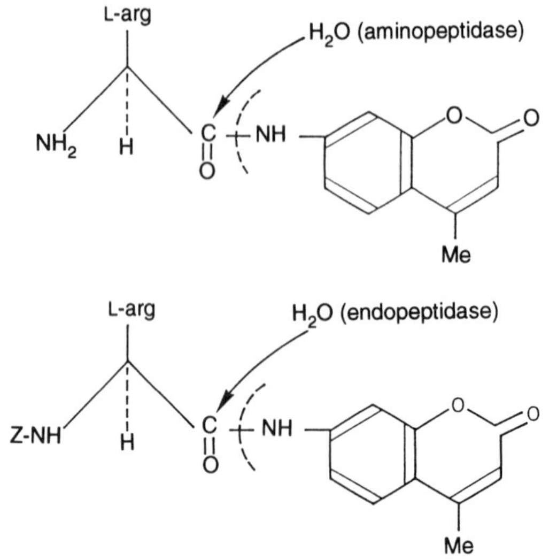

Figure 13.6 Action of peptidolytic enzymes on two related amino-acyl-7-AMC derivatives.

following:

Glutaryl-Phe-AMC

Glutaryl-Gly-Gly-Phe-AMC

Succinyl-Ala-Ala-Phe-AMC

These are all relatively water-soluble substrates of good sensitivity (Figure 13.7). Similar considerations apply to other peptidase substrates. Solubility can be a problem and is influenced by the salt form. Thus, trifluoroacetate (TFA) salts are generally more soluble than the corresponding hydrohalides, which in turn are better than the free base. Likewise, peptide sequences containing basic amino acids such as arginine are more soluble than those rich in hydrophobic amino acids such as phenylalanine or valine. A useful procedure for dealing with problematic substrates is to dissolve 5 mg of the compound in 200 μl of reagent grade dimethylsulphoxide or dimethylformamide and to dilute with 4.8 ml buffer. For more strictly comparative studies a molar basis is recommended, employing 10 μmol of substrate as above.

As with 4-MU substrates, peptidyl derivatives of 7-AMC could be conveniently extended. Greater knowledge of the biochemistry and molecular biology of microbial processes should allow the design of suitable

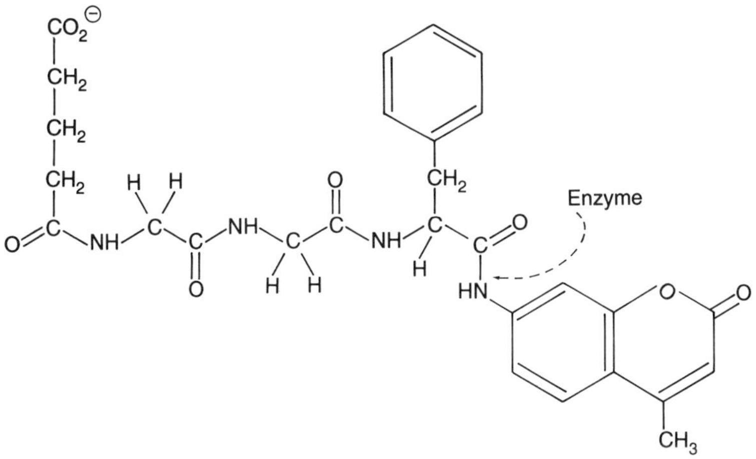

13. *Enzymes in Taxonomy and Diagnostic Bacteriology* 489

Figure 13.7 Structural formula for glutaryl-glycyl-glycyl-phenylalanyl-7-AMC showing positions of hydrolysis by chymotrypsin-like endopeptidase.

substrates. This has already been shown in the case of substrates developed for endotoxin assay by the *Limulus* lysate method and the design of a substrate for staphylocoagulase. The variety of unusual amino acids and peptides encountered in the microbial world should permit synthesis of many additional substrates of taxonomic value.

The fluorescence generated by the two coumarins, 7-AMC and 4-MU, is intense and easily detected in most cases. Fluorescence in the blue region has, however, certain disadvantages, particularly where organisms produce pigments which themselves fluoresce. Such endogenous fluorescence is generally in this blue region. Alternative fluorophores have been employed in analytical and other procedures which fluoresce in other regions of the visible spectrum. These include 6-aminoquinoline, resorufin and trifluoromethyl coumarins. These fluoresce in the yellow, orange-red and green regions, respectively. Whilst the number of substrates based on such fluorophores is limited and their cost prohibitive, it is possible that improved methods of synthesis and greater demand may lead to greater availability.

To date, most enzymes used for taxonomic purposes have been hydrolytic in nature. There are good reasons for this. Hydrolases are frequently, although not invariably, extracellular/extracytoplasmic, no cofactors are normally required other than metal ions, rates of reaction are usually relatively rapid and high water concentration favours a unidirectional process. Furthermore a wide range of substrates exist, both chromogenic and fluorogenic, for ready detection and assay of such enzymes.

13.5.2 Non-hydrolytic enzymes

The hydrolases are only one, albeit large, group of enzymes. But whilst many instances of non-hydrolytic biotransformations of organic molecules by microorganisms have been reported, there has been little attempt to use the enzymes associated with such processes for taxonomic purposes. One spin-off of such an approach could well be the demonstration of enzymes having industrial or commercial importance (Peczyńska-Czoch & Mordarski, 1988; Goodfellow & O'Donnell, 1989).

The importance of detecting putative industrially significant enzymes has been highlighted recently with respect to microbial enzymes catalysing transformations of nitriles (Thompson *et al.*, 1988; Nagasawa & Yamada, 1989). Enzymes such as the nitrile hydratases and nitrilases have great potential for converting nitriles to the corresponding higher value amides or acids under mild conditions, often with high selectivity. Two nitrilase substrates have been examined by Ball (1993). These were anthranilonitrile (2-aminobenzonitrile) and salicylonitrile (2-hydroxybenzonitrile). Both substrates are non-fluorescent but on incubation with appropriate actinomycetes yielded strong blue fluorescence due to conversion to the corresponding amides or acids. Within the family Pseudonocardiaceae, anthranilonitrile was particularly valuable in distinguishing the various species of *Amycolata* from *Amycolatopsis*, *Saccharopolyspora* and related genera (Whitehead, 1990). All *Amycolata* strains tested yielded an intense blue fluorescence. The test would appear to be a valuable adjunct to other fluorescence methods.

By appropriate choice of molecular structure, it should be possible in a similar manner, to survey the distribution of other enzymes such as nitroaryl reductases or aryl hydroxylases. Such enzymes are likely to be less widely distributed than the more ubiquitous hydrolases and hence be capable of distinguishing smaller groups of organisms, as in the case of *Amycolata*.

REFERENCES

Ball, L.G. (1993). Characterisation of the Genus *Microbispora* and Related Actinomycetes. PhD Thesis, University of Newcastle upon Tyne.

Bascomb, S. (1976) Enzymatic activities of bacteria and their survival during treatments affecting cell integrity. *Proc. Soc. Gen. Microbiol.* **3**: 87.

Bascomb, S. (1980) Identification of bacteria by measurements of enzyme activities and its relevance to the clinical diagnostic laboratory. In *Microbiological Classification and Identification* (Goodfellow, M. and Board, R.G., eds), pp. 359–73. Academic Press: London.

Bovill, R. (1991) Rapid methods in staphylococcal systematics. PhD Thesis, University of Northumbria, Newcastle upon Tyne.

D'Amato, R.F., Eriques, L.A., Tomfohrde, K.A. and Singerman, E. (1978) Rapid identification of *Neisseria gonorrhoeae* and *Neisseria meningitidis* by using enzymatic profiles. *J. Clin. Microbiol.* **3**: 77–81.

Ewing, W.H., Davis, B. and Edwards, P. (1960) The decarboxylase reactions of *Enterobacteriaceae* and their value in taxonomy. *Publ. Hlth. Lab.* **18**: 177–83.

Facklam, R.R. (1972) Recognition of group D streptococcal species of human origin by biochemical and physiological tests. *Appl. Microbiol.* **23**: 1131–9.

Feng, P.C.S. and Hartman, P.A. (1982) Fluorogenic assays for the immediate confirmation of *Escherichia coli*. *Appl. Environ. Microbiol.* **43**: 1320–9.

Godsey, J.H., Matteo, M.R., Shen, D., Tolman, G. and Goheke, J.R. (1981) Rapid identification of *Enterobacteriaceae* with microbial enzyme activity profiles. *J. Clin. Microbiol.* **13**: 483–90.

Goodfellow, M. and O'Donnell, A.G. (1989) Search and discovery of industrially-significant actinomycetes. In *Microbial Products: New Approaches* (Baumberg, S., Hunter, I.S. and Rhodes, P.M., eds), pp. 343–83. Cambridge University Press: Cambridge.

Goodfellow, M., Thomas, E.G. and James, A.L. (1987) Characterisation of rhodococci using peptide hydrolase substrates based on 7-amino-4-methylcoumarin. *FEMS Microbiol. Lett.* **44**: 349–55.

Grange, J.M. (1977) A fluorogenic substrate for the rapid differentiation of *Mycobacterium fortuitum* and *Mycobacterium chelonei* on the basis of heat stable esterase activity. *Tubercle* **58**: 147–50.

Grange, J.M. and Clark, K. (1977) Use of umbelliferone derivatives in the study of enzymatic activities of mycobacteria. *J. Clin. Path.* **30**: 151–3.

Grange, J.M. and McIntyre, G. (1979) Fluorogenic glycoside substrates. Their use in the identification of some slow growing mycobacteria. *J. Appl. Bacteriol.* **47**: 285–8.

Humble, M.W., King, A. and Phillips, I. (1977) API-ZYM: A simple rapid system for the detection of bacterial enzymes. *J. Clin. Pathol.* **30**: 275–7.

Ison, C., Glynn, A.A. and Bascomb, S. (1982) Acquisition of new genes by oral *Neisseria*. *J. Clin. Pathol.* **35**: 1153–7.

James, A.L. and Yeoman, P. (1987) Detection of specific bacterial enzymes by high contrast metal chelate formation. Part 1. 8-Hydroxyquinoline β-D-glucoside, an alternative to aesculin in the differentiation of members of the family *Enterobacteriaceae*. *Zbl. Bakt. Hyg.* **A267**: 188–93.

James, A.L. and Yeoman, P. (1988) Detection of specific bacterial enzymes by high contrast metal chelate formation. Part II. Specific detection of *Escherichia coli* on multipoint-inoculated plates using 8-hydroxyquinoline-β-D-glucuronide. *Zbl. Bakt. Hyg.* **A267**: 316–21.

James, A.L., Mistry, M. and Yeoman, P. (1986a) A sensitive method for the demonstration of decarboxylase activities amongst *Enterobacteriaceae* without the use of pH indicators. *Zbl. Bakt. Hyg.* **A262**: 455–61.

James, A.L., Yeoman, R., Rasburn, J.W. and Ng, M. (1986b) Sensitive reagents for detection of indole production by bacteria. *Zbl. Bakt. Hyg.* **A626**: 195–202.

Kersters, K. and De Ley, J. (1971) Enzymatic tests with resting cells and cell-free extracts. *Methods in Microbiology* **6A**: 44–65.

Kilian, M. (1978) Rapid identification of *Actinomycetaceae* and related bacteria. *J. Clin. Microbiol.* **8**: 127–33.

Kilian, M. and Bulow, P. (1976) Rapid diagnosis of *Enterobacteriaceae*. I. Detection of bacterial glycosidases. *Acta Pathol. Microbiol. Scand.* (Sect. B) **84**: 245–51.

Kovacs, N. (1928) Eine vereinfachte Methode zum Nachweis der Indolbildung durch Bakterien. *Z. Immunoforsche Exp. Ther.* **55**: 311–14.

Laughon, B.E., Syed, S.A. and Loesche, W.J. (1982) API-ZYM system for identification of *Bacteroides* spp., *Capnocytophaga* spp. and spirochaetes of oral origin. *J. Clin. Microbiol.* **15**: 97–102.

Lazdunski, V., Busuttel, J. and Lazdunski, A. (1975) Purification and properties of periplasmic aminoendopeptidases from *Escherichia coli*. *Euro. J. Biochem.* **60**: 363–9.

Le Minor, L. and Ben Hamida, F. (1962) Advantages de la recherche de la β-galactosidase sur cell du la fermentation du lactose en milieu complexe dans le diagnostic bactériologique, en particulier des *Enterobacteriaceae*. *Ann. Microbiol.* (Paris) **102**: 267–77.

MacFaddin, J.F. (1980) *Biochemical Tests for the Identification of Medical Bacteria*, 2nd edn. Williams & Wilkins: Baltimore.

McIntyre, J.L., Huber, D., Kuc, J. and Williams, B. (1975) Aminopeptidase profiles of virulent and avirulent *Erwinia amylovora* and *Erwinia herbicola*. *Phytopathology* **65**: 106–21.

Maddocks, J.L. and Greenan, M.J. (1975) A rapid method for identifying bacterial enzymes. *J. Clin. Pathol.* **28**: 686–7.

Miller, J.M. and Wright, J.W. (1982) Spot indole test: Evaluation of four reagents. *J. Clin. Microbiol.* **15**: 589–92.

Muftic, M. (1967) Application of chromogenic substrates to the determination of peptidases in mycobacteria. *Folia Microbiol* (Praha) **12**: 500–7.

Nagasawa, T. and Yamada, H. (1989) Microbial transformations of nitriles. *TIBTECH* **7**: 153–8.

Peczyńska-Czoch, W. and Mordarski, M. (1988) Actinomycete enzymes. In *Actinomycetes in Biotechnology* (Goodfellow, M., Williams, S.T. and Mordarski, M., eds), pp. 219–83. Academic Press: London.

Schofield, G.M. and Schaal, K.P. (1980) Rapid methods for detection of deamination and decarboxylation of amino acids, indole production and reduction of nitrate and nitrite by facultative anaerobic actinomycetes. *Zbl. Bakt. Hyg.* **A247**: 383–91.

Slifkin, M. and Gil, G.M. (1983) Rapid biochemical tests for the identification of groups A, B, C, F and G streptococci from throat cultures. *J. Clin. Microbiol.* **18**: 29–32.

Slots, J. (1981) Enzymatic characterization of some oral and non-oral Gram-negative bacteria with the API-ZYM system. *J. Clin. Microbiol.* **14**: 288–94.

Tharagonnet, D., Sisson, P.R., Roxby, C.H., Ingham, H.R. and Selkon, J.B. (1977) The API-ZYM system in the identification of Gram-negative anaerobes. *J. Clin. Pathol.* **30**: 505–9.

Thompson, L.A., Knowles, C.J., Linton, E.A. and Wyatt, J.M. (1988) Microbial biotransformation of nitriles. *Chemistry in Britain* **24**: 900–2.

Waitkins, S.A., Ball, L.C. and Fraser, C.A.M. (1980) Use of the API-ZYM system in rapid identification of the alpha and non-haemolytic streptococci. *J. Clin. Pathol.* **33**: 53–7.

Wasilauskas, B.L. (1971) Preliminary observations on the rapid differentiation of the *Klebsiella-Enterobacter-Serratia* group on bile-esculin agar. *Appl. Microbiol.* **21**: 162–3.

Watson, R.R. (1976) Substrate specificities of aminopeptidases; specific method for microbial differentiation. *Methods in Microbiology* **9**: 1–14.

Whitehead, D.H. (1990) New strategies for the Isolation and Characterization of Novel Actinomycetes. PhD Thesis, University of Newcastle upon Tyne.

Williams, R.A.D. and Shah, H.N. (1980) Enzyme patterns in bacterial classification and identification. In *Microbiological Classification and Identification* (Goodfellow, M. and Board, R.G., eds), pp. 299–318. Academic Press: London.

14
Analysis of Electrophoretic Whole-Organism Protein Fingerprints

B. Pot, P. Vandamme and K. Kersters
Universiteit Gent, Gent, Belgium

14.1 INTRODUCTION

A microbial cell expresses some 2000 different proteins which form a rich source of information for the characterization, classification and identification of microorganisms. Polyacrylamide gel electrophoresis (PAGE) of cellular proteins yields complex banding patterns which can be considered as highly specific fingerprints of the strains investigated. These protein electrophoregrams are highly reproducible provided strains are cultivated under reproducible conditions and standardized techniques are used. In one-dimensional (1D) gels each protein band usually consists of a number of structurally different proteins having identical electrophoretic mobility.

The value of protein electrophoresis in microbial systematics has been established for over 20 years (for reviews see Jackman, 1985, 1987; Kersters & De Ley, 1980; Kersters, 1985; Vauterin *et al.*, 1993). The rationale for the application of electrophoresis of cellular proteins in microbial systematics is that bacterial strains with 90 to 100% DNA relatedness display only minor differences in their protein fingerprints; strains with at least 70% DNA homology tend to have similarities in their protein electrophoregrams (Owen & Jackman, 1982; Kersters, 1985). Hence, the electrophoretic separation of cellular proteins is a sensitive technique that mainly provides information on the similarity of strains within the same species or subspecies. Depending on the protein electrophoretic variation within a given taxon individual strains can often be recognized by small, but specific and reproducible differences in part of their protein patterns.

A number of different protein electrophoretic techniques are currently

---

Chemical Methods in Prokaryotic Systematics. Edited by M. Goodfellow and A.G. O'Donnell
© 1994 John Wiley and Sons Ltd.

used in microbial systematics:

(a) A mixture of soluble cytoplasmic proteins of a bacterial strain, or proteins solubilized by treatment with a denaturing agent, e.g. sodium dodecyl sulphate (SDS), is submitted to PAGE and stained with a general staining compound for polypeptides. The resultant banding patterns are compared without any attempt to characterize individual protein bands (Figure 14.1).
(b) Native proteins of bacterial strains are submitted to electrophoresis in a suitable support matrix (e.g. starch gel, polyacrylamide gel, or polyacrylamide-agarose mixtures) and the gels stained for specific enzymes. These zymograms are used to compare electrophoretic mobility variants of enzymes between different strains.

Several types of electrophoretic separations in polyacrylamide gels are applied in bacterial systematics: continuous PAGE, discontinuous or disc PAGE, isoelectric focusing, two-dimensional PAGE and gradient PAGE. The following types of bacterial proteins can be compared by PAGE: soluble, native proteins; proteins solubilized by e.g. SDS; cell-envelope and ribosomal proteins; radiolabelled proteins. The present chapter deals only with the most common technique of SDS-PAGE of whole-cell bacterial proteins. Some emphasis will be placed on computer-assisted comparisons of protein electrophoretic fingerprints.

14.2 SDS-PAGE OF WHOLE-CELL BACTERIAL PROTEINS

The SDS-PAGE procedure considered here is based on the techniques described by Laemmli (1976), Jackman (1987), Costas *et al.* (1989) and on our own experience (Pot *et al.*, 1989; Vauterin & Vauterin, 1992). All percentages are expressed as weight per volume (w/v), except when indicated otherwise.

14.2.1 Cultivation of bacteria

The cultivation medium chosen should support good growth of the entire range of bacteria under study. Cells should be harvested in the late exponential or the early stationary phase. It is often convenient to grow bacteria on solid media in large Petri dishes or in Roux flasks, which should be inoculated with fresh precultures. Rich media are recommended and it is important that the pH of the medium remains fairly constant during growth.

14. Electrophoretic Whole-Organism Protein Fingerprints

It is a good practice to check the reproducibility of the technique by recultivating some strains and to compare their protein patterns. A wet weight of 50 to 100 mg of cells is sufficient for the SDS-PAGE technique. For comparative studies, the growth medium, the cell growth stage at harvest, and the protein extraction procedure should of course be standardized to reduce variation in protein electrophoretic patterns.

14.2.2 Preparation and electrophoresis of protein samples

Cells should be harvested by centrifugation at $12\,000 \times g$ and the resulting pellet washed two or three times with sodium phosphate buffer (0.01 M) containing 0.8% NaCl (pH 7.3; sodium phosphate-buffered saline). With gum-producing strains (e.g. xanthomonads), polysaccharides (e.g. xanthans) can usually be removed by suspending and stirring the cells in a relatively large volume of phosphate-buffered saline (100 mg cells (wet weight) in approximately 20 ml buffer).

Samples are prepared using the techniques described in Method 14.1. The procedure for electrophoresis is detailed in Method 14.2.

Method 14.1 Preparation of protein samples.

14.1.A Gram-negative bacteria

1. Suspend washed cells (50 mg wet weight) in 0.9 ml of sample treatment buffer (0.062 M Tris-HCl buffer containing 5% (v/v) mercaptoethanol and 10% (v/v) glycerol; final pH 6.8). Sample preparation can be performed in small (1.5 ml) conical plastic centrifuge tubes (e.g. Eppendorf tubes). Thoroughly mix the suspension.
2. Add 0.1 ml 20% SDS and mix again.
3. Heat the cell suspension at 95 to 100 °C for 10 min (e.g. in an Eppendorf thermostat).
4. Centrifuge samples at $10\,000 \times g$ for 10 min and store the supernatant at −20 °C. Protein samples should preferably be kept at −60 to −80 °C for long-term storage (2 months or longer).

14.1.B Gram-positive bacteria

When cells cannot be lysed by heating in SDS, treatment with lysozyme may be tried.

1. Suspend the bacterial cells in sample treatment buffer (without SDS) containing 0.05 M ethylene diamine tetra-acetate.
2. Add 10 μl lysozyme solution (10 mg lysozyme/0.1 ml 0.05 M Tris-HCl buffer, pH 7.0) per ml suspension, and incubate the mixture for 15 to 30 min (or longer if necessary) at 37 °C.
3. Add SDS as usual and heat for 10 min at 95 °C.

Should lysozyme treatment fail to lyse the cells, methods such as sonication, shaking with glass beads or pressure techniques can be tried prior to SDS treatment.

Such methods usually require a larger amount of cell mass, although a sonicator equipped with a needle-type probe can be used for volumes as small as 1 ml.

Method 14.2 Electrophoresis of protein samples.

Discontinuous gels are run in a vertical slab apparatus (e.g. Hoefer SE600 or LKB 2001); gels are usually 1.0 or 1.5 mm thick. The separation gel is 12.6 cm long and contains 10 or 12% total acrylamide (10 or 12% T) with 2.67% cross-linking (2.67% C). The final concentration of other components in the separation gel are: 0.375 M Tris-HCl (pH 8.8) and 0.1% SDS. The stacking gel is 12 mm long and contains 5% T with 2.67 % C, 0.125 M Tris-HCl (pH 6.8) and 0.1% SDS.

14.2.A Stock solutions for electrophoresis

It is important to dissolve chemicals of the highest possible purity (e.g. 'electrophoresis grade', Bio-Rad; 'Electran grade', BDH) in double-distilled water or in deionized water of high quality, e.g. MilliQ water (resistance 10 MΩ cm^{-1}). We refer to this type of water as 'distilled water'.

1. Acrylamide solution. Acrylamide 29.2 g; N,N'-methylene-bisacrylamide 0.8 g; distilled water to 100 ml. The solution is stable for at least 2 weeks when stored in a dark bottle at 4 °C. Note: acrylamide and its dimer bisacrylamide are toxic compounds.
2. Separation gel buffer. Dissolve 18.15 g Tris in 50 ml distilled water and adjust to pH 8.8 with HCl. Make up to 100 ml with distilled water. Conductivity measurements can be made or volumetric grade acid can be used to enhance the accuracy.
3. Stacking gel buffer. Dissolve 6 g Tris in 50 ml distilled water. Adjust to pH 6.8 with HCl and make up to 100 ml with distilled water.
4. 10% SDS solution. 10 g SDS in 100 ml distilled water.
5. Ammonium persulphate. Make up a 10% solution by adding 1 ml of water to 0.1 g of ammonium persulphate. Always make a fresh solution before use.
6. Tank buffer. Weigh out 12 g Tris, 57.6 g glycine and 4 g SDS. Make up to 4 l with distilled water in a volumetric flask. The pH should be between 8.3 and 8.6. The tank buffer of the upper buffer chamber is prepared freshly before each electrophoretic run and is discarded after each experiment. The tank buffer for the lower buffer chamber should be used for approximately 1 week.

The separation gel buffer and stacking gel buffer should be stored at 4 °C. They are stable for at least 2 weeks. The 10% SDS solution should be kept at room temperature.

14.2.B Preparation of separation gel (12% T, 2.67% C)

1. Assemble cassettes, use clean glass plates and add a small paper label with a unique gel number to e.g. the bottom left corner of each gel cassette. Spacers of 1 mm or 1.5 mm are recommended. Polymerization should be done at a constant temperature which should be identical to the temperature used for the electrophoretic separation (e.g. 10 °C or 20 °C).
2. Add to a clean flask with magnetic stirrer: 26.8 ml distilled water, 20 ml separation gel buffer, 32 ml acrylamide solution and 0.8 ml 10% SDS. Add

14. Electrophoretic Whole-Organism Protein Fingerprints

40 µl N,N,N',N'-tetra-methylethylene diamine (TEMED) and 0.28 ml 10% ammonium persulphate (freshly prepared). Mix thoroughly by stirring. Pour the solution immediately between the plates, and avoid the inclusion of air bubbles. Cassettes should always be filled to constant height (e.g. 12.6 cm). Overlay with 2 ml water-saturated isobutanol (upper layer) to obtain a flat surface. Polymerization should be visible after 10 to 15 min at 20 °C (the speed of polymerization can be adjusted by slightly modifying the amount of TEMED).

3. After 1 h, discard the overlying water-saturated isobutanol; rinse the gel surface at least three times with distilled water. Overlay with 1.6 ml of 1/4 diluted separation gel buffer containing 0.01% SDS.
4. Let the polymerization proceed for 24 h (overnight) in a thermostat-controlled cabinet at e.g. 20 °C. Cover the set (e.g. with a second casting stand) to avoid evaporation.

14.2.C Preparation of stacking gel (5% T, 2.67% C)

Pour the stacking gel at least 1 h before applying the samples.

1. Pour off the liquid layer and rinse the gel surface twice with distilled water. Pour off liquid, invert the casting stand and allow to drain on a paper towel.
2. Mix the following in a clean flask with a magnetic stirrer: 11.3 ml distilled water, 5 ml stacking gel buffer, 3.4 ml acrylamide solution, 0.2 ml 10% SDS, 25 µl TEMED and 0.1 ml 10% ammonium persulphate. Stir thoroughly.
3. Flush the surface of the separation gel with a few millilitres of stacking gel solution, pour off, and fill the cassettes immediately with stacking gel solution. Insert a Teflon comb into each cassette and take care not to trap any air bubbles below the teeth of the combs. Combs forming 15 or 20 slots can be used. Allow the stacking gel to polymerize at room temperature for at least 30 min.
4. Remove the comb and fill the slots with 1/4 diluted stacking gel buffer containing 0.1 % SDS.
5. Just before applying the protein extracts fill the slots with freshly prepared tank buffer, after rinsing them twice with the same buffer.

14.2.D Sample application and electrophoresis

1. Mark the slots with sequence numbers and apply protein extract in each of the slots with the aid of a microsyringe. The volume depends on the concentration of proteins in the extract.
2. The final volume in each slot is adjusted to 15 µl with sample treatment buffer containing 0.001% bromophenol blue (tracking dye).
3. Put the upper buffer chamber in place. Immerse the gel plates in the tank buffer of the lower buffer chamber. Avoid trapping air bubbles under the ends of the cassettes.
4. Pour freshly prepared tank buffer slowly into the upper buffer reservoir taking care not to disturb the loaded slots.
5. Add a spin bar to the lower buffer chamber and place the apparatus on a magnetic stirrer. Circulation of the lower buffer allows the maintenance of a constant temperature during electrophoresis.
6. Connect the cathode to the upper buffer reservoir and run the gel at constant current (10 mA/gel or less depending on cooling facilities).
7. Switch off current when the tracking dye reaches the marker line at e.g. 10 cm from the top of the separation gel.

14.2.E Fixation, staining and destaining

1. Disassemble the cassettes, place the gels in a plastic container and add fresh 3% trichloroacetic acid solution. Shake gently for 30 min.
2. Pour off fixation solution and add staining solution made up of 0.25% Coomassie Blue R-250 in 50% (v/v) methanol and 10% (v/v) acetic acid. Shake gently for at least 1 h.
3. Destain in a solution containing 25% (v/v) methanol and 10% (v/v) acetic acid.
4. Gels can be dried between cellophane sheets on a commercial slab gel dryer. Keep dried gels in a dark, dry place, preferably in a sealed plastic bag.

Silver-staining techniques are beyond the scope of this chapter (Tunon and Johansson, 1984).

14.2.3 Visual comparison of whole-cell protein patterns

Visual comparison is still by far the most frequently used method for the interpretation of bacterial whole-cell protein electrophoretic profiles. Indeed, it is always advisable to compare results of computer analyses with the original gel electrophoretic patterns or their photographs even in cases where computer-assisted analysis is thought to be the most objective method for interpretation. The following factors should be considered for visual interpretation of protein electrophoretic fingerprints.

14.2.3.a Number of protein patterns

It is obvious that a limited number of protein samples applied on one or two polyacrylamide slab gels, which are run in a short period of time using the same buffers and chemicals, are very suitable for visual interpretation. This procedure is a fast and reliable method to pinpoint highly similar patterns. However, comparison of a large number of protein samples is far less objective and often laborious because the reproducibility of the technique has to be checked repeatedly over long periods of time.

14.2.3.b Photographs or plots of the electrophoretic traces

Photographs of protein patterns or plots of densitometric traces can be used to retrieve particular patterns. Such photographs or plots should be normalized for the length of the pattern and each lane should be compared visually against all the others under investigation (Kersters & De Ley, 1975). Groups of similar profiles can thus be delineated and strains can be identified by comparing their electrophoretic fingerprints with those of various reference strains. The use of photographs or plots should always be considered if electrophoretic profiles are to be compared over a long period of time as this will also help to evaluate the level of reproducibility of the electrophoretic system (see Section 14.2.3.e below).

14.2.3.c Type of protein patterns

Visual comparison of protein patterns is very suitable for identification of bacterial strains, but may become laborious when many different groups of bacteria, which produce a variety of protein electrophoretic types, are considered. The objective appreciation of similarities between patterns is extremely difficult by eye, especially when (a) a wide variety of dissimilar patterns are compared and (b) when the profiles contain, despite good overall similarity, a number of dominating protein bands at variable positions. In the latter case, there is a temptation to group patterns according to the matching of the most prominent bands, neglecting or underestimating the overall similarity of the total profile. For these reasons, even very skilled scientists may classify the same set of electrophoretic data in different ways.

14.2.3.d Experience of the investigator

Although it is hard to remember large numbers of protein fingerprints (Feltham & Sneath, 1979), those consistently working with a limited number of electrophoretic types will be able to remember a fairly large number of protein patterns or even to identify a profile at first glance (compare Dijkshoorn et al., 1987a,b with Dijkshoorn et al., 1990). It is therefore advantageous to acquire experience and to include as many reference strains as possible to get acquainted with the natural variation of the protein patterns within the group of bacteria under investigation.

14.2.3.e Reproducibility of the electrophoretic system

It is always advisable to include one or two reference patterns on each slab gel. A mixture of molecular weight markers (containing purified proteins) and/or a reference bacterial protein extract are commonly used as standards to evaluate the reproducibility of the electrophoretic system over a longer period of time. A careful comparison of these standards can reveal major experimental deviations which may distort the results of a visual interpretation. As a final control it is also advisable to group similar profiles on one gel. This will also help to detect patterns which display only minor differences.

14.2.4 Computer-assisted analysis of protein electrophoregrams

Densitometric analysis and computer-assisted comparison of stained banding patterns are necessary for the objective comparison of large numbers of protein extracts. Computer programs have been developed to allow the standardization, normalization, comparison and storage of data (Kersters, 1985; Jackman, 1987; Costas *et al.*, 1989; Pot *et al.*, 1989; Vauterin & Vauterin, 1992). All of these steps will be discussed below and an example will be given of a computer-assisted analysis of 1 D SDS-PAGE protein patterns of 40 strains of the genus *Campylobacter*.

14.2.4.a Densitometry

The first step towards computer-assisted processing of protein electrophoretic patterns is to record the protein concentration along the traces. This is usually done with the aid of a densitometer equipped with a conventional or laser light source (Maurer, 1971; Jackman & Pelczynska, 1986; Jackman, 1987; Costas *et al.*, 1989; Pot *et al.*, 1989). Optical densities are recorded at regular intervals along the traces and the signals are digitized quite often by using an analogue-to-digital (A/D) converting interface linked to a (micro)computer. The number of densitometric values to be recorded per trace depends on the length of the pattern and the resolution of the electrophoretic system. The digitized data are stored on floppy disks or hard disk. The signal can also be fed into a recorder and/or integrator which gives a visual interpretation of the trace on paper with the indication of peak positions, peak numbers and peak areas.

We use an LKB 2202 Ultroscan Laser Densitometer (LKB, Bromma, Sweden), with a resolution of 0.1 mm, which can be either (a) connected to an Apple IIe microcomputer equipped with a Transwarp Accelerator card (Applied Engineering, Carrolton, Texas, USA) via an A/D converter card (LKB, Bromma, Sweden); or (b) linked to a Commodore MS-DOS compatible PC equipped with a 10 kilobyte A/D converter interface card. Raw digitized data are captured and stored with their respective descriptive information (see below) on floppy disk by an adapted version of the program GELSCAN (in Apple Pascal 1.2; LKB, Bromma, Sweden) or by the program GELCOMPAR (in Turbo Pascal; Vauterin & Vauterin, 1992).

Although laser scanners are frequently used, it can be expected that improved high-resolution video apparatus, coupled to more powerful computer systems (Albritton *et al.*, 1988), will become a fast and reliable alternative in the near future.

14.2.4.b Standardization

The use of computers for data capture determines part of the necessary standardization involving the format of (a) the stored digitized traces,

(b) the stored experimental conditions, and (c) the stored sample information. We register patterns of a maximum of 105 mm length as vectors of approximately 1000 measured points. On the computer screen the operator is asked to fill in the following information for each lane to be analysed: name and strain number, indication of type strain and experimental conditions (date of experiment, growth time, growth temperature, growth medium, investigator). This descriptive information is automatically stored with the respective digitized protein profile. The information is labelled with a unique experiment number composed of the gel number and the lane number; this number is also used as the filename.

14.2.4.c Normalization

Normalization involves the standardization of the length of protein patterns to compensate for inevitable small discrepancies within and between gels. The latter are caused by small but notable variations in factors such as the composition of solutions and the conditions of polymerization. For normalization, a reference bacterial extract or a mixture of molecular weight marker proteins has to be included on each slab gel. Digitized traces are brought to equal length by three-point quadratic interpolation techniques (Feltham & Sneath, 1979). At present, two main categories of normalization procedures can be distinguished.

The first is *operator-dependent normalization*. In this process the operator chooses, for every given set of gels, a common profile (a sample pattern or a molecular weight marker profile) which should be included on each gel as an internal reference. One of these references, the standard profile, is continuously kept in the computer memory for comparison with all subsequent reference lanes. Calibration is achieved by operator-directed shrinking and/or stretching of each gel according to the deviation, detected on the computer screen by the operator, between the standard profile and every reference lane separately. Although used with good results (Costas *et al.*, 1987a,b,c, 1989; Costas & Owen, 1987; Holmes *et al.*, 1988; Jackman, 1985, 1987; Owen *et al.*, 1988a,b, 1989a,b), this operator-dependent procedure becomes quite laborious for large numbers of gels or when gels are to be compared over a long period of time.

Besides being group-dependent (for every new set of strains, often a new standard is chosen), this procedure is also operator-dependent as no two operators consistently designate the same standard pattern from all reference patterns, or perform consistently identical shrinking and stretching procedures. Moreover, it is unlikely that one operator, on repetition, will produce identical sets of normalized profiles.

For long-term applications this technique may lead towards lower reproducibility levels and thereby reduce the portability of the system. Consequently, the use of *automated, operator-independent normalization*

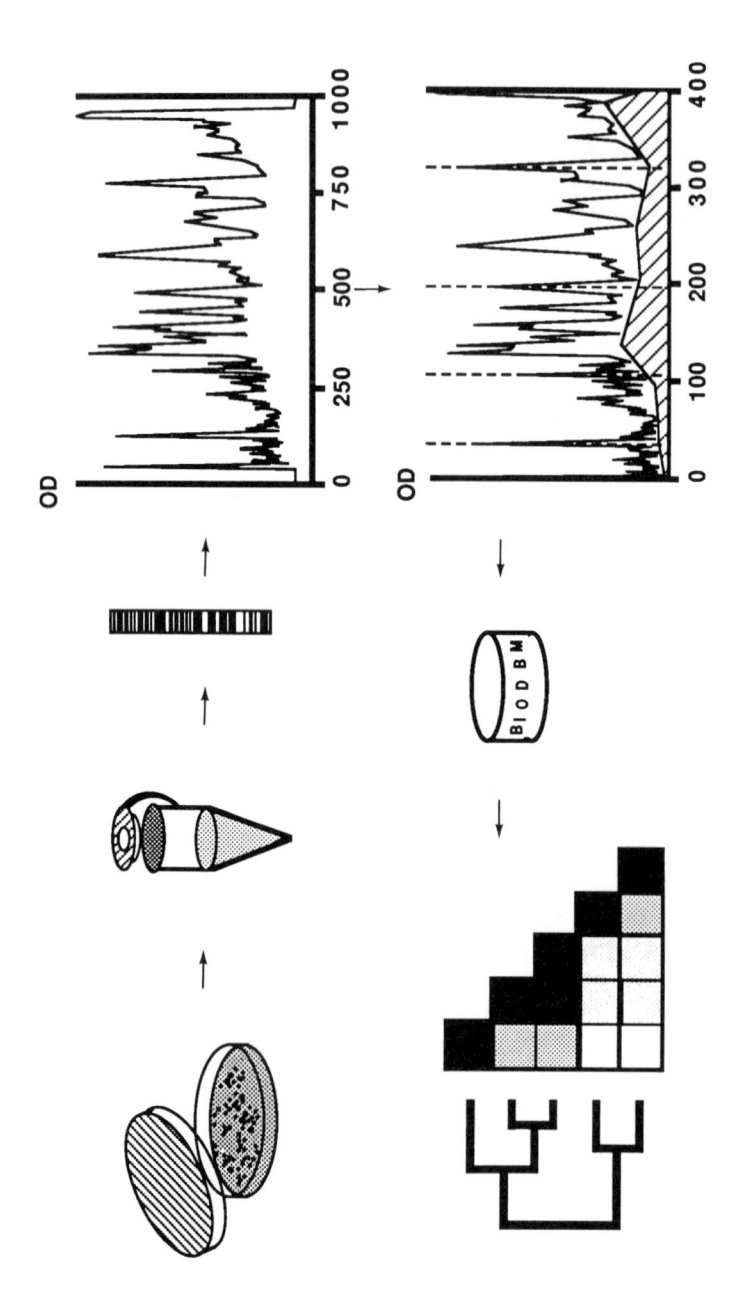

Figure 14.1 Schematic representation of the SDS-PAGE procedure of whole-cell bacterial proteins. Bacterial cells are harvested from Petri dishes and whole-cell protein extracts are prepared. After electrophoretic separation by SDS-PAGE the resulting fingerprint is digitized using a laser densitometer. The 1000 points raw trace (reference strain LMG 1125) is automatically standardized to a shift-corrected 400 points trace on which a background subtraction procedure can be performed. The position of the four reference protein peaks are indicated by dashed lines. The resulting traces can be used for the construction of databases (BIODBM) and for numerical analyses, yielding shaded similarity matrices and dendrograms

procedures is recommended. We developed the programs INTFILE (Pot *et al.*, 1989) and GELCOMPAR (Vauterin & Vauterin, 1992), which use raw digitized data with their respective information as input, and which produce a sequential file of normalized electrophoretic traces that are either transferable to a mainframe computer for further processing (e.g. Siemens 7570-C) or that can be treated on a MS-DOS-compatible personal computer (GELCOMPAR; Vauterin & Vauterin, 1992).

Normalization is achieved by using the bacterial whole-cell protein extract of *Psychrobacter immobilis* LMG 1125 (Pot *et al.*, 1989) as the reference profile. The pattern of this strain consists of a number of well-separated, sharp protein peaks which are well-spread over the entire protein profile in a 12% (T) polyacrylamide gel (Figures 14.1 and 14.2). For use by the program INTFILE, the peak positions of four well-chosen protein bands (Figure 14.1) have been registered for more than 100 electrophoretic runs of the extracts of strain LMG 1125 spread over a long period of time (>1 year) and for eight independent preparations of the bacterial extract. Mean peak positions and respective standard deviations of these reference protein peaks have been calculated and stored in terms of a theoretical (mean) profile in the computer program. This profile is regarded as the standard profile and is used for all further comparisons and alignments. A single electrophoretic run of the extract of strain LMG 1125 is used as the standard with the program GELCOMPAR (Vauterin & Vauterin, 1992). The protein extract of strain LMG 1125 is applied at least four times on each gel to be analysed in order to correct for possible deviations from the standard profile at different locations in the gel. A typical gel for densitometric analysis, containing 18 lanes including the four LMG 1125 lanes, is shown in Figure 14.2.

Normalization is accomplished following the flow chart of the INTFILE program (Figure 14.3) or as described by Vauterin and Vauterin (1992). INTFILE reads the sequence of the traces to be normalized from the screen and fetches the first uncorrected reference profile from the floppy disk. After the determination of the begin and end positions of the reference trace, the program calculates the theoretical peak positions of the four reference protein peaks used for correction (Figure 14.1). A search is then performed within a preset range to determine the real peak positions which will delineate, together with the begin and end position, the five interpolation intervals (Figure 14.1). Any deviation to the theoretical position of the reference peaks will be evaluated. Above a preset value, no interpolation will be performed, the program will warn the user and stop; the defective gel should then be discarded. If the reference trace is found to be acceptable, the program will (a) reduce the 1000 points curve to a 400 points trace using the three-point interpolation algorithm of Lagrange and (b) use this algorithm to perform a linear shrinking or stretching procedure

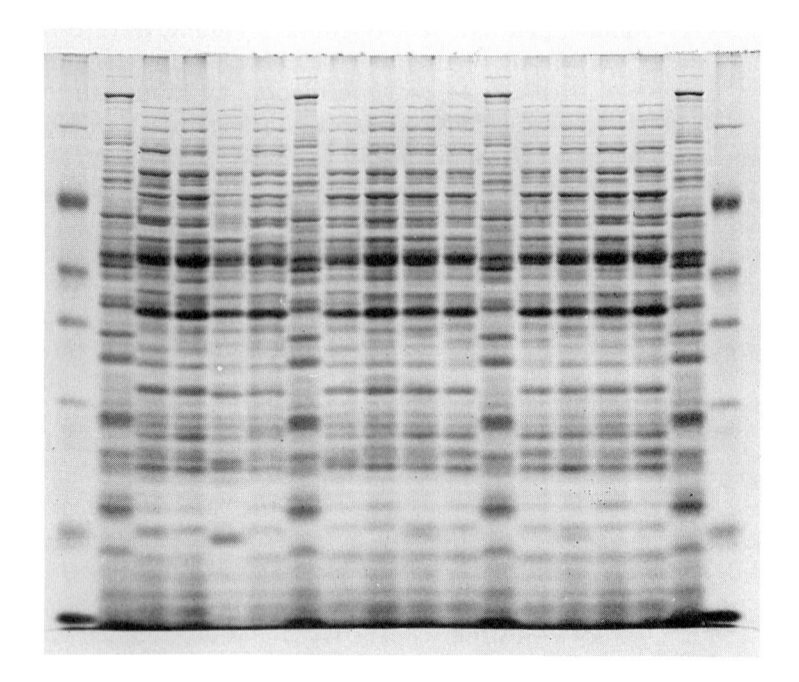

Figure 14.2 Photograph of a typical gel for densitometric analysis. Lanes 1 and 18 contain a mixture of molecular weight markers (from top to bottom: β-galactosidase, 116 000; bovine albumin, 66 000; egg albumin, 45 000; glyceraldehyde-3-phosphate dehydrogenase, 36 000; carbonic anhydrase, 29 000; trypsin inhibitor, 20 100; and lysozyme, 14 500). Lanes 2, 7, 12 and 17 are loaded with the protein extract of the reference strain *Psychrobacter immobilis* LMG 1125. The other lanes contain protein extracts of a selection of highly related *Lactococcus lactis* ssp. *cremoris* strains, which were grouped after visual comparison.

within each of the five interpolation intervals by rigorously maintaining the same number of points per interval. As the total number of points over the five intervals is always 400, the resulting traces are fully normalized and can be compared numerically. Following the interpolation sequence, the neighbouring data lanes of the gel are normalized with the same algorithm using the intervals delineated on the nearest reference trace. After each interpolation, INTFILE performs a linear background subtraction procedure giving the lowest optical density of each trace a value of 1. These data, together with the descriptive information (see above), are then written on a sequential text file and transferred to a mainframe computer for further processing and storage in database format (see below).

The storage of a highly standardized set of uniformly normalized 400 point traces, with the linear background removed (Kersters & Pot, 1988), allows the largest possible flexibility for further processing of the data.

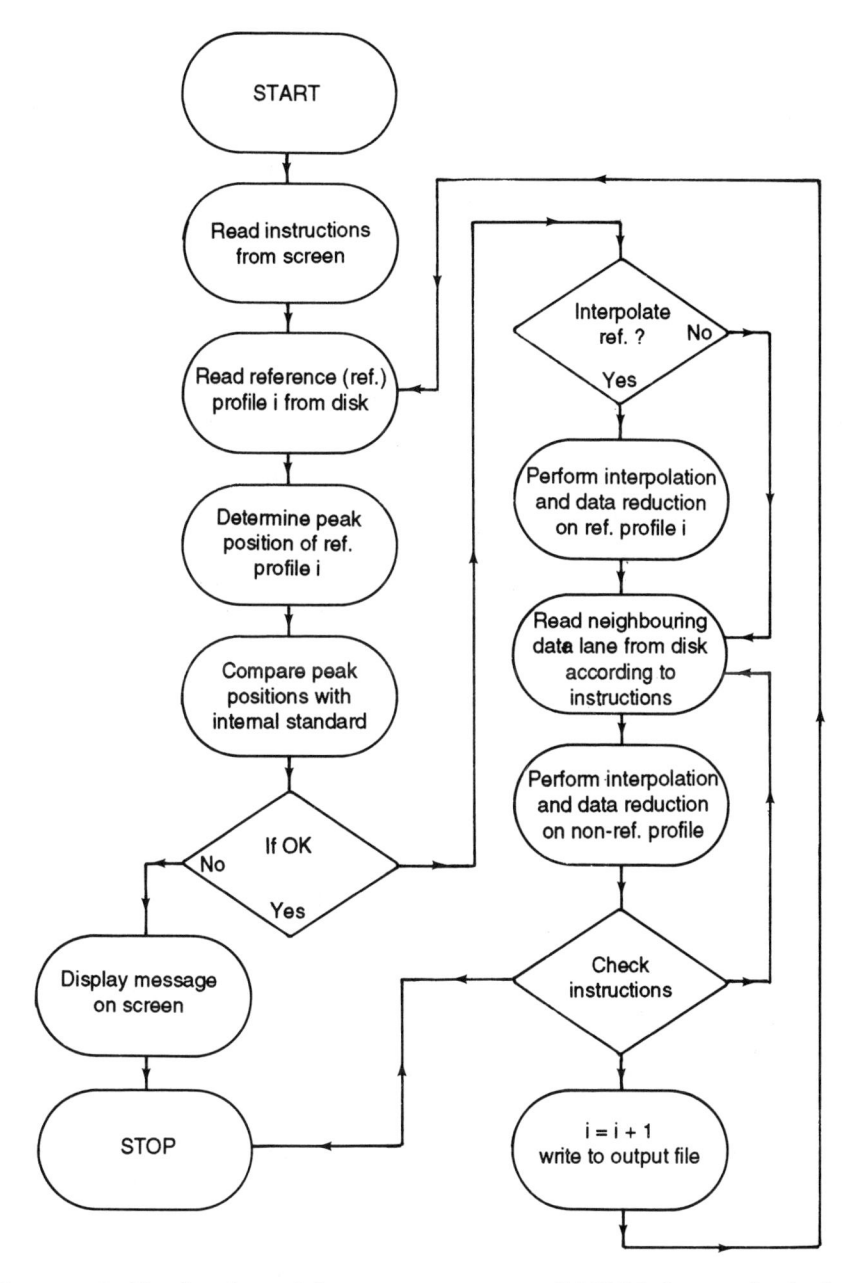

14. *Electrophoretic Whole-Organism Protein Fingerprints* 505

Figure 14.3 The flowchart of the computer program INTFILE. See text for further explanation.

It also enables the level of reproducibility of the electrophoretic system to be continuously measured, even over a long period of time. Therefore, the newly accessed reference profiles of strain LMG 1125 are compared with those already stored in the database. As a high degree of standardization of the electrophoretic technique is required, this type of normalization may seem troublesome and time-consuming in the beginning. However, in the long run it will prove to be time-saving as identification of patterns of various strains can then rely on the taxonomic reference strains stored in the database without the necessity to perform experiments where the taxonomic reference strains have to be run along with the unknowns on slab gels. This property may be an advantage especially in time-limited clinical and epidemiological applications.

In theory a greater precision in alignment may be obtained by increasing the number of intervals, e.g. to 10 or 20. The wide availability of relatively cheap and fast microcomputers has allowed the development of such a program (GELCOMPAR; Vauterin & Vauterin, 1992). Given the inevitable experimental variation in the raw data due to small experimental fluctuations in the preparation of samples, electrophoresis and densitometry, even the finest possible normalization algorithm cannot under all circumstances produce perfectly aligned data. Further improvements of the reproducibility can be obtained by a lateral shift procedure carried out during the calculation of the similarity between individual pairs of traces (see below).

14.2.4.d Computation of similarity

A large number of potentially useful similarity and dissimilarity coefficients have been described (Sneath & Sokal, 1973; Bousfield et al., 1983; Jackman et al., 1983; Sackin, 1987). The most widely used coefficient for comparison of 1 D SDS PAGE protein profiles is the Pearson product moment correlation coefficient (r) which takes into account both peak position and peak intensity (Sneath & Sokal, 1973). The Dice coefficient (Dice, 1945) on the other hand is a similarity coefficient for binary data that are usually derived from the matching of peak positions without taking into account peak intensity. A coefficient has been described that includes both a peak intensity function and the Dice position match function (Albritton et al., 1988). We use routinely the Pearson product moment correlation coefficient (r) (or simply 'correlation coefficient') for the classification and identification of bacteria by means of their protein electrophoretic fingerprints.

The choice of coefficient largely depends on the resolving power and reproducibility of the electrophoretic system employed. The resolution and reproducibility of the technique has increased considerably with the

14. Electrophoretic Whole-Organism Protein Fingerprints

introduction of laser densitometers, high-resolution video apparatus and advanced computing. If the resolution and the reliability of electrophoresis keeps step with this evolution towards the discrimination of larger numbers of bands it is to be expected that classifications and identifications based on protein patterns will tend to rely more on band matching coefficients.

Besides the use of different coefficients, a number of procedures have been developed to increase discrimination between profiles. *Background subtraction* is based on the premise that each bacterial protein fingerprint consists of a series of theoretically non-overlapping peaks superimposed on a non-specific 'background'. A high background level can reduce the discriminatory strength of the protein electrophoretic technique, hence different methods have been described to subtract the background noise. Feltham and Sneath (1979) regarded background as a power series polynomial trend and developed a computer program to calculate trend analyses up to the sixth order. Jackman *et al.* (1983) described a method based on linear regression by subtracting a fraction (0.5–0.8) of a grossly averaged trace consisting of a limited number of points (10% of the total number). Lasters *et al.* (1985) developed an algorithm for the estimation of background trend based on a Fourier smoothing of a selected set of concave kernels of the total profile; although illustrated for the analysis of bacterial whole-cell protein patterns this method has never been tested for the analysis of a large number of strains.

In our laboratory, no 'background trend' is removed systematically. The program INTFILE compensates for small differences in overall absorbance of the gels (different 'background colour') by subtracting the lowest absorbance value minus 1 for the total length of the normalized trace. It is our experience that a simple background removal procedure during data capture or data processing is convenient since it produces uniformly processed traces without loss of information. For some protein profiles, namely ones containing a broad smear of protein bands, the 'background trend' may be relatively high and consequently may disturb the numerical analysis (Pot *et al.*, unpublished data). Removal of this 'background trend' often allows the most discriminative parts of the profile to be detected. For this type of profile, numerical analyses based on background-corrected traces (GEL-COMPAR, version 2.0) yield more reliable results (Vauterin *et al.*, Pot *et al.*, unpublished results). In an interlaboratory comparison of SDS-PAGE whole-cell protein patterns of *Campylobacter* strains (Costas *et al.*, 1990), it was shown that despite the use of two different background subtracting procedures the final groupings of the strains were maintained although there were differences in the overall similarity levels.

A further technique for increasing discrimination is the *removal of major protein bands*. Classification analyses based on the Pearson product moment correlation coefficient have been found to be defective if protein patterns were included which contained one or a few major protein bands that accounted for up to 20% or more of the total protein content (Costas et al., 1987a, 1990; Holmes et al., 1988; Owen et al., 1988a, 1989a). Although identification at the strain level mostly remains possible there is no clear differentiation into the respective species delineated by methods such as e.g. DNA:DNA hybridization. This problem has been overcome by excluding the zone which contains (most of) these heavy bands from the calculation of similarities. This procedure may sound artificial, but its potential usefulness has been demonstrated.

Dijkshoorn et al. (1990) investigated 120 clinical isolates and taxonomic reference strains of the genus *Acinetobacter* by (a) DNA:DNA hybridization and by (b) numerical analysis of SDS-PAGE patterns of cell envelope proteins which displayed a number of major protein bands. It was shown that the total protein pattern was useful for epidemiological purposes. In addition the authors found an acceptable correlation with DNA:DNA hybridization results on the same set of strains by performing a numerical analysis on the first 150 points of the full 400 points trace. Vandamme et al. (1990b) published a comparison of dendrograms derived from analyses of the full protein profile and a restricted part of the pattern. They found that the analysis of the complete protein profiles was an interesting tool for epidemiological purposes as the distribution of the dense bands in the principal band region characterized several intraspecific groups of strains. A perfect correlation was obtained between the electrophoretic and genotypic groups (the latter delineated by DNA:DNA hybridization) when the dense protein band region was deleted. Whether or not major protein bands should be removed from numerical analyses clearly depends on the aims of the study.

The technique of *lateral shifting (best fit on the X-axis)* is widely used (Jackman et al., 1983; Albritton et al., 1988; Costas et al., 1989) to enhance the matching of two normalized traces. The best fit between each pair of traces is obtained by stepwise shifting one normalized trace with respect to the other for up to five positions on either side of the initial alignment (Costas et al., 1989). The drastic rise in calculation time warrants some optimization. We found remarkably improved clustering at the correlation coefficient level $r > 0.80$ based on 750 normalized protein patterns of *Xanthomonas* strains (400 points per trace; Vauterin, unpublished data). An optimal lateral shift was measured by shifting the profiles over a maximum of two points to the left or to the right of the initial alignment. At the lower similarity levels, an optimal shift of three points was more often encountered; this had little or no influence on the final groupings (Vauterin, unpublished results). Optimalization can thus be modified in two ways: (a) depending on the mean correlation level of the analysis, a minimum value of the correlation coefficient at initial alignment can be preset before the lateral shift routine is switched on; (b) the number of points to be shifted should be restricted as a large deviation from the initial alignment position can introduce false 'matches'.

Most of the results above corroborate, in a simplified way, the work of Feltham and Sneath (1979) who were the first to thoroughly investigate the compensation of x-axis distortions using multiple trend analyses. We would like to emphasize that numerical analyses obtained by using one or more of the procedures described above, ought to be published with an appropriate description of the parameters used.

14.2.4.e Grouping of protein patterns (cluster analysis)

The calculation of the similarity between all possible pairs of n patterns yields a matrix of $n(n-1)/2$ similarity values. This matrix is used for grouping electrophoregrams by various clustering techniques (Bergan, 1971) yielding dendrograms or ordered (dis)similarity matrices which are often shaded to enhance readability. Cluster analysis is an explicit way to identify groups in a pool of randomly organized data and helps to find structure in the data (van Tongeren, 1987). However, even if there is a continuous structure in the data, cluster analysis may impose a group structure: a continuum is then arbitrarily partitioned into a discontinuous system of (electrophoretic) types (van Tongeren, 1987). Every user should be aware of this effect when evaluating the results of cluster analyses.

The average linkage clustering technique is most widely used in systematics. With this algorithm the distance between two clusters is given by the average (dis)similarity between all possible pairs of members (one of each group). The unweighted-pair group method using average linkages (UPGMA) (Sokal & Michener, 1958) is generally applied. Many variants using weighted averages, or using single or complete linkage are known. The dendrogram displayed in Figure 14.4 was drawn using UPGMA. Ordination techniques such as principal component analysis (PCA) have been used for the comparison of enzymatic electrophoretic fingerprints (Goullet & Picard, 1987). A heuristic approach to cluster analysis is given by Hogeweg and Hesper, (1981). We use either the CLUSTAN IC program of Wishart (1978) implemented on a Siemens 7570-C mainframe, running under the BS2000 operating system or the program GELCOMPAR. Both software packages allow UPGMA and PCA analyses.

14.2.4.f Construction of protein electrophoretic databases for identification

The construction of protein electrophoretic databases requires a high degree of standardization in experimental methodology. Despite the fact

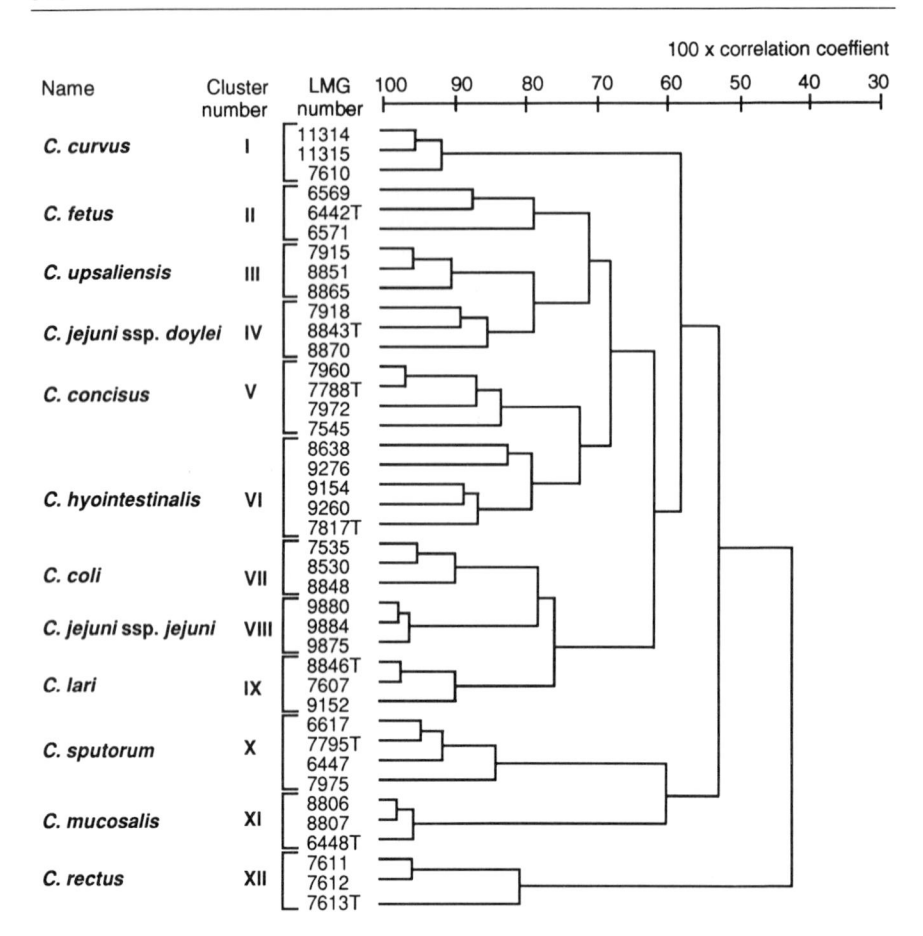

Name	Cluster number	LMG number
C. curvus	I	11314 11315 7610
C. fetus	II	6569 6442T 6571
C. upsaliensis	III	7915 8851 8865
C. jejuni ssp. doylei	IV	7918 8843T 8870
C. concisus	V	7960 7788T 7972 7545
C. hyointestinalis	VI	8638 9276 9154 9260 7817T
C. coli	VII	7535 8530 8848
C. jejuni ssp. jejuni	VIII	9880 9884 9875
C. lari	IX	8846T 7607 9152
C. sputorum	X	6617 7795T 6447 7975
C. mucosalis	XI	8806 8807 6448T
C. rectus	XII	7611 7612 7613T

Figure 14.4 Numerical analysis of the SDS-PAGE protein profiles of 40 strains from eleven different *Campylobacter* species. The resulting dendrogram was calculated using the unweighed pair group method with average linkage of the correlation coefficients (r). Roman numerals indicate cluster numbers.

that many researchers have applied the Laemmli (1970) system, there probably exist as many electrophoretic variants in experimental procedures as there are research groups. Even within a single laboratory, reproducibility and compatibility between the gels of the individual researchers have to be checked continuously and corrected if necessary. If the number of data exceeds a critical mass the construction of a database should be seriously contemplated. The following parameters should be kept in mind.

(a) *Compatibility*. All data should be obtained with one single experimental procedure, processed with a single normalization procedure—

preferably operator—independent (see above)—and have the same format. This implies that traces with different lengths, with different background trends removed, or normalized against different reference strains cannot be included in a single database.

(b) *Consistency.* The digitized fingerprints should be derived from bacterial cells cultivated in highly standardized conditions with respect to growth medium, growth time and growth temperature. If a uniform growth procedure is not possible, the influence of the growth conditions should be estimated by studying duplicated strains to determine overlap. Final classification or identification is only reliable if the variations caused by differences in growth conditions are found to be within the limits of the normal reproducibility of the electrophoretic procedure.

(c) *Reproducibility.* Reference protein patterns have to be included to monitor the reproducibility of the electrophoretic system. These can be molecular weight markers, a standard reference pattern (e.g. strain LMG 1125) or the fingerprint of any other bacterial extract. It is our experience that molecular weight marker lanes tend to lower the reproducibility estimates due to the special shape of their profile. The opposite is true for the standard reference strains, which are individually processed to give a maximum possible match. It is therefore advisable to run a non-reference bacterial extract (mostly this is a type strain of the group under investigation) on every gel that is to be stored in database format. A numerical analysis of all the reference profiles in the database will yield information about overall reproducibility over a long period of time; a run of all multiple non-reference extracts will give a reliable estimate of the reproducibility level for every specific group under investigation. Depending on the nature of the electrophoretic profiles, this latter reproducibility may show some variation. The interlaboratory comparative study of Costas *et al.* (1990) has shown that it is not possible to give precise values for the minimum levels of reproducibility which should be attained as too many parameters are involved. If databases are constructed, the highest possible reproducibility level should be pursued in order to be useful for classification and identification.

Identification of unknown strains can be completely automated once a reliable database is constructed. Most of the literature on identification refers to databases of phenotypic features (Bascomb *et al.*, 1973; Lapage, *et al.*, 1973; Willcox, *et al.*, 1973, 1980; Lapage, 1974; Walczak & Krichevsky, 1982). For 2D electrophoretic protein patterns (O'Farrell, 1975) (rarely used in taxonomy), a number of commercial database systems with identification potentials are available (PDQUEST^{TM}, Protein Databases

Inc., New York; Garrels *et al.*, 1984; Visage™-series, Bio Image, Michigan, USA) but they are beyond the scope of this chapter.

There is little experience with identifications based on 1D electrophoretic protein patterns. As 1D SDS-PAGE is a fingerprinting technique, valuable only at the finest taxonomic levels, a very large number of reference patterns should be available to allow identification at a statistically acceptable level. The large number of reference patterns involved can be handled in two different ways. The first involves the construction of a large number of small databases comprising strains of one genus or even one species. This solution is quite acceptable if (a) unknown profiles encountered are from a restricted origin and if one specific type of microorganism is statistically predominant, or if (b) pre-identification, e.g. to the genus level, is possible using other taxonomically reliable methods. Identification is then achieved by simple comparison of the unknown profile with the profiles of the statistically most relevant database. The use of smaller databases is widespread and the delineation of the database content often depends on the results of taxonomic studies performed on a number of unidentified organisms for which a relationship was found with well-known taxonomic reference strains.

The second procedure involves the construction of a single large database comprising all reference patterns and all unknown profiles. This implies the use of a more powerful computer since data storage capacity needs to be larger and data manipulation time will be longer. The main advantages of this solution are (a) a higher flexibility, since every unknown profile is compared to every profile in the database, including previously unsuccessfully identified patterns, and (b) no pre-identification is needed, shortening the time between arrival and identification.

In our laboratory a Siemens 7570-C implemented FORTRAN 77 program extension (Kindt & Vermeir, 1988) was written based on BIODBM (Biological Database Manager, Casier *et al.*, 1987) for data storage, retrieval and modification of 1D SDS-PAGE protein profiles. At present the database contains approximately 3500 different profiles of yeasts and Gram-positive and Gram-negative bacteria. A UCSD-like PASCAL subprogram (Krols, 1989) was developed to determine specific characteristics of the protein profiles (number of peaks, peak areas, etc.) which are stored along with the original full-length data. These specific characteristics allow fast and reliable identification of protein profiles of unnamed bacterial strains by iterative comparison of a (limited) number of preset characteristics and not by (extensive) calculation of similarities. A final comparison of the strain to be identified with the highest related strains is reported in a number of ways, allowing the user to determine the reliability of the identification.

The program package GELCOMPAR offers the possibility of data storage and allows the construction of specific library entries for well-delineated electrophoretic groups. These libraries are useful for fast identification of unknown electrophoretic patterns (Vauterin & Vauterin, 1992).

14.3 TAXONOMIC CONSIDERATIONS AND APPLICATIONS

Computer-assisted comparison of digitized SDS-PAGE protein patterns is a widely accepted technique for comparison of various biological materials. In microbiology and biotechnology, microbial protein electrophoregrams have been used for:

(a) rapid classification and identification of strains
(b) rapid verification of strain authenticity
(c) detection of contaminants and colony variants of pure cultures
(d) monitoring of protein compositions
(e) ecological and epidemiological studies
(f) screening for the selection of biotechnologically interesting strains
(g) construction of digitized identity cards.

Since the review by Kersters (1985) a large number of bacterial genera and species have been thoroughly studied by electrophoresis of proteins. Examples include *Acidovorax* (Willems *et al.*, 1990), *Acinetobacter* (Dijkshoorn *et al.*, 1987a,b, 1990), *Alcaligenes* (Kiredjian *et al.*, 1986), *Azospirillum halopraeferens* (Reinhold *et al.*, 1987), *Bacillus licheniformis* and *B. polymyxa* (Raspoet *et al.*, 1990); *Clostridium difficile* (Mulligan *et al.*, 1986, 1988), *Corynebacterium* (Jackman & Pelczynska, 1986), *Fusobacterium ulcerans* (Adriaans & Shah, 1988), *Hydrogenophaga* (Willems *et al.*, 1989), *Lactobacillus* (Johnson *et al.*, 1987), *Leuconostoc* (Dicks *et al.*, 1990), *Listeria* (Plikaytis *et al.*, 1986), *Mycoplasma mycoides* (Costas *et al.*, 1987b; Leach *et al.*, 1989) and *M. hominis* (Andersen *et al.*, 1987), *Oceanospirillum* (Pot *et al.*, 1989), *Providencia rustigianii* and *P. alcalifaciens* (Costas *et al.*, 1987a; Holmes *et al.*, 1988), *Saccharomyces* (van Vuuren & Van der Meer, 1987), *Staphylococcus aureus* (Costas *et al.*, 1989), *Streptobacillus moniliformis* (Costas & Owen, 1987), *Streptococcus* (Whiley & Hardie, 1989) and *Streptomyces* and *Streptoverticillium* (Manchester *et al.*, 1990). For some genera there has been an exponential growth of interest: *Campylobacter, Wolinella, Bacteroides* (Ferguson & Lambe, 1984; Pearson *et al.*, 1984; Regbie & Stewart, 1984; Mégraud *et al.*, 1985; Tanner, 1986; Taylor *et al.*, 1987; Ohya *et al.*, 1988; Flores *et al.*, 1989; Vandamme *et al.*, 1989, 1990a,b,c; Goossens *et al.*, 1990) and *Xanthomonas* (Vera Cruz *et al.*, 1984; Thaveechai & Schaad, 1986; Van den Mooter *et al.*, 1987a,b,c; Vauterin *et al.*, 1990, 1993).

An example of the use of SDS-PAGE of whole-cell protein extracts for the classification of members of the genus *Campylobacter* is given below.

Table 14.1 *Campylobacter* strains investigated.

Organism	Strain number	Other strain designations	Strain origin
Campylobacter coli	LMG 8848	NCTC 11353	Porcine placenta (1978)
	LMG 7535	CCUG 10369	Porcine placenta (1978)
	LMG 8530	CCUG 8320	Porcine faeces (1979, Sweden)
Campylobacter concisus	LMG 7788^T	CCUG 13144^T	Gingival sulcus, human (1974, USA)
	LMG 7960	CCUG 20535	Gingival sulcus, human (1974, USA)
	LMG 7972	CCUG 10376	Blood, man with carcinoma of the bronchus (1980, UK)
Campylobacter curvus	LMG 7610	CCUG 11644	Human dental root canal (1981, Sweden)
	LMG 11314	Lauwers CA4118	Human faeces (1991, Belgium)
	LMG 11315	Lauwers CA4125	Human faeces (1991, Belgium)
Campylobacter fetus	LMG 6442^T	NCTC 10842^T	Brain, sheep foetus
	LMG 6571	NIDO 2125/4	Genital, bull (1985, Belgium)
	LMG 6569	NIDO 7572	Bovine foetus (Belgium)
Campylobacter hyointestinalis	LMG 7817^T	CCUG 14169^T	Porcine intestine (1980, USA)
	LMG 8638	ADRI 1048	Faeces, healthy calf (1982, USA)
	LMG 9154	CCUG 20825	
Campylobacter jejuni ssp. *jejuni*	LMG 9880	CCUG 12078	Chicken
	LMG 9884	CCUG 15114	Human
	LMG 9875	CCUG 12067	Human
Campylobacter jejuni ssp. *doylei*	LMG 8843^T	NCTC 11951^T	Faeces, child with diarrhoea (1982–1983, Australia)
	LMG 7918	Goossens M373	Faeces, child with bloody diarrhoea (1986, Belgium)
	LMG 8870	NCTC A613/87	Blood, child (1986, South Africa)
Campylobacter lari	LMG 8846^T	NCTC 11352^T	Cloacal swab, *Larus argentus*
	LMG 7607	CCUG 12774	Faeces, child
	LMG 9152	CCUG 25266	Equine intestine (1981, Sweden)

Campylobacter mucosalis	LMG 6448[T]	CCUG 6822[T]	Porcine jejunum; serotype A (1972, UK)
	LMG 8806	Lawson 1331/76	Porcine; serotype A (1976, UK)
	LMG 8807	Lawson 1075/78	Ileum of pig with adenomatous lesions; serotype A (1978, UK)
Campylobacter rectus	LMG 7611	CCUG 11640	Human dental root canal (1981, Sweden)
	LMG 7612	CCUG 11642	Human dental root canal (1981, Sweden)
	LMG 7613[T]	FDC 371[T]	Human periodontal pocket (1974, USA)
Campylobacter sputorum			
bv. *sputorum*	LMG 7795[T]	CCUG 9728[T]	Human mouth
bv. *bubulus*	LMG 6447	CCUG 11289	Bull sperm (Belgium)
bv. *faecalis*	LMG 6617	CCUG 12015	Ovine faeces
Campylobacter upsaliensis	LMG 7915	CCUG 20818	Blood, child with bronchiolitis (1983, USA)
	LMG 8851	NCTC 11840	Faeces, patient with diarrhoea (UK)
	LMG 8865	NCTC A692/88	Blood, child
Campylobacter sp.	LMG 9260	Goossens K169	Human faeces (1986, Belgium)
	LMG 9276	Goossens G199	Human faeces (1986, Belgium)
	LMG 7545	CCUG 17580	Faeces, child with persistent diarrhoea (1985, Sweden)
	LMG 7975	CCUG 20706	Porcine intestine (UK)

[T] Type strain.

Abbreviations: ADRI, Animal Diseases Research Institute, Nepean, Ontario, Canada; CCUG, Culture Collection of the University of Göteborg, Department of Clinical Microbiology, University of Göteborg, Göteborg, Sweden; FDC, Forsyth Dental Center, Boston, USA; LMG, Culture Collection Laboratorium voor Microbiologie Gent, Universiteit Gent, Gent, Belgium; NCTC, National Collection of Type Cultures, Central Public Health Laboratory Services, London, UK; NIDO, Nationaal Instituut voor Diergeneeskundig Onderzoek, Brussels, Belgium; H, Goossens, Department of Microbiology, University Hospital, Antwerp, Belgium; G, Lawson, Department of Veterinary Pathology, University of Edinburgh, Edinburgh, UK; S. Lauwers, Microbiology Department Academisch Ziekenhuis—Vrije Universiteit Brussel, Brussels, Belgium.

The strains listed in Table 14.1 were chosen from a database consisting of about 800 different *Campylobacter* strains (Goossens *et al.*, 1990; Vandamme *et al.*, 1989, 1990a,b,c; P. Vandamme, unpublished results). The genus *Campylobacter* comprises eleven genuine species (*C. coli, C. concisus, C. curvus, C. fetus, C. hyointestinalis, C. jejuni* ssp. *jejuni* and ssp. *doylei, C. lari, C. mucosalis, C. rectus, C. sputorum,* and *C. upsaliensis*). According to Thompson *et al.* (1988) and Vandamme *et al.* (1991) they are all members of a single rRNA homology group. DNA : DNA hybridization (Roop *et al.*, 1984, 1985; Steele & Owen, 1988) revealed significant DNA binding values only between *C. jejuni* and *C. coli, C. fetus* and *C. hyointestinalis,* and between *C. jejuni* ssp. *jejuni* and *C. jejuni* ssp. *doylei*.

The test strains included four unidentified or misnamed *Campylobacter* strains (Table 14.1) and three representatives of each taxon. Twelve clusters were obtained (Figure 14.4). Clusters I through IV consisted of all of the strains of *C. curvus, C. fetus, C. upsaliensis* and *C. jejuni* ssp. *doylei,* respectively. Cluster V contained the three *C. concisus* strains and strain LMG 7545 which was received as *Campylobacter* sp. Vandamme *et al.* (1991) found that strain LMG 7545 showed 68% DNA binding with the type strain of *C. concisus* thereby confirming the grouping by SDS-PAGE. The three *C. hyointestinalis* strains grouped in cluster VI together with strains LMG 9260 and LMG 9276 (both received as *Campylobacter* sp.). In the corresponding DNA : DNA hybridization studies, both strains were shown to be highly related with the type strain of *C. hyointestinalis* (97 and 94% DNA binding, respectively). Clusters VII, VIII and IX encompassed the representative strains of *C. coli, C. jejuni* ssp. *jejuni,* and *C. lari,* respectively. *Campylobacter coli* was found to be closest ($r=0.78$) to *C. lari* and *C. jejuni* ssp. *jejuni* ($r=0.80$). The *C. mucosalis* strains formed a homogeneous cluster XI, except strain LMG 7975, which was also received as *C. mucosalis* but grouped at $r=0.83$ with the *C. sputorum* strains investigated (cluster X). The DNA : DNA hybridization data confirmed that strain LMG 7975 belongs to *C. sputorum* (83% DNA binding). Finally, cluster XII was formed by the three *C. rectus* strains.

These results prove that identification of unnamed bacteria, or the detection of misnamed organisms, is possible with a restricted set of well-known taxonomic reference strains. For classification purposes, however, a more comprehensive set of taxonomically well-characterized reference strains is required. Generally a good correlation exists between groupings obtained by SDS-PAGE and DNA : DNA hybridization techniques, at least above ±70% DNA binding. In the *Campylobacter* example lower but significant DNA homology values were reported between e.g. *C. jejuni* and *C. coli* but also between *C. fetus* and *C. hyointestinalis*. Such genotypic relationships are clearly not always reflected in the numerical analysis of the electrophoretic protein profiles (Figure 14.4).

ACKNOWLEDGEMENTS

Part of this research was carried out in the framework of contracts BAP-0138-B, BIOT-CT91-0263 and BIOT-CT91-0294 of the Commission of the European Communities. PV is indebted to the Nationaal Fonds voor Wetenschappelijk Onderzoek (NFWO), Belgium, for a position as senior research assistant. KK is grateful to the FGWO for research and personnel grants.

REFERENCES

Adriaans, B. and Shah, H. (1988) *Fusobacterium ulcerans* sp. nov. from tropical ulcers. *Int. J. Syst. Bacteriol.* **38**: 447–8.
Albritton, W.L., Chen, X.P. and Khanna, V. (1988) Comparison of whole-cell protein electrophoretic profiles of *Haemophilus influenzae*: Implementation of a microcomputer mainframe linked system and description of a new similarity coefficient. *Can. J. Microbiol.* **34**: 1129–34.
Andersen, H., Birkelund, S., Christiansen, G. and Freundt, E.A. (1987) Electrophoretic analysis of proteins from *Mycoplasma hominis* strains detected by SDS-PAGE, two-dimensional gel electrophoresis and immunoblotting. *J. Gen. Microbiol.* **133**: 181–91.
Bascomb, S., Lapage, S.P., Curtis, M.A. and Willcox, W.R. (1973) Identification of bacteria by computer: Identification of reference strains. *J. Gen. Microbiol.* **77**: 291–315.
Bergan, T. (1971) Survey of numerical techniques for grouping. *Bacteriol. Rev* **35**: 379–89.
Bousfield, I.J., Smith, G. L., Dando, T. R. and Hobbs, G. (1983) Numerical analysis of total fatty acid profiles in the identification of coryneform, nocardioform and some other bacteria. *J. Gen. Microbiol.* **129**: 375–94.
Casier, P., Meire, P., Develter, D. and Hermy, M. (1987) BIODBM-Biological Database Manager, een voorstudie. Internal report of 'Laboratorium voor oecologie der dieren, zoogeografie en natuurbehoud—Laboratorium voor morfologie, systematiek en ecologie van de planten. Faculteit Wetenschappen; Universiteit Gent.'
Costas, M. and Owen, R. J. (1987) Numerical analysis of electrophoretic protein patterns of *Streptobacillus moniliformis*. *J. Med. Microbiol.* **23**: 303–11.
Costas, M., Holmes, B. and Sloss, L.L. (1987a) Numerical analysis of electrophoretic protein patterns of *Providencia rustigianii* strains from human diarrhoea and other sources. *J. Appl. Bacteriol.* **63**: 319–28.
Costas, M., Leach, R.H. and Mitchelmore, D.L. (1987b) Numerical analysis of PAGE protein patterns and the taxonomic relationships within the "*Mycoplasma mycoides* cluster"'. *J. Gen. Microbiol.* **133**: 3319–29.
Costas, M., Owen, R.J. and Jackman, P.J.H. (1987c) Classification of *Campylobacter sputorum* and allied campylobacters based on numerical analysis of electrophoretic protein patterns. *Syst. Appl. Microbiol.* **9**: 125–31.
Costas, M., Cookson, B.D., Talsania, H.G. and Owen, R.J. (1989) Numerical analysis of electrophoretic protein patterns of methicillin-resistant strains of *Staphylococcus aureus*. *J. Clin. Microbiol.* **27**: 2574–81.
Costas M., Pot, B., Vandamme, P., Kersters, K., Owen, R.J. and Hill, L.R. (1990) Interlaboratory comparative study of the numerical analysis of one-dimensional sodium dodecyl sulphate-polyacrylamide gel electrophoretic protein patterns of *Campylobacter* strains. *Electrophoresis* **11**: 467–74.
Dice, L.R. (1945) Measure of the amounts of ecological association between species. *Ecology* **26**: 297–302.
Dicks, L.M.T., van Vuuren, H.J.J. and Dellaglio, F. (1990) Taxonomy of *Leuconostoc* species, particularly *Leuconostoc oenos*, as revealed by numerical analysis of total soluble cell protein

patterns, DNA base composition, and DNA : DNA hybridizations. *Int. J. Syst. Bacteriol.* **40**: 83–91.
Dijkshoorn, L., Michel, M.F. and Degener, J.E. (1987a) Cell envelope protein profiles of *Acinetobacter calcoaceticus* strains isolated in hospitals. *J. Med. Microbiol.* **23**: 313–19.
Dijkshoorn, L., van Vianen, W., Degener, J.E. and Michel, M.F. (1987b) Typing of *Acinetobacter calcoaceticus* strains isolated from hospital patients by cell envelope protein profiles. *Epidem. Infect.* **99**: 659–67.
Dijkshoorn, L., Tjernberg, I., Pot, B., Michel, M.F., Ursing, J. and Kersters, K. (1990) Numerical analysis of cell envelope protein profiles of *Acinetobacter* strains classified by DNA-DNA hybridization. *Syst. Appl. Microbiol.* **13**: 338–44.
Feltham, R.K.A. and Sneath, P.H.A. (1979) Quantitative comparison of electrophoretic traces of bacterial proteins. *Comp. Biomed. Res.* **12**: 247–63.
Ferguson, D.A., Jr. and Lambe, D.W., Jr. (1984) Differentiation of *Campylobacter* species by protein banding patterns in polyacrylamide slab gels. *J. Clin. Microbiol.* **20**: 453–60.
Flores, B.M., Fennell, C.L. and Stamm, W.E. (1989) Characterization of *Campylobacter cinaedi* and *C. fennelliae* antigens and analysis of the human immune response. *J. Infect. Dis.* **159**: 635–40.
Garrels, J.I., Farrar, J.T. and Burwell IV, C.B. (1984) The QUEST system for computer-analyzed two-dimensional electrophoresis of proteins. In *Two-dimensional Gel Electrophoresis of Proteins—Methods and Applications* (Celis, J.E. and Bravo, R., eds), pp. 37–91. Academic Press: London.
Goossens, H., Pot, B., Vlaes, L., Van den Borre, C., Van den Abbeele, R., Van Naelten, C., Levy, J., Cogniau, H., Marbehant, P., Verhoef, J., Kersters, K., Butzler, J.-P. and Vandamme, P. (1990). Characterization and description of '*Campylobacter upsaliensis*' isolated from human feces. *J. Clin. Microbiol.* **26**: 1039–46.
Goullet, Ph. and Picard, B. (1987) Differentiation of *Shigella* by esterase electrophoretic polymorphism. *J. Gen. Microbiol.* **133**: 1005–17.
Hogeweg, P. and Hesper, B. (1981) Oligothetic characterisation of clusters. *Pattern Recog.* **14**: 131–6.
Holmes, B., Costas, M. and Sloss, L.L. (1988) Numerical analysis of electrophoretic protein patterns of *Providencia alcalifaciens* strains from human faeces and veterinary specimens. *J. Appl. Bacteriol.* **64**: 27–35.
Jackman, P.J.H. (1985) Bacterial taxonomy based on electrophoretic whole-cell protein patterns. In *Chemical Methods in Bacterial Systematics* (Goodfellow, M. and Minnikin, D.E., eds), pp. 115–29. Academic Press: London.
Jackman, P.J.H. (1987) Microbial systematics based on electrophoretic whole-cell protein patterns. *Methods in Microbiology* **19**: 209–25.
Jackman, P.J.H. and Pelczynska, S. (1986) Characterization of *Corynebacterium* group JK by whole-cell protein patterns. *J. Gen. Microbiol.* **132**: 1911–5.
Jackman, P.J.H., Feltham, R.K.A. and Sneath, P.H.A. (1983) A program in BASIC for numerical taxonomy of microorganisms based on electrophoretic protein patterns. *Microb. Lett.* **23**: 87–98.
Johnson, M.C., Ray, B. and Bhowmik, T. (1987) Selection of *Lactobacillus acidophilus* strains for use in 'acidophilus products'. *Antonie van Leeuwenhoek* **53**: 215–31.
Kersters, K. (1985) Numerical methods in the classification of bacteria by protein electrophoresis. In *Computer-assisted Bacterial Systematics* (Goodfellow, M., Jones, D. and Priest, F.G., eds), pp. 337–68. Academic Press: London.
Kersters, K. and De Ley, J. (1975) Identification and grouping of bacteria by numerical analysis of their electrophoretic protein patterns. *J. Gen. Microbiol.* **87**: 333–42.
Kersters, K. and De Ley, J. (1980) Classification and identification of bacteria by electrophoresis of their proteins. In *Microbiological Classification and Identification* (Goodfellow, M. and Board, R.G., eds), pp. 273–97. Academic Press: London.
Kersters, K. and Pot, B. (1988) Electrophoresis of proteins: Data capture, analysis and construction of databanks. In *Biotechnology Action Programme (B.A.P.) Progress Report*, vol. 2. (Magnien, E., ed), pp. 9–17. ECSC-EEC-EAEC: Brussels.
Kindt, G. and Vermeir, J.C. (1988) Databanksysteem poor het beheren van eiwitprofielen. Master Thesis in Informatics, State University Gent, Belgium.

Kiredjian, M., Holmes, B., Kersters, K., Guilvout, I. and De Ley, J. (1986) *Alcaligenes piechaudii*, a new species from human clinical specimens and the environment. *Int. J. Syst. Bacteriol.* **36**: 282–7.

Krols, P. (1989) Programmatuur ter ondersteuning van de identificatie van microorganismen. Master Thesis in Informatics, State University Gent, Belgium.

Laemmli, U.K. (1970) Cleavage of structural proteins during the assembly of the head of bacteriophage T4. *Nature* (London) **227**: 680–5.

Lapage, S.P. (1974) Practical aspects of probabilistic identification of bacteria. *Int. J. Syst. Bacteriol.* **24**: 500–7.

Lapage, S.P., Bascomb, S., Willcox, W.R. and Curtis, M.A. (1973) Identification of bacteria by computer: General aspects and perspectives. *J. Gen. Microbiol.* **77**: 273–90.

Lasters, I., Leyns, F. and Jackman, P.J.H. (1985) Background estimation in one-dimensional electrophoregrams of whole-cell protein extracts. *Electrophoresis* **6**: 508–11.

Leach, R.H., Costas, M. and Mitchelmore, D.L. (1989) Relationship between *Mycoplasma mycoides* subsp. *mycoides* ('large-colony' strains) and *M. mycoides* subsp. *capri*, as indicated by numerical analysis of one-dimensional SDS-PAGE protein patterns. *J. Gen. Microbiol.* **135**: 2993–3000.

Manchester, L., Pot, B., Kersters, K. and Goodfellow, M. (1990) Classification of *Streptomyces* and *Streptoverticilium* species by numerical analysis of electrophoretic protein patterns. *Syst. Appl. Microbiol.* **13**: 333–7.

Maurer, H.R. (1971) *Disc Electrophoresis and Related Techniques of Polyacrylamide Gel Electrophoresis*, 2nd edn. W. de Gruyter: Berlin/New York.

Megraud, F., Bonnet, F., Garnier, M. and Lamouliatte, H. (1985) Characterization of '*Campylobacter pyloridis*' by culture, enzymatic profile and protein content. *J. Clin. Microbiol.* **22**: 1007–10.

Mulligan, M.E., Halebian, S., Kwok, R.Y.Y., Cheng, W.C., Finegold, S.M., Anselmo, C.R., Gerding, D.N. and Peterson, L.R. (1986) Bacterial agglutination and polyacrylamide gel electrophoresis for typing *Clostridium difficile*. *J. Infect. Dis.* **153**: 267–71.

Mulligan, M.E., Peterson, L.R., Kwok, R.Y.Y., Clabots, C.R. and Gerding, D.N. (1988) Immunoblots and plasmid fingerprints compared with serotyping and polyacrylamide gel electrophoresis for typing *Clostridium difficile*. *J Clin. Microbiol.* **26**: 41–6.

O'Farrell, P.H. (1975) High resolution two-dimensional electrophoresis of proteins. *J. Biol Chem.* **250**: 4007–21.

Ohya, T., Kubo, M. and Watase, H. (1988) Electrophoretic protein patterns in *Campylobacter* species with special reference to *Campylobacter mucosalis* and *Campylobacter hyointestinalis*. *Jap. J. Vet. Sci.* **50**: 692–8.

Owen, R.J. and Jackman, P.J.H. (1982) The similarities between *Pseudomonas paucimobilis* and allied bacteria derived from analysis of deoxyribonucleic acids and electrophoretic protein patterns. *J. Gen. Microbiol.* **128**: 2945–54.

Owen, R.J., Costas, M. and Sloss, L.L. (1988a) Electrophoretic protein typing of *Campylobacter jejuni* subsp. '*doylei*' (nitrate-negative *Campylobacter*-like organism) from human faeces and gastric mucosa. *Eur. J. Epidem.* **4**: 277–83.

Owen, R.J., Costas, M., Sloss, L.L. and Bolton, F.J. (1988b) Numerical analysis of electrophoretic protein patterns of *Campylobacter laridis* and allied thermophilic campylobacters from the natural environment. *J. Appl. Bacteriol.* **65**: 69–78.

Owen, R.J., Morgan, D.D., Costas, M. and Lastovica, A. (1989a) Identification of '*Campylobacter upsaliensis*' and other catalase-negative campylobacters from pediatric blood cultures by numerical analysis of electrophoretic protein patterns. *FEMS Microbiol. Lett.* **58**: 145–50.

Owen, R.J., Costas, M., Morgan, D.D., On, S.L.W., Hill, L.R., Pearson, A.D. and Morgan, D.R. (1989b) Strain variation in *Campylobacter pylori* detected by numerical analysis of one-dimensional electrophoretic protein patterns. *Antonie van Leeuwenhoek* **55**: 253–67.

Pearson, A.D., Bamforth, J., Booth, L., Holdstock, G., Ireland, A., Walker, C., Hawtin, P. and Millward-Sadler, M. (1984). Polyacrylamide gel electrophoresis of spiral bacteria from the gastric antrum. *Lancet* **i**: 1349–50.

Plikaytis, B.D., Carlone, G.M. and Plikaytis, B.B. (1986) Numerical analysis of normalized whole-cell protein profiles after sodium dodecyl sulphate-polyacrylamide gel electrophoresis. *J. Gen. Microbiol.* **132**: 2653–60.

Pot, B., Gillis, M., Hoste, B., Van de Velde, A., Bekaert, F., Kersters, K. and De Ley, J. (1989) Intra- and intergeneric relationships of the genus *Oceanospirillum*. *Int. J. Syst. Bacteriol.* **39**: 23–34.

Raspoet, D., Pot, B., De Deyn, D., De Vos, P., Kersters, K. and De Ley, J. (1990) Differentiation of 2,3-butanediol producing *Bacillus licheniformis* and *B. polymyxa* strains by fermentation product profiles and whole-cell protein electrophoretic patterns. *Syst. Appl. Microbiol.* **14**: 1–7.

Regbie, R. and Stewart, C.S. (1984) Polyacrylamide gel electrophoresis of *Bacteroides succinogenes*. *Can. J. Microbiol.* **30**: 863–6.

Reinhold, B., Hurek, T., Fendrik, I., Pot, B., Gillis, M., Kersters, K., Thielemans, S. and De Ley, J. (1987) *Azospirillum halopraeferens* sp. nov., a nitrogen-fixing organism associated with roots of Kallar grass (*Leptochloa fusca* (L.) Kunth). *Int. J. Syst. Bacteriol.* **37**: 43–51.

Roop, R.M., II, Smibert, R.M., Johnson, J.L. and Krieg, N.R. (1984) Differential characteristics of catalase-positive campylobacters correlated with DNA homology groups. *Can. J. Microbiol.* **30**: 938–51.

Roop, R.M., II, Smibert, R.M., Johnson, J.L. and Krieg, N.R. (1985) DNA homology studies of the catalase-negative campylobacters and '*Campylobactier fecalis*', an emended description of *Campylobacter sputorum*, and proposal of the neotype strain of *Campylobacter sputorum*. *Can. J. Microbiol.* **31**: 823–31.

Sackin, M.J. (1987) Computer programs for classification and identification. In *Methods in Microbiology* (Colwell, R.R. and Grigorova, R., eds), pp. 460–94. Academic Press: London.

Sneath, P.H.A. and Sokal, R.R. (1973) *Numerical Taxonomy: The Principles and Practice of Numerical Classification*. W. H. Freeman: San Francisco.

Sokal, R.R. and Michener, C.D. (1958) A statistical method for evaluating systematic relationships. *University of Kansas Scientific Bulletin* **38**: 1409–38.

Steele, T.W. and Owen, R.J. (1988) *Campylobacter jejuni* subsp. *doylei* subsp. nov., a subspecies of nitrate-negative campylobacters isolated from human clinical specimens. *Int. J. Syst. Bacteriol.* **38**: 316–8.

Tanner, A.C.R. (1986) Characterization of *Wolinella* spp., *Campylobacter concisus*, *Bacteroides gracilis*, and *Eikenella corrodens* by polyacrylamide gel electrophoresis. *J. Clin. Microbiol.* **24**: 562–5.

Taylor, A.J., Costas, M. and Owen, R.J. (1987) Numerical analysis of electrophoretic protein patterns of *Bacteroides ureolyticus* clinical isolates. *J. Clin. Microbiol.* **25**: 660–6.

Thaveechai, N. and Schaad, N.W. (1986) Serological and electrophoretic analysis of a membrane protein extract of *Xanthomonas campestris* pv. *campestris* from Thailand. *Phytopathology* **76**: 139–47.

Thompson, L.M., III, Smibert, M.R., Johnson, J.L. and Krieg, N.R. (1988) Phylogenetic study of the genus *Campylobacter*. *Int. J. Syst. Bacteriol.* **38**: 190–200.

Tunon, P. and Johansson, K.-E. (1984) Yet another improved silver staining method for the detection of proteins in polyacrylamide gels. *J. Biochem. Biophys. Meth.* **9**: 171–9.

Vandamme, P., Falsen, E., Pot, B., Hoste, B., Kersters, K. and De Ley, J. (1989) Identification of EF group 22 campylobacters from gastroenteritis cases as *Campylobacter concisus*. *J. Clin. Microbiol.* **27**: 1775–81.

Vandamme, P., Falsen, E., Pot, B., Kersters, K. and De Ley, J. (1990a) Identification of *Campylobacter cinaedi* isolated from blood and feces of children and adult females. *J. Clin. Microbiol.* **28**: 1016–20.

Vandamme, P., Pot, B., Falsen, E., Kersters, K. and De Ley, J. (1990b) Intra- and interspecific relationships of veterinary campylobacters revealed by numerical analysis of electrophoretic protein profiles and DNA : DNA hybridizations. *Syst. Appl. Microbiol.* **13**: 294–302.

Vandamme, P., Pot, B. and Kersters, K. (1990c) Differentiation of campylobacters and *Campylobacter*-like organisms by numerical analysis of one-dimensional electrophoretic protein patterns. *Syst. Appl. Microbiol.* **14**: 57–66.

Vandamme, P., Falsen, E., Rossau, R., Hoste, B., Segers, P., Tytgat, R. and De Ley, J. (1991) Revision of *Campylobacter*, *Helicobacter*, and *Wolinella* taxonomy: emendation of generic descriptions and proposal of *Arcobacter* gen. nov. *Int. J. Syst. Bacteriol.* **41**: 88–103.

Van den Mooter, M., Maraite, H., Meiresonne, L., Swings, J., Gillis, M., Kersters, K. and De Ley, J. (1987a) Comparison between *Xanthomonas campestris* pv. *manihotis* (ISPP List 1980) and *X. campestris* pv. *cassavae* (ISPP List 1980) by means of phenotypic, protein electrophoretic, DNA hybridization and phytopathological techniques. *J. Gen. Microbiol.* **133**: 57–71.

Van den Mooter, M., Steenackers, M., Martens, C., Gosselé, F., De Vos, P., Swings, J., Kersters, K. and De Ley, J. (1987b) Differentiation between *Xanthomonas campestris* pv. *graminis* (ISPP List 1980), pv. *phleipratensis* (ISPP List 1980) emend., pv. *poae* Egli and Schmidt 1982 and pv. *arrhenatheri* Egli and Schmidt 1982, by numerical analysis of phenotypic features and protein gel electrophoregrams. *J. Phytopath.* **118**: 135–56.

Van den Mooter, M., Swings, J., Gosselé, F., Kersters, K. and De Ley, J. (1987c) The taxonomy of the genus *Xanthomonas* Dowson 1939. In *Plant Pathogenic Bacteria* (Civerolo, E.L., Collmer, A., Davis, R.E. and Gillaspie, A.G., eds), pp. 795–6. Martinus Nijhoff: Boston.

van Tongeren, O.F.R. (1987) Cluster analysis. In *Data Analysis in Community and Landscape Ecology* (Jongman, R.H., ter Braak, C.J.F. and van Tongeren, O.F.R., eds), pp. 174–212. Centre for Agriculture Publishing and Documentation (PUDOC): Wageningen, The Netherlands.

van Vuuren, H.J.J. and Van der Meer, L. (1987) Fingerprinting of yeasts by protein electrophoresis. *Am. J. Enol. Vitic.* **38**: 49–53.

Vauterin, L. and Vauterin, P. (1992) Computer-aided objective comparison of electrophoresis patterns for grouping and identification of microorganisms. *European Microbiol.* **1**: 37–41.

Vauterin, L., Vantomme, R., Pot, B., Hoste, B., Swings, J. and Kersters, K. (1990) Taxonomic analysis of *Xanthomonas campestris* pv. *begoniae* and *X. campestris* pv. *pelargonii* by means of phytopathological, phenotypic, protein electrophoretic and DNA hybridization methods. *Syst. Appl. Microbiol.* **13**: 166–76.

Vauterin, L., Swings, J. and Kersters, K. (1993) Protein electrophoresis and classification. In *Handbook of New Bacterial Systematics* (Goodfellow, M. and O'Donnell, A.G., eds), pp. 251–80. Academic Press: London.

Vera Cruz, C.M., Gosselé, F., Kersters, K., Segers, P., Van den Mooter, M., Swings, J. and De Ley, J. (1984) Differentiation between *Xanthomonas campestris* pv. *oryzae*, *Xanthomonas campestris* pv. *oryzicola* and the bacterial 'brown blotch' pathogen on rice by numerical analysis of phenotypic features and protein gel electrophoregrams. *J. Gen. Microbiol.* **130**: 2983–99.

Walczak, C.A. and Krichevsky, M.I. (1982) Computer-aided selection of efficient identification features and calculation of group descriptors as exemplified by data on *Capnocytophaga* species. *Current Microbiology* **7**: 199–204.

Whiley, R.A. and Hardie, J.M. (1989) DNA-DNA hybridization studies and phenotypic characteristics of strains within the '*Streptococcus milleri* group'. *J. Gen. Microbiol.* **135**: 2623–33.

Willcox, W.R., Lapage, S.P., Bascomb, S. and Curtis, M.A. (1973) Identification of bacteria by computer: Theory and programming. *J. Gen. Microbiol.* **77**: 317–30.

Willcox, W.R., Lapage, S.P. and Holmes, B. (1980) A review of numerical methods in bacterial identification. *Antonie van Leeuwenhoek* **46**: 233–99.

Willems, A., Busse, J., Goor, M., Pot, B., Falsen, E., Jantzen, E., Hoste, B., Gillis, M., Kersters, K., Auling, G. and De Ley, J. (1989) *Hydrogenophaga*, a new genus of hydrogen-oxidizing bacteria that includes *Hydrogenophaga flava* comb. nov. (formerly *Pseudomonas flava*), *Hydrogenophaga palleronii* (formerly *Pseudomonas palleronii*), *Hydrogenophaga pseudoflava* (formerly *Pseudomonas pseudoflava* and '*Pseudomonas carboxydoflava*') and *Hydrogenophaga taeniospiralis* (formerly *Pseudomonas taeniospiralis*). *Int. J. Syst. Bacteriol.* **39**: 319–33.

Willems, A., Falsen, E., Pot, B., Jantzen, E., Hoste, B., Vandamme, P., Gillis, M., Kersters, K. and De Ley, J. (1990) *Acidovorax*, a new genus for *Pseudomonas facilis*, *Pseudomonas delafieldii*, EF group 13, EF group 16, and several clinical isolates, with the species *Acidovorax facilis* comb. nov., *Acidovorax delafleldii* comb. nov. and *Acidovorax temperans* sp. nov. *Int. J. Syst. Bacteriol.* **40**: 384–98.

Wishart, D. (1978) *Clustan User Manual, Version 1C, Release 2, 3rd edn.* Edinburgh University Program Library Unit: Edinburgh.

15
Analytical Fingerprinting Methods

John T. Magee
University Hospital of Wales, Cardiff, UK

15.1 INTRODUCTION

Fingerprint techniques used for microbial characterization have many common features. They rely on physico-chemical processes that, in a single analysis, yield data on many parameters reflecting aspects of overall cell composition. However, these data cannot be interpreted directly in terms of macromolecular composition. Infrared spectrometry, for example, yields data on the relative proportions of simple functional groups, such as amino, carboxy and hydroxy residues, in the sample. Cell composition is reflected in the data, but as statistically complex, multiple stacked responses. The infrared absorption at wavelengths characteristic of the amino group reflects the amounts of amino acids, amino sugars and the like as stacked variables. With such techniques, few or none of the parameters are uniquely characteristic of a single macromolecular cell constituent when considered alone. To deduce the amount of any one cell constituent would require comparison of the ratios of multiple variables.

However, for most microbiological applications, direct comparison of fingerprint data gives sufficient information on strain similarities, with no necessity for detailed chemical inferences. Work on the chemometric strategies that allow deduction of macromolecular composition, although promising, is in its infancy (Windig *et al.*, 1981a,b, 1983; Windig & Meuzelaar, 1987). The working philosophy of fingerprint techniques is that one uses a convenient and fast method that yields large amounts of data, but accepts that the parameters measured are complex functions of macromolecular composition. Nature appears to allow either simple, rapid and inexpensive analyses yielding complex data, or complex, slow and expensive analyses that allow direct chemical interpretation.

Having chosen a fingerprint approach, the problems are twofold. The analytical methods, although simple in principle, are often unfamiliar and

---

Chemical Methods in Prokaryotic Systematics. Edited by M. Goodfellow and A.G. O'Donnell
© 1994 John Wiley & Sons Ltd.

difficult to understand in detail, and the mass of pattern data that they yield must be corrected for statistical effects and reduced to inter-strain similarities. Both these problems require some understanding of areas unfamiliar to most microbiologists.

Fingerprint techniques have mostly been developed in multidisciplinary environments, and only pyrolysis mass spectrometry is at a stage where it can be applied by microbiologists with a passing knowledge of other sciences, and a willingness to learn. Others are at an early stage of development, and it would be presumptuous to define a final method protocol. The principles of such methods will be outlined, and a bibliography of work to date given.

15.2 PYROLYSIS TECHNIQUES

15.2.1 Principles

When organic materials are heated to >450 °C in the presence of oxygen, combustion occurs, yielding simple oxides. However, if the material is heated in a chemically inert environment, pyrolysis occurs, with no net oxidation. The thermal energy is sufficient to break covalent bonds, and the compounds shatter and rearrange to low molecular weight, predominantly volatile molecules—the pyrolysate. Bond breakage occurs in an ordered fashion, with weak bonds, such as disulphide bridges, breaking first. The subsequent molecular rearrangement reactions are complex, but reproducible. Provided the pyrolysis conditions are carefully standardized, the qualitative and quantitative composition of the pyrolysate reflects the composition of the sample. Separation and quantification of the pyrolysate components yields data that reflect the composition of the sample.

Pure biological macromolecules yield pyrolysates containing many distinct chemical compounds (Boon *et al.*, 1981), and pyrolysates of distinct macromolecules usually contain several common compounds. When bacteria, comprising many mixed macromolecules, are pyrolysed, the amount of any single pyrolysate component is likely to represent the sum of contributions of several distinct macromolecules. Also, the amount of any single macromolecule in the sample will be reflected in many distinct pyrolysate components. The analytical data for the pyrolysate therefore comprises multiple, stacked, covariant measurements.

15.2.2 History

Pyrolysis techniques are regularly used in chemical analysis. They are particularly suited to the analysis of polymers that are not amenable to 'wet' chemical analysis because of their insolubility. The first published pyrolysis mass spectrometry (Py-MS) study was in 1949 (Madorsky et al., 1949), but much of the early biological work explored the use of pyrolysis gas-liquid chromatography (Py-GLC). In Py-GLC the pyrolysis products are separated and quantified by gas-liquid chromatography. Pyrolysis mass spectrometry was largely neglected, because of the high capital costs, and the lack of purpose-built commercial apparatus.

Pyrolysis gas-liquid chromatography studies clearly showed that high levels of discrimination could be attained. Large inter-species pattern differences were seen in surveys of a broad range of bacteria (Gutteridge & Norris, 1979; Drucker, 1981; Magee, 1993). However, the microbiological application of Py-GLC foundered for three reasons. The highly reactive pyrolysis products combine with, or contaminate the stationary phase of the GLC column ('column poisoning'), changing its separation characteristics, and causing problems in long-term reproducibility (Quinn, 1974; Needleman & Stuchberry, 1977). Interpretation was subjective, because the computing power and statistical programs required for objective interpretation only became widely available in the late 1980s. In addition, the GLC separation is inherently slow (20–45 min/specimen), and the methods devised to overcome 'column poisoning' (Stack et al., 1978; Sekhon & Carmichael, 1973; Quinn, 1974; Magee et al., 1993) further limited the throughput of specimens. Pyrolysis gas-liquid chromatography is a difficult technique, unsuited to routine applications involving large numbers of specimens (French et al., 1989; Magee, 1993). This technique will not be considered here, but several reviews that detail the methodology and give a comprehensive bibliography are available (Quinn, 1976; Gutteridge & Norris, 1979; Drucker, 1981; Magee, 1993).

The breakthrough came in 1973 when a group of Py-GLC workers modified a mass spectrometer to allow automated processing of pyrolysis specimens (Meuzelaar & Kistemaker, 1973). This approach had many advantages over Py-GLC. A complete analysis took <5 min; no column was necessary, eliminating the problem of column poisoning; and the results of the analysis were captured directly on a computer that had sufficient power to allow complex statistical analysis. The problem of long-term and inter-apparatus reproducibility remained, albeit in a different form.

The new statistical analyses could readily detect small differences in processing, caused by, for example, batch-to-batch variation in the growth medium. To this day, Py-MS studies rely on processing samples in a single batch, prepared from organisms grown concurrently on the same batch of medium, under identical conditions. Clearly, medium and incubation variation, drift in the electronics for a single instrument, or differences between instruments may all cause variation in analytical results. Whether these differences are sufficient to prevent valid inter-batch comparison at species or strain level remains unknown. Workers have preferred to explore the many microbiological applications where batch-to-batch comparisons are irrelevant.

In subsequent studies (Meuzelaar *et al.*, 1976a,b, 1982; Huff *et al.*, 1981, 1986; Windig & De Hoog, 1982; Windig & Haverkamp, 1982; Wieten *et al.*, 1981, 1983), Dutch workers explored many applications, and laid the foundations for the sophisticated statistical pattern analyses that are essential to the method. The main thrust of applications shifted to Py-MS, but other pyrolysis techniques are still used (see Section 15).

Two purpose-built commercial mass spectrometers, the Pyromass 8–80 (VG Gas Analysis, Middlewich, Cheshire, UK) and the Extranuclear 5000 (Extranuclear Laboratories, Pittsburgh, PA, USA) were produced soon after, but neither proved popular, probably because of cost. The cost of mass spectrometers fell in the late 1980s, and the latest instruments, the PYMS 200X and RAPyD 400 (Horizon Instruments, Heathfield, Sussex, UK) are beginning to come within the capital resources of large laboratories. These instruments are described here.

In microbiology, Py-MS has been applied in classification, identification, typing and antibiotic action studies (Gutteridge *et al.*, 1985; Shute *et al.*, 1985; Magee, 1993). It is applicable, with little species-specific modification, to any organism that can be grown in pure culture on solid or liquid medium. Specimen preparation is minimal, and further processing is fully automated. Usually, samples are prepared in triplicate or quadruplicate, and can be processed at 40 per hour. The apparatus can be left unattended, allowing up to 400 samples to be processed every 24 h. Although instrument costs are high by microbiologists' standards (£48 000), running costs are low—less than £1 per sample (Magee *et al.*, 1989b).

15.3 PYROLYSIS MASS SPECTROMETRY

Pyrolysis mass spectrometric analysis comprises four steps: coating a small sample (~ 10–100 µg dry weight) onto a metal carrier; pyrolysis, in which the carrier and sample are heated rapidly in a vacuum to a fixed temperature in the range 400–1000 °C, thermally degrading the sample to a mixture of volatile compounds; mass spectrometry, in which the volatile compounds are separated by molecular weight and quantified; and computation, in which the spectra are compared by pattern-recognition statistical methods.

15.3.1 Method

15.3.1.a Experimental design

Workers have adopted the principle that cultures should be blind-coded, and remain so until completion of the mathematical analysis (Magee et al., 1988; Freeman et al., 1990). This adherence to the scientific principle of untainted, objective interpretation has much to commend it. The coded series should contain at least two duplicate cultures, and, for typing studies, the inclusion of at least one coded strain that is clearly epidemiologically distinct is recommended.

Workers diverge a little on replication formats. Most take three replicate samples from each of two replicate cultures of all the blind-coded series (e.g. Freeman et al., 1990). The advantages in blanket re-estimation of inter-plate variation seem small, compared to the cost, so a system in which four replicate samples are taken for each culture, with one in ten of the blind-coded series cultured and analysed in duplicate has been adopted in my laboratory.

It is vital that a reasonable number of spectra contribute to the statistical processing. My practice is to analyse a minimum of 50 spectra. For small collections this is achieved by extended replication of the coded cultures; for typing studies, batch analysis of several independent collections of the same species is a reasonable alternative.

15.3.1.b Foil and tube preparation

Solvent-cleaned pyrolysis foils, 10×2.5 mm thin rigid sheets of iron-nickel alloy folded longitudinally in a V shape, are purchased ready for use (~ 10 p/foil). These are inserted so as to protrude from the mouth of a clean 34×4 mm glass tube, using no-touch technique with flamed forceps (Figure 15.1). Care must be taken to ensure that only a single foil is inserted, as foils tend to adhere to one another. Decontamination by pre-heating foils to 180 °C in a stream of inert gas or hydrogen has been advocated (Windig et al., 1979). My own group formerly heated the assembled tubes to 180 °C for 90 min in air to remove organic contaminants (Magee et al., 1988), but we have lately found little advantage in this step provided a careful no-touch assembly technique is used. After use, the tubes are de-foiled and washed for 24 h in 10% 7X-PF detergent (ICN Biomedicals, Irvine, Scotland), 3×30 min in water, 3×30 min in industrial methylated spirit and $1 \times$ overnight in acetone. Since the narrow lumen of the tubes prevents simple filling and emptying, used tubes are stacked vertically in a wire basket, immersed in the wash in a Koplin jar, and a vacuum drawn

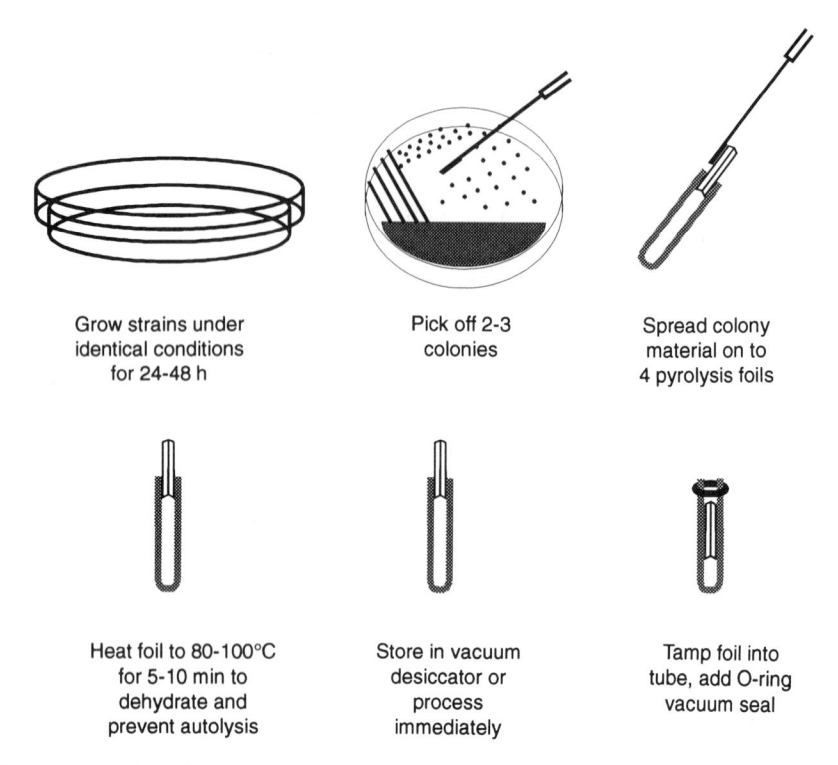

Grow strains under identical conditions for 24-48 h

Pick off 2-3 colonies

Spread colony material on to 4 pyrolysis foils

Heat foil to 80-100°C for 5-10 min to dehydrate and prevent autolysis

Store in vacuum desiccator or process immediately

Tamp foil into tube, add O-ring vacuum seal

Figure 15.1 Sample preparation.

to ensure that the wash penetrates into the tube. The wash solvent can be emptied by shaking, or by standing the inverted basket on paper towel and drawing a vacuum. Detergent and solvent treatment is less hazardous than decontamination in chromic acid used in early work, and appears equally effective.

15.3.1.c Sample preparation

The organisms to be examined are grown on the same batch of medium under identical conditions. Normally, a sugar-free medium, such as blood agar, is used, and the plates incubated for a period that allows luxuriant growth. The cultures should be examined for purity and equality of medium depth. It is wise to reincubate plates after sampling to examine them later for cryptic contamination. Early work with infrared spectroscopy (Kenner *et al.*, 1958) suggested that prolonged incubation, or the use of sugar-rich, selective or nutritionally unbalanced media, results in poorly reproducible cell composition. This has been linked with the accumulation of glycogen or poly-β-hydroxybutyrate storage granules (Kenner et al., 1958; Haynes et al., 1958; Norris & Greenstreet, 1958).

Colonies are taken on a straight wire, carefully avoiding contamination of the sample with the growth medium, and smeared on to each of three or four pyrolysis foils. For liquid cultures, an equivalent amount of washed centrifugate is used. A broad sweep of colonies should be taken, especially for organisms showing colonial polymorphism, e.g. *Clostridium difficile*.

The coated foils are heated to dehydrate the sample, and destroy enzyme activity that might otherwise alter sample composition. The exact temperature and heating time protocols vary somewhat (80–100 °C for 5–10 min) but the heat treatment must be consistent for all samples, and must be applied within 20 min of coating. Alternatively, for highly infectious organisms, the samples may be autoclaved (Wieten et al., 1983). Prepared foils can be stored for long periods (3 months is often quoted) over phosphorus pentoxide in a vacuum desiccator. A set of strain-replicate foils can be prepared within 1–2 min for cultures grown on solid medium.

For processing, the foil is pushed to the centre of the tube with a tamping tool, an O-ring vacuum seal is placed around the tube mouth and the tubes are placed on the instrument sample ring. Further processing with batches of up to 150 sample tubes is automated, controlled by a microcomputer on the instrument, and requires no human intervention.

The first few spectra of a batch are frequently atypical, so it is common practice to prepare an excess of foils, process five to eight at the beginning of the batch, and discard these spectra. The reasons for this effect are obscure. It may reflect saturation of binding sites in the airlock, stabilization of the residual free product mix in the mass spectrometer chamber, or warm-up of the electronics.

Drift in the mass scale can be avoided by allowing the mass spectrometer to stabilize in residual gas analysis mode for 30–60 min. The spectrum of the first spare foil is examined for drift, and the mass scale recalibrated against a volatile standard such as perfluorokerosene, if necessary. Best discrimination is usually obtained in the mass range 50–150, and limiting the mass scan to this range allows more scans during the pyrolysate release phase. The manufacturer's advice should be sought if these scan parameters are to be altered.

15.3.1.d Automated processing (Aries et al., 1986; Magee et al., 1989a)

The tube is picked from the sample ring in the bore of a sample handling rod that clamps the O-ring firmly against a circular knife edge on the airlock face, making a vacuum-tight seal (Figure 15.2). The airlock plunger withdraws to the B position, evacuating the tube via the bypass line, then withdraws to the C position, closing the bypass line and opening the molecular beam tube to the mass spectrometer.

The foil is then heated by passing a high-frequency, high-amperage oscillating current through the pyrolysis coil for a set period, usually 3 s. This produces an intense magnetic field that penetrates the foil causing

AIRLOCK INSERTION

Position A
Sample rod inserts tube into pyrolysis coil and seals O-ring to airlock front face

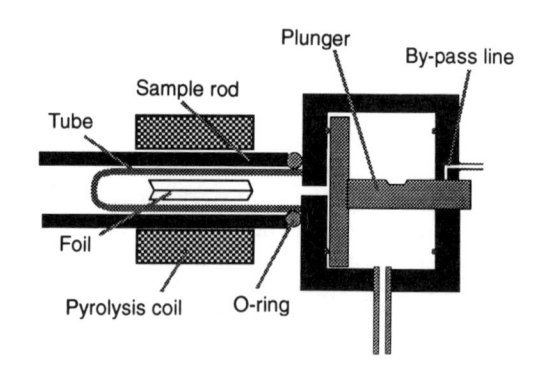

Position B
Plunger withdraws to middle of airlock, opening by-pass line to evacuate tube

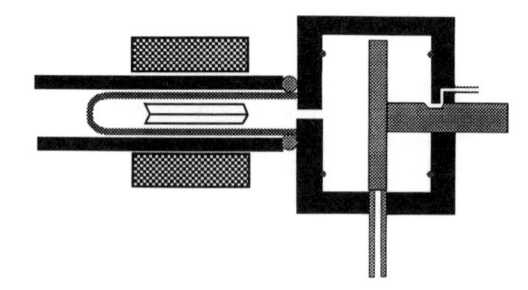

Position C
After evacuation, plunger withdraws to rear position, sealing by-pass line and opening tube to mass-spectrometer

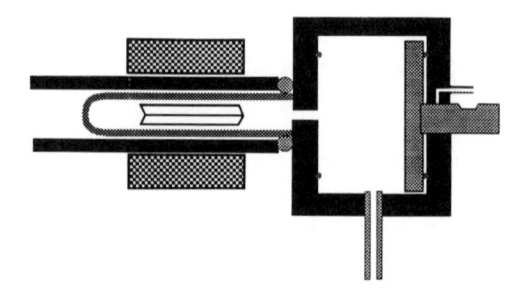

Figure 15.2 Pyrolysis-mass spectrometry (Py-MS) sample insertion.

induction heating. When the temperature of the foil reaches the Curie point of the iron-nickel alloy, the alloy ceases to be ferromagnetic, and the field no longer induces heating. The foil cools slightly, reverts to ferromagnetism, and is heated back to the Curie point. This natural thermostatic effect controls the temperature of the foil within tight limits.

Usually, the alloy has a Curie point of 530 °C, but foils with Curie points of 300 °C to 1000 °C can be purchased. Pyrolysis at low temperatures yields pyrolysates with a high proportion of high-boiling point tarry products. These tend to condense in the tube and airlock, giving contamination problems. High-temperature pyrolysis gives high pyrolysate yields, but low molecular weight products predominate, and these are often poorly characteristic.

The thin sample film rapidly reaches the temperature of the foil. If oxygen were present, combustion would occur, yielding carbon dioxide and water as major products. However, in a vacuum, as here, pyrolysis occurs with no net oxidation. The products are formed by breakage of covalent bonds and intramolecular rearrangements as described above.

The products expand into the airlock chamber, designed to optimize flow mixing, so that early, low molecular weight products and late, high molecular weight products of the pyrolysis mix evenly. The airlock also acts as an expansion chamber, slowing the rate of egress of products down the molecular beam tube and into the mass spectrometer (Figure 15.3). This prevents saturation of the later ionization and detection stages. The sample tube, airlock and molecular beam tube are heated to 100–140 °C to prevent condensation of the products (Windig *et al.*, 1979). Temperatures > 180 °C can induce thermal fragmentation of the products (Windig *et al.*, 1979).

As the products leave the molecular beam tube (Figure 15.3), they encounter a crossing beam of low-energy (30 eV) electrons. Collisions with these electrons result in the formation of ions. This is a complex process; the desire is to remove a single electron from the molecule, leaving a molecular ion with a single positive charge. However, even at the low electron energy employed, some molecules fragment. It is not unusual to see multiple ions when a volatile pure chemical is ionized under these conditions. For example, pyridine yields the molecular ion of mass/charge ratio (m/z, or, colloquially, 'mass') 79 and a hydrogen loss product at mass 78. It appears, however, from examination of spectra at 0.1 mass intervals, that ions with multiple charges are rare.

Un-ionized products freeze on a trap cooled with liquid nitrogen to preserve the high vacuum essential for mass spectrometry. The ions are accelerated out of the electron beam by a positively charged repeller plate, and are focused by an electrostatic lens. They then separate in a quadrupole

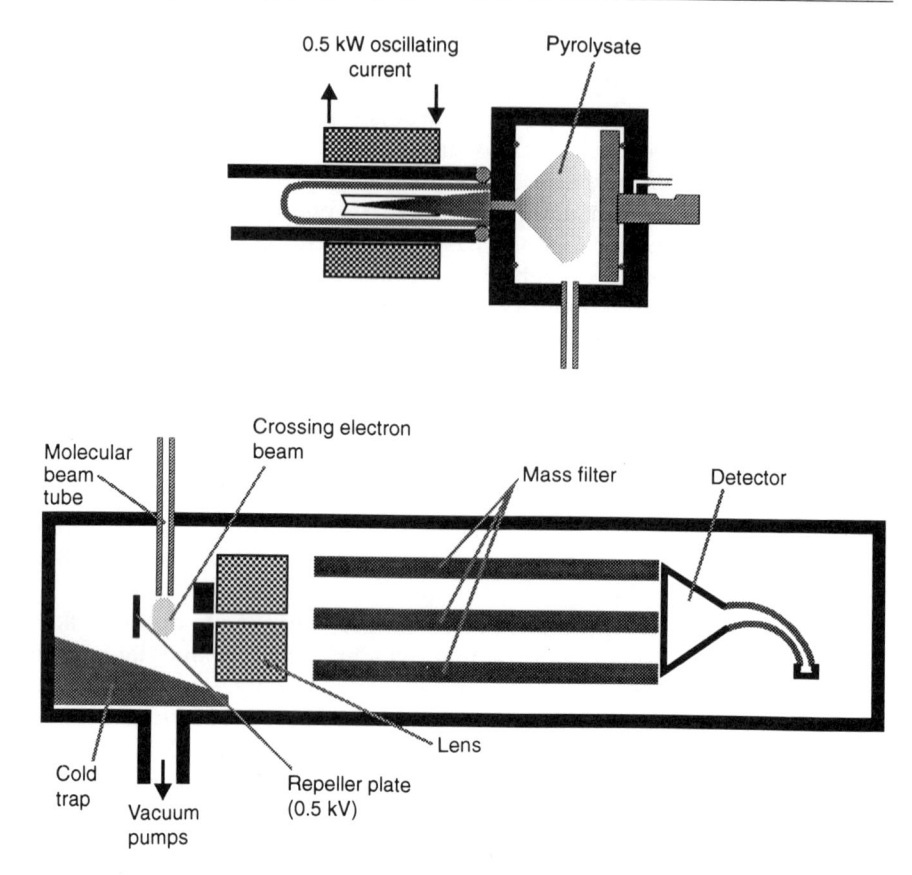

Figure 15.3 Py-MS pyrolysis and product separation.

mass spectrometer, which delivers ions of a specific mass to the detector. The computer records the number of ions detected at this mass, then alters the electromagnetic field in the quadrupole, allowing ions of the next mass through. In the minute following pyrolysis, the instrument can repeatedly scan from mass 11 to 200 (PYMS 200X) or 11 to 400 (RAPyD-400), accumulating ion counts at 0.1 mass intervals.

The spectrum is presented as a histogram of ion counts vs. mass at unit mass intervals. The spectrum is automatically recorded to hard disk and the airlock closes. The sample tube is removed from the airlock face and replaced in the sample ring, and the next sample tube is loaded. The entire process takes about 90 s.

15. Analytical Fingerprinting Methods

15.3.2 Data analysis

15.3.2.a Initial processing

Visual examinations of the spectra are rarely profitable; the eye is drawn to high-intensity masses that are often the least reproducible features. Normally, data for masses 11 to 50 are discarded; these are small, nonspecific products, some of which reflect the quantity of residual oxygen in the sample tube, or residual water in the sample (Magee, 1993). Masses beyond 150 usually have such low intensities that statistical sampling effects minimize their reproducibility.

Correction of the spectra for background by subtraction of mean ion counts for a series of uncoated foils is possible, but this is rarely performed. The 'background' spectra from uncoated foils are of such low intensity compared to sample spectra that they can reasonably be ignored. However, 'background' intensities for masses 11–50 can be significant, and corrections may be necessary if these masses are to be included in the analysis (Windig et al., 1980).

Each spectrum is first normalized, by multiplying the intensity of each mass by a standard total spectrum intensity (say 1×10^6 ions) and dividing by the actual total intensity (sum of ion counts for all masses for the spectrum). This corrects for variations in sample quantity, reducing all spectra to the standard intensity so that they can be directly compared. The statistical complexity of the data ensures that this is not simple. High-intensity masses that contribute disproportionately to the estimate of the total ions are usually excluded (Gutteridge et al., 1985); often these are found to be among the least quantitatively reproducible features.

One would prefer to eliminate such poorly reproducible masses from the estimate of total intensity. However, to do this requires an estimate of within-strain reproducibility for each mass, which can only be calculated for normalized spectrum sets. The solution is to normalize the spectra based on all masses, calculate the reproducibility for each mass, eliminate those masses that show poor reproducibility from the normalization set, and renormalize based on the reduced mass set. This is repeated until no further masses are eliminated. Although mathematically complex, this process can be performed within a few seconds by an 'iterative normalization' program (Huff et al., 1981).

The normalization program also determines the within-strain reproducibility and inter-strain discrimination for each mass. Here, approaches diverge; the manufacturers' software selects the most discriminatory masses for further analysis, while my own software selects the most reproducible masses. The former approach is probably best suited to typing studies, and the latter to classification (Magee, 1993). However, these different approaches often yield similar results.

15.3.2.b Multivariate statistical analysis

Subsequent analysis reduces the statistically significant inter-strain differences to coordinates on a series of derived axes (canonical discriminant or canonical variate axes). These axes represent the statistically significant pattern differences between spectra in a greatly simplified form. Effectively, each spectrum is regarded as the sum of a series of discriminatory sub-spectra; each axis represents a sub-spectrum, and the strain coordinate on the axis is a measure of the contribution of that sub-spectrum to the observed mean-strain spectrum. The axes are numbered sequentially, in order of decreasing discriminating ability. The mathematics and programs involved are dealt with in Appendix 15.1.

This approach has analogies in human perception. Intuitive recognition of underlying pattern differences is basic to many visual processes. As in the statistical analysis of spectrum data, we intuitively interpret a perceived image into derived complex variables, analogous to the sub-spectrum axes. One might describe a person by such derived variables as stoutness, ranginess, or stockiness, thereby conveying an estimate of the ratios of many measurements of their build. These variables are analogous to the sub-spectrum or, more correctly, canonical variate axes derived from the statistical spectrum comparisons.

The results are often presented as dendrograms or ordination diagrams. These show a proportion of the inter-strain discrimination, but the full data can only rarely be represented fully in diagrams. The problem is similar to that encountered in atlases; the network of distances between the cities of the world cannot be preserved completely in a two-dimensional diagram, because the world is a sphere and the distance structure of this three-dimensional structure cannot be crushed into two dimensions without distortion. In Py-MS, the many distinct pattern differences between strains can often only be expressed adequately in more than three dimensions.

Interpretation should therefore include careful examination of the similarity matrix. This tabulates the true statistical distances (χ^2 significance of inter-strain spectrum difference) for each strain pair. Dendrograms, scatter diagrams and similarity matrices are readily produced from the manufacturer's software, or by processing through standard statistical software suites. Annotated examples of processing through the SPSS-PC and CLUSTAN software suites can be obtained from the author.

15.4 OTHER PYROLYSIS-BASED TECHNIQUES

15.4.1 Laser pyrolysis mass spectrometry

15.4.1.a Principles

In the laser microprobe analyser (LAMMA, Leybold-Heraeus, Köln, Germany; Figure 15.4) the sample is introduced into the vacuum chamber of a mass spectrometer on a coated copper grid, as in electron microscopy. The grid is scanned visually by light microscopy, and an area of ~1 μm diameter is selected by the operator for pyrolysis. This area is centred in

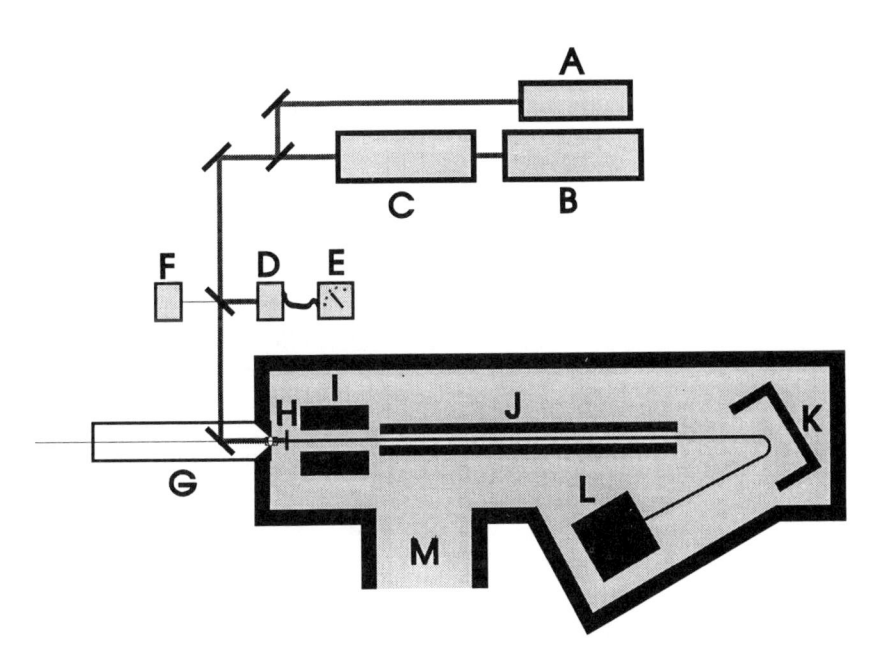

Figure 15.4 Laser pyrolysis in the LAMMA apparatus (simplified from Böhm et al., 1985). The specimen grid (H) is illuminated by the incident light source (F) or a transmitted light source that can be moved into the position of the electrostatic lens (I). It is inspected through the microscope (G) and a suitable area moved into the laser target area, indicated by light from the pilot laser (A). In pyrolysis, light from the high-energy laser (B) passes through a frequency converter (C) and along the same path as the pilot laser beam. The energy of the laser pulse is measured by a photodiode (D) and meter (E). The target area pyrolyses, giving ions that are extracted through the electrostatic lens (I) into the drift tube (J). The magnetic reflector (K) corrects transit time differences due to variation in initial velocity and delivers the ions to the detector (L) in order of mass : charge ratio. The high vacuum in the spectrometer is maintained by pumps (M).

the beam of a low-intensity pilot laser, then pyrolysed by a focused high-energy laser beam giving a target energy intensity of $\sim 10^9$ W/cm^2. The material at the centre of the target decomposes into plasma. Positively and negatively charged pyrolysate ions are produced in the periphery of the plasma cloud by complex and poorly understood reactions (Hercules, 1983); either set of ions can be extracted electrostatically for subsequent mass spectrometry. Spectra from positive ion extraction usually differ from those from negative ion extraction.

The extracted ions move down a drift tube into the field of a magnetic reflector. This turns the ion beam, directing the ions to a detector. The time of flight of the ions, from formation in the laser beam to collision with the detector, is proportional to the mass : charge ratio of the ion. Differences in the transit time in the drift tube due to variation in initial ion velocity are corrected during reflection of the ion beam. Unlike conventional Py-MS, the ion yield in the high-mass range is low, reflecting the considerable differences in energy input during pyrolysis.

15.4.1.b History

This approach offers the possibility of analysing single cells, directly in a specimen, with no culture and incubation steps (Böhm et al., 1985). The potential implications are enormous, particularly in medical microbiology. However, the LAMMA is costly, and much of the published work on its use involves non-microbiological applications. Those microbiological studies that have been performed show that spectra of single vegetative cells of *Bacillus anthracis*, *B. cereus* and *B. thuringiensis* can be distinguished, although there was some overlap between these taxa (Böhm et al., 1985). Also, single cells of *Escherichia coli* that had been subjected to antibiotic treatment differed from untreated cells in the ratio of masses 21 to 43, indicating changes in the sodium : potassium ratio of the cells (Lindner & Seydel, 1983). Detailed methods are given in the papers cited.

15.4.2 Pyrolysis gas-liquid chromatography-mass spectrometry

15.4.2.a Principles

In pyrolysis gas-liquid chromatography-mass spectrometry (Py-GLCMS), strain samples are pyrolysed in the carrier gas flow at the head of a capillary GLC column. The products are separated in the column and the effluent analysed in a mass spectrometer. Usually, an ion trap mass spectrometer is used, as these are less expensive than quadrupole instruments and more readily interfaced with the GLC column. The technical problems in GLC separation of pyrolysis products have been mentioned above. Py-GLCMS has a low throughput, and is slower than Py-MS. However, the technique yields considerably more information on the chemistry of the pyrolysate and hence of the sample.

Unlike Py-MS, the pyrolysate is separated into distinct chemical products which elute from the column at different times. This allows the use of high-energy (70–80 eV) electron impact ionization, giving a high yield of molecular fragment ions. The relative intensities and masses for these fragments, and for the molecular ion, form a pattern that is characteristic of the molecular structure. As each product elutes from the column, the spectrum rises in intensity, proportional to the amount of that product, and shows a characteristic ion pattern. The pattern can be compared to libraries of spectra obtained for pure chemicals under identical ionization conditions and the nature of the product deduced. By contrast, in conventional Py-MS, the distinct chemical components of the pyrolysate are delivered to the mass spectrometer simultaneously; if high ionization energies were used, then the complex fragment spectra for each product would be superimposed, adding still greater complexity to the already complex spectra.

Consequently, Py-GLCMS allows the deduction of the chemical nature of the pyrolysate components. Comparison with products from pyrolysis of pure compounds and knowledge of the likely thermal degradation reactions allows the nature of the macromolecular cell components from which these components originate to be deduced.

15.4.2.b History

Work in this field has concentrated on a search for products that are uniquely characteristic of bacteria in general, or of particular species (Morgan et al., 1991). If such products exist, then Py-GLCMS could be used to detect bacteria directly in specimens, rather than relying solely upon culture techniques. Indeed, if the product were sufficiently distinctive it might be detected by specific ion monitoring in Py-MS.

This search has identified associations of several pyrolysis products with particular bacterial macromolecules (Simmonds, 1970; Hudson et al., 1982; Eudy et al., 1985), and Py-GLCMS has been particularly useful in chemotaxonomic studies of mycolic acids (Etémadi, 1967; Athalye et al., 1984; Kusaka & Mori, 1986). Acetamide is produced primarily from peptidoglycan (Simmonds, 1970; Hudson et al., 1982; Eudy et al., 1985); pyridine is formed by decarboxylation of dipicolinic acid, and is produced in large amounts when sporulating cultures are analysed (Morgan et al., 1991); 2-butenoic acid is a major product of poly-β-hydroxybutyrate pyrolysis (Watt et al., 1991). However, uniquely characteristic products have yet to be found. Methods and results have been reviewed by Morgan *et al.* (1991).

A modification of Py-GLCMS termed pyrolytic-on-line derivatization GLCMS has proved particularly useful in analysis of bacterial lipids (Dworzanski & Meuzelaar, 1991). This is closely similar to Py-GLCMS, but tetramethyl ammonium hydroxide is added to the coated sample before pyrolysis. This procedure results in the methylation of cell fatty acids during pyrolysis. Separation of these fatty acid methyl ester products (FAMEs) in capillary GC, followed by molecular structural characterization in mass spectrometry, gives FAME patterns that are often characteristic of bacterial species, and are closely comparable with those obtained by other, slower methylation methods. The technique, which has also been used in characterization of mycobacterial lipids, has been reviewed by Dworzanski and Meuzelaar (1991).

15.4.3 Pyrolysis tandem mass spectrometry

15.4.3.a Principles

Pyrolysis tandem mass spectrometry (Py-MS-MS) is the most capital intensive adaptation of pyrolytic analysis. It is closely similar to conventional Py-MS in the initial stages of pyrolysis and low-energy electron ionization, and separation of ions. However, from this stage, ions of a specific mass-charge ratio can be selected for redirection past the detector. The redirected ions pass through a high-energy collision and molecular fragmentation stage, and on to a second mass spectrometer, which separates the fragments. As in the MS stage of Py-GLCMS, this secondary fragmentation and separation step gives spectra characteristic of the molecular structure of the redirected ion, allowing chemical identification of the pyrolysis product. Like Py-GLCMS, the pyrolysis products are separated and characterized, but the initial MS separation is rapid compared to capillary GC. Few microbiological studies have been performed on these expensive and complex instruments (Voorhees *et al.*, 1988; DeLuca *et al.*, 1990).

15.5 INFRARED SPECTROMETRY

15.5.1 Principles

Chemical compounds each have a unique pattern of absorption (or emission) across the various wavelengths of electromagnetic radiation. In the infrared band, the absorption characteristics are predominantly dictated by the energy of rotation (only observed in the gaseous phase), or bond-stretching vibration of specific functional groups. These are quantum effects, i.e. a single, specific change in energy accompanies the intramolecular conformational change. Infrared (IR) radiation of that specific energy, and no other, will be absorbed, and promote the molecule through the specific conformational change. Each possible conformation change has a specific energy, and infrared light of the corresponding energies, or wavelengths, will be absorbed strongly. This gives an IR spectrum with narrow absorption bands at wavelengths corresponding to the various conformation change energies. The spectrum abscissa can be expressed in several units; the current standard is to use wave number, the reciprocal of the wavelength measured in centimetres; older papers use wavelength; the absorption frequency multiplied by Planck's constant gives the excitation energy.

The energies of vibration tend to be similar within each of the individual functional groups that account for most of the IR absorption spectrum. In pure chemicals, hydroxyl groups show closely similar energy transitions, affected only slightly by neighbouring residues of the molecule. Ketone groups absorb at another, distinct set of wavelengths. The IR spectrum can therefore be viewed as reflecting the relative proportions of simple functional groups in a sample. Infrared spectrometry (IRS) yields the stacked, covariant data typical of fingerprint analyses.

The main problems of IRS are the broad, intense absorption band of water, which obscures much of the spectrum of hydrated samples, and the presentation of the specimen on a suitable carrier. The choice of carrier materials is tightly restricted by the requirements for transparency throughout the IR band, insolubility in water, and suitable mechanical properties. Silver chloride or zinc selenide plates are the most suitable insoluble carriers, but most organic chemical analysis work is performed with anhydrous specimens, mixed with potassium chloride or bromide, and compressed to a transparent disc.

15.5.2 History

The microbiological applications of IRS were investigated intensively between 1950 and 1958. The conclusion was that IRS could discriminate at species and subspecies level if cultures were grown under tightly reproducible conditions (Norris, 1959). However, the dispersive IR spectrometers used were slow, gave low resolution, and showed great instability of wavelength calibration. Infrared spectrometry of bacteria was possible and useful, but the analysis required considerable expertise and the instruments available did not have the specimen throughput required for routine applications. Further, the spectra were complex, and their comparison at the level of detail required was labour-intensive in this pre-computer era.

As with Py-GLC, dispersive IRS gave the discrimination required but was too difficult and slow to be a practicable routine method. Also, as for Py-GLC, this early work highlighted many of the pitfalls of the method and laid the groundwork for later, more advanced adaptations of the method. Microbiological applications of IRS turned from analysis of whole cells to a tool for investigation of specific biochemical problems in macromolecular structure.

Infrared spectrometry improved greatly with the introduction of Fourier transform IR (FTIR) spectrometers. These stable, computer-controlled instruments were capable of giving high-resolution spectra within a few minutes, and they required much less expertise than the dispersive instruments. This advance was taken up in 1983 by a group of German workers who have published several promising studies (Giesbrecht et al., 1984; Horbach et al., 1988; Helm et al., 1991; Naumann et al., 1991). The most recent advance is the use of microscopes with an adaptor that can direct a beam of IR light through a ~20–100 μm diameter area selected in visible light microscopy, and direct this back to an FTIR spectrometer for analysis (Naumann et al., 1991).

15.5.3 Methods

15.5.3.a Specimen preparation

Most studies describe simple preparation techniques in which a sample of colony material from an agar culture, or the centrifugate from a liquid culture, is deposited on a carrier (see below) and rapidly dried as a thin film. Neumann et al. (1991) suggest taking a standard 1 mm loop of material into 30 μl distilled water, depositing 25 μl onto the carrier and drying in a vacuum desiccator at ~25 torr. Riddle et al. (1956) tilted the carrier during drying, to produce a wedge-shaped film, and selected a portion showing 50% absorption at a wavelength of 8.5 μm. In a few studies (Goulden & Sharpe, 1958), freeze-dried samples were investigated by the potassium bromide disc method (see above). More recently, for IRS-microscopy, microcolonies from plates incubated for 6–8 h have been sampled by drying a contact impression of the plate onto a barium fluoride slide (Naumann et al., 1991).

Early studies established that growth should be on the same batch of medium with concurrent incubation under identical conditions if fine comparisons were to be made (Bolduan et al., 1952; Levine et al., 1953). Further, prolonged incubation or growth on sugar-rich media gave poor reproducibility, which was attributed to the accumulation of glycogen or poly-β-hydroxybutyrate storage granules (Kenner et al., 1958). The analogies with Py-MS preparations are clear, and it may be that inclusion of a heating step, which neutralizes autolytic enzyme activity in Py-MS preparation, may be advantageous.

15.5.3.b Spectrometry

The range of methods for presentation of the specimen in FTIR is so broad as to be confusing. In the simple transmission cell, the IR beam is directed through the sample and carrier plate. In the attenuated total reflectance cell, the beam passes through a zinc selenide prism and is reflected from the rear face, which is coated with the specimen. Complex effects occur at the prism-sample interface; the effective path length through the specimen is dictated by the angle of reflection, which can be tightly defined and adjusted. This may be an advantage over simple transmission where the sample path length is dictated by the film thickness, and is variable. These two presentations seem the most appropriate for the examination of macroscopic bacterial samples (Figure 15.5), and purpose-built multi-sample cuvette systems are available for both (Boden Maltzan Werkstatt, Berlin, Germany).

Spectra are taken in the mid-IR range (4000–5000 cm^{-1}) under standardized instrument parameters. For FTIR instruments, Naumann et al. (1991) suggest 256 scans at 8 cm^{-1} resolution, giving a spectrum within ~40 s with a nitrogen-cooled mercury-cadmium-telluride (MCT) detector or a little longer with a deuterated triglycine sulphate detector. In FTIR microscopy, the low absorption intensity of the tiny sample makes MCT detection essential.

15.5.3.c Numerical analysis

The resolution of the initial spectrum, acquired directly on the control computer in FTIR instruments, can be enhanced by two mathematical techniques, spectral derivatization and Fourier self-deconvolution. These detect and resolve the multiple fused absorption peaks and inflections into their component lines. Programs to perform resolution enhancement, subtract specimen-free background spectra, and normalize the spectrum to a standardized overall intensity are usually supplied with the instrument.

The chemical origin of the spectrum components is known, and broad regions can be assigned as follows. Lines for CH_3, CH_2 and CH predominate from 3000 to 2800 cm^{-1} and from 1500 to 1400 cm^{-1}. With the $C=O$ absorption complex at 1740 cm^{-1}, these probably give information on fatty acids and esters. From 1700 to 1500 cm^{-1}, amide absorption bands predominate, giving information on the ratios of α helices (1655 cm^{-1}), and β pleating (1637 cm^{-1}) in proteins. From 1250 to 1200 cm^{-1}, various phosphodiester absorption lines predominate giving information on DNA,

Fourier transform infrared spectrometer

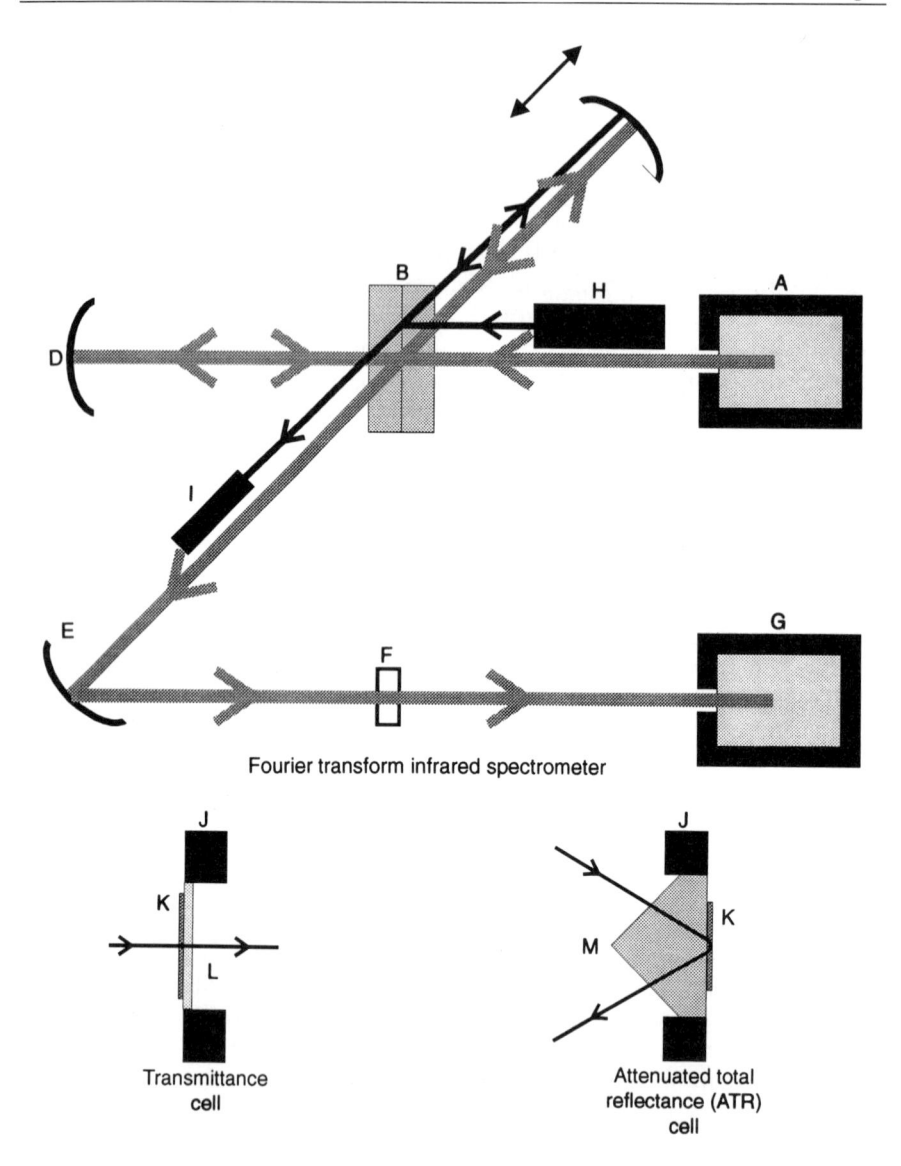

Transmittance
cell

Attenuated total
reflectance (ATR)
cell

RNA and phosphate-containing polysaccharides. From 1200 to 900 cm^{-1}, C—O—C and C—O absorption lines typical of sugar rings occur giving information on polysaccharides. From 900 to 600 cm^{-1}, the spectral features are weak, but discriminatory, and their origin is still unknown.

Spectral comparison in the early studies with dispersive instruments was usually by eye. A few of the later studies of this era used objective univariate mathematical comparisons (Riddle *et al.*, 1956) and the

15. Analytical Fingerprinting Methods

Figure 15.5 (*opposite*) FTIR spectrometry. Focusing and collimating reflectors have been omitted for simplicity. IR light from the source (A) passes through the beam splitter (B), which passes portions of the beam to a fixed reflector (D) and a travelling reflector (C). These beams recombine after passing back through the splitter, and are reflected through the specimen. The movement is monitored by the fringe detector (I) which registers fringe patterns in the visible laser light beam. This beam is generated by the laser (H) and is also split and reflected by the fixed and moving reflectors. One fringe cycle represents movement of reflector C of one wavelength of the laser light. The detector (G) registers an interferogram during the movement scan of C, resulting from the combined effects of interference and absorption in the specimen. This interferogram is recorded by the control computer, averaged over several scans, and converted to an infrared spectrum by a mathematical technique called the Fourier transform. The transmittance cell comprises a holder (J), and a zinc selenide or silver chloride sheet (N) coated with the specimen (K). The attenuated total reflectance cell is similar, but light propagates through the specimen film via a zinc selenide prism (M); the depth of propagation depends upon the angle of incidence of the IR beam.

---

possibility of computerized statistical analysis was mentioned (Norris, 1959). The latter development, however, only came with the development of FTIR, with its computerized data acquisition, and the microcomputer era. The early FTIR work used a difference coefficient, D (Horbach *et al.*, 1988), defined as:

$$D = (1 - \alpha) \times 1000$$

where α is Pearson's product moment correlation coefficient, which essentially measures the non-overlapping area between the two spectrum contours. D values vary from a mean of 0.4, standard deviation (SD) 0.4 for replicate films from the same cell suspension, through 2 (SD = 1) for the same strain grown on different plates of the same batch of media, to 7 (SD = 6) for the same strain on different batches of media. Broadly, within-species D values are usually < 100 and inter-species differences from 100 to 1000 (Naumann *et al.*, 1991). However, later FTIR work (Naumann *et al.*, 1991) has moved to the multivariate strategies outlined briefly for Py-MS above, and considered in greater detail in Appendix 15.1.

15.6 ULTRAVIOLET-RESONANCE RAMAN SPECTROSCOPY

15.6.1 Principles

Physics dictates that the light scattered at right-angles from the intersection of an intense beam of monochromatic light (excitation beam) with a sample is no longer monochromatic. Interaction with the sample molecules has effectively robbed these photons of energy corresponding to the conformational transition energies mentioned above. In effect, the right-angle scattered light is frequency modulated with the infrared spectrum of the sample. This effect is used in Raman spectroscopy (RS), a technique complementary to IRS, because some conformational changes that are inactive in the IR range become evident in RS. The main problem of RS is fluorescence, particularly for excitation wavelengths of 240 to 790 nm.

However, if the excitation beam is at a wavelength that is strongly absorbed by a molecule, the modulation effect is particularly intense for that substance, dominating the RS spectrum even in mixed samples. This is the principle of resonance Raman spectroscopy, where the Raman spectrum of particular components of a mixed sample can be selectively taken by judicious adjustment of the excitation wavelength. In microbiological fingerprint studies, these components are usually macromolecules with strong absorption lines in the ultraviolet (UV) band, from 190 to 250 nm. Although these spectra are selective, the macromolecules contain many functional groups and the typical pattern data of the fingerprint analyses are obtained. As with IRS, microscopy samples can be examined, and, because of the shorter wavelengths involved, spectra of single cells can be taken (Daltiero et al., 1987a).

15.6.2 History

Resonance RS was first applied to examination of carotenoid pigments in chromobacteria, using selective excitation at 488 nm (Howard et al., 1980), which gave spectra showing good species discrimination (Daltiero et al., 1987a). Ultraviolet resonance Raman spectrometry (UVRRS) with excitation at 244 nm (Daltiero et al., 1987b) gave strong nucleic acid spectra, with some contribution from proteins via absorption by tryptophan or tyrosine residues, and strong lines corresponding to dipicolinic acid for sporing cultures (Monoharan et al., 1990). Excitation at 251 nm gave nucleic acid spectra with minimal protein contributions, while excitation at 231 or 222 nm gave protein spectra, with the latter showing some nucleic acid contribution (Britton et al., 1988; Monoharan et al., 1990). Higher resolution spectra with excitation at 231 or 222 nm showed substantial inter-generic spectrum differences.

However, spectra also differ with the growth phase of the culture, particularly for spectra at 242 nm excitation (Nelson & Sperry, 1991). These spectra seem to reflect RNA and protein composition, and can be expected to vary with the marked changes in cell ribosome content that occur between log and stationary phase growth. By contrast, spectra from excitation at 222 nm (protein resonance) show low growth-phase variation and good inter-generic discrimination, but growth phase-dependence reappears for excitation at 218 nm, along with a stronger nucleic acid contribution.

Despite the overall growth phase-dependence at 242 nm excitation, it is possible to estimate DNA base pair (GC/AT) ratios from two lines in spectra obtained at this excitation wavelength. The intense line at 1485 cm^{-1} represents absorption by adenine and guanine (G) residues and the less intense line at 1530 cm^{-1} represents absorption by cytosine (C). Plotting the ratio of intensities for these lines against mole % G + C for a range of well-characterized organisms gave an obvious linear regression, although with a shallow slope.

In summary, the results for UVRRS are promising, but the technique is still in development and routine method protocols have yet to be formulated. Fuller information can be obtained from a recent review article (Nelson & Sperry, 1991) and from the papers cited.

15.7 OVERVIEW

The fingerprint methods described above offer a variety of rapid, automated, high-throughput instrument-based techniques for investigating whole cell composition. These methods tend to be capital-intensive, but with low disposable costs. Sample preparation is minimal, and several of the methods are known to be useful in discrimination at species and subspecies level. Three of the methods, laser pyrolysis, ultraviolet resonance Raman spectroscopy (UVRRS) and Fourier transform infrared (FTIR) spectroscopy can be applied to microscopic samples, offering some prospect that time-consuming incubation cycles could be avoided.

All of the described methods, however, yield complex pattern data. The multivariate statistical processing strategies required to reduce these patterns to similarities have been used for many years in other sciences, but their power is only just being realized in microbiology. Neural network analysis (see Appendix 15.1) may well be a significant addition to these mathematical strategies. Other methods, notably SDS-PAGE, immunoblotting and restriction endonuclease digest analysis, also yield pattern data. These methods might also benefit from the use of the mathematical strategies developed to handle the data generated by fingerprint methods.

Fingerprint techniques are part of the newly developing method infrastructure that may eventually allow laboratories to process samples with a speed more appropriate to patient treatment, cross-infection investigations and contamination and spoilage problems.

REFERENCES

Aries, R.E., Gutteridge, C.S. and Ottley, T.W. (1986) Evaluation of a low-cost, automated pyrolysis-mass spectrometer. *J. Analyt. Appl. Pyrol.* **9**: 81–98.

Athalye, M., Noble, W.C., Mallet, A.I. and Minnikin, D.E. (1984) Gas chromatography-mass spectrometry of mycolic acids as a tool in the identification of medically important coryneform bacteria. *J. Gen. Microbiol.* **130**: 513–19.

Blomquist, G., Johansson, E., Söderström, B. and Wold, S. (1979) Data analysis of pyrolysis-chromatograms by means of SIMCA pattern recognition. *J. Analyt. Appl. Pyrol.* **1**: 53–65.

Böhm, R., Kapr, T., Schmitt, H.U., Albrecht, J. and Wieser, P. (1985) Application of the laser microprobe mass analyser to the differentiation of single bacterial cells. *J. Analyt. Appl. Pyrol.* **8**: 449–61.

Bolduan, O.E.A., Muth, C.F. and Orlando, M.D. (1952) Differentiation of microorganisms by their infrared spectra. *Bacteriol. Proc.* **52**: 32.

Boon, J.J., De Boer, W.R., Kruyssen, F.J. and Wouters, J.T.M. (1981) Pyrolysis mass spectrometry of whole cells, cell walls and isolated cell wall polymers of *Bacillus subtilis* var. *niger* WM. *J. Gen. Microbiol.* **122**: 119–27.

Britton, K.A., Daltiero, R.A., Nelson, W.H., Britt, D. and Sperry J.F. (1988) Ultraviolet resonance Raman spectra of *Escherichia coli* with 225–251 nm pulsed laser excitation. *Appl. Spectrosc.* **42**: 782–8.

Daltiero, R.A., Bask, M., Nelson, W.H., Britt, D., Sperry, J.F. and Purcell, F.J. (1987a) The resonance Raman microprobe detection of single bacterial cells from a chromobacterial mixture. *Appl. Spectrosc.* **41**: 241–4.

Daltiero, R.A., Nelson, W.H., Britt, D. and Sperry, J.F. (1987b) An ultraviolet (242 nm excitation) resonance Raman study of live bacteria and bacterial components. *Appl. Spectrosc.* **41**: 417–22.

DeLuca, S., Sarver, E.W., De, B., Harrington, P. and Voorhees, K.J. (1990) Direct analysis of bacterial fatty acids by Curie-point pyrolysis tandem mass spectrometry. *Anal. Chem.* **62**: 1465–72.

Drucker, D.B. (1981) *Microbiological Applications of Gas Chromatography*. Cambridge University Press: Cambridge.

Dworzanski, J.P. and Meuzelaar, H.L.C. (1991). Rapid detection and identification of microorganisms via pyrolytic on-line derivatisation-gas chromatography-mass spectrometry. In *Modern Techniques for Rapid Microbiological Analysis* (Nelson, W.H., ed), pp. 19–42. VCH Publishers: New York.

Etémadi, A.H. (1967) The use of pyrolysis gas chromatography and mass spectroscopy in the study of the structure of mycolic acids. *J. Gas Chromatogr.* **5**: 447–56.

Eudy, L.W., Walla, M.D., Hudson, J.R., Morgan, S.L. and Fox, A. (1985) Gas chromatography-mass spectrometry studies on the occurrence of acetamide, propionamide, and furfuryl alcohol in pyrolysates of bacteria, bacterial fractions and model compounds. *J. Analyt. Appl. Pyrol.* **7**: 231–47.

Freeman, R., Goodfellow, M., Gould, F.K., Hudson, S.J. and Lightfoot, N.F. (1990) Pyrolysis mass-spectrometry for the rapid epidemiological typing of clinically significant bacterial pathogens. *J. Med. Microbiol.* **32**: 283–6.

French, G.L., Talsania, H. and Phillips, I. (1989) Identification of viridans streptococci by pyrolysis-gas chromatography. *J. Med. Microbiol.* **29**: 19–27.

Giesbrecht, D., Naumann, D., Labischinski, H. and Barnickel, G. (1984) A new method for the rapid identification and differentiation of pathogenic microorganisms using Fourier transform infrared spectroscopy. In *Rapid Methods and Automation in Microbiology and Immunology* (Habermahl, K-O., ed), pp. 198–206. Springer-Verlag: Berlin.

Goodacre, R. and Kell, D.B. (1993) Rapid and quantitative analysis of bioprocesses using pyrolysis mass spectrometry and neural networks: Application to indole production. *Analyt. Chim. Acta* **279**: 17–26.

Goodacre, R., Kell, D.B. and Biachi, G. (1992) Neural networks and olive oil. *Nature* **359**: 1992.

Goulden, J.D.S. and Sharpe, M.E. (1958) The infra-red absorption spectra of lactobacilli. *J. Gen. Microbiol.* **19**: 76–86.
Gutteridge, C.S. and Norris, J.R. (1979) The application of pyrolysis techniques to the identification of microorganisms. *J. Appl. Bacteriol.* **47**: 5–43.
Gutteridge, C.S., Vallis, L. and MacFie, H.J.H. (1985) Numerical methods in the classification of microorganisms by pyrolysis mass spectrometry. In *Computer-Assisted Bacterial Systematics* (Goodfellow, M., Jones, D. and Priest, F.G., eds), pp. 369–401. Academic Press: London.
Haynes, W.C., Melvin, E.H., Locke, J.M., Glass, C.A. and Senti, F.R. (1958) Certain factors affecting the infrared spectra of selected microorganisms. *Appl. Microbiol.* **6**: 298–304.
Helm, D., Labischinski, H., Schallehn, G. and Naumann, D. (1991) Classification and identification of bacteria by Fourier-transform infrared spectroscopy. *J. Gen. Microbiol.* **137**: 69–79.
Hercules, D.M. (1983) Organic mass spectrometry using the laser microprobe. *Pure Appl. Chem.* **55**: 1869–85.
Hindmarch, J.M., Magee, J.T., Hadfield, M.A. and Duerden, B.I. (1990) A pyrolysis mass spectrometry study of *Corynebacterium* spp. *J. Med. Microbiol.* **30**: 137–49.
Horbach, I., Naumann, D. and Fehrenbach, F.J. (1988) Simultaneous infections with different serogroups of *Legionella pneumophila* investigated by routine methods and Fourier transform infrared spectroscopy. *J. Clin. Microbiol.* **26**: 1106–10.
Howard, W.F., Nelson, W.H. and Sperry, J.F. (1980) A resonance Raman method for rapid detection and identification of bacteria in water. *Appl. Spectrosc.* **34**: 72–5.
Hudson, J.R., Morgan, S.L. and Fox, A. (1982) Quantitative pyrolysis gas chromatography-mass spectrometry of bacterial cell walls. *Anal. Biochem.* **120**: 59–65.
Huff, S.M., Meuzelaar, H.L.C., Pope, D.L. and Kjeldsberg, C.R. (1981) Characterisation of leukaemic and normal white blood cells by Curie point pyrolysis mass spectrometry. 1. Numerical evaluation of the results of a pilot study. *J. Analyt. Appl. Pyrol.* **3**: 95–110.
Huff, S.M., Matsen, J.M., Windig, W. and Meuzelaar, H.L.C. (1986) Pyrolysis mass spectrometry of bacteria from infected human urine. *Biomed. Environ. Mass Spectrom.* **13**: 277–86.
Kenner, B.A., Riddle, J.W., Rockwood, S.W. and Bordner, R.H. (1958) Bacterial identification by infrared spectrophotometry. II. Effect of instrumental and environmental variables. *J. Bacteriol.* **75**: 16–20.
Kusaka, T. and Mori, T. (1986) Pyrolysis gas chromatography-mass spectrometry of mycobacterial mycolic acid methyl esters and its application to the identification of *Mycobacterium leprae*. *J. Gen. Microbiol.* **132**: 3403–6.
Levine, S., Stevenson, H.J.R., Chambers, L.A. and Kenner, B.A. (1953) Infrared spectrophotometry of enteric bacteria. *J. Bact.* **65**: 10–15.
Lindner, B. and Seydel, U. (1983) Mass spectrometric analysis of drug-induced changes in Na^+ and K^+ contents of single bacterial cells. *J. Med. Microbiol.* **129**: 51–5.
Madorsky, S.L., Strauss, S., Thompson, D. and Williamson, L. (1949) Pyrolysis of polyisobutene (Vistanex), polyisoprene, polybutadiene, GR-S, and polythene in a high vacuum. *J. Res. Nat. Bur. Stand.* **42**: 499–509.
Magee, J.T. (1993) Whole-organism fingerprinting. In *Handbook of New Bacterial Systematics* (Goodfellow, M. and O'Donnell, A.G., eds), pp. 383–427. Academic Press: London.
Magee, J.T., Hindmarch, J.M., Duerden, B.I. and Mackenzie, D.W.R. (1988) Pyrolysis mass spectrometry as a method for inter-strain discrimination of *Candida albicans*. *J. Gen. Microbiol.* **134**: 2841–7.
Magee, J.T., Hindmarch, J.M., Bennett, K.W., Duerden, B.I. and Aries, R.E. (1989a) A pyrolysis mass spectrometry study of fusobacteria. *J. Med. Microbiol.* **28**: 227–36.
Magee, J.T., Hindmarch, J.M., Burnett, I.A. and Pease, A. (1989b) Epidemiological typing of *Streptococcus pyogenes* by pyrolysis mass spectrometry. *J. Med. Microbiol.* **30**: 273–78.
Magee, J.T., Hindmarch, J.M. and Nicol, C.D. (1991) Typing of *Streptococcus pyogenes* by pyrolysis mass spectrometry. *J. Med. Microbiol.* **35**: 304–6.
Magee, J.T., Hindmarch, J.M. and Meechan, D.F. (1983) Identification of staphylococci by pyrolysis gas-liquid chromatography. *J. Med. Microbiol.* **16**: 483–95.
Manley, B.F.J. (1986) *Multivariate Statistical Methods, A Primer*. Chapman & Hall: London.
Meuzelaar, H.L.C. and Kistemaker, P.G. (1973) A technique for fast and reproducible fingerprinting of bacteria by pyrolysis mass spectrometry. *Anal. Chem.* **45**: 587–90.
Meuzelaar, H.L.C., Kistemaker, P.G., Eshuis, W. and Engel, H.W.B. (1976a) Progress in

automated and computerised characterisation of microorganisms by pyrolysis mass spectrometry. In *Rapid Methods and Automation in Microbiology* (Johnston, H.H. and Newsom, S.W.B., eds), pp. 225–30. Learned Information: Oxford.

Meuzelaar, H.L.C., Kistemaker, P.G., Eshuis, W. and Boerboom, H.A.J. (1976b) Automated pyrolysis mass-spectrometry: Application to the differentiation of microorganisms. *Adv. Mass Spectrom.* **7B**: 1452–6.

Meuzelaar, H.L.C., Haverkamp, J. and Hileman, F.D. (1982) *Pyrolysis Mass Spectrometry of Recent and Fossil Biomaterials. Compendium and Atlas.* Elsevier: Amsterdam.

Monoharan, R., Ghiamati, E., Daltiero, R.A., Britton, K.A., Nelson, W.H. and Sperry J.F. (1990) UV resonance Raman spectra of bacteria, bacterial spores, protoplasts and calcium dipicolinate. *Microbiol. Meth.* **11**: 1–15.

Morgan, S.L., Fox, A., Rogers, J.C. and Watt, B.E. (1991) Identification of chemical markers for microbial differentiation and detection by gas chromatography-mass spectrometry. In *Modern Techniques for Rapid Microbiological Analysis* (Nelson, W.H., ed), pp. 1–18. VCH Publishers: New York.

Naumann, D., Helm, D., Labischinski, H. and Giesbrecht, P. (1991) The characterization of microorganisms by Fourier-transform infrared spectroscopy. In *Modern Techniques for Rapid Microbiological Analysis* (Nelson, W.H., ed), pp. 43–96. VCH Publishers: New York.

Needleman, M. and Stuchberry, P. (1977) The identification of micro-organisms by pyrolysis gas-liquid chromatography. In *Analytical Pyrolysis* (Jones C.E.R. and Cramers, C.A., eds), pp. 77–88. Elsevier: Amsterdam.

Nelson, W.H. and Sperry, J.F. (1991) UV resonance Raman spectroscopic detection and identification of bacteria and other microorganisms. In *Modern Techniques for Rapid Microbiological Analysis* (Nelson, W.H., ed), pp. 97–143. VCH Publishers: New York.

Norris, K.P. (1959) Infra-red spectroscopy and its application to microbiology. *J. Hyg.* **57**: 326–45.

Norris, K.P. and Greenstreet, J.E.S. (1958) On the infrared absorption spectrum of *Bacillus megaterium*. *J. Med. Microbiol.* **19**: 566–80.

Quinn, P.A. (1974) Development of high resolution pyrolysis gas chromatography for the identification of microorganisms. *J. Chromatogr. Sci.* **12**: 796–806.

Quinn, P.A. (1976) Identification of microorganisms by pyrolysis: The state of the art. In *Rapid Methods and Automation in Microbiology* (Johnston, H.H. and Newsom, S.W.B., eds), pp. 178–86. Learned Information: Oxford.

Riddle, J.W., Kabler, P.W., Kenner, B.A., Bordner, R.H., Rockwood, S.W. and Stevenson, H.J.R. (1956) Bacterial identification by infrared spectrophotometry. *J. Bacteriol.* **72**: 593–603.

Saddler, G.C., Falconer, C. and Sanglier, J.J. (1989) Preliminary experiments for the selection and identification of actinomycetes by pyrolysis mass spectrometry. *Actinomycetologia* **2**: S3–S4.

Sekhon, A.S. and Carmichael, J.W. (1973) Column variation affecting a pyrolysis gas-liquid chromatographic study of strain variation in two species of *Nannizzia*. *Can. J. Microbiol.* **19**: 409–11.

Shute, L.A., Berkeley, R.C.W., Norris, J.R. and Gutteridge, C.S. (1985) Pyrolysis mass spectrometry in bacterial systematics. In *Chemical Methods in Bacterial Systematics* (Goodfellow, M., Jones, D. and Priest, F.G., eds), pp. 94–114. Academic Press: London.

Simmonds, P.G. (1970) Whole microorganisms studied by pyrolysis-gas chromatography-mass spectrometry: Significance for extraterrestrial life detection systems. *Appl. Microbiol.* **20**: 567–72.

Söderström, S., Wold, S. and Blomquist, G. (1982) Pyrolysis-gas chromatography combined with SIMCA pattern recognition for classification of fruit-bodies of some ectomycorrhizal *Suillus* species. *J. Gen. Microbiol.* **128**: 1773–84.

Stack, M.V., Donoghue, H.D. and Tyler, J.E. (1978) Discrimination between oral streptococci by pyrolysis gas-liquid chromatography. *Appl. Environ. Microbiol.* **35**: 45–50.

Voorhees, K.J., Durfe, S.L., Holtzclaw, J.R., Enke, C.G. and Bauer, M.R. (1988) Pyrolysis-tandem mass spectrometry of bacteria. *J. Analyt. Appl. Pyrol.* **14**: 7–15.

Watt, B.E., Morgan, S.L. and Fox, A. (1991) 2-Buteneoic acid, a chemical marker for poly-β-

15. Analytical Fingerprinting Methods

hydroxybutyrate identified by pyrolysis-gas chromatography mass-spectrometry in analyses of whole microbial cells. *J. Analyt. Appl. Pyrol.* **20**: 237–50.

Wieten, G., Haverkamp, J., Engel, H.W.B. and Berwald, L.G. (1981) Application of pyrolysis mass spectrometry to the classification and identification of mycobacteria. *Rev. Infect. Dis.* **3**: 871–7.

Wieten, G., Haverkamp, J., Groothuis, D.G., Berwald, L.G. and David, H.L. (1983) Classification and identification of *Mycobacterium africanum* by pyrolysis mass spectrometry. *J. Gen. Microbiol.* **129**: 3679–88.

Windig, W. and De Hoog, G.S. (1982) Pyrolysis mass spectrometry of selected yeast species: II. *Sporidiobolus*. *Studies in Mycology* **22**: 60–4.

Windig, W. and Haverkamp, J. (1982) Pyrolysis mass spectrometry of selected yeast species: I. *Rhodosporidium*. *Studies in Mycology* **22**: 56–59, 69–74.

Windig, W. and Meuzelaar, H.L.C. (1987) Numerical extraction of components from mixture spectra by multivariate data analysis. In *Computer-Enhanced Analytical Spectroscopy* (Meuzelaar, H.L.C. and Isenhour, T.L., eds), pp. 67–102. Plenum: New York.

Windig, W., Kistemaker, P.G., Haverkamp, J. and Meuzelaar, H.L.C. (1979) The effects of sample preparation, pyrolysis and pyrolyzate transfer conditions on pyrolysis mass spectra. *J. Analyt. Appl. Pyrol.* **1**: 39–52.

Windig, W., Kistemaker, P.G. and Haverkamp, J. (1980) Factor analysis of the influence of changes in experimental conditions in pyrolysis-mass spectrometry. *J. Analyt. Appl. Pyrol.* **2**: 7–18.

Windig, W., Kistemaker, P.G. and Haverkamp, J. (1981a) Chemical interpretation of differences in pyrolysis-mass spectra of simulated mixtures of biopolymers by factor analysis with graphical rotation. *J. Analyt. Appl. Pyrol.* **3**: 199–212.

Windig, W., De Hoog, G.S. and Haverkamp, J. (1981b) Chemical characterization of yeasts and yeast-like fungi by factor analysis of their pyrolysis-mass spectra. *J. Analyt. Appl. Pyrol.* **3**: 213–20.

Windig, W., Haverkamp, J. and Kistemaker, P.G. (1983) Chemical interpretation of sets of pyrolysis mass spectra by discriminant analysis and graphical rotation. *Anal. Chem.* **55**: 81–7.

APPENDIX: THE MATHEMATICS OF MULTIVARIATE ANALYSES

The detailed concepts of multivariate analysis are explained in several texts; Manley's 'Primer' (Manley, 1986) is particularly readable. Briefly: the intensities measured for individual masses differ in within-replicate reproducibility and show covariance, i.e. inter-mass correlations. A simple matrix formula can correct the spectra for these statistical effects:

$$D^2 = [A - B] [M]^{-1} [A - B]'$$

D^2 is a measure of statistical significance of pattern differences, the Mahalanobis distance between spectra A and B. $[A - B]$ is a column of figures (column vector), each representing the difference in intensity for a single mass between spectra A and B. $[A - B]'$ is a row vector, the $[A - B]$ column, rearranged as a row. $[M]^{-1}$ is the inverse of a covariance matrix that lists the variance within-masses, and covariance between-masses. D^2 is an approximation of the χ^2 significance of difference between spectra A and B on $n_g - 1$ degrees of freedom, where n_g is the number of replicate spectrum sets. This is the multivariate equivalent of:

$$d^2 = (a-b)^2/v$$

where d is the difference between two measurements, a and b, in standard deviation units, and v is the variance of the measurement.

With this formula, multivariate analyses plot a scatter of spectrum points in n_m dimensional space (where n_m is the number of masses analysed). The axes are corrected for replicate variance and covariance and the origin of the new axis system is placed at the centre of the point scatter: the overall mean-spectrum point. The first axis is an eigenvector, or best fit regression line (i.e. the line giving best discrimination) through the corrected point swarm. Variation along this axis is eliminated, a second regression line is calculated, and so on, repeatedly until $n_g - 1$ axes have been calculated. The position of a spectrum on an axis is given by a formula:

$$C_i = K_{i,1} \times I_1 + K_{i,2} \times I_2 + \ldots\ldots + K_{i,m} \times I_m + k_i$$

where C_i is the coordinate of the spectrum on axis i, m is the number of masses analysed, $K_{i,1\ldots m}$ is the constant for axis i and masses $1 \ldots m$, $I_{1\ldots m}$ is the intensity for mass $1 \ldots m$ and k_i is a constant for axis i that brings the origin to the overall mean spectrum point. In effect, the axis represents a discriminatory sub-spectrum component, the K terms represent the contributions of the masses to the component, and the coordinate of a spectrum on the axis represents the contribution of this sub-spectrum to the overall spectrum pattern. The axes are variously termed principal components, canonical variates or canonical discriminant variates. The Mahalanobis distance is the sum of the squared differences between spectrum coordinates:

$$D^2 = (C_{a1} - C_{b1})^2 + (C_{a2} - C_{b2})^2 \ldots\ldots + (C_{an} - C_{bn})^2$$

The proportion of discrimination represented on each axis can be calculated from its eigenvalue, and is usually tabulated in printouts. Higher numbered axes often represent mere random, sample-specific variation, and some programs give a significance figure for the axes, based on a Wilks' lambda test. Axes with a significance $>1\%$ can justifiably be eliminated from further processing.

Inversion of the variance-covariance matrix $[M]$ presents a firm limitation on experimental design. In univariate analysis a standard deviation cannot be defined from a single figure; in multivariate analysis the n_m^2 elements of the matrix cannot be defined for matrix inversion, unless the

15. Analytical Fingerprinting Methods

number of masses analysed (n_m) is < n_1, the number of spectra analysed minus the number of replicate spectrum sets.

This limit usually requires pruning of the spectrum data, either by selecting the most discriminatory masses (Horizon software), or by selecting the masses showing good within-culture reproducibility (my own approach). The former is probably more appropriate to typing, and the latter to classification (Magee, 1993). Whichever selection method is adopted, it is wise to leave a good margin between n_m and n_1. The univariate analogy is that although a standard deviation can be calculated from two figures, it is preferable to calculate from a larger body of data.

There are several multivariate strategies based on these formulae, most of which are available in the software provided by the manufacturer (based on Genstat 5), and in common PC-based statistical programs such as SPSS-PC and SAS.

In principal components analysis (PCA), the [M] matrix represents the variance and covariance calculated over the entire spectrum set; PCA is useful for a preliminary rough view of data and elimination of grossly aberrant spectra. It has also been used to compress spectrum data into principal component coordinates for processing in discriminant or canonical variate analysis (DA, CVA; see below). The main problem of this stacked PCA-CVA approach is that PCA assumes the data form a single multivariate normal distribution. If PCA shows distinct groups, then this assumption is invalid. The statistical corrections applied in PCA will be exaggerated by inclusion of inter-group comparisons in the variance estimate, and discrimination will be sub-optimal. Therefore, the stacked PCA-CVA strategy, in which the spectrum coordinates from PCA are passed to CVA, has been discarded by most current workers.

Discriminant analysis and CVA, as used in Py-MS, are almost identical, the main operational difference being that DA includes statistical methods to eliminate less discriminatory masses from the analysis, but CVA does not. The most common current approach to Py-MS data analysis is to define the replicate spectrum sets from single culture plates as the replicate groups, and obtain the mean-culture canonical variate coordinates. The inter-culture statistical differences are then visualized as scatter diagrams of points on the first two or three canonical axes, or coordinates on the significant axes are submitted to cluster analysis giving a similarity matrix and dendrogram.

In general, only partial visualizations of the discrimination can be obtained. Scatter diagrams rarely represent >80% of the statistical differences, and dendrograms rarely have cophenetic correlations >0.85. Scatter diagrams tend to be most useful in representation of the group structure, and dendrograms in representation of differences between similar individual cultures. Both should be viewed.

Close examination of the similarity matrix is essential; this tabulates the inter-culture Mahalanobis distances for each pair of cultures, showing the full discrimination. These distances are approximations of the χ^2 probability that the cultures are distinct on the number of cultures minus one degrees of freedom. However, their presentation in printouts is often unwieldy and difficult. In similarity matrices generated from CLUSTAN 2 with the Euclidean distance^2 similarity measure, the distance values are divided for the number of axes and must be multiplied by the number of axes before comparison to χ^2.

In typing studies, D^2 thresholds for indistinguishable strains can be established by examining the range of D^2 for duplicate cultures of the same strain in this table, exemplified in an early typing study (Magee et al., 1988). Alternatively, examination for values $<\chi^2$ on $n_g - 1$ degrees of freedom at $p = 0.001$ has been successful (Magee et al., 1991). The low significance level was adopted because DA and CVA inherently tend to exaggerate the significance of differences.

Identification study data are best analysed in DA, which includes routines to compare 'unknowns' with standard strains. SPSS-PC yields two identification probabilities. One, termed P(G/D), shows the probabilities that the unknown is a member of the first- and second-choice assigned groups, given that the standard strains include all possible groups. This is effectively a measure of the equivocality of identification. The other, termed P(D/G), is the probability that a member of the first-choice group might give a spectrum with the same pattern as the unknown. This is best regarded as a measure of 'strangeness'.

Identification is a powerful confirmatory adjunct to classification. The strains of the proposed groups are divided equally between a teaching and blind challenge set, and analysed in DA. Identification results for each spectrum of the challenge set are noted, the sets reversed, and the former teaching set identified based on the characteristics of the former challenge set. In this way, all spectra are identified 'blind', giving a much better estimate of the prospective identification rate than identification from a uniform teaching set, which gives grossly optimistic estimates. High prospective identification rates indicate homogeneous, well-differentiated groups. Low rates indicate poor homogeneity or differentiation. However, low rates are expected for small groups, where identification is based on comparison of the unknown with only a few strains and spectra (Hindmarch et al., 1990).

Other approaches include SIMCA analysis (Blomquist et al., 1979; Söderström et al., 1982; Saddler et al., 1989), an identification method that has yielded good results, but is not included in common statistical packages; this is a modified form of PCA. Univariate analyses have been largely discarded, but are discussed elsewhere (Magee, 1991). An exciting

15. Analytical Fingerprinting Methods

development is the use of neural network systems for identification and quantitative differentiation from Py-MS data (Goodacre *et al.*, 1992; Goodacre & Kell, 1993). Studies with this approach have proved remarkably successful. The ability to predict quantitative differences in a single component, exemplified by a study of indole production (Goodacre & Kell, 1993), opens a wide range of possible new applications. Also, with only a single group in the teaching set, identification of unknowns is impossible in DA and CVA. However, network analysis handled this type of problem well in a study that compared blind-coded olive oil samples to authentic standards (Goodacre *et al.*, 1992). Adulterated samples were successfully differentiated, and unadulterated samples were assigned to the authentic group.

Index

Note: page numbers in *italics* refer to figures and tables

m-A$_2$pm, 24–5
Acetobacter pasteurianus, difference spectra, *333*
acetylation of carotenoid hydroxy groups, 437
Acinetobacter, 508
Actinomyces pyogenes demethylmenaquinone, 273
actinomycetes, 5, 6
  chemical markers, 15
  fatty acids, 123
  outer membrane, 9
  phosphatidylcholines, 142
  phosphatidylinositol mannosides, 8
acylglycoses, 144
adenine, 545
adsorption HPLC *see* normal-phase high-performance liquid chromatography (HPLC)
aerobic bacteria
  bacteriochlorophyll *a*, 371
  freshwater, 372
  photosynthetic, 371, 372
aglycone portion of glycosidase, 477
D-alanine, 475
L-alanine, 475
alditol glycosides, 77
algae
  hopanoid extraction, 169–70
  major sterols, *192*
  saponification, 169–70
  sterol extraction, 169–70
  sterol taxonomy, 191
allomerization, 347
allylic hydroxy group dehydration of carotenoids, 438
alumina, carotenoid adsorption, 416, 417

D-amino acid oxidase, 474
amino acids
  C-terminal removal, 487
  chiral, 475
  composition of Gram-positive bacteria peptidoglycan, 72
  N-terminal removal, 487
7-amino–4-methylcoumarin (7-AMC), 480, 481, 482, 484
  aminopeptidase demonstration, 486
  endopeptidase demonstration, 486
  fluorescence detection, 489
  peptidyl derivatives, 488
amino-hopanoid derivatives, 173
aminobacteriohopanetetrol, *166*
aminopeptidases, 479
  7-AMC demonstration, 486
aminophosphoglycolipids, 234
aminophospholipids, 234
6-aminoquinoline, 489
amino sugars, 43–4
aminotransferases, 474
aminouronic acids, 80, 81
ammonium persulphate oxidation, 321
amphipathic polar lipids, 7
Amycolata, 490
amygdalin, 477
anaerobic bacteria, cytochromes, 339
analytical instrumentation, 2
anhydrorhodovibrin, 424
anhydrorhodovibrin, absorption spectrum, *433*
anionic polymers, Gram-positive bacteria, 74–81
anoxygenic phototrophs, carotenoids, 407–8
anthranilonitrile, 490
API 20E, 471–2

arabinogalactan, 123
arachaea
  lipids, 10
  plasma membrane, 10
arbutin, 477
archaea, 6–12
  cell envelope-less, 114
  cell wall analysis, 5
  classification, *86–9*
  cytochromes, 339–40
  glycerol ethers, 198
  glycoprotein S-layers, 114
  halophilic, 197, 200
  heteropolysaccharide, 113
  membrane lipids, 198
  methanochondroitin, 113
  methanogenic, 197, 199
  molecular components, 197–8
  murein lack, 112
  protein S-layers, 113
  pseudomurein, 112–13
  S-layers, 113–14
  SDS-resistant layers, 114
  taxonomy, 111–14
archaeal cell envelope analysis, 85–114
  electron microscopy, 85, 90, *91*, 92–3
  isolation, 93–5
  profiles, 90
  S-layer, 90, 92
  specimen preparation, 92–3
archaeal lipids, 197–8
  acid methanolysis, 204
  chromatographic analyses, 200–3
  chromatographic procedures with diethers, 209–11
  core, 206–31
  degradative procedures, 204–6
  diether polar, 241
  elution solvents, 202
  extraction, 198–200
  fractionation by column chromatography, 201
  glycolipid detection, 203
  halophilic, 231, *232*, 233–4
  isopranyl diethers, 207, *208*
  isoprenoid ethers, 227–8
  mass spectrometry, 213, *214*, 215–16
  methanogenic, 234, *235*, 236–7
  neutral, 200
  NMR spectroscopy, 212–13
  overall composition, 203–4
  phospholipid detection, 203
  polar, 200, 202, 231, *232*, 233–4, *235*, 236–8, *239*, 240–9
  purification of fractions, 202
  rapid screening of core, 206
  structural characterization, *205*
  taxonomic significance of diether/tetraether occurrence, 228, *229–30*, 231
  tetraethers, 216, *217*, 218–20, *221*, 222–4, *225*, 226
  thermophilic, 237–8, *239*, 240–1
  uniqueness, 198
  yield, 200, *201*
archaeal quinones, 250–1, *252–3*, 254, *255-6*, 257, *258*, 259
  chromatography, 254
  distribution, 257, *258*, 259
  mass spectrometry, 254, *256*, 257
  NMR spectroscopy, 254, *255*
  structural types, 250–1, 254
  ultraviolet spectroscopy, 257
  visible spectroscopy, 257
Archaeoglobus fulgidus, menaquinones, 267
argentation chromatography, isoprenoid quinones, 286–9, *290*, *291*
arginine dihydrolase (ASH) test, 472
Arthrobacter, 6
aryl polyenes, 444
arylcarotenoids, 410, 444
Athiorhodaceae, 368
average linkage clustering technique, 509
Azotobacter, cytochromes, 328
A. vinelandii, nitrogenase protection, 339

Bacillus anthracis, laser pyrolysis, 536
B. cereus, laser pyrolysis, 536
B. subtilis, cytochromes, 337
B. thuringiensis, laser pyrolysis, 536
bacteria
  cell wall analysis, 5
  chemical techniques in classification/identification, 4
  cultivation, 494–5
  DNA homology of strains, 493
  evolutionary branching events, 387–8
  hopanoid extraction, 169–70
  new systematics, 3

Index

solvent extraction of lipids, 169
sterol extraction, 169–70
bacterial markers, 50–1
bacterial proteins
  automated operator-independent normalization, 501, 503
  background trend, 507
  cluster analysis, 509
  electrophoresis, 495–8
  electrophoretic traces of patterns, 498–9
  fingerprints, 499
  lateral shifting, 508–9
  operator-dependent normalization, 501
  pattern types, 499
  preparation, 495–8
  removal of major bands, 507–8
  reproducibility of electrophoretic system, 499
  stacking gel preparation, 497
  standardization of length of patterns, 501
  taxonomic considerations, 513, *514–15*, 516
  visual comparison of whole-cell patterns, 498–9
  whole-cell SDS-PAGE, 494–501, *502*, 503–4, *505*, 506–13
bacteriochlorin *see* bacteriochlorophyll *a*
bacteriochlorophyll, 12, 345–88
  analytical methods, 349–66
  characterization, 355–9, *360*, 361–3
  in chlorosomes of green bacteria, 378
  chromatography, 351, 353–5
  colour of bacteria, 371
  estimation, 349–51
  green bacteria, 367
  HPLC purification, 354–5
  isolation, 349–51
  light-harvesting function, 348
  organic solvent absorption maxima, *352*
  photosynthetic symbiotic nitrogen-fixing bacteria, 373
  purification, 351, 354
  purple bacteria, 367
  RCs of purple bacteria, 376
  reversed-phase chromatography, 353
  reversed-phase HPLC, 354
  solvent systems, 350
  structure, *346*
  taxonomic considerations, 366–9, *370*, 371–88
bacteriochlorophyll 663, 365
bacteriochlorophyll *a*, 345, *346*, 347, 348
  aerobic bacteria, 371
  aggregation studies, 356–7
  allomerization, 347
  biosynthesis intermediates, 357
  ^{13}-NMR, 357
  characterization, 356–7
  chemical oxidation, 356
  epimerization, 347
  esterifying alcohol, 364
  evolution from chlorophyll, 386
  extinction coefficient, 351
  ^{1}H-NMR, 356–7
  HPLC, 355
  ligation state, 357
  light instability, 347
  pheophytination, 347
  preparation method, 350–1
  structure, 356–7
  synthesis, 386
bacteriochlorophyll *b*, *346*, 347–8
  characterization, 357–8
  esterifying alcohol, 364–5
  ^{1}H-NMR, 357
  mass spectrometry, 357
  photosensitivity, 348
bacteriochlorophyll *c*, *346*, 348–9
  characterization, 358–9, *360*, 361–3
  chromatography, 353, 355
  countercurrent distribution, 353–4
  esterifying alcohols, 365
  extinction coefficient, 351
  green bacteria, 378–9, 380
  ^{1}H-NMR, 358
  δ-methyl groups, 359, 361
  structures, 359, *360*, 361
bacteriochlorophyll *d*, *346*, 348–9
  characterization, 358–9, *360*, 361–3
  chromatography, 353
  conversion to phylloporphyrins, 361
  countercurrent distribution, 353–4
  esterifying alcohols, 365
  extinction coefficient, 351
  green bacteria, 378–9
  structures, 359, *360*

bacteriochlorophyll e, 346, 348–9
  characterization, 358–9, 360, 361–3, 363
  esterifying alcohols, 365
bacteriochlorophyll f, 348 349
bacteriochlorophyll g, 346, 347–8
  characterization, 357–8
  esterifying alcohols, 366
  evolution, 386
  ^{1}H-NMR, 357
  mass spectrometry, 358
  photosensitivity, 348
bacteriochlorophyllides, 349
bacteriohopanepolyols, 163
bacteriohopanetetrol, 166
bacteriopheophorbide, 349
bacteriopheophytin, 347
bacterioviridin, 356
Bacteroides, cytochromes, 339
benzoquinones, 266, 279
benzothiophenquinones
  archaeal, 251, 252
  ultraviolet/visible spectroscopy, 257
BIODBM, 512
blood clotting factor, 265
brown bacteria, 345, 347
  photosynthetic prokaryotes, 367
  taxonomy, 376–81
Brucella abortus, lipid A, 28
butanol, LPS extraction, 33

C24-sterol epimer separation, 189
^{13}C-NMR
  archaeal polar lipids, 243–7
  archaeal quinones, 254
  bacteriochlorophyll a, 357
  carotenoid structure, 442–3
  chemical shifts with archaeal polar lipids, 246
  isopropanid diether analysis, 212–13
  tetraethers, 220, 222–4
caldariellaquinone, 279–80
  mass spectrometry, 297
  NMR, 303
  ultraviolet spectroscopy, 291, 293
caloxanthin, 407
Campylobacter, 507
  cytochromes, 339
  naphthoquinones, 274
  numerical analysis of SDS-PAGE protein profiles, 510
  strains investigated, 513, 514–15, 516
C. jejuni
  constant neutral loss (CNL) scan, 151, 154
  FAB spectra, 151, 152
  methyl-substituted menaquinone, 274
canonical variate analysis (CVA), 551–2, 553
canthaxanthin
  anoxygenic phototrophs, 408
  spectral fine structure, 432
capillary gas chromatography
  fatty acid methyl ester analysis, 131–3, 134, 135–6
  hot needle method, 132–3
capsular polysaccharides, 23–4
carbon monoxide
  cytochrome oxidase identification, 321
  difference spectrum, 321–2
carbonyl group reduction by sodium borohydride, 437–8
carboxy-lyases, 475
β-carotene, 404, 405, 444
  anoxygenic phototrophs, 408
  cyanobacteria, 407
carotene
  gradient procedure HPLC, 422–3
  spectral fine structure, 432
carotenoid glycosides, liberation and identification of sugar residues, 438–9
carotenoid pigments, 10, 12–13
carotenoids, 403–47
  acetylation/silylation of hydroxy groups, 437
  acid condition effects, 413
  acyclic, 412, 424
  adsorbents for chromatography, 416–17
  alkali condition effects, 413
  analysis, 411
  anoxygenic phototrophs, 407–8
  applications of analysis in practice, 443–4
  biosynthesis, 405–6, 410
  bis-allylic fragmentation, 441–2
  carbonyl group reduction by sodium borohydride, 437–8
  chirality, 443
  chromatography, 413, 415–21

Index

chromophore extension, 426–7
chromophore wavelengths of maximal absorption, *428–31*
circular dichroism, 443
cis-peak, 433
column chromatography, 416
composition determination, 411–12
cyanobacteria, 407
cyclic, 427
dehydration of allylic hydroxy groups, 438
derivative formation, 436–9
diagnostic chemical reactions, 436–9
distribution, 407–8, *409*, 410–11
distribution in photosynthetic bacteria, *408–9*
double bonds, 404
end-groups, 404, 440–2
Erythrobacter longus, 371
extraction, 411, 414–15
fragmentations, 440
functional groups, 440–2
geometrical isomers, 433
heat protection, 412–13
HPLC, 411, 421–6
identification, 439
impurities, 414
iodine-catalysed geometrical isomerization, 436–7
isolation, 415–21
isometry, 404–5
lability, 418–19
in lake sediments, 443–4
light protection, 412–13
lyophilization of cell material, 414
molecular mass, 439–40
NMR spectroscopy, 439, 442–3
nomenclature, 403–4
non-photosynthetic prokaryotes, 408, 410–11
normal-phase HPLC, 424–6
oxidation prevention, 412
paper chromatography, 420
plasticizer contamination, 414
polyene chain fragmentations, 440
properties, 412
purification, 411, 415–21
purification for mass spectrometry/ NMR spectroscopy, 420–1
quantitative determination by HPLC, 434–6

reagent purity, 413–14
reversed-phase HPLC, 422–4
saponification, 415
second extraction, 414
semi-systemic names, *449–61*
separation, 411
solvents, 413–14, 415, 417–18
spectral fine structure, 427, *432*
spectrophotometry, 434, *435*
spectroscopy, 426–7, *428–32*, 433
storage conditions, 412
structure, 403–5, *449–61*
TLC, 416, 417, 418–20, 443, *444*
catalase, 331, 474
Cathepsin B, 487
cell envelope
 Gram-negative bacteria, 21
 prokaryote, 5
 proteins, 25–6
 structure of archaea, 85
cell wall polymers, anionic, 64–5
cetyl alcohol, 365
Chattonella antiqua
 24-ethylcholesterol content, 176, 177
 gas chromatogram, *176*
chemical constituent resolution, 3
chemical fingerprinting, 13
chemical ionization, 148
chemical markers, 15–16
chemolithotrophic bacteria, cytochromes, 338–9
chemometric strategies, 523
chemosystematics, 3, 16
chemotaxonomy, 5
chitin hydrolysis, 472
chloramphenicol acetyltransferase, 474
Chlorella, sterol components, 192
chlorins, 349
Chlorobiinae, 376
Chlorobium
 bacteriochlorophyll c/d, 379
 bacteriochlorophyll extraction, 358
 lake sediments, 444
 C. limicola f. *thiosulfatophilum*, 361, 379, 381
 C. phaeovibrioides, bacteriochlorophyll e, 363
 C. vibrioforme f. *thiosulfatophilum*, 379
 bacteriochlorophyll d multiple esters, 365

chlorobiumquinone, 271–2
Chloroflexus aurantiacus, 376, 377
  bacteriochlorophyll *c*, 363
  bacteriochlorophyll *c* esterifying alcohols, 365
  LH complex of bacteriochlorophyll *a*, 380
  RC complex, 380
Chloroflexus-like organisms, 385, 387
chloroform, protein denaturing, 464
Chloronema, 377–8
chlorophyll
  bacteriochlorophyll *a* evolution, 386
  photosynthetic prokaryotes and evolution, 385–8
chlorophyll *a*
  bacterial evolution, 388
  esterifying alcohol, 364
chlorophyll *b*
  esterifying alcohol, 364
  formation from chlorophyll *a*, 388
chlorophyllide, 385
chloroplasts
  evolution of green plant, 384
  evolutionary origin, 384
  phylogenetic origin, 383
  16S rRNA sequence analysis, 383
chlorosomes, 376, 377–8
5α-cholestanol, *185*
Chromatium, 444
C. vinosum, 328
chromatography
  handling precautions for carotenoids, 413
  isoprenoid quinones, 283
chromophore
  carotenoid wavelengths of maximal absorption, *428–31*
  extension of in carotenoids, 426–7
chromophytes, 388
chymotrypsin-like activity, 487–8
circular dichroism in carotenoids, 443
cis-peak for carotenoids, 433
classification, 1, 2
Clostridium difficile, colonial polymorphism, 529
co-chromatography, carotenoid identification, 421
coenzyme Q, 11, 276
collision-induced decomposition (CDI) spectrum for menaquinones, 296, 297

column chromatography
  carotenoids, 416, 417–18
  sterol separation, 171, 172
constant neutral loss (CNL) scan, 151, *154*
coprostanol, 182, *183*
core lipids, archaeal, 206–31
  acid methanolysis, 206
coumarins, 489
cryptophytes, 388
cyanobacteria, 345
  carotenoids, 12, 407
  evolutionary origins, 386
  hopanoids, 168
  lake sediments, 444
  major sterols, *192*
  photosynthetic prokaryotes, 367
  phylogenetic group with chloroplasts, 383
  pigment composition screening, 422
  RCs, 387
  16S rRNA sequence analysis, 383
  sterols, 168
  taxonomy, 382–3
Cyanophora paradoxa cyanelle, 388
cycloartenol, 163, 164, *165*
cycloheptyl fatty acids, 123
cyclohexyl fatty acids, 123
cyclopropane fatty acids, 123
  extraction, 135
cytochrome, 11
  a_1-like pigments, 332, 334
  α-bands, 332
  absorption coefficients, *336*
  ammonium persulphate oxidation, 321
  anaerobic bacteria, 339
  analysis, 311–41
  analytic methods, 316–27
  assignment of visible absorption bands, 327–32, *333*, 334–5, *336*, 337
  β-bands, 332
  Bacillus subtilis, 337
  carbon monoxide difference spectrum, 321–2
  cellular respiration, 311–12
  chemolithotrophic bacteria, 338–9
  classes, 313, 315
  CO complexes, 334
  CO-reduced minus reduced difference spectra, 334–5, 337
  complex preparations, 315

Index   561

distribution, 311
double-beam spectrophotometry, 318
dual wavelength spectrophotometry,
    318–19
extinction coefficients, 341
fluorescence techniques, 326–7
γ-bands, 335
Gram-positive bacteria, 337–8
growth conditions of bacteria, 338
infrared spectroscopy, 325
interpretation of spectra, 331–2, 333,
    334–5
low-intensity absorbance, 332
low-temperature spectra recording,
    319
nomenclature, 312–16, 314, 337
optical spectroscopy, 312
oxidase test, 325–6
oxidation, 320–1
pattern distribution, 337
patterns, 311
peroxidase test, 326
phototrophic bacteria, 339
protein concentration of cell
    suspension, 335
quantification, 335, 336, 337
reduced minus oxidized spectra, 332,
    333, 334
reduction of sample, 320
single beam spectrophotometry, 319
Soret region, 332
spectra of ligand-bound forms, 321–2
spectra requirements, 317–19
spectral band width, 322–4
spectrophotometric techniques, 320–4
spectrophotometry, 316–24
split-beam spectrophotometry, 318
structure, 312–16
taxonomic implications, 337–40
taxonomic status of prokaryotes, 311
types, 327–30
wavelength measurement, 336
wavelength range for
    spectrophotometry, 322
wavelength requirements for
    spectrophotometry, 316
cytochrome a_1, 329
cytochrome α_3-containing oxidases, 329
cytochrome b
    bacterial respiratory chains, 327
    CO difference spectrum, 335
cytochrome $bcaa_3o$, 338

cytochrome c
    CO difference spectrum, 335
    electron transfer, 327–8
    fluorescence techniques, 326–7
    oxidases, 329
    spectral band, 332
    types, 328
cytochrome c-deficient bacterial
    mutants, 325–6
cytochrome cd_1, 330
cytochrome d, 330
    deficiency, 326
    spectrum nomenclature, 334
cytochrome d-type oxidases, 321
cytochrome o, 329–30
cytochrome oxidase, 474
    bacterial, 328–9
    carbon monoxide identification, 321
    CO difference spectra, 341
    electron transfer, 328
    Gram-negative bacteria, 338
    selective media, 326
cytochrome oxidase-deficient bacterial
    mutants, 326
Cytophaga, 445
cytoplasmic membrane, Gram-
    negative bacteria, 21
cytosine, 463, 468–9, 545

database construction
    1D SDS-PAGE, 512
    2D electrophoretic patterns, 511
    FORTRAN 77 program extension,
        512
    protein electrophoresis, 509–13
deaminase, 475
    plates, 485
decarboxylase tests, 485
demethylmenaquinone, 272–3
densitometry, protein electrophoresis,
    500, 504
6-deoxyhexoses, 43
Desulfotomaculum
    cytochrome b, 327
    cytochromes, 339
Desulfovibrio
    cytochrome b, 327
    cytochromes, 339
    menaquinones, 268
Desulfurococcus lipids, 240, 241
Desulfurolobus lipids, 240
trans–22-deyhdroxysterol, 185

diagnostic bacteriology, enzymes in, 471–90
diamino acid, 6
2,3-diamino-2,3-dideoxyglucose (DAG), 28
diaminopimelic acid (DAP), 6
 detection in Gram-positive bacteria, 68, 69, 70–1
Dice coefficient, 506
3,6-dideoxyhexoses, 43
diethers of archaeal lipids, taxonomic implications, 228, 229–30, 231
digitonin precipitation, sterols, 171, 172
dimethylmenaquinone, 276
 NMR, 302
 ultraviolet spectroscopy, 294
dinoflagellate taxonomy, 191–2
dinosterol chemotaxonomy, 192
dipeptidyl aminopeptidase, 487
diphosphatidylglycerol, 142, 143
 Gram-negative bacteria, 45
diplopterol, 166
diplotene, 165
discriminant analysis (DA), 551–2, 553
diversity, 2, 4
DNA:DNA hybridization, 508
 correlation with SDS-PAGE, 516
DNA
 homology of bacterial strains, 493
 HPLC, 465, 466, 467
 hydrolysis, 467
 isolation, 463–6
 melting curve, 468
 prokaryotic, 13
DNA base composition, 466–8
 determination, 463–9
 taxonomic implications, 468–9
DNA-RNA hybridization, 53
dual RC progenitor, 387, 388

echinenone, 407, 444
elaidic acid, 486
electrodialysis, LPS purification, 35
electron impact mass spectrometry (EI-MS)
 carotenoids, 439, 440
 diether core lipids of archaea, 213, 214, 215–16
 tetraethers of archaeal lipids, 224, 225, 226

electron spin resonance spectroscopy, 324–5
electronic spectroscopy, 324–5
electrophoretic whole-organism protein fingerprints, 493–516
endopeptidases, 7-AMC demonstration, 486
endotoxin assay, 489
Enterobacteriaceae
 enzyme characterization, 475
 β-glucuronidase, 478
Enterococcus faecalis,
 demethylmenaquinone, 272
enzymatic treatment, LPS purification, 34
enzymes
 advantages as taxonomic indicators, 472–3
 cleavage point, 487
 colour development, 477
 detection methods, 479–86
 in diagnostic bacteriology, 471–90
 drawbacks of fluorescence techniques, 484
 fluorescence detection, 489
 fluorescence techniques, 480–4
 fluorigenic substrates, 480–1, 482–4
 identification value, 473–6
 known taxonomic value, 473–6
 multipoint techniques, 484–6
 naphthol/naphthylamine based substrates, 479
 non-hydrolytic, 490
 substrate design, 476–7
 substrate solubility, 488
 in taxonomy, 471–90
 visual observation, 479–80
2,3-epoxy menaquinones, 272
epoxyubiquinone, 278
ergosterol, HPLC analysis, 174
Erythrobacter longus, 371, 372
 classification, 374
 LH complexes, 375
Escherichia coli
 aminotransferases, 474
 capsular polysaccharides, 24
 cytochrome bd, 314
 β-glucuronidase, 478
 laser pyrolysis, 536
 linkage map, 473
 peptidoglycan, 21

Index

esculin, 477
  plates, 485
esterases, 476, 478
24-ethyl cholesterol, 176, 177, 188
24-ethyl-5β-cholestanol, 182, *183*
eubacteria
  fatty acids, 121, *122*, 123, *124*, 125
  photosynthetic ancestry, 366–7, 381
  structural lipids, 121–55
  triterpenoids, 163
Eubacterium lentum, 276
eukaryotes
  non-photosynthetic, 163, *165*
  photosynthetic, 163, *165*
  triterpenoids, 163
evolution, envelope chemistry of Gram-negative bacteria, 52–3
Extranuclear 5000, 526

farnesol, 365
trans,trans-farnesol, 365
fast atom bombardment mass spectrometry (FAB-MS), 148–9
  archaeal polar lipids, 232, 235, 247–8
  Bligh–Dyer extraction, 149, 151, *152*
  polar lipids, 144, 149–50
fatty acid methyl ester products (FAMEs), 538
fatty acid methyl esters, 136, *137*, 138–9
  hydroxylated, 138–9
  unsaturation, 139
fatty acid picolinyl esters, 139, *140*, 141
  cyclopropane rings, 141
fatty acids
  acid methanolysis, 125, 126
  alkaline hydrolysis, 125, 126
  amide-linked, 43
  analysis, 8, 9, 43
  analysis methods, 125–7, *128*, 129–33, *134*, 135–6, *137*, 138–9, *140*, 141
  capillary chromatograms, 123
  double bond verification, 135
  equivalent chain lengths, 135
  ester-linked, 43
  eubacteria, 121, *122*, 123, *124*, 125
  extraction, 125–7
  Gram-negative bacteria whole-cell analysis, 48–9
  growth conditions for biomass preparation, 125
  mass spectrometry identification, 136, *137*, 138–9, *140*, 141
  methyl ester analysis by capillary gas chromatography, 131–3, *134*, 135–6
  methyl ester purification, 129–30
  structures in eubacteria, *122*
  TLC, 127, *128*, 129–31
  transesterification, 125
  trimethylsilylation, 135–6
FCRG 51 alga, 177, 182
  sterol gas chromatogram, *177*
FeII, 312
  distinctive cytochrome absorption bands, 320
FeIII, 312
field desorption mass spectrometry (FD-MS)
  archaeal polar lipids, 247
  diether core lipids of archaea, 215
fingerprinting methods, analytical, 523–53
  chemometric strategies, 523
  direct comparison, 523
  infrared spectrometry, 538–41, *542*, 543
  laser pyrolysis, 535–6
  Py-GLCMS, 536–8
  pyrolysis tandem mass spectrometry, 538
  pyrolysis techniques, 524–6
  ultraviolet-resonance RAMAN spectroscopy, 543–5
Flavobacterium, 445
flavoprotein enzymes, 474
Flexibacter, 445
flexirubins, 444–5, 446
fluorescence emission, 480
fluorigenic compounds, 484
fluorophores, 489
FORTRAN 77 program extension, 512
Fourier self-deconvolution, 541
Fourier transform infrared (FTIR) spectroscopy, 540, 541, *542*, 545
fractionation on columns, LPS purification, 34–5
fucosterol, 188

β-galactosidase, 478

galactoside, fluorogenic, 478
galacturonides, 486
gas chromatography, 3
  fatty acid analysis, 8–9
  hopanoids, 175–7, 182–8
  polysaccharides, 81–2
  sterols, 175–7, 182–8
  teichoic acid degradation products, 79–80
gas chromatography-mass spectrometry, conditions for fatty acids, 136
gas chromatography-mass spectroscopy
  hopanoids, 175–7, 182–8
  selected ion monitoring (SIM), 188
  sterols, 175–7, 182–8
gas-liquid chromatography (GLC)
  diether core lipids of archaea, 211
  LPS sugar and fatty acid combined analysis, 39, 40
  with mass spectrometry for bacterial cell direct detection, 50–1
  monosaccharide analysis, 49
  neutral sugar of pseudomurein identification, 97
  tetraethers, 220
  whole-cell fatty acids, 48–9
  whole-cell sugar analysis, 49
gas-liquid chromatography-chemical ionization mass spectrometry, archaeal lipids, 216
GELCOMPAR, 503, 506, 507, 509
  data storage, 512–13
geranylgeraniol, 365, 366
geranylgeranyl diphosphate, 405
glass capillary gas chromatography, 211
β-glucuronidase, 478
Gly-L-Pro-7-AMC, 487
glycan chain length, 5–6
glycerol tetraether, 216, 217
glycerol-dialkyl-glycerol tetraethers (GDGT), 216, 217, 218
  C_{40} chain, 224
  chromatography, 218–20, 221
  stereochemistry, 226
glycerol-dialkyl-nonitol tetraethers (GDNT), 217, 218
  chromatography, 218–20, 221
  stereochemistry, 226
glycerol-trialkyl-glycerol tetraether, 227

glycerophospholipids, 141–2
  FAB spectra, 150
  structures, 142
glycolipids, 45, 144
  archaeal methanogen, 234, 236
  detection of archaeal, 203
  NMR of archaeal polar, 244–5
  thermophilic archaea, 238, 242
N-glycolyl substituent detection, Gram-positive bacteria, 68, 69–70
glycoprotein S-layers, 105–6, 107, 108–10
  archaea, 114
glycosalditols, 77
glycosidases, 476, 477, 478
glycosides in classification, 477
Gram-negative bacteria, 7
  apolar lipids, 46
  capsular polysaccharides, 23–4
  cell envelope, 21
  cell wall, 7, 22
  chemotaxonomy, 53–4
  cytochromes, 338
  cytoplasmic membrane, 21
  demethylmenaquinone, 273
  envelope chemistry and evolution, 52–3
  envelope constituents, 23
  extractable lipids, 45–6
  fatty acids of lipopolysaccharides, 138
  glycolipids, 45
  lipid A, 27, 28, 30–1
  lipopolysaccharides, 9, 22, 27–9, 30–1, 32–9, 40, 41–4
  outer-membrane, 9
  outer-membrane protein (OMP), 22, 25–6
  peptidoglycan, 21, 24–5
  phospholipids, 45
  polysaccharides, 46–7
  rapid whole-cell analysis methods, 47–52
  SDS-PAGE analysis, 22
  sphingolipids, 45
  sulphonolipids, 45
  taxonomy, 52–4
  whole-cell fatty acid analysis, 48–9
  whole-cell sugar profiles, 49
Gram-negative eubacteria
  methylmenaquinones, 274, 275
  ubiquinones, 277

Index 565

Gram-positive bacteria
 anionic polymers, 74–81
 anionic wall polymers, 64–5
 cell wall analysis, 63–6
 cell wall preparation, 66–8
 covalent linked components, 63–4
 cytochromes, 337–8
 diaminopimelic acid (DAP) detection, 68, 69, 70–1
 N-glycolyl substituent detection, 68, 69–70
 peptidoglycan, 64, 68–74
 peptidoglycan isolation, 71–2
 peptidoglycan type determination, 73–4
 polysaccharides, 65, 81–2
 rapid methods, 82–3
 teichoic acids, 74–80
 teichuronic acids, 75–6
 wall autolysis prevention, 67
 wall purification, 67
 wall-associated proteins, 65–6
 whole bacteria analysis, 68–71
 whole-cell detection of teichoic acid components, 82–3
Gram-positive eubacteria
 dimethylmenaquinones, 276
 methylmenaquinones, 274, *275*
green bacteria, 345, 347
 bacteriochlorophyll, 367
 bacteriochlorophyll *c/d*, 378–9
 bacteriochlorophyll synthesis, 380
 chlorosomes, 376, 377–8
 LH complex, 378
 nonsulphur, 376
 photosynthetic prokaryotes, 367
 RC complex, 380–1
 16S rRNA sequences, *370*, 377, 378
 sulphur, 376
 taxonomy, 376–81
guanine, 463, 468–9, 545
gulosaminuronic acid, isolation and identification, 101

¹H-COSY, isoprenoid quinones, 303, *304*
¹H-NMR
 bacteriochlorophyll *a*, 356–7
 carotenoid structure, 442–3
 isoprenoid quinones, 298, *300–1*, 302, 303
 tetrapyrroles, 355

haem, 312–13
 bacterial cytochrome oxidase, 329
 cytochrome classification, 315
 cytochrome nomenclature, 314
 iron in, 324–5
 RAMAN spectra, 325
haem iron, redox reactions, 327
haem proteins, carbon monoxide binding, 321
Haemophilus influenzae, capsular polysaccharides, 24
H. parainfluenzae, demethylmenaquinone, 272–3
haemoproteins, 312
 assignment of visible absorption bands, 327–32, *333*, 334–5, *336*, 337
 classification, 313, 315
 high-spin, 313
Halobacterium, 85
 carbohydrate moiety, 108
 envelope content isolation, 109
 glycolipids, 233
 glycopeptides, 108
 glycoprotein, 109
 glycoprotein S-layers, 106, *107* 108–9, 114
 lipids, 233
 S-layer protein, 109
Halococcus, 85, *91*
H. morrhuae, 100
 chemical composition, *101*
 heteropolysaccharide, 113
halophiles, 197
 cytochromes, 340
halophiles, archaeal
 acidic lipids, 231, *232*
 glycolipids, 231, *232*, 233
 lipids, 231, *232*, 233–4, 241
 quinone content, 257, *258*, 259
haemoproteins
 CO-binding *b*-type, 330
 non-cytochrome, 330–1
Heliobacillus, antenna-RC complexes, 382
heliobacteria
 photosynthetic prokaryotes, 367
 taxonomy, 381–2
Heliobacterium chlorum, 381
 bacteriochlorophyll *g*, 347–8, 357
Heliothrix oregonensis, 378

heptose analysis, 43
heteropolysaccharide, 113
  isolation/analysis, 100–2
  structure, 100
high pH anion exchange chromatography (HPAEC), 41, 44
high-performance liquid chromatography (HPLC), 3
  bacteriochlorophyll purification, 354–5
  p-bromophenacyl esters of mycolic acid, 131, 132
  carotenoids, 411, 421–6
  diether core lipids of archaea, 211
  DNA, 465, 466, 467
  isoprenoid quinones, 285
  peptidoglycan analysis, 25
  quantitative determination of carotenoids, 434–6
  reverse-phase for C24 epimer separation, 189
  silica-based columns, 354
  sterol analysis, 174
  sterol separation, 171, 172
  teichoic acid degradation products, 77–9
  tetraethers, 219–20, 221
hopanepolyols, 164
hopanoids, 10–11
  analysis, 163–93
  bacterial, 191
  biosynthesis, 164–6, 168
  in cyanobacteria, 168
  distribution, 163, 166
  extraction methods, 169–71
  extraction and treatment for chromatography, 173
  gas chromatography, 175–7, 182–8
  gas chromatography-mass spectroscopy analysis, 175–7, 182–8
  GC-MS analysis, 191
  geochemical studies, 192–3
  in microbial systematics, 191–3
  positive identification, 185
  primitiveness of biosynthesis, 165–6
  purification methods, 171–3, 173
  structural diversity, 163
  structure, 166
  substitution, 163–4
hopanols, silylation, 175

hot aqueous phenol, LPS extraction, 29, 32
hydrolases, 474–5, 476
hydrolytic enzymes, 486–89
8-hydroxyquinolone ÅÅ-D-glucoside, 485, 486
hydroxyspheroidene, mass spectrum, 443

identification, 1, 2
iduronides, 486
immunoblotting, 545
indophenol blue, 325
indoxyles, 479
infrared spectrometry, 3
  analytical fingerprinting methods, 538–41, 542, 543
  carriers, 539
  microbiological applications, 539–40
  numerical analysis, 541, 543
  specimen preparation, 540–1
  spectrometry, 541
infrared spectroscopy, 14, 325
  carotenoid characterization, 439
inoculation techniques, multipoint, 484–6
INTFILE, 503–4, 505, 506, 507
iodine-catalysed geometrical isomerization of carotenoids, 436–7
iron, 312–13
  in haem, 324–5
isomerases, 475
isopranoid ether lipids, degradation of archaeal, 204–6
isopranyl diethers
  archaeal lipids, 207, 208
  chromatographic procedures, 209–11
  hydriodic acid-LiAlH$_4$, 209
isopranyl ether lipids, distribution, 229–30
isopranyl glycerol ether lipids, 206
isopranyl tetraethers, 216, 217
isoprenoid ethers of archaeal lipids, 227–8
isoprenoid quinones, 10, 11, 265–305
  archaeal, 250–1, 252–3, 254, 255–6, 257, 258, 259
  argentation chromatography, 286–9, 290, 291
  chromatographic analysis, 283–9, 290
  chromatography, 281–2

Index

extraction, 280–2
¹H-COSY, 303, *304*
¹H-NMR, 298, *300–1*, 302, 303
mass spectrometry, 294, *295*, 296–7, *299*
nuclear magnetic resonance spectroscopy, 298, *300–1*
physicochemical analysis, 290–1, *292–3*, 294, *295*, 296–8, *299–301*, 302–3
purification, 280–2
respiratory, 290–1, *292*, 305
silica cartridge purification, 282
structure, 266–9, *270*, 271–4, *275*, 276–80
supercritical fluid chromatography (SFC), 289, *292*
taxonomic value, 265
ultraviolet spectroscopy, 290–1, *292*, *293–4*
isopropyl thiogalactoside (IPTG), 478

2-keto-3-deoxyoctonate (KDO), 9, 27, 42–3, 46

lactate, 474
Lactobacillus rhamnosus, cyclopropane fatty acid extraction, 135
LAMMA apparatus, 535
lanosterol, 163, 164, *165*
laser pyrolysis, 545
  mass spectrometry, 535–6
lecithinase production, 472
Legionella, fatty acid patterns, 29
Legionellaceae, phosphatidylcholines, 142
leucine aminopeptidase, 487
L-leucyl-7-AMC, 481, 487
ligases, 476
light-harvesting (LH) complex
  green bacteria, 378
  purple bacteria, 374–5
light-harvesting (LH) pigments, 386–7, 388
Limulus lysate method, 489
lipases, 478
lipid A, 10, 16
  aminosugars, 43–4
  analysis, 35–6, 38, 43
  fatty acid substitution pattern, 28–9
  Gram-negative bacteria, 27, 28, 30–1

hydroxy fatty acids in Gram negative bacteria, 123
separation from polysaccharide, 36–7
lipid analyses in microorganism classification/identification, 153, 155
lipid extraction, modified Bligh–Dyer technique, 198–9, 200
lipo-oligosaccharide, 27–8
lipoamino acids, 45
lipopolysaccharide (LPS), 9
  acetyl groups, 44
  analysis of constituents, 38–9, *40*, 41–4
  analysis of intact, 35–6
  antigenic properties, 22–3
  cleavage, 36–7
  endotoxic effects, 22
  ethanolamine, 44
  extraction, 29, 32–4
  Gram-negative bacteria, 22, 27–9, *30–1*, 32–9, *40*, 41–4
  impurities, 35
  phosphate analysis, 44
  purification, 34–5
  R-form, 27, 32–3
  S-form, 27
  sugars and fatty acids combined analysis, 38–9, *40*
lipoquinone, 266
  ultraviolet spectroscopy, 290, 291
lutein, 425
lyases, 475
lycopene, 404, *405*
  anoxygenic phototrophs, 408
  spectral fine structure, *432*
lysine decarboxylase test, 475

magnesium
  dihydroporphyrin, 347, 348
  oxide *see* MgO-kieselguhr G
  removal from bacteriochlorophyll *b/g*, 348
  tetrahydroporphyrin, 345
mannitol, 76, 79
mannurinides, 486
mass spectrometry
  archaeal lipids, 213, *214*, 215–16
  archaeal polar lipids, *232*, *235*, 247–9
  carotenoids, 439–42
  fatty acid identification, 136, *137*, 138–9, *140*, 141

mass spectrometry (cont.)
  fatty acid methyl esters, 136, 137, 138–9
  fatty acid picolinyl esters, 139, 140, 141
  flexirubins, 446
  isoprenoid quinones, 294, 295, 296–7, 299
  polar lipid identification, 148–51, 152
  polysaccharide structure, 81
  pyrolysis, 524, 525–6
  tetraethers of archaeal lipids, 224, 225, 226
  xanthomonadins, 446
mass spectroscopy
  purification of carotenoids, 420–1
  sterols, 185, 186, 187
MAST system, 484–5
mavioquinone, 279
menaquinones, 11, 267–9, 270, 271–2
  archaeal, 251
  argentation chromatography, 287–8, 289, 290
  dihydrogenated, 268
  Gram-positive eubacteria, 268
  ^{1}H-COSY, 304
  hexahydrogenated, 269
  HPLC, 285–6
  mass spectrometry, 294, 295, 296
  methylated, 274, 275
  NMR, 302–3
  with partially saturated side-chains, 269, 270
  RPHPLC separation, 286, 288
  side-chain saturation, 296
  tetrahydrogenated, 268–9
  ultraviolet spectroscopy, 290–1, 293
meningococcal septicaemia, GLC-MS plasma profile, 50
meningococcal vesicle vaccine, outer-membrane proteins, 26
metalloproteins, 325
Methanobacteriales, molecular ratios of cell wall components, 96
Methanobacterium, pseudomurein, 112
M. uliginosum, 91
M. wolfei, 93
Methanobrevibacter, pseudomurein, 112–13
methanochondroitin, 113
  isolation/analysis, 98–9

  structure, 99
  uronic acid determination, 99
methanococci, S-layer isolation, 102, 103–5
Methanococcus, protein S-layers, 113
Methanogenicum, protein S-layers, 113
methanogens, 197
  cytochromes, 339
  lipids, 228
methanogens, archaeal, 234, 235, 236–7
  aminolipids, 242
  lipids, 242
  polar lipids, 242
  taxonomic value of lipids, 237
Methanolobus, 107
Methanopyrus, pseudomurein, 113
Methanosarcina, 91, 98, 99
  methanochondroitin, 113
M. barkeri, 228
Methanosphaera, pseudomurein, 113
Methanospirillum, 90, 107
  S-layer, 110–11
  SDS-resistant layers, 114
  sheath isolation, 111
Methanothermus, 105–6, 107
  glycoprotein S-layers, 105–6, 114
  pseudomurein, 113
Methanothermus fervidis, 91
Methanothrix, 90, 91, 107
  lipids, 241
  S-layer, 110–11
  SDS-resistant layers, 114
  sheath isolation, 111
methionaquinone, 280
  mass spectrometry, 294, 295, 296
methyl bacteriopheophorbide, 349
methyldiplopterol, 166
methylimidazole, 82
methylmenaquinones, 274, 275
  mass spectrometry, 294, 295
  NMR, 302
  ultraviolet spectroscopy, 291, 294
10-methyloctadecanoic acid extraction, 135
Methylomonas rubra, ubiquinones, 278
methylotrophic bacteria, 372–3
  pink-pigmented, 372–3
  16S rRNA sequences, 372–3
4-methylumbelliferone (4-MU), 480, 481, 482, 484, 486
  fluorescence detection, 489

MgO-kieselguhr G, 416, 417, 419–20
Microcell C, 417
Micromonospora, muramic acid, 6
microorganisms, diversity, 4
monosaccharides, Gram-negative bacteria whole-cell analysis, 49
multi-haem purified complexes, 315
multivariate analysis mathematics, 549–53
murein, 112
Mycobacterium, muramic acid, 6
M. avium, mavioquinone, 279
mycolic acids
  actinomycetes, 123
  p-bromophenacyl ester HPLC, 131, 132
  p-bromophenacyl ester preparation, 127
  chemotaxonomic studies, 537
  extraction for analysis, 125, 126–7
  methyl ester TLC, *130*, 131
  structure, *123*, *124*
  TLC, 127

naphthoquinones, 266, 274, 280
  mass spectrometry, 294, *295*, 296
  ring substitution, 290
  ultraviolet/visible spectroscopy, 257
Natronobacterium gregori, 276
Neisseria meningitides
  capsular polysaccharides, 24
  polysaccharides, 46
neisserial LPS marker analyis, 51–2
neural network analysis, 545
neurosporene, spectral fine structure, 432
neutral lipids, archaeal, 200
neutrophiles, archaeal, polar lipids, 233–4
nitrate reductase, 474
nitrilases, 490
nitrile columns for carotenoid HPLC, 425–6
nitrile hydralases, 490
p-nitroanilides, 479
nitrophenols, 479
Nocardia
  menaquinones, 269, *270*, *271*, 272
  muramic acid, 6
N. asteroides, 10-methyloctadecanoic acid extraction, 135

nomenclature, 1–2
non-photosynthetic prokaryotes, carotenoids, 408, 410–11
non-sulphur purple bacteria, 368
normal-phase high-performance liquid chromatography (HPLC), carotenoids, 424–6
nostoxanthin, 407
nuclear magnetic resonance (NMR) spectroscopy, 212–13
  archaeal polar lipids, *239*, 243–7
  carotenoid characterization, 439
  carotenoid structure, 442–3
  flexirubins, 446
  isoprenoid quinones, 298, *300–1*
  polysaccharide structure, 81
  purification of carotenoids, 420–1
  tetraethers, 220, 222–4
  xanthomonadins, 446
nucleic acid
  conserved regions, 15
  hybridization, 53
  sequencing, 15

okenone, *425*, 444
oleic acid, 486
oleyl alcohol, 365
ONPG test, 472
L-ornithine, 6
ornithine decarboxylase test, 475
Oscillochloris, 377–8
outer-membrane protein (OMP) composition, 22
  Gram-negative bacteria, 25–6
  meningococcal vesicle vaccine, 26
  taxonomic potential, 26
oxidase test, cytochromes, 325–6
oxidoreductases, 473–4, 474
2,3-oxidosqualene, *165*
1'-oxomenaquinone, 272

paper chromatography, carotenoids, 420
PDQUEST, 511–12
Pearson product moment correlation coefficient, 506–7, *507–8*
pectin hydrolysis, 472
peptidases, 476, 478–9
peptidoglycan, 5, 6, 24–5
  amino acid composition determination, 72
  determination of type, 73–4

peptidoglycan (cont.)
  Gram-negative bacteria, 21
  Gram-positive bacteria, 64, 71–2
  N and O acetylation of amino sugars, 71
  structure variation, 6
peroxidase, 474
  test for cytochromes, 326
phenol, protein denaturing, 464
phenol/chloroform/petroleum ether (PCP) LPS extraction method, 32–3
phenotypic characteristics, 15
pheophytin a, 348
phorotrophic bacteria, reversed phase HPLC, 423
phosphatases, 475, 476, 478
phosphatidyl ethanolamine, 142
  Gram-negative bacteria, 45
phosphatidylcholine, 142
  FAB spectra, 149, 150
phosphatidylglycerol, 142
  archaeal halophiles, 231, 233
  Gram-negative bacteria, 45
phosphatidylglycerophosphate, archaeal halophiles, 231, 233
phosphatidylglycerylsulfate, 231, 233
phosphatidylinositol, 143
phosphoglycolipids, 234, 236
phospholipase $A_2$1, 150
phospholipids, 8
  archaeal methanogen, 234
  detection of archaeal, 203
  digestion with phospholipase A_2, 150
  FAB-MS, 149
  negative ion spectra, 150
  thermophilic archaeal, 238, 240
phospholucomutase, 475
photosynthesis, cyanobacterial, 386
photosynthetic bacteria
  aerobic, 371, 372
  carotenoid distribution, 408–9
photosynthetic intracytoplasmic membranes (ICM), 373, 374
photosynthetic prokaryotes, 345
  anoxygenic, 345
  classes, 367
  diversity, 366
  evolution of chlorophylls, 385–8
  oxygenic, 345
  subclasses, 367

phototrophic bacteria, 12
  carotenoids, 407
  cytochromes, 339
phototrophic eubacteria
  halophilic, 366
  thermophilic, 366
phototrophic prokaryotes
  anoxygenic, 366
  carotenoids, 403
  oxygenic, 366
phthalates, 414
phycobiloproteins, 388
phylloquinones, 266–7
  ultraviolet spectroscopy, 290
phylogenetic characteristics, 15
3-O-phytanyl-sn-glycerol, 227
tri-O-phytanylglycerol, 227
phytoene, 406
  spectral fine structure, 432
  TLC, 418
phytofluene
  spectral fine structure, 432
  TLC, 418
phytol, 364, 365
picolinyl ester preparation, 139
plasma membrane, 7
plasmalogens, 143
plasmaspray, 148
plastoquinones, 279
  ultraviolet spectroscopy, 291, 293
polar lipids
  analysis methods, 144–5, 146, 147–51, 152
  analysis of mixtures, 151, 152
  archaeal, 200, 231, 232, 233–4, 235, 236–8, 239, 240–9
  archaeal diether, 241
  bacterial, 141–4
  extraction, 144
  FAB technique for mixtures, 151, 152
  fatty acids, 8
  integrated extraction with isoprenoid quinones, 154
  mass spectrometric identification, 148–51, 152
  methanogenic archaea, 242
  rapid analysis of archaeal, 202
  spray reagents for characterizing individual components, 147
  two-dimensional TLC, 145, 146, 147

Index

polar lipids, archaeal
  mass spectrometry, 232, 235, 247–9
  methodologies, 242–9
  NMR spectroscopy, 239, 243–7
polyacrylamide gel electrophoresis
    (PAGE), 493
  types, 494
polyene
  aryl, 444
  bacterial pigmentation, 403
  chain fragmentations, 440
  conjugated system, 412
  non-carotenoid, 444–7
polyethylene columns for
    bacteriochlorophyll
    chromatography, 353, 355
polysaccharides
  acid hydrolysis, 81
  chemical relatedness, 47
  detergent complexes, 46–7
  Gram-negative bacteria, 46–7
  Gram-positive bacteria, 65, 81–2
  sugar analysis, 41–3
principal component analysis (PCA),
    509, 551
Prochloron didemni, 384
prochlorophytes, 345, 384–5
  photosynthetic prokaryotes, 367
  16S-rRNA sequence analysis, 384
Prochlorothrix hollandica, 384
prokaryotes, isoprenoid quinones, 265
proteases, 476
protein
  electrophoregrams, 493
  S-layers, 102, 103–5, 107, 113
  wall-associated of Gram-positive
    bacteria, 65–6
  see also bacterial proteins
protein A, *Staphylococcus aureus*, 63
protein electrophoresis, 493
  average linkage clustering technique,
    509
  background subtraction, 507
  computation of similarity, 506–9
  computer-assisted analysis, 500–1,
    503–4, 505, 506–14
  data compatibility, 510–11
  data consistency, 511
  data reproducibility, 511
  database construction, 509–13
  densitometry, 500, 504

grouping of protein patterns, 509
  lateral shifting, 508–9
  methods, 493–4
  normalization, 501, 503–4, 505, 506
  principal component analysis, 509
  removal of major protein bands,
    507–8
  standardization, 500–1
  unweighted-pair group method using
    average linkages (UPGMA), 509
protein gel electrophoresis, 13
  whole organism, 14
Proteobacteria, 369
protochlorophyllide, 385
Pseudomonas, fatty acid capillary gas
    chromatogram, 133–4
pseudomurein
  amino acid/amino sugar
    identification, 95
  antigenicity, 93
  archaea, 112–13
  biosynthesis, 93
  constituent analysis, 95, 97–8
  determination of type, 98
  hydrolysis of components, 95
  isolation and analysis, 93–5, 97–8
  mechanical disintegration and
    digestion, 94–5
  phosphate determination, 98
  sacculi structure, 92
  structure, 93, 94
Psychrobacter immobilis, 503
pulsed amperometric detection (PAD),
    78–9
purple bacteria, 345, 347, 368–9, 370,
    371–6
  bacteriochlorophylls, 367
  classification on photosynthetic
    membranes, 373
  LH complexes, 374–5
  photosynthetic intracytoplasmic
    membranes (ICM), 373, 374
  photosynthetic prokaryotes, 367
  phylogeny, 368–9
  reaction centre (RC) complexes,
    375–6
  16S rRNA sequences, 368, 369–70
Pyrococcus lipids, 237
pyrolysates, 524
  chemical nature of components, 537
  Py-GLCMS, 537

pyrolysis
  gas chromatography, 3, 13
  gas-liquid chromatography
    (Py-GLC), 525
  gas-liquid chromatography-mass
    spectrometry (Py-GLCMS),
    536–8
  product associations, 537
  tandem mass spectrometry
    (Py-MS-MS), 538
  techniques for fingerprinting, 524–6
pyrolysis mass spectrometry, 3, 13,
    14, 524, 525–6, 526–34
  applications, 526
  automated processing, 529–32
  data analysis, 533–4
  dendrograms, 534
  drift in mass scale, 529
  foil preparation, 527–8
  foils, 529, 530–1
  iterative normalization program,
    533
  method, 527–32
  multivariate statistical analysis, 534
  normalization program, 533–4
  processing data, 533–4
  product separation, 531, 532
  replication formats, 527
  reproducibility, 533
  sample insertion, 529, 530
  sample preparation, 528–9
  similarity matrix, 534
  spectrum presentation, 532
  steps, 526
  tube preparation, 527–8
pyrolytic-on-line derivization 538
  GLCMS, 538
Pyromass 8-80, 526

quinones
  extraction, 280–2
  purification, 280–2
  sulphur-containing, 279–80
quinones, archaeal
  chromatography, 254
  distribution, 257, 258, 259
  mass spectrometry, 254, 256, 257
  NMR spectroscopy, 254, 255
  structural types, 250–1, 254
  ultraviolet spectroscopy, 257
  visible spectroscopy, 257

racemases, 475
RAMAN
  spectra of haem groups, 325
  spectroscopy, 3, 13, 14, 544
rapid enzyme profiling, 13
reaction centre (RC) complex
  cyanobacteria, 387
  green bacteria, 380–1
  heliobacteria, 382
  purple bacteria, 375–6
recapitulation hypothesis, 385
redox potentials, 316
redox reactions, haem iron, 327
redox states, 311, 312
  spectra, 320–1
relative retention times, 183–4, 188
resonance RAMAN spectroscopy, 544
resorufin, 489
respiratory chains, bacterial, 327, 328
restriction endonuclease digest
  analysis, 545
reverse-phase chromatography,
  bacteriochlorophylls, 353
reverse-phase high-performance liquid
  chromatography (HPLC)
  bacteriochlorophylls, 354
  carotenoids, 422–4
reverse-phase partition chromatography
  isoprenoid quinones, 283–6
  ready-made plates, 284
  use, 286–7
Rhizobium, photosynthetic symbiotic
  nitrogen-fixing, 373
Rhodobacter sulfidophilus,
  photosynthesis conditions, 372
Rhodocyclus, cell membrane, 374
rhodophytes, 388
rhodopin, anoxygenic phototrophs, 408
Rhodopseudomonas viridis, 328
  bacteriochlorophyll *b*, 347
  esterifying alcohol for
    bacteriochlorophyll *b*, 364
rhodoquinone, 277, 278–9
Rhodospirillaceae, 444
Rhodospirillum rubrum, 277
  esterifying alcohols for
    bacteriochlorophylls, 364
  rhodoquinone, 278
ribitol, 76, 79
ribonuclease (RNase), 465
RNA hydrolysis, 465

Index

16S rRNA analysis, 53
16S rRNA sequences
  green bacteria, *370*, *377*, *378*
  purple bacteria, 368, *369–70*
  Roseobacter, 408
  R. denitrificans, 371
    photosynthetic apparatus, 373–4

Saccharomonospora viridis, two-dimensional TLC, *146*
Saccharomyces, enzymatic activities in, 471
Saccharopolyspora hirsuta, fatty acid capillary gas chromatogram, *133–4*
salicin, 477
salicylonitrile, 490
Salmonella, outer membrane LPS, 52
S. typhimurium, linkage map, 473
saponification, carotenoids, 415
SDS-PAGE
  analytical fingerprinting methods, 545
  cell envelope protein patterns, 508
  correlation with DNA:DNA hybridization techniques, 516
  Gram-negative bacteria analysis, 22
  Gram-positive bacteria protein analysis, 66
  LPS analysis, 35–6
  one-dimensional, 512
  protein profile comparison, 506
  protein S-layer, 104–5
  taxonomic applications, 513, *514–15*, 516
  whole-cell bacterial proteins, 494–501, *502*, 503–4, *505*, 506–13
sediments, lipid extraction, 171
Serratia rubidaea surface active lipids, 45
sewage contamination, *183*
  biomarker, 182
Shigella sonnei, β-glucuronidase, 478
silica
  carotenoid adsorption, 416, 417, 419
  normal-phase HPLC, 425
silylation of carotenoid hydroxy groups, 437
SIMCA analysis, 552
sirohaem, 331
sodium dodecyl sulphate (SDS)
  Gram-positive bacteria cell wall preparation, 67

polyacrylamide gel electrophoresis *see* SDS-PAGE
soils, lipid extraction, 171
sorbitol dehydrogenase, 474
spectral derivatization, 541
spectrophotometer
  absorption spectrum, 317–18
  calibration, 324
  components, 316–17
  design, 316–20
  difference spectrum, 318
  double-beam, 318
  dual-wavelength, 318–19
  low-temperature accessories, 319–20
  optical configuration, *317*
  reference spectrum, 318
  split-beam, 318
  stray light, 319
  wavelength requirements for cytochromes, 316
spectrophotometry
  carotenoids, *434*, *435*
  spectral band width for cytochromes, 322–3
  spectral peak height, 323–4
spectroscopy
  absorption maxima position for carotenoids, 426–7
  carotenoids, 426–7, *428–32*, 433
spheroidene, 424, 444
spheroidenone, 444
sphingolipids, 45
  hydroxy fatty acids, 123
sphingophospholipids, 144
  FAB spectra, 149
spirilloxanthin pathway, 407
spirochaetes, peptidoglycan, 6, 24
squalene, 164, *165*
5β-stanols, 182
staphylococci, taxonomy, 483
Staphylococcus aureus, protein A, 63
stearic acid, 486
stearyl alcohol, 365
steranes, C24 epimer separation, 189
stereochemistry
  diether core lipids of archaea, 216
  tetraethers of archaeal lipids, 226
sterol biosynthesis, 163
sterol-TMS ether derivatives, characteristic ion fragments, *188*

sterol-TMS ethers, 175, *177*
  mass spectral and gas chromatographic retention data, *178-81*
sterols, 10, 11
  algal taxonomy, 191
  analysis, 163-93, 173-7, 182-9
  biosynthesis, 164, 164-6, 168
  biosynthetic pathways, 192
  C24 epimer analysis, 190
  culture conditions, 190
  in cyanobacteria, 168
  determination of total, 173-4
  digitonin precipitation, 171, 172
  distribution, 163, 166
  extraction efficiency, 190
  extraction methods, 169-71
  gas chromatography, 175-7, 182-8
  gas chromatography-mass spectroscopy analysis, 175-7, 182-8
  GC-MS analysis, 191
  geochemical studies, 192-3
  HPLC analysis, 174
  HPLC purification, 171, 172
  in microbial systematics, 191-3
  polycyclic nuclei, *167-8*
  positive identification, 185
  purification methods, 171-2
  relative retention times, 183-4, 188
  separation of C24 epimer, 189
  side-chains, *167-8*
  structural diversity, 163
  structure confirmation, 185
  substitution, 163-4
  TLC purification, 171, 172
  unsaturated determination, 174
stromatolites, 385, 387
structural lipids of eubacteria, polar lipids, 141-4, 144
succinate dehydrogenase, 474
sucrose columns for bacteriochlorophyll chromatography, 351, 353
sugar analysis
  whole-cell of Gram-negative bacteria, 49
  whole-cell of Gram-positive bacteria, 81
Sulfolobales
  benzothiophenquinones, 251, *252*
  lipids, 228

Sulfolobus, 85, *107*
  envelope content isolation, 110
  glycoprotein S-layers, 114
  lipids, 240, 241
  S-layer, 110
S. acidocaldarius, cytochromes, 340
S. ambivalens
  menaquinones, 367
  sulfolobusquinone, 280
S. solfataricus, 228
sulfolobusquinone, 280
  mass spectrometry, 297
  ultraviolet spectroscopy, 291
sulphatases, 476, 478
sulphonolipids, 45
sulphur purple bacteria, 368
supercritical fluid chromatography (SFC), isoprenoid quinones, 289, 292
systematics, 1

talosaminuronic acid isolation, 97
taxonomy, 2
  chemical, 5
  polyphasic, 15
teichoic acid, 7, 64, 65
  acid hydrolysis of hydrofluoric acid degradation products, 77, 78
  classification in Gram-positive bacteria, 76-80
  GC of degradation products as alditol acetates, 79-80
  GC of degradation products as trifluoroacetyl derivatives, 80
  Gram-positive bacteria, 74-5
  hydrofluoric acid degradation, 76-7, 78
  measurement as organic phosphorus, 75-6
  whole-cell detection of components in Gram-positive bacteria, 82-3
teichuronic acid, 7, 64
  analysis, 80-1
  Gram-positive bacteria, 75-6
  measurement as uronic acid, 76
tetraethers of archaeal lipids, 216, *217*, 218-20, *221*, 222-4, *225*, *226*
  C$_{40}$ hydrocarbons, 222
  ^{13}C chemical shift values, 223
  chromatography, 218-20
  degradative procedures, 218

Index 575

mass spectrometry, 224, 225, 226
polar, 241
stereochemistry, 226
taxonomic implications, 228, 229–30, 231
tetrapyrrole
^{1}H-NMR, 355
photosynthetically active pigments, 367
tetraterpenes, 403 405
thermal melting point (T_m) of DNA, 465, 467
Thermococcus lipids, 237
Thermofilum, S-layer, 111
Thermomicrobium roseum, 377
thermophiles, 197
thermophiles, archaeal
  glycolipids, 242
  lipids, 237–8, 239, 240–1
  quinone content, 257, 258, 259
  tetraethers, 237
Thermoplasma
  cell envelope-less, 114
  naphthoquinones, 274
  T. acidophilum, 85, 91
  cytochromes, 340
  lipids, 238
thermoplasmaquinone, 274
Thermoproteus, 107
  glycolipids, 238
  lipids, 240, 241
  S-layer, 111
  T. tenax, 91, 218
    menaquinones, 267
    phylloquinones, 266, 267
thermospray, 148
thin-layer chromatography (TLC)
  bacteriochlorophyll purification, 351, 353
  carotenoids, 416, 417, 418–20
  diether core lipids of archaea, 210
  fatty acids, 127, 128, 129–31
  isoprenoid quinones, 281–2
  lipid A analysis, 36, 38
  LPS analysis, 36, 38
  one-dimensional of methanolysis products, 129
  polar lipid analysis, 144, 202
  reverse phase, 283, 284–5
  sterol separation, 171
  tetraethers, 218–19

two-dimensional of mycolic acid methyl esters, 130, 131
two-dimensional for polar lipids, 145, 146, 147
Thiorhodaceae, 368
transacetylases, 474
transferases, 474
Treponema pallidum peptidoglycan, 24
trifluoroethyl coumarin, 489
trimethylsilyl ethers (TMS ethers), 175
triose phosphate isomerase, 475
triterpenoids, 163
  phylogenetics, 192
  taxonomic markers, 191
tryptophan deaminase test, 475
tryptophan-2,3-dioxygenase, 330, 331

ubiquinone, 11, 276–9
  argentation chromatography, 288, 291
  HPLC, 285
  log retention time, 285, 286
  mass spectrometry, 295, 296
  NMR, 303
  reverse-phase partition chromatography, 283
  RPHPLC separation, 286, 287
  saturated side-chains, 277
  supercritical fluid chromatography (SFC), 289, 292
  ultraviolet spectroscopy, 291, 293
ultracentrifugation, LPS purification, 34
ultraviolet spectroscopy, isoprenoid quinones, 290–1, 292, 293–4
ultraviolet visible light absorption spectroscopy, 426–7, 428–32, 433
ultraviolet-resonance RAMAN spectroscopy, 543–5
  nucleic acid spectra, 544–5
unweighted-pair group method using average linkages (UPGMA), 509
urease tests, 485
uronic acids, 80, 81

Vibro succingenes, 327 327
vitamin K$_2$, 11, 267
vitamin K, 265, 266, 276
Vitreoscilla haemoglobin (VHb), 331

wax esters, 46
whole-organism fingerprinting, 13–14

xanthomonadins, 445, 446
Xanthomonas, 445, 508

xanthophylls, 403
  gradient procedure HPLC, 422–3
  properties, 412

yeast
  brewing, 182–3

GC sterol identification, 184
hopanoid extraction, 170–1
saponification, 171
sterol extraction, 170–1

zeaxanthin, 407, 425
Zinzadze reagent, 203

Index compiled by Jill Halliday